Lost Sex

Isa Schön · Koen Martens · Peter van Dijk
Editors

Lost Sex

The Evolutionary Biology of Parthenogenesis

Springer

Editors
Isa Schön
Royal Belgian Institute of
Natural Sciences
Freshwater Biology
Vautierstraat 29
1000-Brussels
Belgium
isa.schoen@naturalsciences

Koen Martens
Royal Belgian Institute of
Natural Sciences
Freshwater Biology
Vautierstraat 29
1000-Brussels
Belgium
koen.martens@naturalsciences

Peter van Dijk
Keygene N.V.
Agro Business Park 90
6708 PW Wageningen
The Netherlands
peter.van-dijk@keygene.com

ISBN 978-90-481-2769-6 e-ISBN 978-90-481-2770-2
DOI 10.1007/978-90-481-2770-2
Springer Dordrecht Heidelberg London New York

Library of Congress Control Number: 2009933886

Printed on acid-free paper

Springer is part of Springer Science+Business Media (www.springer.com)

Foreword

The idea to compile this book was first raised during an Exploratory ESF workshop in Wageningen in 2001. At that meeting, it was thought that it should provide the long-needed update on parthenogenesis and its genetic and ecological consequences, but that it should also look at the paradox of sex from an asexual perspective. Already then, it was decided to only focus on eukaryotes, as this would avoid discussions on whether bacteria have sex or not. . .The book would thus deal with these evolutionary opportunists that had given up or lost sexual reproduction. We also wanted to make sure that both animals and plants were included and to help the zoological and botanical communities in understanding each other. When ESF funded the proposal of PVD and KM for a scientific network on parthenogenesis, (PARTNER) we started hunting for authors and contributions during the four PARTNER workshops in 2003 (Wageningen), 2004 (Valencia and Münster) and 2005 (London). This is why most authors of this book have also been participants of one or several PARTNER workshops, and this book can be seen as an ESF deliverable. However, there are also contributions from autors outside of PARTNER, which have added additional expertise to this book. Their contributions also illustrate that the field of research on parthenogenesis is currently moving very fast.

The book contains five parts. The first one consists of the introduction to the theme (see Chapter 1) to catch the reader's interest and its history by concentrating on four major scientists who have significantly contributed to the field (Chapter 2). The general part also comprises the necessary descriptions of the cytology and the mechanisms related to the different asexual and sexual modes of reproduction in plants (Chapter 3) and animals (Chapter 4). In Chapter 3, great care has been taken to explain the botanical terminology of parthenogenetic reproduction and potential pitfalls are pointed out, which have greatly hampered the discussion between zoologists and botanists on asexual reproduction in the past.

The second part of the book discusses the main hypotheses behind the paradox of sex and how modern science regards them. Although there are more than 25 hypotheses attempting to explain the paradox of sex, we have decided to concentrate on four major ones. Is the accumulation of mutations (Muller's ratchet (Muller 1964) and Kondrashov's hatchet (Kondrashov 1988)) still regarded as one of the major problems of long-term asexuality and if so, under which conditions might it apply to the real biological world (see Chapter 5)? What is the reaction norm of asexuals? Can the general-purpose genotype and frozen niche hypotheses indeed be found in biological reality (Chapter 6)? Is the Red Queen still in the running as one of the major theories to explain the prevalence of sex (Chapter 7)? And if so, what are the conditions that have to be matched? Chapter 8 deals with the phenomena of geographic parthenogenesis and presents a new theory on how hybridisation could have caused the patterns in both plants and animals.

The third part of the book deals with more philosophical questions, which have nonetheless a practical impact on working with asexuals: how can clones (Chapter 9) and asexual species (Chapter 10) be defined? It is our hope that readers will find workable solutions in both chapters.

The fourth part comprises numerous case studies, illustrating the enormous variety of mechanisms by which asexual plants and animals reproduce in the living world. The first set of case studies includes examples of putative ancient asexual scandals (following the terminology of Judson and Normark 1996) from the animal world, namely darwinulid ostracods (Chapter 11), oribatid mites (Chapter 12) and bdelloid rotifers (Chapter 13). The surprising conclusion seems to be that none of these so-called scandals has followed the same route in avoiding the long-term disadvantages of asexuality, and that each of them might have found a novel way to circumvent the paradox of sex. There appears to be a general consensus that there are no ancient asexual plants (see Chapter 3).

The three scandalous chapters are followed by a number of items on invertebrates with mixed reproduction. These include examples of cyclic parthenogenesis, namely monogont rotifers and cladocerans (Chapters 14 and 15, respectively), which combine the best of both worlds and provide important insights into the factors triggering the switch between sex and asex. Also, stick insects with an amazing variety of reproductive modes (Chapter 16) are described, followed by examples of Hymenoptera, where microbial infections are one of the most common causes of asexuality (Chapter 17). The last invertebrate case study is on hermaphrodite planarians with asexual reproduction, but also with a little bit of sex (Chapter 18). The invertebrates are followed by vertebrate examples of parthenogenesis, covering fish (Chapter 19), waterfrogs (Chapter 20) and reptiles (Chapter 21). All three chapters provide good explanations on the mechanisms of vertebrate parthenogenesis and why it is less common than in invertebrates.

Two case studies on apomictic plants, the common dandelion (Chapter 22) and Böcher's rock cress (Chapter 23) close the section on case studies. Both author teams provide new, unpublished data on the evolutionary age of an apomixis gene (Chapter 22) and the Meselson effect, respectively (Chapter 23).

Last comes a section on applied aspects of parthenogenesis, a novel and important part. The section starts with examples of asexual diseases and why their reproductive mode poses a major threat to humanity (Chapter 24). The next chapter introduces aphids as agricultural pest and links their detrimental effects with their asexual reproduction. Chapter 26 explains why cloning of mammals is still so unsuccessful and how it can be improved. The last Chapter, 27, deals with clonal grape vines and its implication on wine making.

This last chapter provides the perfect excuse to raise the glass to all authors and readers and to wish them a very fruitful and enjoyable reading – may the bouquet be rich, the flavour be crisp and the finish be lasting!

We express our thanks to the European Science Foundation for funding the PARTNER network as well as the publication of this book.

References

Judson OP, Normark BB (1996) Ancient asexual scandals. Trends Ecol Evol 11: 41–46
Kondrashov AS (1988) Deleterious mutations and the evolution of sexual reproduction. Nature 336: 435–440
Muller HJ (1964) The relation of recombination to mutational advance. Mutat Res 1: 2–9

Ghent, Perth and Wageningen

Contents

Contributors

Philip Agnew Génétique et Evolution des Maladies Infectieuses, Equipe Evolution des Systèmes Symbiotiques, UMR IRD/CNRS 2724, BP 64501, 911 Av. Agropolis, 34394 Montpellier Cedex 5, France, agnew@cepm.mpl.ird.fr

Timothy G. Barraclough Division of Biology, Imperial College London, Silwood Park Campus, Ascot, Berkshire SL5 7PY, UK, t.barraclough@imperial.ac.uk

Bengt O. Bengtsson Genetics, Department of Cell and Organism Biology, Lund University, Sölvegatan 29, SE-223 62 Lund, Sweden, Bengt_olle.bengtsson@cob.lu.se

Andrej Benjak Department of Applied Plant Sciences and Plant Biotechnology, University of Natural Resources and Applied Life Sciences, Vienna, Peter-Jordan Str. 82, A-1190 Vienna, Austria, andrej.benjak@boku.ac.at

Leo W. Beukeboom Evolutionary Genetics, Centre for Ecological and Evolutionary Studies, University of Groningen. P.O. Box 14, NL-9750 AA Haren, The Netherlands, l.w.beukeboom@rug.nl

Arjen Biere Department of Multitrophic Interactions, Netherlands Institute of Ecology, NIOO-KNAW, P.O. Box 40, 6666 ZG Heteren, The Netherlands, a.biere@nioo.knaw.nl

C. William Birky, Jr. Department of Ecology and Evolutionary Biology, The University of Arizona, Tucson, AZ 85745 USA, birky@u.arizona.edu

Jose M. Corral Apomixis Research group, Department of Cytogenetics and Genome Analysis, Institut für Pflanzengenetik und Kulturpflanzenforschung (IPK), D-06466 Gatersleben, Germany, corral@ipk-gatersleben.de

Ellen Decaestecker Laboratory of Aquatic Ecology and Evolutionary Biology, K.U.Leuven, Ch. Debériotstraat 32, B-3000 Leuven, Belgium; Laboratory of Aquatic Biology, Interdisciplinary Research Center, K.U. Leuven Campus Kortrijk, E. Sabbelaan 53, B-8500 Kortrijk, Belgium, Ellen.Decaestecker@kuleuven-kortrijk.be

Hans de Jong Laboratory of Genetics, Wageningen University, Arboretumlaan 4, 6703 BD Wageningen, The Netherlands, Hans.deJong@wur.nl

Luc De Meester Laboratory of Aquatic Ecology and Evolutionary Biology, K.U.Leuven, Ch. Debériotstraat 32, B-3000 Leuven, Belgium, Luc.DeMeester@bio.kuleuven.be

Thierry De Meeûs Génétique et Evolution des Maladies Infectieuses, Equipe Evolution des Systèmes Symbiotiques, IRD, UMR 177 IRD-CIRAD "Trypanosomoses" Centre International de Recherche-Développement sur l'Elevage en zone Subhumide (CIRDES), OI BP 454, Bobo-Dioulasso 01, Burkina-Faso, demeeus@mpl.ird.fr

Gerard Driessen Institute of Ecological Science, Faculty of Earth and Life Sciences, VU University of Amsterdam, De Boelelaan 1085, NL-1081 HV Amsterdam, The Netherlands, gerard.driessen@falw.vu.nl

Thomas G. D'Souza Faculty of Biology, Institute for Evolution and Ecology, Animal Evolutionary Ecology, University of Tuebingen, Auf der Morgenstelle 28, D-72076 Tuebingen, Germany, thomas.dsouza@uni-tuebingen.de

Robert Feil Institute of Molecular Genetics, CNRS and University of Montpellier, Montpellier, France, robert.feil@igmm.cnrs.fr

Astrid Forneck Department of Applied Plant Sciences and Plant Biotechnology, University of Natural Resources and Applied Life Sciences, Vienna, Peter-Jordan Str. 82, A-1190 Vienna, Austria, astrid.forneck@boku.ac.at

Urban Friberg Department of Ecology, Evolution and Marine Biology, University of California, Santa Barbara, CA 93106-9610, USA, friberg@lifesci.ucsb.edu

Matthew K. Fujita Department of Integrative Biology, Museum of Vertebrate Zoology, University of California, 3101 Valley Life Sciences Building, Berkeley, CA 94720, USA, mkfujita@berkeley.edu

Gaston-Denis Guex Institute of Zoology, University of Zürich, Winterthurerstrasse 190, 8057 Zürich, Switzerland, guex@zoolmus.unizh.ch

Michael Heethoff Abteilung für Evolutionsbiologie der Invertebraten, Institut für Evolution und Ökologie, Eberhard Karls Universität Tübingen, Auf der Morgenstelle 28 E, 72076 Tübingen, Germany, heethoff@gmx.de

Elvira Hörandl University of Vienna, Department of Systematic and Evolutionary Botany, Rennweg 14, A-1030 Vienna, Austria, elvira.hoerandl@univie.ac.at

Hansjürg Hotz Institute of Zoology, University of Zürich, Winterthurerstrasse 190, 8057 Zürich, Switzerland, hotz@zoolmus.unizh.ch

Michael Kearney Department of Zoology, The University of Melbourne, Victoria 3010, Australia, mrke@unimelb.edu.au

Britt Koskella Department of Zoology, University of Oxford, South Parks Road, Oxford, OX1 3PS, UK, britt.koskella@zoo.ox.ac.uk

Dunja K. Lamatsch Institute for Limnology, Austrian Academy of Sciences, Mondseestr. 9, A – 5310 Mondsee, Austria, dunja.lamatsch@oeaw.ac.at

Pasqualino Loi Faculty of Veterinary, Department of Comparative Biomedical Sciences, University of Teramo, Piazza Aldo Moro, 45, 64020 Teramo, Italy, ploi@unite.it

Hugh D. Loxdale Institute of Ecology, Friedrich Schiller University, Dornburger Str. 159, 07743 Jena, Germany; Department of Entomology, Max Planck Institute for Chemical Ecology, Hans-Knoell-Strasse 8, D-07745 Jena, Germany, Hugh.Loxdale@uni-jena.de; hloxdale@ice.mpg.de

David B. Mark Welch Josephine Bay Paul Center for Comparative Molecular Biology and Evolution, Marine Biological Laboratory, Woods Hole MA 02543, USA, dmarkwelch@mbl.edu

Mark Maraun J.F. Blumenbach Institute of Zoology and Anthropology, Animal Ecology, Georg August University of Goettingen, Berliner Str. 28, 37073 Goettingen, Germany, mmaraun@gwdg.de

Koen Martens Royal Belgian Institute of Natural Sciences, Freshwater Biology, Vautierstraat 29, B-1000 Brussels, Belgium, koen.martens@naturalsciences

Irene Mateo Leach Evolutionary Genetics, Centre for Ecological and Evolutionary Studies, University of Groningen. P.O. Box 14, NL-9750 AA Haren, The Netherlands, I.Mateo-Leach@rug.nl

Stephanie Meirmans Centre for the Studies of the Sciences and the Humanities, University of Bergen, PO Box 7800, 5020 Bergen, Norway, stephane.meirmans@svt.uib.no

Joachim Mergeay Laboratory of Aquatic Ecology and Evolutionary Biology, K.U.Leuven, Ch. Debériotstraat 32, B-3000 Leuven, Belgium, Joachim.mergeay@bio.kuleuven.be

Matthew Meselson Josephine Bay Paul Center for Comparative Molecular Biology and Evolution, Marine Biological Laboratory, Woods Hole MA 02543, USA; Department of Molecular and Cellular Biology, Harvard University, Cambridge MA 02138, USA, meselson@mbl.edu

Nico K. Michiels Faculty of Biology, Zoological Institute, Animal Evolutionary Ecology, University of Tuebingen, Germany, nico.michiels@uni-tuebingen.de

Maurine Neiman Department of Biology, University of Iowa, Iowa City, IA, 52242, USA, maurine-neiman@uiowa.edu

Roy A. Norton SUNY College of Environmental Science & Forestry, 1 Forestry Drive, Syracuse, 13210 New York, USA, ranorton@esf.edu

Bart A. Pannebakker Institute of Evolutionary Biology, School of Biological Sciences, University of Edinburgh, Mains Road, Edinburgh EH9 3JT, Scotland,

UK, Bart.Pannebakker@ed.ac.uk; Evolutionary Genetics, Centre for Ecological and Evolutionary Studies, University of Groningen, P.O. Box 14, NL-9750 AA Haren, The Netherlands, B.A.Pannebakker@rug.nl

E. Davis Parker, Jr. Department of Ecology and Genetics, Aarhus University, Ny Munkegarde, DK 8000, Aarhus C, Denmark, dave.parker@biology.au.dk

Marcin Piwczynski Department of Plant Taxonomy and Geography, Nicolaus Copernicus University, Gagarina 9, PL-87-100, Torun, Poland, Marcin.Piwczynski@umk.pl

Franck Prugnolle Génétique et Evolution des Maladies Infectieuses, Equipe Evolution des Systèmes Symbiotiques, UMR IRD/CNRS 2724, BP 64501, 911 Av. Agropolis, 34394 Montpellier Cedex 5, France, prugnoll@mpl.ird.fr

Grazyna Ptak Department of Comparative Biomedical Sciences, University of Teramo, Piazza Aldo Moro, 45, 64020 Teramo, Italy, gptak@unite.it

Heinz-Ulrich Reyer Institute of Zoology, University of Zürich, Winterthurerstrasse 190, 8057 Zürich, Switzerland, ulireyer@zool.uzh.ch

William R. Rice Department of Ecology, Evolution and Marine Biology, University of California, Santa Barbara, CA 93106-9610, USA, rice@lifesci.ucsb.edu

Claudia Ricci Department of Biology, Università degli Studi di Milano, via Celoria 26, 20133 Milan, Italy, claudia.ricci@unimi.it

Jessica Ridenour Department of Zoology, The University of Melbourne, Victoria 3010, Australia, jessridenour@gmail.com

Giampaolo Rossetti Dipartimento di Scienze Ambientali, Universitá di Parma, Parco Area delle Scienze 11A, Edificio di Biologia, I-43100 Parma, Italy, giampaolo.rossetti@unipr.it

Ernst Rühl Geisenheim Research Centre, Section for Grapevine Breeding and Grafting, von Lade-Str. 1, D-65366 Geisenheim, Germany, E.Ruehl@fa-gm.de

Anssi Saura Department of Molecular Biology Umeå University, SE-901 87 Umeå, Sweden, Anssi.Saura@molbiol.umu.se

Valerio Scali Department of Biologia Evoluzionistica Sperimentale, University of Bologna, Via Selmi 3, 40126 Bologna, Italy, valerio.scali@unibo.it

Stefan Scheu J.F. Blumenbach Institute of Zoology and Anthropology, Animal Ecology, Georg August University of Goettingen, Berliner Str. 28, 37073 Goettingen, Germany, sscheu@gwdg.de

Dirk S. Schmeller Station d'Ecologie Experimentale du CNRS à Moulis, 09200 Saint Girons, France, dirk.schmeller@ecoex-moulis.cnrs.fr

Maria Victoria Schneider European Bioinformatics Institute, Wellcome Trust Genome Campus, Hinxton, Cambridge, CB10 1SD, UK, vicky@ebi.ac.uk

Isa Schön Royal Belgian Institute of Natural Sciences, Freshwater Biology, Vautierstraat 29, B-1000 Brussels, Belgium, isa.schoen@naturalsciences.be

Manuel Serra Institute of Biodiversity and Evolutionary Biology, Universitat de València, A.O. 2085; E46061 Valencia, Spain, Manuel.Serra@uv.es

Timothy F. Sharbel Apomixis research group, Department of Cytogenetics and Genome Analysis, Institut für Pflanzengenetik und Kulturpflanzenforschung (IPK), D-06466 Gatersleben, Germany, sharbel@ipk-gatersleben.de

Terry W. Snell School of Biology, Georgia Institute of Technology, Atlanta, GA 30332-0230, USA, terry.snell@biology.gatech.edu

Per Stenberg Umeå Center for Molecular Pathogenesis, Umeå University, SE-901 87 Umeå, Sweden, Per.Stenberg@molbiol.umu.se

Matthias Stöck University of Lausanne (UNIL), Department of Ecology and Evolution (DEE) Biophore, CH-1015 Lausanne, Switzerland, matthias.stoeck@unil.ch

Louis van de Zande Evolutionary Genetics, Centre for Ecological and Evolutionary Studies, University of Groningen. P.O. Box 14, NL-9750 AA Haren, The Netherlands, Louis.van.de.Zande@rug.nl

Peter Van Dijk Keygene N.V., Agro Business Park 90, 6708 PW Wageningen, The Netherlands, peter.van-dijk@keygene.com

Kitty Vijverberg Keygene N.V., Agro Business Park 90, 6708 PW Wageningen, The Netherlands, Kitty.vijverberg@keygene.com

Christoph Vorburger Institute of Zoology, University of Zürich, Winterthurerstrasse 190, 8057 Zürich, Switzerland, christoph.vorburger@zool.uzh.ch

Robert C. Vrijenhoek Monterey Bay Aquarium Research Institute, Moss Landing, 7700 Sandholdt Road, Moss Landing, CA, 95039, USA, vrijen@mbari.org

Chapter 1
Asex and Evolution: A *Very* Large-Scale Overview

Bengt O. Bengtsson

Abstract Asexuals come in all sorts. In this personal overview, I identify asexual organisms with eukaryotes that do not regularly go through the meiotic cycle. Such organisms may be asexual in many different ways and of many different reasons. The spread of asexuality is therefore always a unique process, and any notion of a general evolutionary advantage for asexuality is at best misleading. In discussions on the evolution of asexuality, ideas about genetic conflicts are often more helpful than notions about "costs". Many asexuals are associated with different fitness problems, and most of them are not particularly good at being asexual either. Their absence of long-term evolutionary success follows from their lack of recombination, leading to complex effects involving drift and selection that we are just beginning to understand. The interest in asexual organisms comes not from what they say about sex, but from what they say about living as a eukaryote.

1.1 Eukaryote Reproduction and the Meiotic Cycle

The asexual organisms that will be discussed in this chapter have one thing in common: they have all dropped out of the regular meiotic (sexual) cycle. In this they differ from the majority of eukaryotes in which meiosis and fertilization occur regularly. Among the large multi-cellular organisms that make up our visual natural environment – plants, animals and fungi with large fruiting bodies – the cycle is often easy to follow and study, with its prominent outer signs of sexual and reproductive activities. But the meiotic cycle predominates also among the small unicellular eukaryotes about whose lives we often know very little.

Regular meiotic sexuality evolved once, a long time ago. According to recent phylogenetic information, there are strong reasons to believe that the meiotic cycle

B.O. Bengtsson (✉)
Genetics, Department of Cell and Organism Biology, Lund University, Sölvegatan 29, SE-223 62 Lund, Sweden
e-mail: Bengt_olle.bengtsson@cob.lu.se

I. Schön et al. (eds.), *Lost Sex*, DOI 10.1007/978-90-481-2770-2_1,

developed at the origin of eukaryotes or very soon thereafter (Cavalier-Smith 2002). It cannot, of course, be excluded that one day we will find a eukaryote that departed from the "ur-eukaryote" lineage before meiosis evolved, but no such organism is currently known. Meiosis must have had a complex evolutionary history leading up to it, and some more or less alternative variants for decreases in ploidy-numbers have been characterized and discussed (Pontecorvo 1958; Haig 1993; Hurst and Randerson 2000). However, meiosis constitutes by far the most common mechanism for the separation of homologous genomes, and it functions in basically the same way wherever found, down to its most intricate cellular and molecular details (John 1990; Raikov 1995; Cnudde and Gerats 2005; Ramesh et al. 2005).

If the meiotic cycle is common to almost all eukaryotes, this does not mean that it always looks the same. Its basic structure of fertilization (with cell and nuclear fusions), diploid cellular life, meiosis, haploid cellular life, back to fertilization and a new round of the cycle, has taken on an almost infinite number of forms during eukaryote evolution. Careful investigations are often needed to ascertain exactly when, where and how the key events in the process occur. In humans, the length of the diploid state is measured in decades while the haploid state passes in hours. In mosses, the haploid state dominates and the diploid state is an ephemeral outgrowth extending from the top of a long-lived haploid structure. In aphids, after meiosis and fertilization there are rounds of asexual reproduction before meiosis occurs again. In the unicellular organism *Paramecium*, meiosis/fertilization occurs separately from reproduction/multiplication, which comes about via ordinary mitotic cell division. The last example is rare, however, since in most sexual organisms reproduction and dispersal is closely associated with the events of meiosis, gamete formation and fertilization.

Sometimes, the sexual cycle has become elaborated with new structural elements, as illustrated by the evolution in higher plants of a system for *double* fertilizations, where the sole use of the endosperm is to help regulate and support the development of the embryo.

Another way sexual organisms differ is in the appearance of the haploid cells produced by meiosis and intended for fertilization. In animals and plants, a clear division has evolved between small motile "male" gametic cells and large nutrient "female" gametic cells. This morphological distinction is often extended back to the individuals that produce these cells – thus many animals and some plants are segregated into males or females also during their diploid stage. Among the organisms that produce haploid gametes of both kinds (called hermaphrodites), yet other variants become possible. Fertilizations may occur between gametes produced by the same individual (selfing), or between gametes formed by different individuals. In general it appears as if outbreeding, of one kind or another, is the most common mode of reproduction in eukaryotes, and many mechanisms – ranging from cellular surface markers to the manipulation of pollination vectors – have evolved to promote it.

Thus, the underlying structure of the meiotic cycle seems not to have limited the morphological and behavioural evolution in sexual eukaryotes to any noticeable extent; instead it appears as an underlying cause of many new and surprising inventions.

With "asexuals", I will in this chapter mean eukaryotic organisms that do not regularly go through the full sexual cycle, and I will treat asexuality as an alternative reproductive mode to the normal eukaryotic one. This usage is convenient (it falls close to how Gustafsson (1946) and earlier authors used the term "apomixis"), but some comments must be made on what the definition leaves aside. First of all, it does not include bacteria and archebacteria among the asexuals, despite the fact that these organisms do not go through regular cycles of meiosis. Thus, "asexuals" is here used to denote organisms that have *stopped* having regular meiosis and sex, rather than to denote all organisms that do not practice regular sex. Secondly, among the asexuals, those organisms will not be included that have regular meiosis but where some individuals (e.g. males in animals with haplodiploid sex-determination) or life stages (e.g. aphids in summer) are formed without fertilization (parthenogenetically). Thirdly, also excluded from being called "asexual" are a wide number of phenomena where genetic material does not undergo regular DNA-recombination (e.g. Y chromosomes, mitochondrial genomes). The evolution of such genetic material is interesting and sometimes relevant for this overview, but the dichotomy sexual-asexual will here only be applied to the level of organisms.

Even with this restricted definition, asexuals turn out to be a very wide bunch indeed. In particular, it should be noted that with the definition used, meiotic events may sometimes occur in asexual organisms, as well as other processes that lead to DNA-recombination. Without this possibility, most "asexual organisms" as we know them would otherwise have been excluded from our discussion. Among the asexuals, I will also include organisms that reproduce parthenogenetically but retain remnants of meiosis (even if this may lead to some genetic variation among offspring), organisms that commence meiosis but where the process does not run to its normal completion, and organisms where functional meiotic cells of one kind are produced (most often sperm or pollen) but not the other and meiosis thereby becomes irrelevant for the normal transmission of genetic material between generations.

The view of asexuals presented here is by necessity personal. The text is not a review but an *overview*, where some interesting and widely discussed topics and questions are treated very briefly when I believe this improves the flow of the larger argument. Few references are given; for more information, an interested reader may look into Suomalainen et al. (1987), Bell (1988), Asker and Jerling (1992), Mogie (1992), and the other chapters in the present book. The literature on asexuality is highly dispersed, and interesting articles on its different aspects can be found in almost every biological journal.

The aim of this chapter is to summarize some important points valid for all asexuals in a language as free from technical jargon as possible. It is my hope that it will function as a useful introduction, particularly for non-specialists, to the fascinating field of asexual biology.

1.2 Asex Is Often Associated with Polyploidy and Hybridity

The sexual cycle is complicated and so is its key process – meiosis. What an incredible feat it is to pair many centimetre-long DNA-strings down to their exact base pairs within the confines of a nuclear membrane! Also fertilization can be highly complicated, with a lot of interactions occurring before the cellular and nuclear partners may fuse (many fungi, for example, have a rich repertoire of mating signals but also of systems for checking and manipulating the haploid genomes before nuclear fusion).

The term "asexual" is somewhat unfortunate in that it seems to imply that the key element in the phenomenon is the abolishment of sexes. But asexuality occurs among organisms with as well as without any sexual differentiation between individuals. From a cellular point of view, asexuality is not about getting rid of the sexes, but about not going through regular shifts in ploidy.

Thus, one of the most characteristic properties of asexuals is that their cells always remain at the same ploidy level. This may be the haploid level, as in vegetatively reproducing mosses or in asexual "fungi imperfecti" (whether this represents the true haploid level or just appear as such after earlier polyploidization events, is in itself an interesting question). In animals or vascular plants it is, however, never the lower level in the ploidy cycle that takes over, but always the higher, i.e. the diploid or possibly polyploid level (except in the false spider mite *Brevipalpus phoenicis*, where a cytoplasmic bacterium manipulates potentially haploid male embryos so that they become parthenogenetic females; Weeks et al. 2001).

There is a well-known positive correlation between polyploidy and asexuality (Suomalainen et al. 1987; Asker and Jerling 1992; see also Chapter 4), which is easy to understand from many points of view, but not from all. Meiosis often fails in polyploids, particularly in those with an odd number of genomes, due to difficulties with pairing more than two chromosomes of each kind. It is therefore not surprising that asexual reproduction is a great bonus to polyploids – without it many polyploids would not be able to reproduce and would never have been seen in nature. Another explanation for the association between asex and polyploidy is that the complex genotype in polyploids may signal to the control points of meiosis and gamete formation in such a disturbed but precise way, that the formation of unreduced gametes and spontaneous embryogenesis naturally follow.

Less easy to understand is why the well-studied phenomenon of *apomixis* – here used to denote asexual reproduction via seeds in angiosperms – never occurs in diploids, or at least only very rarely so. After some earlier candidates have been excluded by strict genetic tests, only one species is today assumed to be a natural diploid apomict, namely *Boechera holboelli* (a close relative to the well-studied species *Arabidopsis thaliana*; see also Chapter 23.) Interestingly, apomixis may here be caused by a chromosome additional to the normal genomic set (Sharbel et al. 2005), making not even these asexuals strictly diploid. The lack of apomixis among diploid plants is probably due to the complex reproductive machinery used by angiosperms and the special nature of the factors causing apomixis rather than to asexuality itself (as discussed by Grossniklaus et al. 2001), since diploid apomicts can be experimentally produced (example: Nogler 1982). The existence of many diploid asexual animals is also a demonstration that there are no deep genetic or evolutionary difficulties for asexuality to function at the diploid level.

Another common correlate to asexuality is hybridity (nice examples chosen from different organism groups in the Australian desert are given by Kearney 2003). Why asexually reproducing hybrids are common, can be understood from the same two arguments as given above for the positive association between asexuality and polyploidy. The selective enrichment explanation says that when hybrid formation leads to catastrophic meiosis, then it is only those hybrids that circumvent the process that are seen in nature. The disturbed regulation explanation claims that hybridity sometimes leads directly to asexual reproduction by its mismatch of developmental signals. The last scenario may sound contrived, but has repeatedly been proven relevant in experimental studies (example: Wetherington et al. 1987).

Hybridity and polyploidy often happen at the same time, since both are processes that come about by the fusion of non-standard gametic cells. Sometimes they join with incomplete asexuality in a wild evolutionary dance – to taxonomists' horror and delight. This happens when hybrids are repeatedly formed and neither normal meiosis nor asexuality functions perfectly. A range of gametes will then be produced, resulting in individuals with different ploidy numbers that mix reproductive behaviours and morphological characteristics in a multitude of ways. The spontaneous idea that asexuallity leads to organism uniformity is in such cases proven wrong to an almost comic degree. With the aid of molecular and cytogenetic tools, some form-rich examples have recently been analytically characterized, for example the Iberian fresh-water minnow fish *Leuciscus alburnoides* (Alves et al. 2001). The resulting flowchart of genomic relationships and fertilization dependences/independences is astounding. Similar results have been obtained for the predominantly asexual hawkweeds, *Hieracium* subgen. *Pilosella*, making it understandable why Mendel failed when he tried to deduce the basis of their form-variation from simple crossing experiments (Krahulec et al. 2004; Nogler 2006).

Already this first point illustrates that the evolution of asexuality is a surprisingly complicated process. Only rarely, if ever, do eukaryote organisms leave the sexual cycle in a well-ordered fashion.

1.3 Asex Can Have Many Different Immediate Causes

Asexual reproduction can be achieved in a number of ways. All specially adapted systems for the production of sexual offspring may be lost, turning available resources over to "vegetative spread" alone. Or – what appears like a reverse case – all outer aspects of sexual reproduction may be retained both structurally and functionally, with the exception that the fertilized egg is replaced by a suitably prepared diploid cell from the mother. When in such cases male haploid cells are needed to trigger embryo development (many such examples of pseudogamy are known from plants as well as animals, see Asker and Jerling 1992 and Beukeboom and

Vrijenhoek 1998 or Chapters 19 and 20 for further examples), it becomes very difficult to know for certain that reproduction is asexual, unless good genetic markers are available to analyse the offspring. Intermediary situations between these two extremes are often found. They are particularly common among plants, when asexuals develop specialized "propagule" structures, for example runners or bulbils. The outwards signs of asexuality may, thus, vary from situation to situation, just as expected when one considers in how many different ways it must be possible to leave the meiotic cycle.

There are also many different immediate causes of asexuality (Simon et al. 2003). In angiosperms, recent genetic analyses have proved the existence of chromosomally carried determinants for asexuality, sometimes few, sometimes many (examples: van Dijk and Bakx-Schotman 2004; Matzk et al. 2005). Some of them look like "supergenes", i.e. discrete chromosomal regions with a number of genes and functions under close linkage (Grossniklaus et al. 2001). The nuclear factors may interact with the ploidy-level, making also this a factor in the causation of asexuality (Quarin et al. 2001). As discussed before, hybridity may in itself cause asexuality due to the conflicting signalling it gives rise to (it has been suggested to underlie the existence of *all* parthenogenetic vertebrates; Avise et al. 1992; see also Chapters 19, 20 and 21). In the plant *Boechera holboelli* a B chromosome is suspected to be the cause of asexuality (Sharbel et al. 2005; see also Chapter 23). Cytoplasmically transmitted parasites are the determinants of asexuality in many animals. In particular the bacterium *Wolbachia* causes a wide range of reproductive curiosities in insects, arthropods and nematodes (Stouthamer et al. 1999; see also Chapter 17). But also other bacteria may induce parthenogenesis; thus, it has recently been found that tetracycline cures the parasitoid *Neochrysocharis formosa* from parthenogenesis-inducing *Rickettsia* bacteria (Hagimori et al. 2006). Given that most of these causes of asexuality prevent the characterisation of themselves by standard genetic crosses, it is not surprising that very little is as yet known about how they function at the molecular level.

It should be noted that many of the immediate factors for asexuality may well be transmitted via sexual or sexual-like, processes – making asexuality very much like an "infectious" phenomenon. This occurs, for example, in angiosperms when nuclear factors causing apomixis are transmitted via functional pollen and thereby lead to a range of different asexual clones in the progeny. In a narrow perspective, such infectious asexuality may look similar to situations were asexuals are repeatedly formed by the hybridization of sexual species, but in the long run, the difference between the processes is considerable, since in the first case there is a continued selection for improved transmission of the factor(s) causing asexuality that does not exist in the second case (for a well-informed discussion of how genetic systems for asexuality in plants may evolve, see Ozias-Akins and van Dijk 2007).

While the variation in outer signs of asexuality and in its immediate causes often is great, clear lineage effects are nevertheless common, i.e. apparently similar variants of asexuality occur in phylogenetically related organisms. No one will, for example, find it surprising that aphids – whose normal life cycle contains rounds of parthenogenetic reproduction during the "good" part of the year – often give rise to asexual lineages by losing the propensity to reproduce sexually (Simon et al. 2002). In similar ways, but presently unknown to us, other organisms may have evolved specific ways to regulate their meiotic processes that today make them predisposed ("pre-adapted") to particular kinds of asexual evolution – or make such evolution next to impossible.

1.4 Asex Is the Outcome of a Darwinian Process with Special Properties

To understand why a specific instance of asexuality exists, we must regard it as any other biological trait and submit it to a modern Darwinian analysis. With this I mean that we should consider what is known about the transmission of the genetic factor(s) causing it and the ecological reactions of the genotypes involved, and then see if they jointly promote an increase in the trait when rare. If they do, then we may claim that we understand why this particular organism has turned asexual. (The separate but related question of the evolutionary value of the trait in the long term perspective is discussed at the end of the chapter.)

Since there are many different immediate causes of asexuality, there can be no unified explanation for its evolution. All factors discussed in the last section are associated with their own specific transmission rules and they will affect the evolutionary dynamics of the trait in their own particular way. In addition, one can be certain that every case of asexuality is associated with its own specific phenotypic effects that lead to unique ecological reactions and selective processes (example: Weinzierl et al. 1999). In most instances, information about these genetic and ecological factors is at best fragmentary, which means that we can often hardly more than speculate why a specific lineage of asexuality has evolved.

Is it even more difficult to understand why lineages do *not* become asexual? Discussions about the existence of sex have often turned around this question. By postulating a constant recurrence of mutations for asexuality with highly favourable transmission behaviours and weak deleterious selective effects, claims about how easy it is for asexuality to spread have become generally accepted (see the huge literature following the books by Williams 1975 and Maynard Smith 1978) – leading to the impression that it perhaps is the existence of sexual, and not asexual, reproduction that is in need for explanation. However, after decades of impressive empirical research it now stands clear that factors causing organisms to leave the meiotic cycle are generally messy and associated with strange transmission behaviours and complex phenotypic effects, which make them far from the assumptions in the simple didactic models. The Darwinian explanation of asexuality must therefore be considered anew for every fresh case, as there is no simple explanation for its evolution, just as there is no simple general explanation for its lack of bursting forth.

This implies that it is scientifically fruitless to talk about a general "cost of sex". It is, first of all, a misleading expression since the model situations motivating its use are relevant only for a limited set of meiotic organisms, primarily those that have well differentiated sexual functions and use them to outbreed. For very many eukaryotes the expression is therefore not valid. Secondly, the postulated cost is an abstract one that becomes realised only when certain kinds of mutations appear. Since such mutations almost never occur, neither with respect to their genetic nor their ecological properties, the idea of a *general* cost of sex must be considered untenable (see Uyenoyama 1984 for an early, forceful statement of this point).

This does not imply that the effects demonstrated by the theoretical models lack all relevance. It is still true that a lineage consisting of parthenogenetic females under suitable circumstances will outcompete an ecologically parallel one consisting of both males and females (the ecological advantage of asex); the assumed circumstances include resource competition between the organisms, limited deleterious effects on viability and fertility of the mutation, and a balanced sex ratio among the sexuals. Likewise, a mutation in a hermaphroditic organism that causes eggs to develop without fertilization but retains the production of outbreeding male gametes will be associated with a transmission advantage that will spread the trait, if only the deleterious fitness effects associated with the mutation are small enough (the genetic advantage of asex). However, only seldom do these results apply directly to any biological situation. Instead, the specific interactive processes involved must in every specific case be integrated into the full evolutionary analysis.

The feeling that there is something *special* about the evolution of sexuality/asexuality, and which the notion of a "cost of sex" is intended to capture, can be better approached in a different way. The unusual phenomenon in this process is not a unique trait cost, but the segregation distortion involved in the transmission of most immediate causes of asexuality. By this I mean that their inheritance systems function in such a way, that they – by their very nature – favour their own spread, or, in other words, that they act like selfish genes. This is well illustrated by two common examples: the "unfair" transmission pattern of dominant nuclear factors for apomixis in plants that continue to produce functioning pollen, and the induction of feminization and parthenogenesis by parasitic *Wolbachia* bacteria in insects and other animals. With such genetic variation, evolution will not generally move in the direction of higher mean fitness, and this makes all heuristic evolutionary thinking based on individual adaptation difficult. Regarding these "unfair" transmission processes as special individual fitness components (which talk of a cost of sex *de facto* does) strikes me as a very round-about way to approach evolution. Every biologist must be prepared to accept that some complex properties, such as for example sex and asex, find their evolutionary explanations through competitive processes between heritable factors (genes, chromosomes, intracellular parasites) rather than between individuals.

Does this mean that all thinking about asexual evolution must be reduced to the study of complicated recursion relationships between genotype frequencies? I do not think so. The language of "genetic conflicts" (Hurst et al. 1996; Burt and Trivers 2006) retains the strict Darwinian perspective that evolution is understood by the spread of new genetic variants, but drops the insistence on a concomitant increase in individual fitness. And many causes of asexuality – apomixis-inducing nuclear factors, B chromosomes, intracellular parasites – can very easily be seen as selfish genomic intruders and be discussed accordingly. This way of looking upon asexual evolution has the advantage that it helps us see that the spread of a factor for asexuality may provoke the later spread of new factors that "restores" the organism to its initial state. Not much research has as yet been done on such conflicts over the control of the sexual cycle. I am, however, convinced that is in this light some strange asex-related phenomena ought to be analysed, for example the meiosis in dogroses, where some genomes are asexually inherited while others are not (Nybom et al. 2006), the quasi-asexual permanently hybridogenetic system in the waterfrog *Rana esculenta* (Hellriegel and Reyer 2000; see also Chapter 20), and the bizarre cases of androgenesis in *Bacillus* stick insects and the rare tree *Cupressus dupreziana* (Mantovani et al. 2001; Pichot et al. 2001; McKone and Halpern 2003; see also Chapter 16). A nice worked-out example of a conflict between different genetic elements has been described for the parasitoid wasp *Trichogramma kaykai*, where a B chromosome has spread that causes females to produce sons despite the presence of parthenogenesis-inducing *Wolbachia* bacteria (Stouthamer et al. 2001).

1.5 Asex Is Almost Always Associated with Some Sex

Asexuality is seldom perfect. This has for long been known to biologists, who nevertheless prefer to call populations and species asexual even when sexual fertilizations of more or less complex and successful kinds are known to occur occasionally in them. Sometimes the term "facultative (a)sexuality" is used to describe such unclear biological situations, but this term can be misleading in its implication that a well-determined control over the degree of sexuality exists – a presupposition that rarely or never is proven. Terms like "remnant sexuality" or "leaky asexuality" may be more useful with their implication that asexuality is here neither developmentally nor evolutionarily strictly determined (for a study of how difficult it is for a balance between sexual and asexual reproduction to be evolutionarily stable, see Bengtsson and Ceplitis 2000).

Whatever the background to the occurrence of sexuality among asexuals may be, its effects are considerable. When asexuality is complete in a lineage, homologous gene copies will slowly but unboundedly diverge (Birky 1996), until the copies within individuals become more divergent than alleles in sexual populations normally are (the effect has been widely discussed in connection with the anciently asexual bdelloids; see Mark Welch and Meselson 2000 and Chapter 13). However, it turns out that already very little sex in an otherwise asexual population will change this. A few sexual events per generation are sufficient to obliterate the effect of there being more allelic variation within individuals than between them (for studies of this process, see Balloux et al. 2003; Bengtsson 2003; Ceplitis 2003), and the wide divergence between homologous gene copies has generally not been found when looked for in suspected ancient asexuals (example: Schaefer et al. 2006; see also Chapters 11 and 12). With respect to the bdelloid rotifers, it now appears that they are tetraploids, which may have confused the genetic interpretation of their pattern of sequence variance (Mark Welch et al. 2008 and Chapter 13).

When some sexual reproduction occurs in an otherwise asexual population, it cannot be taken for granted that a distinctly clonal distribution of genetic variation will be seen within it. Various statistical methods have been developed to estimate the ratio between sexual and asexual reproduction in populations, normally based on the multi-locus associations between genetic markers that build up whenever recombination (and/or back-mutations) is not there to break them down (Halkett et al. 2005).

Rare sexual events become particularly important when they facilitate the representation of a dominant factor for asexuality in new genetic backgrounds. The factor will thereby escape its original genetic setting and become tested by natural selection in different genotypes. Not only will this transmission via crosses increase its probability of evolutionary spread, but it will also make it possible for the mutation to become associated with better adapted genetic variation during periods of environmental change. With a low but sufficient frequency of sexual fertilisations, chromosomal regions for asexuality may live on in ever renewed genetic backgrounds, making the mutation for asexuality much older than any of the lineages within which it presently is found. (Similar arguments can, of course, be made for intra-cellular bacteria and other potentially infectious agents causing asexuality.)

In general, the closeness between asexual and sexual organisms in many of the natural situations in which they are studied implies that biologists must be prepared for unexpected ecological and genetic interferences between them, the occasional occurrence of sex in apparently well-established asexual taxons, and the retained presence of at least some recombination mechanisms (such as gene conversions or mitotic crossing-over; see also Chapters 11 and 15) in otherwise asexual organisms. All these processes make it difficult to calculate the age of asexuality from the clonal variation currently existing in a population or species, but it also makes such studies important if one really wants to know what has happened in evolution.

1.6 Most Asexuals Are Genetically Variable

A closely related point is that asexual organisms often turn out to be widely genetically variable (Suomalainen et al. 1987; Ellstrand and Roose 1987; Asker and Jerling 1992), a fact that at least spontaneously seems strange since asexuality is supposed to faithfully reproduce the same genotype from generation to generation.

Cases are known where an asexual organism appears to lack any genetic variation and therefore presumably represents the descendents of a recent, single asexual individual (example: Kraft et al. 1996 describe very restricted molecular variation in some blackberry, *Rubus*, "species"). But in most investigated cases, extensive genetic variation is found among studied asexuals. Sometimes the variation is due to the presence of a limited number of well-separated clones, but this is not always the case.

The variability in an asexual organism may derive from its history at the time it became asexual, or from processes which occurred later. Let us first consider variation related to the origin of asexuality. The organism will be genetically variable if it has been and perhaps continues to be repeatedly created, a phenomenon that is particularly common when hybridity is the cause of asexuality (example: Delmotte et al. 2003). Another possibility is that the asexual unit originated from the spread of a mutation for asexuality in, say, a hermaphroditic plant (example: Barcaccia et al. 2006). Computer simulations show that a considerable amount of the original genetic variation will in such situations be retained by the asexuals after the shift in breeding system (Adolfsson and Bengtsson 2007).

What happens after asexuality is established? Drift and natural selection will immediately start causing a gradual loss of any initial variation, at rates depending on the size of the population. Drift purifies small populations most rapidly, while minor differences in fitness become effective only in large populations. Meanwhile, new mutations will appear and sometimes be retained. Occasional sexual events may also occur (for examples of what huge release of variation such crosses may lead to, see e.g. Deng and Lynch 1996, and Ceplitis and Bengtsson 2004). These mutations and recombination events cause the build up of new, complexly structured variation. In a fragmented environment with a high patch-turnover and selection favouring local adaptation, an asexual organism with a low frequency of sex may well present the investigator with the surprising picture of (almost) no genotypic variation existing within patches, at the same time as there are large differences between them that cannot be explained by mutation alone (Balloux et al. 2003; Bengtsson 2003).

Genetically homogeneous clones are interesting, since their distributions can be informative about the role genetic variation plays in the build-up of an organism's niche. However, given the opportunistic nature of natural selection, it seems hard to believe that successful asexual clones will *in general* turn out to have either broader or more narrowly specialized relationships with the environment than standard sexual genotypes (on "general-purpose genotypes", see Baker 1965 and Parker et al. 1977; on "frozen niche variation", see Vrijenhoek 1979; discussions of this topic can be found in e.g. Lynch 1984, Vrijenhoek 1998 and in Chapter 6). Neither do I think that it would be easy to export the demonstration of any such pattern from one particular situation to another. Molecular studies have just begun to add new vital information to these questions by their ability to identify clones and (re)estimate their ages from their neutral genetic background variation (example: Robertson et al. 2006). What ultimate generalizations will come out of this important interface between genetics and ecology seems to me too early to judge today.

Empirically, it has been found that evolutionarily close sexual and asexual lineages often have different ranges, a phenomenon called geographical parthenogenesis after Vandel (1928; see also Chapter 8). That they may tend to exclude each other by reproductive and/or ecological competition and interference is not surprising, since it follows from their close evolutionary relationships (for a detailed small-scaled analysis of sexual and asexual dandelions, see Verduijn et al. 2004).

That the difference in breeding system may have a direct effect on their respective ranges is also possible, in particular if the asexuals become liberated from some important mating constrain such as lack of reliable pollinators. However, the various empirical generalizations that have been reached – for example that asexuals are particularly common in the Australian desert or in the Eurasian land parts that have undergone glaciation – seems mostly to follow from the biological correlates of asexuality rather than from asexuality itself. Thus, there has recently been a discussion whether hybridity or polyploidy is the most important underlying factor of geographical parthenogenesis (Kearney 2005; Lundmark 2006; see also Hörandl 2006). My own reaction is that most asexual organisms have such unique biological characteristics that it becomes difficult to ascribe general properties to them. Polyploidy may explain one case of geographical parthenogenesis and hybridity another; and if among hundred analysed cases the first factor explains 32 and the other factor 47 while we cannot explain the rest – to what use is then a generalization?

1.7 Asexual Lineages Are Comparatively Short-Lived

I have repeatedly stressed how difficult it is to state something general about asexuals that with reasonably accuracy will hold for all or at least most of them. One such generalization is, however, of great importance. It says that asexual lineages are, on average, shorter lived in evolution than sexual lineages.

This conclusion is based on the fact that asexuality appears constantly among sexual eukaryotes, while no reverse evolution – from full asexuality to sexuality – is normally to reckon with. Nevertheless, the eukaryotic tree remains dominated by sexuality. This situation would not hold unless asexual lineages have a shorter lifespan on average than sexual lineages.

The relative disadvantage of asexuality is directly seen in the eukaryote tree in all its immensity, where sexuality dominates all major branches. The same logic applies, however, also to every derived lineage, down to for example the water-fleas *Daphnia pulex* that generate asexuals at a considerable rate and still retains sexuality as their basic reproductive mode (Paland et al. 2005; see Chapter 15).

That asexual lineages on average have a shorter existence time than sexual lineages, is not contradicted by the fact that *some* asexual lineages have existed for considerable periods (Judson and Normark 1996). In fact, there are two reasons for why the distribution of life-times of asexual lineages should have particularly long tails. The first is that "success breeds success". The most likely cause of extinction in the nearby future for any lineage is always a small population size, and an asexual evolutionary unit that has reached some evolutionary success (for whatever reason), measured by its number of individuals, will thereby both be buffered against purely stochastic catastrophic events and be able to obtain new mutations that will continue to keep it adapted to its environment. The probability for a successful asexual lineage to go extinct in the near future is therefore expected to be smaller than the probability for a rare, recent one. The second argument has to do with the disadvantage of asexuals in their competition vis-à-vis sexual lineages. R. A. Fisher explained already in 1930 that any long-term disadvantage of asexuals must be measured relative to sexual, competing forms. No noticeable disadvantage will necessarily be suffered by some asexual lineage that has captured for itself (or been dumped into) a suitable niche where it does not interact with sexual congeners. The disadvantage that in the long term may follow from being asexual (see next section) is not of the kind that it with necessity dooms asexual lineages of reasonable size to mutational melt-down or any similar disaster. Natural selection favouring new mutations arising at random can rejuvenate any organism and constantly does so – no one expects for example bacteria to go extinct just because they do not practice regular sex. An asexual lineage that monopolizes a suitable niche can well live on for a very long time.

Thus, the existence of some long lived asexuals does not affect the conclusion that asexual lineages in general are more short-lived than sexual ones. The importance of ancient asexuals derives instead from their ability to shed light on the long-term dynamics of asexual biological units and what specialized adaptive processes they may evolve to support their particular mode of life.

1.8 The Lack of Recombination Gives Asexual Lineages a Long-Term Disadvantage

Let me recapitulate: New asexual lineages may in principle appear in any sexual organism. They do so rarely – almost never in some branches of the tree of life, more often in others. Due to the mode of inheritance of the factor(s) causing asex and the ecological functioning of the asexual type(s), the trait may spread in nature. Almost never, if ever, is asexuality associated with fitness effects and transmission biases that give it a big selective advantage; the idea of a "two-fold cost of sex" is an interesting theoretical one which does not normally materialize in nature. The meiotic cycle is such a fundamental part of eukaryote life that disturbing it seems to cause inevitable adaptive problems.

But sometimes, the genetic and ecological forces are stacked in such a way that a newly formed asexual lineage thrives and spreads. The immediate cause of this asexuality may be a unique configuration of mutations, but may also follow from parasitic micro-organisms, complementary interactions between diverged hybridizing genomes, or a specialized gene-set that "infects" via more or less normal fertilizations. Positive fitness effects may come from polyploidy and/or heterosis following hybridity.

The most likely fate of any such asexual lineage is rapid extinction, even when it happens to be selectively favoured. This applies to all new traits in stochastic evolutionary processes. Only if the lineage grows rapidly in number and can establish a suitable niche for itself, is it likely that it will live on for some considerable evolutionary time. An asexual lineage may even come to dominate a wide geographic space opened up, for example, after a glaciation. Such windows of opportunity may, however, just as easily close again with continuing environmental change. Similarly, a relatively successful clone may after some time be hit by a parasite particularly adapted to its specific genotype (example: Lively et al. 2004; see also Chapter 7 on the Red Queen).

It seems to me that this summary well explains why so many of the asexual organisms we see in nature have such comparatively short lineage histories. What remains to be considered is why also lineages that have been asexual for fairly long appear to be more prone to extinction than similar sexual lineages. This general disadvantage of asexuality is not self-evident.

One direction such explanations may take is to concentrate on specialized adaptive processes that different organism groups have developed in close association with the meiotic cycle, and which those that leave the cycle thereby miss. For example, the RIP-process (i.e. silencing point-mutations induced and suffered by repeated DNA sequences) that exists in filamentous fungi as part of their regulation of nuclear fusions, is an efficient way to reduce the genetic load caused by transposing elements (Selker 2002). Bypassing the meiotic cycle will normally entail the loss of such derived advantages, which in the long run may lead to trouble.

Nevertheless, it seems necessary to accept that asexuality *in itself* is associated with a selective disadvantage due to its very nature and not just due to effects added on to meiosis later. Since Fisher (1930) and Muller (1932), evolutionary biologists have believed that a lack of recombination between alleles at different loci is the cause of this disadvantage. Today I agree with the view, though I am still interested in what positive effects biased gene conversion may bring to meiosis (and perhaps did at the time of its origin; Bengtsson 1985).

An argument for the existence of a general advantage of allelic recombination is that regular chromosomal DNA-recombination is found in almost all meiotic organisms. That the separation of homologous chromosomes acts best with more than one chromosome per haploid genome may be due to mechanical reasons caused by the functioning of the meiotic spindle, but recombination is generally retained also *within* chromosomes and is distributed over gene-containing regions so comparatively evenly that the process appears selectively favoured. (In the permanent translocation heterozygote *Oenothera biennis*, genomes basically consist of multi-chromosomal non-recombining linkage-groups that unite in sex-determined patterns; see Levy and Winternheimer 1977. This may be the counter-example that proves the rule for the importance of distributed chromosomal recombination.)

Another argument for a long-term advantage of recombination comes from selfing plants. Their patchy phylogenetic distribution is similar to the one that asexuals have, and it is natural to believe that they both suffer from the same lack of "effective recombination" (using a term from Darlington 1939).

It has, however, been difficult to base the belief in a positive effect of recombination on a sound scientific basis, and with time the literature on the evolution of recombination has become enormous. Fisher (1930), Muller (1932) and many with them, believed that recombination is important because it helps create new, balanced polygenic variation (for a review of theoretical work done on the evolution of linkage between interacting genes, see Feldman et al. 1997). However, studies analysing frequency changes in infinitely large populations reached disappointing results: in these models, natural selection tends, if anything, to decrease linkage and recombination. By postulating additional elements in the models, such as positive mutations, changing environments, recurrent mutations, or special fitness interactions, situations may be found in which recombination does indeed increase, but the positive effect of recombination is then only seen in narrow parameter windows.

More promising results have been reached in studies of finite-sized populations, in which deleterious mutations occur at many loci – model assumptions that are valid for all living organisms but which lead to highly complex and analytically intractable interactions. In such situations, random genetic drift in combination with background selection against the deleterious mutations lead to a build-up of chromosomes where normal alleles become associated with deleterious alleles at other loci (Barton and Otto 2005). Modifiers which increase recombination between the loci are then favoured, and in an extensive simulation study Keightley and Otto (2006) have shown that this induced secondary selection may become quite strong (later results by Gordo and Campos 2008 confirm this observation). The advantage of recombination is, in addition, relatively unaffected by the fitness interactions (epistasis) between the deleterious mutations. Particularly important is that the strength of selection for recombination *increases* with population size (at least as long as most genotypes due to stochastic effects are far from their expected values), which implies that recombination is favoured not only in small but also in large populations. In earlier attempts to approach the same effect of drift, selection and lack of recombination by solely focussing on the risk to *fix* deleterious mutations ("Muller's ratchet"; Muller 1964; see also Chapter 5), the selective favouring of recombination could only be found in populations of comparatively small size (Lynch et al. 1995).

Much work remains to be done on this recently demonstrated general advantage of recombination concerning, for example, the expected level of recombination to be reached, but also with respect to questions related to asexuality. In the populations simulated by Keightley and Otto (2006) there are plenty of multi-locus genotypes that would spread if they instantly turned asexual by some mutation without too many associated deleterious effects (in their model there is no intrinsic transmission or ecological advantage of asex). In such asexual lineages, the negative effect of not having recombination would become noticeable only after a standing stock of deleterious mutations in unsuitable linkage associations had been built up within them, and first then would they be outcompeted. It will be interesting to see what time dynamics such ups and downs in asexual lineages will have.

Thus, finally, we seem to be able to conclude that recombination is an evolutionarily favoured trait, and that its absence caused by asexuality accordingly is a handicap. But it must at the same time be recognized that the deleterious effect of having many genes and not having sufficient recombination between them comes from mutations assembled over a long – sometimes very long – time. And that until then the lack of recombination may instead be the key factor behind the spread of asexual clones with particularly good starting genotypes.

1.9 The Paradox of Asex

This overview of asexual biology has stressed its paradoxical aspects. The asexuals constitute a diverse group of organisms with few common features. Most of them are not proper asexuals but reproduce also by other means; some suggested instances turn out not to be asexual at all when more closely studied. The asexuals have been analysed for saying something important about sex, but careful investigations have hardly been able to explain why even some of the most well-known examples of asex exist. Much theorizing around them has been irrelevant at best: asexuals do not lack genetic variation and do not represent evolutionary dead ends, neither are there plenty of potential mutations for asexuality whose lack of spread needs explaining.

So, why go on studying asexuals? The best answer, I think, is because they make such good biology. By questioning the centre in the eukaryotic life cycle, asexuals, even partial ones, open up possibilities to understand interconnections between different levels of organisation – from molecules over tissues and individuals to populations and species – that otherwise are hard to approach. Asexual biology is interesting, not because of what it says about something else, namely sex, but because of what it says about the organisms under study. Many of the organisms that plague us – e.g. parasites and agricultural pests – are asexual (though sometimes with occasional sex), and much can therefore be gained by learning more about their differentiation and evolution (example: de Meeûs et al. 2006; see also Chapters 24 and 25). But the greatest value the study of asexuals brings us is a deeper insight into what it means to be a eukaryote.

Acknowledgements I am grateful to all participants in the European Science Foundation's network on asexuality, PARTNER, for the fascinating and thought-provoking biology that they have taught me. Thanks for comments on the manuscript go to S. Adolfsson, A. Ceplitis, F. B. Christiansen, J. Fehrer, L. Hurst and I. Schön. Support from the Swedish Research Council and the Nilsson-Ehle foundation is gratefully acknowledged.

References

Adolfsson S, Bengtsson BO (2007) The spread of apomixis and its effect on resident genetic variation. J Evol Biol 20: 1933–1940

Alves MJ, Coelho MM, Collares-Pereira MJ (2001) Evolution in action through hybridisation and polyploidy in an Iberian freshwater fish: a genetic review. Genetica 111: 375–385

Asker S, Jerling L (1992) Apomixis in plants. CRC Press, Boca Raton

Avise JC, Quattro JM, Vrijenhoek RC (1992) Molecular clones within organismal clones: mitochondrial DNA phylogenies and the evolutionary histories of unisexual vertebrates. Evol Biol 26: 225–246

Baker HG (1965) Characteristics and modes of origin of weeds. In: Baker HG, Stebbins GL (eds) The genetics of colonizing species. Academic Press, New York, pp. 147–168

Balloux F, Lehmann L, de Meeûs T (2003) The population genetics of clonal and partially clonal diploids. Genetics 164: 1635–1644

Barcaccia G, Arzenton F, Sharbel TF, Varotto S, Parrini P, Lucchin M (2006) Genetic diversity and reproductive biology in ecotypes of the facultative apomict *Hypericum perforatum* L. Heredity 96: 322–334

Barton NH, Otto SP (2005) Evolution of recombination due to random drift. Genetics 169: 2353–2370

Bell G (1988) Sex and death in Protozoa. The history of an obsession. Cambridge University Press, Cambridge, UK
Bengtsson BO (1985) Biased conversion as the primary function of recombination. Genet Res 47: 77–80
Bengtsson BO (2003) Genetic variation in organisms with sexual and asexual reproduction. J Evol Biol 16: 189–199
Bengtsson BO, Ceplitis A (2000) The balance between sexual and asexual reproduction in plants living in variable environments. J Evol Biol 13: 415–422
Beukeboom LW, Vrijenhoek RC (1998) Evolutionary genetics and ecology of sperm-dependent parthenogenesis. J Evol Biol 11: 755–782
Birky CW Jr (1996) Heterozygosity, heteromorphy, and phylogenetic trees in asexual eukaryotes. Genetics 144: 427–437
Burt A, Trivers R (2006) Genes in conflict. The biology of selfish genetic elements. Belknap Press, Cambridge, USA
Cavalier-Smith T (2002) Origins of the machinery of recombination and sex. Heredity 88: 125–141
Ceplitis A (2003) Coalescence times and the Meselson effect in asexual eukaryotes. Genet Res 82: 183–190
Ceplitis A, Bengtsson BO (2004) Genetic variation, disequilibrium and natural selection on reproductive traits in *Allium vineale*. J Evol Biol 17: 302–311
Cnudde F, Gerats T (2005) Meiosis: Inducing variation by reduction. Plant Biol 7: 321–341
Darlington CD (1939) The evolution of genetic systems. Cambridge University Press, Cambridge, UK
Delmotte F, Sabater-Muñoz B, Prunier-Leterme N, Latorre A, Sunnucks P, Rispe C, Simon J-C (2003) Phylogenetic evidence for hybrid origins of asexual lineages in an aphid species. Evolution 57: 1291–1303
de Meeûs T, Lehmann L, Balloux F (2006) Molecular epidemiology of clonal diploids: a quick overview and a short DIY (do it yourself) notice. Infect Genet Evol 6: 163–170
Deng H-W, Lynch M (1996) Change of genetic architecture in response to sex. Genetics 143: 203–212
Ellstrand NC, Roose ML (1987) Patterns of genotypic diversity in clonal plant species. Am J Bot 74: 123–131
Feldman MW, Otto SP, Christiansen FB (1997) Population genetic perspectives on the evolution of recombination. Annu Rev Genet 30: 261–295
Fisher RA (1930) The genetical theory of natural selection. Clarendon Press, Oxford
Gordo I, Campos PRA (2008) Sex and deleterious mutations. Genetics 179: 621–626
Grossniklaus U, Nogler GA, van Dijk PJ (2001) How to avoid sex: the genetic control of gametophytic apomixis. Plant Cell 13: 1491–1497
Gustafsson Å (1946) Apomixis in higher plants. Part I. The mechanism of apomixis. Lunds Universitets Årsskrift NF Avd 2 Bd 42 Nr 3
Hagimori T, Abe Y, Date S, Miura K (2006) The first finding of a *Rickettsia* bacterium associated with parthenogenesis induction among insects. Curr Microbiol 52: 97–101
Haig D (1993) Alternatives to meiosis: the unusual genetics of red algae, microsporidia, and others. J Theor Biol 163: 15–31
Halkett F, Simon J-C, Balloux F (2005) Tackling the population genetics of clonal and partially clonal organisms. Trends Ecol Evol 20: 194–201
Hellriegel B, Reyer H-U (2000) Factors influencing the composition of mixed populations of a hemiclonal hybrid and its sexual host. J Evol Biol 13: 906–918
Hörandl E (2006) The complex causality of geographical parthenogenesis. New Phytol 171: 525–538
Hurst LD, Atlan A, Bengtsson BO (1996) Genetic conflicts. Quart Rev Biol 71: 317–364
Hurst LD, Randerson JP (2000) Transitions in the evolution of meiosis. J Evol Biol 13: 466–479
John B (1990) Meiosis. Cambridge University Press, Cambridge, UK
Judson OP, Normark BB (1996) Ancient asexual scandals. Trends Ecol Evol 11: 41–46

Kearney MR (2003) Why is sex so unpopular in the Australian desert? Trends Ecol Evol 18: 605–607
Kearney M (2005) Hybridization, glaciation and geographical parthenogenesis. Trends Ecol Evol 20: 495–502
Keightley PD, Otto SP (2006) Interference among deleterious mutations favours sex and recombination in finite populations. Nature 443: 89–92
Kraft T, Nybom H, Werlemark G (1996) DNA fingerprint variation in some blackberry species (*Rubus* subg. *Rubus*, *Rosaceae*). Pl Syst Evol 199: 92–108
Krahulec F, Krahulcová A, Fehrer J, Bräutigam S, Plačková I, Chrtek jun. J (2004) The Sudetic group of *Hieracium* subgen. *Pilosella* from the Krkonoše Mts: a synthetic view. Preslia 76: 223–243
Levy M, Winternheimer PL (1977) Allozyme linkage disequilibria among chromosome complexes in the permanent translocation heterozygote *Oenothera biennis*. Evolution 31: 465–476
Lively CM, Dybdahl MF, Jokela J, Osnas EE, Delph LF (2004) Host sex and local adaptation by parasites in a snail-trematode interaction. Am Nat 164: S6–S18
Lundmark M (2006) Polyploidization, hybridization and geographical parthenogenesis. Trends Ecol Evol 21: 9
Lynch M (1984) Destabilizing hybridization, general-purpose genotypes and geographic parthenogenesis. Quart Rev Biol 59: 257–290
Lynch M, Conery J, Bürger R (1995) Mutational meltdowns in sexual populations. Evolution 49: 1067–1080
Mantovani B, Passamonti M, Scali V (2001) The mitochondrial cytochrome oxidase II gene in *Bacillus* stick insects: ancestry of hybrids, androgenesis, and phylogenetic relationships. Mol Phyl Evol 19: 157–163
Mark Welch D, Meselson M (2000) Evidence for the evolution of bdelloid rotifers without sexual reproduction or genetic exchange. Science 288: 1211–1215
Mark Welch DB, Mark Welch JL, Meselson M (2008) Evidence for degenerate tetraploidy in bdelloid rotifers. Proc Natl Acad Sci USA 105: 5145–5149
Matzk F, Prodanovic S, Bäumlein H, Schubert I (2005) The inheritance of apomixis in *Poa pratensis* confirms a five locus model with differences in gene expressivity and penetrance. Plant Cell 17: 13–24
Maynard Smith J (1978) The evolution of sex. Cambridge University Press, Cambridge, UK
McKone MJ, Halpern SL (2003) The evolution of androgenesis. Am Natural 161: 641–656
Mogie M (1992) The evolution of asexual reproduction in plants. Chapman and Hall, London
Muller HJ (1932) Some genetic aspects of sex. Am Nat 66: 118–138
Muller HJ (1964) The relation of recombination to mutational advance. Mutat Res 1: 2–9
Nogler GA (1982) How to obtain diploid apomictic *Ranunculus auricomus* plants not found in the wild state. Bot Helv 92: 13–22
Nogler GA (2006) The lesser-known Mendel: his experiments on *Hieracium*. Genetics 172: 1–6
Nybom H, Esselink GD, Werleman G, Leus L, Vosman B (2006) Unique genomic configuration revealed by microsatellite DNA in polyploid dogroses, *Rosa* sect. *Caninae*. J Evol Biol 19: 635–648
Ozias-Akins P, van Dijk PJ (2007) Mendelian genetics of apomixis in plants. Annu Rev Genet 41: 509–537
Paland S, Colbourne JK, Lynch M (2005) Evolutionary history of contagious asexuality in *Daphnia pulex*. Evolution 59: 800–813
Parker ED Jr, Selander RK, Hudson RO, Lester LJ (1977) Genetic diversity in colonizing parthenogenetic cockroaches. Evolution 31: 836–842
Pichot C, El Maâtaoui M, Raddi S, Raddi P (2001) Surrogate mother for endangered *Cupressus*. Nature 412: 39
Pontecorvo G (1958) Trends in Genetic Analysis. Columbia University Press, New York
Quarin CL, Espinoza F, Martinez EJ, Pessino SC, Bovo OA (2001) A rise of ploidy level induces the expression of apomixis in *Paspalum notatum*. Sex Plant Reprod 13: 243–249

Raikov IB (1995) Meiosis in protists: recent advances and persisting problems. Europ J Protistol 31: 1–7

Ramesh MA, Malik S-B, Logsdon Jr JM (2005) A phylogenomic inventory of meiotic genes: evidence for sex in *Giardia* and an early eukaryotic origin of meiosis. Curr Biol 15: 185–191

Robertson AV, Ramsden C, Niedzwiecki J, Fu JZ, Bogart JP (2006) An unexpected recent ancestor of unisexual *Ambystoma*. Mol Ecol 15: 3339–3351

Schaefer I, Domes K, Heethoff M, Schneider K, Schön I, Norton RA, Scheu S, Maraun M (2006) No evidence for the 'Meselson effect' in parthenogenetic oribatid mites (Oribatida, Acari). J Evol Biol 19: 184–193

Selker EU (2002) Repeat-induced gene silencing in fungi. Adv Genet 46: 439–450

Sharbel TF, Mitchell-Olds T, Dobeš C, Kantama L, de Jong H (2005) Biogeographic distribution of polyploidy and B chromosomes in the apomictic *Boechera holboelli* complex. Cytogenet Genome Res 109: 283–292

Simon J-C, Rispe C, Sunnucks P (2002) Ecology and evolution of sex in aphids. Trends Ecol Evol 17: 34–39

Simon J-C, Delmotte F, Rispe C, Crease T (2003) Phylogenetic relationships between parthenogens and their sexual relatives: the possible routes to parthenogenesis in animals. Biol J Linn Soc 79: 151–163

Stouthamer R, Breeuwer JAJ, Hurst GDD (1999) *Wolbachia pipientis*: Microbial manipulator of arthropod reproduction. Ann Rev Microbiol 53: 71–102

Stouthamer R, van Tilborg M, de Jong JH, Nunney L, Luck RF (2001) Selfish element maintains sex in natural populations of a parasitoid wasp. Proc R Soc Lond B 268: 617–622

Suomalainen E, Saura A, Lokki J (1987) Cytology and evolution in parthenogenesis. CRC Press, Boca Raton

Uyenoyama MK (1984) On the evolution of parthenogenesis: a genetic representation of the "cost of meiosis". Evolution 38: 87–102

Vandel A (1928) La parthénogenèse géographique. Contribution à l'étude biologique et cytologique de la parthénogenèse naturelle. Bull Biol France Belg 62: 164–281

van Dijk PJ, Bakx-Schotman JMT (2004) Formation of unreduced megaspores (diplospory) in apomictic dandelions (*Taraxacum officinale, s.l.*) is controlled by a sex-specific dominant locus. Genetics 166: 483–492

Verduijn MH, van Dijk PJ, van Damme JMM (2004) Distribution, phenology and demography of sympatric sexual and asexual dandelions (*Taraxacum officinale s.l.*): geographic parthenogenesis on a small scale. Biol J Linn Soc 82: 205–218

Vrijenhoek RC (1979) Factors affecting clonal diversity and coexistence. Am Zool 19: 787–797

Vrijenhoek RC (1998) Animal clones and diversity. BioScience 48: 617–628

Weeks AR, Marec F, Breeuwer JAJ (2001) A mite species that consists entirely of haploid females. Science 292: 2479–2482

Weinzierl RP, Schmidt P, Michiels NK (1999) High fecundity and low fertility in parthenogenetic planarians. Invertebr Biol 118: 87–94

Wetherington JD, Kotora KE, Vrijenhoek RC (1987) A test of the spontaneous heterosis hypothesis for unisexual vertebrates. Evolution 41: 721–731

Williams GC (1975) Sex and evolution. Princeton University Press, Princeton

Chapter 2
The Evolution of the Problem of Sex

Stephanie Meirmans

Abstract While numerous reviews on the problem of sex exist, the historical side of the research in this field is often only loosely covered. Here, I provide a more detailed historical overview by analyzing the contributions of four of the most influential biologists: Charles Darwin, August Weismann, Ronald Fisher and John Maynard Smith. More specifically, I discuss why these four biologists became interested in the significance of sex in the first place, describe their respective theories on sexual reproduction and in which context those theories were developed. This approach provides a general overview over the conceptually important changes in the history of the research of sex. Most importantly, it shows that not only the potential answers on the existence of sex have evolved, but also the question itself.

2.1 Introduction

Explaining the widespread occurrence of sexual reproduction in higher organisms is often called the "queen of problems in evolutionary biology" (following Bell 1982), with no consensus regarding the explanation as of today. While numerous reviews of the topic exist (e.g. Barton and Charlesworth 1998; Otto and Lenormand 2002), the historical side of the research on sexual reproduction is usually only loosely covered (see Ghiselin 1988; Mooney 1993, 1995 for exceptions). Traditionally, it is stated that Weismann (1889) first proposed a theory for sex, which was then put into population genetic terms by Fisher and Muller in the 1930s. This "Fisher-Muller" theory was the accepted view until the 1960s and 1970s, when it became clear that certain severe costs were connected to sexual reproduction. The recognition of these costs made the existence of sex so paradoxical that it became the "queen of problems" in the 1980s. Subsequently, numerous other theories were proposed to solve this problem. These theories can be classified into two different types: the "mutational"

S. Meirmans (✉)
Centre for the Studies of the Sciences and the Humanities, University of Bergen, PO Box 7800, 5020 Bergen, Norway
e-mail: stephane.meirmans@svt.uib.no

I. Schön et al. (eds.), *Lost Sex*, DOI 10.1007/978-90-481-2770-2_2,

theories (sex helps to get rid of deleterious mutations) and the "ecological" theories (sex creates good traits).

Though such an overview of the research on sex might be roughly correct, it is clear from its obvious gaps that it does not provide an exhaustive historical representation. For example, one might ask why Weismann or Fisher were interested in sexual reproduction in the first place, given the fact that the idea of a "paradox of sex" did not exist until the 1970s. Was sex problematic for them at all, and if yes, was it for similar or for different reasons than today? In the latter case, can their theories be easily used in modern times, either as single theories or in a pluralistic approach? Furthermore, it might be interesting to know why the problem of sex was not recognized prior to the 1970s.

Here, I address these questions and provide a more detailed historical overview by analyzing the contributions to the topic of sexual reproduction of four of the most influential biologists in the field of the evolution of sex: Charles Darwin, August Weismann, Ronald Fisher and John Maynard Smith. More specifically, I will discuss why these four biologists became interested in the significance of sex in the first place, describe their respective theories on sexual reproduction and in which context those theories were developed. This approach provides a general overview over the conceptually important changes in the history of the research of sex since it presents the development of the theories in steps of roughly 30 years. I will end my historical overview with the publication of Maynard Smith's book on the evolution of sex in 1978, because his book provides much of the current conceptual background for the problem of sex. The remaining development of the ideas on sex, including the "mutational" versus "ecological" theories, will be discussed elsewhere in the present book (see Chapters 5, 6 and 7).

2.2 Darwin: The Effects of Cross-Fertilization

Already Charles Darwin (see Box 2.1) was interested in finding out why sexual reproduction exists (see also Gishelin 1988; Mooney 1995). Darwin's thoughts on sex were, however, quite diverse and changed over time. More specifically, he successively investigated three different aspects that he found relevant to sexual reproduction: (1) sex provides variation, (2) sex unites a species, and (3) sex provides vigorous and fertile offspring.

Box 2.1: *Charles Darwin* (1809–1882) was a remarkable and careful observer of nature (both biological and geological). Since his early youth, he wanted to explain whatever he observed through general laws (Darwin 1958). Nearly everybody is familiar with his long voyage on the "Beagle", during which he observed and collected field data, and with his subsequent development of the theory of evolution by natural selection.

Darwin was also influenced by his immediate surroundings. In his own family and also the wider English society, it was widespread to practice animal breeding and horticulture (Moore 2005). This influence makes itself apparent throughout his books. "On the origin of species" (Darwin 1859) started with a couple of chapters that described the presence of variations in both domesticated and wild organisms. These chapters were followed by a suggestion for how to explain these observations: breeding and artificial selection for domesticated organisms and evolution through natural selection for wild organisms.

Darwin had extensive contacts with other scientists, and developed his ideas and theories continuously, both by collecting more data as well as by performing experiments. Throughout his life, he published an immense number of scientific articles and several books. It is also important to mention that he was suffering from illness, and that also his ten children were subjects to a variety of diseases. It was custom in the Darwin family to marry among cousins. Also Charles Darwin married a first cousin, Emma Wedgewood. Darwin suspected that close inbreeding in his family might have been the major cause for the health problems of his children, and this suspicion reinforced his research on inbreeding (Moore 2005).

2.2.1 The Early Notebooks: Becoming Interested in the Significance of Sex

After disembarking from the "Beagle" in 1836, Darwin spent more than twenty years to develop and refine his theory of evolution before publishing "On the origin of species" in 1859. It is widely known that he pondered much about variation in natural and domestic species during those years. It is perhaps less known that in connection with this, he also thought about the significance of sexual reproduction (Darwin 1958, p. 119).

In fact, Darwin started his investigations into the transmutation of species with considerations on sexual reproduction (published posthumously; Darwin 1960, pp. 41, 42). His thinking about sexual reproduction was thus one of the conceptual fundaments of his theory of evolution. More specifically, in his first notebook on the transmutation of species, he cited his grandfather, Erasmus Darwin, on this topic. His grandfather had noted that sexual species get variable offspring while asexual species produce identical offspring (Darwin 1796). Charles Darwin (1960, pp. 41, 42) now considered that the variation provided by sex might be a means to adapt to a changing world. Interestingly, he had (at this point) not yet developed a mechanism for how this should work. Furthermore, he asked himself whether such varieties wouldn't be endless. How and why can we recognize species? Darwin suggested that the "beautiful law of intermarriages", i.e. sex, could also be important as a means to maintain species.

In order to understand how Darwin further developed his early ideas on the significance of sexual reproduction, it is important to realize what he actually knew of the physiological processes happening during sex. Darwin's grandfather was a physician. He had proposed that in sexual species, the male provides a "living filament", while the female provides nourishment and oxygenation to the developing embryo. In his eyes, this influence from the female on the embryo could explain the variation resulting from sex. It is doubtful, however, whether Darwin shared his grandfather's opinion on how sex generates variable offspring. After all, in Charles Darwin's lifetime, more details about fertilization were available than had been for his grandfather. Most importantly, it had by then been observed that sperm penetrates the egg. Darwin himself thought that both mother and father provide hereditary material for the offspring (Darwin 1876, p. 353). He also thought that up to several sperm can fertilize an egg and that the exact quantity influences the specifics of heredity (ibid, pp. 355–357; see also Darwin 1862a, p. 355 for a similar view on fertilization in plants).

It was common among scholars to assume that the hereditary material includes the germs from all the earlier ancestors. After all, it was known that during fertilization, sperm-cell(s) and germ-cell fuse, and so, they concluded, does the hereditary material (Darwin 1876, p. 457). Offspring generally show traits that are intermediate between parents, so it was further assumed that the hereditary material blended (blending inheritance). It was also acknowledged that offspring could either be more like the father, or more like the mother – so it was thought that the ancestral germs could be transmitted in different quantities, or be differently expressed in the offspring (ibid, pp. 382, 413).

2.2.2 *Sex and Variation*

In his early notebooks, dating from 1837, Darwin had considered sexual reproduction as being one of the fundaments of evolution because it provides the variation that can ultimately result in adaptation (Darwin 1960, pp. 41, 42). However, in his actual publications he dismissed this role of sexual reproduction. At that point, he had gathered crucial information on the nature of domestic variations from breeders and horticulturists, and made two important observations. First, some forms of asexual reproduction (budding) can result in variations. Second, domesticated species are generally more variable than natural species are (discussed in his essay in 1842, see Darwin 1909, pp. 1, 4). Darwin concluded from those observations that sexual reproduction per se cannot be the ultimate source of variation since also asexual organisms can vary. Instead, he proposed that external conditions cause variations. He deduced this explanation first and foremost from the observation that external conditions are generally more variable for domesticated than for natural species.

Despite denying sexual reproduction as the ultimate source for variation, Darwin (1875, pp. 353, 354) nevertheless stressed that sexually produced offspring are generally more liable to vary than asexual ones. He proposed that this feature could be explained because the environment primarily induces variations in the reproductive organs (e.g. Darwin 1859, pp. 131, 132). Darwin had gathered evidence from several different sources that crossing is a very delicate and fragile process, and that it can be easily disturbed by the influence of the environment. For example, breeders had observed that breeding under confinement can be extremely difficult, and that the resulting offspring are often infertile. Darwin concluded that in particular the substance residing in the reproductive organs, the "sexual elements", could change. These modifications in the sexual elements again would lead to huge changes in the resulting offspring because they would be affected so early in development. In Darwin's eyes, the connection between sex and variation was thus the result of physiological/developmental processes rather than genetic processes (see also Ghiselin 1969). It is, however, less clear to which degree the later Darwin regarded the variation resulting from sex as a necessary condition for evolution.

2.2.3 Sex Unites a Species

Already in his notebooks, Darwin had proposed that sexual reproduction might act as a means to unite a species, and he developed this thinking further in his actual publications. His train of thought is not difficult to comprehend when considering that Darwin assumed blending inheritance: While on an individual level, blending inheritance meant that the characteristics of the parents are combined in the offspring (Darwin 1960, pp. 41–42), its consequence on a population level is that varieties occurring in the parental generation are swamped out in the offspring. To Darwin, "intercrossing plays a very important part in nature in keeping the individuals of the same species, or of the same variety, true and uniform in character" (Darwin 1859, p. 103). In short, sex solved for Darwin the problem of endless variations with no species boundaries.

Of course, the swamping effect of sexual reproduction can also be seen as a problem for the evolutionary theory. It makes the inheritance, and thus selection, of favorable varieties difficult. This was one of the main problems of Darwin's theory, and his critics were quick to point that out. From early on, Darwin was aware of this problem but apparently was not too troubled by it (see also Ghiselin 1969). After all, domestic breeding showed that the selection of varieties is possible, despite the presumed effects of blending inheritance. In addition, Darwin suggested that several factors can influence and potentially reduce the swamping effect: for example dispersal, rate and frequency of crossings, and isolation (Darwin 1859, pp. 102, 103). Later in life, Darwin doubted whether the unification of species could explain the origin of sexual reproduction (Darwin 1875, p. 355). At that point, he suggested a different explanation for the widespread occurrence of sex (see below).

2.2.4 Cross-Fertilization: A Law of Nature

Already in the first edition of the "Origin of species", Darwin claimed that all organisms cross-fertilize, and he called this a law of nature. While it was obvious that crossing is a necessity for bisexual organisms, Darwin was convinced that also hermaphrodites occasionally or habitually cross-fertilize. He thought that continuous selfing is deleterious (e.g. Darwin 1859, pp. 96, 97). Darwin derived this law from earlier studies, in which he recognized certain adaptations as ensuring crossing between individual organisms. Originally, he became interested in this topic because he was for a long time convinced that sex was important to keep a species uniform (Darwin 1958, p. 127). More specifically, he investigated flowers as adaptations for cross-fertilization by insects (Darwin 1862a). He detected that some hermaphrodite plant species show flower polymorphisms in order to reduce self-fertilization (Darwin 1958, pp. 128, 129).

It is crucial to understand that with "cross-fertilization", Darwin generally meant crossings between individual organisms, or sex. Darwin was unaware of the existence of homo- and heterozygosity. To him, the effects of self-fertilization, and to a lesser degree, inbreeding, were similar to those of parthenogenesis. Darwin knew about the existence of parthenogenesis since at least 1850 (Darwin, letter to R. Owen, 23 March 1850).

Darwin's early studies on plants did, however, not show why cross-fertilization generally has such an importance. In fact, he was aware that self-fertilization should be advantageous because it ensures reproductive assurance (Darwin 1862a, p. 359). He even noted that in sexual species males are produced, which cannot by themselves produce any offspring (Darwin 1876, p. 461). In 1862, he thus stated that "We do not even in the least know the final cause of sexuality; why new beings should be produced by the union of the two sexual elements, instead of by a process of parthenogenesis" (Darwin 1862b). In his later publications, Darwin specified that selfing species have descended from cross-fertilizing species and that they occur taxonomically isolated. He concluded that selfing represents only a temporary state that can evolve under certain circumstances. More specifically, he deduced that it could be an adaptation to avoid species extinction in cases of limited mating opportunities (Darwin 1877, pp. 290–292).

2.2.5 Sex and Hybrid Vigor

Darwin already early guessed that sex and cross-fertilization might be important for species because it results in vigorous offspring. In animal breeding, it was common knowledge that offspring derived from cross-fertilizations generally are very healthy while inbred progeny often are sickly. The breeders commonly called this feature "hybrid vigor" (Darwin 1859, p. 96). In Darwin's family, there was a tradition of breeding animals, and at this point in time, Darwin had begun to be interested and practice animal breeding. Furthermore, he was interested in inbreeding due to personal interests: his own (inbred) family was sickly, and his favorite daughter died in 1851 (Moore 2005).

Darwin initiated a correlative and experimental study that would last for 11 years, in which he showed the deleterious effects of self-fertilization and investigated its causes (finally published in Darwin 1876). The study corroborated "hybrid vigor";

however, this effect was restricted to cases where the parents (or other ancestors) had been subject to different conditions or had spontaneously varied. Darwin believed that in both cases, the ancestor's "sexual elements have been in some degree differentiated" (ibid, p. 443). As explained earlier, Darwin assumed that external conditions could produce variations in particular in the reproductive system of an organism (ibid, p. 446). Darwin further believed that a differentiated male element would, during and after fertilization, provide a form of different environment for the female germ. In his eyes, this process would ultimately lead to great benefits.

Rather than thinking of more modern benefits of sex such as producing variation, Darwin had benefits from immediate physiological processes during fertilization in mind: he argued that differentiation between the sexual elements can either enhance or interfere with the union of the sexual elements, resulting in partial or whole sterility (ibid, pp. 446, 447). More specifically, Darwin thought that changes in the sexual elements can change the (physiological) way in which pollen and stigma interact with each other, as well as how the contents of pollen and ovules interact (ibid, p. 456). His observations on plants showed that sterility generally appeared in two situations: First, to prevent self-fertilization, and second, to prevent crossing between species. In the first case (self-fertilization), differentiation of the respective sexual elements was too small to allow union. In the second case (crossing between species), differentiation had proceeded too far (ibid, 456). Most fertile were the crosses when differentiation was intermediate, i.e. crossing between distinct varieties from the same species. Darwin compared this process to observations made in chemistry that different molecules show the highest affinity to each other if slightly different (ibid, pp. 456, 457).

Concerning the significance of sexual reproduction, Darwin deduced from his studies that its benefit lies in the advantageous fusion of differentiated individual organisms, which only sexuality makes possible (ibid, pp. 461, 462). Darwin thought that this advantage, resulting in "hybrid vigor", would be "amply sufficient to account for the development of the sexual elements, that is, for the genesis of the two sexes." (ibid, p. 462).

In summary, in his early private notebooks, Darwin considered sex to be advantageous due to its effects on variation and its power to unify species. In fact, these ideas were important to the development of his evolutionary theory as a whole. However, further investigations led him to discard these aspects of sex as reasons for why sexual reproduction exists. Instead, he proposed that sex exists due to physiological advantages during fertilization – resulting in vigorous offspring.

2.3 Weismann: The Significance of Sexual Reproduction in the Theory of Natural Selection

Even though Darwin proposed a reason for why sex exists, Weismann (see Box 2.2) is usually referred to as the biologist who formed the first theory for why sex is advantageous (Burt 2000). This is probably due to the fact that Weismann was the first biologist who was convinced that sex is advantageous because it provides

Box 2.2: *August Weismann* (1834–1914) was one of the first proponents of the evolutionary theory in Germany. He himself claimed to have been the third German to publish in the Darwinian spirit, after Fritz Müller and Ernst Häckel. Even though he was interested in nature since his early childhood, Weismann first became a physician and also practiced several years as such. Later, he turned to zoology and finally became a professor at the university in Freiburg. During his early career, he made a number of important biological discoveries, many of them as a result of cytological studies. However, he had trouble with his eyes and could during long periods of his working life not use the microscope. Instead, his wife and students did most of the observational work, while Weismann himself focused on theoretical work (Conklin 1915; Petrunkevitch 1964).

In 1861, during the first of several periods of illness, Weismann read Darwin's "Origin of species" and immediately became a convinced Darwinian. All his later work was an effort to implement and strengthen the evolutionary theory that was still quite controversial at that time. More specifically, he aimed at unifying aspects from development, evolutionary biology and his own theory, the constancy of the germ line. As a renowned professor, Weismann had access to newly obtained knowledge from many different sources across Europe, a fact that helped him to achieve his goal (Gaupp 1917).

variable offspring (Weismann 1889, 1892). This idea forms the basis for most modern theories for the significance of sex (Burt 2000). In accordance with Darwin, Weismann argued that variability establishes the basis for adaptation. Thus, the difference between Darwin and Weismann is that Weismann believed sex to exist *because* it provides variation while Darwin himself had discarded this mechanism as the reason why sex exists.

2.3.1 The Principles of Heredity

Between 1882–1890, Strassburger, Roux, Beneden, Weismann, Boveri and others established the cytological mechanisms of fertilization and meiosis, and determined that the hereditary material was situated on the chromosomes in the cell nucleus. Weismann stimulated the cytological research with his theoretical efforts to elucidate the genetic meaning of the discoveries that emerged in this area.

Most famously, he developed the theory of a "continuous germ plasm", representing the idea that the hereditary material (called germ plasm) is inherited unchanged. Expressed in modern terms, Weismann distinguished soma and germ line, and declared that while the soma can change over the lifetime of an organism, the germ does not (proposed in 1885; see Weismann 1893). This concept of "hard heredity" is today one of the fundaments of modern genetics, even though modern genetics accepts the infrequent occurrence of mutations.

2.3.2 Heredity, Variability and Sex

Weismann created problems for himself by advocating both the idea of a constant germ line and the theory of evolution. After all, evolutionary change depends on the occurrence of continuous variations while Weismann refused the notion of acquired inherited changes. Darwin had thought that acquired changes could be inherited, and thus result in variation. At a meeting in 1885, Weismann attempted to reconcile his theory of the constancy of the germ line with evolutionary theory: he proposed that sexual reproduction could be the source of the continuous variations that ultimately fuel evolutionary change (Weismann 1889, Preface).

Weismann also developed a theory on the origin of variability. According to this theory, variability arose in simple unicellular organisms via acquired characters. He knew that some unicellular organisms do not have a distinct nucleus, and from this he deduced also no clear germ/ soma distinction in these organisms. Ultimately, this led him to think that in these simpler unicellular organisms acquired characters could indeed be transmitted to the offspring via fission. In all their (multicellular) descendants, this variability is combined and/or redistributed exclusively through sex (Weismann 1889, pp. 277, 278, 1892, pp. 190–195). Weismann continuously developed his ideas on the significance of sex (see also Winther 2001), following his insights into the processes happening during meiosis. Here, I will present two of his essays (originally dating from 1886 and 1891) because they show Weismann's initial idea for sex, and its later conceptual changes. In addition, I will give an overview of Weismann's explanations for the existence of parthenogenesis.

2.3.3 The Significance of Sex, 1886

In 1886, Weismann published a long essay, written in German (English translation published in 1889). In this essay, he proposed for the first time that sexual reproduction is the source for the existence of individual hereditary differences. Parts of this essay are quite modern. However, in particular his ideas on the mechanisms of how variations are produced, were quite different from our current views.

At this point in time, the reduction division of the germ plasm had not been observed, nor was it known that the hereditary material was situated in the chromosomes. Weismann accordingly believed that during fertilization, the germ plasm inherited from the parents was simply fusing together, thus accumulating in quantity. He actually believed that in the tenth generation, a single germ would contain 1024 different germ plasms (Weismann 1889, p. 276; note that he made an error in this calculation as the correct number should have been 512, since he started his example with two germplasms in the second generation). He was therefore astonished by the apparent age of the embalmed bodies of some Egyptian animals, and noted that "...the growth of the germ-plasm of the Egyptian ibis or crocodile must have been stupendous" (ibid, p. 271).

Weismann thus assumed blending inheritance, and simultaneously did not promote the inheritance of acquired characters. Most of Darwin's critics had pointed out that under blending inheritance, variation should *de*crease due to sex. However, Weismann suggested the exact opposite, namely that variation in a population should generally *in*crease due to sex. His reasoning for this rather "startling" proposition (as he himself called it) was rather complex. First, he argued that two different types of variation exist, which differ in their reactions to the process of sex (ibid, pp. 272, 273): (1) large deviations from the species norm, and (2) variations that are of small effect. Weismann thought that sex would indeed *de*crease the variation that is due to large deviations. These deviations, he thought, occur only rarely and would quickly disappear by blending with the large number of "normal" individuals. Small variations, however, would not decrease, as they would be present in extremely high numbers in large populations. This would make complete inter-crossing between individuals, and thus complete blending of variations, impossible (ibid, pp. 272, 273; see also Darwin 1859, pp. 102, 103 for a similar idea). Importantly, Weismann thought that this latter type of variation would lead to the sort of evolutionary change that he thought to be important: change in small and gradual steps (Weismann 1889, p. 264).

Second, Weismann (1889) proposed that sexual reproduction should instead result in the production of new variations. In every generation, the hereditary material from the parents is merged, and he argued that this merging would result in ever new combinations of the hereditary material. Such combinations of germs could gradually change the hereditary material in the required direction, and ultimately result in new species through the creation of new characters (ibid, pp. 272, 275) In fact, Weismann imagined that the number of different combinations provided through sex could even be enhanced: he assumed that sometimes the hereditary material from the father could be more expressed in the offspring, sometimes from the mother, and sometimes even from other ancestors (ibid, p. 276). It was necessary for Weismann to make this assumption because he realized that under his theory for sex full siblings, although obviously never exactly identical, would receive the same hereditary material.

2.3.4 The Significance of Sex, 1891

Already in 1891 (original German essay; English translation published in 1892), only 5 years after the publication of his first essay on sex, Weismann had significantly changed his ideas. This change was due to a number of cytological discoveries that were made in the meantime (Weismann 1892, Preface). Most importantly, chromosomes were thought to be the bearers of the hereditary material. Chromosomes were then also known to exist in equal numbers in the male and female reproductive cells (ibid, p. 112). It was furthermore discovered that two reduction divisions happen during meiosis. One division followed an initial duplication of the hereditary material, while the other resulted in an ultimate halving of the hereditary material in the mature germ cells. The latter process solved for Weismann his previous puzzle that the germ plasm would be endlessly accumulating over the generations.

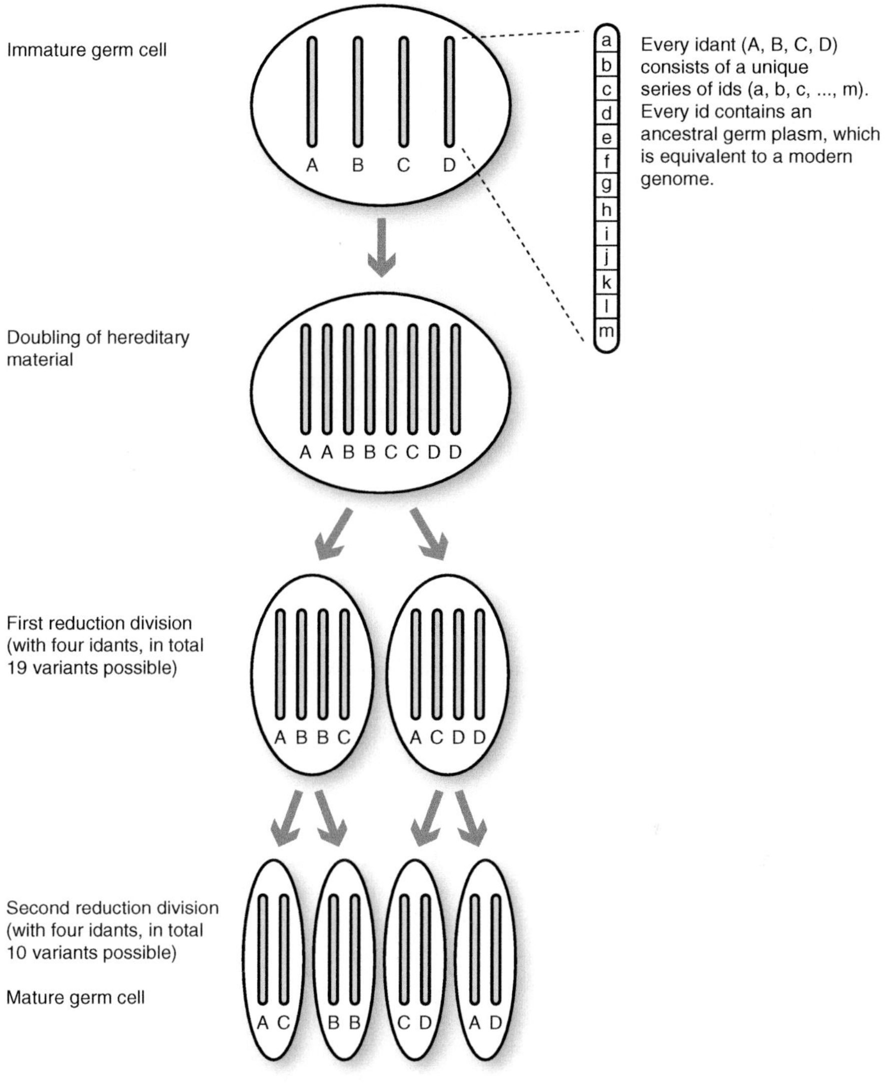

Fig. 2.1 Processes happening during meiosis, according to Weismann in 1891 (see text for details)

According to the cytological observations, Weismann envisioned the nature of the hereditary material differently than only 5 years earlier. Now, he thought that the ancestral germ plasms (called "ids") are lined up linearly on the chromosomes (called "idants") (Fig. 2.1). Weismann believed that each "id" contains the whole hereditary information necessary to build up an organism (Weismann 1892, p. 130). An "id" would be equivalent to our modern concept of a "genome". Information contained herein would be present manifold in each organism. The "ids", occasionally, break up and mix, but usually they were thought to stay fixed in their order on the chromosome (ibid, pp. 130, 131).

As in his previous theory for sex, Weismann reasoned that combinations increase during sex primarily because different individuals combine their germ plasms. However, Weismann now had a sort of "reshuffling" process in mind rather than a simple addition of germ plasm. This change came partly due to the fact that halving of the hereditary material during meiosis had been observed. Furthermore, Weismann deduced that the observed doubling and subsequent reduction of "idants" (chromosomes) during the second reduction division acted to increase the number of possible combinations of chromosomes inherited (see Fig. 2.1 for details; Weismann 1892, p. 133). Weismann, with the help of a mathematician, showed that the resulting potential number of combinations in a species would generally be so high that no individual would be like another, not even its full sibling. The ultimate significance of such variety was then to provide the "material for the operation of natural selection" (ibid, pp. 134, 135).

2.3.5 Cyclical and Obligate Parthenogenesis

Weismann knew the phenomenon of parthenogenesis quite well from his extensive work on daphnids and ostracods (Weismann 1889, pp. 294, 325, 1892, pp. 161ff). He had noticed that parthenogenetic females could quickly produce immense numbers of descendants, and that indeed "the number of egg-producing individuals in all the previous sexual generations would be doubled" (Weismann 1889, p. 290). This wording reminds strongly of the "two-fold cost of sex" as we know today. However, in one case, Weismann argued that the embryos in the thinner-shelled parthenogenetic eggs develop more quickly and hatch earlier than those in sexual eggs (ibid, p. 324). In another case, he probably implied that all parthenogenetically produced eggs could hatch since they do not depend on insemination (Weismann 1889, p. 324, 1892, pp. 216–220).

Weismann concluded that parthenogenetic species have a clear short-term advantage because of their rapid way of reproduction. However, he argued that such species would fail on the long term. His reasoning for such failure changed in accordance with changing insights into the processes of meiosis. Around 1886, when Weismann assumed that only sex could provide variation in metazoa, he contented that parthenogenetic species would ultimately fail since they could never produce any novel characters (Weismann 1889, pp. 273–275). Some years later, observations showed that parthenogenetic species also undergo a doubling of chromosomes with subsequent reduction. Weismann concluded from this that a form of mixis also happens in parthenogenetic species (Weismann 1892, pp. 155, 156). However, he thought this type of mixis to result in limited variation when compared to the variation resulting from sex (ibid, p. 161). More specifically, he argued that there would be an inherent trend in obligate parthenogenetic species towards uniformity of the germ line. This process would occur because parthenogenetic species would lose variants over time, and because this variation could never be re-established (ibid, p. 158). Weismann argued that his theoretical thinking on these issues was empirically corroborated through the rare occurrence of parthenogenesis (Weismann 1892, p. 199). More specifically, he knew that parthenogenesis occurs taxonomically patchy and isolated (Weismann 1889, pp. 290, 291).

Weismann further thought that obligate parthenogenetic species (such as ostracods) could have derived from species reproducing via cyclical parthenogenesis (ibid, pp. 323, 324). Cyclical parthenogenesis, Weismann (1889, p. 289) argued, is an adaptation of the species. Under certain circumstances, it can be vital to reproduce rapidly for a certain amount of time, and thus to reproduce parthenogenetically. In those species, more parthenogenetic offspring ultimately also means more sexually reproducing offspring. Weismann suggested that this advantage could outweigh detrimental effects of temporal parthenogenetic reproduction.

In summary, Weismann presented a theory for sex primarily because he attempted to reconcile his own theory of a constant germ line with the Darwinian theory (see also Mooney 1993). According to Weismann, sexual reproduction serves to provide individual variations. These variations form the basis for evolutionary change and ultimately speciation. Weismann's theory on the significance of sex changed during his career, along with cytological discoveries on meiosis. Around 1886, he believed that the hereditary material is simply merged during sex, thus accumulating in quantity. This process would ultimately result in variation since each individual receives a unique combination of ancestral hereditary material. Later, around 1891, Weismann argued that variation would be established through redistribution of hereditary material. More specifically, he suggested that new combinations could be provided through the mixis of chromosomes. Individual chromosomes, according to Weismann, are made up of a series of ancestral genomes. Ultimately, "genome-packages" would get mixed during sex.

2.4 Fisher: The Contrast Between Sexual and Asexual Reproduction

While Weismann did not incorporate mutations in his theory for sex and actually developed this theory to be able to avoid the acceptance of changes in the hereditary material, Fisher (see Box 2.3) explicitly incorporated mutations in his theory for sex (Fisher 1922,1930). Due to the differing assumptions, the overlap between Fisher's and Weismann's theory for sex is shallow, even though it has often been stated that they basically comprised the same idea (see also Mooney 1995). For Fisher, the ultimate importance of sex was that sexual organisms can both take better advantage of beneficial mutations and get rid of the ones that are deleterious. He presented his idea on the significance of sex in his book on natural selection (1930), but, as Bennett (1983) noted, he outlined the main argument already in 1922, in a short paper on the "Darwinian evolution by mutations".

Box 2.3: *Ronald Fisher* (1890–1962) excelled at school in mathematics, physics and biology. He gained a scholarship in mathematics, reasoning that choosing biology might have been too chancy. However, he often chose biology books as award prizes gained in mathematics. He thus came into the possession of the complete works of Charles Darwin (Bennett 1983). Fisher went on to study mathematics and physics (including statistics) at Cambridge (Mahalanobis 1964). Cambridge was one of the few universities in England that held a chair for genetics. Teaching at this department was mostly influenced by the geneticist Bateson, who had left the university only a couple of years before Fisher entered (Richmond 2006). Bateson believed that the new genetics, based on Mendelism and mutations, had given a deathblow to Darwinism. Fisher also became interested in Mendelism, but instead believed that it provided the grounds for a mathematical treatment of evolutionary theory (Mahalanobis 1964).

Already during those years, Fisher started to write a series of articles that lead up to his later famous book "The genetical theory of natural selection", published in 1930. According to Fisher himself (1930, Preface), mathematics could give a new look on biological problems since it is concerned not only with the *actual* (as a practical biologist would be), but also with a system of *possibilities* that is much wider than the actual. Thus, he explained, if you would want to know why two sexes exist, a mathematician would ask what the consequences are to have three or more sexes. Fisher succeeded in putting the Darwinian theory into a quantitative form; he developed a mathematical theory of evolution. Only later was he recognized as one of the architects of the Modern Synthesis (which combined Darwinism and Mendelism), together with J. B. S. Haldane and S. Wright.

Fisher had been an ardent eugenicist all his life. In 1933, he took over the chair in Eugenics at the London University, as well as the editorial charge of the "Annals of Eugenics". According to Mahalanobis (1964), Fisher developed his new mathematical/statistical theory for evolution in order to "supply a sound basis for eugenics, the science of man".

2.4.1 Mendelism and Darwinism

From 1900 onward, a number of important discoveries changed the way biologists understood heritability. First and foremost, Mendel's laws were rediscovered, and, as Fisher (1922) himself described, the new flowering study of genetics emphasized differences due to single heritable factors. Furthermore, Thomas Hunt Morgan and his group found mutations that they showed to be inherited in a Mendelian manner and linked to a material basis on chromosomes. At first, mendelism seemed to be in disagreement with Darwin's ideas of cumulative, gradual change in small steps. Many paleontologists, who observed just such cumulative changes, therefore opposed themselves to the geneticists. Instead, they believed in the inheritance of acquired characters. Furthermore, some breeders, who had observed pure lines, rejected mutations (Fisher 1922). The consequence of all these discoveries and their reactions was that Darwinism seemed to be dead during those years, and that alternative explanations to evolutionary change abounded (Bennett 1983).

Fisher (1922, 1930) then showed that Mendelism and Darwinism are not opposed but instead can actually reinforce each other. Fisher was an ardent follower of Darwinism, partly due to ideological reasons. He acknowledged that Darwin had wrongly assumed that inheritance is blending. According to Fisher, Darwin had deduced from this form of inheritance that newly appearing variations must be high in order to keep overall variability constant (Fisher 1922, 1930, Chapter 1). However, Hardy and Weinberg had both shown in 1908 that the new form of inheritance, the Mendelian system, instead itself acts to keep the variation constant. Likewise, Fisher (1922) emphasized that most of the variation seen in nature is due to new arrangements of old variation. Only chance and, in particular, selection, can reduce variation, and this is where Fisher (1922) saw the importance of infrequent mutations coming into the picture – first and foremost to restore lost variability.

As is widely known, Fisher used mathematical methods to illustrate how genetics can reinforce Darwinism. More specifically, he described the fate of new mutations supposing a Mendelian system of inheritance, and could show that it can indeed result in Darwinian evolution. It was equally important for Fisher to show that Darwinism results in Mendelism; or, put differently, that a sexual (Mendelian) system should evolve. He thought it likely that asexuality has been the ancestral, "primitive" state, with sexuality having evolved subsequently (Fisher 1930). In fact, he did not believe that any group of organisms today would be entirely asexual. Under asexuality, he understood everything from individual growth to vegetative reproduction by budding.

2.4.2 The Adaptive Significance of Sex

Already in 1922, Fisher argued that a sexual Mendelian system of inheritance should be favored over asexuality. Empirical observations in Fisher's time had shown that mutations, the ultimate source for variability, are largely detrimental. However, already Darwin had established that variation is the essential driving force of evolution. According to Fisher (1922), under this picture, a hereditary system should have evolved that keeps variability high with a minimum rate of (deleterious) mutations. Mendelian inheritance due to sex does, according to Fisher, indeed provide such a system: not only does the Mendelian system allow mutation rates to be kept low because variability is kept high through particulate inheritance, but it also enables every new mutation to be tested in thousands of different genetic backgrounds. Even mutations with only slight effects can thus increase or decrease in frequency, according to whether they are on average beneficial or deleterious. With sex, this process is independent of the specific genetic backgrounds in which these mutations originally arose.

Later, Fisher (1930) chose to turn the argument around, describing instead why being asexual should be disadvantageous. The problem of an asexual group, Fisher argued, is that it cannot always maintain all beneficial mutations that might arise. The problem is that asexual lines can never mix their genes, so that at some point in the future, all individuals will be the descendents of a single individual. However, the probability that a new mutation occurs exactly in this common ancestor is very small. Consequently, an asexual group cannot take advantage of all mutations that are occurring. It should be noted that Muller (1932) independently suggested a similar idea. Muller focused in particular on advantages of sex due to its effect of combining beneficial mutations (for more details on the theories' similarities and differences, see Mooney 1995). This specific theory for sex is therefore often referred to as the "Fisher-Muller theory".

Fisher also argued that the disadvantage of an asexual group is dependent on the mutation rate: if that rate is so low that the normal condition of the group is uniformity, every mutation goes to fixation before the next one occurs. In that case, asexually reproducing organisms can take advantage of all mutations, and there is no disadvantage to asexual reproduction. On the other hand, the low mutation rate means that there is no genetic variation in the group, so that "evolutionary progress will necessarily be almost at a standstill" (Fisher 1930, p. 122). Fisher argued that there should be an optimal mutation rate. However, he failed in calculating this rate, and later referred to this problem as the one problem in the book that puzzled him the most (letter to L. Darwin, 16 July 1930).

But even in the case of optimal mutability, Fisher argued, the rate of progress in asexuals "would still be very inferior to that of a sexual organism placed in the same circumstances" (Fisher 1930, p. 123). He proposed that its relative success, when compared to asexuals, depends on the number of different loci that are freely interchangeable. In his opinion, even a sexual organism with only two loci could thus respond to natural selection at a double rate (ibid, p. 123).

2.4.3 Sex, Time and Levels of Selection

Even though Fisher never published anything else on the significance of sex, it seems clear that he judged the disadvantage of asexuals to increase over time. More specifically, in a letter to another biologist, C. S. Stock, (dating from 24 October 1932) he argued that asexuals would be less and less able to adapt to changed conditions. I take this to mean that an asexual population would be less and less able to make efficient use of mutations. This inefficiency would be due to the fact that an initially diverse asexual population diminishes in diversity over time, due to selection. Thus, mutations could be tested out against less and less genetic backgrounds. Fisher (1922) had previously proposed that sex should be advantageous especially in cases when "complex adaptations have to be made to a slowly changing environment". This statement is in agreement with my interpretation. It is also interesting to note that Fisher (in the letter to Stock; see above) argued that in particular the efficient purging of deleterious mutations would be the ultimate cause for the occurrence of continuous sexual reproduction.

In 1958, Fisher published a second edition of his 1930 book, in which he did not change his arguments for the evolution of sex. However, he had in the meantime thought about the (im)probability of group selection, as can be seen in his letter to K. Mather (16 February 1942) and A. G. Lowndes (23 June 1945). In his second edition, he added a note stating that group selection generally is improbable (in Fisher 1999, pp. 279, 280). He argued that in general, individual selection is stronger than inter-group selection because competition between related groups or species is less strong due to their lower number and longer duration. However, he apparently also acknowledged that his treatment of the evolution of sex involved group selection. Without further explanation, he proposed that sex might have been the only trait that did evolve through group selection.

In summary, Fisher equated sexual reproduction first and foremost with the newly discovered form of Mendelian inheritance. His goal was to modify and strengthen Darwin's ideas on evolution by exploring the consequences of this form of inheritance on evolutionary change. It was equally important for Fisher to show that the sexual Mendelian system is the result of evolution. He argued that sex should evolve because under a Mendelian system, variability can be kept high with a low rate of mutations. This is especially the case because new mutations can be tried out in a variety of different genetic backgrounds. A low rate of mutations is advantageous because most mutations had been shown to be detrimental. Sex provides the opportunity to make the most efficient use of mutations because the fate of each mutation does not crucially depend on the initial placement of the new mutation in a specific genome (see also Mooney 1995).

2.5 Maynard Smith: The Cost of Sex

Contrary to Fisher, Maynard Smith (see Box 2.4) did not present a clear-cut theory that he thought would explain the predominance of sex. Instead, Maynard Smith demonstrated that the Fisher-Muller theory for sex is probably not working. More specifically, he detected problems with the theory connected to newer evidence on the molecular nature of mutations, as well as problems due to the time scale at which the Fisher-Muller theory works. Understanding the adaptive significance of sex turned out to be highly problematic after acknowledging those considerations. Maynard Smith's later work aimed at exploring how to address this "problem of sex".

Box 2.4: *John Maynard Smith* (1920–2004) was a naturalist since his childhood (Szathmary and Hammerstein 2004). However, he actually studied and worked as an engineer for aircrafts several years before entering into the field of evolutionary biology. Maynard Smith was so impressed by J. B. S. Haldane (one of the founders of the Modern Synthesis, along with R. Fisher and S. Wright), that later, at age 27, he went to study evolutionary biology with Haldane as his mentor (Charlesworth 2004). In the beginning of his career as a biologist, Maynard Smith worked experimentally on the genetics of inbreeding and ageing, using *Drosophila subobscura* as a model organism. Only later did he start to approach biological questions in a similar way than Haldane had done, meaning with the help of relatively simple mathematical models (Nee 2004).

Maynard Smith explored a wide variety of issues in evolutionary biology, among which perhaps the most famous are game theory and the evolution of sex. One common thread in his investigations was to take up evolutionary questions that seemed to contradict mainstream Darwinism, and to subject them to rigorous evolutionary analysis (Szathmary and Hammerstein 2004). For a large part, that meant that he was interested in questions that seemed to make sense as a good-for-the-species argument, but seemed to defy adaptations on the individual level (Nee 2004).

2.5.1 The Evolutionary Synthesis and Appearing Complexities

After the 1930s and 1940s, the evolutionary synthesis, the combination of Mendelism and Darwinism, became widely accepted among biologists. Newer evidence on the molecular basis of genes and mutations mostly confirmed the synthesis. However, it also became clear that certain assumptions of classical genetics, such as low genetic diversity and the predominance of additive genetic effects, were too simplistic (Lewontin 1974). With regard to sex, Crow and Kimura (1965) showed that under the Fisher-Muller model, sexual reproduction would be advantageous only under certain conditions such as large population sizes and relatively frequent occurrence of mutations. Furthermore, Crow and Kimura showed that certain genetic complexities, such as gene interactions and linkage, could change the outcome of the Fisher-Muller model for sex. Perhaps most importantly, they demonstrated that in the case of co-adapted gene complexes, sex and recombination could actually be disadvantageous. This has later become known as the "cost of recombination".

In ecology, another general evolutionary problem became apparent, namely the level of selection. This debate was initiated in the early 1960s, when the ecologist Wynne-Edwards (1962) published a book in which he proposed good-for-the-species arguments. Many biologists, including Maynard Smith (1964), reacted to such propositions of group selection. To them, individual selection would generally be more powerful than selection on the group level.

2.5.2 Group Selection and Sex

Around the early 1960s, most population geneticists accepted the Fisher-Muller model for sex. As outlined above, Fisher had stated in 1958 that sex might have been the only trait that evolved via group selection. In connection with the more general debate around group selection, Maynard Smith realized in 1964 that one cannot reject Wynne-Edwards (1962) arguments for group selection and at the same time accept a group selection argument for sex (Maynard Smith 1978, p. 2). Thus, Maynard Smith's first work on sex was to investigate whether the Fisher-Muller model is valid (see details below).

It is important to note, however, that Maynard Smith (1976) later specified that he did not per se reject a group selection argument for sex. More specifically, he argued that at least theoretically, genes could exist, which decrease individual fitness but increase group fitness. However, it is not certain that these genes will be of high evolutionary importance. According to Maynard Smith, the question on group selection is a quantitative one. For sex, the group selection argument would only then be plausible if "the origin of new ameiotically parthenogenetic strains is a sufficiently rare event to be balanced by the extinction of such a strain" (Maynard Smith 1976). In that case, models describing long-term advantages of sex, such as the Fisher-Muller model, can explain the evolution of sex. Maynard Smith was initially suspicious that such conditions could apply in nature, while he later was less opposed to such a scenario.

I identified three different stages during which Maynard Smith contributed to an understanding of the significance of sex: First, he explored the validity of the Fisher-Muller model, second he recognized the problem of the cost of reproducing sexually, and finally he proposed a clear conceptual framework for how one can theoretically and empirically approach the adaptive significance of sex.

2.5.3 The Fisher-Muller Model and a Changing Environment

Maynard Smith's first theoretical work on the evolution of sex aimed at exploring the validity of the Fisher-Muller model for sex. More specifically, Maynard Smith (1968) showed that sex is not advantageous under the Fisher-Muller model if the assumption of mutations being unique events is dropped. He pointed out that empirical observations had shown that many mutations are base substitutions that should not be unique in large populations. However, frequent (unique) mutations and large populations were exactly the conditions that Crow and Kimura (1965) found to be preferable for the adventage of sex. Maynard Smith (1968) suggested that sexual reproduction could instead be advantageous in a changing environment if the genetic variance would be created by the selection for different genotypes in different environments.

2.5.4 The Problem of the Cost of Sex

In 1971, Maynard Smith made obvious that one of the biggest problems is to explain how sexual reproduction is maintained once it has evolved. More specifically, Maynard Smith (1971) detected a severe disadvantage of sex, which has later been called the "cost of males". Compared to sexuals, parthenogenetic organisms can avoid the costly production of males, because usually only females are required for this type of reproduction. In fact, parthenogenetic females arising via mutation from a sexual source population should reproduce twice as many female offspring than sexual females would, resulting in an ultimate "two-fold cost of sex". Since only females contribute directly to the growth rate of a population, a sexually reproducing source population would be quickly outcompeted by parthenogenetic conspecifics if no other mechanisms balance this process. Maynard Smith (1971) suggested that this latter mechanism must be, in some way or another, a long-term advantage to sex, and that "the rarity of parthenogenetic varieties of animals suggests that this long-term selection acts, not by eliminating parthenogenetic varieties when they arise, but by favoring genetic and developmental mechanisms which cannot readily mutate to give a parthenogenetic variety".

Maynard Smith probably became aware of the cost of males because he did earlier experimental studies with parthenogenetic *Drosophila*. During those studies, he noted that parthenogenetic females can lay the same overall number of eggs in their lifetime as sexual females (Maynard Smith 1958). Maynard Smith's 1971 paper strongly influenced another biologist, George Williams. For Williams (1966), group selection in general was a serious threat to evolutionary biology, and he also found group selection models for sex problematic. Maynard Smith's recognition of a cost of sex on the individual level made the widespread occurrence of sex truly paradoxical in Williams' eyes (Williams 1975).

2.5.5 Conceptual Framework

Maynard Smith was in turn inspired by Williams' (1975) treatment of the paradox of sex, and in 1978 he wrote a book on the problem of sex. This book was simply titled "The evolution of sex". Rather than providing any ultimate answers for the problem of sex, Maynard Smith (1978) established in this book a clear framework showing what kind of issues are at stake. In the preface to this book, he stated that he hoped to encourage empirical research directed towards finding an adaptive significance of sex. This conceptual framework laid much of the grounds for the present research on the question of sex. In this section, I will present this framework, but do not intent to go more deeply into the variety of different models for sex that Maynard Smith presented and discussed in his book. Many of these models will be described elsewhere in this book (see, for example, Chapter 5).

In the first chapter of his book, Maynard Smith (1978, p. 6–9) introduced a number of important conceptual distinctions: First, he distinguished between the *origin* of sex and the *maintenance* of sex, because he argued that the two processes might not necessarily involve the same mechanisms. He proposed to direct investigations towards the maintenance of sex rather than the origin, because the investigations on its origin would necessarily be speculative whereas the maintenance of sex might be testable. Second, he distinguished between the evolution of *sex* and the evolution of *recombination*. He argued that while it is not yet clear whether the evolution of sex is maintained by group selection or by individual selection processes, any explanation for the evolution of recombination needs to involve individual selection processes. More specifically, Maynard Smith (1978) reviewed evidence that there is genetic variation within populations for the frequency and location of chiasmata. He then pointed out that this variation provides the grounds for natural selection to reduce crossing-over, and that it therefore indicates that there must be a short-term advantage to recombination. For sex, the evidence is less clear, leaving the possibilities of group selection open.

In his book, Maynard Smith first presented the problem of the cost of sex, similar to his 1971 paper. After that, he discussed some of the long-term consequences of sex. More specifically, he reviewed the Fisher-Muller model and other long-term models for sex that had been suggested, and summarized under which circumstances they would operate. He concluded that generally the models lead to accelerated evolution due to sex if they incorporate the possibility of chance events in finite populations (ibid, pp. 31, 32).

Next, Maynard Smith (1978, pp. 37–71) discussed whether sex could be maintained by group selection. Like in his paper from 1976, his strategy was to evaluate whether group selection in this specific case is quantitatively plausible. He assessed the available empirical evidence and found that it could be plausible: Parthenogenetic lineages are generally found to be evolutionary dead ends, while new parthenogenetic lineages arise rarely. Therefore, parthenogenetic lineages eventually seem to go extinct while sexual species speciate. This process would keep the ultimate number of sexual species constant, and could be counted as maintaining sex through group selection. In this case, most species would, at any given time, be sexual. Finally, Maynard Smith reviewed a variety of models that suggested possible short-term advantages for the maintenance of sex and recombination. Many of those models, however, seemed to suffer from severe shortcomings as general models for the maintenance of sex (ibid, p. 123).

In summary, Maynard Smith became interested in the problem of sex because the leading explanation for sex, the Fisher-Muller model, involved group selection mechanisms, and he rejected the idea that such mechanisms play an important role in evolution. Later, he demonstrated that sexual reproduction might in addition be extremely costly over the short term: He showed a "cost of males", in which case the maintenance of sex appears highly problematic. Maynard Smith finally established a clear framework demonstrating the issues at stake, and what kind of empirical evidence is needed to solve the problem.

2.6 Discussion

Clearly, not only the potential answers on the existence of sex have evolved, but also the question itself. Understanding sex has not been one single "persistent problem": Darwin, Weismann, Fisher and Maynard Smith had different backgrounds and as a result, were interested in the existence of sex for very different reasons (Table 2.1). Prior to developing his theory of evolution, Darwin was interested in sex because his grandfather had suggested that sex results in variable offspring, and Darwin had a general interest in the causes and consequences of variation. Later, he wanted to understand the reasons of inbreeding depression (which he believed were the cause for his own and his family's illnesses), and he thought selfing and parthenogenesis to be related phenomena. Weismann, on the other hand, initially became interested in the adaptive significance of sex because it provided an opportunity to save his theory of a constant germ line from being anti-Darwinian. Later, he wanted to unravel the meaning of the meiotic processes.

Fisher's main concern was to re-adjust the Darwinian theory according to newer observations on the immediate effects of sex – namely the occurrence of particulate (Mendelian) inheritance. Exploring the advantages of sex (the Mendelian system) itself was thus an obvious task for Fisher. Finally, Maynard Smith became interested in sex as he was skeptical of group selection, and the widely accepted Fisher-Muller theory for sex seemed to presuppose group selection. Only later, Maynard Smith became aware that sex itself could be quite costly. Williams subsequently suggested that the widespread occurrence of sex is paradoxical. In essence, Darwin, Weismann, Fisher, and initially Maynard Smith were not interested in the significance of sex due to the recognition of a paradox of sex, but rather because they explored other evolutionary themes.

The ideas on sex presented by the four biologists are fundamentally different because they deeply reflect the knowledge status of their respective times. More specifically, they crucially depended on the empirical observations and assumptions made in their times (Table 2.1). For example, Darwin operated with the assumption of blending inheritance, while Weismann argued for chromosome rather than gene-assortment during sex. It is therefore not evident that older theories can be easily used today. If done so, they need to be re-evaluated against newer observations, methods and assumptions. In the case of Weismann, a re-interpretation has often unconsciously been done (e.g. Burt 2000). After all, nobody would take Weismann's full theory for sex seriously today, including either its merging of germ plasm or its mixis of "idants".

Sex was not seen as a paradox by Darwin, Weismann or Fisher. This was partly due to what they regarded to be the opposite of sex. The paradox arises most clearly when sexual organisms are compared with parthenogenetic ones that are equally fecund, and less clearly if a sexual strategy is more generally compared to asexual reproduction, such as budding. Furthermore, an understanding of what the terms "asexuality" and "parthenogenesis" entail has changed over time, together with the changing knowledge on details concerning those phenomena (in my text, I have used the terminology consistent with the terminology used by the four scientists).

Table 2.1 Type of background assumptions made by different biologists, why they were interested in sex and their theories for sex

	Cytological knowledge of sexual processes	Form of inheritance assumed	Changes in hereditary material assumed	Reasons for interest in sexual reproduction	Theory for sex
Darwin (early)	Union of germ cells	Blending of ancestral hereditary material	?	Erasmus Darwin had proposed that sex results in variation	Production of variation; keeping species constant
Darwin (late)	Union of germ cells	Blending of ancestral hereditary material	Environmentally acquired changes	Explaining hybrid vigor and inbreeding depression	Differentiated germ cells interact better (physiologically)
Weismann (early)	Union of nuclei	Blending of ancestral hereditary material	Environmentally acquired changes (only in unicellular organisms)	Difficulty to fuse theory of constant germ line with evolutionary theory	Unique fusions of ancestral germs, resulting in variation and ultimately enabling adaptation
Weismann (late)	Reduction divisions during meiosis	Particulate inheritance of chromosomes (containing several ancestral genomes)	Environmentally acquired changes (only in unicellular organisms)	Explaining occurrence of reduction divisions	Higher degree of new combinations of chromosomes, thus better ability to adapt
Fisher	Crossing-over	Particulate inheritance of genes	Spontaneous variations of genes (mutations)	Synthesis of Mendelian (sexual) system with evolutionary theory	Sex evolved because it enables best use of mutations
Maynard Smith	Molecular basis of recombination	Molecular basis of genes	Molecular basis of different types of mutations	Refusal of group selection; later problem of sex	No single theory; establishes conceptual framework for how to address the problem of sex

For example, Darwin recognized that sex is disadvantageous because males do not produce offspring by themselves. However, since he never did experiments with parthenogenetic organisms himself, he probably was not aware of the much higher growth rate of parthenogens and thus did not realize the extent of the problem. Furthermore, he found cross-fertilization to have an immediate positive effect on the vigor of offspring. As he thought selfing and parthenogenesis to have similar effects, he took this vigor to be a sufficient advantage to counterbalance the disadvantage of producing males.

In contrast to Darwin, Weismann did extensive experiments and observations on parthenogenetic species and recognized that these species can reproduce at double speed. However, he probably did not realize how quickly sexual species would go extinct under such a scenario since he had limited mathematical skills. Instead, his knowledge of taxonomy suggested to him that parthenogenetic species eventually go extinct. Consequently, their higher growth rate solved for Weismann the problem of how parthenogenetic organisms can exist at all, rather than making him realize the serious problems concerning the existence of sexual reproduction. It is possibly the most astonishing that Fisher did not recognize a paradox of sex. He had the necessary mathematical skills, and eventually also realized that his explanation for sex seemed to presuppose group selection. However, Fisher compared sexual groups to hypothetical asexual groups. He thought asexuality to be the ancestral, "primitive", state, and that any entirely asexual groups no longer existed. Under "asexual", he understood phenomena such as growth and budding. Finally, Maynard Smith did recognize the paradox of sex due to three simultaneously occurring circumstances: First, there was an extensive debate on the levels of selection around this time. Second, he had done experiments where he had noticed the fecundity of parthenogenetic organisms, and third, he had the necessary mathematical skills to recognize the problem.

Despite the tremendous differences between the respective ideas on sex developed by the four biologists presented here as well as their initial interest in sex, there is an important common theme: In all four cases, thinking about the significance of sexual reproduction helped exploring and understanding wider issues in evolutionary biology. In Darwin's case, it even resulted in the development of the evolutionary theory itself. This is perhaps not surprising since all four biologists considered sex to influence variability, even though the understanding of exactly *how* sex influences this variability differed between them. Already Darwin had established variability to be one of the major ingredients of evolutionary theory. Therefore, it is not surprising that "understanding the evolution of sex requires the synthesis of every important process in evolutionary biology" (Otto and Lenormand 2002). Acknowledging that our understanding of sex has so many aspects allows us to recognize why the question of sex is generally deemed so important in evolutionary biology – and why it simultaneously is so complex.

Acknowledgements I thank Patrick Meirmans for his many valuable comments to the manuscript, as well as Richard Burian, Lawrence Kirkendall and Curt Lively for interesting discussions on the history of the research on sex. The manuscript improved further through the constructive criticism by two anonymous reviewers. I am also indebted to the resources providing online access to the books, articles and letters by Fisher (http://digital.library.adelaide.edu.au/coll/special/fisher/), Darwin (http://darwin-online.org.uk/; http://www.darwinproject.ac.uk/) and Weismann (http://www.esp.org/). Much of this research has been done with support from the Norwegian Research Council.

References

Barton NH, Charlesworth B (1998) Why sex and recombination? Science 281: 1986–1990
Bell G (1982) The masterpiece of nature: the evolution and genetics of sexuality. University of California Press, Berkeley
Bennett JH (1983) Introduction. In: Bennett JH (ed) Fisher RA. Natural selection, heredity and eugenics. Oxford University Press, Oxford, pp. 1–50
Burt A (2000) Sex, recombination, and the efficacy of selection – was Weismann right? Evolution 54: 337–351
Charlesworth B (2004) John Maynard Smith: January 6, 1920–April 19, 2004. Genetics 168: 1105–1109
Conklin EG (1915) August Weismann. Proc Am Philos Soc 54: iii–xii
Crow JF, Kimura M (1965) Evolution in sexual and asexual populations. Am Nat 99: 439–450
Darwin CR (1859) On the origin of species by means of natural selection, or the preservation of favoured races in the struggle for life. John Murray, London
Darwin CR (1862a) On the various contrivances by which British and foreign orchids are fertilised by insects, and on the good effects of intercrossing. John Murray, London
Darwin CR (1862b) On the two forms, or dimorphic condition, in the species of Primula, and on their remarkable sexual relations. J Proc Linn Soc Lond (Botany) 6: 77–96
Darwin CR (1875) The variation of animals and plants under domestication, vol II. 2nd edn. John Murray, London
Darwin CR (1876) The effects of cross and self fertilisation in the vegetable kingdom. John Murray, London
Darwin CR (1877) On the various contrivances by which British and foreign orchids are fertilised by insects, and on the good effects of intercrossing, 2nd edn. John Murray, London
Darwin CR (1909) Essay of 1842. In: Darwin F (ed) The foundations of the origin of species. Two essays written in 1842 and 1844. Cambridge University Press, Cambridge, UK, pp. 1–50
Darwin CR (1958) The autobiography. In Barlow N (ed) The autobiography of Charles Darwin 1809–1882. With the original omissions restored. Edited and with appendix and notes by his grand-daughter Nora Barlow. Collins, London, pp. 17–145
Darwin CR (1960) First notebook [B] (July 1837–February 1838). In de Beer G (ed) Darwin's notebooks on transmutation of species. Part I. Bull British Mus (Nat Hist). Histor Ser 2(2): 23–73, pp. 42–73
Darwin E (1796) Zoonomia; or, the laws of organic life, vol I. Johnson, London
Fisher RA (1922) Darwinian evolution by mutations. Eugenics Rev 14: 31–34
Fisher RA (1930) The genetical theory of natural selection. Oxford University Press, Oxford
Fisher RA (1999) The genetical theory of natural selection. Complete variorum edition. Oxford University Press, Oxford
Gaupp E (1917) August Weismann – Sein Leben und sein Werk. Verlag Gustav Fischer, Jena
Ghiselin MT (1969) The triumph of the Darwinian method. University of California Press, Berkley
Ghiselin MT (1988) The evolution of sex: a history of competing points of view. In: Michod RE, Levin BR (eds) The evolution of sex. An examination of current ideas. Sinauer Associates, Sunderland, pp. 7–23
Lewontin RC (1974) The genetic basis of evolutionary change. Columbia University Press, New York

Mahalanobis PC (1964) Professor Ronald Aylmer Fisher. Biometrics 20: 238–252
Maynard Smith J (1958) The effects of temperature and of egg-laying on the longevity of *Drosophila subsobscura*. J Exp Biol 35: 832–842
Maynard Smith J (1964) Group selection and kin selection. Nature 201: 1145–1147
Maynard Smith J (1968) Evolution in sexual and asexual populations. Am Nat 102: 469–473
Maynard Smith J (1971) The origin and maintenance of sex. In: Williams GC (ed) Group selection. Aldine-Atherton, Chicago, pp. 163–175
Maynard Smith J (1976) Commentary: Group selection. Quart Rev Biol 51: 277–283
Maynard Smith J (1978) The evolution of sex. Cambridge University Press, Cambridge, UK
Mooney SM (1993) The evolution of sex: variation and rejuvenescence in the 19th century and today. BioScience 43: 110–113
Mooney SM (1995) H. J. Muller and R. A. Fisher on the evolutionary significance of sex. J Hist Biol 28: 133–149
Moore J (2005) Good breeding: Darwin doubted his own family's "fitness". Nat Hist 114: 45–46
Muller HJ (1932) Some genetic aspects of sex. Am Nat 66: 118–138
Nee S (2004) Professor John Maynard Smith 1920–2004. Trends Ecol Evol 19: 345–346
Otto SP, Lenormand T (2002) Resolving the paradox of sex and recombination. Nature Rev Genet 3: 252–261
Petrunkevitch A (1964) August Weismann. Syst Zool 13: 215–219
Richmond ML (2006) The 'domestication' of heredity: the familial organization of geneticists at Cambridge University, 1895–1910. J Hist Biol 39: 565–605
Szathmary E, Hammerstein P (2004) Obituary – John Maynard Smith (1920–2004). Nature 429: 258–259
Weismann A (1889) The significance of sexual reproduction in the theory of natural selection. In: Poulton EB, Schönland S, Shipley AE (eds) Essays upon heredity and kindred biological problems, vol I. Clarendon Press, Oxford, pp. 255–332
Weismann A (1892) Amphimixis or the essential meaning of conjugation and sexual reproduction. In: Poulton EB, Schönland S, Shipley AE (eds) Essays upon heredity and kindred biological problems, vol II. Clarendon Press, Oxford, pp. 101–222
Weismann A (1893) The germ plasm. A theory of heredity. Charles Scribner's Sons, New York
Williams GC (1966) Adaption and natural selection: a critique of some current evolutionary thought. Princeton University Press, Princeton
Williams GC (1975) Sex and evolution. Princeton University Press, Princeton
Winther R (2001) August Weismann on germ-plasm variation. J Hist Biol 34: 517–555
Wynne-Edwards VC (1962) Animal dispersion in relation to social behavior. Oliver and Boyd, Edinburgh

Chapter 3
Apomixis: Basics for Non-botanists

Peter Van Dijk

Abstract The evolutionary questions studied in apomictic plants and parthenogenetic animals are often the same. This chapter gives a basic introduction to apomixis in flowering plants, in order to make the botanical apomixis literature more accessible to non-specialists. The focus is on the differences and similarities with parthenogenetic animals. The following topics are briefly discussed: 1. apomixis should not include vegetative reproduction, 2. apomixis is a modification of sexual reproduction 3. different mechanisms of apomixis, 4. the role of endosperm development 5. causes of apomixis 6. male function in apomicts 7. intra-clonal variation 8. the phylogenetic distribution of apomixis and 9. constraints in the evolution of apomixis. At the end of the chapter, suggestions for further reading are given.

3.1 Introduction

Evolutionary problems studied in apomictic plants and parthenogenetic animals are often similar. For instance, geographic parthenogenesis is common in both groups (see Chapter 8). It would therefore be very valuable if botanical literature would be easily accessible to non-botanists and *visa versa*. Unfortunately the research field of apomixis in plants is not easy to comprehend for non-botanists. In the first place, this is because apomixis is a modification of sexual reproduction which in itself differs significantly between flowering plants and animals. Therefore, even the basic terminology for asexual reproduction in flowering plants is different from that of animals. For example, in animals, apomixis is a form of parthenogenesis, whereas parthenogenesis is an element of apomixis in plants (see Fig. 3.1). This is because, in contrast to animals, the egg cell is not a direct product of meiosis in plants. Furthermore, the fact that there are several ways how apomictic plants avoid sexual reproduction, adds to the complexity. Finally, the botanical apomixis terminology is often

P. Van Dijk (✉)
Keygene N.V., Agro Business Park 90, 6708 PW Wageningen, The Netherlands
e-mail: peter.van-dijk@keygene.com

I. Schön et al. (eds.), *Lost Sex*, DOI 10.1007/978-90-481-2770-2_3,

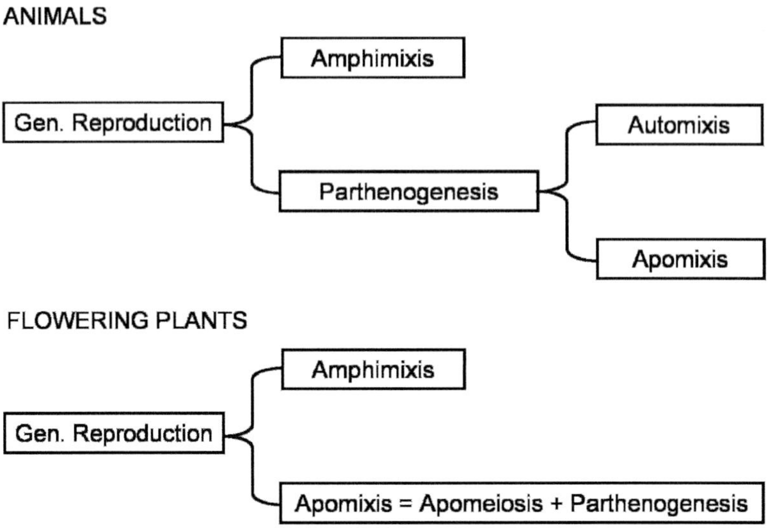

Fig. 3.1 Different uses of the terms apomixis and parthenogenesis in animals and in flowering plants. In animals, apomixis is a type of parthenogenesis, whereas in flowering plants, parthenogenesis is a component of apomixis

redundant and many terms are used for rare exceptions (Nogler 1984b). It is the aim of this chapter to make the field of apomixis in plants more understandable to non-botanists.

With apomixis in plants, I shall specifically mean asexual reproduction through seeds, which has been reported in more than 400 flowering plant taxa (Bicknell and Koltunow 2004). I will restrict this chapter to the angiosperms and not touch upon apomixis in conifers (gymnosperms) and ferns. For these plant groups, the reader may consult Mogie (1992).

Lower animals may have more in common with plants than with higher animals. Apomictic plants are often hermaphrodites, as are parthenogenetic flatworms (see also Chapter 18) and earthworms. The cost of sex in hermaphrodites is a cost of meiosis, which can be up to 1.5, rather than a cost of males in species with two separate sexes, which can be up to 2. Triploidy is common in parthenogenetic animals such as *Potamopyrgus* (fresh water snails, Lively 1987), *Dugesia* (flatworms, Beukeboom et al. 1996), *Poeciliopsis* (gynogenic fish, Schultz 1967; see also Chapter 19) and apomictic plants like *Taraxacum* (dandelions, Van Dijk 2003; see also Chapter 22), *Boechera* (a close relative of *Arabidopsis* Naumova et al. 2001; see also Chapter 23), and *Chondrilla* (skeleton weed, Chaboudez 1994). Apomictic plants are nearly always polyploid, as is the case with many parthenogenetic animals.

Polyploidy, however, also marks a major difference between plants and animals. Whereas most polyploid animals are parthenogenetic, only a small fraction of the polyploid plants are apomictic. In fact, more than 99% of the polyploid plants are sexual. Another important difference is that in most flowering plants there is a double fertilization: besides the egg cell, also the central cell has to be fertilized for the development of the endosperm, a temporary tissue, which surrounds and nourishes the embryo. The endosperm acts like the placenta in mammals and several genes expressed in the endosperm are differently imprinted, depending on whether they are of maternal or paternal origin. Later on, it will be argued that parental genome imprinting may constrain the evolution of apomixis in flowering plants.

The first part of this chapter concerns the mechanisms and causes of apomixis. In the second part, some characteristics of plants will be discussed, which may have special relevance for the evolution of asexuality. However, I will start with making clear what I do not consider to be apomixis.

3.2 Vegetative Reproduction in Plants is Not Apomixis

Some authors include vegetative reproduction in the definition of apomixis. For example, Richards (2003) states that 60% of the British flora is capable of apomictic reproduction. In an article in the same special issue of *Transactions of the Royal Society*, I refer to Mogie (1992), describing that about 0.1% of all flowering plants are apomictic. The special term "agamospermy" is used to describe asexual reproduction by seed. Derived from agamospermy, the term "agamic complex" (Stebbins 1950) is used for species with sexual populations and apomictic clones.

Animals have determinate growth, but plants, with the exception of annual and biennial plants, have a characteristic indeterminate growth. A plant has many meristems, tissues with undifferentiated cells (stem cells), which can undergo mitosis and which can grow and differentiate. Many plant species have specialized structures to produce vegetative daughter plants, such as rhizomes (underground horizontal stem) e.g. sand sedge (*Carex arenaria*) and hop (*Humulus* sp.), stolons (above ground horizontal stem) e.g. strawberry (*Fragaria* sp.), white clover (*Trifolium repens*) and creeping buttercup (*Ranunculusrepens*), tubers e.g. potato (*Solanum tuberosum*) and Dahlias) and bulbils (small bulbs that develop above ground) e.g. wild onion (*Allium vineale*), lesser celandine (*Ranunculus ficaria*) and alpine bistort (*Polygonum viviparum*). Duckweeds (*Lemna* sp.) reproduce vegetatively by budding off from the mother disc-like plant. Probably, all perennial plants can reproduce vegetatively. If we include vegetative reproduction in apomixis, the term becomes almost meaningless.

Although the maternal genotype is maintained with vegetative reproduction, there are good reasons not to include vegetative reproduction in apomixis. Meristems are multicellular, and therefore mutations result in chimeric tissues. Apomicts *sensu stricto* go through a single cell stage, being equivalent to a zygote, which increases the chances of establishment of mutations and restricts the possibility for transmission of viruses to the progeny. Vegetative daughter plants generally establish themselves close to the mother plant whereas seeds are often efficiently dispersed. Seeds are a dry phase, related to dormancy and survival of adverse conditions. Vegetative reproduction in general does not involve dormancy, but establishment probabilities can be much higher in dense competitive vegetations, because the daughter plants are larger and often for some time connected and supported in nutrition by the mother plant. For all these reasons, vegetative reproduction should not be included in apomixis. Fortunately, most authors and textbooks nowadays use apomixis in a strict sense as asexual reproduction through seeds.

3.3 Apomixis Is a Modification of Sexual Reproduction

Apomixis *s.s.* in flowering plants is a modification of sexual seed development. It is therefore imperative to explain first normal sexual seed development. The life cycle of flowering plants is an alteration of a sporophytic phase and a gametophytic phase (Fig. 3.2 left). The phases are also confusingly called "generations". Whereas in animals, the haploid egg cell is one of the four products of female meiosis, in plants, one of the four products (spores) of female meiosis gives rise to the haploid gametophyte generation (also called the embryosac), which produces an egg cell after three mitotic divisions. In both animals and plants, the other three meiotic spores degenerate. Although the gametophyte in flowering plants is much reduced compared to ferns and mosses, meiosis and fertilization in flowering plants are processes that are clearly separated in time and by cell type.

Because of the life cycle phases, botanists have adopted a strict terminology to describe the products of meiosis and fertilization (and of apo-meiosis and parthenogenesis). "2n" is the sporophytic chromosome number, which is reduced by meiosis

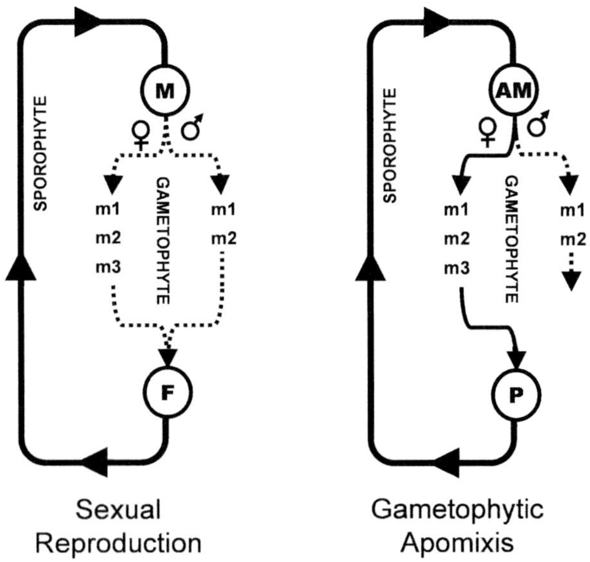

Fig. 3.2 The flowering plant life cycle and sexual reproduction (*left*) versus gametophytic apomixis (*right*). The *solid line* represents the sporophytic generation (2*n*), the *dashed line* the reduced (*n*) male and female gametophytic generation. M = Meiosis; AM = apomeiosis; m1, m2 and m3 = successive mitotic divisions; F = fertilization; P = parthenogenesis. The female gametophyte is reduced (*n*) in sexual reproduction but unreduced (2*n*) in apomixis

to "n" in the gametophyte. Botanists distinguish the gametophytic chromosome number "n" from "x" which is the basic chromosome number of the species. Thus, "x" is not necessarily equal to "n". For example, n equals x in diploids (2n =2x), but n is 2x in tetraploids (2n = 4x). The gametophyte produces the gametes (n), which after fertilization (n + n), form a new sporophyte (2n).

Returning to female meiosis – one spore survives and undergoes three successive mitotic divisions to produce the gametophyte. The gametophyte consists in most species of eight nuclei and seven cells – one cell is binuclear (fused nuclei: n + n), the central cell. One of the other cells is the egg cell. In sexual flowering plants, there is a so-called double fertilization: the pollen grain tube contains two nuclei (both n), of which one fertilizes the egg cell (n + n = 2n) and the other fertilizes the central cell of which the two nuclei have fused ((n + n) + n = 3*n*). The fertilized egg cell forms the embryo and the fertilized central cell forms the endosperm. Below it will be explained that it is important that the maternal to paternal genome ratio in the endosperm is 2:1. Together with the maternal seed coat the embryo and the endosperm make the seed.

3.4 Types of Apomixis in Flowering Plants

Apomixis *s.s.* can be divided into two main types 1. gametophytic apomixis and 2. sporophytic apomixis (Fig. 3.3). In gametophytic apomixis, a gametophytic phase in the development of the embryos takes place (Fig. 3.2 right). Gametophytic apomixis can be further subdivided into 1. diplospory and 2. apospory. In the case of diplospory, a normal reductional meiosis is replaced by a non-reductional division, which can be a mitotic-like division (mitotic diplospory, e.g. some hawkweed species – *Hieracium* spp., some catsfeet/pussytoes – *Antennaria* sp. or a first

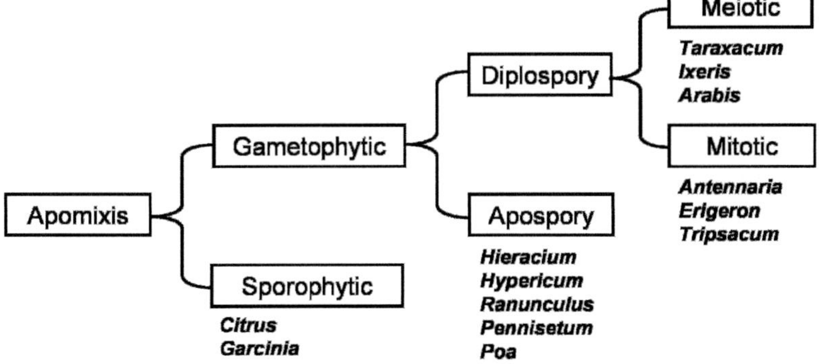

Fig. 3.3 Classification of apomixis types in flowering plants. In the case of sporophytic apomixis and apospory, there is a parallel sexual developmental pathway in the same ovule, in the case of diplospory, the sexual pathway is modified. Examples of apomictic genera are given

division restitution (meiotic diplospory – e.g. dandelion sp. *Taraxacum* sp., eastern gama grass (*Tripsacum deltoides*; Nogler 1984b). In both types of diplospory, two unreduced megaspores (2*n*) are produced, of which one degenerates and the other develops into an unreduced gametophyte with an unreduced egg cell. In some onion (*Allium*) species, unreduced megaspores are produced by a pre-meiotic endomitotic doubling of the somatic chromosome number followed by a normal meiosis with preferentially pairing of doubled chromosomes. In all cases, the unreduced egg cells develop through parthenogenesis into an embryo (2*n*+0; terminology of Harlan and De Wet 1975) that is genetically identical to the mother plant.

In the case of apospory within the ovule, in addition to the normal reduced megagametophyte (n), a second but unreduced (2n) megagametophyte is formed from a non-spore cell (aposporous initial). Within a single ovule, there is competition between the two developing gametophytes. The 2n aposporous gametophyte forms an unreduced egg cell, which develops parthenogenetically into an embryo being genetically identical to the mother plant. Examples of aposporous species are: some hawkweeds – *Pilosella* spp., St John's wort – *Hypericum perforatum* and Kentucky blue grass – *Poa pratensis*. In most cases, the reduced, fertilization-dependent gametophyte is suppressed and an apomictic seed is formed. A low percentage of egg cells (n and 2n), however, can be fertilized producing low frequencies of n + n and 2n + n progeny (also called B_{II} and B_{III} hybrids, according to the terminology of Rutishauser 1948). Especially in aposporous plants, facultative apomixis is common, although the frequencies of sexual offspring plants are generally only a few per cent (see Table 3.1).

The avoidance of meiosis in gametophytic apomicts, either by diplospory or by apospory is often referred to as apomeiosis. Gametophytic apomixis is strongly correlated with polyploidy. Nearly all gametophytic apomicts are polyploid. A classical example of diploid apomixis, *Potentilla agentatum*, was shown to a case of selfing by the use of genetic markers (Holm et al. 1997). Nevertheless, *Boechera holboelli*, a close relative of *Arabidopsis*, is a proven diploid apomict (Naumova et al. 2001; Kantama et al. 2007).

Table 3.1 Four types of offspring that can arise from apomictic plants, because of a residual sexual function, or because of uncoupling of apomeiosis and parthenogenesis

	Parthenogenesis	Fertilization
Meiosis	n + 0 *haploid parthenogenesis* 0.94%	n + n *B_{II} hybrids* 1.91%
Apomeiosis	2n + 0 *Apomixis* 97.05%	2n + n *B_{III} hybrids* 0.10%

Terminology according to Harlan and De Wet (1975) and Rutishauser (1948). The percentages are the actual rates measured in a clone of *Hieracium piloselloides* by Bicknell et al. (2003).

In the other main type of apomixis, sporophytic apomixis, somatic embryos are formed within the sporophytic tissue that surrounds the gametophyte. In contrast to apospory, these cells do not enter a gametophytic phase (no egg cell thus is formed), but remain sporophytically and produce an embryo directly (somatic embryo). Because the surrounding tissue is called nucellus, this type of apomixis is also referred to in the literature as nucellar embryony. Because additional somatic embryos are formed next to the sexual embryo, the term adventitious embryony is also used. In sporophytic apomixis, no parthenogenesis is involved. It occurs in for example *Citrus* and Orchid species (Asker and Jerling 1992; Naumova 1993). In contrast to gametophytic apomixis, there is no association with the ploidy level. In general, sporophytic apomixis is less well investigated than gametophytic apomixis and less is known about the biological basis of genetics and development. Since this is additional rather than alternative to sexual reproduction, it can be argued that this is not a form of apomixis at all. The remainder of this chapter therefore deals with gametophytic apomixis only.

Through the mechanisms outlined above, maternal heterozygosity is maintained, although crossing-over in meiotic diplospory may be possible, causing homozygosity at the chromosomal regions distal of the cross over, which may cause a more than 2-fold cost of sex (Archetti 2004). Automixis, which is a form of parthenogenesis in animals (see also Chapters 4 and 12), does not exist in plants. A second meiotic division restitution, which causes genomic homozygosity similar to terminal fusion in animal parthenogenesis, has been found in plants, but not in combination with parthenogenesis. In contrast, autogamy or selfing, fusion of egg cells and pollen grains produced by the same individual but products from different meioses, is very common in flowering plants. In extreme cases, the selfing occurs in the bud stage of the flower, preventing any cross-fertilization (cleistogamy).

3.5 Pseudogamy and Autonomous Endosperm Development

Above, the double fertilization in flowering plants was described. The function of the endosperm, which arises from the fertilized central cell, is to nourish the developing embryo. The endosperm is temporarily; in some species it is already gone when the seeds are ripe, and in others, it is still present, providing nutrition for the germinating seedling. Also in species, in which the endosperm is very transient, it is crucial for the development of the embryo.

In most apomictic plants, the endosperm develops only after fertilization of the central cell. This is called pseudogamy or pseudogamous apomixis. In animals, the terms pseudogamy and its synonym, gynogenesis, are used for the triggering of the development of the inactivated egg by sperm, without any genetic contribution (see for examples Chapters 16, 19 and 20). It is possible that in plants in addition to sexual endosperm development also triggering the egg cell plays a role. Crossing studies have shown that for proper endosperm development, a 2:1 maternal – paternal genome ratio is essential. Otherwise, the endosperm will collapse, and the embryo will starve, resulting in seed abortion. Haig and Westoby

(1989, 1991) have suggested that this could be due to parent specific gene expressions, which have evolved because of a parental conflict during sexual reproduction. There is evidence for parental imprinting of several genes in the endosperm (Huh et al. 2007). This is comparable to parental imprinting of genes in mammalian embryos (see also Chapter 26). In contrast to the endosperm, there is little evidence for parental imprinting of genes in flowering plant embryos (Gehring et al. 2004). Parental gene imprinting in the endosperm may have constrained the evolution of autonomous endosperm development in most apomicts, although some apomictic species (notably belonging to the Asteraceae family) have developed autonomous apomixis (development of the endosperm without fertilization of the central cell). However, parental imprinting of the endosperm also creates problems for pseudogamy, since pollen grains have a reduced ploidy level compared to egg cells, because apomeiosis mechanisms are specific for the female sex function. Apomictic plant species therefore often have also developed modifications of the double fertilization to maintain the 2m: 1 p ratio, such as no fusion of central cell nuclei, unreduced pollen grains or a 4 nucleated embryo sac.

3.6 Phenotyping Apomixis

Demonstrating pseudogamous apomixis can be tricky because fertilization is necessary for the development of the endosperm. Morphological or molecular markers are needed to prove apomixis beyond doubt. Apomixis can also be demonstrated microscopically by the development of embryos without fertilization. Nomarski Differential Interference Contrast microscopy of methylsalicylate cleared ovules avoids the difficult and laborious sectioning and staining of embedded tissues.

In the case of autonomous apomixis, seed set after removal of the style and anthers (decapitation of the top part of flower buds) are proof for apomixis. High seed set in triploids is a strong indication for apomixis (e.g. *Taraxacum*, *Chondrilla*, Van Dijk 2003), because aneuploid gametes would cause high sterility in case of sexual reproduction. Sometimes, the correlation with apomixis is so strong that triploidy can be used as an indicator for apomixis. Pollen of triploid apomictic dandelions is highly irregular of size. Therefore, it is possible to infer the ploidy level from herbarium material, and indirectly also the breeding systems. This makes it possible to study the distribution of sexuals and apomicts in great detail (Roetman et al. 1988, for *Taraxacum*).

3.7 Facultative Apomixis

Apomictic flowering plants often produce a mixture of different progeny types, because apomeiosis and parthenogenesis can be uncoupled and suppression of sexual reproduction is often not absolute. Therefore, the classifications of Rutishauser

(1948) and Harlan and De Wet (1975) are very useful (see Table 3.1). Bicknell et al. (2003) quantified the frequency of the four different types of offspring in two clones of *Hieracium* species using positive and negative selectable transgenic markers, showing that the frequency of fertilization was 2.0 and 2.4%, respectively. In Table 3.1, the frequencies of all 4 possible progeny classes in the *H.piloselloides* clone are indicated. Facultative apomixis can clearly contribute to clonal diversity commonly observed in natural population of apomictic flowering plants. Matzk et al. (2000) developed a Flow Cytometric Seed Screening method, which allowed the fast reconstruction of the reproductive developmental pathway of individual seeds by the combination of the embryo and the endosperm ploidy levels in dry seeds. They investigated 21 different apomictic flowering plant species and detected facultative apomixis in 17 of them.

3.8 Causes of Apomixis

The genetic basis of apomixis has long been very enigmatic. Experimental research on apomixis in plants dates back to Mendel's crosses in *Hieracium*. Mendel was not aware of the fact that *Hieracium* was facultative apomictic, and the results were very confusing to him. Nevertheless, Mendel was convinced that there was some kind of a general law underlying these results (Nogler 2006). Improved phenotyping methods and molecular markers have significantly contributed to the understanding of the genetic control of apomixis. Mapping of gamma irradiation induced deletions strongly suggests that apomixis in *Hieracium* is controlled by two dominant loci, one controlling apospory and one controlling parthenogenesis (Catanach et al. 2006). Clonal reproduction makes apomicts very suitable to deletion mapping. Because most apomictic plants are hermaphrodites, it is also possible to use them as pollen donors in crosses with sexuals. This way, independent dominant loci have also been shown to control apomixis in two other Compositae apomicts; *Taraxacum* (Van Dijk et al., Chapter 22) and *Erigeron* (Noyes et al. 2007). In contrast, crosses in grasses suggest that apomixis is often controlled by a single dominant locus (e.g. *Panicum maximum,* Savidan 1980; *Pennisetum* sp., Dujardin and Hanna 1989). In all cases described here, the dominant apomixis allele is present in a single dose, suppressing sexual reproduction. Therefore, loss of this dominant allele could result in a reversal to sexuality as long as the recessive sexuality alleles are not yet corrupted by mutations since purifying selection was relaxed during apomictic reproduction. Apomixis Specific Chromosomal Regions (ASCRs) are often characterized by suppression of recombination with many co-segregating markers (reviewed in Ozias-Akins and Van Dijk 2007). It is therefore possible that apomixis loci have a complex architecture and could contain several genes. Some research groups have initiated the map-based cloning of ASCRs. Recently, the first sequences of ASCRs in apomictic grasses have been published (Calderini et al. 2006; Conner et al. 2008). However, the suppression of recombination severely hinders the positional cloning of the apomixis genes themselves.

Crossing diploid sexual and polyploid apomictic *Ranunculus auricomus* plants, Nogler (1984a) discovered that the dominant apomixis factor could only be transmitted through diploid pollen grains, and not through haploid pollen grains. Using dominant genetic markers, Noyes and Rieseberg (2000) showed that this was also the case in *Erigeron annuus*. In Chapter 22, Van Dijk et al. describe the same phenomenon in apomictic *Taraxacum*. Non-transmission of apomixis genes may be a general explanation for the scarcity of diploid apomictic flowering plants. Van Dijk et al. argue that this could be due to a genetic load linked to apomixis genes, accumulated over many previous clonal generations.

There are alternative hypotheses for the control of apomixis. Already in 1918, Ernst suggested that apomixis would be caused by hybridization and polyploidization. However artificial hybridization and polyploidization have not resulted in *de novo* apomixis. Carman (1997) proposed that apomixis might be caused by asynchronous expression of duplicate genes. Hybridization between ecotypes with differently timed developmental programs could lead to the simultaneous execution of developmental steps, which normally would be executed in succession. Polyploidization would then stabilize the hybrids. Although intuitively appealing, so far no convincing evidence has been provided to indicate that that Carman's hypothesis is correct (But see Carman 2007).

3.9 Most Apomictic Plants Are Hermaphrodites

Apomictic plants produce asexual seeds, but most have a sexual male function. Therefore, they can act as pollen donors in crosses with related sexuals or with facultative apomicts and transmit apomixis genes to the offspring. This will generate large arrays of clones, consistent with high clonal diversity commonly found in populations of apomictic flowering plants (e.g. in *Chondrilla juncea* Chaboudez (1994), *Taraxacum officinale s.l.* Van der Hulst et al. (2000), *Ranunculus auricomus s.l.* (Paun and Hörandl 2006), *Pilosella officinarum* Chapman and Brown (2001); however, see LeRoux et al. (2007) for an example of a pan-global single apomictic clone in *Pennisetum setaceum*. Genetically controlled apomixis can thus be infectious (or contagious), as long as the male sexual function remains functional (see Chapter 22, Van Dijk et al.). Lower animals are also often hermaphroditic, for example nematodes, flatworms and earthworms. In such species, apomixis may also be infectious. In some animals like *Daphnia* (see Chapter 15), males can be induced in asexual clones by certain environmental triggers, thus allowing the spread of parthenogenesis genes. Contagious asexuality has also been documented in *Daphnia pulex* in North America (Paland et al. 2005).

3.10 Somatic Mutations

In clonal lineages lacking recombination, genetic variation can only be generated by mutation. In animals, the germline is set apart from the soma at an early embryological stage. Therefore, somatic mutations will not become incorporated in the germline and will not contribute to intra-clonal variation (see also Chapter 9). In contrast, in plants, the gametes are produced very late from somatic cells and often continuously, strongly increasing the possibilities for incorporation of somatic mutations into the gametes. Somatic mutations have been found in apomictic dandelion lineages for rDNA intergenic spacers and the *Adh1* gene (King and Schaal 1990), and for Amplified Fragment Length Polymorphisms (AFLPs), isozymes and microsatellites (Mes et al. 2002). Somatic microsatellite mutations within asexual lineages have also been reported in apomictic lineages of *Ranunculus auricomus s.l.* (Paun and Hörandl 2006).

3.11 The Phylogenetic Distribution of Apomixis in the Flowering Plants

The first angiosperms arose in the Early Jurassic (190 million years ago) to Early Cretaceous (140 mya) (Sanderson and Doyle 2001). It is thought that the sexual flower reproductive biology with specialized pollination systems highly contributed to their present-day diversity. To date, apospory has been reported in 91 genera and diplospory in 51 genera. Assuming that these different forms of apomixis have no common ancestry, this suggests that gametophytic apomixis evolved at least 140 times during the history of the angiosperms (Van Dijk and Vijverberg 2005).

Almost 70% of all flowering plant genera with apomixis belong to three families: the true grasses (Poaceae), the daisy family (Asteraceae) and the rose family (Rosaceae). The first two belong to the largest flowering plant families and the high frequency of apomixis may be less remarkable than often thought. The Rosaceae, however, is a rather small family, and the high frequency of apomixis is certainly noteworthy. Equally remarkable is the absence of apomixis in the large legume family (Leguminosae). Within the Poaceae and the Asteraceae there is a clear clustering of apomixis at the (sub)-tribal level, suggesting pre-adaptation for apomixis or common-ancestry of apomixis (Van Dijk and Vijverberg 2005).

3.12 Constraints on the Evolution of Apomixis in Flowering Plants

Engelstädter (2008) recently reviewed the constraints on the evolution of asexual reproduction in animals. As wide-spread genetic constraints, he listed: 1. the need of egg activation 2. paternal centriole inheritance 3. inbreeding depression in the case of automixis and 4. sex determination. Parental genome imprinting is a constraint specific for mammals (see also Chapter 26) and dry-resistant sexual eggs for aphids (Moran 1992) and *Daphnia* (Hebert 1987) In flowering plants, endosperm development puts a similar constraint on apomixis as egg activation in animals, which in both groups is solved by pseudogamy. In animals, centrosomes with centrioles are essential for cell division. Centrosomes are the main micro-tubule organizing centres and play a role in cell cycle progression. Centrosomes are paternally inherited and this may be a constraint in the evolution of all female reproduction (but see Chapter 16). Plants do not have centrioles and micro-tubule assembly is organized in a different manner. Therefore, this does not pose a constraint on the evolution of apomixis. Cytoplasmic organelles (chloroplasts and mitochondria) are generally maternally inherited in flowering plants. However, in gymnosperms, like the conifers, chloroplasts are paternally inherited through pollen. Apomixis is rare in the gymnosperms and this could be a consequence of non-maternal chloroplast inheritance. A rare case of gymnosperm apomixis is found in *Cupressus dupreziana*. Interestingly, this is a form of rare paternal apomixis: the apomictic embryos are derived from unreduced pollen grains and are thus genetically identical to the pollen donor (Pichot et al. 2001). Automixis does not occur in flowering plants and thus inbreeding is not a constraint in plants. The great majority of plants are hermaphrodites hence apomixis generally does not interfere sex-determination. However in dioecious plants apomixis does occur (genus *Antennaria*; Bayer and Chandler 2007). Whereas parental imprinting is a specific constraint in mammals (see also Chapter 26), parental imprinting of endosperm genes is more or less a universal constraint in plants resulting into pseudogamous apomixis, although in some taxonomical groups, autonomous endosperm development evolved. Sexual and apomictic seeds in plants are ecologically equivalent and there seem to be no intrinsic ecological constraints in the evolution of apomixis in plants, comparable to resting eggs in water fleas.

However, a major constraint in the evolution of apomixis in plants is probably that several independent mutations, at least one for apomeiosis and one for parthenogenesis, are essential for apomixis. Separately, these mutations will have a negative fitness effect, causing the rapid elimination of the mutation. A mutation for apomeiosis alone generates unreduced egg cells which need fertilization in order to develop into an embryo. However, this embryo will have an elevated ploidy level. Thus, a diploid mutant will produce triploid offspring, which in turn will produce tetraploid offspring, and so on. The continuous increase of ploidy levels will sooner or later lead to inviability. An apomeiotic lineage will thus "polyploidize itself out of existence" (Stebbins 1950). Similarly, a separate parthenogenesis mutant will have a low fitness since its haploid offspring will be highly sterile. Only if both mutations occur simultaneously in a population and are combined by crossing, an apomixis genotype with high fitness will be formed. Given strong selection against the individual mutations, the chances that both mutations fitness will occur in the same population are low.

3.13 Ancient Apomicts in Flowering Plants

With perhaps the exception of the genus *Houttuynia*, all apomictic plants have close sexual relatives. It is often possible to use apomicts as pollen donors in crosses with sexual relatives. This suggests that apomicts in flowering plants are relatively recent. Ancient asexual animals, such as the bdelloid rotifers and *Darwinulid* ostracods (see also Chapters 11 and 13), have not been reported in flowering plants. Stebbins

(1950) mentions the small genus *Houttuynia* – with one species, *Houttuynia cordata*, a well known garden plant and used as vegetable in SE Asia – as the only possibly completely apomictic genus. Recent studies have shown high levels of genetic diversity and ploidy level variation in Chinese populations (Wu et al. 2005). Although this is not necessarily in conflict with apomixis (chromosome and gene mutations generating intra-clonal variation), studies on the occurrence of apomixis in this species were conducted in the early 1930's and need to be extended.

3.14 Further Reading

This book chapter only aims to give a brief introduction to the field of apomixis in flowering plants. For further reading, a number of books and articles are recommended. Nogler (1984b) gives a very clear classification of the types of apomixis which has helped many newcomers, not at least because it makes a distinction between essential and redundant terms. Good general books on apomixis are Asker and Jerling (1992): *Apomixis in Plants* and Mogie (1992): *Evolution of Asexual Reproduction in Plants*, although these books are now somewhat outdated, especially the parts concerning the genetics and developmental biology of apomixis. Useful new additions are Savidan et al. (2001): *The Flowering of Apomixis: From mechanisms to genetic engineering* and Hörandl et al. (2007): *Apomixis: evolution, mechanisms and perspectives.* A good review book about sporophytic apomixis is Naumova (1993): *Apomixis in Angiosperms. Nucellar and Integumentary Embryony*. The developmental biology of apomixis is treated in Koltunow and Grossniklaus (2003) while Ozias-Akins and Van Dijk (2007) review the genetics of apomixis. Savidan (2000) and Spillane et al. (2004) consider apomixis from a genetical and an agricultural perspective. Whitton et al. (2008) discuss the evolution of apomixis in the angiosperms and Hörandl (2006) describes the complex causes of geographic parthenogenesis in plants. Fine accounts of the history of apomixis research are given by Nogler (2006) dealing with Mendel's experiments on *Hieracium* and Nogler (2007) about the discovery of parthenogenesis in plants. Finally, a race for the filing of a patent application on an apomixis gene forms the plot in a botanical thriller: *Day of the Dandelion* (Pringle 2007).

Glossary

Adventitious embryony: The formation of somatic next to sexual embryos.
Agamospermy: asexual reproduction by seed.
Apomixis: (in plants):asexual reproduction through seeds.
Apospory: In addition to the normal reduced megagametophyte (*n*), a second but unreduced (2*n*) megagametophyte is formed from a non-spore cell (aposporous initial).
Autogamy: Also called selfing. The fusion of egg cells and pollen grains produced by the same individual.

Autonomous apomixis: The evolution of autonomous endosperm development in some apomictic plants.
Diplospory: a normal reductional meiosis is replaced by a non-reductional division. Two unreduced megaspores (2*n*) are produced, of which one degenerates and the other develops into an unreduced gametophyte with an unreduced egg cell.
Facultative apomixis: the production of a mixture of different progeny types in apomictic plants which is possible because apomeiosis and parthenogenesis can be uncoupled.
Gametophytic apomixis: can consist of diplospory and apospory and is strongly correlated with polyploidy.
Nucellar embryony: see Sporophytic apomixis.
Pseudogamy, pseudogamous apomixis: the endosperm develops only after fertilization of the central cell.
Sporophytic apomixis: Somatic embryos are formed within the sporophytic tissue that surrounds the gametophyte. These cells do not enter a gametophytic phase but remain sporophytically and produce an embryo directly (somatic embryo).

References

Archetti M (2004) Recombination and loss of complementation: a more than two-fold cost for parthenogenesis. J Evol Biol 17: 1084–1097
Asker SE, Jerling L (1992) Apomixis in plants. CRC press, Boca Raton
Bayer RJ, Chandler GT (2007) Evolution of polyploid agamic complexes: a case study using the Catipes group of *Antennaria*, including the *A. rosea* complex (Asteraceae: Gnaphalieae). In: Hörandl E, Grossniklaus U, van Dijk PJ, Sharbel TF (eds) Apomixis: evolution, mechanisms and perspectives. ARG Gantner Verlag KG, Lichtenstein, pp. 317–336
Bicknell RA, Lambie SC, Butler RC (2003) Quantification of progeny classes in two facultatively apomictic accessions of *Hieracium*. Hereditas 138: 11–20
Bicknell RA, Koltunow AM (2004) Understanding apomixis: recent advances and remaining conundrums. Plant Cell 16: S228–S245
Beukeboom LW, Weinzierl RP, Reed KM, Michiels NK (1996) Distribution and origin of chromosomal races in the freshwater planarian *Dugesia polychroa* (Turbellaria: Tricladida). Hereditas 124: 7–15
Calderini O, Chang SB, de Jong H, Bustil A, Paolocci F, Arcioni S, de Vries SC, Abma-Henkens MHC, Klein Lankhorst RH, Donnison IS, Pupilli F (2006) Molecular cytogenetics and DNA sequence analysis of an apomixis-linked BAC in *Paspalum simplex* reveal a non pericentromere location and partial microcolinearity with rice. Theor Appl Genet 112: 1179–1191
Carman JG (1997) Asynchronous expression of duplicate genes in angiosperms may cause apomixis, bispory, tetraspory, and polyembryony. Biol J Linn Soc 61: 51–94
Carman JG (2007) Do duplicate genes cause apomixis? In: Hörandl E, Grossniklaus U, van Dijk PJ, Sharbel TF (eds), Apomixis: evolution, mechanisms and perspectives. ARG Gantner Verlag KG, Lichtenstein, pp. 169–194
Catanach AS, Erasmuson SK, Podivinsky E, Jordan BR, Bicknell R (2006) Deletion mapping of genetic regions associated with apomixis in *Hieracium*. Proc Natl Acad Sci USA 103: 18650–18655
Chaboudez P (1994) Patterns of clonal variation in skeleton weed (*Chondrilla juncea*), an apomictic species. Austr J Bot 42: 283–295
Chapman H, Brown J (2001) 'Thawing' of 'frozen' variation in an adventive, facultatively apomictic, clonal weed. Plant Species Biol 16: 107–118

Conner JA, Goel S, Gunawan G, Cordonnier-Pratt MM, Johnson VE, Liang C, Wang H, Pratt LH, Mullet JE, Debarry J, Yang L, Bennetzen JL, Klein PE, Ozias-Akins P (2008) Sequence analysis of bacterial artificial chromosome clones from the apospory-specific genomic region of *Pennisetum* and *Cenchrus*. Plant Physiol 147: 1396–411

Dujardin M, Hanna WW (1989) Developing apomictic pearl millet – characterization of a BC3 plant. J Genet Breed 43: 145–51

Engelstädter J (2008) Constraints on the evolution of asexual reproduction. Bioassays 30: 1138–1150

Ernst A (1918) Bastadierung als Ursache der Apogamie im Pflanzenreich. Fischer, Jena. (Hybridization as cause of apogamy in the plant kingdom)

Gehring M, Choi Y, Fischer RL (2004) Imprinting and seed development. Plant Cell 16: S203–S213

Haig D, Westoby M (1989) Parent-specific gene expression and the triploid endosperm. Am Nat 134: 147–155

Haig D, Westoby M (1991) Genomic imprinting in endosperm: its effect on seed development in crosses between species, and between different ploidies of the same species, and its implications for the evolution of apomixis. Philos Trans R Soc B 333: 1–13

Harlan JR, De Wet JMJ (1975) On Ö. Winge and a prayer: the origins of polyploidy. Bot Rev 41: 361–390

Hebert PDN (1987) Genotypic characteristics of cyclic parthenogens, their obligately asexual derivatives. In: Stearns SC (ed) The evolution of sex and its consequences. Birkhäuser, Basel, pp. 175–195

Holm S, Ghatnekar L, Bengtsson BO (1997) Selfing and outcrossing but no apomixis in two natural populations of diploid *Potentilla argentea*. J Evol Biol 10: 343–352

Hörandl E (2006) The complex causality of geographical parthenogenesis. New Phytol 171: 525–538

Hörandl E, Grossniklaus U, van Dijk PJ, Sharbel TF (eds) (2007) Apomixis: evolution, mechanisms and perspectives. ARG Gantner Verlag KG, Lichtenstein

Huh JH, Bauer MJ, Hsieh T-F, Fischer R (2007) Endosperm gene imprinting and seed development. Curr Opin Genet Dev 17: 480–485

Kantama L, Sharbel TF, Schranz ME, Mitchell-Olds T, de Vries S, De Jong JH (2007) Diploid apomicts of the *Boechera holboellii* complex display large-scale chromosome substitutions and aberrant chromosomes. Proc Natl Acad Sci USA 104: 14026–14031

King LM, Schaal BA (1990) Genotypic variation within asexual lineages of *Taraxacum officinale*. Proc Natl Acad Sci USA 87: 998–1002

Koltunow AM, Grossniklaus U (2003) Apomixis: a developmental perspective. Annu Rev Plant Biol 54: 547–574

Lively CM (1987) Evidence from a New Zealand snail for the maintenance of sex by parasitism. Nature 328: 519–521

LeRoux JJ, Wieczorek AM, Wright MG, Tran CT (2007) Super-genotype: global monoclonality defies the odds of nature. PLoS ONE 2: e590

Matzk F, Meister A, Schubert I (2000) An efficient screen for reproductive pathways using mature seeds of monocots and dicots. Plant J 21: 97–108

Mes THM, Kuperus P, Kirschner J, Stepanek J, Storchova H (2002) Detection of genetically divergent clone mates in apomictic dandelions. Mol Ecol 11: 253–265

Mogie M (1992) The evolution of asexual reproduction in plants. Chapman and Hall, London

Moran NA (1992) The evolution of aphid life-cycles. Annu Rev Entomol 37: 321–348

Naumova TN (1993) Apomixis in angiosperms. Nucellar and integumentary embryony. CRC Press, Boca Raton

Naumova TN, Van der Laak J, Osadtchiy J, Matzk F, Kravtchenko A, Bergervoet J, Ramulu KS, Boutilier K (2001) Reproductive development in apomictic populations of *Arabis holboellii* (Brassicaceae) Sex Plant Reprod 14: 195–200

Nogler GA (1984a) Genetics of apospory in apomictic *Ranunculus auricomus*. V. Conclusion. Bot Helv 92: 411–423

Nogler GA (1984b) Gametophytic apomixis. In: Johri BM (ed) Embryology of angiosperms. Springer, Berlin, pp. 475–518

Nogler GA (2006) The lesser-known Mendel: his experiments on *Hieracium*. Genetics 172: 1–6

Nogler GA (2007) The discovery of parthogenesis: a long journey to the truth. In: Hörandl E, Grossniklaus U, van Dijk PJ, Sharbel TF (eds) Apomixis: evolution, mechanisms and perspectives. ARG Gantner Verlag KG, Lichtenstein, pp. 25–35

Noyes RD, Baker R, Mai B (2007) Mendelian segregation for two-factor apomixis in *Erigeron annuus*(Asteraceae). Heredity 98: 92–98

Noyes RD, Rieseberg LH (2000) Two independent loci control agamospermy (apomixis) in the triploid flowering plant *Erigeron annuus*. Genetics 155: 379–390

Ozias-Akins P, Van Dijk PJ (2007) Mendelian genetics of apomixis in plants. Annu Rev Genet 41: 509–537

Paland S, Colbourne JK, Lynch M (2005) Evolutionary history of contagious asexuality in *Daphnia pulex*. Evolution 59: 800–813

Paun O, Hörandl E (2006) Evolution of hypervariable microsatellites in apomictic polyploid lineages of *Ranunculus carpaticola*: directional bias at dinucleotide loci. Genetics 174: 387–398

Pichot C, El Maâtaoui M, Raddi S, Raddi P (2001) Surrogate mother for endangered *Cupressus*. Nature 412: 39

Pringle P (2007) Day of the dandelion. Simon and Schuster, New York

Richards AJ (2003) Apomixis in flowering plants: an overview. Philos Trans R Soc B Biol Sci 358: 1085–1093

Roetman E, Den Nijs JCM, Sterk AA (1988) Distribution and habitat range of diploid, sexual dandelions (*Taraxacum* section *Vulgaria*), a Central European flora element in the Netherlands. Acta Bot Neerl 37: 81–94

Rutishauser A (1948) Pseudogamie und Polymorphie in der Gattung *Potentilla*. Julius Klaus Stiftung für Vererb Forsch 23: 267–424

Sanderson M, Doyle JJA (2001) Sources of error and confidence intervals in estimating the age of angiosperms from rbcL and 18S rDNA data. Am J Bot 88: 1499–1516

Savidan Y (1980) Chromosomal and embryological analyses in sexual X apomictic hybrids of *Panicum maximum* Jacq. Theor Appl Genet 57: 153–156

Savidan Y (2000) Apomixis: genetics and breeding. Plant Breed Rev 18: 13–85

Savidan Y, Carman JG, Dresselhaus T (eds) (2001) The flowering of apomixis; from mechanisms to genetic engineering. Mexico, DF: Cimmyt, IRD European Commission DG VI (FAIR)

Schultz RJ (1967) Gynogenesis and triploidy in the viviparous fish *Poeciliopsis*. Science 157: 1564–1567

Spillane C, Curtis MD, Grossniklaus U. (2004) Apomixis technology development – virgin births in farmers' fields? Nat Biotechnol 22: 687–691

Stebbins GL (1950) Variation and evolution in plants. Columbia University Press, New York

Van der Hulst RGM, Mes THM, Den Nijs JCM, Bachmann K (2000) Amplified fragment length polymorphism (AFLP) markers reveal that population structure of triploid dandelions (*Taraxacum officinale*) exhibits both clonality and recombination. Mol Ecol 9: 1–8

Van Dijk PJ (2003) Ecological and evolutionary opportunities of apomixis: insights from *Taraxacum*and *Chondrilla*. Philos Trans R Soc B Biol Sci 358: 1113–1121

Van Dijk PJ, Vijverberg K (2005) The significance of apomixis in the evolution of the angiosperms: a reappraisal. Regnum Vegetabile 143. Bakker FT, Chatrou LW, Gravendeel B and Pelser PB (eds) Plant species-level systematics: new perspectives on pattern and process. Koeltz Scientific Books, Koeningstein, pp. 101–116

Whitton J, Sears CJ, Baack EJ, Otto SP (2008) The dynamic nature of apomixis in the angiosperms. Int J Plant Sci 169: 169–182

Wu W, Zheng YL, Chen L, Wei YM, Yan ZH (2005) Genetic diversity among the germplasm resources of the genus *Houttuynia* Thunb. in China based on RAMP markers. Gen Res Crop Evol 52: 473–482

Chapter 4
Cytology of Asexual Animals

Per Stenberg and Anssi Saura

Abstract We review the cytological mechanisms underlying asexual reproduction, i.e. reproduction without fertilization, in animals. Asexuality or parthenogenesis has evolved many times and the cytological mechanisms to restore the parental chromosome number can vary between and even within species. In automictic or meiotic parthenogenesis, meiosis takes place but the chromosomal constitution of the mother is restored through one or several different mechanisms. Some of these mechanisms enforce homozygosity at all loci while some other mechanisms pass the genome of the mother intact to the offspring. In apomictic or mitotic parthenogenesis the eggs are formed through what is essentially a set of mitoses. Polyploidy, is in general incompatible with chromosomal sex determination and is a rare condition in animals. However, many asexual and hermaphroditic forms are polyploid to various degrees. Polyploidy is divided into allo- and autopolyploidy. In the former mode the chromosome sets are derived from two or more different species while in autopolyploidy the multiplication has taken place within one species. We discuss the evolutionary consequences of the different cytological mechanisms involved in asexual reproduction.

4.1 The Importance of Cytology

In parthenogenesis or *thelytoky*, a female animal produces female progeny through a process that does not involve fertilization (von Siebold 1856). Fertilization is here defined as involving the fusion of gametes from two different individuals. The definition of parthenogenesis is based on cytological observations. The roles of cells and chromosomes in sexual reproduction and parthenogenesis were worked out in the first half of the 20th century. The growth of animal cytology is covered in "Animal Cytology and Evolution" by White (1973), a *magnum opus* that evidently will be

---

P. Stenberg (✉)
Umeå Center for Molecular Pathogenesis, Umeå University, SE-901 87 Umeå, Sweden
e-mail: Per.Stenberg@molbiol.umu.se

I. Schön et al. (eds.), *Lost Sex*, DOI 10.1007/978-90-481-2770-2_4,

also the last of its kind since nobody is expected to master the entire field of the role of cytology in animal evolution like its author.

Cytology had, like e.g. *Drosophila* and population genetics, been an important branch of the new science of genetics. It was overtaken by the spectacular rise of molecular genetics, the tools of which were soon used in *Drosophila* and other model organisms. Chromosomes, the objects of cytogenetics, are large and complex, and the application of molecular methods to cytology has been neither easy nor rapid. Chromosomes still represent a challenge that is not solved as easily as the reductionist exploits of molecular biology thus far. We see, however, that progress is being made again by methods such as fluorescent microscopy and image capturing while epigenetic phenomena stress the importance of cytogenetics once more.

4.2 Cytological Mechanisms of Animal Parthenogenesis

Mendelian genetics, diploidy, meiosis and fertilization are intertwined phenomena. In normal meiosis, the chromosome pairs, chiasmata, are formed and the resulting gametes are haploid. Diploidy is restored at fertilization. In *automictic* or meiotic parthenogenesis, meiosis is present. Since there is no fertilization, the parental level of ploidy has to be restored in one way or another. This can be accomplished through several different mechanisms, the genetic and evolutionary consequences of which differ extensively.

In *apomictic* or mitotic parthenogenesis, egg cells are produced through mitosis. Only one cell division takes place in the eggs. The number of chromosomes is not reduced. As a consequence, the offspring are true *clones*, genetically identical to their mother, save for mutations (see also Chapter 9).

Parthenogenesis has arisen repeatedly and independently during the evolution of animals. As expected, cytological mechanisms do not show any clear evolutionary patterns. A single species or related species can reproduce through different mechanisms. As an example, lepidopterans make use of several automictic mechanisms (Suomalainen et al. 1987) as do strains and species of the crustacean *Artemia*, while parthenogenetic weevils all reproduce through apomictic parthenogenesis (Saura et al. 1993; Stenberg et al. 2003).

Polyploidy is a rare condition in sexually reproducing animals. In contrast, it is common in parthenogenetic forms (Lewis 1980), even though asexuality and polyploidy need not be directly related to each other (Lundmark and Saura 2006). In the following, we describe the different cytological mechanisms of animal parthenogenesis, assess briefly their evolutionary significance and discuss the role of polyploidy in parthenogenesis.

4.2.1 Automictic Parthenogenesis

In this mode of parthenogenesis, the early stages of meiosis are in most cases unaffected. The chromosomes pair and crossing-over takes place. The chromosome number is halved during the formation of the egg and polar bodies. The original diploid chromosome number is restored thereafter through a fusion of the products of meiosis, the details of which are variable. In fact, about all imaginable mechanisms have been described. The main difference to normal fertilization is that the cells involved in fusion are derived from a single individual and a single meiosis.

The evolutionary consequences of automictic parthenogenesis are often poorly understood. It is true that certain modes enforce homozygosity but even in these cases, there may be exceptions. Natural selection can sort out these homozygous lineages and evolution can still take place. Other modes of automixis transfer the genotype of the mother intact to the offspring. The confusion seems to stem mainly from the rather short description of automixis in White (1973, pp. 705–709) and in particular from his essay on heterozygosity (White 1970), where he downplayed the evolutionary potential of parthenogens. In addition, there are cases where members of a single population reproduce simultaneously through several automictic mechanisms (Suomalainen et al. 1987).

4.2.1.1 Gamete Duplication

In this mode, the haploid egg cell divides and produces cleavage nuclei. These nuclei fuse with each other producing a diploid nucleus. Alternatively, the divided chromosomes remain in the same nucleus, which becomes diploid (see Fig. 4.1).

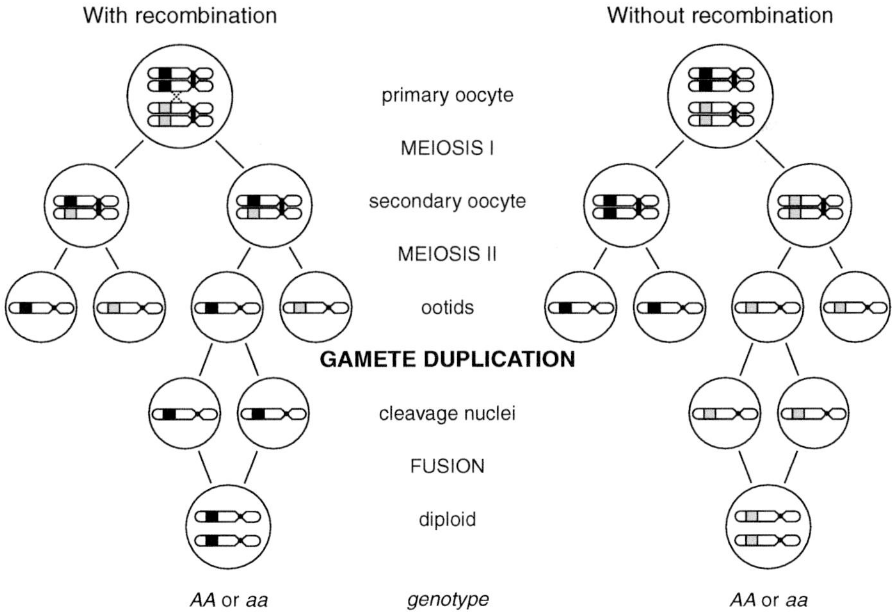

Fig. 4.1 The genetic consequences of gamete duplication (modified after Asher 1970; Suomalainen et al. 1987)

Gamete duplication enforces homozygosity at all loci, irrespective of crossing-over at meiosis. If the founding lineage has been polymorphic, lineages derived from it can differ genetically from each other. They will be subject to evolution being independent from each other. Gamete duplication has been observed in the *Artemia salina* complex (Crustacea), among insects in phasmids, aleyrodids, coccids, lepidopterans, in some parthenogenetic *Drosophila*, cynipids and in certain mites (see Suomalainen et al. 1987 and Norton et al. 1993 for details).

Plantard et al. (1998) demonstrated that *Wolbachia* is responsible for the transition from arrhenotoky (and haplodiploidy) to parthenogenesis through gamete duplication in the cynipid *Diplolepis spinossissimae*. It seems that *Wolbachia* invariably gives rise to parthenogenesis through gamete duplication in hymenopterans (Bordenstein and Werren 2007; see also Chapter 17). This topic is reviewed in van Wilgenburg et al. (2006) and it appears that other microbes, e.g. *Cardinium*, (Bacterioidetes) can have effects.

4.2.1.2 Terminal Fusion

In this mode, the second polar nucleus fuses with the egg nucleus. The genetic consequences of terminal fusion are shown in Fig. 4.2.

Given that the mother is heterozygous for two alleles but there is no crossing-over, terminal fusion enforces homozygosity. If, on the other hand, there is crossing-over between the locus and the centromere, heterozygosity can be maintained (Asher 1970; Suomalainen et al. 1987). A population reproducing through terminal fusion can, at least to a certain extent, maintain polymorphism and selection can operate on the different genotypes.

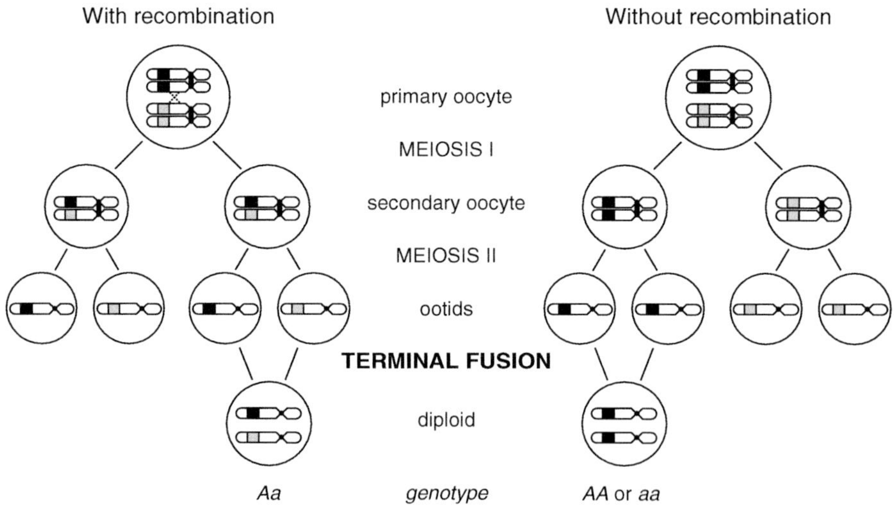

Fig. 4.2 The genetic consequences of terminal fusion (modified after Asher 1970; Suomalainen et al. 1987)

Nematodes, enchytraeids, some *Artemia,* isopods, among insects: acridids, coccids, thrips, certain *Drosophila* and other dipterans, some hymenopterans, some mites and tardigrades reproduce through terminal fusion (see Suomalainen et al. 1987 and Norton et al. 1993 for details; see also Chapter 12).

4.2.1.3 Central Fusion

Here, the two central polar nuclei fuse and give rise to the embryo. If the mother is heterozygous, all her offspring will be heterozygous, given that there is no crossing-over. If there is crossing-over between a locus and the centromere, there will be segregation among the offspring so that a heterozygote *A a* will produce 1/4 *AA* individuals, 1/2 *Aa* individuals and 1/4 *aa* individuals (see Fig. 4.3). In cases where heterozygosity is advantageous, selection will favour loci that are tightly linked to the centromere. This is evidently the case in lepidopterans and in *Drosophila mangabeirai*, which reproduce through central fusion. Lepidopteran females have achiasmate oogenesis that guarantees the maintenance of heterozygosity; *D. mangabeirai* is heterozygous for three inversions that inhibit crossing over (see Suomalainen et al. 1987 for details).

Many hymenopterans combine persistent heterozygosity under single locus complementary sex determination with thelytoky, which implies that the sex locus must be located in a region without recombination (e.g. a centromere) (Beukeboom and Pijnacker 2000; see also Chapter 17).

Animals that reproduce through central fusion include the psychid moth *Dahlica (Solenobia) triquetrella,* the obligately parthenogenetic *Drosophila mangabeirai*

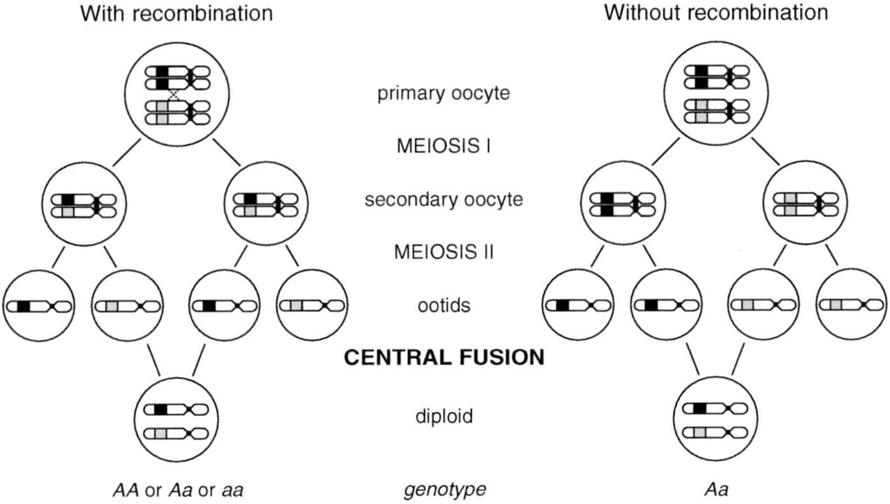

Fig. 4.3 The genetic consequences of central fusion (modified after Asher 1970; Suomalainen et al. 1987)

and strains of other accidentally parthenogenetic *Drosophila* (among other automictic mechanisms), the fly *Lonchoptera dubia*, the parthenogenetic strains of the honey bee (see Seiler 1963 and Suomalainen et al. 1987 for details) and also some other hymenopterans (Beukeboom and Pijnacker 2000).

4.2.1.4 The First Polar Nucleus Fuses with the Nucleus of the Secondary Oocyte

This can either happen directly after the first meiotic division or alternatively, the nuclei derived from the above nuclei fuse (see Fig. 4.4).

A heterozygous mother *Aa* will produce 1/6 *AA* homozygotes, 4/6 *Aa* heterozygotes and 1/6 *aa* homozygotes. Linkage will not affect this erosion of heterozygosity. Narbel-Hofstetter (1964) has shown that several psychid moths reproducing through this mechanism keep the meiotic metaphase plates together. There is one mode, in which the two halves of the first anaphase spindle collapse on one another and form the spindle of the second division. In another mode, second meiotic spindles come together side by side and fuse (White 1973, p. 722). This ensures that the genotype of the mother is kept intact without any segregation of alleles (Narbel-Hofstetter 1964). In addition to the psychids, the liver fluke *Fasciola hepatica* and some strains of the crustacean *Artemia* reproduce through this mechanism (see Suomalainen et al. 1987 for details). The collembolan *Folsomia candida*

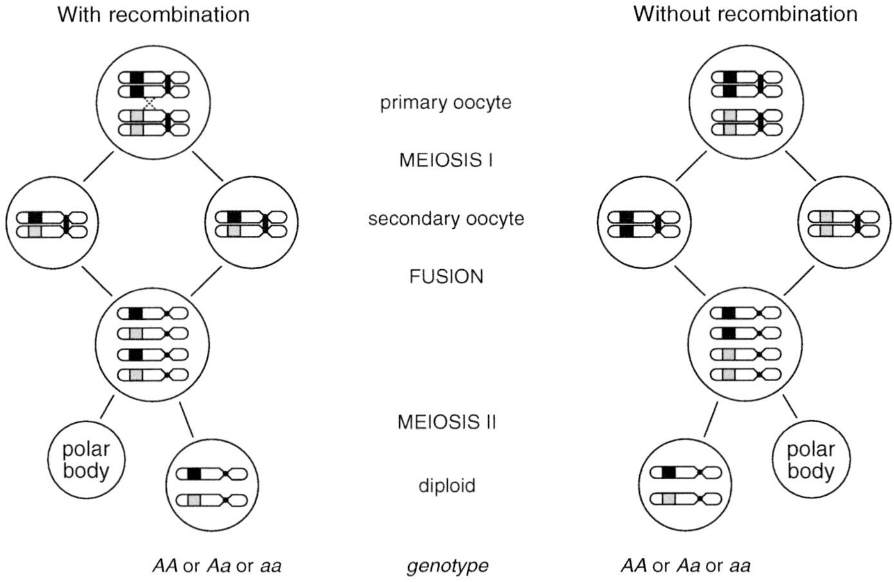

Fig. 4.4 The genetic consequences of a case where the first polar nucleus fuses with the secondary oocyte

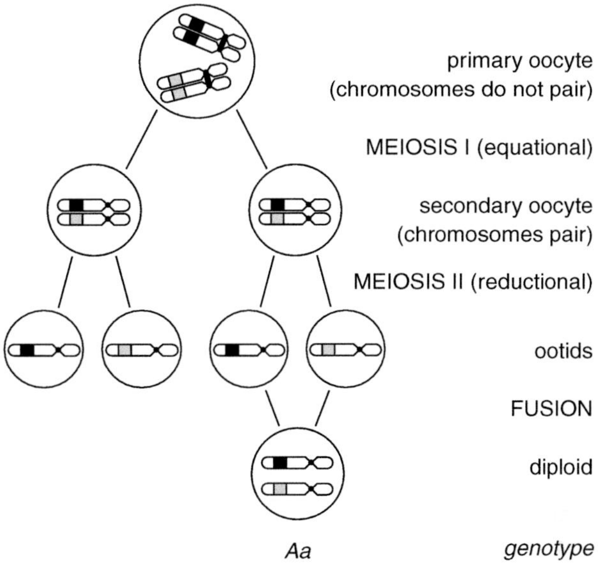

Fig. 4.5 The genetic consequences of gonoid thelytoky (modified after Nur 1979; Suomalainen et al. 1987)

seems, at least in part, to also use this mechanism; Fountain and Hopkin (2005) put evidence forward that *Wolbachia* is responsible for parthenogenesis in *Folsomia.*

4.2.1.5 Gonoid Thelytoky

In this mode, the parthenogenetic egg undergoes two meiotic divisions; the chromosomes pair at the second meiotic metaphase (see Fig. 4.5). Certain scale insects (coccids) reproduce this way (Nur 1979).

The genotype of the mother is passed on unchanged to the offspring (Nur 1979, Suomalainen et al. 1987). Spitzer (2006) has studied the evolutionary potential of different strains of the pest *Saissetia coffeae* that differ extensively in adaptability.

4.2.1.6 Premeiotic Doubling

This mode of automictic parthenogenesis involves an endomitotic process: a premeiotic doubling of chromosome numbers is reduced through meiosis. The resulting daughter chromosomes pair in the first meiotic prophase. All chromosomes pair with their genetically identical counterpart, and the original even number of chromosomes is maintained (see Fig. 4.6).

As a result, the genotype of the mother is passed on to the offspring unchanged.

Premeiotic doubling is a common mechanism used by turbellarians, most parthenogenetic earthworms, several insects, mites, tardigrades and parthenogenetic fishes, amphibians and reptiles (Suomalainen et al. 1987; Cuellar 2005).

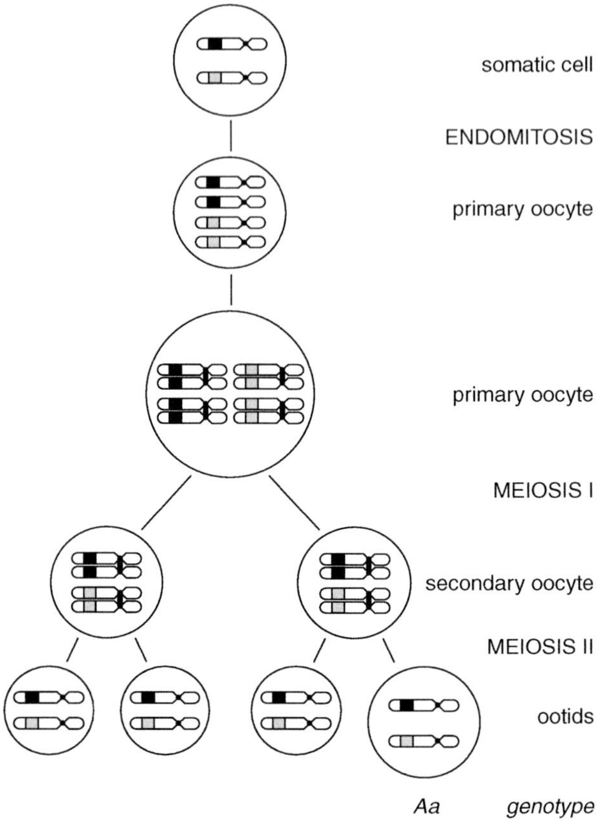

Fig. 4.6 The genetic consequences of premeiotic doubling

4.2.2 Apomictic Parthenogenesis

In apomictic parthenogenesis, the essential features of meiosis are lacking. The chromosomes do not pair and the oocyte undergoes a single maturation division, which is in most cases indistinguishable from mitosis. Consequently, the genotype of the mother is passed on to the offspring without changes. Apomictic parthenogenesis is the most common cytological mode of parthenogenesis and it is found in many animal groups (see Suomalainen et al. 1987 for details).

The parthenogenetic phase of animals with cyclical parthenogenesis is apomictic. Such groups include cladocerans (see Chapter 15), monogonont rotifers (see Chapter 14), aphids, the beetle *Micromalthus*, cecidomyid midges and cynipid wasps. Note that hymenopterans produce males parthenogenetically from haploid unfertilized eggs but these males participate in normal sexual reproduction. Other animals with verified apomictic parthenogenesis (following Suomalainen et al. 1987) include coelenterates, certain tubellarians, trematodes, nematodes, gastrotrichs, bdelloid rotifers (they have two equational maturation divisions in the oogenesis; Hsu 1956; Pagani et al 1993; see also Mark Welch et al. 2004 and

Chapter 13), parthenogenetic gastropods and the earthworm *Dendrobaena octaedra*. The latter record needs to be rechecked, since the distribution of genotypes does not support apomixis and rearing experiments show a segregation of alleles (Terhivuo and Saura 2006; Simonsen and Holmstrup 2008). This variation can be explained either through occasional sex or gene conversion. Also the enchytraeid *Lumbricillus lineatus* and amongst crustaceans, some members of the *Artemia salina* complex, parthenogenetic ostracods (see also Chapter 11) and certain isopods reproduce apomictically. Among the insects, blattids, phasmids, tettigonids, psocids, certain coccids, dipterans, chrysomelids (including *Calligrapha*), the large group of parthenogenetic weevils and certain hymenopterans are apomictic, as are, mites, tardigrades and the fish *Poecilia formosa* (Monaco et al. 1984; Balsano et al. 1989; see also Chapter 19). Finally, Weeks and Braeuwer (2001) have shown that *Wolbachia* gives rise to apomictic parthenogenesis in a mite.

4.3 Evolutionary Consequences

Even though the evolution of asexual animals is discussed in depth elsewhere in this book, we wish to stress the importance of knowing the cytological mechanism of parthenogenesis. Many modes of automixis result in offspring being genetically identical to their mother, and can not be distinguished from offspring being produced through apomictic parthenogenesis. A single morphological species can have several modes of automixis, and a population may be composed of females using different mechanisms in the oogenesis. Finally, a single female can give rise to offspring through different automictic mechanisms. Obviously, the evolutionary consequences will be radically different in such cases.

Most of the experimental cytology on parthenogenesis is old. When molecular results contradict the predictions of cytology, the latter has to be rechecked. This requires learning the techniques described in handbooks and consulting practising cytologists doing routine work, e.g., in hospitals. There is no other way.

4.4 Polyploidy in Association with Parthenogenesis

Polyploidy characterizes angiosperm plants, to the extent that the proportion of polyploids is estimated to range from 43 to about 58% among angiosperm species, but the proportion of species that are polyploid in some sense, so that there are traces of paleopolyploidy, may be as high as 70–80% (Lewis 1980). Virtually all apomictic angiosperms are polyploid (Asker and Jerling 1992; see also Chapters 3 and 8). In contrast, animals are seldom polyploid. Chromosomal sex determination is apparently the main reason for the absence of polyploidy in animals (White 1946). This argument still stands, even though there is evidence for rounds of polyploidization in the history of vertebrates (Carroll et al. 2005). In addition, there are several sexually reproducing frogs and fishes that are polyploid (e.g. Ma and Gustafson 2005; Zhu et al. 2006; Holloway et al. 2006; see also Chapters 19 and 20).

Polyploid formation often involves hybridization between two species (allopolyploidy). The alternative, autopolyploidy (where the multiple chromosomes come essentially from the same genome), has been relegated to the background but has experienced a revival among botanists (e.g. Soltis and Soltis 2000). This distinction needs, however, not be sharp since it depends on the definition of species. There may well be genetic and chromosomal differences that do not allow for a normal segregation of homologous chromosomes in crosses involving forms that belong to what is called a species.

Parthenogenetic animals and the ones with apomictic parthenogenesis in particular, lack the obstacles to polyploidy, i.e. chromosomal sex determination. Polyploidy is indeed very common among parthenogenetic animals (Suomalainen et al. 1987). To give an example, among the 52 morphological species of parthenogenetic weevils that have been cytologically studied, only four cytological races are diploid, while 43 are triploid, 18 tetraploid, 6 pentaploid, three are hexaploid and one is decaploid (Saura et al. 1993). In addition, the polyploid races have a much wider distribution in comparison with diploid sexuals or parthenogens, a pattern called *geographical polyploidy* by Stenberg et al. (2003). Lundmark and Saura (2006) provide a list of insect taxa with geographical polyploidy (see also 8 on geographic parthenogenesis in general).

White (1970, 1973) had a rather negative view on the evolutionary potential of polyploid parthenogens. Even though they can be highly heterozygous, they can not respond to challenges from the environment and are, accordingly, blind alleys of evolution. The merger of different genomes in allopolyploids leads to differently adapted offspring that will have difficulties to reproduce sexually due to improper pairing of chromosomes. It will suffice that at least one asexual lineage out of many will be well adapted for the benefits of asexual reproduction to become manifested (Lundmark and Saura 2006). The tetraploid race of *Dahlica triquetrella* has been shown to be autopolyploid (Seiler 1961). The fecundity of diploid sexual females exceeds that of diploid parthenogens, but the tetraploid females produce more offspring than either diploid race (Seiler 1961). Since the genome of an autopolyploid is duplicated without addition of any new genes, a superior fitness of a polyploid female should then in that case show a positive effect of polyploidy per se.

Earthworms are another interesting example. Up to 40% of the cytologically examined species among Palaearctic earthworms are polyploid. In the fashion of plants, many are hermaphrodites, which shows that animals can attain high proportions of polyploids once they rid themselves of chromosomal sex determination (Lundmark and Saura 2006). Interestingly, there are both sexually reproducing and asexual forms among the polyploids, and both sexuals and asexuals show geographical polyploidy. Lundmark and Saura (2006) argue that geographical polyploidy may be attributable to the adaptive effects of polyploidy rather than asexuality. If so, the two are separate phenomena that need to be studied individually.

D'Souza and Michiels (Chapter 18) show that parthenogenesis is often linked to hermaphroditism in flatworms and their polyploidy is not due to hybridization between species. Flatworm parthenogens can also have occasional sex. We note that the earthworm *Dendrobaena octaedra* mentioned above may be a similar case.

4.5 Conclusions

The evolutionary potential of an asexual lineage depends on the cytological mechanism underlying asexuality. Cytological data are therefore needed in order to understand the evolutionary potential of an asexual animal. As molecular methods have reached the level of sophistication that allows studying chromosome structures, we expect a revival of the interest in cytology of asexual animals. Studying the role of polyploidy in asexual reproduction may be methodologically even more demanding, but it is, nevertheless, important for its own sake. Many asexual animals are polyploid while about all sexuals are diploid.

References

Asher J Jr (1970) Parthenogenesis and genetic variability. PhD thesis. University of Michigan, Ann Arbor

Asker SE, Jerling L (1992) Apomixis in plants. CRC Press, Boca Raton

Balsano JS, Rasch EM, Monaco PJ (1989) The evolutionary ecology of *Poecilia formosa* and its triploid associate. In: Meffe GK, Snelson FF (eds) Ecology and evolution of livebearing fishes (Poeciliidae). Prentice Hall, Englewood Cliffs, pp. 277–297

Beukeboom LW, Pijnacker LP (2000) Automictic parthenogenesis in the parasitoid *Venturia canescens* (Hymeoptera: Ichneumonidae) revisited. Genome 43: 939–944

Bordenstein SR, Werren JH (2007) Bidirectional incompatibility among divergent *Wolbachia* and incompatibility level differences among closely related *Wolbachia* in *Nasonia*. Heredity 99: 278–287

Carroll SB, Grenier JK, Weatherbee SD (2005). From DNA to diversity: molecular genetics and the evolution of animal design. Blackwell, Oxford

Cuellar O (2005) Reproduction and the mechanism of meiotic restitution in the parthenogenetic lizard *Cnemidophorus uniparens*. J Morphol 133: 139–165

Fountain MT, Hopkin SP (2005) *Folsomia candida* (Collembola): a "standard" soil arthropod. Annu Rev Ent 50: 201–222

Holloway AK, Cannatella DC, Gerhart HC, Hillis DM (2006) Polyploids with different origins and ancestors from a single sexual polyploid species. Am Nat 167: E88–E101

Hsu WS (1956) Oogenesis in *Habrotricha tridens* (Milne). Biol Bull 111: 364–374

Lewis WH (ed.) (1980) Polyploidy/biological relevance. Plenum, New York

Lundmark M, Saura A (2006) Asexuality does not explain the success of clonal forms in insects with geographical parthenogenesis. Hereditas 143: 24–33

Ma XF, Gustafson JP (2005) Genome evolution in allopolyploids: a process of cytological and genetic diploidization. Cytogenet Genome Res 109: 236–249

Mark Welch JL, Mark Welch DB, Meselson M (2004) Cytogenetic evidence for asexual evolution of bdelloid rotifers. Proc Natl Acad Sci 101: 1618–1621

Monaco PJ, Rasch EM, Balsano JS (1984) Apomictic reproduction in the Amazon molly, *Poecilia formosa*, and its triploid hybrids. In: Turner BJ (ed.) Evolutionary genetics of fishes. Plenum Press, New York, pp. 311–328

Narbel-Hofstetter M (1964) Les altérations de la meiose chez les animaux parthénogénétiques. Protoplasmatologia VI. F2. Springer-Verlag, Wien

Norton RA, Kethley JB, Johnston DE, O'Connor BM (1993). Phylogenetic perspectives on genetic systems and reproductive modes of mites. In: Wrensch DL, Ebbert MA (eds) Evolution and diversity of sex ratio in insects and mites. Chapman and Hall, New York, pp. 8–99

Nur U (1979) Gonoid thelytoky in soft scale insects (Coccoidea: Homoptera). Chromosoma 72: 89–104

Pagani M, Ricci R, Redi CA (1993) Oogenesis in *Macrotrachela quadricornifera* (Rotifera, Bdelloidea). Hydrobiologia 255–256: 225–230

Plantard O, Rasplus J-Y, Mondor G, Le Clainche I, Solignac M (1998) *Wolbachia*-induced thelytoky in the rose gallwasp *Diplolepis spinosissimae* (Giraud)(Hymenoptera: Cynipidae), and its consequences on the genetic structure of its host. Proc R Soc Lond B 265: 1075–1090

Saura A, Lokki J, Suomalanen E (1993) Origin of polyploidy in parthenogenetic weevils. J Theor Biol 163: 449–456

Seiler J (1961) Untersuchungen über die Entstehung der Parthenogenese bei *Solenobia triquetrella* F.R. (Lepidoptera, Psychidae) III. Die geographische Verbreitung der drei Rassen von *Solenobia triquetrella* (bisexuell, diploid und tetraploid parthenogenetisch) in der Schweiz und in angrenzenden Ländern und die Beziehungen zur Eiszeit. Bemerkungen über die Entstehung der Parthenogenese. Z Vererbungsl 92: 261–316

Seiler J (1963) Untersuchungen über die Entstehung der Parthenogenese bei *Solenobia triquetrella* F.R. (Lepidoptera, Psychidae). IV. Wie besamen begattete diploid und tetraploid parthenogenetische Weibchen von *S. triquetrella* ihre Eier? Schicksal der Richtingskörper im unbesamten und besamten Ei. Vergleich der Ergebnisse mit F1-Aufzuchten und Beziehungen zur Genese der Parthenogenese. Z Vererbungsl 94: 29–66

Simonsen V, Holmstrup M (2008). Deviation from apomictic reproduction in *Dendrobaena octaedra*? Hereditas 145: 212–214

Soltis PS, Soltis DE (2000) The role of genetic and genomic attributes in the success of polyploids. Proc Natl Acad Sci USA 97: 7051–7057

Spitzer B (2006) Local maladaptation in the soft scale insect *Saissetia coffeae* Hemiptera: Coccidae. Evolution 60: 1859–1867.

Stenberg P, Lundmark M, Knutelski S, Saura A (2003) Evolution of clonality and polyploidy in a weevil system. Mol Biol Evol 20: 1626–1632

Suomalainen E, Saura A, Lokki J (1987) Cytology and evolution in parthenogenesis. CRC Press, Boca Raton

Terhivuo J, Saura A (2006) Dispersal and clonal diversity of North-European parthenogenetic earthworms. Biol Inv 8: 1205–1218

Van Wilgenburg E, Driessen G, Beukeboom LW (2006) Single locus complementary sex determination in Hymenoptera: an "unintelligent" design? Frontiers Zool 3: 1

von Siebold C (1856) Wahre Parthenogenesis bei Schmetterlingen und Bienen. Engelmann, Leipzig

Weeks AR, Braeuwer JAJ (2001) *Wolbachia*-induced parthenogenesis in a genus of phytophagous mites. Proc R Soc Lond B 268: 2245–2251

White MJD (1946) The evidence against polyploidy in sexually reproducing animals. Am Nat 80: 610–619

White MJD (1970) Heterozygosity and genetic polymorphism in parthenogenetic animals. In: Hecht MK, Steere WC (eds) Essays in evolution and genetics in honour of Theodosus Dobzhansky, suppl. Evol Biol. Appleton-Century-Crofts, New York

White MJD (1973) Animal cytology and evolution, 3rd ed. Cambridge University Press, Cambridge, UK

Zhu HP, Ma DM, Gui JF (2006) Triploid origin of the gibel carp as revealed by 5S rDNA localization and chromosome painting. Chromosome Res 14: 767–776

Chapter 5
A Graphical Approach to Lineage Selection Between Clonals and Sexuals

William R. Rice and Urban Friberg

Abstract Theories for the evolutionary advantages and disadvantages of sex address two fundamentally different questions: (i) Why does the genome of sexual lineages not "congeal," (i.e., move toward a lowered recombination rate)?, and (ii) When there is a mixture of reproductively isolated clonal and sexual lineages, why do the clonals not accumulate and lead to a predominance of asexual reproduction within a clade? Here, we focus on the latter question. The relevant theory in this case is necessarily based on a special form of "lineage" selection between sexuals and clonals that do not share a common gene pool. We first briefly review the major genetic costs and benefits of clonal reproduction and conclude that the extant assemblage of theories provides an essentially complete description of the phenomenon. We next set out to combine and simplify these seemingly disparate theories by graphically representing the frameworks previously developed by Felsenstein (Genetics 78: 737–756, 1974) and Kimura and Maruyama (Genetics 54: 1337–1351, 1966) to show that the all of the proposed disadvantages to clonal reproduction can be expressed by a single factor: a decreased efficiency of natural selection in non-recombining lineages. This reduced efficiency derives from two distinct processes that only operate in clonal lineages: (i) background-trapping and (ii) the compensatory linkage disequilibrium that accrues in response to epistatic selection.

5.1 Introduction

The adaptive significance of sex has two components: why did sex initially evolve and why is it maintained in extant lineages. Here, we focus on the maintenance of sex. Similarly, the question of why sexual recombination is maintained within the lineages of most multicellular organisms is actually two distinct questions. The first question is: Why do genomes recombine (through independent assortment of

W.R. Rice (✉)
Department of Ecology, Evolution and Marine Biology, University of California, Santa Barbara, CA 93106-9610, USA
e-mail: rice@lifesci.ucsb.edu

I. Schön et al. (eds.), *Lost Sex*, DOI 10.1007/978-90-481-2770-2_5,

non-homologous chromosomes and crossing over between homologous chromosomes)? There is substantial variation in recombination rate along the length of chromosomes, and also between the sexes for the same chromosomal region (Bell 1982; Burt et al. 1991). One way to achieve clonal reproduction is the gradual evolution of reduced recombination until it reaches a zero level. The traditional approach to studying this phenomenon is the development of models – based on selection operating at the individual level – that trace the fate of mutations that modify the level of recombination at specific regions within the genome (see for review Otto and Lenormand 2002 and Keightley and Otto 2006 for a recent advance in this area).

Another route to clonal reproduction is produced by the elimination of the sexual phase of species like monogont rotifers, *Daphnia* and aphids (cyclical parthogens; see also Chapters 14, 15, 25) that alternate between bouts of clonal and sexual reproduction. Because the reproductive machinery for clonal reproduction is already in place, many strictly clonal lineages have evolved within each of these clades (White 1978; Maynard Smith 1978). We also know, however, that clonal reproduction can evolve instantaneously in other groups that are not capable of transient asexual reproduction, through processes such as major mutations, hybridization and infection with cytoplasmic endosymbionts (Stebbins 1950; White 1978; Werren 1997; Lattorff et al. 2005).

Once clonal lineages are formed (gradually or instantaneously), a second question arises concerning the maintenance of sex: Why do established clonal lineages fail to persist over long periods of evolutionary time while their closely related sexual lineages endure? It is this latter question – based on lineage selection between competing sexuals and clonals – upon which we will focus in this chapter.

One final caveat concerns the issue of our lineage-selection approach to the persistence of clonals vs. sexuals. The "knee-jerk" reaction of most evolutionary biologists to evolutionary theory that is based on any form of group selection is that it is somehow inferior, owing to the relatively weak force of group selection when pitted against individual selection. However, as originally pointed out by Fisher (1930), when attempting to understand why reproductively isolated clonal lineages do not out-compete closely related sexual lineages and eventually replace them, a special form of group selection is the only relevant type of selection to consider. This is the case because there is no gene flow between the clonally and sexually reproducing lineages – hence individual selection does not apply because by definition there is not a common gene pool shared by the clonal and sexual lineages. For this reason we will approach the question of why sexual but not clonal lineages persist by comparing the advantages and disadvantages (emergent properties) associated with these two modes of reproduction.

5.2 Costs and Benefits of Sexual vs Clonal Reproduction

5.2.1 Short-Term Costs of Sexual Reproduction

There are two distinct "two-fold" costs of sex (Lively and Lloyde 1990): the demographic cost of producing males (Maynard Smith 1978), and the reduced relatedness cost of meiosis (Williams 1975). The latter cost occurs when an organism is capable of producing a mixture of clonal and sexual offspring (e.g., some citrus species; Westwood 1993), or both inbred and outbred offspring (e.g., many self-compatible plants; Stebbins 1950), and hence when distinct, genetically isolated sexual and clonal lineages do not exist. For example, consider a species in which males help their mates and thereby double a female's net reproductive rate, so that the "cost of producing males" (see below) does not apply. If a mutation caused a female to produce some (but not all) of her eggs via parthenogenesis, or by selfing, then all else being equal, she would place more genes per offspring into the next generation's gene pool. The mutation would be selectively favored because each parthenogenetically produced offspring delivers two-times as many gene copies into a shared gene pool compared to a sexually produced offspring. Because we will focus below exclusively on distinct sexual and clonal lineages, the cost of meiosis will not apply.

The two-fold "cost of producing males," in its simplest between lineages form, occurs when males do not provide resources to their offspring (Maynard Smith 1978) and there are distinct, genetically isolated sexuals and clonals. In this case the net reproductive rate of clonal lineages is double that of outbred sexual lineages because clones are composed entirely of females, whereas the sexual lineages are composed of 50% males which do not contribute to the fecundity of this group. The cost of producing males is a surprisingly potent factor favoring clonal reproduction. To illustrate why, suppose that, by any of a number of possible mechanisms, a clonal individual is formed in a sexual population. Each generation the clonals are expected to contribute double the number of offspring to the collective pool of all offspring used to produce the following generation. Following the logic of Maynard Smith (1978) and assuming that the clonal lineage is not lost early-on by sampling error, some simple calculations demonstrate that, in general, the expected proportion of clonal individuals ($P_{clonal,G}$) after G generations will be,

$$P_{clonal,G} \approx \frac{2^G P_o}{1 + 2^G P_o}, \tag{5.1}$$

where P_o is the initial proportion of clonals (generally assumed to be $1/N$, where N is the population size), the scalar "2" represents the two-fold fecundity advantage of clonals (because they are all female, rather than 50% female), and G is the number of generations since the clonal lineage originated. Solving for the generation (G^*) when the clonals are virtually fixed (i.e., when $P_{clonal,G*} = x = [N\text{–}1]/N$) gives,

$$G^* \approx \frac{\ln(x) - \ln(1-x) - \ln(P_0)}{\ln(2)}. \tag{5.2}$$

If the demographic advantage is less than two-fold, say R-fold, the 2^G is replaced with R^G in equation 5.1, and ln(2) is replaced with ln(R) in equation 5.2. In Fig. 5.1, we plot the number of clonal individuals over time (generations) assuming that a single clonal individual is formed in generation-1, the clonals collectively have double the reproductive output of the sexuals, and that the population size is constant at

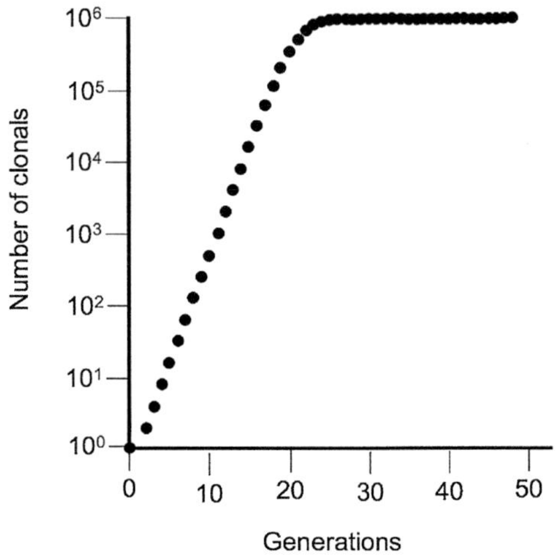

Fig. 5.1 The number of clonal individuals in a population of 1,000,000 after *G* generations

10^6 individuals. Counterintuitively, it takes only about 40 generations for the clonals to completely displace the sexuals even at this very large population size! Any advantage to sex must forestall and eventually overcome this remarkably strong, potential advantage to clonal reproduction.

Of course, for the two-fold benefit of clonal reproduction to be as strong as indicated in Fig. 5.1, the clonals must truly have doubled the reproductive output compared to the sexuals. However, as has been pointed out previously (Maynard Smith 1978), the demographic advantage of clonals can be greatly reduced in certain species. For example, (i) in species like socially monogamous birds in which males provision females and thereby augment their fecundity, (ii) in hermaphroditic organisms when the investment of resources in male function is far less than that in female function, and (iii) in species in which females alternate between long stretches of clonal reproduction and rarer bouts of sexual reproduction (e.g., rotifers and aphids). Even when these ameliorating conditions are not met, the full two-fold demographic advantage of clonals may be far from fully realized because the clonals are recently derived from sexual lineages (Stebbins 1950; White 1978; Maynard Smith 1978; Bell 1982). In this case, the cytological mechanisms used by clonals to bypass the ploidy-reduction process during egg production are frequently inefficient modifications of the sexual process (e.g., leading to instantaneous, genome-wide homozygosity), which can substantially reduce the fecundity of the clonals (Stebbins 1950; White 1973). Nonetheless, even when clonals achieve a more modest demographic advantage, this would be sufficient to provide them with a potent short-term advantage that must somehow be compensated for by counterbalancing advantages to sexual reproduction. For example, if the clonals had only a 1.1-fold reproductive advantage (instead of 2-fold) it would take only about 300 generations for the clonals to displace the sexuals.

In addition to the cost of producing males, another short-term cost of sexual reproduction is interlocus sexual conflict, in which males evolve traits that increase their fertilization success at a cost to the fecundity of their mates (Parker 1979; Rowe et al. 1994; Rice and Holland 1997). While these costs have been documented in a wide diversity of species (reviewed in Arnqvist and Rowe 2005), they have been studied most extensively in the *Drosophila melanogaster* laboratory model system (e.g., Fowler and Partridge 1989; Chapman et al. 1995; Rice 1996a). In this system, the net cost to females due to male-female interaction is estimated to be about a 22% reduction in their lifetime fecundity (Rice et al. 2005, 2006). If these costs are representative of most species under natural conditions, then interlocus sexual conflict may be an important, yet underappreciated, factor contributing to the cost of sex.

Clonal reproduction also has two major long-term benefits. First, selfish elements can impose a substantial genetic load in sexual lineages (Burt and Trivers 2006). Because the fitness of selfish elements is perfectly coupled with that of the clonal lineage within which they reside, there will be strong selection to ameliorate any harmful effects of selfish elements within clonal lineages. Second, intra-locus sexual conflict due to the expression of sexually antagonistic alleles may create a substantial gender load in sexual species that reduces their net reproductive rate (reviewed in Rice and Chippindale 2001; Arnqvist and Rowe 2005). In all-female clonal lineages that are derived from sexual progenitors, male-benefit/female-detriment sexually antagonistic alleles gradually will be purged by selection from the gene pool – thereby increasing the net reproductive rate of clonal compared to sexual lineages.

5.2.2 Long-Term Costs of Clonal Reproduction

Although some ecological models suggest that sex can provide benefits in a single generation (e.g., parent-offspring pathogen transmission, Rice 1983; and resistance to rapidly evolving pathogens, Hamilton 1980), most of the genetic benefits to sexual recombination occur over longer periods of time. One long-term disadvantage to clonal reproduction is a higher mutational load (Kimura and Maruyama 1966; Kondrashov 1988), which is expected to accumulate over a period of a few hundred to a few thousands of generations after a clonal lineage is generated–but this disadvantage only applies when epistatic interactions between mutations at different loci are common, and of a prescribed type (see below). A second long-term disadvantage of clonal reproduction occurs because new mutations are trapped in their original genetic background (background-trapping, Fisher 1930; Rice 1987, 1996b; Charlesworth 1994) which strongly influences the operation of natural selection (Felsenstein 1974). One manifestation of background-trapping in clonal populations is the Muller's ratchet process (Muller 1964; the stochastic, recurrent loss of the least-mutated class, causing mildly deleterious mutations to accumulate over time in the genomes of clonal organisms). However, a conceptually related process – drift decay (the stochastic, recurrent fixation of mildly deleterious mutations due to random genetic drift overpowering natural selection) – operates in sexual lineages, and as illustrated below, Muller's ratchet can be re-expressed as the same process.

Nonetheless, the accumulation of harmful mutations (retrogressive evolution) in finite populations operates more rapidly in clonal compared to sexual lineages (see below).

A second manifestation of background-trapping in clonal populations is that most beneficial mutations originate in low-fitness genetic backgrounds, causing them to be purged from the gene pool by natural selection (the "Ruby in the Rubbish" process, Fisher 1930; Manning 1983; Peck 1994; Charlesworth 1994). In sexual populations these same mutations can escape from an initially low-fitness genetic background and this greatly increases their probability of fixation (Barton 1995; Charlesworth 1994). The increased efficiency of recruiting new beneficial mutations in sexual lineages causes adaptive evolution to be faster and this effect is the genetic foundation for the benefits of sex due to the Red Queen process (Van Valen 1973; Jaenike 1978; Maynard Smith 1978; Hamilton 1980; Bell 1982) and greater genetic diversity in sexual lineages (Burt 2000). An additional consequence of background-trapping in the context of adaptive evolution is hitchhiking decay (Rice 1987). In this case, a beneficial mutation that fixes in a clonal lineage will drag along with it any mildly deleterious mutations that were in its genetic background of origin – so adaptation at one locus occurs at the expense of retrogressive evolution at other loci in the same genome.

Two extensions of the fact that sexual recombination removes background-trapping are: (i) In order for two different beneficial mutations to become fixed in a clonal population they must be sequentially produced in the same lineage (whereas in a sexual population these same mutations could both become fixed more rapidly because they can come together in the same genome by syngamy; (Fisher 1930; Muller 1932); and (ii) If two beneficial mutations are produced in different clonal lineages, and if each lineage has very similar net fitness, the rate of accumulation of both mutations will be slowed due to competition with the other clone (clonal interference – Clarke et al. 1994; Gerrish and Lenski 1998). Although there may be additional undiscovered potential advantages to sex that have not been described, we suspect that the extant tally of costs and benefits is essentially complete.

The faster retrogressive evolution and the slower adaptive evolution in finite clonal populations are collectively called the "Hill-Robertson effect" (Felsenstein 1974) because both can be expressed as a consequence of the smaller effective population size of clonal populations compared to their close sexual relatives. The Hill-Robertson effect is a reduction in the efficiency of natural selection that occurs because finite populations accumulate non-random gene associations that interfere with the process of natural selection.

From the above brief summary of the long-term disadvantages of clonal reproduction, it would appear that there are many unrelated long-term advantages to sexual recombination (see for a more complete review of all of the models for the disadvantage of clonal reproduction: Maynard Smith 1978; Kondrashov 1993; West et al.1999). In the following section, we create a general framework to illustrate why we believe that there is only one fundamental long-term advantage to sex in the context of competition between sexual and clonal lineages: sexual recombination increases the efficiency of natural selection – and the various models of unique advantages of sex in this context are all different ramifications of this single, fundamental advantage.

5.3 Fundamental Benefits to Sex

Recombination only has a net population genetic effect when there is linkage disequilibrium (a non-random association between alleles at different loci). Linkage disequilibrium can be produced deterministically by epistatic selection and stochastically by sampling error in finite populations (because all possible combinations of segregating genes within genomes, and new mutations, are not present in the proportions expected by chance). To illustrate, consider two favored SNPs that, by chance, only occur in coupling (i.e., only $A^+B^+$and A^-B^-haplotypes are present; positive linkage disequilibrium, LD). In this case, selection on one SNP reinforces that on the other, causing fixation of the beneficial haplotype to be relatively fast. But when chance leads to the occurrence of only the disfavored repulsion SNPs ($A^+B^-$and A^-B^+; negative LD), selection on each SNP interferes with that on the other, which causes fixation of the favored haplotype to be relatively slow, and hence the non-neutral polymorphism is expected to persist longer. In sum, the joint operation of sampling error and natural selection cause negative disequilibrium to predominate because it endures longer (Hill-Robertson interfering disequilibrium, or Hill-Robertson interference), which reduces the efficacy of natural selection. Below we extend this logic to graphically illustrate how genome-wide linkage disequilibrium strongly interferes with natural selection, and therefore why recombination – by reducing linkage disequilibrium – acts to facilitate the process of natural selection.

Our purpose in writing this chapter was not to break new theoretical ground concerning the adaptive significance of sex. We think that that the extant theory is essentially complete. Instead we tried to re-express existing theoretical findings with two graphical models so that they all can be expressed as a simple, single advantage to sex: an increased efficacy of selection when recombination is present. The two graphical models described below follow the logic from the analytical work of Kimura and Maruyama (1966) on mutational load, that of Felsenstein (1974) on selection in finite populations (the Hill-Robertson effect), and numerous other studies that have built upon these two fundamental papers.

5.3.1 Equilibrium Mutational Load in Clonal Lineages

Mutational load (L_{mut}) is the reduction in a population's mean fitness (W_{mean}) due to the accumulation of mutations in the gene pool. If the fitness of a mutation-free genotype is scaled to unity, then $L_{mut} = 1 - W_{mean}$. To solve for the equilibrium mutational load, suppose that a new population is begun with a single clonally reproducing organism that carries no harmful mutations. As the population grows over time, mildly deleterious mutations will be introduced at an average rate of U_{del} per genome per generation, creating a spectrum of fitness (mutational) classes upon which natural selection can operate. Beneficial mutations will also occur – but at a much lower rate – and these will be temporarily ignored until the following section on background-trapping. As harmful mutations accumulate, the variance in fitness among individuals increases, which causes the collective strength of selection at removing them to also increase (Fisher 1930). Eventually, sufficient genetic variation accrues to cause selection to remove deleterious mutations at the same rate that recurrent mutation introduces them and the proportion of each fitness class remains constant (mutation-selection equilibrium). For simplicity, we focus on the proportion of individuals in the fittest class ($P_{fittest}$), but the same result could be obtained by focusing on any fitness class (Kimura and Maruyama 1966).

At equilibrium, the proportion of individuals in the fittest class does not change, i.e., $P'_{fittest} = P_{fittest}$. Some of the progeny of the fittest class will receive new harmful mutations and will therefore be recruited to lesser fitness classes rather than the fittest class (Fig. 5.2). Because there are thousands of mutable loci in a genome and because, at most, only relatively few randomly located loci are expected to be mutated each generation, the number of new mutations per genome is expected to follow an approximate Poisson distribution. This distribution is the limit of the binomial distribution when the number of Bernoulli trials (a Bernoulli trial is a mutation being present or absent at each gene locus in the genome) is very large and the probability of a "success" – i.e., a mutation – is very small. Because $e^{-\mu}$ (also expressed here as $exp[-\mu]$, where μ is the mean number of "successes") is the expected zero-class of a Poisson distribution, $exp(-U_{del})$ is the expected proportion of offspring from the fittest class that carries no new mutations. All other offspring are "dead by mutation" with respect to the reproduction of the fittest class because they do not contribute to create this class anew. Each generation, the proportion of individuals in the fittest class is increased by natural selection by a factor of $W_{fittest}/W_{mean}$ (where $W_{fittest}$ is the fitness of the best class; see Crow and Kimura (1970, p. 179

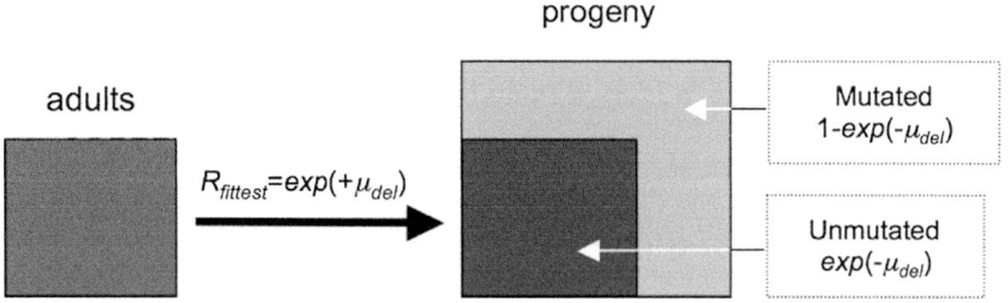

Fig. 5.2 "Death by mutation." Mutation causes a fraction of $1-exp(-\mu_{del})$ offspring from the fittest class to be newly mutated and therefore to be recruited to other, more mutated classes of the distribution of the number of deleterious mutations/genome. As a consequence, for the fittest class to reproduce itself ($R_{fittest} = 1$, after discounting for newly mutated offspring) its net reproductive rate must be $1 = R_{fittest} * exp(-\mu_{del}) \Rightarrow R_{fittest} = exp(+\mu_{del})$

for derivation) and it loses a fraction 1–$exp(-U_{del})$ of its fecundity to "death by mutation" (Fig. 5.2). At equilibrium,

$$P'_{fittest} = P_{fittest} * \text{(selective gain)} * \text{(fraction of offspring not lost by mutation)}$$

$$= P_{fittest} * \left(\frac{W_{fittest}}{W_{mean}}\right) * e^{-\mu_{del}}.$$

If we scale the fitness of the best class to be one, set $P'_{fittest} = P_{fittest}$, and solve for W_{mean},

$$W_{mean} = e^{-\mu_{del}}$$

So, as originally shown by Kimura and Maruyama (1966), the mean fitness of a clonal lineage will evolve to be a fraction $exp(-U_{del})$ of the fitness of the best class. When U_{del} is substantial, this constraint may prevent low fecundity groups such as vertebrates from persisting as clones. To see why, we can express fitness in the currency of the net reproductive rates (R), and then solve for the value of $R_{fittest}$ needed to achieve equilibrium, i.e., the requisite net reproductive rate of the best class needed to prevent mutations from accumulating in an open-ended fashion (note that R_{mean}= 1 at equilibrium; see Rice 1998 and Fig. 5.2),

$$R_{fittest} = e^{+\mu_{del}}.$$

If for example, when U_{del} =3, then the requisite net reproductive rate of the fittest class would need to be about 20, a value too large for organisms like humans and many other large vertebrates.

In summary, the equilibrium mutational load of a clonal population is 1–$exp(-U_{del})$ and the net reproductive rate of the best class needed to stop open-ended mutation accumulation – and eventual extinction – is $exp(+U_{del})$. These analyses about mutational load in clonal lineages made no assumptions about the shape of the distribution of fitness classes, nor the presence or absence of epistasis, so the results are independent of these factors – something that will not be true for sexual lineages (see below). We did assume, however, that there were no compensatory nor unconditionally beneficial mutations – which is obviously not true. When unconditionally beneficial mutations are rare relative to harmful mutations, they are expected to have little effect on the mutational load (Kimura and Maruyama 1966). Compensatory mutations, however, can be an important factor preventing open-ended mutation accumulation in both clonal and sexual populations, as described later in Section 5.3.3.

5.3.2 Contrasting Equilibrium Mutational Load in Sexual and Clonal Lineages

Dominance, epistasis, and the cost of removing harmful mutations: The mutational load of a population is determined by the cost of removing an average of U_{del}

mutations/genome each generation. In this section, we describe why interactions between mutations (dominance and epistasis) can strongly influence mutational load. Recall that for simplicity we have assumed that selection acts via differential mortality of genomes with different numbers of mutations – but our results will be unchanged when fitness is also mediated by differences in the fecundity of surviving individuals (see for discussion, Rice 1998). Since at equilibrium an average of U_{del} mutations/genome must be removed each generation, the more mutations that are removed per selective death the lower the mutational load. To illustrate this point, consider selection on a single locus in a diploid sexual population that is initially mutation-free. Next, let recurrent mutation cause mutations to accumulate until mutation-selection equilibrium is achieved. If all mutations were dominant and lethal, then one mutation would be removed per selective death (ignoring rare double mutants at the same locus in the same individual). At the other extreme, if the mutations were lethal and completely recessive, initially they would be hidden from selection in heterozygotes, so they would accumulate further – compared to the case of dominant mutations – until sufficient homozygotes were produced and culled by selection. However, now two mutations would be removed per selective death, meaning that only half as many individuals would die per generation. Selection is thus twice as efficient when the mutations are recessive and the mutational load is half of what it was when the mutations were dominant (Crow and Kimura 1970, p. 300).

Epistasis is an extension of the concept of dominance (masking of the expression of one allele by another at the same locus) to the case of multiple loci (masking the expression of alleles at one gene locus by those at one or more different loci). Epistasis, like dominance, can strongly influence mutational load by creating genome-wide linkage disequilibrium that strongly affects the efficiency of selection (the number of mutations removed per selective death).

A simple, graphical measure of genome-wide linkage disequilibrium: Before comparing the mutational load in a sexual vs. clonal lineage, we will build a graphical measure of genome-wide linkage disequilibrium. In an effectively infinite population, harmful mutations will occur at low frequency at all gene loci. In a single genome, there are a very large number of gene loci and at each gene locus the probability of carrying a deleterious mutation is very small. If there is a random association of mutations at each locus, then the total number of mutations in a genome represents the sum of a large collection of independent Bernoulli trials (one Bernoulli trial – a mutation present or absent – for each gene locus in the genome) in which the probability of success (a mutation being present at a specific locus) is very small. This relationship does not require all gene loci to have the same equilibrium frequency of mutations since the sum of different Poisson distributions also follows a Poisson distribution. Using the same logic as we did in the above section on the distribution of new mutations, the distribution of numbers of mutations per genome also is expected to follow a Poisson distribution. Therefore, the shape of the distribution of mutations per genome can be used as a metric of linkage disequilibrium in a population, the larger the deviation from a Poisson distribution, the greater the genome-wide linkage disequilibrium.

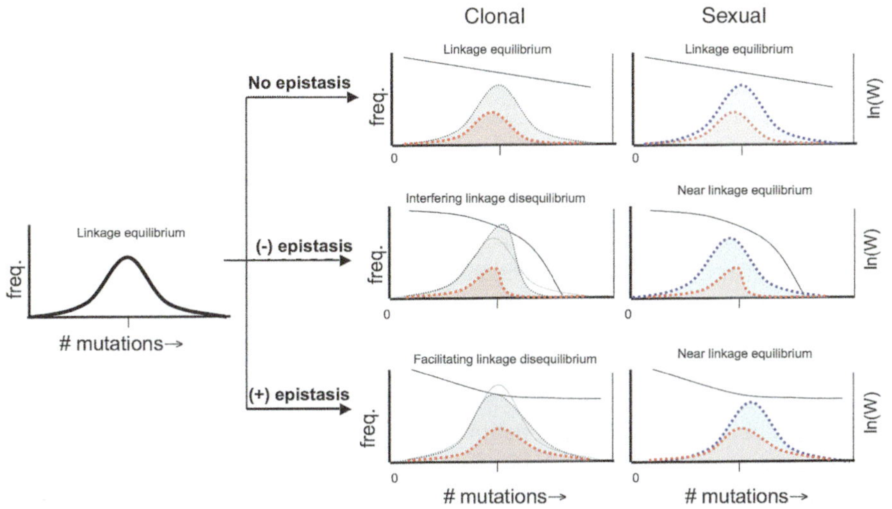

Fig. 5.3 We begin (*left*) with a population that is in mutation-selection balance and linkage equilibrium (a Poisson distribution of mutations/genome, left ordinate). Next, the effect of the form of the fitness function (*rows*) and reproductive mode (*columns*) on the shape of the distribution across two generations is plotted. *Black lines/curves* above the distributions represent the fitness functions (*right* ordinate with log-scale). Selection operates first (*red-dotted* distributions; surviving individuals), followed by mutation and clonal reproduction (*grey-dotted* distributions) or mutation and sexual reproduction (*blue-dotted* distributions). For reference, when reproduction is clonal, the linkage equilibrium distribution (Poisson, with μ set to the mean number of mutations in the population) is shown by a *grey-solid outline*. With clonal reproduction, (i) positive epistasis among mutations (less-steep than linear fitness function) creates facilitating linkage disequilibrium (fatter right-hand tail of distribution) in the next generation that offsets the disadvantage of this less efficient form of selection, and (ii) negative epistasis (steeper than linear fitness function) creates interfering linkage disequilibrium (shallower right-hand tail) in the next generation that offsets the advantage of this more efficient form of selection. Over multiple generations, the "compensating" linkage disequilibrium created by an epistatic fitness function is completely counterbalancing, so there is no net effect on equilibrium mutational load (Kimura and Maruyama 1966). In contrast, linkage disequilibrium created by an epistatic fitness function in one generation is strongly reduced in the next generation by recombination in sexual lineages. The reduction of linkage disequilibrium in sexuals prevents the counterbalancing effects seen in clonals and thereby permits the mutational load to change in response to epistasis among mutations

Selection builds linkage disequilibrium when there is epistasis among mutations: Consider a clonal population at the earliest embryo stage – a point in time just before the operation of natural selection begins. For simplicity we will again assume that all mutations are harmful, that each mutation reduces fitness by the same small amount (s), and that selection acts via differential survival. Lastly, as a simple starting point of reference, we further assume that the population is initially in linkage equilibrium, in which case the shape of the distribution of number of mutations per genome will be Poisson (see above section and Fig. 5.3, left).

Next, we explore the consequences of the operation of selection and mutation across one generation with and without epistasis among mutations at different loci.

First, consider the case of no epistasis. This condition implies that selection acts independently on different mutations. Accordingly, the fitness (W_n) of an individual carrying n mutations is multiplicative (i.e., $W_n = [1–s]^n$) because of the general result from probability theory that the probability of a collection of n independent events (not dying due to the expression of a mutation at each of the n mutated loci in the genome, with probability at each locus being $1–s$) is the product of the probability of each event (i.e., $[1–s]^*[1–s)^* \ldots = [1–s]^n$). This form of independent selection across loci is referred to as "multiplicative fitness". A semi-log plot of fitness (W_n) vs. number of mutations in the genome is expected to be linear (y = intercept + slope*x) because $\ln[W_n] = 0 + n^*\ln[1–s]$ (Fig. 5.3, top row). Because selection and mutation act independently of each other, and independently across loci in the genome, and because at equilibrium selection removes an average of U_{del} mutations per genome and mutations add this same value back, the distribution remains in linkage equilibrium after the joint action of mutation and selection – irrespective of the presence or absence of recombination (Kimura and Maruyama 1966; Fig. 5.3, top row).

When selection does not act independently across mutations at different loci, there is epistasis and the joint action of selection and mutation will act to change the shape of the distribution of mutations per genome. First consider a clonal population. When fitness declines in a steeper than linear fashion (e.g., when multiple mutations in the same genome reduce fitness more than would be predicted by the independent effects of each mutation), then genomes with more mutations are removed by selection at a higher rate than predicted by multiplicative fitness (negative epistasis: a steeper than linear slope, Fig. 5.3, second column, middle row). The subsequent effect of mutation and reproduction then causes the starting distribution in the next generation to deviate from a Poisson distribution (thinner right-hand tail) such that genomes with higher numbers of mutations per genome are underrepresented in the next generation, i.e. there is linkage disequilibrium (Fig. 5.3, second row of middle column). Lastly, a different type of linkage disequilibrium is created when the cost of a mutation diminishes as mutations accumulate in the genome (positive epistasis: a less steep than linear slope, Fig. 5.3, bottom row, middle column). In this case, the joint action of selection and mutation thicken the right-hand tail of the distribution in the next generation such that genomes with higher numbers of mutations are overrepresented compared to a Poisson distribution (Fig. 5.3, bottom row of middle column). In sum, in clonal populations, epistatic selection builds linkage disequilibrium, and this is manifested by a change in the shape of the distribution of mutations/genome, compared to a Poisson distribution. When recombination is present, it will act to breakdown the linkage disequilibrium that is built up by epistatic selection, and thereby influence the efficiency of selection in the next generations, as described in the next section.

A graphical depiction of genome-wide epistasis, selection and compensating linkage disequilibrium in clonals: When there is no epistasis, the joint action of mutation and selection creates no linkage disequilibrium in an infinitely large population, so recombination has no effect (Fig. 5.3, top row of right column) and the equilibrium load is expected to be the same with and without sex. Epistatic selection creates linkage disequilibrium – which is manifested through changes in the shape of the distribution of mutations/genome – predominantly by increasing or decreasing the thickness of the right-hand tail of the distribution compared to a Poisson distribution. When the epistasis is negative, the number of genomes with many mutations is lower, relative to linkage equilibrium, so it takes more selective deaths per generation to achieve mutation/selection equilibrium (Fig. 5.3, right column and middle row). Therefore, negative epistasis builds "interfering linkage disequilibrium," which will reduce the efficiency of selection in the subsequent generations. When epistasis is positive, selection creates linkage disequilibrium that thickens the right-hand tail of the distribution, so it takes fewer selective deaths per generation to achieve mutation/selection equilibrium (Fig. 5.3, right column, bottom row). In this case, positive epistasis builds "facilitating linkage disequilibrium," which will increase the efficiency of selection in the subsequent generations.

In the previous section on load in asexual populations we saw that the equilibrium mutational load must be $L_{mut} = 1\text{–}exp(-U_{del})$, irrespective of the presence or absence of epistatic selection. The reason that the load is unchanged by epistatic selection is due to its two counteracting effects when recombination is absent. With negative epistasis, selection culls genomes that are more highly encumbered with harmful mutations at a higher rate compared to multiplicative selection, and this acts to reduce the mutational load. However, this type of selection also creates interfering disequilibrium which makes highly encumbered genomes less common for selection to act upon in subsequent generations (thinner right-hand tail of the distribution of mutations/genome, Fig. 5.3 middle column, middle row). Because we know that the load is unchanged by epistasis in clonal lineages (i.e., $L_{mut} = 1 - W_{mean} = 1\text{–}\exp[-U_{del}]$, see earlier section), it must be true that these two effects are exactly counterbalancing. The same offsetting effects - but in the opposite direction – occur in the case of positive epistasis in clonal lineages. In this case, selection less effectively culls high fitness genotypes (compared to multiplicative fitness), but the facilitating disequilibrium created by positive epistasis (thicker right hand tail of the distribution of mutations/genome, Fig. 5.3, middle column, bottom row) makes these genomes more available for selection to act upon in subsequent generations –and the two effects are exactly counterbalancing. In sum, with clonal reproduction "compensating linkage disequilibrium" accrues in response to epistatic selection, and this causes epistasis to have no net effect on the mutational load of the population.

Recombination breaks down compensating linkage disequilibrium in sexuals: When recombination is present, there is a decoupling between the offsetting effects of epistatic selection and the linkage disequilibrium that it produces. With negative epistasis, selection is more effective at culling genomes that are highly encumbered with mutations, but recombination moves the distribution of mutations back toward linkage equilibrium (Poisson shape). As a consequence, selection in one generation interferes less with selection in the next generation, compared to the case of clonals described above because at least some of the interfering linkage disequilibrium is removed. When there is free recombination between all mutations in the genome, and all mutations are rare so that virtually all are heterozygous, then recombination removes virtually all linkage disequilibrium each generation. But when at least some mutations are linked (i.e., have a map distance less than 50 centiMorgans) and/or there is an appreciable frequency of mutational homozygotes, then recombination substantially reduces – but does not totally eliminate – linkage disequilibrium in each generation. In sexual lineages, the mutational load is decreased in response to negative epistasis because more mutations are removed per selective death, and recombination prevents compensating linkage disequilibrium from fully counterbalancing this effect in subsequent generations (Fig. 5.3, right column, middle row). With positive epistasis – in contrast – selection is less effective at culling genomes that are highly encumbered with mutations since recombination moves the distribution of mutations back toward linkage equilibrium (Poisson shape) so that selection in one generation does not augment selection in the next by creating facilitating linkage disequilibrium. In this case, the mutational load is increased because positive epistasis causes fewer mutations to be removed per selective death and recombination prevents compensating disequilibrium from fully accruing. Whether recombination increases or decreases the mutational load depends on the relative occurrences of epistatic vs. multiplicative fitness, and on the levels of positive vs. negative epistasis when it is present – a topic we return to in Sections 5.4 to 5.5.

5.3.3 A Graphical Depiction of Background-Trapping and the Accumulation of Favourable and Harmful Mutations

Genetic polarization in clonals: Consider a clonal population and for simplicity assume that the distribution of fitness classes is approximately normally distributed (Fig. 5.4; the shape of the distribution will not be important in what follows). The value of U_{del} determines the width (variance) of the distribution – the larger the value the broader the distribution. Because the genomes are non-recombining, they are analogous to multiple alleles at a single haploid locus. Whenever there is variation among genotypes for net fitness, this will give rise to a selective sweep by the fittest class and eventually all individuals in the population will be descended from the fittest class. Note that this type of selective sweep by the fittest class does not depend on the presence of new beneficial mutations. Such a selective sweep is not expected to purge all genetic variation, however, because as the sweep ensues, new harmful mutations accumulate and the distribution in Fig. 5.4 will therefore represent a dynamic steady-state in which the fittest lineages are recurrently sweeping through the population –accumulating new harmful mutations as they do so. As a consequence, the distribution is "polarized" with a net movement of mutated genotypes over generations from right to left ("genetic polarization," Rice 1987, 1996b). In effect, genetic polarization causes a stream-like flow of genotypes across the distribution from right to left, with the flow generated by the recurrent accumulation of new mutations by non-recombining genomes.

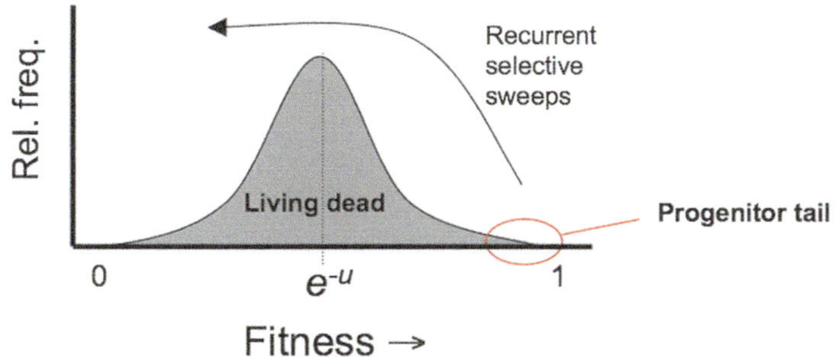

Fig. 5.4 Recurrent selective sweeps by the fittest genomes in a clonal population (the "progenitor-tail") create a dynamic steady-state, in which lesser fit genomes (the "living-dead", which constitute the majority of the population when U_{del} is substantial) are continually being replaced by recurrently mutated descendents of the fittest genotypes. Genomes in the living-dead (most of the population) are therefore marching toward eventual extinction. Only mutations (beneficial or harmful) that originate in the progenitor tail (or that move their recipient genome to this subpopulation) can become fixed in the population, so the size of the progenitor tail is the effective size of a clonal population. As a consequence, and compared to sexuals, (i) fewer beneficial mutations accumulate because they are trapped in the living-dead (slower progressive evolution), and (ii) more harmful mutations accumulate because they need only drift to fixation in the small progenitor-tail (trapped in high-fitness genotypes) before they rapidly spread to the remainder of the population due to selection on their superior genetic background (faster retrogressive evolution)

Clonal reproduction causes nearly all lineages in the population to be marching toward eventual extinction because any lineages that are not a member of the fittest class will eventually be competitively displaced (Fisher 1930). Therefore, the only lineages with an evolutionary future are the minority residing in the fittest class. When both beneficial and harmful mutations occur, and when selection coefficients are variable, then there is recurrent change in the rank order among lineages with similar fitness values (neighbor-shuffling). Because of neighbor-shuffling and recurrent selective sweeps of the fittest class, genotypes residing in the extreme right-hand tail of the distribution (the "progenitor-tail", rather than the single fittest class; Rice 1987, 1996b) have an evolutionary future. The vast majority of lineages, residing in the remaining part of the distribution, are marching toward eventual extinction, and hence they are collectively the "living-dead" (Fig. 5.4). In sum, a clonal population can be bisected into two functional components: the living-dead and the progenitor-tail – the greater the standing variance in heritable fitness, the greater the disparity in size between the two components. When the genome-wide deleterious mutation rate is large, the number of individuals residing in the progenitor tail will be far fewer than the number residing in the living-dead.

Background-trapping and favorable mutations: Consider the fate of a new beneficial mutation in clonal vs. sexual lineages (see Fisher 1930; Manning 1983 and Peck 1994; for the original development of these ideas). We can compare the efficiency of selection in clonal and sexual lineages by calculating the probability of fixation of the mutation. The arguments below implicitly assume the organism is haploid. However, nothing is changed if the organism is diploid except that the term "fixation" of a new mutation in a clonal organism means that it achieves a temporary maximal possible frequency of 50% (rather than 100%). This distinction is necessary because a new mutation initially becomes fixed only on one homolog in a diploid clonal genome, with fixation on the other homolog occurring only after a tandem mutation, or a gene conversion, generates the mutation on this chromosome.

As a starting point of reference, suppose that there was no variation in fitness among genomes in a finite population. Under these conditions, the probability of fixation is approximately equal to $2s$, where s is the selective advantage of the allele in the heterozygous state (Crow and Kimura 1970, p. 426). Most individual beneficial mutations are expected to be lost by sampling error, even in very large populations, because they start out as single copies and are susceptible to drifting to loss by sampling error while their total number of copies is small – the smaller the population size the greater the level of sampling error. When the selection coefficient (s) of a beneficial mutation is larger, it accumulates faster to a copy number suitably large to protect it from loss via sampling error. When s is small relative to the reciprocal of N_e (the effective population size – the smaller this value, the greater the sampling error in gene frequency across generations), sampling error overpowers natural selection and the beneficial mutation is no more likely to fix than a neutral mutation (Crow and Kimura 1970, p. 426). More specifically, and to a close approximation, selection is only effectual on a new beneficial mutation (i.e., it overpowers drift) when $s>1/N_e$ (Kimura 1983, p. 44; Li 1978).

With clonal reproduction, any beneficial mutation entering the living-dead is destined to extinction by virtue of it being trapped in a lineage that is marching toward eventual extinction (unless the new mutation increases the fitness of the recipient genetic background to the point where it moves it into the progenitor-tail). When a beneficial mutation enters the progenitor-tail, selection will overpower drift only when $s > 1/N_{e(clonal)} = 1/N_{prog\text{-}tail}$. In a recombining population mutations freely move between genetic backgrounds and selection on genetic backgrounds, on average, has only a small influence on the probability of fixation of a new mutation (Barton 1995). In this case, selection is expected to overpower drift when $s>1/N_e$(Kimura 1983, p. 44; Li 1978). Since the carrying capacity of sexual and clonal populations is expected to be similar, the N_e of a sexual population is expected to generally be far larger than $N_{prog\text{-}tail}$ of an asexual population (assuming U_{del} is not trivially small – so there is substantial heritable variation in fitness). The probability of fixation (and hence the rate of accumulation) of beneficial mutations will thus be far greater in recombining lineages (Fig. 5.5).

Background-trapping and deleterious mutations: In a finite population with no variation in fitness among genetic backgrounds, sampling error will overpower selection and permit the mutation to fix by drift whenever $|s|<1/N_e$ (Li 1978; Kimura 1983, p. 44). Because recombination allows mutations to escape from their initial genetic background (i.e., it removes linkage disequilibrium that is generated by chance), this same constraint ($|s|<1/N_e$) is also a suitable approximation – on average – in finite sexual populations when there is substantial variation among individuals in their genetic backgrounds (Barton 1995). In a clonal population, the effective size of the population is approximately the size of the progenitor-tail and

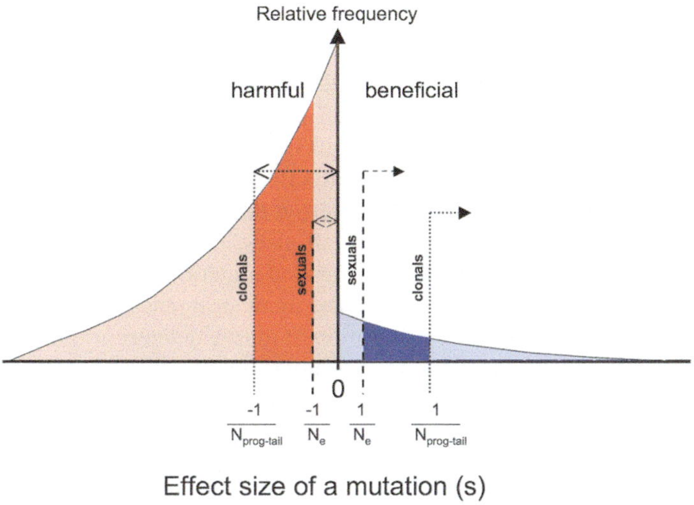

Fig. 5.5 In order for selection to overpower drift, the $|s|$ must be greater than the reciprocal of the effective population size ($1/N_e$ in a sexual population and $1/N_{prog\text{-}tail}$ in a clonal population, with $N_e \gg N_{prog-tail}$ when U_{del} is substantial). As a consequence, a smaller proportion of the beneficial mutation spectrum (*dark blue area*) and a larger portion of the harmful mutation spectrum (*dark red area*) will accumulate in clonal populations. The actual shapes of the mutation spectra are unknown, but the shapes shown are meant to reflect the much higher frequency of harmful compared to beneficial mutations, and the higher frequency of minor effect compared to major effect harmful mutations

harmful mutations can accumulate when $|s|<1/N_{prog-tail}$. Because N_e of a sexual population is expected to be much larger than $N_{prog-tail}$ of a clonal population (assuming U_{del} is not trivially small –so there is substantial heritable variation in fitness), a much larger proportion of the spectrum of harmful mutations will drift to fixation in a clonal compared to a sexual lineage (Fig. 5.5). Moreover, the time to fixation of harmful mutations can be substantially faster in clonal populations. In a sexual population, a harmful mutation must spread through the entire population by stochastic drift – a very slow process unless the population size is quite small. In a clonal population of the same size, however, the mutation needs only accumulate by chance to fixation in the relatively small progenitor-tail, after which it will spread to the remainder of the population by deterministic selection on its superior genetic background – a relatively fast process.

Due to the factors described in the above paragraph, the fittest class of a clonal population can drift to loss – and be replaced by the next best class – whenever $|s|<1/N_{prog-tail}$. So the Muller's ratchet process can be described as genetic drift within the progenitor-tail. Once a gene fixes in the progenitor-tail, it will ultimately fix in the entire population due to selection on its high-fitness genetic background (Rice 1987, 1996b). In the past, Muller's ratchet process has been repeatedly (an erroneously) described to operate only when the number of individuals in the best class is small. This is not true. The more accurate description is that open-ended mutation accumulation occurs in clonal populations whenever $|s|$ is small relative to the size of the progenitor-tail ($N_{prog-tail}$), no matter how large the size of the best class. Past models of Muller's ratchet have assumed for analytical tractability that s is constant and equal to the estimated average effect of a harmful mutation. When this over-simplification is removed by permitting s to have a distribution of effects with a large proportion of mutations with very small effect (Eyre-Walker et al. 2006), there always will be mutations satisfying the criterion that $|s| < 1/N_{prog-tail}$, so Muller's ratchet will turn in clonal populations of all sizes, but at a slower rate in larger populations. This same phenomenon (open-ended mutation accumulation) occurs in sexual populations (drift-decay), but with the more stringent constraint: $|s| < 1/N_e$. To escape from the open-ended accumulation of mildly deleterious mutations, compensatory mutations must continually be recruited in both sexual and clonal lineages – a process that is less efficient due to background-trapping in clonal lineages.

Because a larger range of harmful and a smaller range of beneficial mutations accumulate in clonal compared to sexual lineages, the rate of progressive evolution is reduced and the rate of retrogressive evolution is increased in clonal lineages (Fig. 5.5). These processes will cause the competitive ability of clonal lineages to erode over time – relative to their sexual competitors – and hence the cost of clonal reproduction will increase over evolutionary time. The magnitude of this effect depends primarily on U_{del}: the greater its magnitude, the broader the distribution of fitness, and hence the smaller the size of the progenitor-tail relative to that of the living-dead. However, the cost of clonal reproduction also depends on the population size (N). The difference in the range of mutations that accumulate in sexual vs. clonal populations depends on the difference, $\Delta = 1/N_e - 1/N_{prog-tail}$ (Fig. 5.5). When N is very large (e.g., $N > 10^9$, as occurs in many micro-invertebrates), then $N_{prog-tail} = P_{fittest} * N$ can also be a large value. In this case, Δ is trivially small, and hence the efficiency of selection is similar in sexual and clonal populations.

5.4 Clonal vs Sexual Lineages

The demographic cost of producing males, in addition to any costs of male-induced harm to their mates, potentially gives clonal lineages a substantial short-term advantage. This potentially large advantage has led many researchers to conclude that sex must provide a short-term counterbalancing advantage that compensates sexual lineages (for review, see Maynard Smith 1978; Kondrashov 1993; West et al. 1999). However, one simple observation makes it clear that the advantage of sex is frequently long-term rather than short-term: it is common to observe clonal lineages (with extant, closely related sexual lineages) that have persisted for many hundreds to tens of thousands of generations (Butlin and Griffiths 1993). If the main advantage of sex were short-term, then one would not expect clonals to persist so long in the face of competition with sexuals.

In order for sexual lineages to persist in the face of competition from clonals there needs to be: (i) a much less than two-fold advantage to clonal reproduction (this will lengthen the time available for sexuals to gain from their long-term advantages),

(ii) some sort of spatial refuge that prevents the clonals from displacing the sexuals over their entire range, or (iii) a combination of these two circumstances. Once one or more of these condition is met, the fitness of clonal lineages is expected to continuously decline, relative to the sexuals, due to their slower progressive evolution and faster retrogressive evolution, and possibly due to an higher deterministic mutational load if the genome-wide deleterious mutation rate is high and negative epistasis is common relative to positive epistasis. This cumulative decay in the relative fitness of the clonals will eventually offset the short-term advantages to clonal reproduction and the sexuals are expected to eventually prevail. If clonals are recurrently generated from sexuals, however, as occurs in the dandelion (*Taraxacum officinale*; Vijverberg et al. 2004; see also Chapter 22), a steady-state mixture of sexuals and clonals should occur, but the clonals will always be closely related to the sexuals and the clonals will not give rise to major adaptive radiations.

The advantage of sex depends critically on the genome-wide mutation rate to deleterious mutations (U_{del}). The larger the value of U_{del}, the broader the distribution of heritable variation in fitness and the smaller the proportion of individuals in the progenitor-tail - and hence the smaller the effective population size of a clonal ($N_{prog-tail}$) compared to a sexual population (N_e). Recent evidence indicates that the genome-wide mutation rate to harmful mutations is substantial ($U_{del} = 1.2$ in *D. melanogaster*, Haag-Liautard et al. 2007). Laboratory studies of the same species support the conclusion that multicellular species harbor substantial standing heritable variation for fitness, and hence there appears to be substantial scope for the operation of background-trapping (Rice and Chippindale 2001). As a consequence, the effective population sizes of closely related sexual and clonal lineages are expected to differ substantially, and this will cause progressive evolution to be faster and retrogressive evolution to be slower in sexual lineages (the Hill-Robertson effect, Felsenstein 1974). One offsetting factor, however, is the observation that clonal lineages typically have substantially less standing genetic variation compared to their sexual relatives (Bell 1982). Assuming this reduced genetic variation translates into less standing genetic variation for fitness, the cost of background-trapping in clonals will be diminished, although we doubt that it will ever be removed altogether.

In multicellular organisms, there is still no consensus of the relative levels of positive vs. negative epistatic selection (recently, but briefly, reviewed in Beerenwinkel et al. 2007), so the importance of mutational load in favoring sexual lineage is still unknown. However, the most exhaustive study on this topic (based on only 27 doubly mutated genomes) indicates that both positive and negative epistasis were about equally common (Elena and Lenski 1997; Beerenwinkel et al. 2007), at least in one species of bacteria reared under laboratory conditions. Until we have a better measure of the relative levels of positive and negative epistasis, the importance of mutational load in selecting for sexual vs. clonal lineages will remain obscure.

Assuming that negative epistasis is sufficiently common, the value U_{del} is also of critical importance in evaluating the importance of mutational load in favoring sex – the larger the value, the larger the potential benefits to sex. The value $exp(U_{del})$ is the requisite net reproductive rate of the fittest class to achieve an equilibrium distribution of mutations per genome. If this value cannot be achieved (because the intrinsic rate of increase is low, such as occurs in many vertebrates) then the fittest class will be recurrently lost by the deterministic operation of recurrent mutation pressure and a clonal linage will experience open-ended fitness decay. Even when this requisite net reproductive rate can be achieved, the magnitude of the sexuals' advantage with respect to mutational load increases with the value of U_{del}.

We have characterized the advantage of sex in the currency of recombination's capacity to break down linkage disequilibrium. The more intuitive idea is that sex is favored because it creates heritable phenotypic diversity (Burt 2000). In most cases creating heritable genetic diversity and reducing the linkage disequilibrium are interchangeable concepts. However, we prefer to describe the advantage of recombination in the currency of reduced linkage disequilibrium because it more accurately describes the capacity for recombination to prevent background-trapping and the "interfering disequilibrium" that contributes to increased mutational load, which are the core functional advantages of sex.

No discussion of the advantages and disadvantages of clonal reproduction would be complete without considering the major exception to the rule: the clonal lineages of bdelloid rotifers, which have persisted over long periods of evolutionary time and given rise to a major adaptive radiation (Welch and Meselson 2000; see also Chapter 13). How has this clade escaped the disadvantages of asexual reproduction, which have prevented its spread in nearly all other clonal lineages? We think that the existence of the bdelloid rotifer clade is the exception that proves the rule. So long as U_{del} is not so large that the requisite net reproductive rate to prevent open-ended mutation accumulation is unachievable by a species, theory does not predict that clones cannot persist – only that the clonals will be susceptible to competitive displacement by close sexual relatives. If bdelloid rotifers occupy a unique ecological niche (so that there are no ecological equivalents that are sexual), and if early on the short-term benefits of clonality permitted all sexual lineages of bdelloid rotifers to be competitively driven to extinction, then theory predicts that the remaining clonal lineages could persist and undergo adaptive radiation. Progressive evolution would be slowed and retrogressive evolution speeded, compared to a sexual lineage, but if sexual competitors were absent, this would pose no insurmountable problem.

One might argue that theory predicts an eventual demise of all-clonal clades like the bdelloid rotifers due to open-ended deleterious mutation accumulation (i.e., due to Muller's ratchet, or more generally $|s| < 1/N_{prog-tail}$). However, there is also open-ended deleterious mutation in sexual species due to drift-decay (i.e., for mutations with $|s| < 1/N_e$). Both sexuals and clonals can escape this downward trajectory in mean fitness by compensatory evolution when one compensatory mutation ($s > 1/N_e$ or $s > 1/N_{prog-tail}$) can compensate for multiple mildly deleterious mutations ($|s| < 1/N_e$ or $|s| < 1/N_{prog-tail}$) (see Rice 1998 for fuller discussion). Another reason why bdelloids may be exceptional is their very large population size (many billions of individuals). Because the advantage of sex via background-trapping is proportional to the magnitude of the difference ($\Delta = 1/N_e - 1/N_{prog-tail}$, Fig. 5.5), in bdelloids, both $1/N_e$ and $1/\ 1/N_{prog-tail}$ are both expected to be very small values, and therefore the advantage of sex should be small. In summary, once the bdelloids successfully made the transition to an all clonal clade, and assuming they have adapted to a unique ecological niche, the persistence and adaptive radiation of the bdelloids is fully consistent with theory.

5.5 Conclusions

Although there have been a wide diversity of models developed to describe the disadvantage of clonal reproduction (reviewed in Maynard Smith 1978; Kondrashov 1993; West et al. 1999), we think that they can all be expressed as a single disadvantage: the efficacy of natural selection is lowered in clonal genomes due to the buildup of linkage disequilibrium that interferes with natural selection. In the context of an increased mutational load of clonals, the linkage disequilibrium is generated by natural selection (Kimura and Maruyama 1966, and graphically illustrated in Fig. 5.3). It is not yet clear whether recombination is expected to substantially increase the competitive ability of sexual lineages with respect to mutational load because we do not yet have a firm empirical determination of whether negative epistasis is sufficiently predominant to cause the load of sexuals to be substantially smaller than that of clonals (Beerenwinkel et al. 2007). We do know, however, that the mutation rate to harmful mutations is probably large enough in most macroscopic species to make this advantage potentially large (Haag-Liautard et al. 2007). In the context of selection on new mutations, the higher linkage disequilibrium of clonals is built up by chance (Hill-Robertson effect, Felsenstein 1974, and illustrated graphically in Fig. 5.4). Because most natural populations of macroscopic organisms are expected to harbor substantial standing genetic variance for fitness because of a substantial harmful deleterious mutation rate (Haag-Liautard et al. 2007), the effective population size of clonal populations will be typically far smaller than that of their sexual competitors ($N_{e(clonal)} = N_{prog-tail} << N_{e(sexual)}$). This reduced $N_{e(clonal)}$ will cause the clonals' competitive ability to continually wane with time, compared to their sexual relatives, due to the faster accumulation of harmful mutations and the slower accumulation of beneficial and compensatory mutations. Theory, however, does not preclude the persistence of clonals when they have very large N, high intrinsic reproductive potential ($R_{fittest} \geq e^{+\mu_{del}}$), and do not have closely matched sexual competitors.

Acknowledgements WRR was supported by grants from the National Science Foundation (DEB-0128780 and DEB-0111613) and UF by a stipend from Lennader's foundation and a fellowship from the Wenner-Gren Foundations. We thank K. Schoenrock for comments on the manuscript.

References

Arnqvist G, Rowe L (2005). Sexual conflict. Princeton University Press, Princeton
Barton NH (1995) Linkage and the limits to natural selection. Genetics 140: 821–841

Bell G (1982) The masterpiece of nature: the evolution and genetics of sexuality. Croom Helm and University of California Press, London and Berkeley
Beerenwinkel N, Pachter L, Sturmfels B, Elena SF, Lenski RE (2007) Analysis of epistatic interactions and fitness landscapes using a new geometric approach. BMC Evol Biol 7: 60
Burt A (2000) Perspective: sex, recombination, and the efficacy of selection – was Weismann right? Evolution 54: 337–351
Burt A, Bell G, Harvey PH (1991) Sex differences in recombination. J Evol Biol 4: 259–277
Burt A, Trivers R (2006) Genes in conflict: the biology of selfish genetic elements. Belknap Press, Cambridge, MA
Butlin RK, Griffiths HI (1993) Ageing without sex? Nature 364: 680
Chapman T, Liddle LF, Kalb JM, Wolfner MF, Partridge L (1995) Cost of mating in *Drosophila melanogaster* females is mediated by male accessory gland products. Nature 373: 241–244
Charlesworth B (1994) The effect of background selection against deleterious mutations on weakly selected, linked variants. Genet Res 63: 213–227
Clarke DK, Duarte EA, Elena SF, Moya A, Domingo E, Holland JJ (1994) The Red Queen reigns in the kingdom of RNA viruses. Proc Natl Acad Sci USA 91: 4821–4824
Crow JF, Kimura M (1970) An introduction to population genetic theory. Harper and Row, New York
Elena SF, Lenski RE (1997) Test of synergistic interactions among deleterious mutations in bacteria. Nature 390: 395–398
Eyre-Walker A, Woolfit M, Phelps T (2006) The distribution of fitness effects of new deleterious amino acid mutations in humans. Genetics 173: 891–900
Felsenstein J (1974) Evolutionary advantage of recombination. Genetics 78: 737–756
Fisher RA (1930) The genetical theory of natural selection. Oxford University Press, Oxford
Fowler K, Partridge L (1989) A cost of mating in female fruitflies. Nature 338: 760–761
Gerrish PJ, Lenski RE (1998) The fate of competing beneficial mutations in an asexual population. Genetica 102–103: 127–144
Haag-Liautard C, Dorris M, Maside X, Macaskill S, Halligan DL, Charlesworth B, Keightley PD (2007) Direct estimation of per nucleotide and genomic deleterious mutation rates in *Drosophila*. Nature 445: 82–85
Hamilton WD (1980) Sex versus non-sex versus parasite. Oikos 35: 282–290
Jaenike J (1978) A hypothesis to account for the maintenance of sex within populations. Evol Theory 3: 191–194
Keightley PD, Otto SP (2006) Interference among deleterious mutations favours sex and recombination in finite populations. Nature 443: 89–92
Kimura M (1983) The neutral theory of molecular evolution. Cambridge University Press, Cambridge, UK
Kimura M, Maruyama T (1966) Mutational load with epistatic gene interactions in fitness. Genetics 54: 1337–1351
Kondrashov AS (1988) Deleterious mutations and the evolution of sexual reproduction. Nature 336: 435–440
Kondrashov AS (1993) Classification of hypotheses on the advantage of amphimixis. J Hered 84: 372–387
Lattorff HMG, Moritz RFA, Fuchs S (2005) A single locus determines thelytokous parthenogenesis of laying honeybee workers (*Apis mellifera capensis*). Heredity 94: 533–537
Li W-H (1978) Maintenance of genetic variability under joint effect of mutation, selection and random drift. Genetics 90: 349–382
Lively CM, Lloyde DG (1990) The cost of biparental sex under individual selection. Am Nat 135: 489–500
Manning JT (1983) The consequences of mutation in multiclonal asexual species. Heredity 50: 15–19
Maynard Smith J (1978) The Evolution of sex. Cambridge University Press, Cambridge, UK
Muller HJ (1932) Some genetic aspects of sex. Am Nat 66: 118–138

Muller HJ (1964) The relation of recombination to mutational advance. Mutat Res 1: 2–9
Otto SP, Lenormand T (2002) Resolving the paradox of sex and recombination. Nat Rev Genet 3: 252–261
Parker GA (1979) Sexual selection and sexual conflict. In: Blum MS and Blum NA (eds) Sexual selection and reproductive competition in insects. Academic Press, New York, pp. 123–166
Peck JR (1994) A ruby in the rubbish: beneficial mutations, deleterious mutations and the evolution of sex. Genetics 137: 597–606
Rice WR (1983) Parent-offspring pathogen transmission: a selective agent promoting sexual recombination. Am Nat 121: 187–203
Rice WR (1987) Genetic hitch-hiking and the evolution of reduced genetic activity of the *Y* sex chromosome. Genetics 116: 161–167
Rice WR (1996a) Sexually antagonistic male adaptation triggered by experimental arrest of female evolution. Nature 361: 232–234
Rice WR (1996b) Evolution of the Y sex chromosome in animals. BioScience 46: 331–343
Rice WR (1998) Requisite load, pathway epistasis, and deterministic mutation accumulation in sexual versus asexual populations. Genetica 102–103: 71–81
Rice WR, Chippindale AK (2001) Sexual recombination and the power of natural selection. Science 294: 555–559
Rice WR, Holland B (1997) The enemies within: intergenomic conflict, interlocus contest evolution (ICE), and the intraspecific Red Queen. Behav Ecol Sociobiol 41: 1–10
Rice WR, Linder JE,Friberg U, Lew TA, Morrow EH, Stewart AD (2005) Inter-locus antagonistic coevolution as an engine of speciation: assessment with hemiclonal analysis. Proc Natl Acad Sci USA 102 (Suppl. 1): 6527–6534
Rice WR, Stewart AD, Morrow EH, Linder JE, Orteiza N, Byrne PG (2006) Assessing sexual conflict in the *Drosophila melanogaster* laboratory model system. Philos Trans R Soc Biol 361: 287–299
Rowe L, Arnqvist G, Sih A, Krupa JJ (1994) Sexual conflict and the evolutionary ecology of mating patterns: water striders as a model system. Trends Ecol Evol 9: 289–293
Stebbins GL (1950) Variation and evolution in plants. Columbia University Press, New York
Van Valen L (1973) A new evolutionary law. Evol Theory 1: 1–30
Vijverberg K, Van der Hulst RGM, Lindhout P, Van Dijk PJ (2004) A genetic linkage map of the diplosporous chromosomal region in *Taraxacum officinale* (common dandelion; Asteraceae). Theor Appl Genet 108: 725–732
Welch DM, Meselson M (2000) Evidence for the evolution of bdelloid rotifers without sexual reproduction or genetic exchange. Science 288: 1211–1215
Werren JH (1997) Biology of Wolbachia. Annu Rev Entomol 42: 587–609
West SA, Lively CM, Read AF (1999) A pluralist approach to sex and recombination. J Evol Biol 12: 1003–1012
Westwood MN (1993) Temperate-zone pomology, physiology and culture. Timber Press, Portland, Oregon
White MJD (1973) Animal cytology and evolution, 3rd ed. Cambridge University Press, Cambridge, UK
White MJD (1978) Modes of speciation. WH Freeman, San Francisco
Williams GC (1975) Sex and evolution. Princeton University Press, New Jersey

Chapter 6
Geographical Parthenogenesis: General Purpose Genotypes and Frozen Niche Variation

Robert C. Vrijenhoek and E. Davis Parker Jr.

> *"It is not entirely clear, however, how forms whose genetic system must be very inflexible manage to become adapted to new environments when they* do *get transported to them: the apparent ecological versatility in* space *seems to be at variance with their lack of ecological versatility in* time*" (Original italics; MJD White, 1973, p. 748).*

Abstract Clonally reproducing all-female lineages of plants and animals are often more frequent at higher latitudes and altitudes, on islands, and in disturbed habitats. Attempts to explain this pattern, known as geographical parthenogenesis, generally treat the parthenogens as fugitive species that occupy marginal environments to escape competition with their sexual relatives. These ideas often fail to consider the early competitive interactions with immediate sexual ancestors, which shape alternative paths that newly formed clonal lineages might follow. Here we review the history and evidence for two hypotheses concerning the evolution of niche breadth in asexual species – the "general-purpose genotype" (GPG) and "frozen niche-variation" (FNV) models. The two models are often portrayed as mutually exclusive, respectively viewing clonal lineages as generalists versus specialists. Nonetheless, they are complex syllogisms that share common assumptions regarding the likely origins of clonal diversity and the strength of interclonal selection in shaping the ecological breadth of asexual populations. Both models find support in ecological and phylogeographic studies of a wide range of organisms, and sometimes generalist and specialist traits (e.g., physiological tolerance, microspatial preference, seasonal abundance, food habits, etc.) are found together in an asexual organism. Ultimately, persistent natural clones should be viewed as microspecies in ecological models that consider spatial and temporal heterogeneity rather than multi-locus genotypes in simplistic population models.

R.C. Vrijenhoek (✉)
Monterey Bay Aquarium Research Institute, Moss Landing, 7700 Sandholdt Road, Moss Landing, CA, 95039, USA
e-mail: vrijen@mbari.org

I. Schön et al. (eds.), *Lost Sex*, DOI 10.1007/978-90-481-2770-2_6,

6.1 Introduction

All-female reproduction offers twice the generative potential of biparental sex, but strictly asexual species are rare in the plant and animal kingdoms. The overwhelming prevalence of biparental sex is paradoxical, given the cost of producing males and risks associated with finding and obtaining mates. With few exceptions, biparental sex persists in taxa that can produce all-female clones (Williams 1975); therefore, everything else is not equal between sexual and asexual lineages. Considerable discussion has focused on identifying immediate benefits that can compensate for the demographic costs of sex (West et al. 1999; West and Peters 2000). Comparative studies have focused on ecological circumstances that might favor asexual lineages, because exceptions to the 'rule of sex' can provide insight into reasons for the rule (Vrijenhoek 1989a). Geographical parthenogenesis is one such an exception (see also Chapter 8). Vandel (1928, 1940) first recognized that parthenogenetic arthropods tend to have wider and more northern distributions than their sexual relatives in Europe. Since his initial observations based mostly on comparisons of glaciated versus non-glaciated distributions, many other examples have been reported of parthenogens being more prevalent at higher altitudes, in anthropogenically disturbed environments, at the margins of a species range, on islands versus the mainland, and in successional versus climax communities (Glesener and Tilman 1978; Bell 1982; Lynch 1984; Parker 2002).

With these diverse examples, a number of hypotheses have been proposed to explain geographical parthenogenesis: (1) all-female reproduction provides superior colonizing ability and reproductive assurance (Baker 1965; Tomlinson 1966); (2) biotic uncertainty favors genotypically diverse sexuals in species-packed central environments (Glesener and Tilman 1978; Hamilton et al. 1981); (3) destabilizing hybridization favors displacement of parthenogens from the sexual range (Lynch 1984; cf. Paulissen et al. 1988); (4) heterozygosity assurance due to cloning favors asexual lineages in subdivided metapopulations (Vrijenhoek 1985; Haag and Ebert 2004); and so forth. As opposed to hypotheses that parthenogens simply escape from competition with their immediate sexual ancestors, we consider two commonly invoked hypotheses that focus on selection pressures occurring while parthenogenetic clones first arise from their sexual progenitors.

The "general-purpose genotype" (GPG) and "frozen niche-variation" (FNV) hypotheses appear at first glance to be mutually exclusive explanations for the success of parthenogenetic lineages. The GPG model proposes that individual clones (see Chapter 9 on clone definition) have broader environmental tolerances than their sexual relatives, whereas the FNV model proposes that individual clones are specialists. We consider selective pressures that might favor generalist versus specialist clones in central versus marginal environments and examine mechanisms for clonal origin that might shape interclonal selection. Finally, we address the possibility that these hypotheses are not mutually exclusive. Hereafter, we use the term parthenogenesis for all-female clonal reproduction in a broad sense. We restrict the term thelytoky for obligate parthenogenesis that does not require sperm. Gynogenesis and hybridogenesis, on the other hand, are sperm-dependant forms parthenogenesis

(Beukeboom and Vrijenhoek 1998). Gynogens are strictly clonal and require sperm only to initiate embryogenesis. Hybridogens are hemiclonal and incorporate sperm and express paternal genes, but transmit only the maternal genome to ova (for examples, see Chapters 16, 19 and 20).

6.1.1 Adaptation at the Margins

To consider factors that might contribute to geographic parthenogenesis, we must first review some conditions that are expected to occur on the edge of a species range. Mayr (1954) was among the first to indicate that adaptation to marginal environments might be limited by gene flow from central populations. This idea implies that the genetic composition of marginal populations differs in some way from populations at the center of a species range, and that gene flow contributes to outbreeding depression (Wallace 1959). Partially isolated marginal populations are expected to exhibit reduced genetic diversity due to population fluctuations, genetic drift and possibly directional selection for local adaptations (da Cuhna et al. 1950; Carson 1968). Nevertheless, adaptive gains may be swamped by episodic gene flow, unless marginal populations evolve means to increase their isolation (García-Ramos and Kirkpatrick 1997). Therefore, selfing, parthenogenesis and chromosomal rearrangements that impede recombination will be favored in marginal populations by closing the genome to introgression (Wallace 1959). Selection that favors genetic diversity and effective recombination in central populations is replaced by selection that favors closure of the genome at the periphery of a species range (Stalker 1954; Templeton 1982).

The dynamics of marginal populations have been considered in several attempts to explain geographical parthenogenesis. Metapopulation simulations by Ladle et al. (1993) suggest that dispersal advantages and Red Queen processes (see also Chapter 7) might favor the accumulation of parthenogens in low-density marginal habitats to which they escaped during a parasite-free stage in the life cycle. Peck et al. (1998) modeled parthenogenesis along a gradient in annual reproductive output that decreased with latitude. In their simulations excess immigration from more productive southern populations diluted the open genomes of locally adapted, northern, sexual populations. In contrast, locally adapted clones were protected from dilution and tended to accumulate in the north due in part to their dispersal advantages. However, Horne and Martens (1999) argue that the absence of sexual freshwater ostracods in northern Europe is not just a consequence of the superior colonization abilities of clones. Fossil evidence indicates that sexual ostracods also inhabited northern Europe during post-glacial times and were replaced as climates gradually became more stable. Consequently, they argue that modern climatic stability favored the replacement of sexual lineages by competitively superior clones. Though these arguments were intended to explain geographical parthenogenesis, they do not explore the nature of clonal competitive superiority, and thus do not exclude aspects of either the GPG or FNV models.

A putatively "new" idea about geographical parthenogenesis was recently outlined by Haag and Ebert (2004). Sexual populations inhabiting subdivided marginal habitats are subject to metapopulation processes that should lead to genetic drift and contribute to inbreeding depression. Apomixis, however, protects clones from inbreeding depression associated with founder events. This potential benefit of parthenogenesis was clearly articulated in earlier publications (Vrijenhoek and Lerman 1982; Vrijenhoek 1985; Niklasson and Parker 1994) and called "heterozygosity assurance" (Beukeboom and Vrijenhoek 1998; Vrijenhoek 1998b; Kearney and Shine 2004). Evidence for the temporary benefits of heterozygosity assurance is clearly documented in studies of sexual and clonal forms of *Poeciliopsis* (Vrijenhoek and Lerman 1982; Vrijenhoek 1989b; Lively et al. 1990).

6.2 General Purpose Genotype (GPG)

Vandel (1928) first associated broad environmental tolerance with geographical parthenogenesis. He reported that parthenogenetic races of the terrestrial isopod, *Trichoniscus*, are distributed in cooler and drier habitats than their sexual relatives. White (1973), who studied the wingless Australian grasshopper *Warramaba*, also associated parthenogenesis with broadly tolerant genotypes, noting that "Natural selection may possibly tend to favor genotypes that are especially plastic, phenotypically, in the case of thelytokous populations (replacing genetic polymorphism by physiological adaptation).

The concept of a general-purpose genotype was defined by Baker (1965) to describe the life history syndrome associated with weedy species of plants. He defined it as a genotype characterized by the ability to grow in a multitude of climates and edaphic situations, i.e. a genotype with broad environmental tolerance. Baker's original description was explicitly comparative: colonizing species of weeds were more tolerant of physical stresses, more plastic in flowering phenology, more likely to have vigorous vegetative growth, and more likely to be self-compatible or apomictic than their closest non-colonizing relatives.

Soon after, allozyme studies revealed that many parthenogenetic taxa are composed of multiple clones (Solbrig 1971; Hebert 1974; Lokki et al. 1975, 1976; Parker and Selander 1976; Saura et al. 1976a, b; Vrijenhoek et al. 1977, 1978). How did these findings bear on the concept of a GPG?

A simple model (Fig. 6.1) for the evolution of GPGs was articulated by Parker et al. (1977). While selection in sexual populations acts primarily on individual genes with small additive effects, selection among clones acts on epistatic interactions within composite genotypes. The continuous generation of new clones from sexual ancestors would favor the persistence of clones that can survive in all environments – i.e. genotypes with broad ecological tolerance. The key idea in this model was that the geometric mean of fitness (replacement rate) had to be greater than zero for clonal lineages to persist. It was not necessary that a successful clone should have the highest fitness in all environments, only that it can survive in the most environments. Lynch (1983) advanced a similar argument, but the model

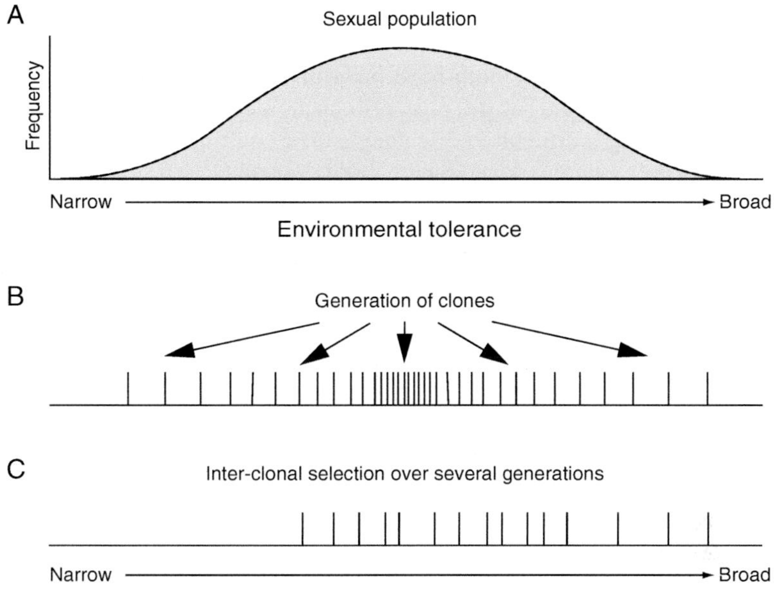

Fig. 6.1 The origin of general purpose genotypes in colonizing parthenogenetic taxa (adapted from Niklasson 1995). **(a)** A sexual population that varies in the range of tolerances (*narrow* to *broad*) that individual genotypes have for a particular form of environmental stress. **(b)** A range of genotypes is frozen among clones produced by the sexual ancestor. **(c)** Natural selection over several generations fixes clones with wider than average tolerance

focused on mutations within clonal lineages as the primary source variation (Lynch 1985; Lynch and Gabriel 1987) rather than multiple origins of clones from sexual progenitors. The race to improve fitness within a clonal lineage may be compromised over the long term, however, by mutational meltdown (Lynch et al. 1993). Multiple origins of clones from genetically variable sexual ancestors, on the other hand, have been demonstrated experimentally to create new genotypic combinations that exhibit considerable fitness variance and opportunities for interclonal selection (Annest and Templeton 1978; Wetherington et al. 1989b). Templeton (1982) emphasized a radical shift in the unit of selection that accompanies clonal origins – from individual genes in sexual progenitors to whole genomes in clonal populations – which could result in distinct evolutionary trajectories for closely related parthenogenetic and sexual lineages under identical environmental conditions. Multiple origins of clones from genetically variable sexual ancestors are now well established for many asexual taxa (White 1978; Parker 1979; Vrijenhoek 1979; Suomalainen et al. 1987; Hebert et al. 1989; Moritz et al. 1989). Occasionally "hopeful monsters" (*sensu* Goldschmidt 1940) with wide environmental tolerances may arise among new clones. Consequently, the capacity of parthenogens to generate robust and flexible physiologies "cheats the system," because polyphyletic clones generated by recombination in the immediate sexual ancestors creates the variation on which inter-clonal selection acts to produce GPGs.

6.2.1 Elevated Ploidy and Hybridity

Elevated ploidy and hybridity both have been invoked as rapid means to evolve broadly tolerant genotypes, but these genetic phenomena are often confounded in parthenogens, making it difficult to disentangle their contributions to fitness (Parker and Niklasson 2000; Lundmark and Saura 2006). Vandel (1940) suggested that geographical parthenogenesis was a direct consequence of selection for larger and more robust polyploids in extreme environments. Polyploid races of the terrestrial isopod, *Trichoniscus*, are distributed in cooler and drier habitats than their diploid sexual relatives (Vandel 1928). Polyploid races of the curculionid weevil, *Otiorrynchus*, are more cold tolerant than their smaller sexual relatives (Lindroth 1954). Apparently, the larger body sizes of polyploids allow them to occupy more extreme environments in the north of Europe, but many of these polyploids also are hybrids, so heterosis might be involved. A few appear to be autopolyploids, however, lending credence to Vandel's idea that elevated polyploidy alone may be a sufficient explanation for geographical parthenogenesis (Lundmark and Saura 2006).

Kearney (2005) argued that selection for the stabilization of hybrid genotypes is more important than polyploidy or all-female reproduction per se. Essentially all parthenogenetic vertebrates and many parthenogenetic insects are inter-specific or inter-populational hybrids; thus clonal reproduction has been viewed as a means for preserving heterotic or intermediate genotypes (White 1970; Schultz 1971). Elevated heterozygosity is evident in many parthenogenetic animals (reviewed in Vrijenhoek 1990), and the physiological breadth of some unisexual-hybrid fish (*Poeciliopsis monacha-lucida* and *Phoxinus eos-neogaeus*) and a frog (*Rana esculenta* = *R. ridibunda-lessonae*) was interpreted by several researchers as evidence for heterosis (Bulger and Schultz 1979; Tunner and Nopp 1979; Schlosser et al. 1998). Nevertheless, experimental studies with *Poeciliopsis monacha-lucida* did not reveal evidence for "spontaneous heterosis" in laboratory-synthesized hybrids (Wetherington et al. 1987). However, enhanced larval growth rates and size at metamorphosis in laboratory-synthesized *Rana ridibunda-lessonae* hybrids was interpreted as evidence for heterosis (Hotz et al. 1999). It is difficult to generalize from so few experimental studies. More studies are obviously needed, but caution should be exercised in interpreting the results from such studies, because enhanced performance of hybrids for some life history and somatic traits does not necessarily translate into enhanced fitness (euheterosis) – witness the sterile mule.

Parthenogenesis has also been viewed as a means to preserve hybrid phenotypes that are optimally adapted to transitional habitats (ecotones) between the ranges of the sexual progenitors (Wright and Lowe 1968; Moore 1977). Hybrid intermediacy differs from the heterosis hypothesis, because it does not view hybrids as superior; instead, it views hybrids as inferior competitors within the ranges of their progenitors. Thus, parthenoforms are thought to persist as ecological fugitives that escape competition by occupying marginal habitats, which are likely to be intermediate for hybrids (Moore 1984). The geographical distribution of allodiploid and allotriploid *Poeciliopsis* within and between river systems is broadly consistent with this view (Moore et al. 1970; Thibault 1978). Nonetheless, the hybrid intermediacy hypothesis fails to account for stable coexistence of multiple clones in geographically intermediate habitats (see Moore 1984 for a discussion of this issue).

6.2.2 Evidence for General Purpose Genotypes

Evidence for and against the GPG model comes from two sources: (1) field and experimental studies that compare parthenogenetic taxa with their closest sexual relatives (summarized in Table 6.1); and (2) comparative population genetic and phylogeographic studies of parthenogenetic taxa that focus mainly on local differentiation versus wide geographic ranges of clones (summarized in Table 6.2). We present cases that illustrate the range of conclusions derived from different taxa and from varying criteria used to define generalist versus specialist clones. Many of the cited authors produced earlier papers on this subject, which can be accessed from the papers we cite.

Experimental and observational studies (Table 6.1) offer mixed support for the GPG model. Acute tolerance to stress and performance across a range of environments (breadth of tolerance) are both criteria for GPG's (Baker 1965). Ideally, direct ecological or physiological comparisons should be made between parthenogens and their closest sexual relatives, but in some cases (*Artemia*, *Octolasion*, *Sitobion*) this was not possible, due to the lack of a known sexual ancestor. In some comparisons, males were found to be less tolerant, but sexual and parthenogenetic females races showed no consistent differences (*Pycnoscelus*, *Nemasoma*). The studies on *Antennaria* illustrate one of the most powerful experimental designs to test the GPG hypothesis. Ramets of sexual and apomictic genotypes were reared under several temperature and desiccation regimes, and performance was plotted against mean performance of all the genotypes in these environments. This method of joint-regression (Bierzychudek 1989; Michaels and Bazzaz 1989) is optimal for testing the GPG model, but it requires replicated sexual genotypes, which limits the design to plants and a few animal groups in which researchers can experimentally clone sexually produced genotypes. Genotypes that show a "flatter" response to environmental sources of variance than the mean of all genotypes can then be argued to be "generalists".

Two problems often occur with claims that geographically widespread parthenogens are GPGs (Table 6.2). First, parthenogens should be examined genetically to determine whether their distribution is comprised of many independent clonal genotypes or a single clonal lineage. The second problem lies in the evolutionary ages of clonal lineages. A geographically widespread clone might have evolved very recently to occupy a narrow but universally available niche and depend on human transport or habitat disruption (Vrijenhoek 1979). Many of the cases in Table 6.2 involve aphid pests of crop plants (see also Chapter 25). These pests have spread with human assistance, even to new continents, during the last few hundred years. In these cases, wide distributions of particular clones could be random consequences of serendipitous introductions (cf. Fenton et al. 1998; Delmotte et al. 2002;

Table 6.1 Experimental and observational evidence for and against the general purpose genotype hypothesis in parthenogenetic taxa

Order: family	Taxon	Mode[a]	Character	Evid[b]	Comments	References
Annelida						
Haplotaxida Lumbricidae	*Octolasion tyrtaeum*	P	Distribution across soil types and forest types	+	Two dominant clones distributed in all habitats. Frequencies not correlated with soil texture or pH. Rare clones at a few localities. No evidence for differentiation between dominant clones	Jaenike et al. (1980)[c]
Arthropoda						
Anostraca Artemiidae	*Artemia pathenogenetica*	P	Salinity and thermal tolerance	–	Thelytokes have narrower tolerance than sexual species for salinity and temperature. Differences between 2N and 4N thelytokes from the same population. Sexual ancestors are unknown	Browne and Wanigasekera (2000)
Cladocera Daphniidae	*Daphnia pulex*	P	Salinity and thermal tolerance	±	Permanent ponds have reduced clonal diversity implying selection for generalist clones. No significant differences between obligate clones and sexually produced clones in sensitivity to salinity stress. Sexuals less sensitive to thermal stress	Lynch (1983) and Weider (1993)
Acariformes Penthaleidae	*Penthaleus major*	P	Spatial distribution	–	Clonal diversity decreases toward population margin, but no evidence that a general-purpose clone increases in frequency at margins	Robinson et al. (2002)
Julida Nemasomatidae	*Nemasoma varicorne*	P	Desiccation resistance	–/?	Males of sexual race less tolerant than females. Differences between sexual and clonal females inconclusive	Enghoff (1976) and Hoy Jensen et al. (2002)
Coleoptera Curculionidae	*Otiorrhynchus dubius*	P	Thermal tolerance	+	3N thelytokes are larger and more cold tolerant than 2N sexuals	Lindroth (1954)
Dictyoptera Blaberidae	*Pycnoscelus surinamensis*	P	Life history traits and desiccation resistance	–	Fertility and viability differences among clones consistent with FNV hypothesis. Sexual *P. indicus* males less tolerant than females. Highly tolerant sexual race in lab for over 30 years	Niklasson and Parker (1994), Parker and Niklasson (1995) and Gade and Parker (1997)

Table 6.1 (continued)

Order: family	Taxon	Mode[a]	Character	Evid[b]	Comments	References
Hemiptera Aphididae	*Myzus persicae*	P	Host plant associations and performance	±	Predominant clones do not segregate by host plants in Scotland. No difference in geometrical mean fitness between obligate and cyclical parthenogens across experimental host plants	Fenton et al. (1998), Vorburger et al. (2003a)
	Sitobion avenae	P	Host plant performance	+	Two dominant clones on all host plants in two successive years in France. Agricultural practices may have favored generalist clones	Haack et al. (2000)[c]
Ostracoda Cyprididae	*Heterocypris incongruens*	M	Salinity and thermal tolerance	–	Four clones identified across two Belgian population samples. No evidence for GPG clones, but may be consistent with FNV model	Van Doninck et al. (2002)
Darwinulidae	*Darwinula stevensoni*	P	Salinity and thermal tolerance	+	Several clones from European sites have broad tolerance to environmental stresses, though some differences exist among geographical clones	Van Doninck et al. (2002)
	Penthesilenula brasiliensis	P	Salinity and thermal tolerance	+	Widespread species with broad salinity tolerance	Van Doninck et al. (2003)
	P. aotearoa	P	Salinity and thermal tolerance	–	Limited distribution and salinity tolerance	Van Doninck et al. (2003)
	Vestalenula molopoensis	P	Salinity and thermal tolerance	–	Limited distribution and salinity tolerance	Van Doninck et al. (2002)
Chordata						
Anura Ranidae	*Rana esculenta*	H	Asymmetric competition, GxE interactions	±	Crowding and competition affect hybridogenetic *R. esculenta* less than sexual *R. lessonae*, but fitness of individual hemiclones is strongly habitat dependent	Semlitsch (1993) and Semlitsch et al. (1997)
Cypriniformes Cyprinidae	*Phoxinus eos-neogeneus*	G/H	Spatial distribution, tolerance of hypoxia	±	Widespread single clone outperforms related sexuals under hypoxic conditions, leading to partial niche segregation. Situation is complicated by 2N/3N mosaicism	Elder and Schlosser (1995), Schlosser et al. (1998), and Doeringsfeld et al. (2004)

Table 6.1 (continued)

Order: family	Taxon	Mode[a]	Character	Evid[b]	Comments	References
Cyprinodontiformes Poeciliidae	*Poeciliopsis monacha-lucida* complex	H/G	Thermal tolerance	±	2N hemiclones exceed parental sexual forms, but 3N gynogenetic clones vary among one another in survival of thermal stresses	Bulger and Schultz (1979)
	P. monacha-occidentalis	H	Thermal tolerance	–	Hybridogens do not exceed parental sexual species for thermal tolerance	Bulger and Schultz (1982)
Squamata Teiidae	*Aspidoscelis tesselatus*	P	Physiological	–	Parthenogens do not exceed sexual relatives for burst speeds and endurance	Cullum (1997)
Gekkonidae	*Heteronotia binoei*	P	Physiological, reproductive, geographical	–	Parthenogens do not exceed sexual relatives for burst speeds but have greater endurance. Parthenogens more cold tolerant but have greater water loss, a higher standard metabolic rate, and lower fecundity than sexuals. Geographic range of parthenogens is broader than sexuals but "environmental niche" may be narrower	Kearney and Porter (2004), Kearney and Shine (2005) and Kearney et al. (2005)
Mollusca						
Neotaenioglossa Hydrobiidae	*Potamopyrgus antipodarum*	P	Salinity tolerance, feeding rate, growth and reproduction	±	Introduced European populations exhibit wide tolerance and relatively constant fitness ranks across a range of environments. No sexuals present	Jacobsen and Forbes (1997) and Dybdahl and Kane (2005)
Magnoliaphyta						
Asterales Asteraceae	*Antennaria parlinii*	A	Light and nutrient gradients	±	Apomicts respond more evenly across range of nutrients, but sexuals better at some treatments	Michaels and Bazzaz (1989)
	A. parvifolia	A	Desiccation and thermal tolerance	+	Apomicts less sensitive to variation in environmental conditions than sexuals	Bierzychudek (1989)
	Erigeron annuus	A	Competition and shading	–	Apomict performed no better across experimental treatments than sexual relative *E. philadelphicus*	Kenny (1996)

[a] Reproductive mode: A = apomictic; G = gynogenesis; H = hybridogenesis; P = thelytoky or obligate parthenogenesis; M = mixed sexual and apomictic parthenogenesis.
[b] Evidence: + is positive; – is negative; and ? is equivocal.
[c] No direct comparisons made with a sexual ancestor (see text).

Table 6.2 Phylogeographic evidence for and against the GPG hypothesis for parthenogenetic taxa

Order: family	Taxon	Mode[a]	Marker(s)	Evid[b]	Comments	References
Arthropoda Hemiptera Aphididae	*Diuraphis noxia*	P	Microsatellites	–	No widespread clones across 38 localities in Iran. High population structure, with $F_{ST} = 0.23$	Dolatti et al. (2005)
	Myzus persicae	P	Microsatellites	+	Two clones among obligate parthenogens in Victoria, Australia. Low population structure, with $F_{ST} = 0.021$	Vorburger et al. (2003b)
	Rhopalosiphum padi	P	Allozymes, microsatellites	+/?	Widespread clones persist over 4 years in S. France. Clones sampled from various host plants, but no data on host range	Delmotte et al. (2002)
	Sitobion miscanthi, *S. fragariae*	P	Microsatellites, SSCP	+/?	Both geographic partitioning and widespread clones in New Zealand	Wilson et al. (1999)
	*S.*miscanthi, *S. fragariae*	P	Microsatellites	+	Single dominant clones within 4 chromosomal races of *S. miscanthi,*one clone of *S. fragariaei* in Australia	Sunnucks et al. (1996)
Julida Nemasomatidae	*Nemasoma varicorne*	P	AFLPs	+	Single monophyletic clone in Denmark, England and Poland; >30% sequence divergence from closest potential sexual ancestor (see text)	Hoy Jensen et al. (2002)
Ostracoda Cyprididae	*Heterocypris incongruens*	M	Allozymes	+/?	Multiple clones are found, but a single clone dominates 66% of the localities (mostly rice fields) sampled in N. Italy. Is widespread clone a rice field specialist?	Rossi et al. (2006)
Darwinulidae	*Darwinula stevensoni*	P	mtDNA, ITS1	+	Ancient worldwide asexual species with a single ITS1 sequence but 3.8% mtDNA sequence divergence among lineages. Lives in wide range of aquatic habitats	Schön et al. (1998)
	Microdarwinula zimmeri	P	Clonal diversity not reported	+/?	Associated with mosses on four continents	Van Doninck et al. (2003)

Table 6.2 (continued)

Order: family	Taxon	Mode[a]	Marker(s)	Evid[b]	Comments	References
	Penthesilenula brasiliensis	P	Clonal diversity not reported	+/?	Found on four continents in springs, ponds and lakes.	Van Doninck et al. (2003)
	P. aotearoa	P	Clonal diversity not reported	–	Only known from type locality in Brazil	Van Doninck et al. (2003)
	Vestalenula molopoensis	P	Clonal diversity not reported	–	Restricted to dolomite springs in S. Africa	Van Doninck et al. (2003)
Chordata						
Squamata Teiidae	*Aspidoscelis tesselata*	P	Allozymes	+	One multi-locus 2N genotype at 12 of 18 localities ranging across 1000 km	Parker and Selander (1976)
			Skin grafts	+	One histocompatibility clone across nearly entire range of the 2N race	Maslin (1967)
Mollusca						
Neotaenioglossa Thiaridae	*Melanoides tuberculata*	P	mtDNA	+	Two haplotypes co-occur in all freshwater habitats on French Polynesian Islands	Myers et al. (2000)
Cnidaria						
Actiniaria Edwardsiidae	*Nematostella vectensis*	P	RAPDs, AFLPs	±	Single dominant clone (= 61% of population) in S.E. England, but regional population structure in New England, USA	Pearson et al. (2002) and Darling et al. (2004)

[a] Reproductive mode: A = apomictic; G = gynogenesis; H = hybridogenesis; P = thelytoky or obligate parthenogenesis; M = mixed sexual and apomictic parthenogenesis.
[b] Evidence: + is positive; – is negative; and ? is equivocal.

Vorburger et al. 2003b). Other cases involving "natural" post-Pleistocene dispersals provide more convincing evidence for the long-term persistence of one or a few clonal genotypes (Hoy Jensen et al. 2002). However, wide distributions of individual clones (superclones according to Vorburger et al. 2003b) following recent colonization events suggests that these clones have characteristics of GPGs, because they must have survived temporal fluctuations between generations to be dispersed over such wide areas.

The GPG model for geographic parthenogenesis has considerable appeal, but mixed support in the literature. Given the various ways in which parthenogenesis evolves (hybridization, polyploidization, spontaneously, by infection, etc.) and the differing ages of parthenogenetic lineages, it is not surprising that a diversity of patterns can be found. The prevalence of polyploid and hybrid genotypes among parthenogens does not negate the generality of this model. It only suggests that multiple processes in addition to post-formational mutations can generate the fitness variation that allows natural selection to fix GPG's, but the same is true for specialist genotypes under the FNV model (below). In this sense polyploidy and hybridization should be viewed as sources of variation that can drive the phenotypic evolution of asexual populations, rather than the *raison d'être* of parthenogenesis (cf. Kearney 2005, 2006; Lundmark 2006). Regardless of the ways in which broadly tolerant genotypes may evolve, clonal reproduction provides the additional benefit of reducing fitness costs associated with meiosis and the production of "unbalanced genomes" (Parker and Niklasson 2000).

6.3 Frozen Niche-Variation (FNV)

When organisms colonize open habitats with low species diversity, they often expand their phenotypic distribution and niche breadth, a phenomenon called "ecological release" (MacArthur and Wilson 1967). Sexual reproduction impedes disruptive selection and niche diversification, because recombination forces offspring distributions to regress to the population mean. Selfing, assortative mating, limited dispersal, and chromosomal rearrangements are ways to counteract antagonistic recombination and foster diversification in sexual species (Wallace 1959; Maynard Smith 1962; Antonovics 1968). Parthenogenesis completely eliminates recombination; therefore, White (1970) suggested that new parthenogenetic lineages might be adapted to new ecological niches. Roughgarden (1972, p. 684) concluded from a modeling study that "an asexually reproducing population is capable of very rapid ecological release, ... an important reason why asexual populations should be good colonists." The model partitions niche width into within- and between-phenotype components that contain (1) the breadth of resources used by each phenotype, and (2) the variety of phenotypes in the population. Given an adequate source of clonal variation, an asexual population should rapidly attain an optimal phenotypic distribution, because the between-phenotype component of niche width "is more malleable to the force of natural selection than in sexual populations" (Roughgarden 1972, p. 712).

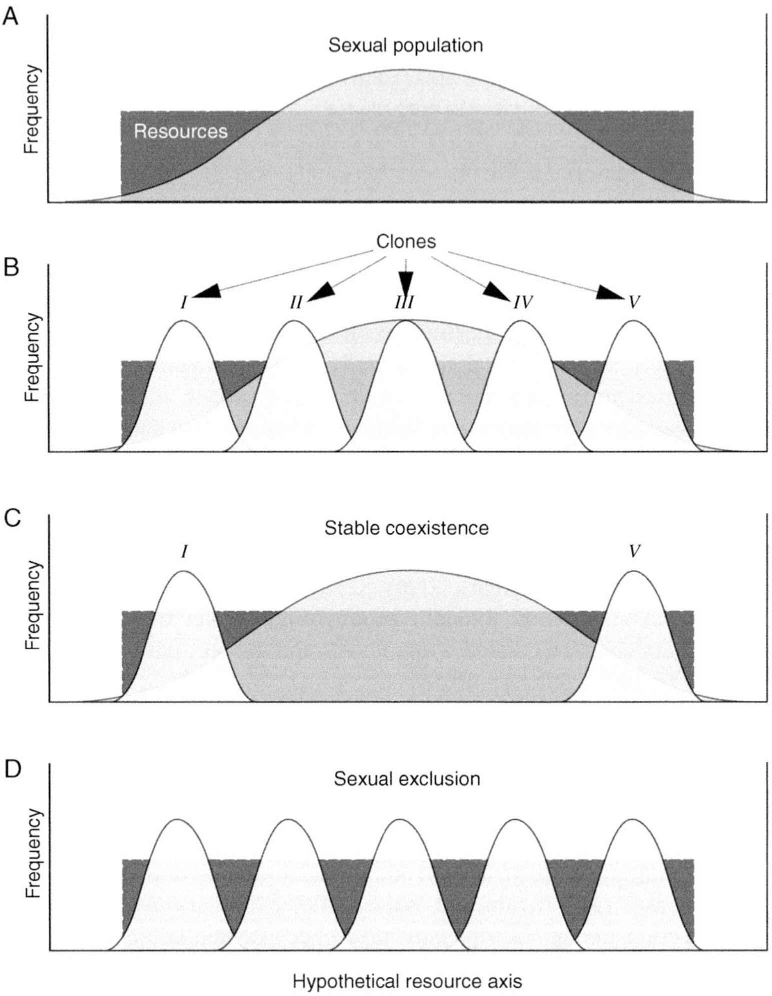

Fig. 6.2 Frozen Niche variation and asymmetric competition between new clones and their sexual progenitors. **(a)** A sexual population (*broad curve*) exhibits genetic variation for the utilization efficiency of a natural resource that is evenly distributed (*dotted line*). **(b)** A range of genotypes is frozen among clones produced by the sexual ancestor. **(c)** Natural selection rapidly eliminates clones that substantially overlap one-another and the centrally distributed sexual phenotypes, leading to stable coexistence if rate of clonal origins is not too high. **(d)** Too high a rate of clonal formation will eclipse resource use by the sexual ancestor and lead to its extinction

The Frozen Niche Variation model (Vrijenhoek 1979, 1984) was directly stimulated by Roughgarden's (1972) ideas and a need to explain the stable coexistence of *Poeciliopsis* fish clones in spatially heterogeneous desert streams (Vrijenhoek 1978). The model, as illustrated in Fig. 6.2, involves three stages: (1) multiple clones arising from genetically variable sexual ancestors provide frozen genotypic variation; (2) interclonal selection eliminates clones that significantly overlap the niches of established clones and the sexual progenitors; and (3) selection fixes an array of specialized clones that efficiently partition underutilized resources. A few specialized clones should be able coexist with a broad-niched sexual ancestor as long as the rate of clonal synthesis is not too high; however, " a broad panel of efficient specialist clones could competitively exclude the sexual host..." (Vrijenhoek 1979, p. 792).

Bell's (1982) Tangled Bank model led to a similar conclusion, but his goal was broader – to explain the maintenance of sex rather than the coexistence of clones. The idea stems from Weisman's (1889) view that the diversity generated by sex increases ecological efficiency. Phenotypic differences among sexual genotypes should decrease competition among siblings (Williams and Mitton 1973) and thereby increase the breadth of resources exploited by the population as a whole (Ghiselin 1974). The Tangled Bank model incorporated ideas about the maintenance of genetic diversity in multi-niche environments (Levene 1953; Strobeck 1974). Individual genotypes are regulated by density-dependent factors and imperfect competition within their respective subniches; therefore, rare genotypes suffer the least competition and fitness becomes frequency-dependent. Sexual recombination makes each individual unique and therefore rare and favored. Bell (1982, p. 131) concluded that a single specialized "clone cannot displace the sexual population from the whole of its ecological range . . . [but] If the number of clones were as great as the number of sexual genotypes the sexual population would have no refuge . . ." His conclusion was anticipated by empirical studies that showed how clonal diversity of *Poeciliopsis* hybridogens correlated with increased abundance of asexual fish relative to the sexual fish (Vrijenhoek 1979); however, a similar association has not been observed in hybridogenetic waterfrogs (Hotz et al. 1994).

Theoretical analyses have explored many aspects of the Tangled Bank and FNV models in fine- vs. course-grained environments and in temporally stable vs. fluctuating environments. Asymmetric competition - in which clones have a greater inhibitory effect on identical clone-mates than on genetically diverse sexual individuals – tends to facilitate coexistence, as long as the sexuals maintain a broader overall niche than that of the combined clones (Ghiselin 1974; Bell 1982; Case and Taper 1986; Koella 1988; Gaggiotti 1994; Lomnicki 2001). Clones lost due to genetic drift or temporal fluctuations in fitness can be replaced with new clones arising from the sexual ancestors, but the rate of clone-formation will influence the persistence time of sexuals and lead to their exclusion if clone-formation is too frequent (Vrijenhoek 1979; Bell 1982).

Computer simulations revealed that invasive clones initially exploit marginal resources, where competition from sexual relatives is weak, but eventually new clones exclude the sexuals from the center of the resource distribution, unless clone-formation is infrequent (Weeks 1993). Invading parthenogens should succeed best in ecologically marginal habitats where inter-specific interactions are weak, which is completely consistent with observations of geographical parthenogenesis (Gaggiotti 1994). Nonetheless, intraspecific competition under high density can prevent clones from realizing their two-fold advantage (Doncaster et al. 2000). Thus, competition with a reproductively isolated sexual species or among distinct clonal genotypes should be considered forms of inter-specific competition (Lomnicki 2001), and basic Lotka-Volterra dynamics apply. So again, we return to the critical role that asymmetrical competition plays in determining the window for coexistence among clones and their sexual progenitors, or among distinct clones (Pound et al. 2002). Treating clones as ecological microspecies with distinct life histories and niches results in different outcomes than the simpler models that assume "all else is equal" between completely replaceable sexual and clonal genotypes, consequently the persistence of sex may require much smaller advantages than previously thought (Tagg et al. 2005b).

6.3.1 Evidence for Frozen Niche Variation

Few experimental studies have tested the first premise of the FNV model – that ecologically relevant genotypic variation can be frozen among new clones when they arise. Schultz (1973) first demonstrated that different trophic morphologies could be frozen among hemiclones of *Poeciliopsis monacha-lucida* that he synthesized in the laboratory via crosses of *P. monacha* × *P. lucida*. Subsequent studies with additional synthetic *P. monacha-lucida* hemiclones demonstrated the capacity to freeze genotypic variation in life-history and reproductive traits (Wetherington et al. 1989b; Lima et al. 1996; Lima 2005) and in predatory behavior (Lima and Vrijenhoek 1996). Similar experiments might also be conducted with laboratory-synthesized strains of hybridogenetic *Rana esculenta* (see for example Hotz et al. 1999), with the pseudogamous leaf hopper *Muellerianella 2-fairmairei-brevipennis* (Drosopoulis 1978), and with tychoparthenogenetic *Drosophila mercatorum* (Annest and Templeton 1978). Laboratory clones generated from cyclically parthenogenetic aphids, cladocerans and rotifers might also be studied in this light, but we are not aware of such studies to date.

The second FNV premise – that inter-clonal selection weeds out new clones that substantially overlap the niche of established clones and the sexual ancestors – is harder to test. Stringent selection eliminates most newly formed tychoparthenogenetic clones of *Drosophila mercatorum* (Annest and Templeton 1978), but this results mostly from fitness penalties associated with complete homozygosity of these automictic strains. Unisexual-hybrid clones may also suffer fitness penalties due to genetic and developmental incompatibilities between the combining genomes (Vrijenhoek 1989a). More than half of the laboratory-synthesized *Poeciliopsis monacha-lucida* hemiclones mentioned previously had poor survival abilities, limited reproductive potential, and reduced developmental stability (Wetherington et al. 1987, 1989b). Nonetheless, a few of these strains performed as well as natural *P. monacha-lucida* strains, and the differences in life history, reproductive, behavioral, and trophic characters among these strains illustrate the scope for inter-clonal selection. Newly synthesized triploid-hybrid clones of *Taraxacum* closely resemble diploid sexuals in leaf size and phenology, which suggests that selection has shifted the ecological characteristics of natural triploid apomicts (de Kovel and de Jong 2000).

Indirect evidence for inter-clonal selection is found in seasonal shifts in the frequencies of cladoceran, ostracod and insect clones (Table 6.3). Generally, obligate parthenogenetic populations of these arthropods show lower clonal diversity than cyclical populations (Hebert 1974; Tomiuk and Wöhrman 1981; Llewellyn et al. 2003), which can be interpreted as evidence that inter-clonal selection has eliminated less fit and ecologically redundant clones. Extant clones in nature "can only be a minute fraction of those that have participated in the contest. The competitive exclusion principle should apply in the extreme... Occasionally [a new clone] succeeds and crowds out one or more of the older clones, wholly or partly. Thereafter the prevailing clones represent an even more select array." (Williams 1975, p. 29).

Many studies have shown that coexisting clones are not ecologically identical (Table 6.3), starting with the classical example in apomictic North American *Taraxacum* (Solbrig and Simpson 1974, 1977). Coexisting dandelion clones flourish under different disturbance regimes. One clone (A) was the majority type in disturbed fields and another clone (D) performed best in undisturbed fields. *Taraxacum* is more complex in Europe (see also Chapter 22), however, where higher clonal diversity results from hybridization between triploid apomicts and diploid sexuals (see Menken et al. 1995; de Kovel and de Jong 2000). Many of the studies listed in Table 6.3 were interpreted as support for the FNV model, whereas other studies sought to compare phenotypic distributions or niche breadths between closely related sexual and asexual forms. Asymmetric competition between narrow clones and ecologically broad sexual relatives has been reported for *Poeciliopsis*, *Aspidoscelis (Cnemidophorus)*, *Rana* and *Alsophila*. Differences in trophic morphology, feeding behavior and diet exist among *Poeciliopsis* and *Aspidoscelis* clones, and life history differences exist among clones of *Heterocypris*, *Poeciliopsis*, *Rana* and *Taraxacum* clones. Host plant associations differ among *Brevipalpus*, *Uroleucon* and *Alsophila*. Clones segregate spatially along environmental gradients in *Lumbricillus*, *Penthaleus*, *Artemia*, *Trichoniscus*, *Poeciliopsis*, *Aspidoscelis*, *Lepidodactylus*, *Potamopyrgus*, and *Asplanchna*, and different clones exhibit unique tolerances to hypoxic and thermal stress in *Daphnia*, *Poeciliopsis*, *Rana*, and *Lepidodactylus*. Finally, temporal frequencies vary among clones of *Daphnia*, *Heterocypris*, *Dipsa*, *Trichoniscus*, and *Asplanchna*.

6.4 Evolution of Generalist Versus Specialist Clones

None of the geographically widespread clones listed as GPGs in Table 6.2 are sperm-dependent hybridogens or gynogens. True parthenogens can escape direct competition with their sexual progenitors, and their ability to disperse and colonize new habitats is greatly facilitated by not having to produce males or find mates. Consequently, some researchers view parthenogens as "weedy" species (sensu Baker 1965) –i.e., inferior competitors that can only thrive as fugitives in marginal or disturbed habitats where competition with sexual relatives is minimal (Lowe and Wright 1966; Cuellar 1977). A generalist should be favored under these conditions

Table 6.3 Experimental and observational evidence for and against the FNV hypothesis for parthenogenetic taxa

Family	Taxon	Mode[a]	Characters	Evid.[b]	Comments	References
Annelida						
Haplotaxida Enchytraeidae	*Lumbricillus lineatus*	G	Spatial segregation	+	Clonal genotypes segregate along salinity gradients in the littoral zone of Danish fjords	Christensen (1980)
Arthropoda						
Acariformes Penthaleidae	*Penthaleus major*	P	Spatial and temporal segregation	+	Environmental heterogeneity effects relative fitness of clonal genotypes. Intense selection maintains clonal diversity	Weeks and Hoffmann (1998)
Tenuipalpidae	*Brevipalpus phoenicis*	P	Host plant associations	+	Clones are specialized. Fitness reduced when clones moved to alternative host plants	Groot et al. (2005)
Anostraca Artemiidae	*Artemia salina* complex	P	Life history, physiology, spatial segregation	+	Extremely high clonal diversity. Sexuals are not known. Clones differ in key life history traits and segregate along salinity gradients	Browne and Hoopes (1990) and Barata et al. (1996)
Cladocera Daphniidae	*Daphnia magna*	P	Temporal distribution, physiology	+	Seasonal changes in clone frequencies; thermal differentiation. Individual clones constrained to narrower niche	Hebert (1974), Hebert and Crease (1980), Carvalho (1987) and Hebert et al. (1988)
	D. obtusa	P	Competitive abilities, experimental evidence	+	Genetically diverse clones have competitive advantage when invading a genetically uniform population	Tagg et al. (2005a)
	D. pulex	P	Temporal and spatial segregation, physiology	+	Physiological differences and spatial segregation among clones. Seasonal changes in clone frequencies. Higher clonal diversity in temporary ponds may favor specialized genotypes	Lynch (1983), Weider and Hebert (1987), Weider (1989) and Weider (1993)
			Competitive abilities, experimental evidence	+	Genetically diverse clones experience competitive release (higher birth rates) when competing with uniform population vs. competing with themselves	Tagg et al. (2005b)
	D. pulicaria	M	Spatial and temporal segregation	+	Habitat partitioning and microhabitat specialization of clones in response to fish predation	Hembre and Megard (2006)

Table 6.3 (continued)

Family	Taxon	Mode[a]	Characters	Evid.[b]	Comments	References
Ostracoda Cyprididae	*Heterocypris incongruens*	M	Temporal distribution, life history	+	Recurrent frequency shifts among clones that differ in growth rates, survival and reproductive characteristics	Rossi and Menozzi (1993)
Coleoptera Curculionidae	*Ips acuminatus dubius*	G	Spatial segregation	+	Niche differentiation between gynogens and sexual hosts	Løyning (2000)
Diptera Lonchopteridae	*Dipsa bifurcata*	P	Temporal distribution	+	Seasonal changes in clone frequencies are synchronous among populations	Niklasson et al. (2004)
	Loncoptera dubia	P	Temporal distribution	+	Seasonal changes in clone frequencies	Ochman et al. (1980)
Lepidoptera Geometridae	*Alsophila pometaria*	G	Asymmetric competition, host plant associations	+	High clonal diversity. Sexuals present. Individual clones have narrower niche than sexuals. Dominant clones associated with particular tree stands	Mitter et al. (1979), Futuyma et al. (1981), Futuyma et al. (1984) and Harshman and Futuyma (1985)
Hemiptera Aphididae	*Uroleucon rudbeckiae*	P	Host plant associations	+	Fitness of clones varies across host-plant species and across host plant phenotypes	Service and Lenski (1982) and Service (1984)
Isopoda Trichoniscidae	*Trichoniscus pusillus*	P	Temporal and spatial segregation, physiology	+	Temporal changes in clone frequencies. Clones differ in moisture tolerance and distribution along moisture gradients	Christensen (1979), Christensen and Noer (1986) and Christensen et al. (1988)
Chordata						
Anura Ranidae	*Rana esculenta*	H	Asymmetric competition, life history, thermal physiology, GxE interactions	±	Life history and thermal differences among hemiclones. GxE interactions among hemiclones. Hemiclonal mixtures outperform single clones in competition experiments. No correlation between clonal diversity and abundance relative to sexuals	Hotz et al. (1994), Rist et al. (1996) and Negovetic et al. (2001)
Caudata Ambystomidae	*Ambystoma jeffersonianum* complex	G	Asymmetric competition	+	Larvae of gynogenetic triploids more adversely affected by competition with sexuals than vice versa	Wilbur (1971)

Table 6.3 (continued)

Family	Taxon	Mode[a]	Characters	Evid.[b]	Comments	References
Atheriniformes Atherinidae	*Menidia clarkhubbsi*	G	Spatial segregation	+	3N gynogens and 2N sexuals may segregate partially on salinity gradients. Differences among 3N clones not assessed	Echelle and Echelle (1997)
Cypriniformes Cyprinidae	*Squalius alburnoides*	H/G	Spatial segregation, morphology, diets	±	3N females in shallower, high velocity, water than 2N sexuals. Differences among clones not assessed, but ploidy levels appear to have same dietary niches	Martins et al. (1998) and Gomes-Ferreira et al. (2005)
Cyprinodontiformes Poeciliidae	*Poeciliopsis monacha-lucida*	H	Spatial segregation, life history, GxE interactions behavior, diet	+	Coexisting hybridogenetic hemiclones differ in trophic and aggressive behavior, diets, disease resistance, microhabitat use, and growth rates	Schenck and Vrijenhoek (1986), Schultz and Fielding (1989), Wetherington et al. (1989a), Weeks et al. (1992) and Leberg and Vrijenhoek (1994)
			Competition experiments	±	Mixed clones outperform single clones in competition with sexuals. Sexuals have wider niche than single clones but not greater than clonal mixture	Weeks (1995)
	P. 2 *monacha-lucida*	G	Asymmetric competition, physiology, diet, spatial segregation	+	3N clones segregate along stream gradients, and differ in survival of thermal and hypoxic stresses. Within-phenotype component of dietary breadth narrower in clones than sexual relatives	Vrijenhoek (1978), Vrijenhoek and Pfeiler (1997) and Gray and Weeks (2001)
Squamata Teiidae	*Aspidoscelis tesselatus*	P	Asymmetric competition, spatial distribution	+	Clones segregate into different habitats. Removal experiments indicate that sexuals have small inhibitory effect on asexuals but not *vice versa*	Scudday (1973) and Price (1986)
	A. sonorae	P	Asymmetric competition, diet	+	Sexual, *A. tigris*, has greater between-individual variation in diet than parthenogens. Both have similar within-individual diet breadth	Case (1990)

Table 6.3 (continued)

Family	Taxon	Mode[a]	Characters	Evid.[b]	Comments	References
Gekkonidae	*Lepidodactylus lugubris*	P	Spatial segregation, physiology	+	Clones differ in thermal preferences and geographical distribution	Bolger and Case (1994) and Radtkey et al. (1995)
Mollusca						
Neotaenioglossa Hydrobiidae	*Potamopyrgus antipodarum*	P	Spatial distribution, morphology, life history	+	High diversity of 3N clones. Sexuals present. Individual clones express narrower phenotypic variation than sexuals. Clones differ in size, morphology and life history traits. Common clones typically found in one habitat	Fox et al. (1995), Jokela et al. (1997) and Jokela et al. (2003)
Platyhelminthes						
Tricladida Planariidae	*Schmidtia polychroa*	G	Spatial segregation	+	Weak niche differentiation between sexuals and parthenogens in Italian lake	Weinzierl et al. (1999)
Rotifera						
Ploima Asplanchnidae	*Asplanchna brightwelli* and *A. girodi*	P	Temporal and spatial segregation, life history	+	Experimental evidence for strong interclonal competition between clones adapted to environments separated in time and space	Snell (1979)
Magnoliaphyta Asterales Asteraceae	*Taraxacum officinale*	A	Life history, GxE interactions	+	Dominant N. American clones differ in competitive and reproductive abilities. Coexistence related to environmental disturbance	Solbrig (1971), Solbrig and Simpson (1974) and Vavrek et al. (1996)
	Erigeron annuus	A	GxE interactions		Differential performance of clones across environments	Stratton (1994)
Myrtales Onagraceae	*Oenothera lanciniata*	H	Asymmetric competition	+	Clonal diversity is higher when asexuals do not compete locally with sexuals	Ellstrand and Roose (1987)

[a] Reproductive mode: A = apomictic; G = gynogenesis; H = hybridogenesis; P = thelytoky or obligate parthenogenesis; M = mixed sexual and apomictic parthenogenesis.
[b] Evidence: + is positive; - is negative; and ? is equivocal.

because fugitives experience a wide range of marginal conditions while dispersing (Parker et al. 1977), and because no premium exists on competitive abilities (Lowe and Wright 1966).

Sperm-dependant parthenogens, on the other hand, must live with a closely related sexual host that is likely to be similar ecologically (reviewed in Beukeboom and Vrijenhoek 1998). Unless the unisexual form is constrained by a unique carrying capacity or by mating behaviors that limit its reproductive potential, the all-female form should eliminate its sexual host and thereby ensure its own demise (Clanton 1934; Moore 1975; Kawecki 1988). Coexistence is greatly facilitated by resource partitioning that diminishes direct competition between the sexual parasite and its host (Stenseth et al. 1985; Kirkendall and Stenseth 1990; Schley et al. 2004). Therefore, it is not surprising that essentially all of the sperm-dependant parthenogens listed in Tables 6.1 and 6.3 exhibit some degree of niche separation from their sexual hosts. Of course, many of the listed examples are hybrids, which may partially shift the ecological niche away from the progenitor it uses as a sexual host. Additionally, many of these sperm-dependant parthenogens (Table 6.3) exhibit evidence for phenotypic variation and resource partitioning among coexisting clones. Strong conspecific mate preferences by males of the sexual host limits sperm availability (Moore and McKay 1971). If sexual mating preferences regulate unisexuals below their carrying capacity, clones would compete for sperm rather than food and spatial resources. Such conditions might favor clonal generalists, but this does not appear to be the case for *Poeciliopsis* where multiple clones negatively impact the sexual population by occupying a wider range of subniches (Vrijenhoek 1979). Sperm-dependence and limitation are not necessary conditions for operation of the FNV model because many true parthenogens also have coexisting specialist clones (Table 6.3).

Sources of clonal variation may also play a role in determining the evolutionary trajectories of asexual populations. Spontaneous mutations affecting life history characters can accumulate in clonal lineages, but each new mutation must express itself against a genetic background that has a history of favorable selection, and thus phenotype effects may be limited (Lynch 1985). The within-phenotype component of niche breadth might slowly evolve in this way, but point mutations are not likely to be as effective as frozen variation in affecting changes in the between-phenotype component (Roughgarden 1972). Polyploidization, on the other hand, should be able to affect both components instantaneously as a result of genome duplications and subsequently as a result of genome reorganization and reduction. Hybrid origins, which produced all known chordate and many arthropod clones, adds another dimension to inter-clonal variation, because genomic interactions and dominance will play strong roles in generating phenotypic variation (Wetherington et al. 1989b). Finally, hybridization and polyploidy together have the capacity to create a tremendous variety of genotypes and corresponding phenotypes. Multiple origins of clones provides sufficient variation to drive both the FNV and GPG models (Parker et al. 1977). To the extent that variation exists in the sexual ancestors for the within-phenotype component of niche breadth, newly arising clones might also freeze differences in their tolerance to environmental conditions (Vrijenhoek

1979). Polyphyletic origins of clones simply increase the scope for natural selection among clones, but the trajectory taken by an asexual population will depend mostly on the competitive regime it encounters. Hard selection that is density- and frequency-independent should favor the evolution of GPGs in regions of low intra- and inter-specific competition; whereas soft selection that is density- and frequency-dependent should favor specialist clones in regions of high competition (Kenny 1996).

6.4.1 GPG and FNV Are Not Mutually Exclusive

Attempts to contrast the FNV and GPG models as mutually exclusive hypotheses ignore the complex interplay between spatial and temporal sources of variation in fitness. Simple contrasts of these models often cast the GPG model as focused only on factors that dampen temporal fluctuations in fitness, and the FNV model as focused only on spatial variance. Yet environments can vary in space and time, and different clones possess complex multivariate phenotypes that respond differentially to these sources of variance in fitness; thus, it is possible to have a clone with narrow trophic preferences and wide thermal tolerances, or *vice versa* (Vrijenhoek 1979, 1998a). A problem with many models for the maintenance of clonal diversity in time and space results from an extrapolation of genetic models to the ecology of clones (Tomiuk et al. 2004). Genetic models reveal that it is difficult to maintain allelic diversity under temporal variation in fitness, unless fitnesses are precisely balanced (Haldane and Jayakar 1963; Hedrick et al. 1976). Spatial heterogeneity is generally more conducive to maintaining diversity (Levene 1953; Strobeck 1970). Nonetheless, seasonal differences in fitness do appear to contribute to the maintenance of clonal diversity in *Taraxacum officinale* (Vavrek et al. 1996) and *Dipsa bifurcata* (Niklasson et al. 2004). Different clones of these organisms appear to be favored during different seasons, which can have profound effects on growth rates and diets that are affected by seasonal temperatures and food availability. As spatial and temporal fluctuations can both affect fitness variation, each clone should be considered a reproductively isolated microspecies with unique demographic properties (birth and death rates, diapause, immigration rates, etc.). Consequently, ecological models (e.g., Chesson 1985; Hedrick 1995; Pound et al. 2002) are likely to be more realistic than simple genetic models for explaining the coexistence of clones and the maintenance of sexual and asexual lineages.

Does the FNV model require recurrent origins of new clones? The *Poeciliopsis* example suggests that the opportunity for recurrent clonal origins drives the demographic success of hybridogenetic populations (Vrijenhoek 1979, 1984). Apparently this is not the case for *Rana esculenta*, as no correlation exists between clonal diversity and abundance (Hotz et al. 1994), but this problem should be addressed in other taxa. Nonetheless, essentially all the well-documented cases of ecological diversification among clones (Table 6.3) involve taxa that have extant sexual progenitors or cyclical progenitors. Elimination of the sexual progenitors or escape from their genetic influence reduces the sources of variation to mutation, which may be insufficient to drive rapid clonal differentiation.

6.5 Conclusions

The FNV and GPG models share a number of features. First, clonal diversity can be frozen during multiple origins of clones from genetically variable sexual ancestors. Second, newly formed clones will likely differ from one another in niche breadth, and depending on the competitive regime, inter-clonal selection will fix generalist or specialist genotypes by expanding the within- or between-phenotype components, respectively (Roughgarden 1972). The competitive regimes faced by new clones will likely be influenced by the ability of clones to escape competition from their sexual progenitors (sperm-dependence versus independence) and by the rate at which new clones are formed (mutation versus recurrent frozen variation). Although the GPG and FNV models are often portrayed as mutually exclusive hypotheses, this view is simplistic because the models, as originally stated, focused on different aspects of environmental variation in time and space. Now, the distinction between environmental heterogeneity in time and space has begun to blur as more robust ecological models are applied to these problems (Chesson 1985; Hedrick 1995). Finally, these ideas must be reconsidered in the light of "pluralistic" models for the maintenance of sex (e.g. West et al. 1999; Pound et al. 2004). Both generalist and specialist genotypes must survive rapidly evolving parasites, mutational deterioration, and an ever-changing environment. It is hard to imagine how genuinely ancient clones can persist with such multifarious threats, but apparently some have (Schön et al. 1998; Mark Welch and Meselson 2000). For the rest, it appears that the recurrent freezing of new clonal genotypes from extant sexual ancestors provides the diversity that allows asexual taxa to refresh their genotypes, spread to new locations and occupy new niches.

Acknowledgments The authors wish to thank the C. Patrick Doncaster and the editors for helpful suggestions and clarifications.

References

Annest JL, Templeton AR (1978) Genetic recombination and clonal selection in *Drosophila mercatorum*. Genetics 89: 193–210

Antonovics J (1968) Evolution in closely adjacent plant populations, V. Evolution of self-fertility. Heredity 23: 219–238

Baker HG (1965) Characteristics and modes of origin of weeds. In: Baker HG, Stebbins GL (eds) Genetics of colonizing species. Academic Press, New York, pp. 147–172

Barata C, Hontoria F, Amat F, Browne R (1996) Competition between sexual and parthenogenetic *Artemia*: temperature and strain effects. J Exp Mar Biol Ecol 196: 313–328

Bell G (1982) The Masterpiece of nature: the evolution and genetics of sexuality. University of California Press, Berkeley

Beukeboom L, Vrijenhoek RC (1998) Evolutionary genetics and ecology of sperm-dependent parthenogenesis. J Evol Biol 11: 755–782

Bierzychudek P (1989) Environmental sensitivity of sexual and apomictic *Antennaria*: do apomicts have general-purpose genotypes? Evolution 43: 1456–1466
Bolger DT, Case TJ (1994) Divergent ecology of sympatric clones of the asexual gecko, *Lepidodactylus lugubris*. Oecologia 100: 397–405
Browne RA, Hoopes CW (1990) Genotypic diversity and selection in asexual brine shrimp (*Artemia*). Evolution 44: 1035–1051
Browne RA, Wanigasekera G (2000) Combined effects of salinity and temperature on survival and reproduction of five species of *Artemia*. J Exp Mar Biol Ecol 244: 29–44
Bulger AJ, Schultz RJ (1979) Heterosis and interclonal variation in thermal tolerance in unisexual fish. Evolution 33: 848–859
Bulger AJ, Schultz RJ (1982) Origins of thermal adaptation in northern vs southern populations of a unisexual hybrid fish. Evolution 36: 1041–1050
Carson H (1968) The population flush and its genetic consequences. In: Lewontin R (ed) Population biology and evolution. Syracuse University Press, Syracuse, New York, pp. 123–137
Carvalho G (1987) The clonal ecology of *Daphnia magna* (Crustacea: Cladocera): II. Thermal differentiation among seasonal clones. J Anim Ecol 56: 469–478
Case T (1990) Patterns of coexistence in sexual and asexual species of *Cnemidophorus* lizards. Oecologia 83: 220–227
Case TJ, Taper ML (1986) On the coexistence and coevolution of asexual and sexual competitors. Evolution 40: 366–387
Chesson PL (1985) Coexistence of competitors in spatially and temporally varying environments: a look at the combined effects of different sorts of variability. Theor Popul Biol 28: 263–287
Christensen B (1979) Differential distribution of genetic variants in triploid parthenogenetic *Trichoniscus pusillus* (Isopoda: Crustacea) in a heterogeneous environment. Hereditas 91: 179–182
Christensen B (1980) Constant differential distribution of genetic variants in parthenogenetic forms of *Lumbricillus lineatus* (Enchytraeidae, Oligochaeta) in a heterogeneous environment. Hereditas 92: 193–198
Christensen B, Noer H (1986) Spatial and temporal components of genetic variation in triploid parthenogenetic *Trichoniscus pusillus* (Isopoda, Crustacea). Hereditas 105: 277–285
Christensen B, Noer H, Theisen BF (1988) Differential response to humidity and soil type among clones of triploid parthenogenetic *Trichoniscus pusillus* (Isopoda, Crustacea). Hereditas 108: 213–217
Clanton W (1934) An unusual situation in the salamander *Ambystoma jeffersonianum* (Green). Occ Pap Mus Zool Univ Michigan 290: 1–15
Cuellar O (1977) Animal parthenogenesis. Science 197: 837–843
Cullum A (1997) Comparisons of physiological performance in sexual and asexual whiptail lizards (genus *Cnemidophorus*): implications for the role of heterozygosity. Am Natur 150: 24–47
da Cuhna A, Burla H, Dobzhansky T (1950) Adaptive chromosome polymorphism in *Drosophila willistoni*. Evolution 4: 212–235
Darling JA, Reitzel AM, Finnerty JR (2004) Regional population structure of a widely introduced estuarine invertebrate: *Nematostella vectensis* Stephenson in New England. Mol Ecol 13: 2969–2981
de Kovel CGF, de Jong G (2000) Selection on apomictic lineages of *Taraxacum* at establishment in a mixed sexual apomictic population. J Evol Biol 13: 561–568
Delmotte F, Leterme N, Gauthier J-P, Rispe C, Simon. J-C (2002) Genetic architecture of sexual and asexual populations of the aphid *Rhopalosiphum padi* based on allozyme and microsatellite markers. Mol Ecol 11: 711–723
Doeringsfeld MR, Schlosser IJ, Elder JF, Evenson DP (2004) Phenotypic consequences of genetic variation in a gynogenetic complex of *Phoxinus eos-neogaeus* clonal fish (Pisces: Cyprinidae) inhabiting a heterogeneous environment. Evolution 58: 1261–1273

Dolatti L, Ghareyazie B, Moharramipour S, Noori-Daloii MR (2005) Evidence for regional diversity and host adaptation in Iranian populations of the Russian wheat aphid. Entomol Exp Appl 114: 171–180
Doncaster CP, Pound GE, Cox SJ (2000) The ecological cost of sex. Nature 404: 281–285
Drosopoulis S (1978) Laboratory synthesis of a pseudogamous triploid "species" of the genus *Mullerianella* (Homoptera, Delphacidae). Evolution 32: 916–920
Dybdahl MF, Kane SL (2005) Adaptation vs. phenotypic plasticity in the success of a clonal invader. Ecology 86: 1592–1601
Echelle AA, Echelle AF (1997) Patterns of abundance and distribution among members of a unisexual-bisexual complex of fishes (Atherinidae: *Menidia*). Copeia 1997: 249–259
Elder JF, Jr, Schlosser IJ (1995) Extreme clonal uniformity of *Phoxinus eos/neogaeus* gynogens (Pisces: Cyprinidae) among variable habitats in northern Minnesota beaver ponds. Proc Natl Acad Sci USA 92: 5001–5005
Ellstrand NC, Roose ML (1987) Patterns of genotypic diversity in clonal plant species. Am J Bot 74: 123–131
Enghoff H (1976) Parthenogenesis and bisexuality in the millipede, *Nemasoma varicorne* C. L. Koch, 1847 (Diplopoda: Blaniulidae). Morphological, ecological and biogeographical aspects. Vidensk Meddr Dansk Naturhistorisk Forening 139: 21–59
Fenton B, Woodford JAT, Malloch G (1998) Analysis of clonal diversity of the peach-potato aphid, *Myzus persicae* (Sulzer), in Scotland, UK and evidence for the existence of a predominant clone. Mol Ecol 7: 1475–1487
Fox JA, Dybdahl MF, Jokela J, Lively CM (1995) Genetic structure of coexisting sexual and clonal subpopulations in a freshwater snail (*Potamopyrgus antipodarum*). Evolution 50: 1541–1548
Futuyma DJ, Cort RP, van Noordwijk I (1984) Adaptation to host plants in the fall cankerworm (*Alsophila pometaria*) and its bearing on the evolution of host affiliation in phytophagous insects. Am Natur 123: 287–296
Futuyma DJ, Leipertz SL, Mitter C (1981) Selective factors affecting clonal variation in the fall cankerworm *Alsophila pometaria* (Lepidoptera: Geometridae). Heredity 47: 151–172
Gade B, Parker ED Jr (1997) The effect of life cycle stage and genotype on desiccation tolerance in the colonizing parthenogenetic cockroach *Pycnoscelus surinamensis* and its sexual ancestor *P. indicus*. J Evol Biol 10: 479–493
Gaggiotti OE (1994) An ecological model for the maintenance of sex and geographic parthenogenesis. J Theor Biol 167: 201–221
García-Ramos G, Kirkpatrick M (1997) Genetic models of adaptation and gene flow in peripheral populations. Evolution 51: 21–28
Ghiselin MT (1974) The economy of nature and the evolution of sex. University of California Press, Berkeley
Glesener RR, Tilman D (1978) Sexuality and the components of environmental uncertainty: clues from geographic parthenogenesis in terrestrial animals. Am Nat 112: 659–673
Goldschmidt RB (1940) The material basis of evolution. Yale University Press, New Haven
Gomes-Ferreira A, Ribeiro F, Moreira da Costa L, Cowx IG, Collares-Pereira MJ (2005) Variability in diet and foraging behaviour between sexes and ploidy forms of the hybridogenetic *Squalius alburnoides* complex (Cyprinidae) in the Guadiana River basin, Portugal. J Fish Biol 66: 454–467
Gray MM, Weeks SC (2001) A comparison of dietary use patterns of clonal and sexual fish (Poeciliidae: Poeciliopsis). Can J Fish Aquat Sci 58: 1313–1318
Groot TVM, Janssen A, Pallini A, Breeuwer JAJ (2005) Adaptation in the asexual false spider mite *Brevipalpus phoenicis*: Evidence for frozen niche variation. Exp Appl Acarol 36: 165–176
Haack L, Simon J-C, Gauthier J-P, Plantegenest M, Dedryver C-A (2000) Evidence for predominant clones in a cyclically parthenogenetic organism provided by combined demographic and genetic analyses. Mol Ecol 9: 2055–2066
Haag C, Ebert D (2004) A new hypothesis to explain geographic parthenogenesis. Ann Zool Fennici 41: 539–544

Haldane JBS, Jayakar SD (1963) Polymorphism due to selection of varying direction. J Genet 58: 237–242

Hamilton WD, Henderson P, Moran N (1981) Fluctuation of environmental and coevolved antagonist polymorphism as factors in the maintenance of sex. In: Alexander RD, Tinkle D (eds) Natural selection and social behavior. Chiron Press, New York, pp. 363–381

Harshman LG, Futuyma DJ (1985) The origin and distribution of clonal diversity in *Alsophila pometaria* (Lepidoptera: Geometridae). Evolution 39: 315–324

Hebert PDN (1974) Ecological differences among genotypes in a natural population of *Daphnia magna*. Heredity 33: 327–337

Hebert PDN, Beaton MJ, Schwartz SS, Stanton DJ (1989) Polyphyletic origins of asexuality in *Daphnia pulex*. I. Breeding-system variation and levels of clonal diversity. Evolution 43: 1004–1015

Hebert PDN, Crease TJ (1980) Clonal coexistence in *Daphnia pulex* (Leydig): another planktonic paradox. Science 207: 1363–1365

Hebert PDN, Ward RD, Weider LJ (1988) Clonal diversity patterns and breeding-system variation in *Daphnia pulex*, an asexual-sexual complex. Evolution 42: 147–149

Hedrick P (1995) Genetic polymorphism in a temporally varying environment: effects of delayed germination or diapause. Heredity 75: 164–170

Hedrick PW, Ginevan ME, Ewing EP (1976) Genetic polymorphism in heterogeneous environments. Annu Rev Ecol Syst 7: 1–32

Hembre LK, Megard RO (2006) Direct and indirect effects of predation on the genetic structure of a *Daphnia* population. J Plankton Res 28: 1129–1141

Horne DJ, Martens K (1999) Geographical parthenogenesis in European non-marine ostracods: post-glacial invasion or Holocene stability? Hydrobiologia 391: 1–7

Hotz H, Beerli P, Guex G-D, Semlitsch RD, Uzzell T (1994) Clonal diversity and hybrid frequency are not correlated in water frogs: is the Frozen Niche Variation model wrong? II International Symposium on Ecology and Genetics of European Water Frogs 39: 513–514

Hotz H, Semlitsch RD, Gutmann E, Guex G-D, Beerli P (1999) Spontaneous heterosis in larval life-history traits of hemiclonal frog hybrids. Proc Natl Acad Sci USA 96: 2171–2176

Hoy Jensen L, Enghoff H, Frydenberg J, Parker ED Jr (2002) Genetic diversity and the phylogeography of parthenogenesis: comparing bisexual and thelytokous populations of *Nemasoma varicorne* (Diplopoda: Nemasomatidae) in Denmark. Hereditas 136: 184–194

Jacobsen R, Forbes V (1997) Clonal variation in life history traits and feeding rates in the gastropod, *Potamopyrgus antipodarum*: performance across a salinity gradient. Funct Ecol 11: 260–267

Jaenike J, Parker ED Jr, Selander RK (1980) Clonal niche structure in the parthenogenetic earthworm *Octolasion tyrtaeum*. Am Natur 116: 196–205

Jokela J, Lively CM, Dybdahl MF, Fox JA (2003) Genetic variation and maintenance of sex. Biol J Linn Soc 79: 165–181

Jokela J, Lively CM, Fox JA, Dybdahl MF (1997) Flat reaction norms and 'frozen' phenotypic variation in clonal snails (*Potamopyrgus antipodarum*). Evolution 51: 1120–1129

Kawecki TJ (1988) Unisexual/bisexual breeding complexes in Poeciliidae: why do males copulate with unisexual females? Evolution 42: 1018–1023

Kearney M (2005) Hybridization, glaciation and geographical parthenogenesis. Trends Ecol Evol 20: 495

Kearney M (2006) Response to Lundmark: Polyploidization, hybridization and geographical parthenogenesis. Trends Ecol Evol 21: 10

Kearney M, Porter WP (2004) Mapping the fundamental niche: physiology, climate, and the distribution of a nocturnal lizard. Ecology 85: 3119–3131

Kearney M, Shine R (2004) Developmental success, stability, and plasticity in closely related parthenogenetic and sexual lizards (Heteronotia, Gekkonidae). Evolution 58: 1560–1572

Kearney M, Shine R (2005) Lower fecundity in parthenogenetic geckos than sexual relatives in the Australian arid zone. J Evol Biol 18: 609–618

Kearney M, Wahl R, Autumn K (2005) Increased capacity for sustained locomotion at low temperature in parthenogenetic geckos of hybrid origin. Physiol Biochem Zool 78: 316–324
Kenny N (1996) A test of the general purpose genotype hypothesis in sexual and asexual *Erigeron* species. Am Midl Nat 136: 1–13
Kirkendall LR, Stenseth NC (1990) Ecological and evolutionary stability of sperm-dependent parthenogenesis: effects of partial niche overlap between sexual and asexual females. Evolution 44: 698–714
Koella JC (1988) The tangled bank: the maintenance of sexual reproduction through competitive interactions. J Evol Biol 2: 95–116
Ladle R, Johnstone R, Judson O (1993) Coevolutionary dynamics of sex in a metapopulation: escaping the Red Queen. Proc R Soc Lond B 253: 155–160
Leberg P, Vrijenhoek RC (1994) Variation among desert topminnows in their susceptibility to attack by exotic parasites. Cons Biol 8: 419–424
Levene H (1953) Genetic equilibrium when more than one ecological niche is available. Am Nat 87: 331–333
Lima NRW (2005) Variations on maternal-embryonic relationship in two natural and six laboratory made hybrids of *Poeciliopsis monacha-lucida* (Pisces, Cyprinodontiformes). Braz Arch Biol Technol 48: 73–79
Lima NRW, Koback CJ, Vrijenhoek RC (1996) Evolution of sexual mimicry in sperm-dependent clonal forms of *Poeciliopsis* (Pisces: Poeciliidae). J Evol Biol 9: 185–203
Lima NRW, Vrijenhoek RC (1996) Avoidance of filial cannibalism by sexual and clonal forms of *Poeciliopsis* (Pisces: Poeciliidae). Anim Behav 51: 293–301
Lindroth CH (1954) Experimentelle Beobachtungen an parthenogenetischem und bisexuellem *Otiorrhynchus dubius* Stroem (Col., Curculionidae). Entomol Tidskr 75: 111–116
Lively CM, Craddock C, Vrijenhoek RC (1990) The Red Queen hypothesis supported by parasitism in sexual and clonal fish. Nature 344: 864–866
Llewellyn K, Loxdale H, Harrington R, Brookes C, Clark S, P S (2003) Migration and genetic structure of the grain aphid (*Sitobion avenae*) in Britain related to climate and clonal fluctuation as revealed using microsatellites. Mol Ecol 12: 21–34
Lokki J, Saura A, Lankinen P, Suomalainen E (1976) Genetic polymorphism and evolution in parthenogenetic animals. VI. Diploid and triploid *Polydrosus mollis* (Coleoptera: Curculionidae). Hereditas 82: 209–216
Lokki J, Suomalainen E, Saura A, Lankinen P (1975) Genetic polymorphism and evolution in parthenogenetic animals. II. Diploid and polyploid *Solenobia triquetrella*(Lepidoptera: Psychidae). Genetics 79: 513–525
Lomnicki A (2001) Carrying capacity, competition and maintenance of sexuality. Evol Ecol Res 3: 603–610
Lowe CH, Wright JW (1966) Evolution of parthenogenetic species of *Cnemidophorus* (whiptail lizards) in western North America. J Arizona Acad Sci 4: 81–87
Løyning MK (2000) Reproductive performance of clonal and sexual bark beetles (Coleoptera: Scolytidae) in the field. J Evol Biol 13: 743–748
Lundmark M (2006) Polyploidization, hybridization and geographical parthenogenesis. Trends Ecol Evol 21: 9
Lundmark M, Saura A (2006) Asexuality alone does not explain the success of clonal forms in insects with geographical parthenogenesis. Hereditas 143: 23–32
Lynch M (1983) Ecological genetics of *Daphnia pulex*. Evolution 37: 358–374
Lynch M (1984) Destabilizing hybridization, general-purpose genotypes and geographical parthenogenesis. Quart Rev Biol 59: 257–290
Lynch M (1985) Spontaneous mutations for life history characters in an obligate parthenogen. Evolution 39: 804–818
Lynch M, Bürger R, Butcher D, Gabriel W (1993) The mutational meltdown in asexual populations. J Hered 84: 339–344
Lynch M, Gabriel W (1987) Environmental tolerance. Am Nat 129: 283–303

MacArthur RH, Wilson EO (1967) The theory of island biogeography. Princeton University Press, Princeton, NJ

Mark Welch D, Meselson M (2000) Evidence for the evolution of bdelloid rotifers without sexual recombination or genetic exchange. Science 288: 1211–1215.

Martins MJ, Colares-Pereira MJ, Cowz IG, Coelho MM (1998) Diploids*v.*triploids of *Rutilus alburnoides*: spatial segregation and morphological differences. J Fish Biol 52: 817–828

Maslin TP (1967) Skin grafting in the bisexual teiid lizard *Cnemidophorus sexlineatus* and in the unisexual *C. tesselatus*. J Expt Zool 166: 137–150

Maynard Smith J (1962) Disruptive selection, polymorphism and sympatric speciation. Nature 195: 60–62

Mayr E (1954) Change of genetic environment and evolution. In: Huxley J (ed) Evolution as a process. Allen and Unwin, London, pp. 157–180

Menken SBJ, Smit E, Nijs HCMD (1995) Genetical population structure in plants: gene flow between diploid sexual and triploid Asexual Dandelions (*Taraxacum* Section Ruderalia). Evolution 49: 1108–1118

Michaels HJ, Bazzaz FA (1989) Individual and population responses of sexual and apomictic plants to environmental gradients. Am Nat 134: 190–207

Mitter C, Futuyma DJ, Schneider JC, Hare JD (1979) Genetic variation and host plant relations in a parthenogenetic moth. Evolution 33: 777–790

Moore WS (1975) Stability of unisexual-bisexual populations of *Poeciliopsis* (Pisces: Poeciliidae). Ecology 56: 791–808

Moore WS (1977) An evaluation of narrow hybrid zones in vertebrates. Q Rev Biol 52: 263–277

Moore WS (1984) Evolutionary ecology of unisexual fishes. In: Turner BJ (ed) Evolutionary genetics of fishes. Plenum Press, New York, pp. 329–398

Moore WS, McKay FE (1971) Coexistence in unisexual-bisexual species complexes of *Poeciliopsis* (Pisces: Poeciliidae). Ecology 52: 791–799

Moore WS, Miller RR, Schultz RJ (1970) Distribution, adaptation and probable origin of an all-female form of *Poeciliopsis* (Pisces: Poeciliidae) in northwestern Mexico. Evolution 24: 789–795

Moritz C, Brown WM, Densmore LD, Wright JW, Vyas D, Donnellan S, Adams M, Baverstock P (1989) Genetic diversity and the dynamics of hybrid parthenogenesis in *Cnemidophorus* (Teiidae) and *Heteronotia* (Gekonidae). In: Dawley R, Bogart J (eds) Evolution and ecology of unisexual vertebrates. Bulletin 466, New York State Museum, Albany, New York, pp. 87–112

Myers MJ, Meyer CP, Resh VH (2000) Neritid and thiarid gastropods from French Polynesian streams: how reproduction (sexual, parthenogenetic) and dispersal (active, passive) affect population structure. Freshw Biol 44: 535–545

Negovetic S, Anholt BR, Semlitsch RD, Reyer H-U (2001) Specific responses of sexual and hybridogenetic European waterfrog tadpoles to temperature. Ecology 82: 766–774

Niklasson M (1995) Clonal diversity and individual adaptability in two colonizing parthenogenetic insects. PhD dissertation, Lund University, Lund, Sweden

Niklasson M, Parker ED Jr (1994) Fitness variation in an invading parthenogenetic cockroach. Oikos 71: 47–54

Niklasson M, Tomiuk J, Parker ED Jr (2004) Maintenance of clonal diversity in *Dipsa bifurcata* (Fallén, 1810) (Diptera: Lonchopteridae). I. Fluctuating seasonal selection moulds long-term coexistence. Heredity 93: 62–71

Ochman HB, Stille B, Niklasson M, Selander RK (1980) Evolution of clonal diversity in the parthenogenetic fly *Lonchoptera dubia*. Evolution 34: 539–547

Parker ED Jr (1979) Ecological implications of clonal diversity in parthenogenetic morphospecies. Am Zool 19: 753–762

Parker ED Jr (2002) Geographic parthenogenesis in terrestrial invertebrates: generalist or specialist clones? In: Hughes RN (ed) Progress in asexual propagation and reproductive strategies. Oxford-IBH, Oxford, pp. 93–114

Parker ED Jr, Niklasson M (1995) Desiccation resistance in invading parthenogenetic cockroaches: a search for the general purpose genotype. J Evol Biol 8: 331337

Parker ED Jr, Niklasson M (2000) Genetic structure and evolution in parthenogenetic animals. In: Singh R, Krimbas C (eds) Evolutionary genetics from molecules to morphology. Cambridge University Press, Cambridge, UK, pp. 456–474

Parker ED Jr, Selander RK (1976) The organization of genetic diversity in the parthenogenetic lizard *Cnemidophorus tesselatus*. Genetics 84: 791–805

Parker ED Jr, Selander RK, Hudson RO, Lester LJ (1977) Genetic diversity in colonizing parthenogenetic cockroaches. Evolution 31: 836–842

Paulissen MA, Walker JM, Cordes JE (1988) Ecology of syntopic clones of the parthenogenetic whiptail lizard, *Cnemidophorus 'laredoensis'*. J Herpetol 22: 331–342

Pearson CVM, Rogers AD, Sheader M (2002) The genetic structure of the rare lagoonal sea anemone, *Nematostella vectensis* Stephenson (Cnidaria; Anthozoa) in the United Kingdom based on RAPD analysis. Mol Ecol 11: 2285–2293

Peck JR, Yearsley JM, Waxman D (1998) Explaining the geographic distributions of sexual and asexual populations. Nature 391: 889–892

Pound GE, Cox SJ, Doncaster CP (2004) The accumulation of deleterious mutations within the frozen niche variation hypothesis. J Evol Biol 17: 651–662

Pound GE, Doncaster CP, Cox SJ (2002) A Lotka-Volterra model of coexistence between a sexual population and multiple asexual clones. J Theor Biol 217: 535–545

Price AH (1986) The ecology and evolutionary implications of competition and parthenogenesis in *Cnemidophorus*. PhD dissertation, New Mexico State University, Las Cruces, NM

Radtkey RR, Donnellan SC, Fisher RN, Moritz C, Hanley KA, Case TJ (1995) When species collide: the origin and spread of an asexual species of gecko. Proc R Soc Lond B 259: 145–152

Rist L, Semlitsch RD, Hotz H, Reyer H-U (1996) Feeding behaviour, food consumption and growth efficiency of hemiclonal and parental tadpoles of the *Rana esculenta* complex Funct Ecol 11: 735–742

Robinson MT, Weeks AR, Hoffmann AA (2002) Geographic patterns of clonal diversity in the earth mite species *Penthaleus major* with particular emphasis on species margins. Evolution 56: 1160–1167

Rossi V, Benassi G, Leonardi S, Piotti A, Menozzi P (2006) Clonal diversity of *Heterocypris incongruens* (Crustacea: Ostracoda) in Northern Italian ricefields. Arch Hydrobiol 166: 225–240

Rossi V, Menozzi P (1993) The clonal ecology of *Heterocypris incongruens* (Ostracoda): life-history traits and photoperiod. Funct Ecol 7: 177–182

Roughgarden J (1972) Evolution of niche width. Am Nat 106: 683–718

Saura A, Lokki J, Lankinen P, Suomalainen E (1976a) Genetic polymorphism and evolution in parthenogenetic animals. III. Tetraploid *Otiorrhynchus scaber* (Coleoptera: Curculionidae). Hereditas 82: 79–100

Saura A, Lokki J, Lankinen P, Suomalainen E (1976b) Genetic polymorphism and evolution in parthenogenetic animals. IV. Triploid *Otiorrhynchus salicis* Strom (Coleoptera: Curculionidae). Entomol Scand 7: 1–6

Schenck RA, Vrijenhoek RC (1986) Spatial and temporal factors affecting coexistence among sexual and clonal forms of *Poeciliopsis*. Evolution 40: 1060–1070

Schley D, Doncaster CP, Sluckin T (2004) Population models of sperm-dependent parthenogenesis. J Theor Biol 229: 559–572

Schön I, Butlin RK, Griffiths HI, Martens K (1998) Slow molecular evolution in an ancient asexual ostracod. Proc R Soc Lond B 265: 235–242

Schlosser IJ, Doeringsfeld MR, Elder J, F. AL (1998) Niche relationships of clonal and sexual fish in a heterogeneous landscape. Ecology 79: 953–968

Schultz RJ (1971) Special adaptive problems associated with unisexual fishes. Am Zool 11: 351–360

Schultz RJ (1973) Unisexual fish: laboratory synthesis of a "species". Science 179: 180–181

Schultz RJ, Fielding E (1989) Fixed genotypes in variable environments. In: Dawley R, Bogart J (eds) Evolution and ecology of unisexual vertebrates. New York State Museum, Albany, New York, pp. 32–38
Scudday JR (1973) A new species of lizard of the *Cnemidophorus tesselatus* group from Texas. J Herpetol 7: 363–371
Semlitsch RD (1993) Asymmetric competition in mixed populations of tadpoles of the hybridogenetic *Rana esculenta* complex. Evolution 47: 510–519
Semlitsch RD, Hotz H, Guex G-D (1997) Competition among tadpoles of coexisting hemiclones of hybridogenetic *Rana esculenta*: support for the frozen niche variation model. Evolution 51: 1249–1261
Service P (1984) Genotypic interactions in an aphid-host plant relationship: *Uroleucon rudbeckiae* and *Rudbeckia laciniata*. Oecologia V61: 271–276
Service PM, Lenski RE (1982) Aphid genotypes, plant genotypes and genetic diversity: a demographic analysis of experimental data. Evolution 36: 1276–1282
Snell TW (1979) Intraspecific competition and population structure in rotifers. Ecology 60: 494–502
Solbrig O (1971) The population biology of dandelions. Am Sci 59: 686–694
Solbrig OT, Simpson BB (1974) Components of regulation of a population of dandelions in Michigan. J Ecol 62: 473–486
Solbrig OT, Simpson BB (1977) A garden experiment on competition between biotypes of the common dandelion (*Taraxacum officinale*). J Ecol 65: 427–430
Stalker HD (1954) Parthenogenesis in *Drosophila*. Genetics 39: 4–34
Stenseth NC, Kirkendall LR, Moran N (1985) On the evolution of pseudogamy. Evolution 39: 294–307
Stratton DA (1994) Genotype-by-environment interactions for fitness of *Erigeron annuus* show fine-scale selective heterogeneity. Evolution 48: 1607–1618
Strobeck C (1970) Haploid selection with *n* alleles in *m* niches. Am Nat 113: 439–444
Strobeck C (1974) Sufficient conditions for polymorphism with N niches and M mating groups. Am Nat 108: 152–156
Sunnucks P, England PR, Taylor AC, Hales DF (1996) Microsatellite and chromosome evolution of parthenogenetic *Sitobion* aphids in Australia. Genetics 144: 747–756
Suomalainen E, Saura A, Lokki J (1987) Cytology and evolution in parthenogenesis. CRC Press, Boca Raton
Tagg N, Doncaster CP, Innes DJ (2005b) Resource competition between genetically varied and genetically uniform populations of *Daphnia pulex* (Leydig): does asexual reproduction confer a short-term ecological advantage? Biol J Linn Soc 85: 11–123
Tagg N, Innes DJ, Doncaster CP (2005a) Outcomes of reciprocal invasions between genetically diverse and genetically uniform populations of *Daphnia obtusa* (Kurz). Oecologia 143: 527–536
Templeton AR (1982) The prophecies of parthenogenesis. In: Dingle H, Hegmann JP (eds) Evolution and genetics of life histories. Springer-Verlag, Berlin, pp. 75–102
Thibault RE (1978) Ecological and evolutionary relationships among diploid and triploid unisexual fishes associated with the bisexual species, *Poeciliopsis lucida* (Cyprinodontiformes: Poeciliidae). Evolution 32: 613–623
Tomiuk J, Niklasson M, Parker ED Jr (2004) Maintenance of clonal diversity in *Dipsa bifurcata* (Fallén, 1810) (Diptera: Lonchopteridae). II. Diapause stabilizes clonal coexistence. Heredity 93: 72–77
Tomiuk J, Wöhrman K (1981) Changes in genotype frequencies at the MDH-locus in populations of *Macrosiphum rosae* (L.) (Hem., Aphididae). Biol Zentralbl 100: 631–640
Tomlinson J (1966) The advantage of hermaphroditism and parthenogenesis. J Theor Biol 11: 54–58
Tunner HG, Nopp H (1979) Heterosis in the common European water frog. Naturwissenschaften 66: 268–269

Van Doninck K, Schön I, De Bruyn L, Martens K (2002) A general purpose genotype in an ancient asexual. Oecologia 132: 205–212
Van Doninck K, Schön I, Maes F, De Bruyn L, Martens K (2003) Ecological strategies in the ancient asexual animal group Darwinulidae (Crustacea, Ostracoda). Freshwater Biol 48: 1285–1294
Vandel A (1928) La parthénogénèse géographique contribution a l'étude biologique et cytologique de la parthénogénèse naturelle. Bull Biol France Belg 62: 164–281
Vandel A (1940) La parthénogénèse géographique. IV. Polyploidie et distribution géographique. Bull Biol France Belg 74: 94–100
Vavrek MC, McGraw JB, Yang HS (1996) Within-population variation in demography of *Taraxacum officinale*: Maintenance of genetic diversity. Ecology 77: 2098–2107
Vorburger C, Lancaster M, Sunnucks P (2003b) Environmentally related patterns of reproductive modes in *Myzus persicae* and the predominance of two 'superclones' in Victoria, Australia. Mol Ecol 12: 3493–3503
Vorburger C, Sunnucks P, Ward SA (2003a) Explaining the coexistence of asexuals with their sexual progenitors: no evidence for general-purpose genotypes in obligate parthenogens of the peach-potato aphid, *Myzus persicae*. Ecol Lett 6: 1091–1098
Vrijenhoek RC (1978) Coexistence of clones in a heterogeneous environment. Science 199: 549–552
Vrijenhoek RC (1979) Factors affecting clonal diversity and coexistence. Am Zool 19: 787–797
Vrijenhoek RC (1984) Ecological differentiation among clones: the frozen niche variation model. In: Wöhrmann K, Loeschcke V (eds) Population biology and evolution. Springer-Verlag, Heidelberg, pp. 217–231
Vrijenhoek RC (1985) Animal population genetics and disturbance: the effects of local extinctions and recolonizations on heterozygosity and fitness. In: Pickett STA, White P (eds) The ecology of natural disturbance and patch dynamics. Academic Press, New York, pp. 265–285
Vrijenhoek RC (1989a) Genetic and ecological constraints on the origins and establishment of unisexual vertebrates. In: Dawley R, Bogart J (eds) Evolution and ecology of unisexual vertebrates. New York State Museum, Albany, New York, pp. 24–31
Vrijenhoek RC (1989b) Genotypic diversity and coexistence among sexual and clonal forms of *Poeciliopsis*. In: Otte D, Endler J (eds) Speciation and Its Consequences. Sinauer Associates, Sunderland, pp. 386–400
Vrijenhoek RC (1990) Genetic diversity and the ecology of asexual populations. In: Wöhrmann K, Jain S (eds) Population biology and evolution. Springer-Verlag, Berlin, pp. 175–197
Vrijenhoek RC (1998a) Animal clones and diversity. Bioscience 48: 617–628
Vrijenhoek RC (1998b) Clonal organisms and the benefits of sex. In: Carvalho G (ed) Advances in molecular ecology. IOS Press, Amsterdam, pp. 151–172
Vrijenhoek RC, Angus RA, Schultz RJ (1977) Variation and heterozygosity in sexually vs. clonally reproducing populations of *Poeciliopsis*. Evolution 31: 767–781
Vrijenhoek RC, Angus RA, Schultz RJ (1978) Variation and clonal structure in a unisexual fish. Am Nat 112: 41–55
Vrijenhoek RC, Lerman S (1982) Heterozygosity and developmental stability under sexual and asexual breeding systems. Evolution 36: 768–776
Vrijenhoek RC, Pfeiler E (1997) Differential survival of sexual and asexual *Poeciliopsis* during environmental stress. Evolution 51: 1593–1600
Wallace B (1959) The influence of genetic systems on geographic distribution. Cold Spr Harb Symp Quant Biol 24: 193–205
Weeks A, Hoffmann A (1998) Intense selection of mite clones in a heterogeneous environment. Evolution 52: 1325–1333
Weeks SC (1993) The effects of recurrent clonal formation on clonal invasion patterns and sexual persistence: a Monte Carlo simulation of the frozen niche variation model. Am Nat 141: 409–427

Weeks SC (1995) Comparisons of life-history traits between clonal and sexual fish (*Poeciliopsis*: Poeciliidae) raised in monoculture and mixed treatments. Evol Ecol 9: 258–274

Weeks SC, Gaggiotti OE, Spindler KP, Schenck RE, Vrijenhoek RC (1992) Feeding behavior in sexual and clonal strains of *Poeciliopsis*. Behav Biol Sociobiol 30: 1–6

Weider LJ (1989) Spatial heterogeneity and clonal structure in arctic populations of apomictic *Daphnia*. Ecology 70: 1405–1413

Weider LJ (1993) A test of the "general-purpose" genotype hypothesis: differential tolerance to thermal and salinity stress among *Daphnia* clones. Evolution 47: 965–969

Weider LJ, Hebert PDN (1987) Ecological and physiological differentiation among low-Arctic clones of *Daphnia pulex*. Ecology 68: 188–198

Weinzierl RP, Beukeboom LW, Gerace L, Michiels NK (1999) Spatial and ecological overlap between coexisting sexual and parthenogenetic *Schmidtea polychroa* (Tricladida; Platyhelminthes) Hydrobiologia 39: 170–185

Weisman A (1889) The significance of sexual reproduction in the theory of natural selection. In: Poulton EB, Schonland S, Shipley AE (eds) Essays upon heredity and kindred biological problems. Oxford University Press, Oxford, pp. 254–338

West SA, Lively CM, Read AF (1999) A pluralistic approach to sex and recombination. J Evol Biol 12: 1003–1012

West SA, Peters AD (2000) Evolution: paying for sex is not easy. Nature 407: 962

Wetherington JD, Kotora KE, Vrijenhoek RC (1987) A test of the spontaneous heterosis hypothesis for unisexual vertebrates. Evolution 41: 721–731

Wetherington JD, Schenck RA, Vrijenhoek RC (1989a) Origins and ecological success of unisexual *Poeciliopsis*: the Frozen Niche Variation model. In: Meffe GA, Snelson FF, Jr (eds) The ecology and evolution of poeciliid fishes. Prentice Hall, Englewood Cliffs, pp. 259–276

Wetherington JD, Weeks SC, Kotora KE, Vrijenhoek RC (1989b) Genotypic and environmental components of variation in growth and reproduction of fish hemiclones (*Poeciliopsis*: Poeciliidae). Evolution 43: 635–645

White MJD (1970) Heterozygosity and genetic polymorphism in parthenogenetic animals. In: Hecht MK, Steere WC (eds) Essays in evolution and genetics in honor of Theodosius Dobzhansky. Appleton-Century-Crofts, New York

White MJD (1973) Animal cytology and evolution. Cambridge University Press, Cambridge, UK

White MJD (1978) Modes of speciation. WH Freeman, San Francisco

Wilbur HM (1971) The ecological relationship of the salamander *Ambystoma laterale* to its all-female, gynogenetic associate. Evolution 25: 168–179

Williams GC (1975) Sex and evolution. Princeton University Press, Princeton

Williams GC, Mitton JB (1973) Why reproduce sexually? J Theor Biol 39: 545–554

Wilson AC, Sunnucks, CP, Hales DF (1999) Microevolution, low clonal diversity and genetic affinities of parthenogenetic *Sitobion* aphids in New Zealand. Mol Ecol 8: 1655–1666

Wright JW, Lowe CH (1968) Weeds, polyploids, parthenogenesis and the geographical and ecological distribution of all-female species of *Cnemidophorus*. Copeia 1968: 128–138

Chapter 7
Sex and the Red Queen

Maurine Neiman and Britt Koskella

> *The essence of sex in our theory is that it stores genes that are currently bad but have promise for reuse. It continually tries them in new combination, waiting for the time when the focus of disadvantage has moved elsewhere – Hamilton et al. (1990).*

Abstract Negative frequency-dependent selection exerted by parasites and pathogens can generate a selective advantage for rare host genotypes. This mechanism, known as the Red Queen, is currently considered to be one of the most likely explanations for the predominance of sexual reproduction in natural populations. Even so, the extent to which the Red Queen can and does provide an advantage to sex in nature is fiercely debated. Here, we survey the history of the development of the Red Queen hypothesis as applied to the maintenance of sex and discuss its theoretical underpinnings. We then review and synthesize the current body of theory and empirical data relevant to assessing whether Red Queen dynamics are likely to contribute to any general explanation for why sex is so common. We conclude that while there are many independent lines of evidence in support of a role for the Red Queen, important theoretical and empirical gaps remain. In particular, there is a need for theory addressing the breadth of conditions under which the Red Queen can favor sex, predictions for the patterns of molecular evolution expected for loci under negative frequency-dependent selection, and empirical research evaluating the strength of parasite-mediated selection in nature and the genetics of susceptibility and infection.

7.1 Sex and the Red Queen – Introduction

Negative frequency-dependent selection exerted by parasites such that rare host genotypes are favored is now considered to be one of the most likely selective

M. Neiman (✉)
Department of Biology, University of Iowa, Iowa City, IA, 52242, USA
e-mail: maurine-neiman@uiowa.edu

I. Schön et al. (eds.), *Lost Sex*, DOI 10.1007/978-90-481-2770-2_7,

forces underlying the persistence of sex and outcrossing[1] in natural systems. Disproportionately high infection in common host genotypes can create a short-term advantage to sex, because sex results in the production of offspring with variable genotypes.

A more general version of this mechanism originated with Van Valen's (1973) observation that species within a taxonomic group tend to go extinct at a constant rate. Van Valen argued that this pattern could be due to "stochastically constant" deterioration of environmental conditions (caused by either biotic or abiotic forces, though he emphasized antagonistic biotic interactions) from the perspective of species adapted to earlier conditions, and termed this mechanism "The Red Queen's hypothesis." While Van Valen left the origin of the name unstated in the main body of the paper, he did cite Lewis Carroll's *Through the Looking Glass*; *"Now here, you see, it takes all the running you can do, to keep in the same place."* This was the Red Queen's explanation to a confused Alice as to why she could run as fast as she could in Wonderland but never get anywhere, a situation analogous to the constant evolutionary pressure exerted by a changing environment. Bell (1982) recognized the parallels between Van Valen's view of extinction rates and the constant adaptation that should characterize host-parasite coevolution and applied the Red Queen moniker to a new hypothesis that invoked pressure from coevolving parasites to generate an advantage for sex.

Conceptual development of the Red Queen hypothesis ultimately began, as noted by Hamilton (1980), with Haldane (1949), and was revisited by Clarke (1976), both of whom identified disproportionately high attack of common host or prey genotypes as a potential explanation for surprisingly high levels of allelic polymorphism. Haldane (1949) was also the first to argue that disease is likely to be a selective force of profound importance for the maintenance of genetic diversity in many species. While selection imposed by biotic factors was increasingly implicated in the persistence of sex and outcrossing in the 1970's (e.g. Levin 1975, Glesener and Tilman 1978), the explicit coupling of the Red Queen with sex originated with Jaenike (1978; also see Bell 1982). He responded to Williams's (1966, 1975) assertion that the maintenance of sex requires immediate, individual advantages by building on models (Maynard Smith 1971; Charlesworth 1976) showing that recombination can persist if environments change rapidly enough to favor different combinations of alleles from one generation to the next (also see Lewontin 1974). In this case, sex is beneficial because it allows for the production of genetically variable offspring and increases the probability of producing offspring with the combination of alleles that happen to be favored at the time. While this mechanism can provide an individual-level advantage to sex, both Maynard Smith and Charlesworth assumed that the agent of selection was abiotic and unlikely to change rapidly enough to allow sex to persist (also see Maynard Smith 1978).

[1] Since sex and outcrossing are similar phenomena, and since Red Queen dynamics may contribute to favoring both in a similar manner, "outcrossing" is also included when "sex" is mentioned, and vice versa, unless stated otherwise.

Jaenike's innovation was to provide "biological realism" for these previously-developed models by postulating that the mechanism of selection was biotic in origin (see also Levin 1975; Glesener and Tilman 1978; Bell 1982). Specifically, he pointed out that if parasites preferentially attack common genotypes to which they are adapted, sex will produce novel or rare genotypes that are temporarily free from infection. Although Levin (1975) suggested that pressure from pests could favor *novel* genotypes and thus recombination in plants, Jaenike's idea was an important step forward in that it provided an advantage to rareness, not just novelty, and provided a more general, intrinsic advantage to sexual reproduction.

These ideas were formalized by Hamilton (1980) (also see Glesener 1979, Hamilton 1982), who provided a model verifying that negative frequency-dependent selection exerted by parasites can create cycles in host and parasite genotypic fitness and, under certain conditions, favor sex. Hutson and Law (1981) obtained a similar result with regard to the frequency of recombination, while Price and Waser (1982) found that an allele for sex will spread under stable conditions when there is strong selection for rare genotypes and when offspring dispersal is limited. The Red Queen gained additional credibility from models presented by Hamilton et al. (1990) that, unlike previous models, demonstrated how the Red Queen can still provide an advantage to sexual reproduction under "challenging" but "realistic" conditions (e.g., with repeated mutations to parthenogenesis and when the full two-fold cost of sex is realized). Despite these major theoretical advances, however, the broad applicability of the Red Queen hypothesis remains a contentious issue due to the potentially restrictive conditions under which sex evolves and is maintained (May and Anderson 1983; Howard and Lively 1994; Otto and Nuismer 2004; Agrawal and Otto 2006; Gandon and Otto 2007). A particular focus of this controversy has been the assumed requirements for high virulence, tight linkage, and rapidly fluctuating epistasis, as discussed below.

7.2 Assumptions and Predictions of the Model

The primary assumption of the Red Queen hypothesis is that parasites that are able to infect common host genotypes achieve a significant selective advantage, which thus favors rare hosts. The theory of negative frequency-dependent selection by parasites predicts that loci involved in host infectivity should be polymorphic both within and between natural populations and that genotypic combinations of alleles at these loci should oscillate in frequency over time as local parasites continually adapt to their host population. The implications are that common host genotypes in natural populations should be or become over-infected by parasites relative to their frequency in the population. This type of "over-reactive" frequency-dependent selection has been shown to lead to cyclical dynamics, where parasites eventually track common host genotypes and, assuming there is a fitness cost associated with infection, drive them to lower frequency (Hamilton 1980). The constant cycle of adaptation predicts that only a subset of common genotypes is expected to be over-infected at any one time. This means that interpretation of data relating infection and genotype frequency can be misleading unless data are collected across many populations and/or across time.

The occurrence and characteristics of the predicted oscillatory cycles have been the focus of numerous theoretical and empirical studies (e.g., Jaenike 1978; Bell 1982; Seger 1988; Hamilton et al. 1990; Dybdahl and Lively 1998; Koskella and Lively 2007). For example, computer simulations of host-parasite coevolution have shown that the period of oscillations (i.e., how quickly parasites drive common clones down in frequency) is determined primarily by parasite generation time, while the amplitude of the oscillations (i.e., the degree of change of individual genotypes over time) is driven mainly by the virulence of the parasite (Lively 1999; Gandon 2002).

A key feature of these dynamics is the time lag between the rise in frequency of a recently rare and resistant host genotype and the subsequent chance introduction of a matching parasite genotype via migration, mutation, or recombination. This means that there will also be a time lag prior to over-infection of the newly common host by the local parasites. This time lag (or phase difference) is essential for driving oscillatory dynamics and has therefore been the focus of much theoretical work (Hutson and Law 1981). For example, host recombination rate, parasite migration rate between host populations, the level of parasite specificity, and the degree of parasite virulence (i.e. the fitness cost of infection for the host) have all been predicted to affect the phase difference (Gandon 2002). These findings suggest that parasite populations that are able to respond more quickly to selection, e.g., more genetically variable populations, will more closely track common host genotypes and more quickly drive changes in the genetic makeup of local host populations.

Where present, these negative frequency-dependent dynamics are predicted to lead to local adaptation of parasites to host populations since parasites are more likely to successfully infect common host genotypes in populations with which they are coevolving. Specifically, if parasites are driving genetic changes in host populations and if geographically distinct populations are evolving independently, each parasite population should be better at infecting hosts from local, sympatric populations than hosts from allopatric populations.

7.2.1 Population Genetics

Population genetic models that explore the Red Queen hypothesis are based on the premise that sex can be advantageous in the face of a changing environment because recombination disrupts non-random association between alleles, i.e., linkage disequilibrium (LD). The idea here is that recombination acts to re-randomize genotypes by breaking up any genotypes that are more frequent in a population than would be expected by random assortment.

Linkage disequilibrium caused by epistasis for fitness can be disadvantageous under two scenarios; first, if there is directional selection and epistasis for fitness is negative; or second, if there is fluctuating selection in which the sign of epistasis

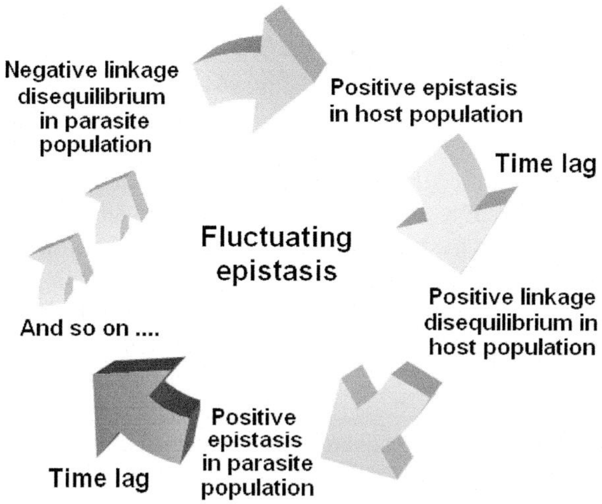

Fig. 7.1 Diagram depicting fluctuating epistasis in which the sign of epistasis in host populations changes in response to the sign of linkage disequilibrium ("LD") in the parasite population, and vice versa. For example, negative LD in the parasite causes positive epistasis in the host population whereby host genotypes that are rare and currently resistant have a disproportionately high fitness. This positive epistasis in the host population then leads to positive LD in the host population after a time lag (as the uninfected host genotype increases in frequency and, eventually, becomes more common than would be predicted based on the individual frequencies of each allele). This positive LD in the host creates positive epistasis in the parasite population since any parasite genotype able to infect this now-common host genotype will have a significant fitness advantage. This would then, after a time lag, lead to positive LD in the parasite population and negative epistasis in the host population, and the cycle continues (Peters and Lively 1999)

changes frequently (Barton 1995; Peters and Lively 1999, 2007; Gandon and Otto 2007). Under the first scenario, negative epistasis is generated any time that the fitness of an individual with two beneficial alleles is less than the product of the fitness for two separate individuals carrying only one of the two alleles. Under weak selection, the build-up of negative epistasis in a population allows for the increased spread of advantageous alleles via recombination and is therefore thought to confer an advantage to sex (Lythgoe 2000). Under the second scenario, the signs of both epistasis and linkage disequilibrium fluctuate over time in a genotype-specific manner (see Fig. 7.1; Maynard Smith 1978; Seger and Hamilton 1988; Barton 1995). This scenario requires that the fluctuations occur over a short and precise time-scale such that epistasis and LD are of opposite signs for a large portion of coevolutionary time (Barton 1995).

While the ability of fluctuating epistasis (see Fig. 7.1) to maintain sex was initially viewed with skepticism because fluctuations had to be so rapid (e.g. Maynard Smith 1978; Barton 1995), new theory has suggested that this particular criticism may be unwarranted (see below). Moreover, the assumption of genotype-specific fluctuations in linkage disequilibrium and epistasis is likely often met during host-parasite coevolution since there is strong evidence for genotype-specific infection patterns and since rare genotypes are favored via negative frequency-dependent selection (i.e., epistasis and LD are of opposite signs).

A good example of theory that supports the case for fluctuating epistasis is presented in Peters and Lively (1999). They used a series of numerical simulations of host-parasite coevolution to show that tight genetic specificity for infection points to fluctuating epistasis rather than directional selection as the more likely candidate for the maintenance of recombination. The same model was later used under a wider range of parameters to show that recombination spreads in a population due to both short-term benefits in the face of fluctuating epistasis and long-term benefits in response to directional selection (Peters and Lively 2007). Lythgoe (2000) used a deterministic model of parasites with acquired immunity and was also able to implicate fluctuating epistasis as a likely mechanism for the maintenance of sex. Furthermore, Gandon and Otto (2007) used a series of deterministic models to isolate the effects of fluctuating epistasis in large populations. They showed that higher levels of host recombination are favored when LD and epistasis are more often out of phase, which happens when parasites are better adapted to their hosts (see also Fig. 7.2).

The amount of time that LD and epistasis are of opposite signs is also a function of the strength of selection, such that the signs differ more often under strong selection when the period of fluctuations is small (Barton 1995). For example, under a model incorporating highly virulent parasites, negative frequency-dependent selection is predicted to result in a time lag of only 1–2 generations between a given change in the sign of epistasis and the corresponding change in LD (Peters and Lively 1999).

However, even with appropriately rapid fluctuating epistasis, the theory remains restricted by requirements of tight linkage (Gandon and Otto 2007; Peters and Lively 2007) and high virulence (Peters and Lively 2007, but see Salathé et al. 2007). Furthermore, recent theoretical work by Kouyos et al. (2007) suggests that small modifications of the standard Red Queen model can lead to dampening of the LD cycles that are critical to the Red Queen. The authors go on to show that the amplitude of the LD oscillations is inversely correlated with population size and that drift counteracts the dampening of the cycles. In light of the current theory, the ability of host-parasite interactions to single-handedly maintain oscillations in natural populations remains unclear.

The strong selection required to maintain sex in many models has also been a source of concern for many theoreticians, because it implies that only very virulent and/or highly prevalent parasites will be able to maintain sex and outcrossing (e.g., Otto and Nuismer 2004). This problem has been circumvented in some models through the incorporation of further parameters in an attempt to more closely approximate biological realism. For example, Hutson and Law (1981) determined that a recombination modifier allele will spread under relatively weak selection if the coevolutionary time lag is relatively long. More recent models found an increased advantage to sex under larger numbers of interacting loci (Hamilton et al. 1990), when there is some vertical transmission of parasites (Agrawal 2006), when recombination is present but infrequent (Peters and Lively 2007), and when

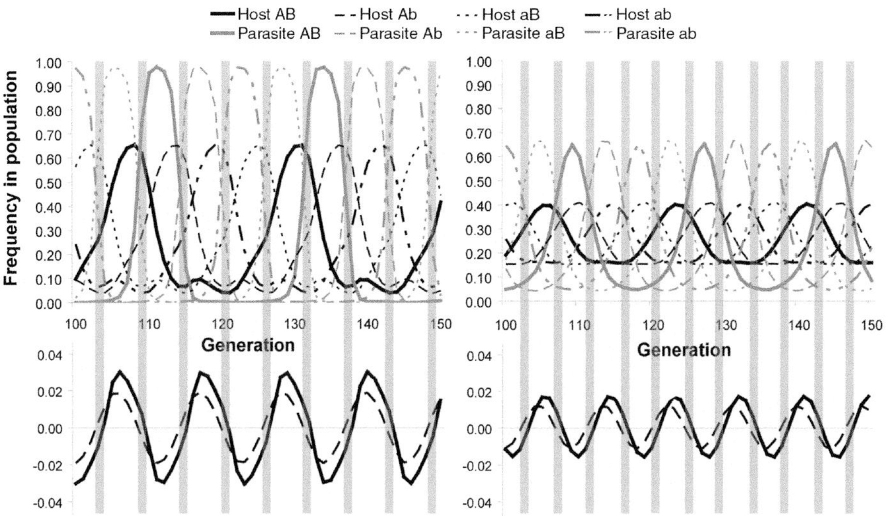

Fig. 7.2 Simulation results using the model of Lively (1999), in which population genetic recursion equations are used to track changes in allele frequencies over time in response to selection, as determined by infection success. The model assumes two diallelic loci with infection of hosts determined through a matching-alleles (MA) framework. The MA model of infection assumes that parasites infect hosts in a genotype-specific manner, such that an Ab parasite is only able to infect Ab hosts. The *top panel* of the figure shows genotype-specific changes in frequency of hosts (*black lines*) and parasites (*grey lines*) over 50 generations of coevolution. The *bottom panels* show the corresponding changes in linkage disequilibrium ("LD") (*solid line*) and epistasis (*dashed line*) over the same period of time. The *grey bars* highlight the periods of time in which epistasis and LD have opposite signs. The *left-hand panels* represent a low rate of parasite migration (0.001), while the *right-hand panels* represent a higher rate of migration (0.05) where parasites are quicker to respond to changes in host genotype frequencies. Because fluctuations in genotype frequencies are occurring more rapidly in the *right-hand panels*, the proportion of time that epistasis and LD are of opposite signs is greater (Barton 1995)

segregation in diploids (as opposed to just recombination in haploids) is considered as a mechanism generating novel combinations of alleles (Agrawal and Otto 2006).

One important consideration is that certain characteristics of parasite community dynamics are likely to play an important role in the realized selective pressure for sex, but are rarely included in laboratory estimates of parasite virulence. As a result, the strength of parasite-mediated selection in nature may often be underestimated. For example, parasite prevalence in a given host population directly influences the strength of parasite-mediated selection, and is thus a critical determinant of the evolutionary response of the host population (Otto and Nuismer 2004). A recent model also shows that virulence may be greater when host condition is density-dependent (Lively 2006). Furthermore, it is rare that only a single parasite species is exerting pressure on a given host population. This means that, to the extent that it is possible, the combined effects of all parasite species in a given community should be considered (Hamilton et al. 1990).

Another way to circumvent the assumption of high parasite virulence was introduced by Salathé et al. (2007), who turn the tables by highlighting the importance of host-mediated selection on parasite populations. The authors created a model that incorporates a biologically realistic fitness cost to parasites that are unable to infect a host in the next generation. They found that when parasites are well-adapted to host populations, as would occur under strong selection for successful infection, recombination in the host population is more likely to create novel and resistant genotypes than to break up currently resistant genotypes. The authors interpreted this result as a consequence of the strong selection for parasites to match local hosts, which results in a greater frequency of susceptible host genotypes. In this case, the primary outcome of recombination would be to break these susceptible host genotypes up and to create rare or novel genotypes, which are temporarily resistant. In addition to the theoretical advances discussed in the previous paragraph, this work is an example of how the incorporation of more biological realism into current working models can uncover the robustness of the Red Queen hypothesis.

7.2.2 Infection Dynamics

One important assumption underlying many host-parasite models is tight genetic specificity for infection such that certain host genotypes are susceptible to a subset of parasite genotypes and resistant to others. Computer simulations have shown that high specificity for infection can lead to oscillations in genotype frequencies and the long-term maintenance of genetic diversity (e.g., Seger 1988). As discussed previously, these oscillatory dynamics are central to many theories regarding host-parasite coevolution, including both local adaptation and the maintenance of sexual reproduction (Hamilton 1980; Bell 1982; Price and Waser 1982; Hamilton et al. 1990).

Two models describing the genetics of host-parasite interactions have received the vast majority of theoretical and empirical attention, though many others exist. The first model described was the "gene-for-gene" model (GFG), which predicts that the interaction between parasite virulence loci and host resistance loci determines successful infection (Flor 1956). The matching alleles model (MA), on the other hand, is based upon a system of self/nonself recognition molecules such that hosts can successfully defend against any parasite genotype that does not match their own (Hamilton 1980; Bell 1982; Peters and Lively 1999).

Agrawal and Lively (2002, 2003) have shown that both GFG and MA models of infection can lead to genotypic oscillations, but that the resulting dynamics are often very different. For example, under the GFG model, the possibility of universally virulent parasites and/or universally resistant hosts exists. In this case, with any migration or mutation, a universally virulent parasite would quickly sweep to fixation and genetic polymorphism within the host and parasite populations would not be maintained. GFG can result in oscillatory dynamics if the model incorporates a fitness cost to resistant host genotypes in an avirulent parasite environment or to virulent parasite genotypes in a mainly susceptible host population (May and Anderson 1983). Polymorphism can also be maintained under the GFG model in the absence of costs if genetic drift is incorporated (Salathé et al. 2005)

A model incorporating a combination of GFG and MA-type interactions led to sustained oscillations, suggesting that movement away from the strict genetic models might more often lead to dynamic cycling (Agrawal and Lively 2003). This result is particularly interesting since many host-parasite systems do not conform to the predictions of any one proposed infection genetic model (e.g., Rolff and Siva-Jothy 2003; Nidelet and Kaltz 2007; Wegner et al. 2007). In other words, it is likely that the true infection genetics of any given system act more as a combination of the current models than as predicted under any one model. The dynamics observed under GFG conditions, however, have been shown to have more damped and less frequent oscillations, which are less likely to favor sex (but see Agrawal and Lively 2002). Thus, while models incorporating GFG can favor sex under certain conditions, negative frequency-dependent selection leading to the maintenance of polymorphism, and perhaps sex, occurs more readily in MA-based models (Frank 1992, 2000; Brunet and Mundt 2000; Agrawal and Lively 2002).

7.3 Does It Work?

While theory suggests that Red Queen dynamics can maintain sex under certain conditions, direct empirical tests incorporating even several of the integral components of the hypothesis are rare (Lively and Apanius 1995; Apanius et al. 1997; Meirmans and Neiman 2006). This is largely due to the complexity of the Red Queen, which involves (but is not excluded to) host-parasite coevolution, local adaptation, negative frequency-dependent selection, risk of infection, and time-lagged oscillations in host and parasite genotype frequency and in the frequency of infection of rare vs. common genotypes (Wuethrich 1998; Lively 2001; Meirmans and Neiman 2006).

As a result, empirical work relevant to evaluating whether the Red Queen contributes widely to the maintenance of sex has focused on examining at best a few of its components (Apanius et al. 1997; Meirmans and Neiman 2006). This means that the majority of these studies are limited in their ability to exclude other potential explanations for results that fit the predictions of the Red Queen (Wuethrich 1998). As reviewed below, while there is mounting empirical evidence for each of the basic components of the theory, several key questions remain unanswered.

7.3.1 Geographical Distribution of Sex and Outcrossing

There is a well-documented pattern of high frequency of outcrossing and sex in undisturbed, biologically complex habitats (Levin 1975; Glesener and Tilman 1978; Barrett and Eckert 1990; Hamilton et al. 1990) where disease and other "natural enemies" are prevalent (Levin 1975; Glesener and Tilman 1978; Lloyd 1980; Bell

1982). In fact, the recognition of the predominance of sex in stable communities led to the devaluing of models hypothesizing that sex is common because it provides a selective advantage when abiotic conditions are unpredictable (Glesener and Tilman 1978; Maynard Smith 1978; Price and Waser 1982). Levin (1975) was the first to propose that the ecological association between sex and stability may result from pathogen-mediated selection for outcrossing in natural populations. The documentation of these patterns set the stage for more direct exploration of the role of Red Queen dynamics in the maintenance of sex.

7.3.2 Frequency of Sex vs Frequency of Infection

If Red Queen dynamics contribute to the maintenance of sex within species, and if variance in the risk of infection is high, primarily sexual/outcrossing populations and species should be found where the risk and prevalence of infection is high, and vice versa (Lively and Apanius 1995; Lively 2001). As noted above, the geographical distribution of sex broadly fits these predictions.

Whether the Red Queen explains the distribution of sex and outcrossing within and between species was first explicitly addressed in two studies from 1987 that employed comparative approaches to determine whether variables assumed to be associated with the intensity of parasite pressure could explain variation in the frequency of sex or recombination. In a study aimed at identifying the determinants of recombination frequency amongst 40+ mammal species, Burt and Bell (1987) showed that ~75% of the variance in recombination above the level required for normal meiosis was explained by variance in generation time. They argued that this result was in line with the expectations of the Red Queen, since longer generation times should favour higher levels of recombination because parasites would have more time to adapt to a particular host genotype prior to host reproduction. Lively (1987) found that a large fraction of the variation in the relative frequency of sexual vs. asexual *Potamopyrgus antipodarum*, a New Zealand snail, was explained by variation in the frequency of infection by virulent parasites. In other words, highly sexual populations were also heavily infected, as expected if sex can only gain an advantage in the face of intense pressure from coevolving parasites (also see Lively 1992; Lively and Jokela 2002; King and Lively 2009). Importantly, later studies established that this link between the frequency of sex and infection was not due to increased susceptibility of sexual *P. antipodarum* (Lively and Jokela 1996; Jokela et al. 1997). Positive associations between parasitism and outcrossing in two other hermaphroditic freshwater snail species have also been documented (Schrag et al. 1994, reviewed in Johnson et al. 1997), but are not a ubiquitous feature of snail species with mixed sexual/asexual populations (Ben-Ami and Heller 2005, 2008).

Covariance between the relative frequency of sex/outcrossing and the frequency of parasitism is now established in a wide range of taxa, providing broad but indirect support for a role for the Red Queen in the maintenance of sex. For example, Kumpulainen et al. (2004) showed that the variation in the relative frequency of parasitoids explained a large fraction of the variance in the relative frequency of sexual psychid moths, while asexual moths tend to predominate in locations that are relatively parasitoid-free. Outcrossing has also been shown to be more prevalent in plant species subject to attack by multiple pathogen and parasite species (Busch et al. 2004).

Associations between the frequency of recombination within a species and the presence of virulent parasites are also expected if the Red Queen can provide a short-term advantage to sex and recombination. Such a pattern was documented in a recent selection experiment conducted by Fischer and Schmid-Hempel (2005). They found evidence for increased frequency of recombination in populations of the flour beetle, *Tribolium castaneum*, after only 8 generations of exposure to a microsporidian parasite, *Nosema whitei*, relative to parasite-free control populations. The results from this experimental coevolution study are a major contribution to the body of empirical data relevant to the Red Queen in that they suggest that parasite-mediated selection can quickly facilitate the spread of recombination in a population.

7.3.3 Susceptibility to Infection

If sex/outcrossing are favored by Red Queen dynamics, individuals that are the product of asexual reproduction or selfing should be, *on average,* easier to infect with local, coevolving parasites than individuals that are the product of outcrossed sex (Lively and Apanius 1995; Agrawal and Lively 2001). Empirical assessment of this prediction is complicated by a number of factors including the likelihood of periodic maladaptation of coevolving parasites (Morand et al. 1996), the extent to which selection for particular mating systems relies upon the genetics of infection and resistance (e.g., Agrawal and Lively 2001, 2002), and the difficulty of accurately measuring the risk of infection in natural populations (Lively 2001).

Direct support for this element of the Red Queen also requires that higher levels of infection in asexuals, relative to sympatric sexuals, are due to increased mean susceptibility of asexuals to infection by coevolving, local parasites. Indirect evidence for this pattern comes from observational studies finding that asexual or selfed individuals have higher rates of infection or more enemy damage than sympatric sexually-produced individuals, though other explanations for this pattern cannot be rejected with such data alone (Mee and Rowe 2006). A good early example of this type of study comes from Burt and Bell (1991), who found that asexually produced American beech (*Fagus grandifolia*) seedlings suffered more herbivore damage when very young than did sympatric sexually-produced seedlings. Similar patterns have since been documented in asexual vs. sexual lizards (Moritz et al. 1991; see also Chapter 21), fish (Hakoyama et al. 2001; Mee and Rowe 2006; see also Chapter 19), flatworms (Michiels et al. 2001; see also Chapter 18), and inbred vs. outbred sheep (Coltman et al. 1999).

A few studies have directly compared the susceptibility of asexual vs. sexual taxa to infection by parasites and pathogens. Hakoyama et al. (2001) examined the metacercarial load and immunocompetence in sexual vs. asexual *Carassius* (crucian carp) living sympatrically in a Japanese river. They found that sexual *Carassius* had a significantly lower load of metacercarian parasites and ~50% higher immune activity than their asexual counterparts. Hakoyama et al. concluded that the reduced immune activity of asexuals could underpin their higher parasite load, but that more data are needed to understand whether this phenomenon actually helps to maintain sex in this species. Another recent study showed that asexual lineages of a freshwater snail (*Potamopyrgus antipodarum*) had ~ 30% lower count of a particular type of immune defense cells than sexuals (Osnas and Lively 2006), though whether this plays a role in the persistence of sexual *P. antipodarum* is unclear.

There still remains no clear and generalizable relationship between mating system and susceptibility to parasites (e.g., Stevens et al. 1997; Haag et al. 2003; Puurtinen et al. 2004); there are systems where parasite prevalence is statistically indistinguishable in sexuals and asexuals (e.g., Tobler and Schlupp 2005) and in selfing vs. outcrossing individuals (e.g., Puurtinen et al. 2004), or even higher in sexuals than in asexuals (e.g., Hanley et al. 1994; Brown et al. 1995). Such results are difficult to interpret definitively in light of potentially complicating factors such as differences in ploidy level between sexual and asexuals (Osnas and Lively 2006) and high among-clone variance in asexual lineage susceptibility (Brown et al. 1995). A meta-analysis that takes these phenomena into account could be quite illuminating. Either way, it is clear that more research into the genetics and physiology of infection and resistance in sexuals vs. asexuals and selfers vs. outcrossers is needed to fully evaluate the role of parasite pressure in the predominance of sex and outcrossing.

7.3.4 Rare Advantage

Antonovics and Ellstrand (1984) used sweet vernal grass (*Anthoxanthum odoratum*) to perform the first empirical study explicitly aimed at testing the Red Queen. Although their research did not focus upon parasites per se, it did test a key component of the Red Queen: that minority genotypes would realize a fitness advantage simply as a function of being rare. Minority *A. odoratum* genotypes did in fact experience a two-fold advantage, leading Antonovics and Ellstrand to conclude that negative frequency-dependent selection by a biotic agent could facilitate the maintenance of sex in natural populations (also see Ellstrand and Antonovics 1985). A follow-up study by Schmitt and Antonovics (1986) found no minority-genotype advantage for *A. odoratum* with regard to probability of infection by aphids, suggesting that negative frequency-dependent selection was not prominent in this particular parasite. They did observe, however, that infected plants had ~30% higher survivorship when surrounded by unrelated vs. related neighbors, which indicates that aphid infection could interact with other mechanisms to favor sex and outcrossing. Kelley (1994) summarized data from this system in order to determine whether an aphid-transmitted virus that infects *A. odoratum* might favor sexually-produced genotypes.

He determined that the virus provided a weak advantage for rare and/or sexually produced *A. odoratum* genotypes, but concluded that more study was needed to clarify the role of this and other pathogens in favoring rare *A. odoratum* genotypes species.

A 1990 study of parasite load in sexual vs. asexual topminnows (*Poeciliopsis* spp.) in several natural populations provided particularly compelling support for negative frequency-dependent selection exerted by parasites favoring rare hosts (Lively et al. 1990). One population, Sandal Pool, contained members of two topminnow clones along with sexual individuals, while another population, Log Pool, contained sexuals but only one of the two Sandal Pool clones. The clone present in both populations was at low frequency relative to the other clone in Sandal Pool. Lively et al. (1990) found that the single Log Pool clone had a significantly higher load of trematode parasites than the coexisting sexuals, while the other, more common clone was more infected than both the other clone and the sexuals in Sandal Pool. This result indicated that the frequency of infection was a function of clonal frequency rather than among-clone variance in susceptibility. This inference was bolstered by a superficially contradictory finding from a third population, where sexual topminnows were infected at higher levels than their sympatric clonal counterparts. However, this pool had recently experienced a drought that resulted in the loss of nearly all of the genetic diversity of the sexual topminnow population. Consequently, the sexual individuals in the pool were very homozygous relative to the asexuals, which maintained permanent high heterozygosity due to their apomictic reproduction (Vrijenhoek and Lerman 1982). Once the genetic diversity of the sexuals was supplemented by an introduction of sexual fish from a different population, the common clone was found to have significantly higher frequency of infection than the sexuals only 2 years following the transplant. The latter result also demonstrates that parasites can respond rapidly to changes in host genotypic frequency, a key component of the Red Queen hypothesis (Lively and Apanius 1995; Peters and Lively 1999).

It is clear, however, that not all interactions between hosts and their coevolving parasites involve negative frequency-dependence. For example, Strauss and Karban (1994) found that, counter to the predictions of the Red Queen, genetic diversity (or rareness, per se) provided no advantage for *Erigeron glaucas* (seaside daisy) subject to infection by a herbivorous thrip, *Apterothrips apteris*. Similar results were documented in another plant species, *Allium vineale*, by Ronsheim (1996), and in a freshwater bryozoan (Vernon et al. 1996). The negative results from plants are perhaps not surprising given that plant-parasite interactions are often characterized by GFG infection genetics that are less likely to provide a direct advantage to sex (Kover and Caicedo 2001).

Clay and Kover (1996) concluded that there were still not sufficient data to determine whether negative frequency-dependent selection against common genotypes is widespread enough to provide a general explanation for the predominance of sex. More than a decade later, there is now evidence for negative frequency-dependent selection exerted by parasites in a diverse array of taxa (e.g. Carius et al. 2001; Bruvo et al. 2007; Mundt et al. 2008), suggesting that this particular component of the Red Queen is a relatively common feature of host-parasite interactions.

7.3.5 *Parasitic Tracking of Common Host Genotypes*

The Red Queen requires not just rare advantage but also evolutionary tracking of common parasite genotypes, or so-called "reciprocal entrainment" (Lively and Apanius 1995). A good example of this comes from Dybdahl and Lively (1998), who found that clonal genotype frequencies of the New Zealand freshwater snail, *Potamopyrgus antipodarum,* were correlated with time-lagged changes in infection by the sterilizing trematode, *Microphallus*. By examining genotype-specific infection rates in natural populations over a period of five years they found that clones that had been rare and under-infected in samples from previous generations tended to become over-infected after they became common. In a follow-up study, Lively and Dybdahl (2000) showed that common clones were infected at ~30% higher levels than sympatric rare clones by local parasites, but that levels of infection between rare and common clones did not differ when exposed to allopatric parasites. This result showed that common genotypes were specifically being targeted by coevolving parasites and were not just inherently more susceptible. Jokela et al. (2009) took this approach a step further by showing that clones from a mixed sexual/asexual population that were both common and resistant 7–10 years ago had since become extremely rare and more susceptible to infection. This is a key finding because it provides definitive evidence that key elements of the Red Queen can operate on the short time scale required to maintain sex, and in a population where sexuals and asexuals compete.

Lythgoe and Read (1998) highlighted the *Potamopyrgus-Microphallus* system, and pointed out that one can test the predictions of Red Queen models by looking backward in time to determine which host genotypes were common and which were rare prior to the infection data at the time of the study. They called this method the "Advice of the Rose" after Lewis Carroll's novel in which Alice is told by the rose to walk backwards in order to find the Red Queen.

This idea was recently tested using experimental coevolution methods comparing infectivity of *Microphallus* populations that were coevolving directly with *Potamopyrgus antipodarum* with infectivity of *Microphallus* populations on hosts that were lagged behind by one generation (Koskella and Lively 2007). After only three generations of experimental coevolution, the coevolving parasites were ~65% more infective to their own, coevolving hosts than to the control hosts, which had not received parasites since the start of the experiment. Furthermore, the experimental host populations were ~32% less susceptible than control hosts to parasites from their original population, indicating that the host populations had responded to selection imposed by the coevolving parasites. These results are indicative of just how powerful such experiments can be for detecting the effects of parasite-mediated selection. Furthermore, the results showed that the lagged host treatment had a consistently higher rate of infection than the coevolving host treatment. Parasites were therefore tracking their host populations in a time-lagged manner such that parasites from any given generation were better able to infect hosts that were lagged behind by one generation than contemporaneous hosts that already had an opportunity to respond to selection by the parasite.

Data examining the ubiquity of parasite-mediated, frequency-dependent selection has focused on uncovering its two major components; time-lagged dynamics in parasite tracking and genotype-specific tracking of host genotypes. However, it is important to note that, although over-infection of common clones is a direct prediction of the Red Queen hypothesis, it is only predicted to be found in roughly half of the cases. This is because the time lag dynamics allow hosts to remain common and under-infected until a suitable parasite genotype arises in the population via either migration or mutation (Dybdahl and Lively 1995; Morand et al. 1996). This means that absence of infection in a common host genotype is not sufficient to reject the Red Queen. Therefore, future studies examining parasite-mediated, negative frequency-dependent dynamics should follow multiple common host genotypes and should test for over- or under-infection in a genotype-specific manner over time.

7.3.6 Parasite Local Adaptation

As mentioned previously, empirical data on parasite local adaptation has been viewed as a key piece of the Red Queen puzzle. Whether a parasite is better able to infect members of its local, sympatric host population than allopatric hosts depends in part upon the degree of genetic specificity for infection and the dynamics of coevolution.

A distinct alternative to the Red Queen is the hypothesis of Arms Race coevolution (Dawkins and Krebs 1979), which describes a situation in which hosts continually build up their armory of resistance to parasitism while parasites respond by becoming more virulent or evolving ever-changing mechanisms of evading host immunity. This form of unidirectional coevolution is characterized by repeated selective sweeps (Woolhouse et al. 2002) and, unlike the Red Queen, does not generate a rare advantage per se. Rather, hosts and parasites are continually evolving novel adaptations in response to one another which, once defeated, will never again incur resistance/virulence against the other player. In addition, the adaptations resulting in resistance of hosts and virulence of parasites are an inherent property of the individual genotype, such that some hosts are more resistant overall and some parasites are more virulent overall. Thus, another major demarcation between Arms Race and Red Queen coevolution is that under the Red Queen, a genotype which is susceptible to one parasite genotype may be resistant to another and vice versa. Since populations with highly virulent parasites would most likely also contain highly resistant hosts, parasite local adaptation is not predicted to be universal because virulent parasites would be more infective to any allopatric hosts that have not yet evolved a similarly high level of resistance. The implication of these differences between the two mechanisms is that Arms Race dynamics are less likely to favor sexual reproduction. Distinguishing between these two theories, and thus, the extent to which parasite pressure is likely to favor sex, requires insight into how host and parasite populations change over time and whether directional selection or negative frequency-dependent selection explains the antagonistic interaction between the species.

As outlined above, one important type of evidence in favor of Red Queen dynamics is parasite tracking of common host genotypes. Systems where dormant stages of both host and parasite are occasionally preserved present a promising avenue for direct empirical tests of parasite tracking. Decaestecker et al. (2007) used this approach to show that *Daphnia magna* were consistently more susceptible to contemporaneous parasites than those from past generations (see also Chapter 15). Since this type of data is very difficult if not impossible to collect for most systems, reciprocal cross-infection experiments testing for local adaptation can provide indirect support for parasite tracking (Lively and Apanius 1995; Johnson et al. 1997, Lively et al. 2004), as well as for tight genetic specificity for infection (e.g., Carius et al. 2001). Assessment of the presence and extent of local adaptation is complicated by the expectation that the time-lag characterizing host-parasite coevolution under negative frequency-dependent selection should result in periodic "local maladaptation" (Morand et al. 1996). Hence, it not surprising that while local adaptation has been documented in many taxa, it does not characterize all host-parasite systems (as reviewed in Greischar and Koskella 2007).

Among the many studies documenting a strong degree of local adaptation is Shykoff and Schmid-Hempel (1991). They exposed uninfected groups of bumblebees (*Bombus terrestris*) to *B. terrestris* infected with a horizontally-transmitted trypanosome parasite, and found that the prevalence of infection following exposure was more than twice as high when the source and target hosts were related. A more recent study examining infection dynamics of the freshwater crustacean, *Daphnia magna,*and its bacterial parasite, *Pasteuria ramosa,* found that host clone/parasite combinations that had been isolated together from natural populations resulted in successful new infections nearly twice as often as novel host-parasite combinations (Carius et al. 2001). This study further showed that no single host clone was more resistant to every parasite genotype, which is suggestive of matching-allele type dynamics rather than gene-for-gene, as discussed previously. There is also strong evidence for local adaptation in the *Microphallus-Potamopyrgus antipodarum* system (Lively et al. 2004; Jokela et al. 2009), and from several other snail-parasite systems (reviewed in Lively 1996; Johnson et al. 1997; Webster and Davies 2001).

Local adaptation has also been demonstrated within experimental coevolution studies. For example, Nidelet and Kaltz (2007) compared the frequency of infection in three lineages of the protozoan *Paramecium caudatum* by "resident," coevolving lines of a bacterial parasite (*Holospora undulata*) to infection by novel, non-coevolving *H. undulata* lines. In all three *P. caudatum* lineages, there was a small but significant increase in infection prevalence for resident vs. novel parasites following $\sim$30 bacterial generations of host-parasite sympatry. These types of laboratory experiments offer a powerful tool for investigating both genotypic specificity for infection and patterns of coevolution.

7.3.7 Molecular Evolution in Disease Resistance Loci

Haldane (1949) was the first to suggest that selection by parasites might favor rare hosts and explain the high intraspecific allelic diversity of vertebrate cell-surface molecules. This possibility remained an abstract concept for decades, requiring the elucidation of the molecular structure of immunological factors and the ability to generate and analyze DNA sequence data. Since then, there has been a growing body of empirical research in support of the idea that negative frequency-dependent selection may underlie evolution at many genes associated with immune function (e.g. Stahl et al. 1999; Bergelson et al. 2001; Tiffin et al. 2004; Lazarro 2005; Bakker et al. 2006; Dionne et al. 2007; Schwensow et al. 2007; reviewed in Bernatchez and Landry 2003; Charlesworth 2006; Piertney and Oliver 2006).

The most compelling data come from studies of the major histocompatibility complex (MHC), a multi-gene region that is the main source of immunological self/non-self recognition in vertebrates. Its unusual structure, co-dominant expression, and high allelic diversity led Bodmer (1972) and Doherty and Zinkernagel (1975) to suggest that heterozygosity or rareness at MHC might confer a fitness advantage mediated by coevolutionary interactions with pathogens (also see Hamilton 1982). Doherty and Zinkernagel (1975) provided initial empirical support for this hypothesis by showing that mice heterozygous for one component of MHC had enhanced immunological surveillance relative to homozygous mice.

Whether MHC diversity is actually maintained by negative frequency-dependent selection exerted by pathogens has been controversial and attracted much attention (e.g., Hughes and Nei 1988, 1992; Takahata and Nei 1990; Slade and McCallum 1992; Potts et al. 1994; Apanius et al. 1997; Edwards and Hedrick 1998; Hughes and Yeager 1998; Wedekind et al. 2005; Milinski 2006; Piertney and Oliver 2006; Knapp 2007), though there is little doubt that some form of selection is involved (Apanius et al. 1997; Bernatchez and Landry 2003; Piertney and Oliver 2006). This subject has been the focus of recent reviews (e.g. Bernatchez and Landry 2003; Milinski 2006; Piertney and Oliver 2006), so we provide only a brief overview.

Hughes and Nei were the first to use DNA sequence data to study MHC evolution by conducting a series of comparisons of the relative amounts of synonymous and non-synonymous polymorphism in class I and class II MHC (Hughes and Nei 1988, 1989a, b). Non-synonymous polymorphism is predicted to be low relative to synonymous polymorphism in genetic regions that are subject to purifying selection (as most genetic regions are assumed to be) because non-synonymous mutations change protein structure. However, if rareness per se is favored, as hypothesized in the antigen recognition site (ARS) of MHC, non-synonymous polymorphisms are predicted to accumulate at relatively high rates.

Hughes and Nei found that this prediction was met in both Classes I and II MHC, with high non-synonymous polymorphism in the ARS relative to other parts of

MHC and to other protein-coding genes not directly involved in pathogen recognition (Hughes and Nei 1988, 1989a, 1989b; also see Schaschl and Wegner). While Hughes and Nei (1988) attributed this result to overdominance, Hughes et al. (1994) concluded that MHC polymorphism is maintained by selection favoring diversity in the ARS, and noted that the association of MHC haplotypes with resistance to malaria implicates pathogens as an important selective agent. Since then, other researchers have documented similar patterns of sequence evolution in other taxa, and have come to similar conclusions (e.g., Bernatchez and Landry 2003; Mayer and Brunner 2007).

The results of Hughes and Nei inspired research more directly aimed at disentangling the selective force(s) underlying the unique pattern of evolution at MHC. An important body of work comes from studies that address whether particular MHC genotypes are more resistant to infection than others. Some studies find that certain MHC genotypes are associated with higher susceptibility to infectious disease (e.g. Briles et al. 1977; Todd et al. 1990; Hill et al. 1991, 1994; Apanius et al. 1997; Ameisen et al. 2002; Wegner et al. 2004; Westerdahl et al. 2004; Milinski 2006; Wedekind et al. 2005; Knapp 2007), or that MHC heterozygotes are more resistant to infection (e.g. Penn et al. 2002; McClelland et al. 2003; Wegner et al. 2004), while other studies find no consistent link between heterozygosity and resistance to infection (e.g. Wedekind et al. 2005; Schwensow et al. 2007; reviewed in Bernatchez and Landry 2003; Milinski 2006). A small body of data (e.g., Schwensow et al. 2007) does find that rare MHC genotypes are infected at low levels relative to more common genotypes, as expected under negative frequency-dependent selection (reviewed in Knapp 2007).

Potts et al. (1994) reviewed the state of the empirical data relevant to whether pathogens directly mediate MHC diversity, and concluded that more data were needed to address this question. Recent reviews have come to the same conclusion (Wedekind et al. 2005; Milinski 2006; Piertney and Oliver 2006; Knapp 2007). One complicating factor is that, as for many complex biological phenomena (West et al. 1999; Meirmans and Neiman 2006), multiple, non-mutually exclusive mechanisms are likely to be involved in the selective maintenance of MHC diversity (Apanius et al. 1997; Piertney and Oliver 2006). A complication that applies more generally is that there are still no clear expectations for the pattern of molecular evolution that will definitively allow identification of negative frequency-dependent selection (Bernatchez and Landry 2003; Tiffin et al. 2004; Nordborg et al. 2005; Bakker et al. 2006; Charlesworth 2006; Piertney and Oliver 2006; Tiffin and Moeller 2006).

7.4 Pluralism

As noted by Kondrashov (1993), there exist at least 20 potential explanations for the predominance of sex (e.g., Kondrashov 1993). However, none of these hypotheses (including the Red Queen) has found sufficient empirical support to consider the sex problem solved. In particular, single mechanisms seem to be limited by narrow applicability; each model can only favor sex/outcrossing under a set of strict, often unrealistic assumptions (West et al. 1999; Meirmans and Neiman 2006). The inability to explain sex via single mechanisms has resulted in a recent shift in focus away from discrimination between mechanistically simple hypotheses towards more complex approaches that can potentially be applied under a broader rage of conditions (Barton and Charlesworth 1998; West et al. 1999; Burt 2000; Meirmans and Neiman 2006; de Visser and Elena 2007). Several recent studies have theoretically explored pluralistic mechanisms for sex, incorporating the Red Queen along with other mechanisms such as mutation accumulation (Howard and Lively 1994, 1998) and mate choice (Howard and Lively 2003).

7.4.1 The Red Queen and Pluralism

As detailed above, the extent to which the assumptions underlying the Red Queen hold is the subject of much controversy. Even if the many requirements for Red Queen are in place, simulations show that when asexual assemblages are diverse, the Red Queen will not drive individual lineages to extinction due to the selective advantage that accrues to a rare asexual genotype under conditions of coevolving virulent parasitism (Howard and Lively 1994, 1998). Furthermore, the host-parasite coevolutionary dynamics that are required for operation of the Red Queen select for genetic diversity rather than sex per se (Lively and Howard 1994). This means that if all else is equal, a genetically diverse array of asexual lineages is as well-equipped to deal with parasitism as a sexual population (Glesener and Tilman 1979; Lively and Howard 1994; Lythgoe 2000).

Recognizing some of these difficulties, Howard and Lively (1994) pointed out that the Red Queen may apply more broadly if it is operating in concert with other mechanisms. For example, they used computer simulations to show that bottlenecking of host lineages caused by intense, coevolving parasitism targeting common host genotypes could accelerate the rate of deleterious mutation accumulation by decreasing the efficacy of purifying selection (Howard and Lively 1994, 1998), posing a further challenge to asexuals. These simulations found that extinction of asexual lineages due to interaction between the Red Queen and mutation accumulation can happen within several hundred generations of the origin of the asexual lineage, rapidly enough to allow sex to persist (Howard and Lively 1994). In contrast, their models suggest that neither the Red Queen nor mutation accumulation alone can give sex enough of a short-term advantage to explain its widespread persistence.

7.4.2 Empirical Tests

Authors have increasingly implicated pluralistic mechanisms such as the model described above as potential explanations for the persistence of sex (e.g. Jokela et al. 2003; Bruvo et al. 2007), though empirical tests remain scarce (West et al. 1999; Meirmans and Neiman 2006). In fact, to the best of our knowledge, no direct empirical tests of pluralist hypotheses for sex exist in mixed sexual/asexual systems in which sex requires an explanation. While two recent studies in asexual bacterial model systems (*Escherichia* and *Pseudomonas*) have documented interactions between parasite pressure and mutations in a manner that is consistent with that postulated in pluralist models (Cooper et al. 2005; Buckling et al. 2006; also see Coltman et al. 1999), widespread empirical application of pluralist models has been hampered by the common perception that these models may be effectively untestable in non-model eukaryotic systems (Meirmans and Neiman 2006).

Instead, early attempts to empirically test pluralistic hypotheses for sex have largely been indirect, investigating whether readily measurable phenomena (e.g., low fertility) that may be linked to an expected direct consequence of asexuality that itself is difficult to measure, high mutation load, are associated with higher rates or more severe consequences of disease (e.g. Bruvo et al. 2007). Other tests have used sexual model eukaryotic systems (e.g. Peters 1999; Haag et al. 2003; Salathé and Ebert 2003; Killick et al. 2006). Many of these studies obtain results consistent with those expected under particular pluralist mechanisms (e.g. Killick et al. 2006; Bruvo et al. 2007), paving the way for more direct and powerful tests of these ideas (Meirmans and Neiman 2006).

7.5 Conclusions

There is increasingly strong empirical support for many of the key elements of the Red Queen hypothesis for sex. However, whether Red Queen dynamics play an important role in the maintenance of sex and outcrossing in many natural populations remains unresolved. While recent theoretical advances have emphasized that host-parasite interactions can favor sex under a broader range of conditions than originally envisioned (especially if pluralism is considered), reasonable doubts remain. In general, the complexity of the mechanisms involved means that a definitive verdict on the utility of the Red Queen as an explanation for sex requires both more theory (e.g., Gandon and Otto 2007) and more data (Peters and Lively 2007). We believe that in particular, data evaluating the strength and extent of parasite-mediated selection in natural populations and studies directed at elucidating the genetics of infection/ resistance and whether/why there are differences in susceptibility to infection in sexual vs. asexuals are needed to give the Red Queen a fair and full test.

Acknowledgments We would like to thank P. Tiffin and D. Moeller for discussion of the molecular population genetics of disease resistance genes, J. Jokela for discussion of ploidy and susceptibility in *P. antipodarum*, C. Lively and D. Taylor for comments on an earlier version of the MS, and S. Bonhoeffer and an anonymous reviewer for thoughtful and helpful reviews.

References

Agrawal AF (2006) Similarity selection and the evolution of sex: revisiting the Red Queen. PLoS Biol 4: 1364–1371

Agrawal AF, Lively CM (2001) Parasites and the evolution of self fertilization. Evolution 55: 869–879

Agrawal AF, Lively CM (2002) Infection genetics: gene-for-gene versus matching-allele models, and all points in between. Evol Ecol Res 4: 79–90

Agrawal AF, Lively CM (2003) Modeling infection genetics as a two-step process combining gene-for-gene and matching-allele genetics. Proc R Soc Lond B 270: 323–334

Agrawal AF, Otto SP (2006) Host-parasite coevolution and selection on sex through the effects of segregation. Am Nat 168: 617–629

Ameisen JC, Lelièvre JD, Pleskoff O (2002) HIV/host interactions: new lessons from the Red Queen's country. AIDS 16: S25–S31

Antonovics J, Ellstrand NC (1984) Experimental studies of the evolutionary significance of sexual reproduction. I. A test of the frequency-dependent selection hypothesis. Evolution 38: 103–115

Apanius V, Penn D, Slev PR, Ruff LR, Potts WK (1997) The nature of selection on the major histocompatibility complex. Crit Rev Immunol 17: 179–224

Bakker EG, Toomajian C, Kreitman M, Bergelson J (2006) A genome-wide survey of *R* gene polymorphisms in *Arabidopsis*. Plant Cell 18: 1803–1818

Barrett SCH, Eckert CG (1990) Variation and evolution of mating system in seed plants. In: Kawano S (ed) Biological approaches and evolutionary trends in plants. Acad Press, London, pp. 229–254

Barton NH (1995) A general model for the evolution of recombination. Genet Res 65: 123–144.

Barton NH, Charlesworth B (1998) Why sex and recombination? Science 281: 1985–1990

Bell G (1982) The masterpiece of nature. Croon Helm, London

Ben-Ami F, Heller J (2005) Spatial and temporal patterns of parthenogenesis and parasitism in the freshwater snail *Melanoides tuberculata*. J Evol Biol 18: 138–146

Ben-Ami, F, Heller J (2008) Sex versus parasitism versus density. Biol J Linn Soc 93: 527–544.

Bergelson J, Kreitman M, Stahl EA, Tian D (2001) Evolutionary dynamics of plant *R*-genes. Science 292: 2281–2285

Bernatchez L, Landry C (2003) MHC studies in nonmodel vertebrates: what have we learned about natural selection in 15 years? J Evol Biol 16: 363–377

Bodmer WF (1972) Evolutionary significance of the HL-A system. Nature 237: 139–145

Briles WE, Stone HA, Cole RK (1977) Marek's disease: effects of B histocompatibility alloalleles in resistant and susceptible chickens. Science 195: 193–195

Brown SG, Kwan S, Shero S (1995) The parasitic theory of sexual reproduction: parasitism in unisexual and bisexual geckos. Proc R Soc Lond B 260: 317–320

Brunet J, Mundt CC (2000) Disease, frequency-dependent selection, and genetic polymorphisms: experiments with stripe rust and wheat. Evolution 54: 406–415.

Bruvo R, Schulenburg H, Storhas M, Michiels NK (2007) Synergism between mutational meltdown and Red Queen in parthenogenetic biotypes of the freshwater planarian *Schmidtea polychroa*. Oikos 116: 313–323

Buckling A, Wei Y, Massey RC, Brockhurst MA, Hochberg ME (2006) Antagonistic coevolution with parasites increases the cost of host deleterious mutations. Proc R Soc Lond B 273: 45–49

Burt A (2000) Perspective: sex, recombination, and the efficacy of selection: was Weismann right? Evolution 54: 337–351

Burt A, Bell G (1987) Mammalian chiasma frequencies as a test of two theories of recombination. Nature 326: 803–805

Burt A, Bell G (1991) Seed reproduction is associated with a transient escape from parasite damage in American beech. Oikos 61: 145–148

Busch J, Neiman M, Koslow JM (2004) Evidence for maintenance of sex by pathogens in plants. Evolution 58: 2584–2590

Carius HJ, Little TJ, Ebert D (2001) Genetic variation in a host-parasite association: potential for coevolution and frequency-dependent selection. Evolution 55: 1136–1145

Charlesworth B (1976) Recombination modification in a fluctuating environment. Genetics 83: 181–195

Charlesworth D (2006) Balancing selection and its effects on sequences in nearby genome regions. PLoS Genet 2: 0379–0384

Clarke B (1976) The ecological genetics of host-parasite relationships. In: Taylor AER, Muller R (eds) Genetic aspects of host-parasite relationships. Blackwell Scientific, Oxford, pp. 87–103
Clay K, Kover PX (1996) The Red Queen hypothesis and plant/pathogen interactions. Annu Rev Phytopathol 34: 29–50
Coltman DW, Pilkington JG, Smith JA, Pemberton JM (1999) Parasite-mediated selection against inbred Soay sheep in a free-living island population. Evolution 53: 1259–1267
Cooper TF, Lenski RE, Elena SF (2005) Parasites and mutational load: an experimental test of a pluralistic theory for the evolution of sex. Proc R Soc Lond B 272: 311–317
Dawkins R, Krebs JR (1979) Arms races between and within species. Proc R Soc Lond B 205: 489–511
Decaestecker E, Gaba S, Raeymaekers JAM, Stocks R, Van Kerckhoven L, Ebert D, De Meester L (2007) Host-parasite 'Red Queen' dynamics archived in pond sediment. Nature 450: 870–873
de Visser JAGM, Elena SF (2007) The evolution of sex: empirical insights into the roles of epistasis and drift. Nat Rev Genet 8: 139–149
Dionne M, Miller KM, Dodson JJ, Caron F, Bernatchez L (2007) Clinal variation in MHC diversity with temperature: evidence for the role of host-pathogen interaction on local adaptation in Atlantic salmon. Evolution 61: 2154–2164
Doherty PC, Zinkernagel RM (1975) Enhanced immunological surveillance in mice heterozygous at the H-2 gene complex. Nature 256: 50–52
Dybdahl MF, Lively CM (1995) Host-parasite interactions: infection of common clones in natural populations of a freshwater snail (*Potamopyrgus antipodarum*). Proc R Soc Lond B 260: 99–103
Dybdahl MF, Lively CM (1998) Host-parasite coevolution: evidence for rare advantage and time-lagged selection in a natural population. Evolution 52: 1057–1066
Edwards SV, Hedrick PW (1998) Evolution and ecology of MHC molecules: from genomics to sexual selection. Trends Ecol Evol 13: 305–311
Ellstrand NC, Antonovics J (1985) Experimental studies on the evolutionary significance of sexual reproduction. II. A test of the density-dependent selection hypothesis. Evolution 39: 657–666
Fischer OM, Schmid-Hempel P (2005) Selection by parasites may increase host recombination frequency. Biol Lett 1: 193–195
Flor HH (1956) The complementary genic systems in flax and flax rust. Adv Genet 8: 275–296
Frank SA (1992) Models of plant–pathogen coevolution. Trends Genet 8: 213–219
Frank SA (2000) Specific and non-specific defense against parasitic attack. J Theor Biol 202: 283–304
Gandon S (2002) Local adaptation and the geometry of host-parasite coevolution. Ecol Lett 5: 246–256
Gandon S, Otto SP (2007) The evolution of sex and recombination in response to abiotic or coevolutionary fluctuations in epistasis. Genetics 175: 1835–1853
Glesener RR (1979) Recombination in a simulated predator-prey interaction. Am Zool 19: 763–771
Glesener RR, Tilman D (1978) Sexuality and the components of environmental uncertainty: clues from geographic parthenogenesis in terrestrial animals. Am Nat 112: 659–673
Greischar M, Koskella B (2007) A synthesis of experimental work on parasite local adaptation. Ecol Lett 10: 418–434
Haag CR, Sakwiska O, Ebert D (2003) Test of synergistic interaction between infection and inbreeding in *Daphnia magna*. Evolution 57: 777–783
Hakoyama H, Nishimura T, Matsubara N, Iguchi K (2001) Difference in parasite load and non-specific immune reaction between sexual and gynogenetic forms of *Carassius auratus*. Biol J Linnean Soc 72: 401–407
Haldane JBS (1949) Disease and evolution. Ric Sci 19 (Suppl): 68–76
Hamilton WD (1980) Sex vs. non-sex vs. parasite. Oikos 35: 282–290
Hamilton WD (1982) Pathogens as causes of genetic diversity in their host populations, In: Anderson RM, May RM (eds) Population biology of infectious diseases. Springer, Berlin, pp. 269–296

Hamilton WD, Axelrod R, Tanese R (1990) Sexual reproduction as an adaptation to resist parasites (A review). Proc Natl Acad Sci USA 87: 3566–3573

Hanley KA, Bolger DT, Case TJ (1994) Comparative ecology of sexual and asexual gecko species (*Lepidodactylus*) in French Polynesia. Evol Ecol 8: 438–454

Hill AVS, Allsopp CEM, Kwiatkowski D, Anstey NM, Twumasi P, Rowe PA, Bennett S, Brewster D, McMichael AJ, Greenwood BM (1991) Common West African HLA antigens are associated with protection from severe malaria. Nature 352: 595–600

Hill AVS, Yates SNR, Allsopp CM, Gupta S, Gilbert SC, Lalvani A, Aidoo M, Davenport M, Plebanski M (1994) Human leukocyte antigens and natural selection by malaria. Philos Trans R Soc Lond B 346: 379–385

Howard RS, Lively CM (1994) Parasitism, mutation accumulation, and the maintenance of sex. Nature 367: 554–557

Howard RS, Lively CM (1998) The maintenance of sex by parasitism and mutation accumulation under epistatic fitness functions. Evolution 52: 604–610

Howard RS, Lively CM (2003) Opposites attract? Mate choice for parasite evasion and the evolutionary stability of sex. J Evol Biol 16: 681–689

Hughes AL, Hughes MK, Howell CY, Nei M, Howard JC, Higgs P (1994) Natural selection at the class II major histocompatibility complex loci of mammals. Philos Trans R Soc Lond B 345: 359–367

Hughes AL, Nei M (1988) Pattern of nucleotide substitution at major histocompatibility complex class I loci reveals overdominant selection. Nature 335: 167–170

Hughes AL, Nei M (1989a) Evolution of the major histocompatibility complex: independent origin of non-classical class I genes in different groups of mammals. Mol Biol Evol 6: 559–579

Hughes AL, Nei M (1989b) Nucleotide substitution at major histocompatibility complex class II loci: evidence for overdominant selection. Proc Natl Acad Sci USA 86: 958–962

Hughes AL, Nei M (1992) Models of host-parasite interaction and MHC polymorphism. Genetics 132: 863–864

Hughes AL, Yeager M (1998) Natural selection at major histocompatibility complex loci of vertebrates. Annu Rev Genet 32: 415–435

Hutson V, Law R (1981) Evolution of recombination in populations experiencing frequency-dependent selection with time delay. Proc R Soc Lond B 213: 345–359

Jaenike J (1978) An hypothesis to account for the maintenance of sex within populations. Evol Theor 3: 191–194

Johnson SG, Lively CM, Schrag SJ (1997) Evolution and ecological correlates of uniparental and biparental reproduction in freshwater snails, In: Streit B, Städler T, Lively CM (eds) Evolutionary ecology of freshwater animals. Birkhäuser Verlag, Basel, pp. 263–291

Jokela J, Dybdahl MF, Lively CM (2009) The maintenance of sex, clonal dynamics, and host-parasite coevolution in a mixed population of sexual and asexual snails. Am Nat 174: S43–S53

Jokela J, Lively CM, Dybdahl MF, Fox JA (2003) Genetic variation in sexual and clonal lineages of a freshwater snail. Biol J Linn Soc 79: 165–181

Jokela J, Lively CM, Fox JA, Dybdahl MF (1997) Flat reaction norms and 'frozen' phenotypic variation in clonal snails (*Potamopyrgus antipodarum*). Evolution 51: 1120–1129

Kelley SE (1994) Viral pathogens and the advantage of sex in the perennial *Anthoxanthum odoratum*. Philos Trans R Soc Lond B 346: 295–302

Killick SC, Carlsson AM, West SA, Little TJ (2006) Testing the pluralist approach to sex: the influence of environment on synergistic interactions between mutation load and parasitism in *Daphnia magna*. J Evol Biol 19(5): 1603–1611

King KC, Lively CM (2009) Geographic variation in sterilizing parasite species and the Red Queen. Oikos: in press (DOI 10.1111/j.1600-0706.2009.17476.x)

Knapp LA (2007) Selection on MHC? A matter of form over function. Heredity 99: 241–242

Kondrashov AS (1993) Classification of hypotheses on the advantage of amphimixis. J Hered 84: 372–387

Koskella B, Lively CM (2007) Advice of the Rose: experimental coevolution of a trematode parasite and its snail host. Evolution 62: 152–159
Kouyos RD, Salathé M, Bonhoeffer S (2007) The Red Queen and the persistence of linkage-disequilibrium oscillations in finite and infinite populations. BMC Evol Biol 7: 211–219
Kover PX, Caicedo AL (2001) The genetic architecture of disease resistance in plants and the maintenance of recombination by parasites. Mol Ecol 10: 1–16
Kumpulainen T, Grapputo A, Mappes J (2004) Parasites and sexual reproduction in psychid moths. Evolution 58: 1511–1520
Lazarro BP (2005) Elevated polymorphism and divergence in the class C scavenger receptors of *Drosophila melanogaster* and *D. simulans*. Genetics 169: 2023–2034
Levin DA (1975) Pest pressure and recombination systems in plants. Am Nat 109: 437–451
Lewontin RC (1974) The genetic basis of evolutionary change. Columbia University Press, New York
Lively CM (1987) Evidence from a New Zealand snail for the maintenance of sex by parasitism. Nature 328: 519–521
Lively CM (1992) Parthenogenesis in a freshwater snail: reproductive assurance versus parasitic release. Evolution 46: 907–913.
Lively CM (1996) Host-parasite coevolution and sex. Bioscience 46: 107–114
Lively CM (1999) Migration, virulence, and the geographic mosaic of adaptation by parasites. Am Nat 153: S34–S47
Lively CM (2001) Trematode infection and the distribution and dynamics of parthenogenetic snail populations. Parasitology 123: S19–S26
Lively CM (2006) The ecology of virulence. Ecol Lett 9: 1089–1095
Lively CM, Apanius V (1995) Genetic diversity in host-parasite interactions. In: Grenfell BT, Dobson AP (eds) Ecology of infectious diseases in natural populations. Cambridge University Press, Cambridge, UK, pp. 421–449
Lively CM, Craddock C, Vrijenhoek RC (1990) The Red Queen hypothesis supported by parasitism in sexual and clonal fish. Nature 344: 864–866
Lively CM, Dybdahl MF (2000) Parasite adaptation to locally common host genotypes. Nature 405: 679–681
Lively CM, Dybdahl MF, Jokela J, Osnas EE, Delph LF (2004) Host sex and local adaptation by parasites in a snail-trematode interaction. Am Nat 164: S6–S18
Lively CM, Howard RS (1994) Selection by parasites for clonal diversity and mixed mating. Philos Trans R Soc Lond B 346: 271–281
Lively CM, Jokela J (1996) Clinal variation for local adaptation in a host-parasite interaction. Proc R Soc Lond B 263: 891–897
Lively CM, Jokela J (2002) Temporal and spatial distributions of parasites and sex in a freshwater snail. Evol Ecol Res 4: 219–226
Lloyd DG (1980) Demographic factors and mating patterns in angiosperms. In: Solbrig OT (ed) Demography and evolution in plant populations. Botanical Monographs. University of California Press, Berkeley, pp. 67–88
Lythgoe KA (2000) The coevolution of parasites with host-acquired immunity and the evolution of sex. Evolution 54: 1142–1156
Lythgoe KA, Read AF (1998) Catching the Red Queen? The advice of the rose. Trends Ecol Evol 13: 473–474
May RM, Anderson RM (1983) Epidemiology and genetics in the coevolution of parasites and hosts. Proc R Soc Lond B 219: 291–313
Mayer F, Brunner A (2007) Non-neutral evolution of the major histocompatibility complex class II gene DRB1 in the sac-winged bat *Saccopteryx bilineata*. Heredity 99: 257–264
Maynard Smith J (1971) What use is sex? J Theor Biol 30: 319–355
Maynard Smith J (1978) The evolution of sex. Cambridge University Press, Cambridge, UK

McClelland EE, Penn DJ, Potts WK (2003) Major histocompatibility complex heterozygote superiority during coinfection. Infect Immun 71: 2079–2086

Mee JA, Rowe L (2006) A comparison of parasite loads on asexual and sexual *Phoxinus*(Pisces: Cyprinidae). Can J Zool 84: 808–816

Meirmans S, Neiman M (2006) Methodologies for testing a pluralist idea for the maintenance of sex. Biol J Linn Soc 89: 605–613

Michiels NK, Beukeboom LW, Pongratz N, Zeitlinger J (2001) Parthenogenetic flatworms have more symbionts than their coexisting, sexual conspecifics, but does this support the Red Queen? J Evol Biol 14: 110–119

Milinski M (2006) The major histocompatibility complex, sexual selection, and mate choice. Annu Rev Ecol Evol Syst 37: 159–186

Morand S, Manning SD, Woolhouse MEJ (1996) Parasite-host coevolution and geographic patterns of parasite infectivity and host susceptibility. Proc R Soc Lond B 263: 119–128

Moritz C, McCallum H, Donnellan S, Roberts JD (1991) Parasite loads in parthenogenetic and sexual lizards (*Heteronotia binoei*): support for the Red Queen hypothesis. Proc R Soc Lond B 244: 145–149

Mundt CC, Brunet J, Sackett KE (2008) Impact of density and disease on frequency-dependent selection and genetic polymorphism: experiments with stripe rust and wheat. Evol Ecol 22: 637–657.

Nidelet T, Kaltz O (2007) Direct and correlated responses to selection in a host-parasite system: testing for the emergence of genotype specificity. Evolution 61: 1803–1811

Nordborg M, Hu TT, Ishino Y, Jhaveri J, Zheng H, Bakker E, Calabrese P, Gladstone J, Goyal R, Jakobsson M, Kim S, Morozov Y, Padhukasahasram B, Plagnol V, Rosenberg NA, Shah C, Wall JD, Wang J, Zhao K, Kalbfleisch T, Schulz V, Kreitman M, Bergelson J (2005) The pattern of polymorphism in *Arabidopsis thaliana*. PLoS Biol 3: 1289–1299

Osnas EE, Lively CM (2006) Host ploidy, parasitism and immune defense in a coevolutionary snail-trematode system. J Evol Biol 19: 42–48

Otto SP, Nuismer SL (2004) Species interactions and the evolution of sex. Science 304: 1018–1020

Penn DJ, Damjanovich K, Potts WK (2002) MHC heterozygosity confers a selective advantage against multiple-strain infections. Proc Natl Acad Sci USA 99: 11260–11264.

Peters AD (1999) The effects of pathogen infection and mutation on life-history characters in *Arabidopsis thaliana*. J Evol Biol 12: 460–470

Peters AD, Lively CM (1999) The Red Queen and fluctuating epistasis: a population genetic analysis of antagonistic coevolution. Am Nat 154: 393–405

Peters AD, Lively CM (2007) Short- and long-term benefits and detriments to recombination under antagonistic coevolution. J Evol Biol 20: 1206–1217

Piertney SB, Oliver MK (2006) The evolutionary ecology of the major histocompatibility complex. Heredity 96: 7–21

Potts WK, Manning CJ, Wakeland EK (1994) The role of infectious disease, inbreeding and mating preferences in maintaining MHC genetic diversity: an experimental test. Philos Trans R Soc Lond B 346: 369–378

Price MV, Waser NM (1982) Population structure, frequency-dependent selection, and the maintenance of sexual reproduction. Evolution 36: 35–43

Puurtinen M, Hytönen M, Knott KE, Taskinen J, Nissinen K, Kaitala V (2004) The effects of mating system and genetic variability on susceptibility to trematode parasites in a freshwater snail, *Lymnaea stagnalis*. Evolution 58: 2747–2753

Rolff J, Siva-Jothy MT (2003) Invertebrate ecological immunology. Science 301: 472

Ronsheim ML (1996) Evidence against a frequency-dependent advantage for sexual reproduction in *Allium vineale*. Am Nat 147: 718–734

Salathé M, Ebert D (2003) The effects of parasitism and inbreeding on the competitive ability in *Daphnia magna*: evidence for synergistic epistasis. J Evol Biol 16: 976–985

Salathé M, Kouyos RD, Regoes RR, Bonhoeffer S (2007) Rapid parasite adaptation drives selection for high recombination rates. Evolution 62: 295–300

Salathé M, Scherer A, Bonhoeffer S (2005) Neutral drift and polymorphism in gene-for-gene systems. Ecol Lett 8: 925–932

Schmitt J, Antonovics J (1986) Experimental studies of the evolutionary significance of sexual reproduction. VI. Effect of neighbor relatedness and aphid infestation on seedling performance. Evolution 40: 830–836

Schrag SJ, Mooeres AO, Ndifon GT, Read AF (1994) Ecological correlates of male outcrossing ability in a simultaneous hermaphrodite snail. Am Nat 143: 636–655

Schwensow N, Fietz J, Dausmann KH, Sommer S (2007) Neutral versus adaptive genetic variation in parasite resistance: importance of major histocompatibility complex supertypes in a free-ranging primate. Heredity 99: 265–277

Seger J (1988) Dynamics of some simple host parasite models with more than two genotypes in each species. Philos Trans R Soc Lond B 319: 541–555

Seger J, Hamilton WD (1988) Parasites and sex. In: Michod RE, Levin BR (eds) The evolution of sex. Sinauer, Sunderland, pp. 176–193

Shykoff JA, Schmid-Hempel P (1991) Parasites and the advantage of genetic-variability within social insect colonies. Proc R Soc Lond B 243: 55–58

Slade RW, McCallum HI (1992) Overdominant vs. frequency-dependent selection at MHC loci. Genetics 132: 861–862

Stahl EA, Dwyer G, Mauricio R, Kreitman M, Bergelson J (1999) Dynamics of disease resistance polymorphism at the *Rpm1* locus of *Arabidopsis*. Nature 400: 667–671

Stevens L, Yan G, Pray LA (1997) Consequences of inbreeding on invertebrate host susceptibility to parasitic infection. Evolution 51: 2032–2039

Strauss SY, Karban R (1994) The significance of outcrossing in an intimate plant-herbivore relationship. I. Does outcrossing provide an escape from herbivores adapted to the parent plant? Evolution 48: 454–464

Takahata N, Nei M (1990) Allelic genealogy under over-dominant and frequency-dependent selection and polymorphism of major histocompatibility complex loci. Genetics 124: 967–978

Tiffin P, Moeller DA (2006) Molecular evolution of plant immune genes. Trends Genet 22: 662–670

Tiffin P, Hacker R, Gaut BS (2004) Population genetic evidence for rapid changes in intraspecific diversity and allelic cycling of a specialist defense gene in *Zea*. Genetics 168: 425–434

Tobler M, Schlupp I (2005) Parasites in sexual and asexual mollies (*Poecilia*, Poeciliidae, Teleostei): a case for the Red Queen? Biol Lett 1: 166–168

Todd JR, West BC, McDonald JC (1990) Human leukocyte antigen and leprosy: study in northern Louisiana and review. Rev Infect Dis 12: 63–74

Van Valen L (1973) A new evolutionary law. Evol Theor 1: 1–30

Vernon JG, Okamura B, Jones CS, Noble LR (1996) Temporal patterns of clonality and parasitism in a population of freshwater bryozoans. Proc R Soc Lond B 263: 1313–1318

Vrijenhoek RC, Lerman S (1982) Heterozygosity and developmental stability under sexual and asexual breeding systems. Evolution 36: 768–776.

Webster JP, Davies CM (2001) Coevolution and compatibility in the snail-schistosome system. Parasitology 123: S41–S56

Wedekind C, Walker M, Little TJ (2005) Major histocompatibility complex (MHC) effects, but no general MHC heterozygote advantage in single-strain infections. Genetics 170: 1427–1430

Wegner KM, Kalbe M, Reusch TBH (2007) Innate versus adaptive immunity in sticklebacks: evidence for trade-offs from a selection experiment. J Evol Ecol 21: 473–483

Wegner KM, Kalbe M, Schaschl H, Reusch TBH (2004) Parasites and individual major histocompatibility complex diversity – an optimal choice? Microbes Infect 6: 1110–1116

West SA, Lively CM, Read AF (1999) A pluralistic approach to sex and recombination. J Evol Biol 12: 1003–1012

Westerdahl H, Hansson B, Bensch S, Hasselquist D (2004) Between-year variation of MHC allele frequencies in great reed warblers: selection or drift? J Evol Biol 17: 485–492

Williams GC (1966) Adaptation and natural selection. Princeton University Press, Princeton

Williams GC (1975) Sex and evolution. Princeton University Press, Princeton
Woolhouse MEJ, Webster JP, Domingo E, Charlesworth B, Levin BR (2002) Biological and biomedical implications of the coevolution of pathogens and their hosts. Nat Genet 32: 569–577
Wuethrich B (1998) Evolution of sex: putting theory to the test. Science 281: 1980–1982

Chapter 8
Geographical Parthenogenesis: Opportunities for Asexuality

Elvira Hörandl

Abstract Asexual organisms often occupy larger and more northern distribution areas than their sexual relatives. These phenomena, summarized under the term "geographical parthenogenesis", seem to confirm a short term advantage of asexual reproduction. Geographical parthenogenesis may be explained by better colonizing abilities of asexual organisms, or by a swamping of sexual populations because of introgression of asexuality. Asexual organisms may perform better in diverse and narrow ecological niches, or may benefit in colder climates from a lower pressure of parasites and predators. The distributional success of asexuals has been also referred to indirect advantages of hybridity and/or polyploidy. Sexual hybrids or polyploids, however, do not show patterns of geographical parthenogenesis. Here, I present a novel model for those asexual organisms that have originated from hybridization. Climatic changes may have triggered interspecific hybridization, which increases frequencies of new origins of asexuality, but decreases fitness of sexual progenitor species. Asexuality is further advantageous for re-colonization of devastated areas. Therefore, frequencies of asexual populations increase relative to those of related sexuals. Glaciations during the Pleistocene may have provided great opportunities for the evolution of asexual organisms, while the Tertiary might have been a period of predominant sexuality. The geologically rather recent wave of asexuality helps to explain that most extant asexual animals and plants are evolutionarily young and appear scattered on the tips of phylogenetic trees.

8.1 Introduction

It has long been recognized that related asexual and sexual taxa have different distributional patterns. Since Vandel (1928) coined the term "geographical parthenogenesis", several authors have described the phenomenon both in animals

---

E. Hörandl (✉)
Department of Systematic and Evolutionary Botany, University of Vienna, Rennweg 14, A-1030 Vienna, Austria
e-mail: elvira.hoerandl@univie.ac.at

I. Schön et al. (eds.), *Lost Sex*, DOI 10.1007/978-90-481-2770-2_8,

and plants (Bell 1982; Bierzychudek 1985; Asker and Jerling 1992; Van Dijk 2003; Haag and Ebert 2004; Kearney 2005; Hörandl 2006). Recent literature surveys summarize cases for animals (Kearney 2005), lichens (Poelt 1970) and flowering plants (Hörandl et al. 2008) and review the main characteristics of biogeographical patterns: (1) asexual organisms have larger distributional ranges, often considerably exceeding that of their sexual relatives, (2) tend to range to higher latitudes and altitudes than their sexual relatives, and (3) tend to colonize previously glaciated areas. A tendency to devastated habitats is also seen in parthenogenetic geckos (*Heteronotia*) and grasshoppers (*Warrambia*) colonizing extremely arid environments in the Australian deserts (Kearney 2003). Some cases in flowering plants suggest that sexual species have distributions centered within much larger ranges of apomictic complexes (sunflower family: *Antennaria*: Bayer 1990; grasses: *Paspalum*: Urbani 2002; *Taraxacum* and *Chondrilla*: Van Dijk 2003; *Erigeron*: Noyes 2007; buttercups: *Ranunculus*: Hörandl and Paun 2007). These characteristics may not necessarily occur in combination, but each taxon has its own specific pattern (Hörandl et al. 2008).

Despite a widespread taxonomic distribution, geographical parthenogenesis is not a ubiquitous feature of asexuality. For example, the phenomenon is not observed in several tropical plants which reproduce via adventitious embryony, a mode of reproduction forming asexual embryos in parallel to sexual ones (Richards 1997). In apomictic taxa with exceptional high frequencies of facultative sexuality, geographical patterns remain often unclear (Hörandl et al. 2008). Such intertwined sexual-asexual systems may lack a distinct differentiation pattern of distribution areas for either mode of reproduction.

The distributional success of asexual organisms is sometimes, but not necessarily correlated to taxonomic diversity of asexual organisms (Hörandl 2006; Hörandl et al. 2008). Geographical parthenogenesis is often described as an infraspecific phenomenon, e.g., the widespread occurrence of asexual populations compared to sexual populations, often connected to a differentiation of cytotypes (ploidy levels) and/or ecotypes. In some apomictic plant complexes, the geographical pattern is rather referred to a high diversity of hundreds or thousands of agamospecies which differ from the sexual counterparts and from each other by ploidy level, morphological and ecological features (e.g. in dandelions, *Taraxacum*; hawkweeds: *Hieracium* subg. *Pilosella*; buttercups: *Ranunculus auricomus*; blackberries: *Rubus fruticosus*). Single agamospecies within such complexes may have rather restricted distributions (e.g. Hörandl 2006). To avoid a bias of different taxonomic concepts and different speciation processes, it is useful to treat geographical parthenogenesis as a quantitative phenomenon on the population level.

In this respect, the distributional success of parthenogenetic organisms seems to support a concept of theoretical disadvantages of sexuality because of two costs of sex: that of males (which do not produce offspring) and that of meiosis which breaks up co-adapted gene combinations (e.g. Maynard Smith 1978; Bell 1982). The main hypotheses to explain the "paradox" of the maintenance of sex suggest that sex would allow for a faster response to environmental changes, avoids the accumulation of deleterious mutations, or would simply be the phylogenetically established mode of reproduction (review by Birdsell and Wills 2003). Geographical parthenogenesis is often seen as a short term success of asexuality (e.g. Van Dijk 2003), but the evolutionary relevance of the phenomenon is still not well understood.

The causality of geographical parthenogenesis is in much dispute and has been comprehensively discussed by many authors (e.g. Bierzychudek 1985; Kearney 2005; Lundmark and Saura 2006; Hörandl 2006). In this review, I will first briefly summarize the main current hypotheses for geographical parthenogenesis. Second, I discuss more comprehensively the difficulty of disentangling effects of the reproductive mode from those of polyploidy and hybridization as potential causal factors. Third, I will discuss the hypothesis that geographical parthenogenesis is due to an increase of opportunities of *de novo* origins of parthenogenetic lineages via hybridization and/or polyploidization of sexual species. In this context, I will outline that shifts from sexual to asexual reproduction have several functional and evolutionary constraints, inferring a "cost" to asexuality. I will further present a hypothesis that climatic changes may provide opportunities for asexuality by increasing frequencies of new origins of asexuality, and at the same time increasing the costs of sexual reproduction. On two examples of buttercups (*Ranunculus*), I will try to exemplify how modes of origins, different intrinsic features of asexual reproduction, and environmental factors may shape patterns of geographical parthenogenesis. Some future prospects are added to stimulate further research in this direction.

8.2 The Main Current Hypotheses

Several non-exclusive hypotheses have been proposed for the causality of geographic parthenogenesis but there is no general line of conclusions. One group of hypotheses relies on intrinsic features of asexuality such as uniparental reproduction (e.g. Stebbins 1950; Baker 1967; Mogie et al. 2007): a single asexual individual can found a population, which provides an advantage for colonization ("Baker's Law", Baker 1967). Reproductive certainty without the need of mating partners and the reduction of density-dependence is an obvious advantage for parthenogenetic organisms. Recent studies on dandelions (e.g. Van Dijk 2007) and on buttercups (Hörandl 2008) indeed support the idea that reproductive assurance is a main advantage of apomictic hermaphroditic plants compared to related sexual species. It may even outweigh cases where fertility as seed set of apomictic plants is lower than that of sexual taxa (Hörandl 2008).

One main problem for hypotheses relying on uniparental reproduction is the occurrence of pseudogamy (sperm-dependent parthenogenesis or pollen-dependent apomixis; see Beukeboom and Vrijenhoek 1998; Hörandl et al. 2008). In the case of pseudogamy, only a hermaphrodite that is capable of self-fertilization would have the ability of uniparental reproduction. In the animal kingdom, hermaphroditism and pseudogamy are not frequently coupled (e.g. in plathyhelminths, nematodes, annelids, and mollusks, but neither in arthropods nor in vertebrates); self-fertilization of hermaphrodites is rare (possible in *Turbellaria*; Beukeboom and Vrijenhoek 1998). Therefore, most pseudogamous animals still have the need of two individuals for reproduction, the costs of producing male organs, and the costs of mate searching. In flowering plants, the great majority (c. 90%) of apomicts are pseudogamous and hermaphroditic (Richards 1997); pollen is needed for the fertilization of the endosperm nuclei to form the endosperm, the nutritious tissue for the embryo (see also Chapter 3). Autonomous (pollen-independent) apomixis, which is most commonly observed in the sunflower family (Asteraceae), seems to be overrepresented in taxonomic surveys of geographical parthenogenesis, which would confirm an advantage of pollen-independent uniparental reproduction (Hörandl et al. 2008). In fact, also many pseudogamous plants show a break-down of self-incompatibility systems, which would otherwise already reject self-pollen on the stigma or in the style (Dickinson et al. 2007; Hörandl 2008). Self-compatibility in apomicts may evolve as follows: clone mates sharing the same genotype would have a self-incompatibility reaction of stigma, pistil or pollen. A self-incompatible apomictic mother plant pollinated by another genotype will be soon surrounded by its own maternal offspring and receive mostly pollen of the same, incompatible genotype. A self-incompatible genotype allocating pollen to the fertilization of the endosperm, but not the egg cell of another individual, will disappear from the population (Noirot et al. 1997). Therefore, selection should strongly favour self-compatible genotypes in apomictic populations with a clonal structure (Hörandl 2008).

Higher plants also offer the possibility of disentangling uniparental reproduction from the reproductive system by comparing sexual self-fertilization and apomixis in hermaphrodites. Self-fertilization of hermaphroditic flowering plants is a widespread phenomenon, as at least 20–40% of angiosperms are facultative selfers (Richards 1997). In flowering plants, sexual self-fertilization appears to be the most efficient mode of reproduction for colonization, despite the fact that selfing rapidly reduces heterozygosity in the offspring and may potentially cause inbreeding depression (e.g. Carr and Dudash 2003). However, under certain circumstances, inbreeding may purge populations from a high load of deleterious recessive mutations and reduce inbreeding depression (Byers and Waller 1999). Colonizing abilities of sexual self-fertilizing hermaphrodites may still be superior because of a higher genetic diversity caused by recombination and segregation of chromosomes during meiosis (Burt 2000). In island floras, a higher proportion of selfing colonizers in the source populations of the continent may also cause higher frequencies of selfers than of apomicts.

Despite obvious advantages, uniparental reproduction is not a requirement for geographical parthenogenesis: the hermaphroditic flatworm *Schmidtea polychroa* is not capable of self-fertilization, but nevertheless shows a typical pattern of geographical parthenogenesis (Pongratz et al. 2003). In such cases, uniparental reproduction *alone* is not a convincing argument for geographical parthenogenesis. A main advantage for colonizing scenarios is that heterozygosity of small founder populations is still maintained (Beukeboom and Vrijenhoek 1998). Moreover, pseudogamy requires the maintenance of male functions and consequently, keeps the potential for facultative sexuality. The possibility of occasional recombination creating genetic variation may be a main advantage of pseudogamy in *Schmidtea* (Beukeboom 2007). The maintenance of a male function provides the potential for "a little bit of sex" which theoretically creates genetic variation as efficiently as regular sex, but with lower costs (Green and Noakes 1995).

Another theoretical approach regards the interactions between sexuals and apomicts in mixed populations as the main cause for geographical parthenogenesis (e.g. Mogie 1992; Carillo et al. 2002; Mogie et al. 2007). If asexuality is a heritable trait, then the maintenance of male functions will always allow the transfer of genetic control factors from asexual individuals to the offspring of sexual individuals, whereas sexuality will not be transferred to parthenogens. This unidirectional introgression should finally result in the loss of sexual individuals from a population. The model is contradicted by co-existing populations, which may perpetuate because of crossing barriers between different cytotypes (e.g. Van Dijk 2007) or by niche differentiation between sexual and asexual lineages (e.g. Verduijn et al. 2004). In flowering plants, so-called mentor effects in mixed sexual-asexual populations may turn sexual species to selfing and prevent at least partly introgression of apomixis into sexuals (Tas and van Dijk 1999; Brock 2004; Hörandl 2006; Hörandl and Temsch 2009). The applicability of introgression models may depend on rather group-specific reproductive and cytological features. Geographical parthenogenesis might be also due to gene flow during immigration processes (Peck 1998). His model predicts that sexual populations may have a disadvantage because of immigration of maladapted genes into a new environment via random mating. Since productivity of sexual populations is higher in southern than in northern regions, there is a higher "immigration load" of maladapted genes in the North. This loss of fitness in sexuals is not effective in apomicts. Empirical evidence for this model, however, is so far not available.

Alternative explanations for geographical parthenogenesis have been sought in environmental factors selecting for asexuality under certain conditions. The idea that parthenogens are generalists (general purpose genotypes; Baker and Stebbins 1965; Lynch 1984) is not well supported from recent empirical evidence. Molecular population genetic studies do not confirm the occurrence of single widespread clones, but rather reveal an unexpected high genotypic diversity (e.g. Loxdale and Lushai 2003; Lushai et al. 2003; Hörandl and Paun 2007). In the majority of cases, genotypes or clones of asexual lineages are restricted to a single locality (reviewed in Hörandl and Paun 2007). Experimental studies on parthenogenetic animals (Gade and Parker 1997; Robinson et al. 2002; Vorburger et al. 2003) and apomictic plants (de Kovel and Jong 1999) do not support the "general purpose genotype" hypothesis. In contrast, the "Frozen Niche Variation Model" (FNV) by Vrijenhoek (1984, 1994; see also Chapter 6) fits the overall high genotypic diversity of most asexuals. The model predicts a better occupation of ecological niches by genetically different clones. The original concept of the model suggests that sexual species with different ecological niches hybridize and create a segregating, genetically and ecologically diverse hybrid progeny. If these hybrids shift to asexual reproduction, each clone "freezes" a part of the ecological spectrum of the parental species. Each clone is in its narrow niche fitter than either parent, and together all clones use the whole resource space more efficiently than their sexual parents. The model is not necessarily restricted to asexual hybrids, but is probably broadly applicable to genetically diverse asexual organisms. The FNV model gets increasing support from experimental data and population genetics (Jokela et al. 1997; Semlitsch et al. 1997; Vavrek et al. 1998; Hörandl et al. 2000; Gray and Weeks 2001; Meirmans et al. 2003; Verduijn et al. 2004; Paun et al. 2006a). The basic assumption of differentiated asexual hybrids is in accordance with the ecological differentiation observed in some sexual hybrids (e.g. Rieseberg et al. 2003; Rieseberg and Willis 2007; Mallet 2007; see also below).

The tendency of parthenogens to occupy areas in higher altitudes and latitudes has often been related to a lower capacity of asexuals to keep pace with coevolving parasites and pathogens (Glesener and Tilman 1987). In cooler climates, there are less biotic interactions, and the disadvantage of asexuality because of lowered genotypic diversity would be decreased. Empirical data raise the question whether such Red Queen mechanisms are generally applicable (see also Chapter 7 of this book). First, biotic interactions of parasites often cause shifts in frequencies of clones and may cause the maintenance of clonal diversity in asexual organisms by negative frequency-dependent selection (e.g. Van Dijk 2003; Jokela et al. 2003). Such data fit the conclusion of Otto and Nuismer (2004) that the Red Queen model explains the maintenance of infrequent sex in typically asexual species rather than the ubiquitous distribution of obligate sex. Recent experimental studies do not support the expectation that frequencies of parasites relate to sexuality (Ben-Ami and Heller 2005; Meirmans et al. 2006). Second, response of asexual organisms to pathogens must be carefully examined. Red Queen mechanisms are based on different levels of genetic diversity in sexual and asexual populations, "all else being equal". In the case of an altered response of asexual organisms to parasites, the theoretical concept of the Red Queen becomes inapplicable. Hybridization may cause different response to parasites, herbivores and pathogens, including additivity, higher susceptibility, but also resistance (Fritz et al. 1999). Hybrid origin plus polyploidy of parthenogens may dramatically alter gene expression patterns and consequently, the response to pathogens (e.g. Adams and Wendel 2005; Comai 2005; Chen 2007). In higher plants, alterations in expression patterns of secondary metabolites can change the response of hybrids and/or polyploids to parasites or predators (Orians 2000; Kirk et al. 2005). A stronger resistance of polyploids to pests and pathogens as compared to their diploid relatives is known from several sexual plants (reviewed in Levin 2002). The widespread herb *Chondrilla juncea* (Asteraceae) has sexual populations in Central Europe and in the Aegean area, and apomictic lineages in the whole Mediterranean area, extending eastwards to Turkey and Central Asia; in Turkey resistance of apomictic lineages against pathogenic fungi has been assessed in several clones (Van Dijk 2003). Third, distribution data do not support a generally higher frequency of asexual organisms in colder climates. Apomictic plants do not occur in extremely cold areas in the arctic and in high elevations of the Alps, where sexual species are still found. Furthermore, absolute numbers of apomictic plants in Fennoskandia decrease from the North to the South, which contradicts the basic assumptions of the model (Asker and Jerling 1992).

Many authors have emphasized that geographical parthenogenesis is probably due to a combination of factors (Haag and Ebert 2004; Pound et al. 2004; Hörandl 2006; Mogie et al. 2007). The factors discussed above may have different importance in different model systems. Here, I will elaborate an aspect already mentioned in Hörandl (2006), namely that geographical parthenogenesis may also relate to frequencies and geographical areas of the origin of asexual reproduction.

8.3 The Connection of Polyploidy, Hybridization and Asexuality

Some authors regard the advantages of hybrid origin, such as genomic novelty and a higher adaptive potential as the main factor causing geographical parthenogenesis (e.g. Vrijenhoek 1994; Kearney 2005). This hypothesis is reasonable for asexual animals, where hybridity and polyploidy have a strong correlation with asexuality. In flowering plants, the widespread occurrence of sexual hybridization and polyploidy allows for disentangling these factors from the reproductive system. Unfortunately, the actual distribution patterns of sexual hybrids do not confirm a general advantage of these features. Around 10% of plant species hybridize regularly, but with unequal taxonomic distributions; in some plant families, no hybrids have been reported (Arnold 1997). Geographical patterns as observed in asexual organisms are not a common feature of sexual hybrids. Homoploid hybrids have the problem of establishing sufficient reproductive barriers against the parents (e.g. Coyne and Orr 2004). Some well-documented examples of sexual hybrid speciation suggest that hybrids may have novel adaptive traits because of recombination and transgressive segregation: in the F_2 and following generations, new genotypes with extreme features may arise in the offspring, which allow for establishment in more extreme habitats compared to their parents (e.g. Rieseberg et al. 2003; Seehausen 2004; Rieseberg and Willis 2007; Mallet 2007). Nevertheless, in well-documented cases of homoploid hybrid speciation in animals and plants, hybrids have *restricted* distribution ranges compared to their sexual parents (Mavárez and Linares 2008). Homoploid hybrids can occupy novel ecological niches, but they do not show the range expansions typical for geographical parthenogenesis.

Polyploidy alone is also not a causal factor for large distributions. Reviews by Stebbins and Dawe (1987) and Hörandl (2006; based on Brochmann et al. 2003) have shown that sexual polyploidy in flowering plants is not correlated to larger distribution areas compared to diploid relatives. In contrast, autopolyploid sexuals often have even much smaller distribution areas within the range of their diploid progenitors (e.g. in *Ranunculus cassubicifolius,* Hörandl and Paun 2007; *Galax urceolata, Heuchera grossifolia, Vaccinium grossifolium*; Soltis et al. 2007). In *Chamerion angustifolium*, which is widespread in North America, diploids occupy the northern part and tetraploids the southern part of the distribution range (Soltis et al. 2007). In none of the cases, a pattern typical for geographical parthenogenesis has been reported. In allopolyploid species, the difficulty arises again to disentangle hybrid origin from effects of polyploidization. Both processes create without any doubt "genomic novelty" (e.g. Comai 2005; Chen 2007). If hybrid biotypes retain some fertility, filial generations and backcrosses may be formed and selection may either stabilize hybrids in novel niches or replace parental species by fitter hybrids. This "evolutionary novel model" by Arnold (1997) may lead to evolutionarily stable lineages. In this respect, hybridization and polyploidization can be seen as a mode of rapid speciation, which may occur within a few generations (e.g. Rieseberg and Willis 2007). Nevertheless, such rapid speciation processes do not necessarily create patterns of geographical parthenogenesis. Geographical parthenogenesis, in turn, is not necessarily correlated to speciation processes, but is mainly a spatial phenomenon.

8.4 Hybridization and Polyploidy, and the Cost of Origins of Asexuality

Hybridization and polyploidy may play a crucial role in new origins of asexuality from sexual progenitors. Asexual organisms usually appear scattered on phylogenies, mainly on the tips of phylogenetic trees (Simon et al. 2003; Van Dijk and Vijverberg 2005), indicating multiple and mostly young shifts from sexual ancestors to asexual lineages. Hybridization and polyploidy are seen as major triggers for parthenogenesis in animals (Simon et al. 2003; Kearney 2005; Gomez-Zurita et al. 2006; Mable 2007). In plants, spontaneous origin of apomixis via single mutations is unlikely because of the complexity of apomictic reproduction (Van Dijk and Vijverberg 2005; see also Chapter 3). The shift from sexuality to apomixis in angiosperms is constrained by the coordination of three steps, which on their own are deleterious: (1) the skip or bypass of meiosis (apomeiosis), (2) the parthenogenetic development of an unreduced egg cell, and (3) modifications in the formation of the endosperm, the nutritious tissues for the embryo. These processes can be uncoupled and are in many apomictic taxa under different genetic controls (Nogler 1995; Noyes and Rieseberg 2000; Van Dijk et al. 1999, 2003). Incomplete steps towards apomixis are deleterious and will be selected against; only the combination of all features will be successful (Van Dijk and Vijverberg 2005). In flowering plants, recent evidence most likely relates natural origins of apomixis to a combination of hybridization and polyploidy. These may cause a "genomic shock" resulting in the spatial or temporal deregulation of gene expression in the regulatory system and synchronously establishing apomeiosis and parthenogenesis (Grimanelli et al. 2001; Koltunow and Grossniklaus 2003; Bicknell and Koltunow 2004; Curtis and Grossniklaus 2007). Carman (1997, 2001, 2007) proposed an elaborated hypothetical model for the natural origin of apomixis via hybridization: sexual species that have differentiated in the timing of meiosis, embryo sac development, and embryogenesis are brought together by range fluctuations caused by environmental changes (e.g. glaciations). In hybrids, rare apomeiotic (unreduced) embryo sacs arise and will cause polyploidization via fertilization and formation of triploid (2n+n) hybrids as a first step. Epigenetic modifications in the polyploids would then lead to a collapse of the timing of developmental pathways, where developmental steps that are normally expressed in sequel now occur simultaneously causing apomeiotic development of the embryo sac and full expression of apomixis.

Carman (2007) occasionally observed apomeiotic embryo sacs in experimentally produced F_1 hybrids of diploid, sexual species of *Sorghum* (Poaceae) and *Antennaria* (Asteraceae) which are heterochronous in the timing of their developmental pathways. Frequencies of apomeiosis partially increase in segregating F_2 generations (J. Carman, personal communication). According to these findings, the formation of F_1 hybrids with an altered timing of reproductive pathways would be the first step towards apomixis. The hybrid origin of natural apomictic plants has been assessed by many cytogenetic studies (reviewed in Asker and Jerling 1992) and by recent analyses based on molecular markers in different plant families and genera of flowering plants (e.g. *Boechera*: Koch et al. 2003; Dobeš et al. 2004ab; *Ranunculus*: Paun et al. 2006a, b; *Limonium*: Palop-Esteban et al. 2007; *Antennaria*: Bayer and Chandler 2007). For animals, (Kearney 2005) reports documented cases of hybrid origin for gastropods, some ostracods, cladocerans, isopods, coleopterans, and vertebrates. Alternative pathways to asexuality are mutagenic origins (e.g. in ostracods, vertebrates, aphids and moths) and infectious origins in insects induced by *Wolbachia*, a Proteobacterium (Simon et al. 2003). For many cases, it is still not known how parthenogenetic lineages originated (Kearney 2005).

The role of polyploidy for new origins of apomixis is still disputed. Whereas in animals, polyploidy is mostly connected to asexuality (Kearney 2005) this correlation does not exist in higher plants. In contrast, almost all apomictic flowering plants and ferns are polyploid. Apomixis is observed in cytotypes of autopolyploid origin (e.g. in *Antennaria*, Bayer 1991; *Paspalum*; Quarin et al. 2001), inferring that hybrid origin is not a prerequisite for asexual reproduction. Genome duplication alone may cause a similar "genomic shock" as hybridization resulting in the deregulation of reproductive pathways (Grimanelli et al. 2001). Alternative explanations suggest that polyploidy is required for inheritance of genetic factors controlling apomixis. Nogler (1984) observed that apomeiosis in buttercups (*Ranunculus auricomus*) is transmitted via a Mendelian factor A- which is dominant for the expression of apomixis, but with lethal recessive effects in the gametes. The wildtyp allele A+ does not contribute to asexual reproduction. Therefore, only diploid gametes being heterozygous for A-A+ can transfer the apomixis factor to their offspring. Non-transmission of apomixis via haploid pollen occurs also in some genera of Asteraceae (see Chapter 22). Nevertheless, the generality of the model is under dispute. The frequent connection of asexuality and diploidy in animals suggests that polyploidy is not a ubiquitous condition for asexuality.

If hybridization and/or polyploidization of sexual species are required for the majority of new origins of asexuality, then it must be considered that both processes are constrained by the frequencies of hybrid formation or polyploidization, and are further by reduced fertility of the F_1 generation. Moreover, apomeiotic and parthenogenetic development of egg cells are two processes that cannot easily be uncoupled. Apomeiosis with continued fertilization would lead to a continuous increase of ploidy levels, which will finally be deleterious (e.g. Comai 2005). Parthenogenetic development of meiotically reduced eggs of previously diploid organisms would express deleterious recessive alleles that have been masked in a diploid, heterozygous genome.

Consequently, new origins of asexuality via hybridization and/or polyploidization from sexual species infer a multiple cost:

(1) Breakdown of existing pre- or postzygotic crossing barriers between the parental sexual species to form hybrids and/or increased frequencies of unreduced gametes for polyploidization,
(2) Overcoming a reduced fertility of the F_1 hybrid generation,
(3) The concerted break-up of the meiosis-outcrossing cycle and establishment of an alternative way of embryo formation (see also Chapter 3).

Altogether, these constraints may help to explain the low frequencies and the scattered phylogenetic distribution of asexuality in animals and seed plants (less than 1% of species). Already step one, the hybrid formation, might be constrained by taxonomic predispositions and mating systems of sexual taxa (e.g. Arnold 1997). Rough estimates suggest that around 10% of plant species regularly hybridize, but with uneven taxonomic distributions among families and orders (Arnold 1997). Most records of hybrids are concentrated in the families Asteraceae, Poaceae, Rosaceae, Cyperaceae, Salicaceae, and Scrophulariaceae (s.l.); in the first three plant families, apomixis is most commonly observed. In animals, hybridization is mostly known from fish, birds, and butterflies (e.g. Mavárez and Linares 2008), but many cases may still need further documentation. Step two will be only possible if at least a few viable and fertile F_1 offspring are produced. Step three is the most complex step in the whole process, and still not fully understood. The diversity of asexual mechanisms in angiosperms suggests that several pathways have evolved to establish the combination of apomeiosis and parthenogenetic development of unreduced egg cells (e.g. Asker and Jerling 1992). Genetic control of apomixis is often complex and can involve up to five different genes (Matzk et al. 2005, 2007). These functional constraints suggest that the shift from a well-established sexual system to asexuality is not a regular process, but requires special opportunities.

8.5 Climatic Changes as Opportunities for Origins and Dispersal of Asexuality

From the reasoning above it is obvious that the costs of the shift to asexuality can be reduced if frequencies of potential shifts, i.e., frequencies of hybridization and polyploidy increase. Unfortunately, the two processes do not automatically result in asexuality. In flowering plants, 30–70% of species is estimated to be polyploid (e.g. Arnold 1997), but less than 1% of species are apomictic (Mogie 1992). Sexual polyploidy and hybridization still outnumbers shifts to apomixis, which may be explained by the constraints listed above. Consequently, costs for sexual reproduction must additionally be increased to balance the low frequencies of asexuality. The most likely opportunity for asexuality arises when costs of sexuality become higher than the costs for shifts to apomixis.

That is, asexuality has the highest chance to spread if (1) frequencies of origin via hybridization and/or polyploidy increase, (2) costs of sexual reproduction increase, and (3) re-colonization of previously devastated areas is more efficient by asexuals than by sexuals. A combination of these conditions is most likely found in the climatic oscillations, glaciations and retreating glaciers in the Pleistocene. During cold periods and glacial maxima (in the northern hemisphere), the distribution range of most species would have pushed southwards or to lower elevations in high mountain regions. In warmer interglacial periods, distribution ranges expanded again northwards and upwards. These fluctuations in distribution ranges have brought previously geographically or ecologically isolated sexual species together (e.g. Hewitt 1996). Prezygotic isolation mechanisms are often regarded as stronger crossing barriers than postzygotic ones (e.g. Coyne and Orr 2004). The breakdown of ecological and geographical isolation of sexual species through range fluctuations should increase frequencies of interspecific hybridization. In higher plants, both hybrid crosses and low temperatures increase frequencies of unreduced gametes, which is the first step towards polyploidy (Levin 2002). Polyploidy may help to establish asexual reproduction (see above). Consequently, more opportunities for new origins of asexuality will arise. As soon as apomeiosis is established, reduced fitness because of the bypass of disturbances of meiosis will be avoided in newly formed hybrids or polyploids. Moreover, asexuality establishes an efficient reproductive barrier against the sexual species, thus avoiding gene flow from the sexual parents (Kearney 2005).

In contrast, hybridization and/or polyploidy will increase the costs of sexuality, at least in the first generations. Heterospecific matings often fail to produce any progeny because of post-pollination or post-insemination crossing barriers (Arnold 1997); such matings infer a loss of gametes or gametophytes. Even after successful heterospecific fertilization, the first hybrid generation is often sterile or has a strongly reduced fertility, mainly because of disturbances of chromosome pairing during meiosis. The reduced fitness of hybrids is a further cost of sexual reproduction. If sexual hybrids are partly fertile and remain sympatric with their parents, then backcrossing and introgression may "swamp" the parental sexual species (Arnold 1997). The main disadvantages of polyploidy are cellular constraints, problems of chromosome pairing during meiosis and conflicts in the gene expression patterns of multiple copies in the genome (Comai 2005). Polyploid cytotypes establish rapidly crossing barriers against the parents, potentially leading to speciation (Arnold 1997). Nevertheless, newly arisen sexual polyploids will have problems to establish themselves within the parental diploid population because of a minority cytotype disadvantage (e.g. Levin 1975, 2002): a newly arisen polyploid within a diploid sexual population is in a minority and will be surrounded by individuals with the wrong ploidy level, generating a lack of appropriate mating partners.

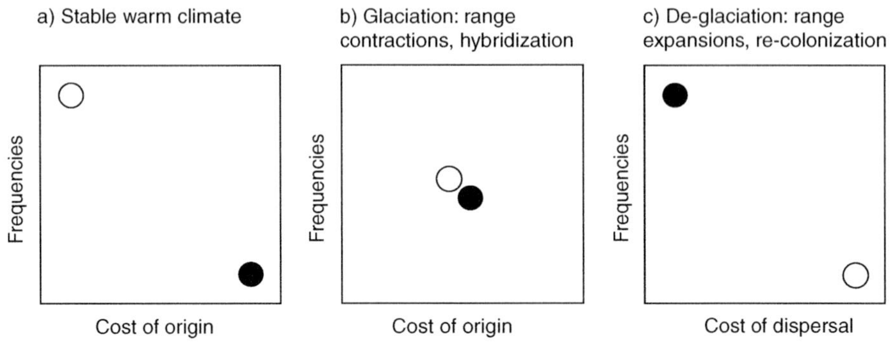

Fig. 8.1 Hypothetical scenario of the "opportunistic model" for geographical parthenogenesis assuming a hybrid origin of asexuality. *Filled circle*: asexual populations; *empty circle*: sexual populations. (**a**) in a stable warm climate, sexuality as the default system is frequent, whereas frequencies of asexuality remain low because of the constraints of new origins (high cost of origin). (**b**) during glaciations (and aridifications), sexual species undergo area fragmentation, drift and interspecific hybridization; asexual populations become more frequent because of higher occurencies of interspecific hybridization of sexual species (lowered cost of origin). Multiple hybrid origins and secondary origins of asexual lineages via backcrossing of asexuals will further lower this cost. (**c**) during re-colonization, asexual populations disperse rapidly because of uniparental reproduction (low cost of dispersal), sexual populations remain restricted because of density-dependence of sex. Frozen niche variation mechanisms may further enhance the spread of asexuality

Glaciations may further cause area fragmentation and geographical isolation of sexual populations (Hewitt 1996). Extinction rates of sexual populations will increase if mating partners become rare because of density-dependence of sexual reproduction. Drift will cause a loss of genetic diversity and heterozygosity will be lost because of inbreeding, resulting in inbreeding depression in small and isolated populations. In contrast, asexuals being capable of uniparental reproduction are not limited by density-dependence, and do not suffer from a loss of heterozygosity in small populations.

During re-colonization, asexuality is advantageous because reproduction without a mating partner is certain, especially if colonization is connected to long distance dispersal. In contrast to sexual selfers (through autogamy or automixis), apomictic individuals maintain heterozygosity and will not suffer from founder events. Sexual reproduction is density-dependent and therefore less efficient for rapid re-colonization. Small sexual founder populations may be hampered by bottlenecks, inbreeding and a loss of heterozygosity (e.g. Lynch 1984). Areas devastated by glaciations or by aridification thus provide open habitats, which can be faster occupied by asexuals. Asexuality consequently reduces the costs of dispersal and spread (Fig. 8.1c).

Once asexual reproduction has arisen, contagious origins of asexuality, i.e., hybridization of asexual lineages with their sexual progenitors may further increase its frequencies. Such backcrossing may add to the frequencies of secondary origins and the genetic diversity of asexuals. In the ostracod *Eucypris virens*, multiple origins of asexual lineages from sexual ancestors could be demonstrated by phylogenetic reconstructions (Schön et al. 2000). Crosses between asexual females and sexual males may have caused multiple origins in parthenogenetic *Schmidtea polychroa* and maintain genetic diversity of asexuals (Pongratz et al. 2003). Allozyme analyses in several plant groups suggest an increase of genetic diversity in sympatric sexual-asexual populations (reviews in Hörandl et al. 2001; Hörandl and Paun, 2007). Backcrossing between sexual and apomictic individuals is probably a main factor for creating new asexual lineages (Hörandl and Paun 2007). Furthermore, facultative sexuality within asexuals may result in occasional crosses among asexual lineages; such processes may have contributed to the high diversity of asexuals of the *Ranunculus auricomus* complex (Hörandl et al. in press). Since constraints of a functional meiosis are lacking, asexual lineages are not under a selective pressure to have crossing barriers against other asexuals. Therefore, a potential crossability remains, and occasional facultative sexuality is likely to result in successful cross-fertilization among asexual lineages. The continued origin of new genotypes may enhance the occupation of different environments via FNV mechanisms and may contribute to the further range expansion of apomictic complexes. Facultative sexuality is the best explanation for the formation of huge and diverse apomictic complexes in various plant families, like blackberries (*Rubus fruticosus*, Weber 1995), hawkweeds (*Hieracium* subg. *Pilosella*; Fehrer et al. 2005, 2007) and goldilocks (*Ranunculus auricomus*, Hörandl and Paun 2007). In these complexes, single agamospecies have restricted distribution areas and narrow ecological niches (e.g. Hörandl 1998), but apomictic complexes as a whole show typical patterns of geographical parthenogenesis.

Thus, geographical parthenogenesis can be explained as a result of a temporary increase of the costs of sexuality paralleled by a huge decrease of the cost for shifting to asexuality, plus advantages of asexuality for re-colonizating devastated areas. This "opportunistic model of geographical parthenogenesis" (Fig. 8.1) would not only resolve the problem of larger distribution areas, but could also explain why geographical parthenogenesis appears in areas influenced by climatic oscillations of the Pleistocene. Further opportunities for asexuality may arise through human influence and the introduction of species into new geographical areas, which may trigger hybridization. In hawkweeds (*Hieracium* subg. *Pilosella*), the introduction of European species to New Zealand resulted in intensive hybridization, secondary origins and dispersal of asexuals (Chapman et al. 2000; Trewick et al. 2004).

The time when evolutionary opportunities arose, may explain why most extant asexual taxa appear to be "young", not older than the Pleistocene (e.g. Johnson and Bragg 1999; Moritz and Heideman 1993; Schön et al. 2000; Pongratz et al. 2003; Dobeš et al. 2004a, b; Strasburg and Kearney 2005; Paun et al. 2006b). The glaciations of the Pleistocene were, in geological dimensions, a rather recent event, which may have dramatically increased frequencies of asexual origins compared to their sexual ancestor. The apparently "young" age of asexuals would then not be a consequence of a short persistence of asexuals because of the accumulation of deleterious mutations, but is simply a consequence of the latest opportunities for asexuality to arise, in geological dimensions, rather recently. In the warmer periods

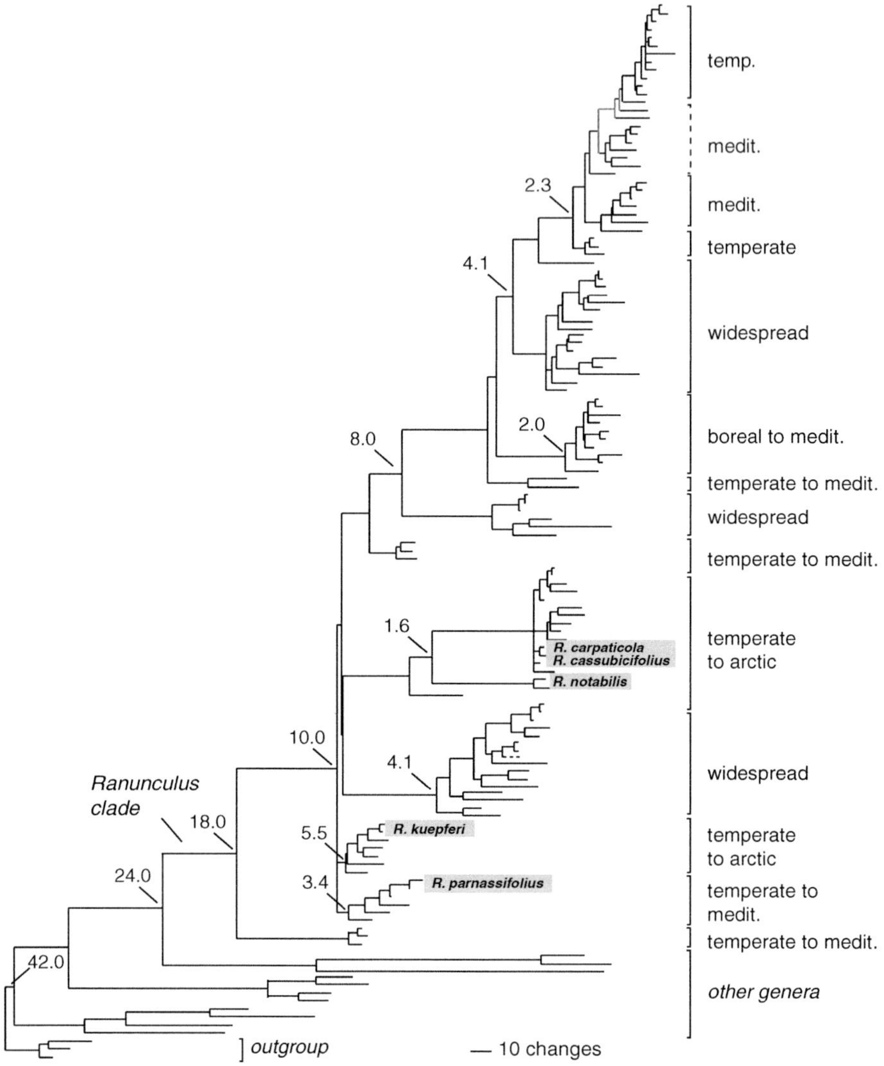

Fig. 8.2 Phylogenetic reconstruction of the genus *Ranunculus* s.l. and multiple origins of apomixis (*grey boxes*) on terminal branches of clades. Phylogram based on a combined ITS and *matK-trnK* dataset, including sexual 140 species. Clades with bootstrap support over 80% are marked in *bold*. Sexual ancestors of apomictic taxa are printed in *bold*. The predominant geographical distribution of clades is indicated on the *right* (medit. = regions with a Mediterranean-like climate). Age estimates (in Mya) using penalized likelihood based on plastid data are given for selected nodes (modified after Paun et al. 2005, with permission of the publisher)

of the Mesozoic and the Tertiary, opportunities for asexuality were probably less frequent and speciation of sexuals has established the basis for the great diversity of animals and plants that we observe today (see e.g. Fig. 8.2). Rare origins may have caused limited numbers of asexuals that would be "ancient" from our present point view. This consideration may help to resolve the paradox of "ancient asexual scandals" such as Darwinulid ostracods, which have been asexual for at least 200 myr (fossil evidence; Martens et al. 2003). In angiosperms, the reproductive mode cannot be inferred from the fossil record. If rare opportunities for asexuality limit the number of asexual lineages, then ancient asexuals are not necessarily an evolutionary "scandal". They are more appropriately regarded as "living fossils" from an older time with more frequent origins of asexuality. Various mechanisms such as the loss of transposable elements or cryptic recombination (gene conversion) may lower the load of deleterious mutations in ancient asexuals (e.g. Butlin 2002), see also Chapters 11, 12, and 13.

8.6 Case Studies in *Ranunculus*

The importance of frequencies and constraints of the origins of apomixis and various patterns of geographical parthenogenesis will be illustrated for two case studies of buttercups from the genus *Ranunculus*. *Ranunculus* is a big and widespread, predominantly sexual genus, the largest in the family Ranunculaceae. Ca. 600 sexual species are known and occur on all continents, but with centers of diversity in the high mountain systems of the world and previously glaciated areas of the Pleistocene. Molecular clock analyses suggest an origin of the genus in the late Oligocene, and several waves of diversification during the late Tertiary and Quaternary (Paun et al. 2005; Fig. 8.2). Sexual polyploidy and hybridization is common in the genus and has shaped the genetic structure and evolution of many clades (Lockhart et al. 2001; Hörandl et al. 2005). Phylogenetic reconstructions of the genus using ITS and plastid sequences (*matk-trnk*) have provided evidence for at least three independent origins of apomixis, each of them showing a pattern of geographical parthenogenesis (Fig. 8.2): (1) the *Ranunculus auricomus* complex, a huge apomictic polyploid complex, distributed mainly throughout temperate to arctic Eurasia with 600–800 agamospecies and four sexual species. The phylogenetic position of three of these sexual species, *R. cassubicifolius*, *R. carpaticola* and *R. notabilis* suggest a close but not a sister relationship (Fig. 8.2). The sexual taxa occur in temperate Europe and the Mediterranean (Hörandl and Paun 2007; Fig. 8.3); (2) *Ranunculus kuepferi*, a single species distributed in the European Alps, with diploid sexuals in the Alps Maritimes, and tetraploid apomicts in previously glaciated parts of the Alps (Huber 1988; Cosendai and Hörandl 2008; Hörandl et al. 2008); and (3) *Ranunculus parnassifolius*, a species with diploid sexuals in the Pyrenees and tetraploids in the European Alps (Jalas and Suominen 1989; Vuille and Küpfer 1985). The three apomictic taxa appear in different main clades of the genus and may have originated in different time periods (Paun et al. 2005; Fig. 8.2). The taxonomic distribution of apomixis in *Ranunculus* fits with the generally scattered pattern of apomixis on the tips of phylogenies. The enormous diversity of sexual species in *Ranunculus* as compared to the three apomictic taxa illustrates a crucial problem of asexual reproduction: origins of apomixis are too rare to compete with frequencies of sexual speciation (Fig. 8.2).

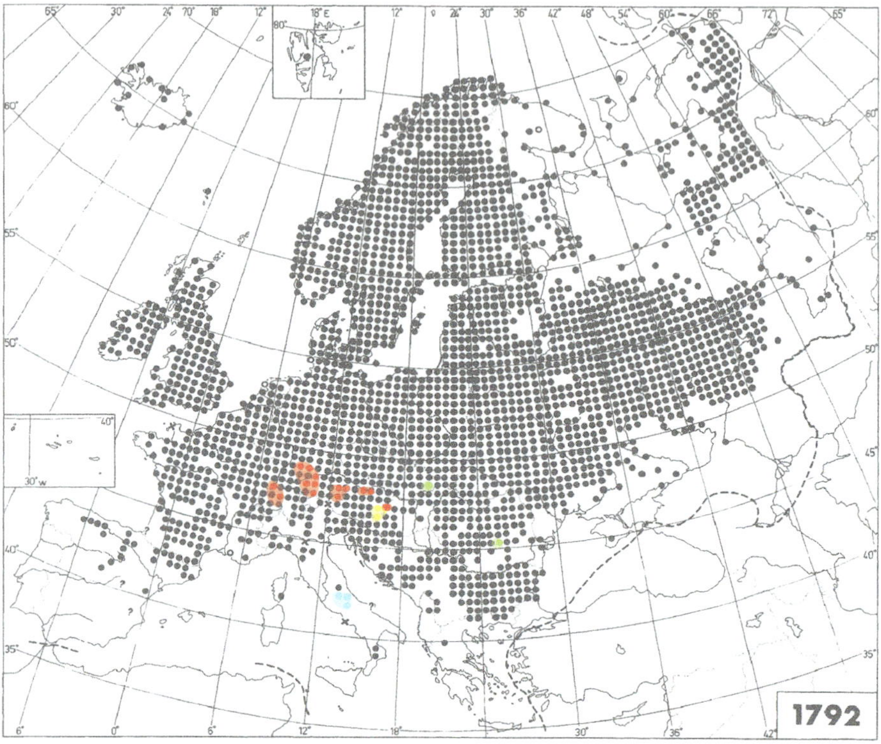

Fig. 8.3 Distribution of the *Ranunculus auricomus* complex in Europe. *Black dots*: general distribution, based on Jalas and Suominen (1989); *colored dots*: sexual species; *red*: *R. cassubicifolius*; *green*: *R. carpaticola*, diploid sexual cytodemes; *yellow*: *R. notabilis*; *orange*: *R. marsicus*. Sexual species are sympatric with agamospecies; outside their distribution areas only apomictic taxa are known so far (see text). Reproduced from Hörandl and Paun (2007) (with permission of the publisher)

Each case of apomictic buttercups tells a different story on evolution and the possible origin of geographical parthenogenesis. The *Ranunculus auricomus* complex is by far the best-studied group of the genus. Intensive work in the 20th century has revealed aposporous apomixis for most polyploids, an unreduced embryo sac developing out of a somatic cell of the nucellus. The unreduced egg cell develops parthenogenetically into an embryo, which is genetically identical to the mother plant. Pollination is necessary for the fertilization of the endosperm (pseudogamy) (Häfliger 1943; Rutishauser 1954; Izmaiłow 1967; Nogler 1971, 1984). The complex is widely distributed and abundant in the temperate to arctic zones of Europe and Asia (Fig. 8.3). The strongly disjunct distribution areas of the sexual species *R. cassubicifolius* and *R. carpaticola* can be best explained by area fragmentation during glacial maxima in the Pleistocene (Hörandl and Greilhuber 2002; Paun et al. 2006b, Hörandl et al. in press). Interestingly, polyploidy also occurs in the sexual taxa but their geographic distribution ranges remain restricted: *R. cassubicifolius* has autotetraploid cytotypes in Lower Austria (Hörandl and Greilhuber 2002), and *R. marsicus* is tetra- to hexaploid. These examples contradict the hypothesis that polyploidy is responsible for the large distribution area of this apomictic complex. The apomictic biotypes are rather easily recognized by bad pollen quality, reduced petals, and partly aborted fruits. Apomicts have been reported throughout the whole range of the complex, and detailed cytological and histological investigations have confirmed apomixis in Fennoskandia (Rousi 1956), Poland (Izmaiłow 1965, 1967), Germany, Austria, and Switzerland (Häfliger 1943; Nogler 1971).

Molecular studies using DNA fingerprinting (AFLPs, microsatellites) and DNA sequence analysis suggest that apomicts have originated from hybridization of sexual taxa. Apomictic populations have elevated values of observed heterozygosity, which are fixed at most loci, confirming the hybrid origin of apomicts (Hörandl et al. 2000, 2001; Hörandl and Greilhuber 2002; Paun et al. 2006a). ITS sequence polymorphisms suggest a reticulate evolution of the whole complex (Hörandl et al. 2005).

Experimental crosses of the sexual taxa *Ranunculus notabilis* and *R. carpaticola* show significantly higher frequencies of seed abortion in the F_1 hybrid generation as compared to the sexual parental species (Hörandl 2008). In apomicts, disturbances of chromosome pairing during meiosis are common causing bad pollen quality and high frequencies of megaspore abortion. On average, in only about 25% of ovules is the degenerating megaspore tetrad replaced by a somatic (aposporous) cell (Izmaiłow 1965, 1967).

The complex evolved probably from hybridizations of sexual species. A case study on hexaploid apomicts in Slovakia suggests a single hybrid origin from diploid *Ranunculus carpaticola* and tetraploid *R. cassubicifolius* (Paun et al. 2006b). Crosses between different ploidy levels are general often sterile in plants because of endosperm imbalance (e.g. Levin 2002); in this case, apomixis has probably stabilized a rare secondary contact event with hybridization. DNA sequence analysis and morphology suggest a hybrid origin of tetraploid apomicts involving *R. notabilis* and *R. carpaticola* (Hörandl et al. in press). Also here, historical range fluctuations and hybridization through secondary contact must be assumed, because at present, the species are geographically separated (see Fig. 8.3). Contagious origins of apomictic lineages via crossings of sexuals and apomicts appear to be rare in this complex (Hörandl and Temsch 2009); in mixed populations, sexuals and apomicts remain genetically distinct (Hörandl et al. 2000, 2001). Taken together, the postulated constraints of (1) hybridization of sexual species and (2) hybrid sterility do apply for shifts towards apomixis (Fig. 8.1a) and may be reduced by fluctuations of distribution ranges and the break-down of geographical barriers (Fig. 8.1b).

The apomictic complex is constituted of several distinct morphotypes with various ecological niches and local distributions, which have often been classified as agamospecies (Hörandl 1998; Jalas and Suominen 1989). Geographical parthenogenesis is here a phenomenon constituting of a high diversity of apomictic lineages, cytotypes and ecotypes. Two of the sexual species, *R. cassubicifolius* and *R. carpaticola*, are widely distributed, but with big disjunctions and isolated populations in small areas. A regional genetic differentiation of various populations is apparent in

AFLP clusters (Paun et al. 2006b), and inbreeding can be inferred from increased levels of homozygosity in iso-enzyme markers (Hörandl and Greilhuber 2002).

Population genetic studies using iso-enzymes and DNA fingerprinting have revealed a reduced genotypic diversity of apomicts as compared to sexual species, but pure clones are rare; most populations show a predominant clone and a few deviating genotypes. Facultative sexuality and occasional crossing between apomictic lineages are likely to be the main engine for the continuous formation of new genotypes (Hörandl and Paun 2007; Hörandl et al. in press). Apomictic populations occur in a broad spectrum of habitats and tend to colonize more frequently man-made habitats such as meadows, pastures and even ruderals, whereas sexual populations preferentially occur in forest floors (Hörandl et al. 2000; Paun et al. 2006a; Hörandl and Paun 2007). Based on these observations, FNV mechanisms may play an important role in the colonization.

The capability of apomictic biotypes for uniparental reproduction has been demonstrated by Hörandl (2008). Enforced self-pollination via pollinator exclusion tests revealed normal seed set in apomicts, but failure of seed set in both diploids and autotetraploids of the sexual species. When pollinators are available, apomicts have a lower reproductive fitness as compared to sexuals because of high frequencies of aborted fruits. This may contribute to maintenance of sexual populations in their distribution range, and supports the importance of uniparental reproduction for founder events. Therefore, also the colonization hypothesis is strongly supported in this case.

To summarize, the pattern of geographical parthenogenesis in *Ranunculus auricomus* is probably shaped by maintenance and continuous origin of clonal diversity, by a more efficient niche exploration and better colonizing abilities of apomicts. The role of single vs. multiple origins needs to be studied further. Apomictic lineages of *R. auricomus* benefit probably mainly from a reduced cost of dispersal during re-colonization (Fig. 8.1c).

Ranunculus kuepferi is a different case of geographical parthenogenesis. Here, geographical parthenogenesis appears to be an infraspecific phenomenon, constituted by the spatial separation of diploid sexuals and tetraploid apomicts. The latter do not show any morphological differences as compared to the former, except for higher frequencies of aborted fruits, aborted pollen, and a lower number of stamens (Huber 1988). A clear ecological differentiation is also missing: both cytotypes grow in alpine grassland vegetation, with no differences in the main ecological indicator values (Huber 1988). Whether the apomicts have originated from hybridization is questionable. DNA fingerprinting profiles do not suggest involvement of another genome in the origin of tetraploids, because of the low percentage of fragments being specific for tetraploid apomictic populations (Cosendai and Hörandl 2008).

Apomicts show similar developmental features as *R. auricomus*, i.e. the production of an unreduced embryo sac, parthenogenetic development of the egg cell into the embryo and pseudogamous endosperm formation (Huber 1988; Hörandl et al. 2008). As in *R. auricomus*, frequencies of pollen and seed abortion are much higher than in sexuals (Huber 1988; Hörandl, unpubl.). Surprisingly, AFLP studies revealed exceptional high levels of genotypic diversity within apomictic populations, with unique genotypes for each individual. Sexual and apomictic populations do not show differences in genetic variation. Apomicts have separated gene pools for each population, suggesting multiple founder events of populations throughout the Alps (Cosendai and Hörandl 2008). Taken together, the geographical success of apomictic *R. kuepferi* is enigmatic, because evidence for a hybrid genome is missing. The equal level of genetic variation in sexual and apomictic populations also questions the applicability of Red Queen mechanisms.

Possible answers come from population genetic and cytological studies: variation in ploidy levels and high frequencies of triploids in the geographical contact zone of diploids and tetraploids in the Alps Maritimes suggest high frequencies of facultative sexuality and also introgression of apomixis into sexuals (Cosendai and Hörandl 2008). Continued hybridization between diploid sexuals and polyploid apomicts and introgression of apomixis into sexual populations is probably important for the formation and maintenance of genotypic diversity. Reproductive systems are highly flexible (Cosendai and Hörandl in prep.). Multiple secondary origins and mixed reproductive systems may be an alternative strategy of geographical parthenogenesis, as already suggested by Schön et al. (2000). Such cases imply an advantage to apomixis by reducing the costs for secondary origins of asexuality (Fig. 8.1b).

8.7 Suggestions for Future Research

The here presented cases studies suggest that geographical parthenogenesis cannot be explained by a single scheme, and many aspects of this hypothetical framework need to be analyzed further. For many organisms, it is still not known how asexual reproduction originated; empirical tests of the hybrid origin of asexuality are required to confirm the basic assumption of the model. Molecular studies and experimental crossings are most promising to clarify how asexual reproduction can originate from hybrids of sexual species. Nevertheless, hybridization is not the only mechanism for creating parthenogenetic lineages, and frequencies of alternative pathways need to be estimated before general conclusions can be drawn. The functionality and genetic control of asexual reproduction needs to be better understood to estimate the proposed "costs" of shifts from sexual systems to asexuality. Control mechanisms may differ considerably in various taxonomic groups and need to be studied for each case.

An additional question is the extent and ecological significance of sexual hybrid speciation, i.e., hybridization of sexual species without a shift to asexuality. Such hybrid lineages may have similar evolutionary forces and are potential competitors of asexual hybrids for novel ecological niches. Sexual hybrids do have a potential for speciation, whereas asexual hybridity may in many cases only result in a short-term spatial expansion. Fertility problems, a major constraint of sexual F_1 hybrids, may be overcome in later hybrid generations by exogenous and endogenous selection (Arnold 1997), and reduce the "costs of dispersal" of hybrids as hypothesized in Fig. 8.1. That is, the consequences of hybridization need to be analyzed in a broader framework to fully understand their role for the evolution of asexuality.

Besides the frequencies of asexual origin, other factors such as uniparental reproduction, FNV and Red Queen mechanisms have to be considered, while synergistic effects are likely. Again, group-specific features have to be taken into account, and the careful examination of reproductive systems of a specific taxon is required before uniparental reproduction can actually be postulated as a general advantage for asexuality. Genetic variation of populations needs to be studied to provide evidence for the genetic basis of frozen niche mechanisms; moreover, experimental approaches are highly desired to actually prove a superior niche occupation of asexual populations. The understanding of Red Queen mechanisms for geographical parthenogenesis needs experimental work and field studies of various diploid and polyploid sexual-asexual model systems.

Modelling approaches will be useful for estimating the long-term consequences of geographical parthenogenesis. Do asexual organisms have the potential to finally replace sexual related species simply by occupying the available resource space? What are the ecological and evolutionary consequences of geographical parthenogenesis? What are the consequences of climatic changes, e.g., global warming, for evolution of sex vs. asexuality? Taken together, multidisciplinary approaches are needed to fully understand the evolutionary significance of geographical parthenogenesis.

Acknowledgements I would like to thank Isa Schön, Peter Van Dijk and one anonymous referee for valuable comments on the manuscript. This work was funded by the Austrian Research Foundation (FWF), P 19006-B03.

Glossary

Agamospecies: apomictic lineages with distinct morphological, ecological and geographical features, which have been classified formally as species.
Apomeiotic: the development of the embryo sac without meiosis in flowering plants (see also Chapter 3).
Aposporous: a form of apomitic reproduction in flowering plants, whereby the embryo sac develops out of a somatic cell (see also Chapter 3).
Homoploid hybrids: hybrids with the same ploidy level as the parents; usually applied to diploid hybrids, which have originated from diploid parental species.
Pseudogamy: sperm-dependent or pollen-dependent asexual reproduction (see also Chapter 3).
Uniparental reproduction: reproduction via only one parental individual (including unisexual reproduction, apomixis, automixis, and autogamy of hermaphroditic organisms; vegetative propagation is not included here; see also Chapter 3).

References

Adams KL, Wendel JF (2005) Polyploidy and genome evolution in plants. Curr Opin Plant Biol 8: 135–141
Arnold ML (1997) Natural hybridization and evolution. Oxford University Press, Oxford

Asker SE, Jerling L (1992) Apomixis in plants. CRC Press, Boca Raton
Baker HG (1967) Support for Baker's law – as a rule. Evolution 21: 853–856
Baker HG, Stebbins GL (1965) The genetics of colonizing species. Academic Press, New York
Bayer RJ (1990) Investigations into the evolutionary history of the *Antennaria rosea* (Asteraceae: Inuleae) polyploid complex. Plant Syst Evol 169: 97–110
Bayer RJ (1991) Allozymic and morphological variation in *Antennaria* (Asteraceae:Inuleae) from the low arctic of northwestern North America. Syst Bot 16: 492–506
Bayer RJ, Chandler GT (2007) Evolution of polyploid agamic complexes: a case study using the Catipes group of *Antennaria*, including the *A. rosea* complex (Asteraceae: Gnaphalieae). In Hörandl E, Grossniklaus U, van Dijk P, Sharbel T (eds) Apomixis: Evolution, mechanisms and perspectives. ARG Gantner Verlag KG, Lichtenstein, pp. 317–336
Bell G (1982) The masterpiece of nature: The evolution and genetics of sexuality. California Press, Berkeley
Ben-Ami F, Heller J (2005) Spatial and temporal patterns of parthenogenesis and parasitism in the freshwater snail *Melanoides tuberculata*. J Evol Biol 138: 138–146
Beukeboom LW (2007) Sex to some extent. Heredity 98:123–124
Beukeboom LW, Vrijenhoek RC (1998) Evolutionary genetics and ecology of sperm-dependent parthenogenesis. J Evol Biol 11: 755–782
Bicknell RA, Koltunow AM (2004) Understanding apomixis: recent advances and remaining conundrums. Plant Cell 16: S228–S245
Bierzychudek P (1985) Patterns in plant parthenogenesis. Experientia 41: 1255–1264
Birdsell JA, Wills C (2003) The evolutionary origin and maintenance of sexual recombination: a review of contemporary models. Evol Biol 33: 27–138
Brochmann C, Gabrielsen TM, Nordal I, Landvik JY, Elven R (2003) Glacial survival or tabula rasa? The history of North Atlantic biota revisited. Taxon 52: 417–450
Brock MT (2004) The potential for genetic assimilation of a native dandelion species, *Taraxacum ceratophorum* (Asteraceae), by the exotic congener *T. officinale*. Am J Bot 91: 656–663
Burt A (2000) Perspective: sex, recombination and the efficacy of selection-was Weismann right? Evolution 54: 337–351
Butlin R (2002) The costs and benefits of sex: new insights from old asexual lineages. Nat Rev Genet 3: 311–317
Byers DL, Waller DM (1999) Do plant populations purge their genetic load? Effects of population size and mating history on inbreeding depression. Annu Rev Ecol Syst 30: 479–513
Carillo C, Britton NF, Mogie M (2002) Coexistence of sexual and asexual conspecifics: a cellular automaton model. J Theor Biol 217: 275–285
Carman JG (1997) Asynchronous expression of duplicate genes in angiosperms may cause apomixis, bispory, tetraspory, and polyembryony. Biol J Linn Soc 61: 51–94
Carman JG (2001) The gene effect: genome collisions and apomixis. In: Savidan Y, Carman JG, Dresselhaus T (eds) The flowering of apomixis: From mechanisms to genetic engineering. CIMMYT, Mexico DF, pp. 95–110
Carman JG (2007) Do duplicate genes cause apomixis? In: Hörandl E, Grossniklaus U, van Dijk P, Sharbel T (eds), Apomixis: Evolution, mechanisms and perspectives. ARG Gantner Verlag KG, Lichtenstein, pp. 169–194
Carr DE, Dudash MR (2003) Recent approaches into the genetic basis of inbreeding depression in plants. Philos Trans R Soc Lond B 358: 1071–1084
Chapman HM, Parh D, Oraguzie N (2000) Genetic structure and colonizing success of a clonal weedy species, *Pilosella officinarum* (Asteraceae). Heredity 84: 401–409
Chen ZJ (2007) Genetic and epigenetic mechanisms for gene expression and phenotypic variation in plant polyploids. Annu Rev Plant Biol. 58: 377–406
Comai L (2005) The advantages and disadvantages of being polyploid. Nat Rev Genet 6: 836–846

Cosendai A-C, Hörandl E (2008) *Ranunculus kuepferi* (Ranunculaceae), an alpine apomict showing geographical parthenogenesis. Plant species concepts and evolution, NCCR Plant Survival International Conference, Jan 30th to 1st Feb 2008, Neuchatel, Switzerland, 36

Coyne JA, Orr HA (2004) Speciation. Sinauer Ass, Sunderland

Curtis MD Grossniklaus U (2007). Amphimixis and apomixis:two sides of the same coin. In: Hörandl E, Grossniklaus U, Van Dijk P, Sharbel T (eds.) Apomixis: Evolution, mechanisms and perspectives. ARG Gantner Verlag KG, Lichtenstein, pp. 37–62

de Kovel CGF, Jong G (1999) Responses of sexual and apomictic genotypes of *Taraxacum officinale* to variation in light. Plant Biol 1: 541–546

Dickinson TA, Lo E, Talent N (2007) Polyploidy, reproductive biology, and Rosaceae: understanding evolution and making classifications. Plant Syst Evol 266: 59–78

Dobeš C, Mitchell-Olds T, Koch MA (2004a) Extensive chloroplast haplotype variation indicates Pleistocene hybridization and radiation of North American *Arabis drummondii*, *A.* × *divaricarpa*, and *A. holboellii* (Brassicaceae). Mol Ecol 13: 349–370

Dobeš C, Mitchell-Olds T, Koch MA (2004b) Interspecific diversification in North American *Boechera stricta* (= *Arabis drummondii*), *Boechera* × *divaricarpa,* and *A. holboellii* (Brassicaceae) inferred from nuclear and chloroplast molecular markers – an integrative approach. Am J Bot 91: 2087–2101

Fehrer J, Krahulcová A, Krahulec F, Chrtek J, Rosenbaumová R, Bräutigam S (2007) Evolutionary aspects in *Hieracium* subgenus *Pilosella*. In Hörandl E, Grossniklaus U, van Dijk P, Sharbel T (eds) Apomixis: Evolution, mechanisms and perspectives. ARG Gantner Verlag KG, Lichtenstein, pp. 359–395

Fehrer J, Šimek R, Krahulcová A, Krahulec F, Chrtek J, Bräutigam E, Bräutigam S (2005) Evolution, hybridisation, and clonal distribution of apo- and amphimictic species of *Hieracium* subgen. *Pilosella* (Asteraceae, Lactucaceae) in a central European mountain range. In: Bakker F, Chatrou L, Gravendeel B, Pelser PB (eds) Plant species-level systematics: New perspectives on pattern and process. ARG Gantner Verlag KG, Lichtenstein, pp. 175–201

Fritz RS, Moulia C, Newcombe G (1999) Resistance of hybrid plants and animals to herbivores, pathogens, and parasites. Annu Rev Ecol Syst 30: 565–591

Gade B, Parker ED (1997) The effect of life cycle stage and genotype on desiccation tolerance in the colonizing parthenogenetic cockroach *Pycnoscelus surinamensis* and its sexual ancestor *P. indicus*. J Evol Biol 10: 479–493

Glesener RR, Tilman D (1987) Sexuality and the components of environmental uncertainty, clues from geographic parthenogenesis in terrestrial animals. Am Nat 112: 169–673

Gomez-Zurita J, Funk DJ, Vogler AP (2006) The evolution of unisexuality in *Calligrapha* leaf beetles: molecular and ecological insights on multiple origins via interspecific hybridization. Evolution 60: 328–347

Gray MM, Weeks SC (2001) Niche breadth in clonal and sexual fish (*Poeciliopsis*), a test of the frozen niche variation model. Can J Fish Aqu Sci 58: 1313–1318

Green RF, Noakes DLG (1995) Is a little bit of sex as good as a lot? Theor Biol 174: 87–96

Grimanelli D, Leblanc O, Perotti E, Grossniklaus U (2001) Developmental genetics of gametophytic apomixis. Trends Genet 17: 597–604

Haag CR, Ebert D (2004) A new hypothesis to explain geographic parthenogenesis. Ann Zool Fenn 41: 539–544

Häfliger E (1943) Zytologisch-embryologische Untersuchungen pseudogamer Ranunkeln der *Auricomus*-Gruppe. Ber Schweiz Bot Ges 53: 317–379

Hewitt GM (1996) Some genetic consequences of ice ages, and their role in divergence and speciation. Biol J Linn Soc 58: 247–276

Hörandl E (1998) Species concepts in agamic complexes: applications in the *Ranunculus auricomus* complex and general perspectives. Folia Geobot. 33: 335–348

Hörandl E (2006) The complex causality of geographical parthenogenesis. New Phytol 171: 525–538

Hörandl E (2008) Evolutionary implications of self-compatibility and reproductive fitness in the apomictic *Ranunculus auricomus* polyploid complex (Ranunculaceae). Int J Plant Sci 169: 1219–1228

Hörandl E, Cosendai A-C, Temsch E (2008) Understanding the geographic distributions of apomictic plants: a case for a pluralistic approach. Plant Ecol Divers 2: 309–320

Hörandl E, Greilhuber J (2002) Diploid and autotetraploid sexuals and their relationships to apomicts in the *Ranunculus cassubicus* group: insights from DNA content and isozyme variation. Plant Syst Evol 234: 85–100

Hörandl E, Greilhuber J, Dobeš C (2000) Isozyme variation and ploidy levels within the apomictic *Ranunculus auricomus* complex: evidence for a sexual progenitor species in southeastern Austria. Plant Biol 2: 53–62

Hörandl E, Greilhuber J, Klimova K, Paun O, Temsch E, Emadzade K, Hodálová I (in press) Reticulate evolution and taxonomic concepts in the *Ranunculus auricomus* complex (Ranunculaceae): insights from morphological, karyological and molecular data. Taxon

Hörandl E, Jakubowsky G, Dobeš C (2001) Isozyme and morphological diversity within apomictic and sexual taxa of the *Ranunculus auricomus* complex. Plant Syst Evol 226: 165–185

Hörandl E, Paun O (2007) Patterns and sources of genetic diversity in apomictic plants: implications for evolutionary potentials. In: Hörandl E, Grossniklaus U, Van Dijk P, Sharbel T (eds) Apomixis: Evolution, mechanisms and perspectives. ARG Gantner Verlag KG, Lichtenstein, pp. 169–194

Hörandl E, Paun O, Johansson J, Lehnebach C, Armstrong T, Lockhart P (2005) Phylogenetic relationships and evolutionary traits in *Ranunculus* s.l. (Ranunculaceae) inferred from ITS sequence analysis. Mol Phyl Evol 36: 305–327

Hörandl E, Temsch E (2009) Introgression of apomixis into sexual species is inhibited by mentor effects and ploidy barriers in the Ranunculus auricomus complex. Ann Bot 104: 81–89

Huber W (1988) Natürliche Bastardierungen zwischen weißblühenden *Ranunculus*-Arten in den Alpen (Natural hybridizations between white-flowered species of *Ranunculus* in the Alps). Veröff Geobot Inst ETH, Zürich 100:1–160 [German with English abstract]

Izmaiłow R (1965) Macrosporogenesis in the apomictic species *Ranunculus cassubicus*. Acta Biol Cracov 8: 183–195

Izmaiłow R (1967) Observations in embryo and endosperm development in various chromosomic types of the apomictic species *Ranunculus cassubicus* L. Acta Biol Cracov 10: 99–111

Jalas J, Suominen J (1989) Atlas Florae Europaeae. 8. Nymphaeaceae to Ranunculaceae. The Committee for Mapping the Flora of Europe and Societas Biologica Fennica Vanamo, Helsinki

Johnson SG, Bragg E (1999) Age and polyphyletic origins of hybrid and spontaneous parthenogenetic *Campeloma* (Gastropoda: Viviparidae) from the southeastern United States. Evolution 53: 1769–1781

Jokela J, Lively CM, Dybdahl MF, Fox JA (2003) Genetic variation in sexual and clonal lineages of a freshwater snail. Biol J Linn Soc 79: 165–181

Jokela J, Lively CM, Fox JA, Dybdahl MF (1997) Flat reaction norms and 'frozen' phenotypic variation in clonal snails (*Potamopyrgus antipodarum*). Evolution 51: 1120–1129.

Kearney M (2003) Why is sex so unpopular in the Australian desert? Trends Ecol Evol 18: 605–607

Kearney M (2005) Hybridization, glaciation and geographical parthenogenesis. Trends Ecol Evol 20: 495–502

Kirk H, Choi YH, Kim HK, Verpoorte R, van der Meijden E (2005) Comparing metabolomes: the chemical consequences of hybridization in plants. New Phytol 167: 613–622

Koch MA, Dobeš C, Mitchell-Olds T (2003) Multiple hybrid formation in natural populations: concerted evolution of the internal transcribed spacer of nuclear ribosomal DNA (ITS) in North American *Arabis divaricarpa* (*Brassicaceae*). Mol Biol Evol 20: 338–350

Koltunow A, Grossniklaus U (2003) Apomixis, a developmental perspective. Annu Rev Plant Biol 54: 547–574

Levin DA (1975) Minority cytotype exclusion in local plant populations. Taxon 24: 35–43

Levin DA (2002) The role of chromosomal change in plant evolution. Oxford University Press, Oxford

Lockhart P, McLechnanan PA, Havell D, Glenny D, Huson D, Jensen U (2001) Phylogeny, dispersal and radiation of New Zealand alpine buttercups: molecular evidence under split decomposition. Ann Missouri Bot Gard 88: 458–477

Loxdale H, Lushai G (2003) Rapid changes in clonal lines: the death of a 'sacred cow'. Biol J Linn Soc 79: 3–16

Lundmark M, Saura A (2006) Asexuality alone does not explain the success of clonal forms in insects with geographical parthenogenesis. Hereditas 143: 23–32

Lushai G, Loxdale HD, Allen JA (2003) The dynamic clonal genome and its adaptive potential. Biol J Linn Soc 79: 193–20

Lynch M (1984) Destabilizing hybridization, general-purpose genotypes and geographic parthenogenesis. Q Rev Biol 59: 257–290

Mable K (2007). Sex in the postgenomic era. Trends Ecol Evol 2: 559–561

Mallet J (2007). Hybrid speciation. Nature 446: 279–283

Martens K, Rossetti G, Horne DJ (2003) How ancient are ancient asexuals? Proc R Soc London B 270: 723–729

Matzk F, Prodanovic S, Bäumlein H, Schubert I (2005) The inheritance of apomixis in *Poa pratensis* confirms a five locus model with differences in gene expressivity and penetrance. Plant Cell 17:13–24

Matzk F, Prodanovic S, Cziha A, Tiedemann J, Arzenton F, Blattner F, Kumlehn J, Altschmied L, Schubert I, Johnston A, Grossniklaus U, Bäumlein H (2007) Genetic control of apomixis – preliminary lessons from *Poa*, *Hypericum* and wheat egg cells. In: Hörandl E, Grossniklaus U, Van Dijk P, Sharbel T (eds) Apomixis: Evolution, mechanisms and perspectives. ARG Gantner Verlag KG, Lichtenstein, pp. 169–194

Mavárez J, Linares M (2008) Homoploid hybrid speciation. Mol Ecol 17: 4181–4185

Maynard Smith J (1978) The evolution of sex. Cambridge University Press, Cambridge, UK

Meirmans PG, Vlot EC, Den Nijs JCM, Menken SBJ (2003) Spatial ecological and genetic structure of a mixed population of sexual diploid and apomictic triploid dandelions. J Evol Biol 16: 343–352

Meirmans S, Skorping A, Lonying MK, Kirkendall LR (2006) On the track of the Red Queen: bark beetles, their nematodes, local climate and geographic parthenogenesis. J Evol Biol 19: 1939–1947.

Mogie M (1992) The evolution of asexual reproduction in plants. Chapman and Hall, London

Mogie M, Britton NF, Stewart-Cox JA (2007) Asexuality, polyploidy and the male function. In: Hörandl E, Grossniklaus U, van Dijk P, Sharbel T (eds) Apomixis, evolution, mechanisms and perspectives. ARG Gantner Verlag KG, Lichtenstein, pp. 195–214

Moritz C, Heideman A (1993) The origin and evolution of parthenogenesis in *Heteronotia binoei* (Gekkonidae): Reciprocal origins and diverse mitochondrial DNA in western populations. Syst Biol 42: 293–306

Nogler GA (1971) Genetik der Aposporie bei *Ranunculus auricomus* s.l. W. Koch. I. Embryologie. Ber Schweiz Bot Ges 81: 139–179

Nogler GA (1984) Genetics of apospory in apomictic *Ranunculus auricomus*: 5. Conclusion. Bot Helv 94: 411–423

Nogler GA (1995) Genetics of apomixis in *Ranunculus auricomus*. VI. Epilogue. Bot Helv 105: 111–115

Noirot M, Couvet D, Hamon S (1997) Main role of self-pollination rate on reproductive allocations in pseudogamous apomicts. Theor Appl Genet 95: 479–483

Noyes RD (2007) The evolutionary genetics of apomixis in *Erigeron* sect. *Phalacroloma* (Asteraceae). In: Hörandl E, Grossniklaus U, van Dijk P, Sharbel T (eds) Apomixis: Evolution, mechanisms and perspectives. ARG Gantner Verlag KG, Lichtenstein, pp. 337–358

Noyes RD, Rieseberg LH (2000) Two independent loci control agamospermy (apomixis) in the triploid flowering plant *Erigeron annuus*. Genetics 155: 379–390

Orians CM (2000) The effects of hybridization in plants on secondary chemistry: implications for the ecology and evolution of plant-herbivore interactions. Am J Bot 87: 1749–1756

Otto SP, Nuismer SL (2004) Species interactions and the evolution of sex. Science 304:1018–1020

Palop-Esteban M, Segarra-Moragues JG, Gonzalez-Candelas F (2007) Historical and biological determinants of genetic diversity in the highly endemic triploid sea lavender *Limonium dufourii* (Plumbaginaceae). Mol Ecol 16: 3814–3827

Paun O, Greilhuber J, Temsch E, Hörandl E (2006a) Patterns, sources and ecological implications of clonal diversity in apomictic *Ranunculus carpaticola* (*Ranunculus auricomus* complex, Ranunculaceae). Mol Ecol 15: 897–910

Paun O, Lehnebach C, Johansson JT, Lockhart P, Hörandl E (2005) Phylogenetic relationships and biogeography of *Ranunculus* and allied genera in the Mediterranean and the European alpine system (Ranunculaceae). Taxon 54: 911–930

Paun O, Stuessy TF, Hörandl E (2006b) The role of hybridization, polyploidization and glaciation in the origin and evolution of the apomictic *Ranunculus cassubicus* complex. New Phytol 171: 223–236

Peck JR, Yearsley JM, Waxman D (1998) Explaining the geographic distributions of sexual and asexual populations. Nature 391: 889–892

Poelt J. (1970) Das Konzept der Artenpaare bei den Flechten. Deutsch Bot Ges 4: 187–198.

Pongratz N, Storhas M, Carranza S, Michiels NK (2003) Phylogeography of competing sexual and parthenogenetic forms of a freshwater flatworm: patterns and explanations. BMC Evol Biol 3: 23

Pound GE, Cox SJ, Doncaster CP (2004) The accumulation of deleterious mutations within the frozen niche variation hypothesis. J Evol Biol 17: 651–662

Quarin CL, Espinoza F, Martinez EJ, Pessino SC, Bovo OA (2001) A rise of ploidy level induces the expression of apomixis in *Paspalum notatum*. Sexual Plant Reprod 13: 243–249

Richards AJ (1997) Plant breeding. Chapman and Hall, London

Rieseberg L, Willis JH (2007) Plant speciation. Science 317: 910–914

Rieseberg LH, Raymond O, Rosenthal DM, Lai Z, Livingstone K, Nakazato T, Durphy J, Schwarzbach AE, Donovan LA, Lexer C (2003) Major ecological transitions in wild sunflowers facilitated by hybridization. Science 301: 1211–1216

Robinson MT, Weeks AR, Hoffmann AA (2002) Geographic patterns of clonal diversity in the earth mite species *Penthaleus major* with particular emphasis on species margins. Evolution 56: 1160–1167

Rousi A (1956) Cytotaxonomy and reproduction in the apomictic *Ranunculus auricomus* group. Ann Bot Soc Zool Bot FennVanamo 29:1–64

Rutishauser A (1954) Die Entwicklungserregung des Endosperms bei pseudogamen Ranunculusarten. Mitt Naturforsch Ges Schaffhausen 25: 1–45

Schön I, Gandolfi A, Di Masso E, Rossi V, Griffiths HI, Martens K, Butlin RK (2000) Persistence of asexuality through mixed reproduction in *Eucypris virens* (Crustacea, Ostracoda). Heredity 84: 161–169

Seehausen O (2004) Hybridization and adaptive radiation. Trends Ecol Evol 19: 198–207

Semlitsch RD, Hotz H, Guex GD (1997) Competition among tadpoles of coexisting hemiclones of hybridogenetic *Rana esculenta*: support for the Frozen Niche Variation model. Evolution 51: 1249–1261

Simon JC, Delmotte F, Rispe C, Crease T (2003) Phylogenetic relationships between parthenogens and their sexual relatives: the possible routes to parthenogenesis in animals. Biol J Linn Soc 79: 151–163

Soltis DE, Soltis PS, Schemske D, Hamcock JF, Thompson JN, Husband BC, Judd WS (2007) Autopolyploidy in angiosperms: have we grossly underestimated the number of species? Taxon 56: 13–30

Stebbins GL (1950) Variation and evolution in plants. Columbia University Press, New York

Stebbins GL, Dawe JC (1987) Polyploidy and distribution in the European flora: a reappraisal. Bot Jahrb Syst 108: 343–354

Strasburg JL, Kearney M (2005) Phylogeography of sexual *Heteronotia binoei* (Gekkonidae) in the Australian arid zone: climatic cycling and repetitive hybridization. Mol Ecol 14: 2755–2772

Tas ICQ, Van Dijk P (1999) Crosses between sexual and apomictic dandelions (*Taraxacum*). I. The inheritance of apomixis. Heredity 83: 707–714

Trewick SA, Morgan-Richards M, Chapman HM (2004) Chloroplast DNA diversity of *Hieracium pilosella* (Asteraceae) introduced to New Zealand: reticulation, hybridization, and invasion. Am J Bot 91: 73–85

Urbani MH (2002) Cytogeography and reproduction of the *Paspalum simplex* polyploid complex. Plant Syst Evol 236: 99–105

Van Dijk P (2003) Ecological and evolutionary opportunities of apomixis: insights from *Taraxacum* and *Chondrilla*. Philos Trans R Soc Lond B 358: 1113–1121

Van Dijk P, Tas ICQ, Falque M, Bakx-Schotman T (1999) Crosses between sexual and apomictic dandelions (*Taraxacum*). II. The breakdown of apomixis. Heredity 83: 715–721

Van Dijk P, van Baarlen P, de Jong JH (2003) The occurrence of phenotypically complementary apomixis-recombinants. Sex Plant Repr 16: 71–76

Van Dijk PJ (2007) Potential and realized costs of sex in dandelions, *Taraxacum officinale* s.l. In: Hörandl E, Grossniklaus U, van Dijk P, Sharbel T (eds) Apomixis: Evolution, mechanisms and perspectives. ARG Gantner Verlag KG, Lichtenstein, pp. 215–233

Van Dijk PJ, Vijverberg K (2005) The significance of apomixis in the evolution of the angiosperms:a reappraisal. In: Bakker F, Chatrou L, Gravendeel B, Pelser PB (eds) Plant species-level systematics: New perspectives on pattern and process. ARG Gantner Verlag KG, Lichtenstein, pp. 101–116

Vandel A (1928) La parthénogenese geographique. Contribution a l'étude biologique et cytologique de la parthénogenese naturelle. Bull Biol France Belg 62: 164–182

Vavrek JMC, McGraw JB, Yang HS (1998) Within-population variation in demography of *Taraxacum officinale* (Asteraceae): differential genotype response and effect on interspecific competition. Am J Bot 85: 947–954

Verduijn MH, Van Dijk P, van Damme JMM (2004) Distribution, phenology and demography of sympatric sexual and asexual dandelions (*Taraxacum officinale* s.l.): geographical parthenogenesis on small scale. Biol J Linn Soc 82: 205–218

Vorburger C, Sunnucks P, Ward SA (2003) Explaining the coexistence of asexuals with their sexual progenitors:no evidence for general-purpose genotypes in obligate parthenogens of the peach-potato aphid, *Myzus persicae*. Ecol Lett 6: 1091–1098

Vrijenhoek RC (1984) Ecological differentiation among clones: the frozen niche variation model. In: Woermann K, Loeschcke V (ed) Population biology and evolution, Springer, Berlin, pp. 217–231

Vrijenhoek RC (1994) Unisexual fish: model systems for studying ecology and evolution. Annu Rev Ecol Syst 25: 71–96

Vuille C, Küpfer P (1985) Aposporie chez le *Ranunculus parnassifolius*. L. I. Etude cytoembryologique. Bull Soc Neuchat Sci Nat 108: 123–134

Weber HE (1995) *Rubus*. In: Weber HE (ed) Gustav Hegi, Illustrierte Flora von Mitteleuropa, vol. IV 2A, ed. 3, Parey, Munich, pp. 284–595

Chapter 9
The Elusive Clone – In Search of Its True Nature and Identity

Koen Martens, Hugh D. Loxdale and Isa Schön

Abstract Sex in eukaryotes involves a combination of meiosis and syngamy; absence of these processes leads to asexual reproduction of which there are different kinds (Box 9.1). Lineages comprising subsequent generations of such asexually reproducing individuals are nearly invariably referred to as clones. In addition, they are mostly identified by genetic techniques using different molecular markers and with differing powers of resolution. Here, we demonstrate that various clonal concepts are being used without clear discrimination, which can significantly impede both the repeatability of, and comparisons between, studies. An example of the resulting dilemma is that, according to one concept, the more than 350 species of obligate asexual bdelloid rotifers all belong to one clone, whereas according to another concept, each of these species comprises several clones.

Box 9.1: Definitions of Sex and Asex

Sex is a combination of genetic processes that involve exchange of genetic material *between individuals*, mostly but not always, associated with reproduction. In Eukaryotes, sex is believed to involve the processes of meiosis and the subsequent fusion of these haploid nuclei, originating from different individuals. *Vegetative reproduction* occurs when one or more new individuals are formed from the soma of the parent organisms. *Parthenogenesis* is the process where new individuals are produced out of unfertilised eggs. In thelytokous parthenogenesis, daughters are produced out of diploid eggs or seeds. In arrhenotokous parthenogenesis, sons are produced through haploid eggs,

K. Martens (✉)
Royal Belgian Institute of Natural Sciences, Freshwater Biology, Vautierstraat 29, B-1000 Brussels, Belgium
e-mail: koen.martens@naturalsciences

I. Schön et al. (eds.), *Lost Sex*, DOI 10.1007/978-90-481-2770-2_9,

while the production of daughters requires fertilisation to obtain a diploid individual.

A whole array of processes intermediate between mitosis and meiosis are known. Some of these are grouped under the term *automixis*. Some automictic processes lead to homozygosity in offspring, other enhance heterozygosity (see Chapter 4 for more details).

Some authors use a different definition of sex, and also consider parthenogenesis to be a form of sex, because it involves procreation from the germline. In that case, only vegetative reproduction is considered to be fully asexual. This definition is here rejected, as it ignores the fundamental consequences of the absence of meiosis and syngamy.

9.1 What Are Clones?

Since the production of Dolly the sheep by synthetic cloning in the late 1990's, several domestic animals have been produced by cloning techniques – mice, cattle, pigs, horses, rats, cats, a dog and a mule. These animals were generated by transferring a nucleus from a somatic cell into an enucleate oocyte. Owing to epigenetic programming of the donor cell, this causes problems with the reproducibility of the technology as well as low productivity and high inefficiency (Pennisi and Vogel 2000; Aldhous and Coghlan 2006; see also Chapter 25). However, less well known is the fact that several of our agricultural crops such as corn, soybean, rapeseed and cotton are grown as transgenic clones, currently covering 81 million hectares in 17 countries (Wenzel 2006). The negative evolutionary consequences of using such crops, including a potentially higher susceptibility to parasites or diseases due to the loss of genetic diversity, are hardly realised by the scientific community in general, let alone a non-scientific audience.

Besides applied aspects, clones and cloning are an important topic in biology and one that relates to many diverse phenomena of considerable importance, both in the understanding of nature per se and issues such as the role of sex in evolution in particular. For sure, cloning is a successful strategy when produced naturally. A majority of phyla (over 50%) have unisexual taxa (Hull 1988; Hughes 1989; Simon et al. 2003; see also Chapter 4). Many groups, including the so-called ancient asexuals, have been reproducing asexually for millions of years (see Chapters 11, 12 and 13), while even animals with asexual phases such as aphids (see also Chapter 25) or *Daphnia* (see also Chapter 15) can be asexual for numerous, perhaps hundreds of generations.

In the case of parthenogenesis (as opposed to vegetative asexuality – see Box 9.1), the complexity of this form of reproduction may differ between taxa in terms of fundamental genetics as well as in life cycle (see several articles in Loxdale and Lushai 2003a and Chapters 11–28 of this book). Indeed, there is not just one type of clone. To further complicate matters, this topic is rife with philosophical complexities that are in large part semantic as discussed below. Many of the current definitions of "clone" are as confusing as they are inadequate. Whilst the general view of *genetic fidelity* is well-established in the literature, little empirical evidence exists for or against this concept and belief. Modern, high resolution molecular (DNA) markers, which seemed capable of answering the question "what exactly is a clone?", may to a large extent confuse the issue, to the point that again, semantics and definitions snatch the concept from us and only an operational model remains, for what that is worth. Here, we consider these questions, including clonal concepts, germ-line versus soma issues and theoretical considerations.

9.2 A Brief History of Clonal Concepts

Asexuality in plants and animals has been known for a very long time. Aristotle (384-322 BC; www.iep.utm.edu/a/aris-bio.htm) was aware of asexuality in some taxa, such as fishes, whilst gardeners have been propagating plant cuttings for many centuries. Only in the late 18th century was the phenomenon of parthenogenesis studied scientifically. For example, Anthony van Leeuwenhoek (1632–1723) showed that aphids could propagate by means of viviparous reproduction, whilst Charles Bonnet (1720–1793) demonstrated such reproduction to be parthenogenetic (Bonnet 1745). In the early 20th century, Hebert J. Webber (1865–1946; Fig. 9.1), a plant physiologist at the Citrus Experiment Station at UC Riverside California, USA, was the first to coin the term "clon". It is derived from the Greek "*klon*", meaning a twig or cutting and Webber used it in relation to vegetative propagation of a single plant (Webber 1903). Since then, many attempts have been made to define clones (see for example McLaren 2000). Initially, clones were defined on morphological characters and in the case of fruits, on aspects such as yield, flavour,

Fig. 9.1 Herbert John Webber (1865 – 1946), plant physiologist and father of the clone

shelf life and other traits that could be quantified and shared as a property of uniformity (but see Gill et al. 1995 and below). Perhaps this was also a reflection of our industrial interest and ability to produce man-made objects "clonally" by the technology of mass production going back to the mid 19th century.

This philosophical construct was rooted in the belief that whatever a clone is , it shares the property of being derived from a common ancestral mother. In the case of eukaryotes, this was the last sexually produced ancestor prior to the proliferation of asexual generations or "vertical clones" (i.e. between generations). Normally, mammalian clones e.g., monozygotic twins, are in effect "horizontal clones" (i.e. within generations). With synthetic mammalian clones, more especially humans, the practice has of course severe ethical ramifications (see Chapter 26).

In the late 20th century, with the onset of molecular markers, the focus centred on the use of these markers to define "genetic identity" by qualifying and quantifying genotypes, initially single locus, later multi-locus. Much important and useful work was done using allozymes and valuable information on clones gained, including their fundamental biology and population dynamics (e.g. *Daphnia*: Hebert et al., 1988; aphids: Loxdale and Brookes 1990; Simon and Hebert1995; see also Loxdale 1994).

With the widespread application of high-resolution DNA markers utilizing mitochondrial DNA and nuclear sequences (ribosomes, RAPDs, microsatellites and AFLPs), the amount of polymorphism at various regions of the eukaryotic genome was further assessed, and the nature and reality of the clonal genome thereby further elucidated. Again, new insights were gained from these novel approaches (e.g. plants: Van der Hulst et al. 2000; Paun et al. 2006; fungi: Gryta et al. 2006; ostracods: Schön et al. 1998, 2000; *Daphnia*: Hebert et al. 2003; monogonont rotifers: Gomez 2005; aphids: Loxdale and Lushai 2007; for the use of molecular markers generally to assess the genetic structure of clonal and partially clonal animals, see Halkett et al. 2005).

However, as became apparent when these various approaches were applied to identify the clone in terms of genotypes, the concept of the clone disappeared in direct proportion to the resolving power of the technique applied ("resolution phenomenon"; see Loxdale and Lushai 2003b). Clearly, the final proof of genetic fidelity is partial or complete sequencing of the nuclear and mitochondrial genome. But here, the nonsense begins as the definition is a victim of the approach used to address it. Furthermore, a clone, whatever it is in terms of its primary DNA sequence (and even here it may vary due to tandem repeat differences, e.g. microsatellite or ribosomal DNA intergenomic spacer repeat numbers; Fenton et al. 2005), may additionally show intraclonal variation due to epigenetic programming and expression. Good examples are the highly insecticide-resistant strains of the Peach-potato aphid, *Myzus persicae* (Sulzer), where expression waxes and wanes depending on pesticide selection pressures (Field and Blackman 2003; see also Chapter 25), or bdelloid rotifers, where the age of the mother influenced protein expression patterns of daughters from the next generation (Ricci 1999; see also Chapter 13). Even single nematode worms, *Caenorhabditis elegans*, (from predominantly hermaphroditic and self fertilising lineages, hence largely homozygous and effectively a "clone") have recently been shown to display heterogeneity for telomere repeat length at chromosome VL (possibly related to telomerase-mediated or end-replication losses, although the exact mechanism/s responsible for such heterogeneity is/are still not clear) (Cheung et al. 2004). So, whilst multilocus genotypes and sequences may provide strong clues as to relatedness and the origin of so-called clones, the entity as such cannot be easily defined.

9.3 Germline Versus Soma: The Weismannian Doctrine Revisited

Whereas Charles Darwin (1809–1882) proposed a largely Lamarckian system of inheritance, August Weismann (1834–1914) postulated a strict division between germline and somatic cell lineages (see also Chapter 2). He was the first to postulate that genetic inheritance between generations is restricted to germline cells, and that all acquired mutations in the soma are lost upon the death of the organism (Löther 1989). His theory was revived almost a century later by Kirkwood and Holliday (1979) and is now commonly known as the "disposable soma theory" (see also Lankenau 2008 and Schön et al. 2008 for recent discussions on the link between germ line and asexuality). In animals, nuclear germline is defined very early on in ontogeny. In worms, insects and all animals up to anurans, the germ plasma is already set aside during oogenesis. In urodeles, reptiles, birds and mammals, germline cells are distinguished after gastrulation (Hilscher 1999), and in preformistic development (Hilscher 1999), by changes in DNA methylation that regulate gene expression of primordial germline-specific genes (Maatouk et al. 2006). Germline cells remain relatively protected with up to 100 times lower mutation frequencies and frequencies of mitotic recombination (e.g. in embryonic stem cells; Hong et al. 2006) and by mechanisms, which eliminate germline cells with mutational burdens (Hong et al. 2006). Metabolic processes in germline cells are furthermore limited compared to specialised cells such as brain or liver cells, and nuclear germline DNA is at the highest risk of accumulating mutations during meiosis when it produces gametes.

The somatic cells of eukaryotes produce proteins (both structural and enzymatic) during the entire ontogeny of the organism and afterwards throughout adult life and continue to undergo regular cell divisions. Therefore, somatic nuclear DNA has potentially higher mutation accumulation rates. In mammalian somatic cells, the in vivo rates have been estimated to be around 10^{-4} per gene per generation (Hong et al. 2006). The different mutation rates between soma and germline also become obvious by comparing female and male mammals with long generation times (Goetting-Minesky and Makova 2006). The male mutation bias in male perissodactyls (odd-toed mammals) is 3.88 (indicating around four times more mutations in the screened regions of the male Y chromosome; Goetting-Minesky and Makova 2006); this bias is mainly due to the higher number of cell divisions during male spermatogenesis, a subsequent process after pro-spermatogenesis that is lacking in females (Hilscher 1991). In addition, somatic mutation rate can also be much more influenced by environmental factors than germline mutation accumulation rates, as germline cells are generally located inside the body, thereby better protecting them against UV or other external mutational sources.

Parthenogenetic reproduction occurs via the germline. Therefore, this mode has a higher chance of maintaining genetic fidelity than vegetative reproduction, which has the potential to pass on somatic mutations over generations (Gill et al. 1995; Fig. 9.2). Plants lack a clearly defined germline during early developmental stages; the cells producing gametophytes only differentiate in the adult plant (Scott and Spielman 2006; see also Chapter 3), so even parthenogenetic reproduction mostly uses seed precursor cells derived from the soma. Bermuda grass, *Cynodon dactylon* (L.), which propagates vegetatively, has been found to have as many as ten somatic mutations per triploid genome per generation (Caetano-Anolles 1999). Vegetative reproduction will thus potentially have higher mutation accumulation rates and parthenogenetic and vegetative clones can therefore have different evolutionary histories. Maybe the lack of germline and the resulting inferior genetic fidelity over

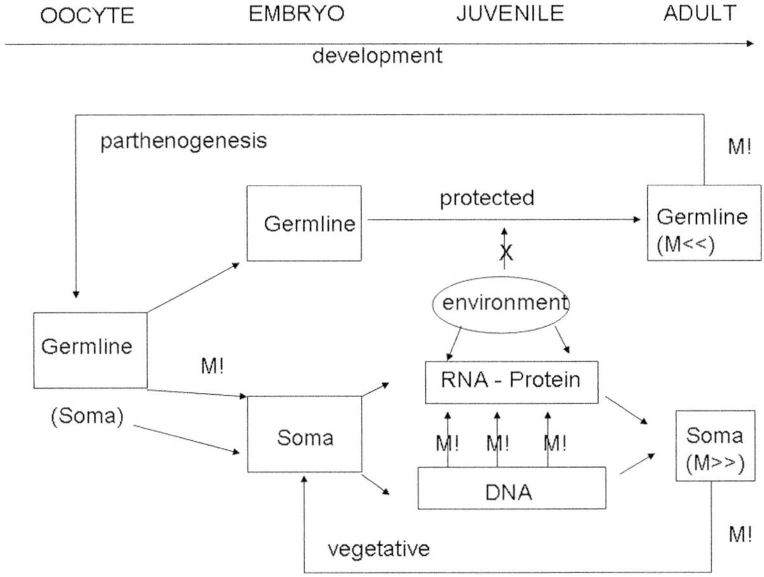

Fig. 9.2 Diagram illustrating different mutation accumulation rates in germline and soma during ontogenetic development in animals. Germline is defined very early on in ontogeny (around the 4th cleavage) and after that remains relatively protected, therefore its mutation accumulation rate is low. Somatic cells produce proteins during the entire ontogeny and also afterwards continuously during the adult life and therefore have a higher mutation accumulation rate. *Parthenogenesis* reproduces through the germline and therefore has a higher chance for genetic fidelity over generations than *vegetative reproduction*, which has the potential to pass on somatic mutations. M! stands for mutation risks

generations are responsible for the fact that thus far, no putative ancient asexual lineages have been identified in plants (Docking et al. 2006; see also Chapter 3). Any models predicting early extinction of clonal lineages through the accumulation of (deleterious) mutations would have to incorporate significantly higher rates in groups with vegetative asexuality, due to the potential incorporation of somatic mutations.

Because somatic mosaicism is common in plants, a genetic mosaicism hypothesis (GMH) has been proposed such "that arborescent plants accumulate spontaneous mutations and become genetically mosaic as they grow" (Gill et al. 1995; Klekowski 2003). According to Gill et al. (1995), spontaneous mutants occur in 0.1–19% of asexual plants. Like plants, many clonal animals (i.e. spread across 19 phyla and including cnidarians, platyhelminths, bryozoans, annelids and entoprocts) violate Weismann's doctrine; they "either do not sequester germline cells at all or do so later in development (epigenesis)". In such animals, it is possible for somatic mutations to find their way into the germ line and be inherited. In these animals lacking preformistic development, the GMH should apply too, but to date no estimates of mutant frequencies or mutation rates within colonies of clonal invertebrate animals are available.

The discovery of Numts or nuclear copies of mitochondrial sequences are especially troublesome to systematics because of their likelihood of confusing phylogenies based on mitochondrial DNA sequence divergence (see for example Sorenson and Quinn 1998; Song et al. 2008). The existence of Numts also demonstrates that even at the DNA level, the nuclear and mitochondrial genome are not necessarily entirely separated, hence again violating Weismann's doctrine to some degree. This could have a bearing on attempts to assess the genetic fidelity of clones using mitochondrial markers.

9.4 Clonal Concepts

Most recent taxonomic revisions indicate which of the more than 25 extant species concepts are applied in their studies (Mayden 1997; George and Mayden 2005). This is important to understand many of the decisions involved in recognising certain specific categories. Sadly, the intrinsic relevance of this practice has not yet been recognised by students of asexual organisms. The term "clone" is nearly invariably used without any discrimination and without clear definition of its content. A review of recent literature on parthenogenesis reveals that at least five very different concepts of clones and clonality are used, and often one paper applies more than one concept, but without explicitly stating so. Box 9.2 provides an overview of these concepts, with brief definitions. In addition, in order to visualise the implications of these concepts, the consequences for the status of clones versus species of the putative ancient asexual group Bdelloidea (with more than 350 morphologically recognised species) is given.

Box 9.2: Brief Description of Extant Clonal Concepts

Most clonal concept names refer to names of species concepts; this does not imply that clonal and species concepts must be completely philosophically congruent.

1. *Genetic identity* or *cohesion* clonal concept

   Definition: A clone consists of genetically identical individuals.

   Prediction: One species of bdelloid rotifers comprises several clones.

   Remarks: This concept sometimes implies genealogical relatedness amongst individuals, but this is not always explicitly stated. The name is inspired by Templeton's (1989) Cohesion concept for species.

2. *Last sex event* or *evolutionary* clonal concept

   Definition: A clone consists of all offspring of a single female since the last sexual event.

   Prediction: All bdelloids (>350 presently recognised species) are comprised into one clone.

   Remarks: The name originates from Wiley's (1978) Evolutionary species concept.

3. *Morphological* clonal concept

   Definition: A clone consists of individuals that are morphologically identical.

   Prediction: Bdelloid clones and species will be congruent.

4. *Phylogenetic* clonal concept

   Definition: A clone is a monophyletic cluster of (genetically and morphologically) (very) similar lineages.

   Prediction: Some bdelloids species might have several clones, some bdelloid species might be distributed over different clones, possibly even widely spaced over the phylogram.

   Remarks: The concept refers to Cracraft's (1983) Phylogenetic species concept.

5. *Molecular* or *genotypic* clonal concept

   Definition: A clone is a group of individuals amongst which no genetic difference can be found with the genetic markers at hand.

   Prediction: depending on which markers are used, bdelloid species will have a variable number of clones, or one clone might comprise several species.

Remarks: Although this is largely technique-based, there is an underlying implied concept that the use of this marker will identify clones.

A 6th concept might be termed the *Ecological concept*, where genotypically-indistinguishable groups of individuals might occupy different ecological niches. However, at present we know of no clear cut example of such a situation, and this therefore remains to date a purely hypothetical possibility (See also http://plato.stanford.edu/entries/species/).

The results are most surprising. Depending on which concept is used, all bdelloids might be seen as forming one clone or each (morphologically determined) bdelloid species might comprise several clones (see also Birky et al. 2005 and Chapter 10). Surely, such discrepancy deserves much attention, if the true meaning of the word "clone" in any primary research paper is to be correctly interpreted. This also appears to apply to medically relevant, host-associated dermatophyte fungi in which, as the authors state "[there] . . . is reason to suggest that clonal dermatophyte lineages are able to maintain ongoing populations and to follow independent evolutionary trajectories" (Graser et al. 2006). Yet, researchers often start by formulating their null hypothesis in which clones are (mostly implicitly) seen as "last sexual event" lineages, after which the hypothesis is tested in a way that uses clones as identified by whatever molecular marker at hand (allozymes, RAPDs, AFLPs, direct sequencing, etc). The incongruence between the concepts of the hypotheses and those used in the tests are crucial, because the null hypotheses rely on inherent *similarity* between individuals, while the tests determine empirical *differences* between individuals, to identify clones. However, the absence of observed differences does not necessarily equate with similarity. In effect, this is the philosophical construct evoked by Karl Popper (1902–1994) in relation to the falsification of hypotheses (Magee 1997).

9.5 Validity of the Clonal Concepts

Table 9.1 applies the three criteria used by Hull (1997) to compare the relevance of extant species concepts. These criteria are *universality* (is the concept applicable throughout the living world, or is it only relevant to a subset?), *applicability* (how easy is it to apply this concept in routine investigations?) and *theoretical content* (is the concept purely pragmatic, or is it underpinned by biological reality?).

It is immediately apparent that none of the five clonal concepts scores high on all three criteria. For example, the *Genotypic clonal concept* is highly applicable, but scores low on theoretical content. The *Cohesion clonal concept*, on the contrary, has the most powerful theoretical content, but is effectively not applicable in the field.

Table 9.1 Test of validity of five clonal concepts using Hull's (1997) criteria

			Applicability	
Clonal concept	Universality	Theoretical	Field	Lab
Cohesion	3	1	1	2
Evolutionary	3	3	1	2
Morphological	1	1	1	1
Phylogenetic	3	3	1	2
Genotypic	2	1	3	3

Scores are between 1 and 3.

The *Cohesion clonal concept*, implying genetic identity, is intuitively the most commonly used concept, and even Stephen J. Gould (1941–2002) probably had this concept in mind when he asked if a clone is "an aggregate of individuals or one gigantic evolutionary body?" (Gould 2002), a question very much related to that of the nature of individuality (see also Janzen 1977 and Chapters 22 and 25 for examples). However, genetic fidelity is a myth, and even direct (first generation) parthenogenetic offspring of one female will most likely already have incorporated one or even several mutations (Loxdale and Lushai 2003b; Vorwerk and Forneck 2007). One might consider relaxing this requirement of strict identity, but how much interclonal variability should than be allowed: one nucleotide? Or one amino acid? Or should functionality of the entire protein be the cut-off point? Even if the theoretical wisdom of such relaxations could be determined, the practicality of such an approach would be minimal.

The *Phylogenetic clonal concept* (a clone is a monophyletic clade) also requires an arbitrary cut-off point to define clones, in this case most often defined as 1, 5 or 10% divergence of a clade, although Birky et al. (2005 and Chapter 10) provide a model, in which population genetic parameters are used to separate clades. In addition, and most importantly, many clones have polyphyletic origins, for example through multiple hybridisation events (e.g. Schön et al. 2000; Vorburger 2001; Delmotte et al. 2003).

The *Evolutionary clonal concept* (the last sex event) runs especially astray in long-term asexuality, where clones will become vast reservoirs of both genetic and morphological variation, and will loose all intuitive and practical content. In addition, the identification of morphological clones can in some cases be useful, but epigenetic and reaction norm effects will ensure that identical genomes might produce very different phenotypes (Yin et al. 1999). This concept is the most vulnerable to such phenotypic effects.

Lastly, the *Genotypic clonal concept* will be as reliable as the method which is applied and unless whole genome sequences are available, even multilocus marking techniques will always be prone to underestimation of clonal variability. More importantly, one will mostly be clueless as to the degree of intraclonal variability allowed by such a method-based concept, while linking clones derived by different genotyping techniques might not be straightforward. In addition, this clonal concept is most prone to suffer from effects of polyploidy.

9.6 Conclusions

Perhaps the simile to draw from clones and clonal studies is the fundamental nature of the atom; the more this has been delved, the more mysterious it has been found to be, to the extent that it becomes almost metaphysical (Harrison and Dunham 1999). The clone resists total understanding, because its fundamental nature is still unclear; it is a multi-facetted phenomenon and it tends to change as each facet is explored in term. When models of speciation are applied to concepts of clonality as shown above, many of these models are applicable to one or a few criteria, but none to them all. Because of this, our strongest take-home message is that for any study using clones, the concept adopted needs to be clearly stated and emphasised. At least by so doing, it will then be possible to determine which particular studies are themselves open to comparison and which are not. At the very least, theoretical (hypothesis tested) and practical (genotyping) clonal concepts should be the same in any paper.

The fact that the origins of clones and the mechanisms of clonality are so diverse means probably that any universal concept of the clone will remain elusive, if not unobtainable. Even so, workable and testable approaches will remain in evidence, since they continue to provide valuable insights into the workings of the natural world.

We have not sought to sink the clone as a concept, only to provide caution to its study and employment in both the laboratory and the field. Whereas renaming clonally derived mutations as "*clonotype*" or "sub-clone" maybe more scientifically accurate (especially in terms of genetic fidelity; Lushai and Loxdale 2002; Fenton et al. 2005), these are not particularly useful terms. The clone celebrated its centenary in 2003. Whether it will live to celebrate another such centenary in an unrevised form seems unlikely in the light of the philosophical and technical aspects outlined here.

Acknowledgements KM and IS wish to thank the EC (Project no. MRTN-CT-2004-512492) and the Belgian OSTC (project no. MO/36/015) for funding. HDL was supported by an EU Marie Curie Senior Research Fellowship (TOK FP6-029983). We thank Marty Nemeth of the University of California Riverside Science Library (Special Collections) for kindly providing the photograph of H. J. Webber.

References

Aldhous P, Coghlan A (2006) Ten years on – has the cloning dream died? New Scientist 2558: 01 July, 8–10

Birky CW, Wolf C, Maughan H, Hebertson L, Henry E (2005) Speciation and selection without sex. Hydrobiologia 546:29–45

Bonnet C (1745). *Traité d'insectologie*. 2 parts. Durand, Paris

Caetano-Anolles G (1999) High genome-wide mutation rates in vegetatively propagated bermudagrass. Mol Ecol 8: 1211–1221

Cracraft J (1983) Species concepts and speciation analysis. Curr Ornithology 1: 159–187

Cheung I, Schertzer M, Baross A, Rose AM, Lansdorp PM, Baird DM (2004) Strain-specific telomere length revealed by single telomere length analysis in *Caenorhabditis elegans*. Nucleic Acids Res 32: 3383–3391

Delmotte F, Sabater-Muñoz B, Prunier-Leterme N, Latorre A, Sunnucks P, Rispe C, Simon J-C (2003) Phylogenetic evidence for hybrid origins of asexual lineages in an aphid species. Evolution 57: 1291–1303

Docking TR, Saade FE, Elliott MC, Schoen DJ (2006) Retrotransposon sequence variation in four asexual plant species. J Mol Evol 62: 375–387

Fenton B, Malloch G, Woodford JAT, Foster SP, Anstead J, Denholm I, King L, Pickup J (2005) The attack of the clones: tracking the movement of insecticide resistant peach–potato aphids *Myzus persicae* (Hemiptera: Aphididae). Bull Entomol Res 95: 483–494

Field LM, Blackman RL (2003) Insecticide resistance in the aphid *Myzus persicae* (Sulzer): Chromosome location and epigenetic effects on esterase gene expression in clonal lineages. Biol J Linn Soc 79: 107–113

George AL, Mayden RL (2005) Species concepts and the Endangered Species Act: How a valid biological definition of species enhances the legal protection of biodiversity. Nat Res J 45: 369–407

Gill DE, Chao L, Perkins SL, Wolf JB (1995) Genetic mosaicism in plants and clonal animals. Ann Rev Ecol Syst 26: 423–444.

Goetting-Minesky MP, Makova KD (2006) Mammalian male mutation bias: impacts of generation time and regional variation in substitution rates. J Mol Evol 63: 537–544.

Gomez A (2005) Molecular ecology of rotifers: from population differentiation to speciation. Hydrobiologia 546: 83–99

Gould SJ (2002) The structure of evolutionary theory. Harvard University Press, Harvard, pp. 807–810

Graser Y, De Hoog S, Summerbell RC (2006) Dermatophytes: recognising species of clonal fungi. Med Mycol 44: 199–209.

Gryta H, Carriconde F, Charcosset JY, Jargeat P, Gardes M (2006) Population dynamics of the ectomycorrhizal fungal species *Tricholoma populinum* and *Tricholoma scalpturatum* associated with black poplar under differing environmental conditions. Environ Microbiol 8: 773–786

Halkett F, Simon J-C, Balloux F (2005) Tackling the population genetics of clonal and partially clonal organisms. Trends Ecol Evol 20: 194–201

Harrison S, Dunham P (1999) Practical inadequacy or inadequate practice? Quantum theory, 'reality' and the logical limits to realism. Trans Inst British Geograph 24: 236–242

Hebert PDN, Witt JDS, Adamowics SJ (2003) Phylogeographical patterning in *Daphnia ambigua*: Regional divergence and intercontinental cohesion. Limnol Oceanogr 48: 261–268.

Hebert PDN, Ward RD, Weider LJ (1988) Clonal-diversity patterns and breeding-system variation in *Daphnia pulex*, an asexual-sexual complex. Evolution 42: 147–159

Hilscher W (1991) The genetic control and germ cell kinetics of the female and male germ line in mammals including man. Hum Reprod 6: 1416–1425

Hilscher W (1999) Some remarks on the female and male Keimbahn in the light of evolution and history. J Exp Zool 285: 197–214

Hong Y, Cervantes RB, Tichy E, Tischfield JA, Stambrook PJ (2006) Protecting genomic integrity in somatic cells and embryonic stem cells. Mut Res 614: 48–55

Hughes R N (1989) A functional biology of clonal animals. Chapman and Hall, London

Hull D (1988) Science as a Process. University of Chicago Press, Chicago.

Hull D (1997) The ideal species concept – and why we can't get it. In: Claridge MF, Dawah HA, Wilson MR (eds) Species: The units of biodiversity, Chapman and Hall, London, pp. 357–377

Janzen DH (1977) What are dandelions and aphids? Am Nat 111: 586–589

Kirkwood TB, Holliday R (1979) The evolution of ageing and longevity. Proc R Soc Lond B 205: 531–546

Klekowski EJ (2003) Plant clonality, mutation, diplontic selection and mutational meltdown. Biol J Linn Soc 79: 61–67

Lankenau D-D (2008) The legacy of the germ line – maintaining sex and life in Metazoans: cognitive roots of the concept of hierarchical selection. In: Egel R, Lankenau D-H (eds) Genomic dynamics and stability, Vol. 2: Meiosis and recombination. Crossing over and disjunction. Springer, Berlin, Heidelberg, pp. 289–339

Löther R (1989) Wegbereiter der Genetik: Gregor Johann Mendel und August Weismann. Urania, Leipzig/Jena

Loxdale HD (1994) Isozyme and protein profiles of insects of agricultural and horticultural importance. In: Hawksworth DL (ed) The identification and characterization of pest organisms, CABI, Wallingford, Oxford, pp. 337–375

Loxdale HD, Brookes CP (1990) Genetic stability within and restricted migration (gene flow) between local populations of the blackberry-grain aphid *Sitobion fragariae* in south-east England. J Anim Ecol 59: 495–512

Loxdale HD, Lushai G (eds) (2003a) Intraclonal genetic variation: Ecological and evolutionary aspects, Proceedings of the joint Royal Entomological Society-Linnean Society Symposium, Biol J Linn Soc, 79: 1–208

Loxdale HD, Lushai G (2003b) Rapid changes in clonal lines: the death of a 'sacred cow'. Biol J Linn Soc 79: 3–16

Loxdale HD, Lushai G (2007) Population genetic issues: the unfolding story revealed using molecular markers. In: van Emden HF and Harrington R (eds) Aphids as crop pests, CABI, Wallingford, Oxford, pp. 31–67

Lushai G, Loxdale HD (2002) The biological improbability of a clone (mini-review). Genet Res 79: 1–9

Maatouk DM, Kellam LD, Mann MR, Lei H, Li E, Bartolomei MS, Resnick JL (2006) DNA methylation is a primary mechanism for silencing postmigratory primoridial cell genes in both germ cell and somatic cell lineages. Development 133: 3411–3418

Mayden RL (1997) A hierarchy of species concepts: the denouement in the saga of the species problem. In: Claridge MF, Dawah HA, Wilson MR (eds) Species: The units of biodiversity, Chapman and Hall, London, pp. 381–424

Magee B (1997) Popper (Modern Masters), Fontana Press, London

McLaren A (2000) Cloning: pathways to a pluripotent future. Science 288: 1775–1780

Paun O, Greilhuber J, Temsch EM, Hörandl E (2006) Patterns, sources and ecological implications of clonal diversity in apomictic *Ranunculus carpaticola* (*Ranunculus auricomus* complex, Ranunculaceae). Mol Ecol 15: 897–910

Pennisi E, Vogel G (2000) Clones: a hard act to follow. Science 288: 1722–1727

Ricci C (1999) Epigenetic inheritance systems in bdelloid rotifers. I. Maternal-age-related biochemical effects. Ital J Zool 66: 333–339

Schön I, Butlin RK, Griffiths HI, Martens K (1998) Slow molecular evolution in an ancient asexual ostracod. Proc R Soc Lond B 265: 235–242

Schön I, Gandolfi A, Di Masso E, Rossi V, Griffiths HI, Martens K, Butlin RK (2000) Long-term persistence of asexuality through mixed reproduction in *Eucypris virens*. Heredity 84: 161–169

Schön I, Lamatsch D, Martens K. (2008) Lessons to learn from ancient asexuals. In: Egel R, Lankenau D-H (eds) Genomic dynamics and stability, Vol. 2: Meiosis and recombination. Crossing over and disjunction. Springer Publishers, Berlin, pp. 341–376

Scott RJ, Spielman M (2006) Genomic imprinting in plants and mammals; how life history constrains convergence. Cytogenet Genome Res 113: 53–67

Simon J-C, Hebert PDN (1995) Patterns of genetic variation among Canadian populations of the bird cherry-oat aphid, *Rhopalosiphum padi* L. (Homoptera: Aphididae). Heredity 74: 346–353

Simon J-C, Delmotte F, Rispe C, Crease T (2003) Phylogenetic relationships between parthenogens and their sexual relatives: the possible routes to parthenogenesis in animals. Biol J Linn Soc 79: 151–163

Song H, Buhay JE, Whiting MF, Crandall KA (2008) Many species in one: DNA barcoding overestimates the number of species when nuclear mitochondrial pseudogenes are coamplified. Proc Natl Acad Sci USA 105:13486–13491

Sorenson MD, Quinn TW (1998) Numts: a challenge for avian systematics and population biology. Auk 115: 214–221

Templeton A (1989) The meaning of species and speciation. In: Otte D, Endler JA (eds) Speciation and its consequences. Sinauer Associates, Sunderland, pp. 3–27

Van der Hulst RGM, Mes THM., den Nijs JCM, Bachmann K (2000) Amplified fragment length polymorphism (AFLP) markers reveal that population structure of triploid dandelions (*Taraxacum officinale*) exhibits both clonality and recombination. Mol Ecol 9: 1–8

Vorburger C (2001) Fixation of deleterious mutations in clonal lineages: evidence from hybridogenetic frogs. Evolution 55: 2319–2332

Vorwerk S, Forneck A (2007) Analysis of genetic variation within clonal lineages of grape phylloxera (*Daktulosphaira vitifoliae* Fitch) using AFLP fingerprinting and DNA sequencing. Genome 50: 660–667

Webber HJ (1903) New horticultural and agricultural terms. Science 18: 501–503, See also: http://lib.ucr.edu/agnic/webber/Vol1/vol1dedication.html

Wenzel G (2006) Molecular plant breeding: achievements in green biotechnology and future perspectives. Appl Microbiol Biotech 70: 642–650

Wiley EO (1978) Evolutionary species concept reconsidered. Syst Zool 27: 17–26

Yin Y, Geiger W, Martens K (1999) Effects of genotype and environment on phenotypic variability in *Limnocythere inopinata* (Crustacea, Ostracoda). Hydrobiologia 400: 85–114

Chapter 10
Asexual Speciation

C. William Birky Jr. and Timothy G. Barraclough

Abstract Population and evolutionary genetic theory shows that asexual organisms can undergo speciation to form inclusive populations that are independent arenas for the evolutionary processes of mutation, selection, and random drift; in other words, evolutionary species. Asexual speciation occurs when members of a species become physically isolated from each other for a long time or undergo divergent selection for adaptation to different niches. Such species form genotypic clusters separated by long-lasting gaps, as opposed to transient gaps due to random genetic drift. Species will often be phenotypic clusters as well, but these clusters may be cryptic; such cryptic species will be detectable only by genetic means. Our theoretical model of the nature and origin of species suggests two different methods of assigning individuals to species using gene sequence data. One combines population genetics and sampling theory to identify clusters that are separated by gaps too deep to be caused by drift. The other method uses the change in branching rates of lineages in a phylogenetic tree to detect the transition from between-species to within-species branching. These species criteria have been used to demonstrate that bdelloid rotifers and oribatid mites have undergone speciation; the two criteria show a reassuring amount of agreement in delimiting species. Not surprisingly, some of the resulting species are cryptic, not distinguishable by phenotype. At least some of these asexual species are adapted to different ecological niches. Our species concept and criteria can be used to test theories about the population genetics and evolutionary diversification of asexual organisms.

10.1 The Importance of Asexual Species

There is a general consensus among biologists that species are real and important units of biological diversity, and understanding the mechanisms of speciation is a hot research topic (Coyne and Orr 2004). Nearly all of that work is focused on species

---

C. William Birky Jr. (✉)
Department of Ecology and Evolutionary Biology, The University of Arizona, Tucson, AZ 85721 USA
e-mail: birky@u.arizona.edu

I. Schön et al. (eds.), *Lost Sex*, DOI 10.1007/978-90-481-2770-2_10,

and speciation in animals and plants that reproduce sexually. But it is also critically important to understand species and their origins in asexual organisms such as parthenogenetic animals and plants. A substantial part of the earth's biological diversity consists of organisms that reproduce only asexually so far as we know, or reproduce sexually very rarely or with extreme inbreeding so that they are effectively asexual or clonal. Asexual organisms include not only the parthenogenetic animals and plants that are the subject of this book, but also many fungi; some green algae, including the large and ancient genus *Chlorella*; and many other eukaryotic protists, including possibly all of the members of one of the five major groups of eukaryotes, the Excavates. In addition to eukaryotes, many bacteria are largely asexual, although levels of homologous recombination vary and may be high enough to confer evolutionary advantages of sex in many groups (Cohan 2004; Ochman et al. 2005). Many asexual eukaryotes and prokaryotes are medically important parasites (see Chapter 24; see also Ayala 1998), agricultural pests (see Chapter 25), or invasive species (Mergeay et al. 2006).

A deep understanding of what asexual species are and how to find them is a prerequisite for testing theories about the evolutionary advantage of sex, i.e. about the evolutionary consequences of losing sex. For example, theory predicts that asexual organisms should accumulate a higher load of detrimental mutations (Muller's ratchet (Muller 1964) and Kondrashov's hatchet (Kondrashov 1982); see also Chapter 5) than do sexual organisms, resulting in a higher extinction rate (reviewed by Otto and Lenormand 2002). It is useful to test for the accumulation of detrimental mutations along evolutionary lineages between species as well as within species. Theory also predicts that asexuals should be less able to fix advantageous mutations (Orr 2000), perhaps resulting in both higher rates of extinction in changing environments and lower rates of speciation. Testing this theory obviously requires models of species (species concepts) and operational definitions (species criteria) that identify comparable evolutionary units in asexuals and sexuals.

Even determining whether a group of organisms is asexual requires knowing species boundaries. Some methods use population genetic analyses to look for linkage disequilibrium, while others look for incongruence in the phylogenetic trees of different genes within a species. Rare cases where gene trees are incongruent could be due to infrequent outcrossing and recombination within a species, or to horizontal transfer between species; only the former is sexual reproduction, although horizontal transfer may play an important role in ecological speciation in bacteria (Cohan 2004) or even some eukaryotes (Gladyshev et al. 2008). Clearly the use of population genetic methods to test whether a species is sexual or asexual is facilitated by having a species concept that is compatible with both asexuals and sexuals.

In spite of the importance of asexual organisms and species, the nature of species is even more controversial in asexual organisms than it is in sexuals; even the existence of species in asexual organisms has been questioned (Maynard Smith and Szathmary 1995). In this review we focus on the question of when and how speciation can occur in parthenogenetic animals and plants and other asexuals and on how species can be identified. We will show that asexual organisms are similar to sexual organisms in that they can undergo allopatric speciation due to physical isolation or sympatric speciation as a result of divergent selection for adaptation to different habitats. We also describe two different criteria for assigning individuals to species, and compare the resulting species to sexual species defined by reproductive isolation or other criteria.

10.2 General Theory of Speciation and Species in Asexuals

Theoretical studies of the evolutionary consequences of losing sex (aka the evolutionary advantage of sex) have shown that asexual organisms should accumulate more detrimental mutations than do sexual species and in consequence should have a higher rate of extinction. Also asexual organisms should be less able to fix advantageous mutations, which can also contribute to extinction in a changing environment, as well as making it more difficult to speciate by adapting to different niches. To test these theories, we need to compare speciation rates in asexual and sexual organisms. This requires a species concept or concepts that identify comparable evolutionary units in asexual and sexual organisms. This is especially difficult because the very existence of species in asexuals is controversial. Many people believe that asexual (and clonal) organisms will form a continuum of genetic variation with no clearly distinguishable species.

To address this problem, we (Barraclough et al. 2003) developed a general theory of speciation for asexual and sexual organisms. We focused on one commonly used criterion, namely the very general observation that many, if not all, organisms fall into discrete clusters of individuals with very similar genotypes and phenotypes, well-separated from other such clusters. Our theory of the nature and origin of these clusters can be developed as follows:

We begin by considering speciation in an established asexual lineage or inclusive population descended from a single asexual individual. That individual could have arisen either by loss of sexual reproduction in an organism that alternates asexual and sexual reproduction, or by hybridization between two obligatory sexual species. Because all the genes in an asexual organism are completely linked and behave as a unit, the origin and ploidy of the asexual lineage are irrelevant (with exceptions described below). We further assume that the asexual lineage is evolving as an independent evolutionary arena, separated physically or by adapting to a different niche from its parent lineage(s).

The diagrams in Fig. 10.1 show phylogenies of a gene in an asexual organism under several different scenarios. Note that because there is no outcrossing and recombination, the phylogeny of one gene is also the phylogeny of all the genes in the organism, and the phylogeny of the individuals as well. This model assumes that all the copies of the gene are orthologous, not the products of gene duplications and losses that can distort phylogenies. Similar problems can arise in diploids that show the Meselson effect or in polyploids (Birky 1996). These problems can be avoided by using mitochondrial or chloroplast genes.

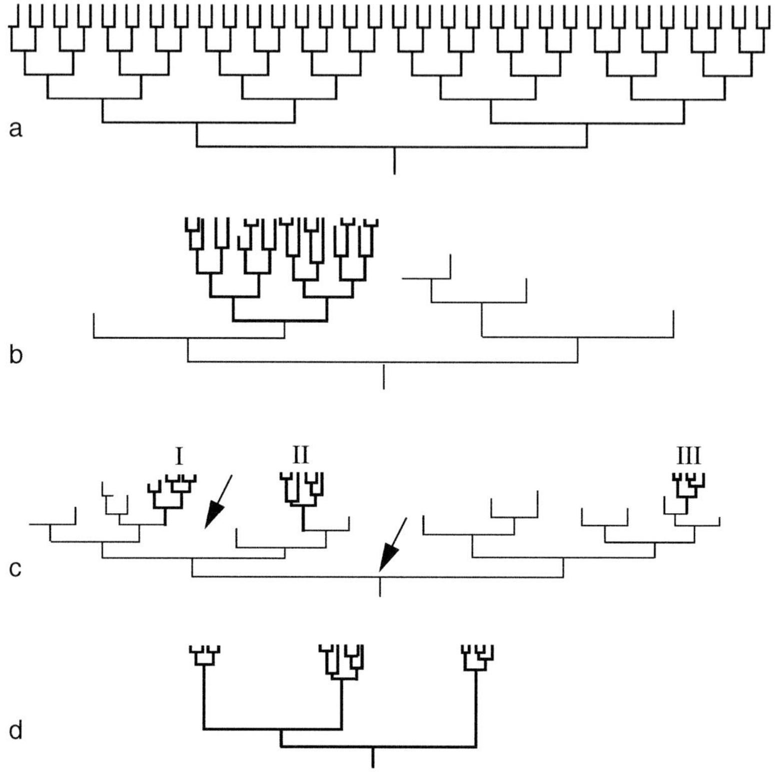

Fig. 10.1 Four models of asexual growth. Each one is a phylogenetic tree of genes or individuals reproducing by binary fission. **a** Unlimited growth, no variation in offsp;ring number. **b** Stochastic model with finite population size limited by carrying capacity of niche, stochastic variation in offspring number (random drift). **c** Stochastic model, with two speciation events (*arrows*) producing the species I, II, and III. Each event could be allopatric spciation by physical isolation of two populations, or sympatric speciation by divergent selection. **d** The tree that would be observed if one could sample all individuals in each species

Figure 10.1a is a naïve model of the phylogeny (or genealogy) of a gene in a clone. For simplicity it is assumed that the organism is reproducing by simple binary fission without limit: each individual leaves two offspring in the next generation. In this model, different lineages acquire different mutations so that different copies of the gene gradually diverge from each other. At any one time the individuals in a clone show a continuous distribution of sequence differences. If this model accurately reflected the genealogy of a clone, there would be no species by any definition. But the model is completely unrealistic, because it assumes that all individuals divide in synchrony and there is no limit to the growth of the clone.

In real life, growth of the clone is limited by finite resources that set an upper limit on the total size of the population (the census size N). Moreover, some individuals reproduce more often than others. This is partly because of stochastic effects, which cause random genetic drift of allele frequencies in population genetic theory, and partly because of natural selection. Figure 10.1b shows a more realistic model in which the population size is limited by the carrying capacity of the organism's habitat. The stochastic loss of lineages has created clusters separated by shallow gaps. These gaps will have a depth of approximately $2N_e$ generations on average, and the clusters they form will be transient with an average lifetime of approximately $2N_e$ generations. The clusters would not be considered different species by population geneticists. They do form clades and might have morphological differences that would allow them to be distinguished in the field. However, they are not adapted to different ecological niches and are not evolving independently. They are not comparable to ecological or sexual species, which can coexist indefinitely even in sympatry.

Longer-lasting clusters can be formed by allopatry or by divergent selection, as illustrated in Fig. 10.1c. In this example, two individuals in the first generation were physically isolated, e.g. by dispersal to distant locations. In the second generation two individuals in one location were selected for adaptation to different ecological niches while remaining sympatric. Consequently the offspring of these three individuals formed independently evolving populations. Each one forms an independent arena for mutation, random drift, and selection. Each one shows transient clusters separated by shallow gaps $2N_e$ generations deep or less. However, they are separated by long-lasting gaps with depths significantly greater than $2N_e$ generations. Figure 10.1d shows the tree with extinct lineages pruned; this is the realized tree that is actually observed using sequence data from extant individuals.

The resulting populations show several characteristics that are widely, although not universally, considered to be desirable traits in species: (i) they are independently evolving lineages, as in the Evolutionary Species Concept (EvSC, Simpson 1951; Hey 2006); (ii) they are reciprocally monophyletic, so that every individual is more closely related to another member of the same species than to any member of another species; and (iii) assuming that phenotypic evolution is neutral or under divergent selection, they will form distinct phenotypic clusters reflecting their significant differences in genotype. In addition, the two populations on the left that are adapted to different niches will be different species by the Ecological Species Concept (EcSC, Van Valen 1976); the population on the right is a different ecological species from at least one of these and over time is likely to diverge ecologically from both.

At least two of the clusters meet an important criterion under the Biological Species Concept (BSC), namely persistence in sympatry. It is unlikely that the two species on the left in Fig. 10.1c can ever merge again, as this would require reversal of the evolutionary steps that adapted them to different niches. In this sense they are equivalent to BSC species, which will not fuse even in sympatry because they are reproductively isolated.

Hey (2006) argued that most or all species concepts are really variants on the EvSC, or are species criteria for the EvSC, working definitions for assigning individuals to evolutionary species. The model of Barraclough et al. (2003) can also be considered a variant of the EvSC. However, unlike Simpson's original description of the EvSC and subsequent versions, our model has an explicit theoretical basis in population and evolutionary genetics. Moreover it generates explicit criteria and associated statistics for identifying species from finite samples of specimens, as described below. In view of these differences, one of us has proposed that it be considered a new model called the Evolutionary Genetic Species Concept (EGSC, Birky in preparation).

Barraclough et al. (2003) justified the above model of speciation and species in terms of coalescent theory, but the same model can be derived from classical population genetic theory where $2N_e$ generations is the approximate expected time to the most recent common ancestor of the entire population. Note that although coalescent theory is frequently considered to be a neutral theory, one can define N_e so that it takes selection, including hitchhiking, into account; this N_e is smaller and the time to the most recent common ancestor is reduced accordingly. The coalescent approach provides an intuitive conceptual linkage to phylogenetic analysis of gene sequences.

10.3 Criteria for Assigning Individuals to Species

Models of the nature of species do not directly address the practical problem of assigning individuals to species. The task of identifying species in a collection of individual organisms, and of assigning newly collected individuals to those species, requires a species criterion. Consideration of expected patterns of clustering outlined above suggests two different approaches to assigning individual asexual organisms to evolutionary genetic species.

10.3.1 The 4X Rule

The first species criterion, called the "4X rule" (Birky et al. 2005), has two characteristics that are uncommon among species criteria: (i) it explicitly recognizes that species are identified on the basis of samples, usually very small samples, of very large populations of individuals; and (ii) the probability that a species assignment is incorrect is set in advance at ≤ 0.05. The 95% confidence limit is widely used in biology but more or less stringent intervals can be used if desired.

The 4X rule works as follows:

1. DNA is extracted and one or more genes are amplified and sequenced from each of a sample of individuals collected from nature. Mitochondrial or chloroplast genes are used because they are nearly always homoplasmic (only one allele in high frequency in each individual); consequently one can sequence amplified mitochondrial genes directly without having to clone them.
2. Phylogenetic trees are made of the resulting sequences and used to identify reciprocally monophyletic groups of sequences. These clades are samples from putative species.

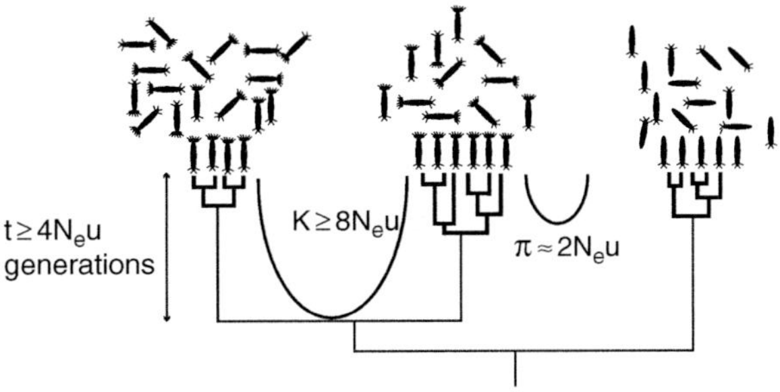

Fig. 10.2 Rationale for the 4X rule. In this imaginary example, samples of 4–6 individuals from each of 3 independently evolving populations will be separated by $K \geq 8N_eu$ sequence differences per site. The mean sequence difference among individuals in a sample π estimates $2N_eu$, so the ratio of K to π is 4

3. The matrix of pairwise sequence differences is examined to find putative species (clades) that obey the 4X rule: the sequence differences between individuals in different clades are $\geq$4X greater than the mean pairwise sequence differences within the clades. More specifically, the mean pairwise sequence difference within a putative species is used as an estimate of the nucleotide diversity in the species.

The theoretical rationale for the rule is illustrated in Fig. 10.2. As Barraclough et al. (2003) noted, gaps formed within asexual species by stochastic processes will be $2N_e$ generations deep on average; the 95% confidence interval is $4N_e$ generations. Consequently, one can infer with 95% confidence that gaps deeper than $4N_e$ generations separate independently evolving populations, i.e. EG species. The sequence difference between individuals in two such populations, corrected for multiple hits, will be $K = 8N_eu$ where u is the mutation rate per site per generation. At the same time the expectation of the mean pairwise sequence difference between individuals within each population will be $\theta \approx 2N_eu$. The parameter θ is approximated by the mean pairwise sequence difference or nucleotide diversity π. The ratio of the sequence difference between populations to the difference between individuals within a population is $K/\theta \geq 8N_eu/2N_eu = 4$.

Of course we cannot get sequences from every individual in a species; species are identified on the basis of samples, usually very small, from populations that are usually very large. Fortunately, sampling theory by Rosenberg (2003) shows that if samples of two or more individuals from two different populations separated by $\geq 4N_e$ generations are reciprocally monophyletic, then the entire populations are reciprocally monophyletic with probability $\geq$95%. If one population is represented by two individuals and the other by only one, the probability is still approximately 94%. This remarkable statistical power is possible because the probabilities are conditioned on the samples being reciprocally monophyletic, and because after $4N_e$ generations it is very likely that the species as a whole consists of a single clade which is necessarily the same as the sample clade.

This criterion will fail to identify populations that have recently begun to evolve independently but have not had time to show the phylogenetic signatures of independent evolution. The proportion of species missed will depend on the ratio of the speciation rate to $2N_e$ generations. Of course very recent speciation events can confound species recognition with any species concept or criterion.

10.3.2 Analysis of Branching Rates

An alternative framework for considering whether a clade has diversified into distinct genetic clusters is to consider branching models (Pons et al. 2006). This approach is more complex than the 4X rule, but allows for a global test of whether the study clade has diversified into separate species or not. Under a null model that the entire sample derives from just a single asexual population, i.e. without divergence into independently evolving or ecologically distinct species, then the pattern of branching is expected to conform to a standard coalescent for a single population (Fig. 10.1b). A large body of theory is available to specify the likelihood of a given pattern of branching under a particular model (Hudson 1991), ranging from a neutral coalescent in a population of constant size, through to populations that have changed size through time or experienced different forms of selection (Rosenberg and Nordborg 2002).

Under the alternative model that the clade has diversified into species, as defined above, then we expect to observe distinct genetic clusters (Fig. 10.1d). Branching events within each cluster will reflect coalescence within populations of the same kind considered in the null model, but in this case with each cluster representing a separate, independently evolving population. Branches connecting the different clusters now represent the timing of divergence events, for example events causing geographic isolation or the onset of adaptive divergence into different niches. In other words, branching within clusters reflects population processes, whereas branching between clusters reflects diversification. Again, a large body of theory is available to specify the likelihood of branching times in a phylogeny depending on different models of speciation, extinction and sampling of species (Nee et al. 1994; Barraclough and Savolainen 2001; Nee 2001).

Pons et al. (2006) combined previous approaches for considering population coalescence and branching times within phylogenies to derive a general expression for the likelihood of branching times under the alternative model that the sample derives from a set of independently evolving populations. The model relies on the waiting intervals between successive branching events as input data (Fig. 10.2, Pons et al. 2006). Under a relatively broad set of circumstances, the net observed branching rate within species is expected to exceed that between species, hence the signature of clusters can be detected as a transition from slow to fast branching rates occurring near the tips of the tree (Fig. 10.1d). The likelihood of models with and without a transition in branching rate can be used to test whether the alternative model provides a better fit to the data than a null model in which the entire sample conforms to a single branching process. The models were modified to allow a range of qualitative departures from standard assumptions, such as constant population size, neutrality or constant speciation rates. Pons et al. (2006) applied the approach to a sample of tiger beetles from salt lakes in Australia and showed that this lineage had diversified into a set of independently evolving species. The method is particularly suitable for asexual clades in which any single locus is expected to reveal the species history, not just the history of that gene.

In principle, the model of speciation and the criterion for finding species that we have described can be applied to any asexual organism. However, in keeping with the theme of this volume, we will focus on parthenogenesis, which is limited to animals and plants. Moreover, we focus on clades that (i) are thought to have been reproducing without sex long enough to have completed speciation, and (ii) have a suitable data base of gene sequence data. Here we consider two groups of animals that meet these criteria, the bdelloid rotifers and oribatid mites.

10.4 Application to Bdelloid Rotifers

The bdelloid rotifers are an ancient asexual group that includes more than 400 species defined by phenotype, principally morphology (see Chapter 13). Birky et al. (2005) applied the 4X rule to 110 specimens of bdelloid rotifers. These specimens were obtained from more than 34 sites in 7 states in the US and 1 site in Italy. Morphological study revealed that they included members of six genera. The collecting sites included permanent and temporary springs, streams, ponds, lakes, soil and moss from a wide range of ecosystems between sea level and 3660 m. A 591 bp segment of the mitochondrial *cox1* (also known as *COI*) gene was amplified and sequenced from clones of females descended from each specimen.

Phylogenetic analysis (Fig. 10.3) revealed the presence of a number of clades, including 21 terminal clades that obey the 4X rule and are simple, i.e. they contain no such clades within them. The mean sequence difference among individuals in a clade ranged 0–2%, similar to other invertebrates (Avise 1994). These are samples from inclusive populations that are evolving independently because they are adapted to different niches or are geographically isolated. Moreover they have been separated long enough to become reciprocally monophyletic; they are species under the evolutionary and evolutionary genetic species concepts. There are also 14 singlets, individuals that are not part of any of these species; most or all of these singlets are separated from each other and from the species by such large sequence differences that they almost certainly represent samples of size one from additional species.

We also analyzed branching rates to find species in this sequence data set. The maximum likelihood solution for the independently evolving clusters model infers the presence of 24 clusters, with the remaining individuals inferred to be singletons (Fig. 10.3). Confidence limits within 2 log likelihood units of the maximum likelihood solution range from 22 to 25 clusters. The pattern of clustering is significantly greater than expected under the null model that the sample derives from a single population (likelihood ratio test, $X^2 = 108.5$, $p < 0.0001$). Overall, the branching rate method resolves clusters more finely than the 4X rule approach: 8 clusters found in the latter are split into more clusters by the former. One possible explanation is that

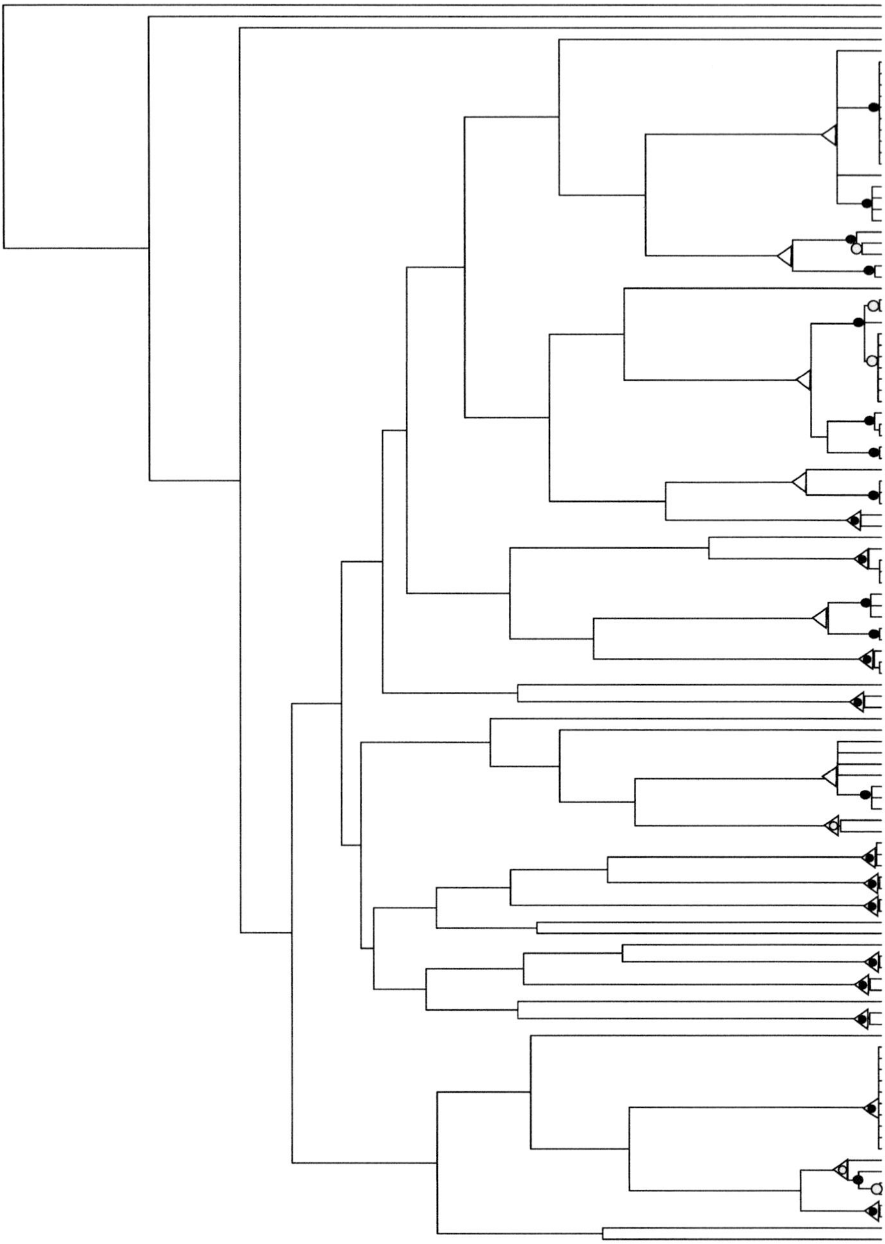

Fig. 10.3 (continue)

if a few populations harbor greater genetic diversity than the predominant trend, then they may tend to be split by the branching rate model because it assumes a single threshold for the transition from between- to within-cluster branching. Future implementations that allow for different genealogical branching rates within different clusters are in development (T.G. Barraclough, unpubl.). An alternative explanation is that, by treating each case individually, the 4X rule may be more conservative: knowing the typical level of within-cluster diversity across the sample may allow the branching rate method to split clusters that haven't diverged by 4X the intraspecific variation. Simulations of known scenarios could be used to compare the power of the two methods. Power aside, the branching rate method has the advantage of adopting a global analysis of the data, which facilitates hypothesis testing of different scenarios, whereas the 4X rule is simple to apply while still founded on robust population genetic theory.

Evidence for genetic clusters is consistent with independently evolving entities, but the nature of those entities remains in question. One possibility is that some or all of them evolve independently solely as a result of geographic isolation. Traditional accounts have tended to exclude such cases from consideration as species, because they cannot explain the existence of alpha-diversity, i.e. coexistence of species in a single area. Our analyses provide clear evidence of distinct clusters present in sympatry over successive years, but it remains possible that their coexistence is transient or a feature of ongoing invasion from source areas. A more direct approach is to look for evidence of ecological divergence between clusters. Birky et al. (2005) found that Pha2 and Pha3 differ in temperature tolerance range; clones of Pha2 survive and reproduce at 36° while clones of Pha3 do not. However, it is not clear whether this difference detected in the lab reflects differences in their habitat.

Fontaneto et al. (2007) tested directly for divergent selection on presumed eco-morphological traits in the genus *Rotaria*. If morphology has evolved neutrally or under constant selection among all lineages, we expect levels of intraspecific and interspecific variation to mirror levels of neutral genetic variation (McDonald and Kreitman 1991). However, if there has been divergent selection on morphology, for example as a result of divergence to occupy different niches, then we expect more eco-morphological divergence between species than within them. This pattern was indeed found in *Rotaria*, consistent with their adaptive radiation rather than simply

Fig. 10.3 Phylogenetic tree of the bdelloid rotifers sampled by Birky et al. (2005), showing clusters inferred using their 4X rule and by the branching rate method by Pons et al. (2006). Names of the specimens are not shown in order to save space, but are available from CWB on request. *Triangles* indicate clusters identified by the 4X rule. *Black circles* indicate clusters recovered in the maximum likelihood solution, *grey dots* indicate those appearing within confidence limits defined by 2 log likelihood units. The tree was reconstructed by neighbour-joining using GTR maximum likelihood distances. An ultrametric tree was obtained using penalized likelihood implemented in the software r8s by Sanderson (2002) with a smoothing parameter of 1. The root was arbitrarily scaled to have an age of 1 unit. The branching rate method was implemented using software in the R statistical programming language (Team 2006) by TGB neutral divergence in geographic isolation. Interestingly, morphologically coherent clades experiencing divergent selection often contained two or more genetic clusters: these could either be solely the result of geographic isolation, or they may have diverged in unmeasured phenotypic traits, such as behaviour, life-history or other morphological traits. Now that the molecular systematics of bdelloids is being established, more work is needed to explore the biological significance of the genetically distinct clusters. All but one of the traditional species in the genus were found to comprise monophyletic clades, but many of them subsumed several genetic clusters: more resolved morphological techniques may be needed to provide non-genetic means for cluster identification.

10.5 Application to Oribatid Mites

The Oribatidae is a diverse and widespread group of soil-dwelling mites that includes a number of apparently obligate parthenogenetic lineages (see Chapter 12). Some of the parthenogenetic lineages are quite ancient and speciose. Heethoff et al. (2007) obtained *cox1* sequences of 65 specimens of the morphological species *Platynothrus peltifer* collected from Europe, Russia, Japan, and the United States. Phylogenetic analyses revealed seven clades within this species; sequence differences within these clades were less than 2%, while the differences between clades averaged 56% after correction for multiple hits. Using their data kindly provided by Michael Heethoff, we found that all seven clades qualify as species under the 4X rule. The branching rate method recognizes 7 clusters plus one for the outgroup (Fig. 10.4). At present, it is not clear if the seven clades of *P. peltifer* can be distinguished by morphology. However, there is some evidence that they have been physically isolated by continental drift and mountain uplifts (Heethoff et al. 2007 and Chapter 12).

10.6 Predictions About Relative Speciation Rates in Sexuals and Parthenogens

We have argued that asexual organisms should diversify into independently evolving and biologically distinct species broadly equivalent to sexual species. Evidence from bdelloid rotifers and oribatid mites supports this claim. But the question remains whether asexuals should diversify to a greater or lesser extent than sexuals. On the one hand, sexuals need to evolve reproductive isolation before diversity can evolve and be maintained in sympatry (Futuyma 1987; Eldredge 1989). Because asexuals have no such requirements, we might expect them to evolve sympatric diversity more easily. Also a single asexual individual can found a new allopatric population, compared to two individuals in outcrossing sexuals. However, sexual organisms are expected to adapt more quickly to changing environments: if the rate of adaptation to new conditions were the limiting step in diversification, sexuals might have the edge over asexuals. Finally, if the events leading to geographic isolation or the onset

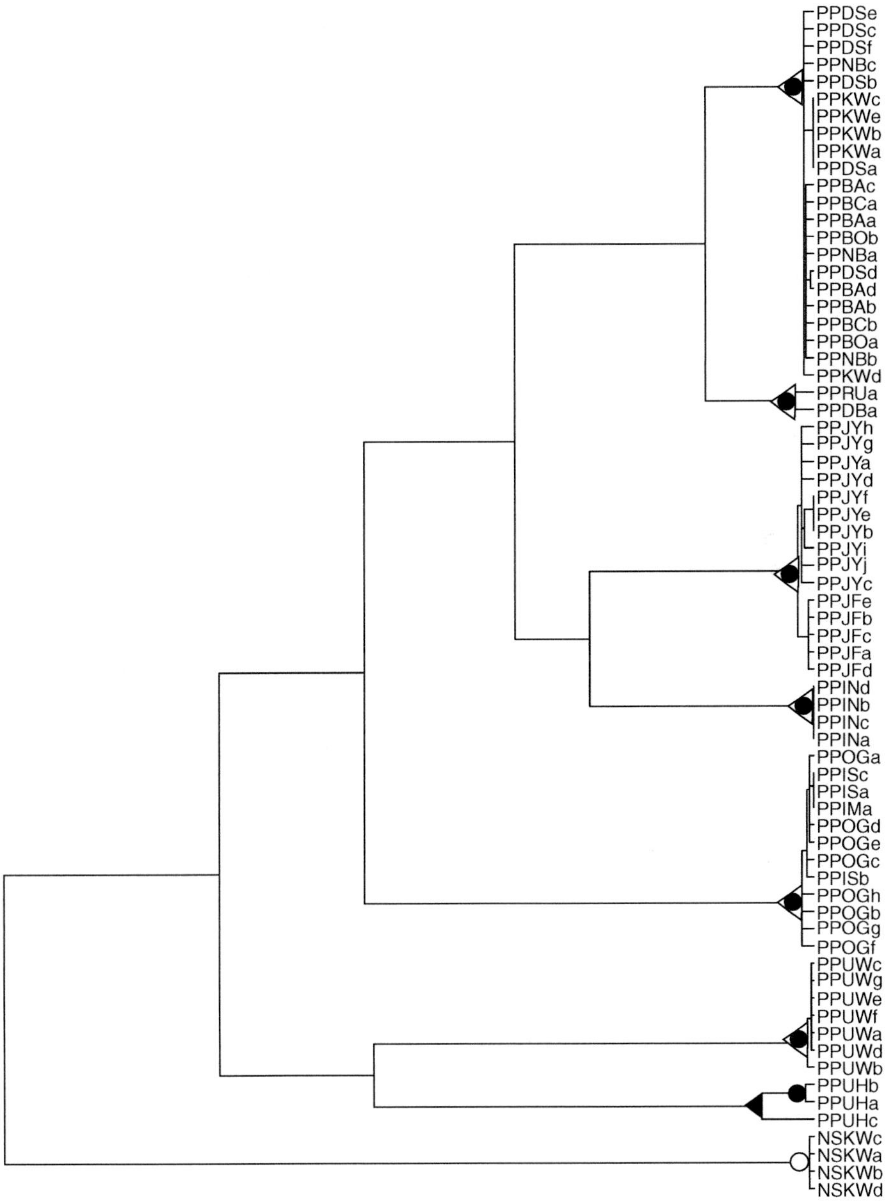

Fig. 10.4 Phylogenetic tree of the oribatid mites sampled by (Heethoff et al. 2007) showing clusters inferred using the 4X rule of Birky et al. (2005) (*triangles*) and the branching rate method of Pons et al. (2006) (*black circles*). The *open circle* indicates the outgroup. All of the branching rate clusters were recovered in the maximum likelihood solution: there were no other solutions within the standard 2 log likelihood units. The tree and clustering analyses were obtained following identical methods to those described for the bdelloid rotifers

of divergent selection were actually the limiting step in diversification, rather than rates of adaptation or the origin of reproductive isolation, then both sexuals and asexuals might have similar expected rates of diversification, all other things being equal (Barraclough et al. 2003). Therefore, the answer to the question could provide insights into which steps in the process of diversification are most important for determining speciation rates within clades.

Unfortunately, there are a number of difficulties in performing a rigorous test of these ideas. By far most origins of asexuality appear to be evolutionary dead-ends. Most asexuals originated from their sexual ancestors very recently and may have had insufficient time to diversify into distinct species. Moreover, in most groups with frequent origins of asexuality, such as land plants, there appear to be repeated origins of asexuality from a single sexual ancestor, meaning that genetic and phenotypic patterns of variation result more from gene flow from the sexual ancestor than as an outcome of asexuality.

Ancient asexual groups such as bdelloids offer the opportunity to study diversification in a group that has been asexual long enough for diversification to occur. Patterns of variation could be compared quantitatively with those in their nearest and most comparable facultatively sexual relatives, the monogonont rotifers. One difficulty here is that other traits aside from obligate versus facultative asexuality could differ between the clades, for example bdelloids often live in more desiccating habitats than monogononts and differ in several key aspects of life-history (e.g. anhydrobiosis). However, bdelloids and monogononts do co-occur in some habitats, and therefore clades could be compared at least in comparable environments.

Ideally, we would have repeated comparisons of sexual and asexual sister clades, but other cases of ancient asexuals are few and far between. Oribatid mites or darwinulid ostracods offer some potential, but again may suffer from being so ancient that their nearest sexual relatives live very different lives. Perhaps more promising, although even less well known, would be fungi, single-celled eukaryotes, or prokaryotes. In fact species can be detected by applying the 4X rule to sequence data from several groups of unicellular eukaryotes (Birky, submitted). However, a large amount of basic knowledge and direct evidence for strict asexuality would need to be accumulated before sexual versus asexual comparisons could be made in any of these groups. Ultimately, a survey of patterns of diversification among clades with differing frequencies of recombination may prove most feasible and most satisfying, rather than striving for the difficult goal of finding unambiguous asexual versus sexual comparisons.

10.7 Conclusions

Population and evolutionary genetic theory shows that asexual organisms can form species that share a number of properties with other species concepts. In particular, they are inclusive populations that are independent arenas for the evolutionary processes of mutation, selection, and random drift. As a result they form genotypic clusters separated by long-lasting gaps due to physical isolation and/or divergent selection, as opposed to transient gaps due to random genetic drift. The genotypic clusters will often be phenotypic clusters as well, but they may not be readily detectable, in which case the clusters will be cryptic species. This theoretical model of the nature and origin of species suggests two different methods of distinguishing species using gene sequence data; one detects the difference in branching rates of lineages between species and within species, while the other identifies clusters that are separated by gaps too deep to be caused by drift. These species criteria have been used to demonstrate that bdelloid rotifers and oribatid mites have undergone speciation. Some of the resulting species are cryptic, not presently distinguishable by phenotype.

Theory also shows that under some conditions, either asexuals or sexuals may show a higher rate of speciation. Which is true is an empirical question, and the answer may differ in different groups of organisms. Our species criteria provide the first step toward answering this question.

References

Avise JC (1994) Molecular markers, natural history and evolution. Chapman and Hall, New York

Ayala FJ (1998) Is sex better? Parasites say no. Proc Natl Acad Sci USA 95: 3346–3348

Barraclough TG, Burt A, Birky CW Jr (2003) Diversification in sexual and asexual organisms. Evolution 57: 2166–2172

Barraclough TG, Savolainen V (2001) Evolutionary rates and species diversity in flowering plants. Evolution 55: 677–683

Birky CW Jr (1996) Heterozygosity, heteromorphy, and phylogenetic trees in asexual eukaryotes. Genetics 144: 427–437

Birky CW Jr, Wolf C, Maughan H, Herbertson L, Henry E (2005) Speciation and selection without sex. Hydrobiologia 546: 29–45

Cohan FM (2004) Concepts of bacterial biodiversity for the age of genomics. In: Fraser CM, Read T, Nelson KE (eds) Microbial Genomes. Humana Press, Totowa, NJ, pp. 175–194

Coyne JA, Orr HA (2004) Speciation. Sinauer Ass., Sunderland, MA

Eldredge N (1989) Macroevolutionary dynamics. McGraw-Hill, New York

Fontaneto DEA, Herniou EA, Boschetti C, Caprioli M, Melone G, Ricci C, Barraclough TG (2007). Independently evolving species in asexual bdelloid rotifers. PLoS Biology 5: 1–8

Futuyma DJ (1987) On the role of species in Anagenesis. Am Nat 130: 465–473

Gladyshev EA, Meselson M, Arkhipova IR (2008) Massive horizontal gene transfer in bdelloid rotifers. Science 320: 1210–1213

Heethoff M, Domes K, Laumann M, Maraun M, Norton RA, Scheu S (2007) High genetic divergences indicate ancient separation of parthenogenetic lineages of the oribatid mite *Platynothrus peltifer* (Acari, Oribatida). J Evol Biol 20: 392–402

Hey J (2006) On the failure of modern species concepts. Trends Ecol Evol 21: 447–450

Hudson RR (1991) Gene genealogies and the coalescent process. Oxf Surv Evol Biol 7: 1–44

Kondrashov AS (1982) Selection against harmful mutations in large sexual and asexual populations. Genet Res 40:325–332

Maynard Smith J, Szathmary E (1995) The major transitions in evolution. Freeman, New York

McDonald JH, Kreitman M (1991) Adaptive protein evolution at the *adh* locus in *Drosophila*. Nature 351: 652–654

Mergeay JD, Verschuren D, De Meester L (2006) Invasion of an asexual American water flea clone throughout Africa and rapid displacement of a native sibling species. Proc R Soc Lond B 273: 2839–2844

Muller HJ (1964) The relation of recombination to mutational advantage. Mutat Res 1: 2–9
Nee S (2001) Inferring speciation rates from phylogenies. Evolution 55: 661–668.
Nee S, May RM, Harvey PH (1994) The reconstructed evolutionary process. Philos Trans R Soc Lond B 344: 305–311
Ochman H, Lerat E, Daubin V (2005) Examining bacterial species under the specter of gene transfer and exchange. Proc Natl Acad Sci USA 102: 6595–6599
Orr HA (2000) The rate of adaptation in asexuals. Genetics 155: 961–968
Otto SP, Lenormand T (2002) Resolving the paradox of sex and recombination. Nat Rev Genet 2: 252–261
Pons J, Barraclough TG, Gomez-Zurita J, Cardoso A, Duran DP, Hazell S, Kamoun S, Sumlin WD, Vogler AP (2006) Sequence-based species delimitation for the DNA taxonomy of undescribed insects. Syst Biol 55: 1–15
Rosenberg NA (2003) The shapes of neutral gene genealogies in two species: Probabilities of monophyly, paraphyly, and polyphyly in a coalescent model. Evolution 57: 1465–1477
Rosenberg NA, Nordborg M (2002) Genealogical trees, coalescent theory and the analysis of genetic polymorphisms. Nat Rev Genet 3: 380–390
Sanderson MJ (2002) Estimating absolute rates of molecular evolution and divergence times: A penalized likelihood approach. Mol Biol Evol 19: 101–109
Simpson GG (1951) The species concept. Evolution 5: 285–298
Team RDC (2006) R: A language and environment for statistical computing. R Foundation for Statistical Computing AD, Vienna, Austria
Van Valen L (1976) Ecological species, multispecies, and oaks. Taxon 25: 93–106.

Chapter 11
Darwinulid Ostracods: Ancient Asexual Scandals or Scandalous Gossip?

Isa Schön, Giampaolo Rossetti and Koen Martens

Abstract Whereas studies of putative ancient asexuals could help solve the paradox of sex, most research on such groups still focuses on consolidating their status. The evidence for the darwinulid ostracods is as yet inconclusive. Recent males have been found in a single species, but their functionality is uncertain and their morphology highlights the erroneous assignment of male status to a single individual of *Darwinula stevensoni*, presently the best candidate for an ancient asexual darwinulid. Previous records of putative fossil males for the past 200 million years have been rejected. Genetic signatures of ancient asexuality are equally inconclusive: there is no Meselson effect in the darwinulids, but neither the presence nor the absence of the Meselson effect does provide conclusive evidence for or against sex. However, it would seem that a combination of a general purpose genotype with powerful homogenising genetic mechanisms (gene conversion, DNA repair) could counter the deleterious effects of the absence of sex in at least a number of darwinulid species.

11.1 Introduction

The debate on whether or not ancient asexuals really exist is still ongoing (see e.g. Martens and Schön 2008). Some researchers of the paradox of sex simply do not accept the existence of any long-term asexuality at all (Little and Hebert 1996). Others make an exception for the bdelloid rotifers, but reject all other claims such as for oribatid mites or darwinulid ostracods (Hayden 2008) (see Box 11.1 for further explanations on ostracods in general and darwinulids in particular).

I. Schön (✉)
Royal Belgian Institute of Natural Sciences, Freshwater Biology, Vautierstraat 29, B-1000 Brussels, Belgium
e-mail: isa.schoen@naturalsciences

I. Schön et al. (eds.), *Lost Sex*, DOI 10.1007/978-90-481-2770-2_11,

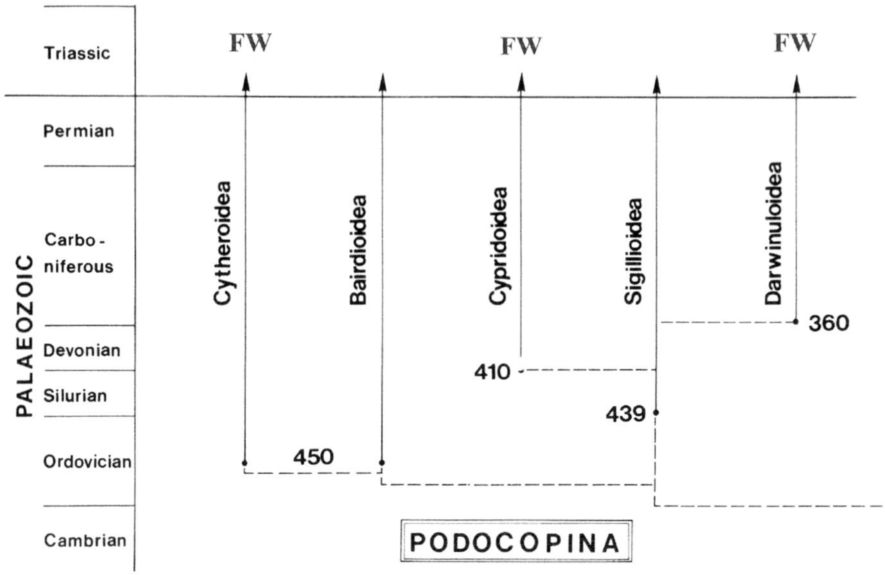

Fig. 11.1 Geological history of non-marine ostracods. Modified from Martens et al. (1998). Most species of the Darwinuloidea were lost after the Permian-Triassic mass extinction

Box 11.1 Ostracoda

Ostracods in General

Ostracods are small, bivalved crustaceans, which occur in almost all aquatic and moist (semi-) terrestrial habitats. The non-marine ostracods belong to three main lineages in the order Podocopida, all of them with marine roots: Cytheroidea (c 450 Myr old), Cypridoidea (c 410 Myr) and Darwinuloidea (c 360 Myr) (see Fig. 11.1; Martens et al. 1998).

Special morphological characteristics of ostracods, important to determine asexual status, are their calcified valves, providing an extensive real-time record of microfossils and the presence of paired hemipenes and (in some groups) giant sperm, prehensile palps and special sperm pumps ("Zenker's organs") in males of sexual species. Darwinulids are smaller than the average non-marine ostracod (adult size 0.6–0.8 mm; *Vestalenula mathilda* <0.4 mm). Darwinulid valves are mostly transparent, smooth and more or less elongated.

Ostracod Diversity

Of the three major ostracod lineages, the Cypridoidea are almost exclusively non-marine and by far the most speciose group, comprising c 1500 described species, about 3/4 of the total number of non-marine ostracods (Martens et al.

2008). The cytheroids are mostly marine and have about 500 non-marine species. The darwinulids are exclusively non-marine, but are at present far less speciose: only about 35 living species in 5 genera (*Darwinula*, *Alicenula*, *Vestalenula*, *Penthesilenula*, and *Microdarwinula*) are presently recognised. The overall number of recent darwinulid species will be higher as we have recently identified five cryptic species in the morphospecies *Penthesilenula brasiliensis*, mainly following continental distributions (Schön et al. submitted). Fossil records indicate that the darwinulids were more diverse up to the Triassic (see below and in Martens et al. 1998). (For discussions on species definitions in ancient asexuals in general, see Chapter 10).

Reproductive Modes in Ostracods

The Cytheroidea are generally sexual, though there is mixed (sexual/asexual) reproduction and even full parthenogenesis in some groups (Horne et al. 1998a). The Cypridoidea have generally mixed reproduction, and are in some cases fully sexual, in others presumed fully asexual. Only the darwinulids are believed to be fully asexual, and to have been so for 200 million years (Martens et al. 2003).

Ecology of Darwinulidae

Darwinulids often occur in so-called marginal (eg. semi-terrestrial) habitats, much like bdelloid rotifers (see Chapter 10). A few species, such as *Penthesilenula brasiliensis*, are more common, also occur in lakes and ponds and have intercontinental distributions. The most common species, *Darwinula stevensoni* (Fig. 11.2), is cosmopolitan (except for Antarctica) and ubiquitous (occurs in a wide range of habitats: rivers, lakes, interstitially etc.).

Life History of Darwinulidae

All darwinulids are brooders, and 4 out of 5 genera have the posterior part of the carapace inflated so as to create a brood pouch (Horne et al. 1998b; see also Fig. 11.2). With an average of 6–8 offspring per female in *Penthesilenula brasiliensis* (Pinto et al. 2007), and 11–15 offspring in *D. stevensoni* (Van Doninck et al. 2003a), darwinulids generally have low fecundity as compared to other ostracods (Geiger 1998) and, especially in higher latitudes, rather long life cycles. *Darwinula stevensoni* takes 4 years in Canada (McGregor 1969) and in Finland (Ranta 1979) to complete its life cycle, although further South,

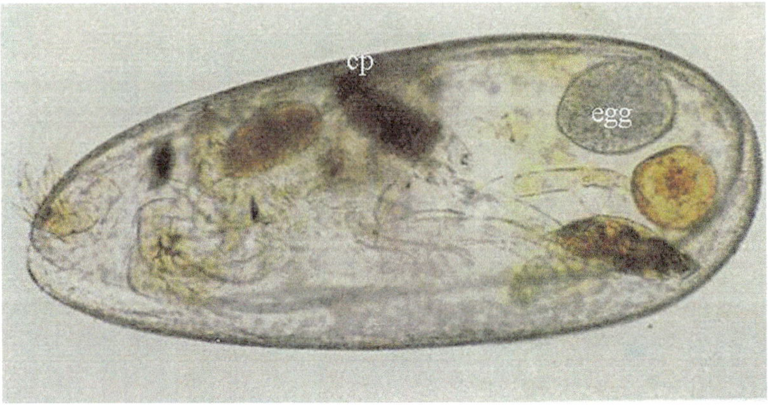

Fig. 11.2 *Darwinula stevensoni*. Note the egg in the brooding pouch. cp = colour patch. From Okubo, (2001)

e.g. Belgium, it takes only 1 year (Van Doninck et al. 2003a), which is still considerably longer than in most other Cypridoidea (e.g. Martins et al. 2008). Darwinulids do not have resting eggs like all of the Cypridoidea and some Cytheroidea (Geiger 1998), and they do not swim but only crawl slowly.

There are several reasons for this scepticism within the scientific community. Most importantly, as several philosophers of science have already shown, science progresses within paradigms (Kuhn 1962) and breaking out of a paradigm requires very solid evidence. Present day evolutionary theory predicts that asexual lineages will be short-lived, because of many reasons, a.o. the inability to prevent, through sex and recombination, their genomes from accumulating deleterious mutations (Muller's ratchet, Muller 1964, and Kondrashov's hatchet, Kondrashov 1988). A closer inspection of asexuality in the animal kingdom (see Chapter 4, 14, 15, 16, 17, 18, 19, 20 and 21) reveals that most asexual taxa are at one point of their life cycle also sexual or have close ties with sexual relatives. Putative ancient asexuals have also been named "ancient asexual scandals" (Judson and Normark 1996), because their mere existence would contradict the relevance of the various hypotheses for the prevalence of sex if they are genuine *ancient* asexuals. This is why it is so important to verify their *ancient* status independently of the patterns and processes in other animal groups (see Chapters 3 and 9 for an putative explanation as to why there are no ancient asexual plants).

Thus, in order to break out of the scientific paradigm around the asexual status of the Darwinulidae (or part thereof), namely that they have occasional or rare sex, although males have not yet been found in most species (e.g. Little and Hebert 1996; Hayden 2008), solid evidence would be required, not just circumstantial indications. We feel that the paradigm should also work the other way round – solid evidence should also be required to convincingly reject ancient darwinulid asexuality. Rejecting sexual reproduction is difficult because it requires negative evidence nl., for the *absence* of sex (Schurko et al. 2009). The three males found by Smith et al. (2006; see below) from a single species amongst thousands of females were recently put on stage in Nature (Hayden 2008) with the premature conclusion that darwinulids are scandalous gossip (and sexual) rather than ancient asexual scandals. However, as long as these males have not been shown to be functional, and to be common enough to produce a meaningful amount of recombination, they are not the smoking gun that shot down the ancient asexual status of this group (Martens and Schön 2008; (see also Schurko et al. 2009) for a broader discussion on how to provide evidence for exclusive asexuality). Smith et al. (2006) themselves never claimed this.

Here, we provide an overview of current knowledge regarding the presumed ancient asexual status of darwinulid ostracods. We review taxonomical, morphological, palaeontological, genetic and genomic evidence for or against ancient asexuality, and difficulties encountered with each of these lines of enquiry and we suggest some possible avenues for future research.

11.2 Demonstrating the Status of Long-Lived Asexuals

Research programmes on ancient asexuals have their main merit as alternative approaches to investigate the "paradox of sex". Indeed, if sex has so many advantages, then which special adaptations – if any – allow long-term survival without it? However, the main task of the research teams dealing with such putative ancient asexuals has thus far been to demonstrate that their respective groups (mainly bdelloids, darwinulids and certain lineages within oribatid mites) indeed merit the status. Below, we outline the main results of these various lines of enquiries for the darwinulids.

11.2.1 Recent Males

One commonly used sign that a certain group might have lived for a long time without sex is obviously the absence (or extreme rarity) of males in extant populations. (In hermaphrodites, this would be the absence of male functions; see Chapter 18 for more details). This was also the first hint of the possible long-term asexuality in darwinulids. Until 2006, only two putative occurrences of males of darwinulids have ever been reported. There is mention of a male of *Darwinula stevensoni* in Turner (1895), but no sexually dimorphic features were illustrated. Turner, however, might have been convinced that males were possible, because with the original description of the species, Brady and Robertson (1870 – see also Brady and Norman 1889) illustrated the so-called hemipenes of a putative male. Maddocks (1973) later made an attempt to homologise parts of this drawing with hemipenes of other ostracod groups. It has also been suggested that the putative hemipenis from Brady and

Robertson might actually be another, misidentified part of the anatomy, namely the head shield plus attached basal segments of Antennulae and Antennae. No fifth-limb prehensile palps (typical of males in many ostracods) were illustrated by Brady and Robertson (1870).

Since then, the discovery of three males, one each in three different populations amongst hundreds of females of *Vestalenula cornelia* by Smith et al. (2006), was most valuable from a morphological and phylogenetic point of view, but did not conclusively solve the discussion on ancient asexuality (see below for more details).

The following topics are of interest here:

1. *V. cornelia* males resemble A-1 females, i.e. smaller than adult females, without a posterior brood pouch and with muscle scars (attachments of the transversal adductor muscles) in the middle of the valves (Fig. 11.3a and b; see below for relevance of this central position);
2. these males do have their fifth pair of appendages transformed into the typical male prehensile palps (see Fig. 11.3b);
3. the hemipenes are relatively small and lack the internal labyrinth present in both Cypridoidea and Cytheroidea, consisting basically only of a small (paired) bladder through which a spermiduct runs (see Fig. 11.3b and c). There are no clear sperm pumps (like a Zenker organ), but there is a structure that could be interpreted as a precursor to a Zenker organ. Note that non-marine Cytheroidea have functional males without Zenker organ (Martens et al. 2008). In those cases, the hemipenes are large and muscular.

Since the morphology of the *V. cornelia* hemipenes and fifth limbs (see Fig. 11.3b) is so different from those illustrated for the putative male of *Darwinula stevensoni* (Smith et al. 2006), the "head shield" hypothesis for *D. stevensoni* gains in momentum, so that it becomes more and more likely that Brady and Robertson (1870) were indeed dealing with an (A-1) juvenile female (which would explain size and shape in dorsal view), and not with a male.

What, then, is the significance of the three males of *Vestalenula cornelia* for the ancient asexual status of darwinulids as a whole? Their discovery certainly strengthens the status of the species *D. stevensoni* as a long term asexual (c 25 myr; Straub 1952) because of the support for the head shield hypothesis and the fact that the putative male of *D. stevensoni* described by Brady and Robertson does not morphologically resemble the males of *V. cornelia* as would be expected from ostracods belonging to the same family.

But it is also far from certain that *V. cornelia* can now be seen as a sexual darwinulid. No copulation has been observed and no sperm was found in the males, nor in sympatric females, which leaves the possibility open that these males are non-functional atavisms.

Instances of rare males in ostracods (and in many other animal groups) are not at all uncommon. Yin et al. (1999) managed to rear a few males in the laboratory from fully parthenogenetic females in the species *Limnocythere inopinata*. Dozens of incidences of singular males in otherwise all-female populations have been reported

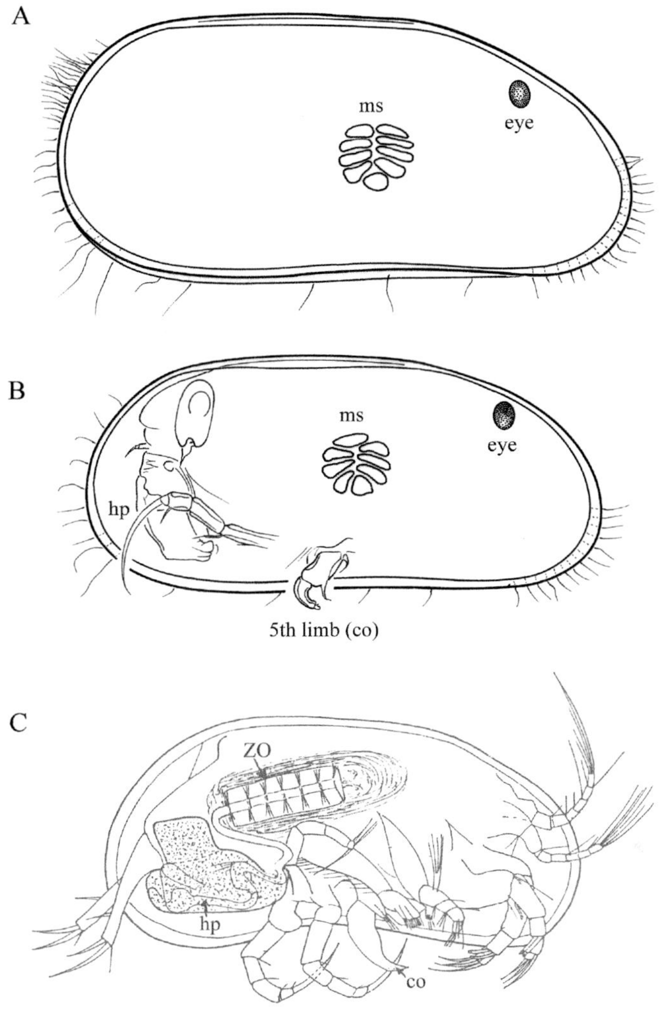

Fig. 11.3 Morphology of *Vestalanula cornelia* as compared to *Candona neglecta*. **a** Valve outline and muscle scar (ms) position of female. **b** Valve outline, muscle scar (ms) position, hemipenis (hp) and fifth limb with hook (co = clasping organ) of male. Modified from Smith et al. (2006). **c** Morphology of male of *Candona neglecta* (modified after Horne et al. 1998b) with more complicated hemipenis (hp), fifth limb with hook (co = clasping organ) and sperm pump (Zenker's organ – ZO)

from other ostracod species. In oribatid mites, rare (non-functional) males have been known for a long time (Taberly 1987; see also Chapter 12). Males are also reported in asexual lineages of *Potamopyrgus antipodarum* (Lively and Jokela 2002, Fig. 1 therein).

In any asexual species, it is possible that occasional males are produced through mutations in the regulatory cascade controlling sex. In XY-controlled systems, males would need the Y chromosome to be functional. Several ostracods of the Cypridoidea are known to have XY-controlled systems (Schön and Martens 1998a). But even if the control occurs through other means than XY, if males are sufficiently rare, the genes coding for any spermatogenesis-related functions would be expected to decay over time through mutations and drift, as selection will be weak because of the rare expression of these genes. "Real evidence for sex would be to show that these males are fertile and sufficiently common to produce a meaningful amount of recombination for the whole population." (W.R. Rice, pers. comm. 18.4.2008).

In order to have a meaningful impact on the population and species level, not only do all the necessary male genes for spermatogenesis and copulation still have to be functional; the same applies for the female counterparts coding for a functional female reproductive system (e.g. spermatoreceptors) but even more so for copulatory behaviour. Therefore, even if rare males can be functional, females must still be receptive to copulatory signals provided by the males and accept them for reproduction. Vandekerkhove et al. (in prep.) found that male *E. virens* copulated significantly shorter with parthenogenetic than with sexual females; there was no evidence yet from these experiments in the lab for insemination or the successful production of offspring.

Although the discovery of the males in *Vestalenula cornelia* is most useful from a phylogenetic point of view because it finally allows to reconstruct ostracod phylogenies with morphological characters of the male appendices from all families, which was not possible before (Horne et al. 2005) and it pays homage to the patience by which their discoverers have worked their way through thousands (R. Smith, pers. comm.) of last instar females, their significance for the status of darwinulids as ancient asexuals depends on the ability to demonstrate their functionality see also Schurko et al. (2009). Preliminary results on genetic signals of recombination in females from these populations (Schön and Martens 2007) seem to indicate that these males are not functional.

11.2.2 Fossil Males

The first recognised darwinulids were found in deposits from the Carboniferous, and possible even as far back as the Devonian (Molostovskaya 2000). By the end of the Permian, hundreds of morphologically distinguishable darwinulid species had evolved in several families, some with recognisable sexual dimorphism (Abushik 1990). It seems therefore clear that at least some groups of the Darwinuloidea had sex in the Palaeozoic. The Permian/Triassic mass extinction, the largest of the Phanerozoic (Bowring et al. 1998), reduced the group dramatically, leaving only a few dozen species in one lineage, the Darwinulidae (Martens et al. 1998).

From the first half of the Mesozoic, Triassic and Jurassic (c 200 myr), several putative males were reported in the literature (Christensen 1963; Styk 1976; Urosevic 1979). Invariably, these records concerned one or two smaller specimens amongst few larger adult females. This difference in size was then interpreted as sexual size dimorphism because the presence of the brood pouch makes the valve imprints of female specimens broader than that of males (see Martens et al. (2003), for more details).

Martens et al. (2003) investigated a large, well-preserved assemblage of the darwinulid species *Alicenula leguminella* (Forbes 1885) from the latest Jurassic (c.145 myr) of England. The specimens were found to be markedly variable in size and shape, and some were sufficiently small to be putative males. Three criteria were used to test whether or not the assemblage was bisexual. Firstly, length and height was measured and the normality of the distributions was tested for a large number of specimens. If there was indeed sexual size dimorphism, these size/frequency distributions should have deviated from normality and fall into two (or more, if juveniles are present) distinct distribution classes. Secondly, the presence/absence of a posterior brood pouch was determined. In males, no such brood pouch should occur. Finally, the position of the central muscle scars was evaluated. Ostracod valves are closed by adductor muscles attached to the approximate centre of each valve, the functionally most efficient position. In adult brooding ostracods (e.g., female darwinuloids), the posterior extension of the carapace to form the brood pouch results in a relatively more anterior position for the adductor muscles (Griffiths and Horne 1998). Anteriorly-positioned adductor muscle scars on the inside of the valves may thus be taken as a female characteristic in darwinulids, whereas males are indicated by more central scars (see Fig. 11.4).

Martens et al. (loc. cit.) found that the length/height distributions did not significantly deviate from normality, that all specimens had externally visible brood pouches and that all specimens had their adductor muscle scars situated well in front of the middle. It was thus concluded that the assemblage consisted of females only, and that the observed size difference was a result of the fact that the assemblage represented dozens to hundreds of generations. It is well known that differences in salinity and temperature can cause substantial changes in size and shape of ostracod valves (Yin et al. 1999 for *Limnocythere inopinata*, and Rossetti and Martens1996 for *Darwinula stevensoni*).

However, no juveniles at all were found in this assemblage, possibly because larval instars have thin/poorly calcified valves that are less easily preserved, or because current or wave action had winnowed out the smaller valves. Since we now know

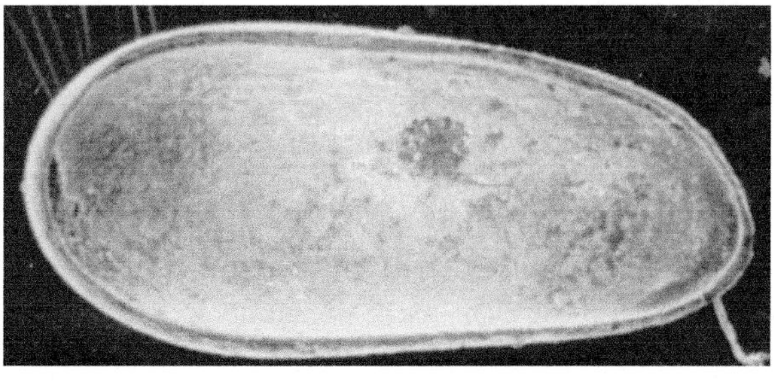

Fig. 11.4 Position of the adductor muscle scar in fossil valve of female *Alicenula* Modified after Martens et al. (2003).

that male darwinulids can resemble penultimate instar females (Smith et al. 2006), the possibility cannot entirely be excluded that also some putative males were not preserved in this particular assemblage. Other assemblages might have to be more thoroughly checked for the presence of males.

Note however, that in the L/H graph of Smith et al. (loc.cit.), the three male specimens do not cluster completely amongst the A-1 females, adult but rather cluster between the largest A-1 and the adult female specimens.

11.2.3 Genetic Signatures of Ancient Asexuality: The Meselson-White Effect

White (1973) was the first to suggest that ancient asexuality could be proven by estimating genetic divergence within individuals, because of the independent accumulation of mutations within the two alleles of a diploid organism. This test has meanwhile become widely known as the Meselson effect and it has been successfully demonstrated for bdelloid rotifers (Mark Welch and Meselson 2000). Meanwhile, new studies on bdelloid rotifers have provided an alternative explanation for the existence of the Meselson effect (Hur et al. 2008; Mark Welch et al. 2008; see also Chapter 13): Bdelloids are most likely allotetraploids resulting from an ancient hybridization event. Consequently, the high allelic diversities were not observed among homologue alleles from the same genome, but from intergenomic comparisons of the ancestors of these hybrids. Thus, the Meselson effect does not exist in bdelloid rotifers and might also not exist in other ancient asexuals. In the plant *Boechera,* some kind of Meselson effect was observed in flanking regions of microsatellites of apomicts; it has not been attributed to an old age of these lineages, but rather to a combination of mutation accumulation, gene duplication and hybridization events (see Chapter 23 for more details).

So far, the Meselson effect could not be confirmed for other investigated cases of putative ancient asexuals such as mites and fungi (Kuhn et al. 2001; Pawlowska and Taylor 2004; Hijri and Sanders 2005; Schäfer et al. 2006) while it revealed sex in the Placozoa (Signorovitch et al. 2005). However, as was pointed out by Butlin (2000) and Schön et al. (2008) – the Meselson effect is asymmetric: its presence may confirm long-term asexuality, but its absence does not necessarily refute it.

The first evidence that *Darwinula stevensoni* is genetically quite peculiar when compared to other ostracods was provided by Rossi et al. (1998) and Schön et al. (1998, 2000). Using allozyme markers, Rossi et al. (loc cit.) identified only 7 clones in 30 European populations of *Darwinula stevensoni* (Table 11.1). In other ostracods such as e.g. *Bradleystrandesia reticulata* (Turgeon and Hebert 1995), *Limnocythere inopinata* (Rossi et al. 1998) and *Eucypris virens* (Rossi et al. 1998) 185, 31 and 211 clones, were found in 29, 12 and 55 populations, respectively. Direct DNA sequencing (Schön et al. 1998) confirmed this low genetic diversity and provided results that were even more puzzling: in European populations of *D. stevensoni*, COI showed only about 2.2% variability and ITS1 none at all, whereas both markers

Table 1 Genetic variability of *Darwinula stevensoni*

Marker	N	Pop	Alleles	Multilocus geno	Private geno	Polyploidy
Allozyme	1822	36	2.5	7	0	0
DNA	**N**	**Pop**	**bp**	**% in pop**	**% in Europe**	**% in species**
ITS	56	14	320	0	0	0
COI	17	19	447	0	2.2	3.8

Allozyme data are from Rossi et al. (1998), DNA data from Schön et al. (1998).
N= total number of genotypes individuals; pop = populations; alleles = average number of alleles per locus. geno = genotype; bp = basepairs. % = % sequence diversity.

have shown substantial divergence in *E. virens* (Schön et al. 1998, 2000). Note that recent mitochondrial analyses has revealed more than 40 cryptic species in European populations of the morphospecies *E. virens* (Bode et al. submitted). A subsequent comparison between the European specimens of *D. stevensoni* and material from the South African Lake Sibaya confirmed this picture: 3.8% variability in COI, and 0% in ITS1 were observed. This very low variability in nuclear DNA sequences could not be owing to a recent selective sweep, because the variability in the COI sequences allowed Schön et al. (loc. cit.) to estimate that the European and South African populations must have diverged about 5–8 million years ago (depending on how the molecular clock was calibrated). The Sibaya population was therefore not a recent introduction by Europeans, which is easy to understand as Lake Sibaya was a holy place in Zululand and no foreigners where allowed near it until recently. The Zulu were fierce warriors ...

The difference between the net mutation rates of COI and ITS is not owing to a speed-up in COI, but to a slow down of ITS evolution in *D. stevensoni* (Schön et al. 1998). This slow-down is not present in the other analysed darwinulid species (Schön et al. 2003), whereas ITS evolved slower than COI in all ostracod superfamilies (Schön et al. 2003). Whether this difference between ITS and COI reflects an overall difference of molecular evolution between nuclear and mitochondrial ostracod genomes remains to be tested more thoroughly.

Even so, it was still possible that these results for *D. stevensoni* were linked to a particularity of its ITS region or the fact that this nuclear region was sequenced directly, in which case small numbers of point mutations might have been overlooked. Therefore, Schön and Martens (2003) used vector-cloning of PCR products to reanalyse three nuclear regions of *D. stevensoni*, the multi-copy ITS1 region and part of the single-copy genes *hsp82* and Calmodulin. In such an approach, single PCR products are sequenced, allowing detection of all possible genetic changes within an individual at that locus. Schön and Martens (2003) demonstrated similar results for all three nuclear regions, namely low levels of genetic diversity both within and between geographically different populations, confirming the earlier data on ITS1 from direct sequencing.

The low levels of genetic diversity within *Darwinula stevensoni* did not provide any evidence of the Meselson effect. The absence of a Meselson effect could be an indication of recombination, but can also be explained by the action of homogenizing mechanisms such as gene conversion or highly efficient DNA repair (Schön and Martens 1998b). Several authors found statistically significant evidence for gene conversion acting in ITS and *hsp82* of *D. stevensoni* and some *Penthesilenula* and *Microdarwinula* species (Schön and Martens 2003; Pinto et al. submitted). It seems doubtful that gene conversion on its own would be sufficient to keep genomes of ancient asexuals almost mutation-free for millions of years. The presence of other mechanisms such as highly efficient DNA repair (Schön and Martens 1998b) or ameiotic recombination (Omilian et al. 2006) are possible and their importance for the Darwinulidae is currently being tested.

11.2.4 Genomics: Transposons

In all genomes screened so far, transposons have been found (Feschotte and Pritham 2007). Two major classes of transposons can be distinguished – retrotransposons with vertical transmission (from parents to off-spring) and DNA transposons with vertical and horizontal transmission (between individuals within populations, even between species). For a long time, evolutionary theory predicted that ancient asexuals should contain fewer functional retrotransposons than sexuals or even none at all, because sex was thought to be necessary to spread these transposons within populations (Hickey 1982; Schön and Martens 2000; Arkhipova and Meselson 2005a). Except for bdelloid rotifers (Arkhipova and Meselson 2000) and yeast (Zeyl et al. 1996), however, this pattern could not be confirmed.

Schön and Arkhipova (2006) found one group of degenerated retrotransposons, *Daphne*, in *Darwinula stevensoni*, and a second transposon group, *Syrinx*, that was still functional and had obviously recently been active. Other asexuals such as the yeast *Candida albicans* (Mattews et al. 1997; Goodwin and Poulter 2000), *Entamoeban* protozoans (Pritham et al. 2005) and Foraminifera (Maumus et al. 2008) have also been shown to harbour functional retrotransposons in high numbers.

Computer simulations have provided us with more insights into which factors shape the relationships between host and transposons in short- and long-term asexuals (Docking et al. 2006; Dolgin and Charlesworth 2006). The latter authors demonstrated that population size is most crucial for determining the fate of retrotransposons in long-term asexual lineages as an equilibrated number of transposons is only reached in large (infinite) populations. If these populations furthermore have the ability to excise elements, transposons will eventually be purged from asexual populations. If there is no horizontal transfer, such asexual populations will be immune against further transposition. The existence of an infinite population size together with excision might resemble the situation in bdelloid rotifers, where only one group of retrotransposons called *Athena* (Arkhipova et al. 2003), but DNA transposons from five different superfamilies have been found (Arkhipova and Meselson 2005b).

The simulations by Dolgin and Charlesworth (2006) provide several scenarios under which active retrotransposons can persist in ancient asexuals: a large initial copy number before sex was abandoned, a small effective population size, low excision or high transposition rates, no synergism between elements or insertion of most transposable elements in neutral sites. Whether any of these assumptions can be put forward to explain the existence of active *Syrinx* retrotransposons in *Darwinula stevensoni* (Schön and Arkhipova 2006; see also above),will have to be revealed by future genomic and genetic research on this and other darwinulid species.

There is also growing evidence that (retro) transposons can be domesticated for beneficial host functions. Examples include *Drosophila* and bdelloid rotifers, where retrotransposons have taken over telomeric functions (Arkhipova et al. 2003; Pardue and DeBaryshe 2003). Research is currently under way to test whether also *Syrinx* elements preferentially insert into telomeric regions of *Darwinula stevensoni*, as their exceptionally long terminal repeats at the 3'-end might indicate.

11.2.5 Chromosomal Evidence: Aneuploidy

As suggested by several authors (Judson and Normark 1996; Schurko et al. 2009), the long-term absence of meiosis can cause aneuploidy (or unequal numbers of chromosomes) because there is no more selection for a functioning meiosis, during which chromosomes would normally pair (see also Chapter 4). There is little information on the chromosomal structure of non-marine ostracods (see Schön and Martens 1998a for a review). Cytogenetic studies on non-marine ostracods date back to the 1960s and 1970s, when, for example, Tétart (1978) studied female karyotypes of about a dozen species. In his diagrams, *Darwinula stevensoni* has 22 dot-like chromosomes. Because of the strange morphology of the chromosomes, which makes them all look alike, no homologue pairs can be attributed. Without further studies, it is also not certain that *D. stevensoni* is indeed diploid as the even number of dot-like chromosomes might indicate. Thus, there is so far no positive evidence for aneuploidy in the only screened darwinulid species, whereas an odd number of chromosomes was found in one out of four screened species of bdelloid rotifers (J. Mark Welch et al. 1998).

11.2.6 Ancient Asexual Status for Darwinulid Ostracods?

The jury is still out on whether the Darwinulidae as a whole are ancient asexuals, as none of the evidence (for or against) available to date is conclusive. The darwinulids do not seem to have the Meselson effect, but much of the presumed power of this effect is now doubtful, while there are good indications of active gene homogenizing mechanisms in several darwinulid species. There are active retrotransposons in at least *Darwinula stevensoni*, but these might be domesticated (e.g. in telomeric regions). No aneuploidy has so far been found, but the chromosomal data on darwinulids are quite old and again available for only one species.

The discovery of the extant males of *Vestalenula cornelia* is a highly interesting event in itself, but offers surprisingly little help in answering the question at hand, unless they can be shown to be functional and to be contributing a meaningful amount of recombination. In addition, now that we know what a male darwinulid looks like, the position of *D. stevensoni* as ancient asexual is strengthened, as the old record of the putative male in this species can now be dismissed. At least part of the Darwinulidae as a whole, and *D. stevensoni* in particular, are still firmly in the race for the title of ancient asexuals.

11.3 Ecological Strategies of Darwinulid Ostracods

If darwinulids have indeed managed to survive for millions of years without sex and recombination, which (if any) special mechanisms did they develop to counter the deleterious consequences? We have already indicated (see above) the potential of non-sex related homogenising mechanisms (gene conversion, DNA repair, mitotic recombination, . . .) to successfully counter the accumulation of deleterious mutations through Kondrachov's hatchet (Kondrashov 1988) and Muller's ratchet (Muller 1964), two of the most cited hypotheses that are expected to explain the existence of sex in the first place. But what about more ecologically oriented hypotheses, such as the Red Queen (Van Valen 1973), the Tangled Bank (Ghiselin 1974; Bell 1982), Fluctuating Selection (Maynard Smith 1980; Roughgarden 1991) and others? Most of these hypotheses deal with the fact that asexuals, in the absence of recombination, have impeded potential to evolve (and thus adapt) in changing environments, be they abiotic (e.g. Fluctuating Selection in fluctuating climatic changes) or biotic (e.g. the Red Queen for host-parasite, competitor and predator-prey relationships; see also Chapter 7).

11.3.1 GPG Versus FNV

Two major hypotheses describe how asexuals will use niches in relation to their sexual ancestors: the General Purpose Genotype (GPG) and the Frozen Niche hypotheses (FNV) (see also Chapter 6; Baker 1965; Vrijenhoek 1979, 1984). The FNV dictates that clonal spin-offs from sexual population will "freeze" the ecological niche of these sexual populations (e.g. with regard to tolerances related to temperature, salinity, oxygen, etc.). Because most sexual populations are adapted to current environmental conditions, their asexual spin-offs will generally inherit these limited tolerance ranges. However, since a species with mixed reproduction can have a large number of different clones spinning off from sexual ancestors (amongst other origins of asexual lineages; see for ostracod examples Schön et al. (2000) and Rossi et al. (1998)), the total ecological tolerance of a set of clones might still cover a wide range of environmental conditions.

In contrast, the GPG postulates that asexuals do not need to adapt at all to changing environments, for example to fluctuating climates, if they have a genotype which allows them to survive in a wide range of ecological conditions. If this genotype produces a phenotype with wide ecological tolerance, then the GPG hypothesis predicts that the resulting phenotype will have (1) a broad tolerance against a wide range of environmental factors; and (2) a very low variance in the tolerances of phenotypes derived from different populations.

Both predictions were confirmed in populations of *Darwinula stevensoni* by Van Doninck et al. (2002): the tested specimens (with similar genotypes; Schön et al. 1998 and Van Doninck et al. 2002) showed a wide tolerance for a mixture of temperature and salinity treatments, while a logit linear model analysis of the "survival" data showed that responses between animals from several freshwater lakes (Ireland, France) were indistinguishable. Only the responses of animals from a (slightly) saline Belgian lake deviated to some extent, which might indicate that there was a maternal effect. Responses from other darwinulid species varied: *Penthesilenula brasiliensis* (a species with intercontinental distribution) has even wider tolerances than *D. stevensoni* for some variables, while endemic darwinulids such as *Vestalenula molopoensis* had much narrower tolerance ranges and *P. aotearoa* is found in-between (Van Doninck et al. 2003b).

One could argue that a GPG can only persist in fully asexual lineages as recombination will almost certainly break-up the allele-combinations required for a GPG. Therefore, a GPG can also only originate and persist in fully asexual lineages. However, the chance that a sexual lineage would have a GPG at the time clones originate from this sexual root (and can thus freeze the GPG in the clonal lineage) is very small indeed (Van Doninck et al. 2003b), which might explain why few taxa seem to have evolved a *real* GPG (see Chapter 6 for examples). This represents something of a paradox, as this would imply that a GPG can only evolve through adaptation in asexual lineages, which by definition have impeded evolvability (but see Chapter 6).

11.3.2 Parasites or No Parasites?

A remaining question is the relevance of parasites in the persistence of ancient asexuals. The action of the Red Queen dynamics in host-parasite relationships has been extensively investigated, with a particular focus on the maintenance of sex (Hamilton et al. 1990; see also Chapter 7). However, it seems that a rotation system between common and rare clones prevents parasites from driving all asexual lineages into extinction (see for example Lively and Jokela (2002) and Jokela et al. (2003) on the *Potamopyrgus*-system): the most common clone is most heavily infected and decimated, and therefore becomes one of many rare clones in a population; a seemingly randomly chosen rare clone can thus rise to supremacy. These dynamics have been documented to take only a few dozen generations or less (Jokela et al. 2003; Decaestecker et al. 2007; see also Chapters 7 and 15).

How putative ancient asexual ostracods can deal with the Red Queen remains still unknown. Preliminar results (S. Adolfsson, unpubl.) seem to indicate that *Eucypris virens* and other non-marine ostracods have very low incidences of parasites, which might have predisposed them as a more likely group in which ancient asexuality may arise. Speculation aside, the relevance of parasites in the survival of long term asexual lineages remains unclear (see Chapter 7).

11.3.3 Marginal Habitats and Long-Term Asexuality

Most darwinulids and bdelloids (see Chapter 13) can be found in what could generally be termed "marginal habitats", such as moist mosses and other (semi-) terrestrial environments while oribatid mites occur almost exclusively in the soil (see Chapter 12). The exceptions to a distribution in marginal habitats are the darwinulid species with GPG, such as *Darwinula stevensoni* and *Penthesilenula brasiliensis*: they are ubiquitous, or nearly so, and can also be found in lakes, rivers, interstitially, etc. But what is the significance of the fact that the three putative ancient asexual groups generally occur in such habitat types?

One possibility is that competition (with sexual species) would be lower, as there are simply fewer species occurring in such highly fluctuating habitats. Another reason could be that asexuals are able to survive in very low densities over many generations, since they do not need to find a mate in order to reproduce (Van Dijk 2007, Hörandl 2008). For all sexual populations, there is a density threshold (mediated by size and mobility of animals, amongst other biological characteristics) below which the probability of finding a mate is too low to ensure sufficient reproduction for the population to remain viable. In marginal habitats, such as semi-terrestrial ones,conditions may vary widely and asexuals would have the advantage over sexuals. If asexuals are sufficiently small so that effective population sizes still remain large in spite of low densities (and all three putative asexual groups have very small body sizes), such low densities may not necessarily lead to genetic bottle necks as sufficient genetic variability might survive. Additionally, this kind of habitats might provide opportunities for ancient asexuals to escape parasites (see above and Chapter 7).

11.3.4 Reduced Mutation Rates

As mentioned above, the accumulation of deleterious mutations is still regarded as one of the major problems of long-term asexuality (Muller 1964; Kondrashov 1988). Although *Darwinula stevensoni* can be found in several types of habitats, most of these have in common that their exposure to potential mutagens may be relatively low: if the species occurs in lakes, it is mostly in rather turbid water with low oxygen levels and at depths of around 1 m where little UV penetrates. The same is true for interstitial habitats, slow flowing rivers etc. This distribution pattern in itself reduces mutation rates. Also, shaded litter in rain forests, mosses near waterfalls etc. generally have reduced UV input. In addition, darwinulid ostracods have transparent valves without colours, except for *D. stevensoni* which has two dark-coloured patches on the valves, approximately near the mid-dorsal position of the gonads, which could perhaps help to keep the germline free of mutations (see also Fig. 11.2). For *Eucypris virens*, preliminary experiments indicate that its greenly pigmented valves provide protection against UV (Van Den Broecke et al. 2007). It is as yet unknown whether the chemical structure of the calcified valves themselves (without pigmentation) can significantly reduce UV damage to the underlying tissue.

Also, the low number of maximal 11–15 offspring in *Darwinula stevensoni* (Van Doninck et al. 2003a) can further reduce overall mutation rates, as can a generally slower metabolism. The latter remains apparent from the slow movement and long developmental times as compared to other ostracods (see Box 11.1). Last but not least, the absence of males is an additional important component of decreasing mutation rates because most mutations arise in the male germline (Redfield 1994). Computer simulations (loc. cit.) show that the cost of male mutations can easily exceed the benefit of recombination.

Schön et al. (2003) have demonstrated that molecular evolution is slow in all investigated ostracods, sexual and asexual. The presence of homogenising mechanisms such as gene conversion, DNA repair and mitotic recombination are most likely the most important factors to reduce mutation rates in darwinulids (Schön and Martens 1998b), although the quantification of their effects still remains to be demonstrated.

11.3.5 Brood Selection or Enhanced Fecundity?

It has been suggested (Lively and Johnson 1994) that brooding asexuals have an additional advantage over non-brooders to counter, or at least slow down, the accumulation of deleterious mutations by actively removing defective embryos from the brood pouch. Such brood selection could indeed purge genomes with a high incidence of deleterious mutations from the gene pool, provided that females have the means to detect such deficient embryos. All darwinulids are brooders and Horne et al. (1998c) showed that *Darwinula stevensoni* has the ability to detect, select and eject unwanted material such as eggs or embryos from the brood pouch. But whether or not the observed ejection of eggs truly constitutes brood selection in the sense of Lively and Johnson (1994) remained equivocal.

Pinto et al. (2007) tested the brood selection hypothesis in the darwinulid *Penthesilenula brasiliensis* (Pinto and Kotzian 1961) by checking for the incidence of early ejection of eggs and first instar juveniles from the brood pouch (here termed *abortion*) and for the viability of the ejected eggs. Also, effects of temperature on the incidence of release and on the rate of early development were checked. *Penthesilenula brasiliensis* can brood its young up to the second instar (third instar in *D. stevensoni*) but almost 30% of the eggs of *P. brasiliensis* were aborted at 17°C (the temperature of the original locality), against less than 10% at 22°C, a presumably less favourable temperature. The majority of the ejected eggs remained viable and hatched and the juveniles moulted to later instars. They survived for several months, but none ever reached adult hood in the experiments of Pinto et al. (loc.cit.). Eggs and embryos developed significantly faster at the higher temperature, but there was no significant difference between brooded and rejected eggs in developmental rate.

There are two major strategies that may lie at the core of these observations. On the one hand, these egg releases might indeed be incidences of brood selection. On the other hand, these early releases of eggs might be a strategy to increase fecundity in favourable circumstances: such release reduces overcrowding in the brood space and allows for higher numbers of offspring. The first hypothesis is corroborated by the fact that 80% of the individuals that had hatched from aborted eggs had died within one year. The second one is supported by the observation that early release is temperature-dependent (and highest at the most favourable temperature) and by the fact that initial mortality was not significantly different between brooded and aborted individuals. A third possibility is that females also abort healthy individuals as an error of judgement, but this seems highly wasteful, and does not explain the temperature dependence of the phenomenon.

Since the juveniles developed about twice as fast at the higher temperature, and since a female brood pouch in this species can generally hold only up to 5–6 offspring, it was concluded that increased fecundity, and not brood selection, was most likely the main explanation for the early release of eggs. The fact that nearly always eggs, not juveniles, were rejected also supports this hypothesis. Nevertheless, incidence of brood selection could not be entirely excluded, as some releases occurred when the brood pouch was only half full. Some exceptionally slowly developing individuals might have been deficient individuals eliminated by brood selection. Both strategies could have occurred simultaneously.

11.4 How Darwinulids Could Have Survived Without Sex for Millions of Years

Given the differences regarding speciosity, individual genetic variability, genetic mechanisms underlying the formation of gametes (apomixis versus automixis), incidence of rare males and prevalence of homogenizing mechanisms between the three putative ancient asexual groups, it is clear that there is no single scenario that will explain the origin and persistence of all three groups. Likewise, it seems wise to at least consider the possibility that even within these groups, different strategies might exist.

Darwinula stevensoni is the most likely candidate within the darwinulids to be a true ancient asexual (Martens and Schön 2008): it has an all-female fossil record going back 25 million years (Straub 1952), no males have thus far been found (functional or atavistic, recent or fossil) for this widespread species, it has a GPG and active gene conversion, amongst other homogenising genetic mechanisms. It is thus possible that this species has managed to survive without sex over longer time frames by simply arresting evolution, as the presence of a GPG eliminates (or at least reduces) the need for fast adaptation to changing environments. The fact that this genus is monospecific (Schön et al. submitted) is congruent with the absence of evolution and speciation. In order to maintain the GPG, the species needed to give up sex altogether (to prevent recombination from destroying the unique combination of alleles that make up the GPG; Van Doninck et al. 2003b), as well as maintain a high activity of homogenising mechanisms, such as gene conversion (Schön and Martens 2003), but possibly also efficient DNA repair systems (Schön and Martens 1998b). Several aspects of this integrated strategy have already been demonstrated (Van Doninck et al. 2002; Schön and Martens 2003; Schön et al. submitted;), others are presently under investigation.

Ecological strategies of species in *Alicenula* and *Microdarwinula* are thus far completely unknown. Neither of these genera is monospecific (Rossetti and Martens 1998) and their apparent rarity seems to exclude the presence of a GPG, but ecological experiments have to verify this hypothetical correlation between distribution and ecological strategies (Van Doninck et al. 2003a). The lack of sufficient genetic data prevents further statements to date.

The genera *Penthesilenula* and *Vestalenula* each comprise about a dozen species (Rossetti and Martens 1998), which seems to contradict the concept of arrested evolution as in *Darwinula*. It is thus likely that these lineages have evolutionary strategies that are at least partly incongruent with those of *D. stevensoni*. This does not necessarily mean that these species are sexual, or have been so very recently. At least *P. brasiliensis* also has a GPG (Van Doninck et al. 2003b) and is seemingly ubiquitous and close to cosmopolitan; also, this species is a good candidate for an ancient asexual, and it seems that similar homogenising mechanisms as in *D. stevensoni* are active (Pinto et al. submitted). Other species in *Penthesilenula* and *Vestalenula* seem to have restricted distributions and narrow ecological tolerances (Van Doninck et al. 2002, 2003b). If our hypothetical link between a GPG and homogenizing, genetic mechanisms is right and they have been asexual for as long as *D. stevensoni* (25 myr; Straub 1952), then they might have developed other genetic strategies to counter the deleterious effects of the absence of sex.

11.5 What Remains to be Discovered...

Given that *Darwinula stevensoni* is thus far the best studied model-darwinulid, further studies on this species appear most logical. Truly efficient homogenizing mechanisms might have turned this species into a functional haploid, which could make its long-term survival more likely, because lower mutational loads can be facilitated by lower ploidy levels (Otto and Whitton 2000). This could be verified by estimating its genome size and discovering the developmental mechanisms for apomictic egg production. It also remains to be thoroughly tested whether this species indeed has highly efficient DNA repair systems. Also, additional ecological studies, for example the response of *D. stevensoni* to abiotic stress (an alternative way to confirm the GPG) or investigating parasitic loads, could provide important insights for testing standing hypotheses on the prevalence of sex. Finally, verifying and obtaining more detail on the fossil record of this species could also be most useful.

General questions needing further research on darwinulids include the importance of valves for the prevention of UV-induced mutations, the general activity of transposons and of meiotic proteins and of course the functionality (or absence thereof) of the males of *Vestalenula cornelia*.

Studies of evolutionary strategies of ancient asexual lineages could go a long way towards helping us to understand exactly which deleterious effects of the long term absence of sex need to be countered. Research on putative ancient asexual groups thus remains a vital part of the concerted efforts to solve the paradox of sex.

Acknowledgements IS and KM acknowledge the Belgian OSTC for funding (MO/36/005 and MO/36/015). This work was also funded by the EU Marie Curie Research Training Network SexAsex (From Sex to Asex: a case study on interactions between sexual and asexual reproduction, contract MRTN-CT-2004-512492).

References

Abushik AF (1990) Palaeozoic Ostracoda. In: Abushik AF (ed) Manual of the Microfauna of the USSR 4, VSE-GEI, NEDRA, pp. 1–35 [In Russian]

Arkhipova I, Meselson M (2000) Transposable elements in sexual and ancient asexual taxa. Proc Natl Acad Sci USA 97: 14473–14477

Arkhipova I, Meselson M (2005a) Deleterious transposable elements and the extinction of asexuals. Bioessays 27: 76–85

Arkhipova IR, Meselson M (2005b) Diverse DNA transposons in rotifers of the class Bdelloidea. Proc Natl Acad Sci USA 102: 11781–11786

Arkhipova IR, Pyatkov KI, Meselson M, Eugen'ev MB (2003) Retroelements containing introns in diverse invertebrate taxa. Nat Genet 33: 123–124

Baker HG (1965) Characteristics and modes of origin of weeds. In: Baker HG, Stebbins GL (eds) Genetics of Colonizing Species, Academic Press, New York, pp. 147–172

Bell G (1982) The Masterpiece of Nature. Croom Helm, London, 635 pp.

Bowring SA, Erwin DH, Jin YG, Martin MW, Davidek K, Wang W (1998) U/Pb Zircon geochronology and tempo of end-Permian mass extinction. Science 280: 1039–1045

Brady GS, Norman AM (1889) A monograph of the marine and freshwater Ostracoda of the North Atlantic and of North-Western Europe. Section I: Podocopa. Sci Trans R Dublin Soc Ser 2, 4: 63–270

Brady GS, Robertson D (1870) The Ostracoda and Foraminifera of tidal rivers. Ann Mag Nat Hist Ser 4, 6: 1–33, 307–309

Butlin RK (2000) Virgin rotifers. Trends Ecol Evol 15: 389–390

Christensen OB (1963) Ostracods from the Purbeck-Wealden beds in Bornholm. Danmarks Geol Unders Ser II 86: 1–58

Decaestecker E, Gaba S, Raeymaekers J, Stoks R, Van Kerckhoven L, Ebert D, De Meester L (2007) Host-parasite Red Queen dynamics archived in pond sediment. Nature 450: 870–873

Docking TR, Saadé FE, Elliott MC, Schoen DJ (2006) Retrotransposon sequence variation in four asexual plant species. J Mol Evol 62: 375–387

Dolgin ES, Charlesworth B (2006) The fate of transposable elements in asexual populations. Genetics 174: 817–827

Feschotte C, Pritham EJ (2007) DNA Transposons and the evolution of Eukaryotic Genomes. Annu Rev Genet 41: 331–368

Geiger W (1998) Population dynamics, life histories and reproductive modes In: Martens K (ed) Sex and Parthenogenesis: Evolutionary Ecology of Reproductive Modes in Non-marine Ostracods, Backhuys Publ, Leiden, pp. 215–228

Ghiselin MT (1974) The economy of nature and the evolution of sex. University of California Press, Berkeley, CA

Goodwin TDJ, Poulter RTM (2000) Multiple LTR retrotransposon families in the asexual yeast *Candida albicans*. Genome Res 10: 174–191

Griffiths HI, Horne DJ (1998) Fossil distribution of reproductive modes in non-marine ostracods. In: Martens K (ed) Sex and parthenogenesis: evolutionary ecology of reproductive modes in non-marine ostracods, Backhuys Publ, Leiden, pp. 101–118

Hamilton WD, Axelrod R, Tanese R (1990) Sexual reproduction as an adaptation to resist parasites (a review). Proc Natl Acad Sci USA 87: 3566–3573

Hayden EC (2008) Scandal! Sex-starved and still surviving. Nature 452: 678–680

Hickey DA (1982) Selfish DNA: a sexually-transmitted nuclear parasite. Genetics 101: 519–531

Hijri M, Sanders IR (2005) Low gene copy number shows that arbuscular mycorrhizal fungi inherit genetically different nuclei. Nature 433: 160–163

Hörandl E (2008) Evolutionary implications of self-compatibility and reproductive fitness in the apomictic *Ranunculus auricomus* polyploid complex (Ranunculaceae). Int J Plant Sci 169: 1219–1228

Horne DJ. Baltanas A and Paris G (1998a) Geographical distribution of reproductive modes in living non-marine ostracods. In: Martens K (ed) Sex and parthenogenesis: evolutionary ecology of reproductive modes in non-marine ostracods, Backhuys Publ, Leiden, pp. 77–99

Horne DJ, Danielopol DL, Martens K (1998b) Reproductive behaviour. In: Martens K (ed) Sex and parthenogenesis: evolutionary ecology of reproductive modes in non-marine ostracods. Backhuys Publ, Leiden, pp. 157–196

Horne DJ, Martens K, Mösslacher F (1998c) A short note: is there brood selection in *Darwinula stevensoni*? In: Crasquin-Soleau S, Braccini E and Lethiers F (eds) What about Ostracoda! Actes du 3e Congrés Européen des Ostracodologistes, Paris-Bierville, France, 8–12 Juillet 1996. Bull Centre Rech Elf Explor Prod Mem 20: 33–35

Horne DJ, Schön I, Smith RJ, Martens K (2005) What are Ostracoda? A cladistic analysis of the extant superfamilies of the subclasses Myodocopa and Podocopa (Ostracoda: Crustacea). Crustac Iss 16: 249–274

Hur JH, Van Doninck K, Mandigo ML, Meselson M (2008) Degenerate tetraploidy was established before bdelloid rotifer families diverged. Mol Biol Evol 26: 375–383

Jokela J, Lively CM, Dybdahl MF, Fox JA (2003) Genetic variation in sexual and clonal lineages of a freshwater snail. Biol J Linn Soc 79: 165–181

Judson PO, Normark BB (1996) Ancient asexual scandals. Trends Ecol Evol 11: 41–46

Kondrashov AS (1988) Deleterious mutations and the evolution of sexual reproduction. Nature 336: 435–440

Kuhn G, Hijri M, Sanders IR (2001) Evidence for the evolution of multiple genomes in arbuscular mycorrhizal fungi. Nature 414: 745–748

Kuhn TS (1962) The Structure of Scientific Revolutions, 1st ed, University of Chicago Press, Chicago, IL

Little T, Hebert PDN (1996) Ancient asexuals: scandal or artefact? Trend Ecol Evol 11: 296

Lively CM, Johnson SG (1994) Brooding and the evolution of parthenogenesis: strategy models and evidence from aquatic invertebrates. Proc R Soc Lond B 256: 89–95

Lively CM, Jokela J (2002) Temporal and spatial distributions of parasites and sex in a freshwater snail. Evol Ecol Res 4: 219–226

McGregor DL (1969) The reproductive potential, life history and parasitism of the freshwater ostracods *Darwinula stevensoni* (Brady and Robertson). In: Neale JW (ed) The Taxonomy, Morphology and Ecology of Recent Ostracoda. Oliver and Boyd, Edinburgh, pp. 194–221.

Maddocks R (1973) Zenker's organ and a new species of *Saipanetta* (Ostracoda). Micropalaentology 19: 193–208.

Mark Welch D, Meselson M (2000) Evidence for the evolution of bdelloid rotifers without sexual reproduction or genetic exchange. Science 288: 1211–1214

Mark Welch DB, Mark Welch JL, Meselson M (2008) Evidence for degenerate tetraploidy in bdelloid rotifers. Proc Natl Acad Sci USA 105: 5145–5149

Mark Welch JL, Meselson M (1998) Karyotypes of bdelloid rotifers from three families. Hydrobiologia 387/388: 403–407

Martens K, Horne D, Griffiths H (1998) Age and diversity of non-marine ostracods. In: Martens K (ed) Sex and Parthenogenesis: Evolutionary Ecology of Reproductive Modes in Non-marine Ostracods , Backhuys Publ, Leiden, pp. 37–55

Martens K, Rossetti G, Horne DJ (2003) How ancient are ancient asexuals? Proc R Soc Lond B 270: 723–729

Martens K, Schön I (2008) Opinion: ancient asexuals: darwinulids not exposed. Nature 453: 587

Martens K, Schön I, Meisch C, Horne DJ (2008) Global diversity of ostracods (Ostracoda, Crustacea) in freshwater. Hydrobiologia 595: 185–193

Martins MJF, Vandekerkhove J, Namiotko T (2008) Environmental stability and the distribution of the sexes: insights from life history experiments with the geographic parthenogen *Eucypris virens* (Crustacea: Ostracoda). Oikos 117: 829–836

Mattews G, Goodwin TJD, Butler MI, Berryman TA, Poulter RTM (1997) pCal, a highly unusual Ty1/copia retrotransposon from the pathogenic yeast *Candida albicans*. J Bact 179: 7118–7128

Maumus F, Allen A, Jabbari K, Vardi A, Bowler C (2008) Pirate transposons in diatom genomes. Abstract band of the International Congress on Transposable Elements, April 20–23 2008, Saint-Malo, France, p. 155

Maynard Smith J (1980) Selection for recombination in a polygenic model. Genet Res 35: 269–277

Molostovskaya II (2000) The evolutionary history of Late Permian Darwinulocopina Sohn, 1988 (Ostracoda) from the Russian Plate. Hydrobiologia 419: 125–130

Muller HJ (1964) The relation of recombination to mutational advance. Mutat Res 1: 2–9

Okubo I (2001) Freshwater Ostracod *Darwinula stevensoni* from Lake Biwa. Bull Biogeogr Soc Jp 56: 39–43

Omilian AR, Cristescu MEA, Dudycha JL, Lynch M (2006) Ameiotic recombination in asexual lineages of *Daphnia*. Proc Natl Acad Sci USA 103: 18638–18643

Otto P, Whitton J (2000) Polyploid incidence and evolution. Annu Rev Genet 34: 401–437

Pardue ML, DeBaryshe PG (2003) Retrotransposons provide an evolutionarily robust non-telomerase mechanism to maintain telomeres. Annu Rev Genet 37: 485–511

Pawlowska TE, Taylor JW (2004) Organization of genetic variation in individuals of arbuscular mycorrhizal fungi. Nature 427: 733–737

Pinto RL, Rocha CEF, Martens K (2007) Early release of eggs and embryos in a brooding ancient asexual ostracod: brood selection or bet-hedging to increase fecundity? Hydrobiologia 585: 249–253

Pritham EJ, Feschotte C, Wessler SR (2005) Unexpected diversity and differential success of DNA transposons in four species of *Entamoeba* Protozoans. Mol Biol Evol 22: 1751–1763

Ranta E (1979) Population biology of *Darwinula stevensoni* (Crustacea, Ostracoda) in an oligotrophic lake. Ann Zool Fenn 16: 28–35

Redfield RJ (1994) Male mutation rates and the cost of sex for females. Nature 369: 145–147

Rossetti G, Martens K (1996) Redescription and morphological variability of *Darwinula stevensoni* (Brady & Robertson, 1870) (Crustacea, Ostracoda). Bull K Belg Inst Natuurw Biol 66: 73–92

Rossetti G, Martens K (1998) Taxonomic revision of the Recent and Holocene representatives of the family Darwinulidae (Crustacea, Ostracoda), with a description of three new genera. Bull K Belg Inst Natuurw Biol 68: 55–110

Rossi V, Schön I, Butlin RK, Menozzi P (1998) Clonal genetic diversity. In: Martens K (ed) Sex and Parthenogenesis: Evolutionary Ecology of Reproductive Modes in Non-marine Ostracods, Backhuys Publ, Leiden, pp. 257–274

Roughgarden J (1991) The evolution of sex. Am Nat 138: 934–953

Schäfer I, Domes K, Heethoff M, Schneider K, Schön I, Norton RA, Scheu S, Maraun M (2006) No evidence for the 'Meselson effect' in parthenogenetic oribatid mites (Oribatida, Acari). J Evol Biol 19: 184–193

Schön I, Arkhipova IR (2006) Two families of non-LTR retrotransposons, *Syrinx* and *Daphne,* from the darwinulid ostracod, *Darwinula stevensoni.* Gene 371: 296–307

Schön I, Butlin RK, Griffiths HI, Martens K (1998) Slow molecular evolution in an ancient asexual ostracod. Proc R Soc Lond B: 265, 235–242

Schön I, Gandolfi A, Di Masso E, Rossi V, Griffiths HI, Martens K, Butlin RK (2000) Persistence of asexuality through mixed reproduction in *Eucypris virens* (Crustacea, Ostracoda). Heredity 84: 161–169

Schön I, Lamatsch D, Martens K. (2008) Lessons to learn from ancient asexuals. In: Egel R, Lankenau D-H (eds) Genomic Dynamics and Stability, Vol. 2: Meiosis and Recombination. Crossing Over and Disjunction. Springer Publishers, Berlin, pp. 341–376

Schön I, Martens K (1998a) Sex determination in non-marine ostracods. In: Martens K (ed) Sex and Parthenogenesis: Evolutionary Ecology of Reproductive Modes in Non-marine Ostracods), Backhuys Publ, Leiden, pp. 25–36

Schön I, Martens K (1998b) Opinion: DNA-repair in ancient asexuals: a new solution to an old problem? J nat Hist 32: 943–948

Schön I, Martens K (2000) Transposable elements and asexual reproduction. Trends Ecol Evol 15: 287–288

Schön I, Martens K (2007) Is there molecular evidence for sex in *Vestalenula cornelia*? Abstract book of the 19th Senckenberg Conference and the European Ostracodologist's Meeting VI, Frankfurt Main, Germany (5–7 september 2007), p. 45

Schön I, Martens K (2003) No slave to sex. Proc R Soc Lond B 270: 827–833

Schön I, Martens K, Van Doninck K, Butlin RK (2003) Evolution in the slow lane: molecular rates of evolution in sexual and asexual ostracods (Crustacea: Ostracoda). Biol J Linn Soc 79: 93–100

Schurko AM, Neiman M, Logsdon JM Jr (2009) Signs of sex; what we know and how we know it. Trends Ecol Evol 24: 208–217

Signorovitch AY, Dellaporta SL, Buss LW (2005) Molecular signatures for sex in the Placozoa. Proc Natl Acad Sci USA 102: 15518–15522

Smith RJ, Kamiya T, Horne DJ (2006) Living males of the 'ancient asexual' Darwinulidae (Ostracoda: Crustacea). Proc R Soc Lond B 273: 1569–1578

Straub EB (1952) Mikropaläontologische Untersuchungen im Tertiär zwischen Ehingen und Ulm a.d. Donau. Geol Jb 66: 433–523

Styk O (1976) Triassic assemblages of ostracods from Brackish marine waters of Poland. Abh Verh Naturwiss Ver Hamburg (NF) 18/19 (Suppl): 275–277

Taberly G (1987) Recherches sur la parthénogenèse thélytoque de deux espèces d'acariens oribatides: *Trhypochthnonius tectorum* (Berlese) et *Platynothrus peltifer* (Koch). IV. Observations sur les males ataviques. Acarologia 29: 95–107

Tétart J (1978) Les garnitures chromosomiques des Ostracodes d' eau douce. Trav Lab Hydrobiol Univ Grenoble 39–70: 113–140

Turgeon J, Hebert PDN (1995) Genetic characterization of breeding systems, ploidy levels and species boundaries in *Cypricercus* (Ostracoda). Heredity 75: 561–570

Turner CH (1895) Freshwater Ostracoda of the United States. Minn Geol Nat Hist Surv Zool Ser 2: 277–337.

Urosevic D (1979) Stratigraphic position of sediments with *Darwinula* in the Rhaetian of Stara Planina Mountain (Yugoslavia). In: Krstic N (ed) Taxonomy, Biostratigraphy and Distribution of Ostracodes, Proc 7th Int Symp Ostracodes. Serbian Geol Soc, Beograd, pp. 109–112 [in Russian]

Van den Broecke L, Schön I, Martens K (2007) DNA repair in the asexual ostracods *Darwinula stevensoni* and *Eucypris virens*. Abstract book of the Open Project Meeting of the SEXASEX EU RTN MCA project "Paradox of sex, theory and data", Zürich, Switzerland (14–15 september 2007)

Van Dijk PJ (2007) Potential and realized costs of sex in dandelions, *Taraxacum officinale* s.l. In: Hörandl E, Grossniklaus U, van Dijk P, Sharbel T (eds) Apomixis: Evolution, Mechanisms and Perspectives. Gantner, Ruggell, pp. 215–233

Van Doninck K, Schön I, De Bruyn L, Martens K (2002) A general purpose genotype in an ancient asexual. Oecologia 132: 205–212

Van Doninck K, Schön I, Martens K, Godderis B (2003a) The life cycle of the ancient asexual ostracod *Darwinula stevensoni* (Brady and Robertson, 1870) (Crustacea, Ostracoda) in a temperate pond. Hydrobiologia 500: 331–340

Van Doninck K, Schön I, Maes F, De Bruyn L, Martens K (2003b) Ecological strategies in the ancient asexual animal group Darwinulidae. Freshwater Biol 48: 1285–1294

Van Valen LM (1973) A new evolutionary law. Evol Theory 1: 130

Vrijenhoek RC (1979) Factors affecting clonal diversity and coexistence. Am Zool 19: 787–797

Vrijenhoek RC (1984) Ecological differentiation among clones: the frozen niche variation model. In: Wöhrmann K, Loeschcke V (eds) Population Biology and Evolution. Springer-Verlag, Heidelberg, pp. 217–231

White MJD (1973) Animal cytology and evolution, Cambridge University Press, Cambridge, UK

Yin Y, Geiger W, Martens K (1999) Effects of genotype and environment on phenotypic variability in *Limnocythere inopinata* (Crustacea: Ostracoda). Hydrobiologia 400: 85–114

Zeyl C, Bell G, Green DM (1996) Sex and the spread of the retrotransposon Ty3 in experimental populations of *Saccharomyces cerevisiae*. Genetics 143: 1567–1577

Chapter 12
Parthenogenesis in Oribatid Mites (Acari, Oribatida): Evolution Without Sex

Michael Heethoff, Roy A. Norton, Stefan Scheu and Mark Maraun

Abstract Oribatid mites (Acari, Oribatida) are an extraordinarily old and speciose group of chelicerate arthropods that probably originated in Silurian times. A high number (∼10%) of oribatid mite species reproduces via parthenogenesis, presumably by terminal fusion automixis with holokinetic chromosomes and an inverted sequence of meiotic divisions. Several of the old taxa of oribatid mites likely have radiated while being parthenogenetic. Many species of those parthenogenetic clusters are morphologically distinct – this distinctness contrasts with high genetic variance, as has been confirmed by molecular studies, e.g. for *Platynothrus peltifer* and species of the genus *Tectocepheus*. *Platynothrus peltifer* comprises at least seven distinct molecular lineages which are geographically separated and may be recognized as cryptic species. Stable isotope ratios ($^{15}N/^{14}N$ and $^{13}C/^{12}C$) of oribatid mite species indicate that they occupy distinct trophic niches; however, the exact nature of these niches is unknown. One of the few microhabitats colonized by specific oribatid mite species is the bark of trees. The tree-inhabiting genus *Crotonia* re-evolved sexual reproduction from parthenogenetic ancestors, potentially while colonizing trees. Understanding the high degree of parthenogenetic reproduction in soil living oribatid mites allows the dissection of the functional role and evolution of sexual reproduction, and the factors responsible for the long-term survival and radiation of parthenogenetic species.

12.1 General Biological Aspects of Oribatid Mites

12.1.1 Overview

Oribatid mites (Acari, Oribatida; Fig. 12.1) are a speciose group: about 10,000 species are now described (Subias 2004) and 100,000 may actually exist (Schatz 2002). While mainly soil-living, several groups also live on trees or in aquatic

M. Heethoff (✉)
Abteilung für Evolutionsbiologie der Invertebraten, Institut für Evolution und Ökologie, Eberhard Karls Universität Tübingen, Auf der Morgenstelle 28 E, 72076 Tübingen, Germany
e-mail: heethoff@gmx.de

I. Schön et al. (eds.), *Lost Sex*, DOI 10.1007/978-90-481-2770-2_12,

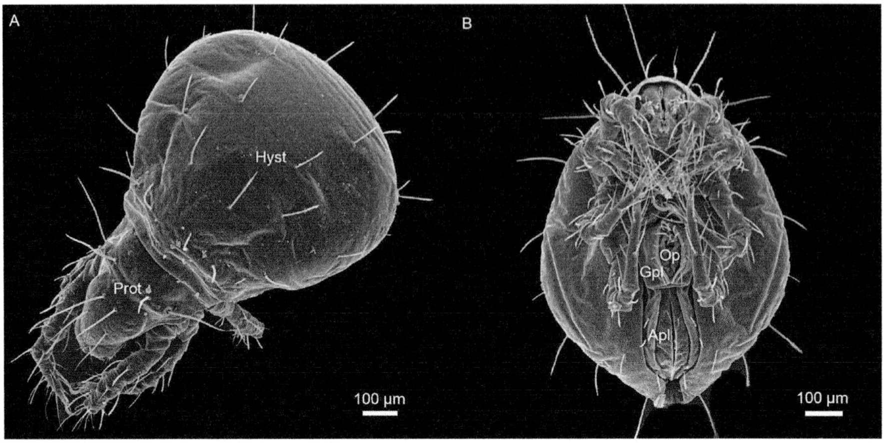

Fig. 12.1 Micrographs (SEM) of *Archegozetes longisetosus* (Trhypochthoniidae). **a**: dorsal view, **b**: ventral view. Apl: anal plate, Gpl: genital plate, Hyst: hysterosoma, Op: ovipositor, Prot: proterosoma.

systems. They range in length mostly between 200 and 800 μm, and are unusual among chelicerates for two reasons. Firstly, they are principally particle-feeding saprophages and microbivores, rather than fluid-feeders (Heethoff and Norton 2009). Secondly, as adults, they display a great range of physical and chemical defense mechanisms (Sanders and Norton 2004; Schmelzle et al. 2008), the most common of which is heavy sclerotization or mineralization of the cuticle that results in the common names "beetle mites", "armoured mites" and "box mites". An interesting feature of oribatid mites is that parthenogenesis is one or two orders of magnitude more common than in most other eukaryotic groups: about 10% of all species reproduce parthenogenetically (Norton and Palmer 1991; Norton et al. 1993).

12.1.2 Geological Age

Oribatid mites have seemingly existed since the first complex terrestrial communities evolved. The first indisputable fossil record is from the Devonian (380 million years; Shear et al. 1984; Norton et al. 1988), although the origin of the group probably dates back to 400–440 million years (Lindquist 1984). Based on specific patterns of occurrence, low dispersal power and genetic distances, the distribution of oribatid mite species seems to largely reflect vicariance associated with continental drift, rather than dispersal (Hammer and Wallwork 1979; Heethoff et al. 2007a). From biogeographic inferences, some species predated the breakup of the great landmass of Pangea about 200 million years ago yet kept their distinct morphology (Hammer and Wallwork 1979).

12.1.3 Population Density

The population density of oribatid mites is predictable within broad ranges (Maraun and Scheu 2000). In acidic boreal forests, they reach densities of up to 400,000 ind/m^2 whereas in calcareous forests, densities are usually somewhere between 20,000 and 40,000 ind/m^2. In tropical lowland and mountain rainforests densities are also rather low, e.g. in a tropical Brazilian lowland about 10,000 ind/m^2 (Badejo et al. 2002) and in a tropical mountain rainforest about 15,000 ind/m^2 (Maraun et al. 2008). There is little seasonal fluctuation of oribatid mite densities, indicating that the communities are in equilibrium conditions.

One of the main factors regulating the density of oribatid mites is the substrate: however, this affects oribatid mites indirectly by triggering, for example, the presence of macrofauna decomposers such as earthworms and diplopods. These reach high density and biomass in calcareous forests, and by feeding and removing litter material they destroy the habitat of litter-living mesofauna, thereby detrimentally affecting oribatid mite communities. The density of oribatid mites in base-rich forests therefore is low. Exceptions are some Canadian forests which are base-rich but virtually devoid of macrodecomposers, in particular earthworms, due to glaciation. In these forests (e.g. in the mountain ranges of western Alberta, Canada), oribatid mite densities are high. However, recent invasion by European earthworm species has transformed these systems by incorporating the ectorganic matter into the mineral soil thereby damaging the habitat of litter-living invertebrates; this in turn has caused the density of oribatid mites to decline strongly (Migge-Kleian et al. 2006; Eisenhauer et al. 2007).

Laboratory and field experiments with European soil showed that mechanical disturbance by earthworms is one of the main factors responsible for low densities of oribatid mites (Maraun et al. 2001). The low densities in tropical mountain rainforests cannot be explained by high densities of macrodecomposers since, at these sites, the density of such macrodecomposers is low; presumably other factors including low resource quality are responsible for these low densities of oribatid mites in tropical forest ecosystems (Maraun et al. 2008). Overall, the low density of oribatid mites in calcareous forest soils presumably is mainly due to macrofauna activity, whereas their density in more acidic forest soils probably is limited by the availability of high quality resources (bottom-up control; Salamon et al. 2006). Predation (top-down control) is likely to be of little importance as a regulatory factor for adult oribatid mites since these are well-defended (enemy free space hypothesis; Peschel et al. 2006). However, predation on more vulnerable juvenile oribatid mites has not been studied and may be an important factor regulating oribatid mite density.

12.1.4 Niche Differentiation and Feeding Biology

Anderson (1975) hypothesized that trophic niche differentiation contributes to the high diversity of soil animal species. However, laboratory feeding choice

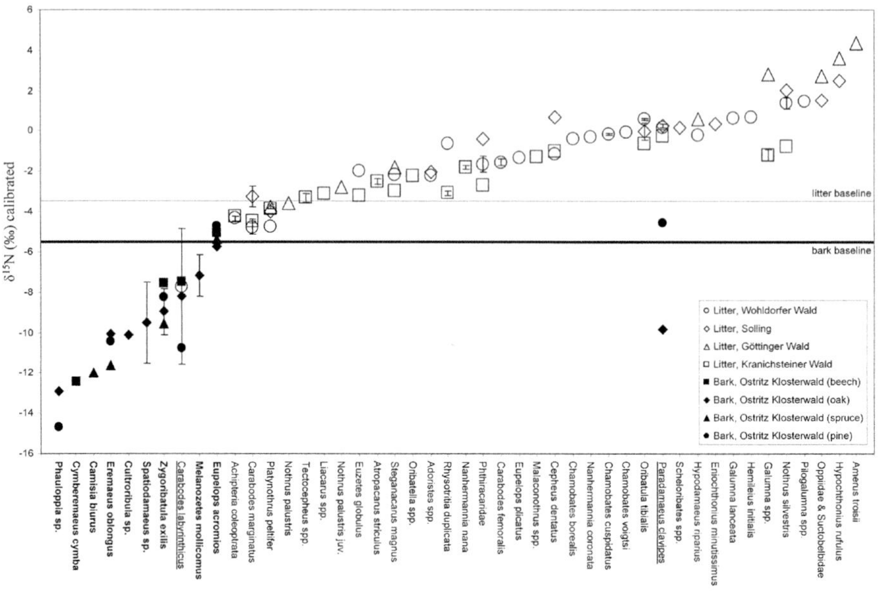

Fig. 12.2 Variation of $\delta^{15}N$ values of oribatid mites from litter (*open symbols*) and from the bark of trees (*filled symbols*; species names in **bold**) of different beech forests. Data for litter are calibrated to the L/F litter material of the Solling (L/F layer); data for bark are calibrated to the average $\delta^{15}N$ signature of the outer bark layer of the four forest types. Single measurements (without SD) and means of 2–5 replicates with SD. Species were ordered according to increasing $\delta^{15}N$ value. Underlined species occurred in the litter and also on the bark of trees (modified after Schneider et al. 2004 and Erdmann et al. 2007).

experiments indicate that soil-living microarthropods such as adult oribatid mites are generalist feeders, suggesting that trophic niche differentiation is poor. They feed on a variety of substrates when offered the choice, including algae, lichens, different species of saprophytic and mycorrhizal fungi and dead organic material (Maraun et al. 1998a, 2003a; Schneider and Maraun 2005; Schneider et al. 2005; Heethoff et al. 2007b). In contrast, natural variations in stable isotope ratios ($^{15}N/^{14}N$, $^{13}C/^{12}C$) of oribatid mites indicate distinct trophic niches that collectively span a wide range of trophic levels, including phytophagous species feeding on algae, primary decomposers feeding on detritus, secondary decomposers feeding on fungi, predators presumably feeding on nematodes or necrophages feeding on small dead invertebrates (Schneider et al. 2004). Stable isotope signatures ($\delta^{15}N$) of soil and bark-living oribatid mite species span over about 20 δ units – equivalent to about six trophic levels (cf. Minagawa and Wada 1984; Post 2002) – which is unique for a single animal taxon (Schneider et al. 2004; Erdmann et al. 2007; Fig. 12.2). The contradiction between field data (which indicate narrow trophic niches) and laboratory data (which indicate generalist feeding) is unresolved and deserves further attention. Possibly, immature oribatid mites are more specialized feeders and change to a wider food spectrum when adult.

The high diversity of soil animals, especially of soil microarthropods, has been considered enigmatic (Anderson 1975; Maraun et al. 2003a), mainly for two reasons. First, the spatial structure of the soil system appears to be rather homogeneous compared with above-ground habitats. Second, co-evolutionary processes between decomposer soil invertebrates and their resources are probably weak; there is no selection pressure for dead organic matter to "defend" itself against detritivores, and co-evolution between microbivorous microarthropods and their prey species may have been prevented by difficulty in physically isolating individual microbial species for ingestion (Scheu and Setälä 2002). Like many soil animal taxa, few oribatid mite species inhabit specific physical niches (microhabitats) such as earthworm burrows or specific litter types (Migge et al. 1998; Maraun et al. 1999; Hansen 2000). While soil-associated microhabitats like mosses and decaying wood might contain oribatid mite species with strong substrate affiliations, these species often are also present at other places. If there is microhabitat specificity it is mostly found in the immature instars, many of which burrow in rather specific types of living (fungal fruiting bodies, lichens) or decomposing (needles, twigs, seeds, wood) substrates. One exception may be arboreal oribatid mite species which, even as adults, almost exclusively occur on the bark or in the canopy of trees (Erdmann et al. 2006, 2007; Lindo and Winchester 2006).

12.1.5 Functioning

As a combined result of feeding on decomposing organic matter and their high density, oribatid mites are important decomposers in forest ecosystems, fallows, fields and meadows (Hansen 1999; Maraun and Scheu 2000). In particular in acidic forests, they are among the major decomposer groups (along with collembolans, enchytraeids and dipteran larvae) and drive mineralization processes and humus formation. The fossil record of oribatid mites indicates that this was already the case in Palaeozoic forests (Labandeira et al. 1997). Despite this perceived ancient importance of oribatid mites for terrestrial ecosystems, their function in the soil system and their role in aboveground – belowground interactions have been little studied. The main reason for this probably is their slow reproduction rate which makes rearing for experimental purposes difficult. The few available studies indicate that oribatid mites affect the composition of microbial communities via dispersal of fungal spores (Maraun et al. 1998b). These can be dispersed in the gut of the animals via their faeces but also attached to the body surface (Behan and Hill 1978; Renker et al. 2005). Besides these indirect effects on decomposition processes via modification of the microbial community, oribatid mites also affect decomposition processes directly by feeding on dead plant material. For example, box mites (Ptyctima) and *Adoristes ovatus* (Koch) feed inside decomposing needles and leaves (Harding and Stuttard 1974; Lions and Gourbiere 1988).

Due to their wide ecological tolerance, their limited reaction after disturbances and their broad feeding habits, oribatid mites are of limited use as bioindicators in forest ecosystems (Lindo and Visser 2004). However, they may be used as bioindicators during ecosystem succession, especially in early successional stages of agroecosystems (Behan-Pelletier 1999).

12.2 Reproductive and Developmental Biology

12.2.1 General Aspects

Mites are without peer among chelicerates with respect to their reproductive potential and the diversity of their reproductive strategies, genetic systems and ontogenies (Norton et al. 1993; Walter and Proctor 1999). Oribatid mites are diplodiploid organisms (usually with $2n = 18$; Heethoff et al. 2006) in which fertilization is usually indirect, via spermatophores deposited by males in the absence of females; alternatively, parthenogenetic development occurs. They retain the presumed ancestral developmental series of Acari: embryological development terminates in a regressive prelarva which is succeeded by the larva, protonymph, deuteronymph, tritonymph and adult. As in most acariform mites, the first active instar is the hexapod larva; the prelarva is inactive (calyptostase) and remains inside the egg shell in all known species, unlike some other groups with active prelarvae in primitive species (Otto 1997).

Egg laying strategies range from iteroparity (repeated production of a few eggs) to semelparity (single production of many eggs). Eggs are usually laid in crevices at an early developmental stage (embryo or prelarva), but larviparity also occurs (Walter and Proctor 1999). With some exceptions, oribatid mites tend to have low reproductive outputs, and long developmental times of 50 weeks or more are common for species of temperate zones (Norton 1994; Walter and Proctor 1999, Sovik and Leinaas 2003; Heethoff et al. 2007b).

12.2.2 Female System and Reproductive Strategies

The morphology of the female reproductive system in the Acari is variable (Evans 1992). Depending on the group, the ovary can be unpaired, paired or divided into germinal and nutritive regions (Evans 1992; Alberti and Coons 1999; Bergmann et al. 2008). Oribatid mites usually have an unpaired ovary from which two oviducts originate and proceed through the opisthosoma until they fuse at the vagina (Alberti and Coons 1999; Heethoff et al. 2007b; Bergmann et al. 2008). Oviposition is accomplished with an extrusible ovipositor (Fig. 12.3), but the mechanism of extrusion is poorly understood. It has been assumed that haemolymph pressure extends the ovipositor and that muscles attached to its wall retract it (Michael 1884; Grandjean 1956). However, more probably, direct muscular action is also responsible for ovipositor extrusion (U. Kurz and M. Heethoff, unpubl.). Wallwork (1977)

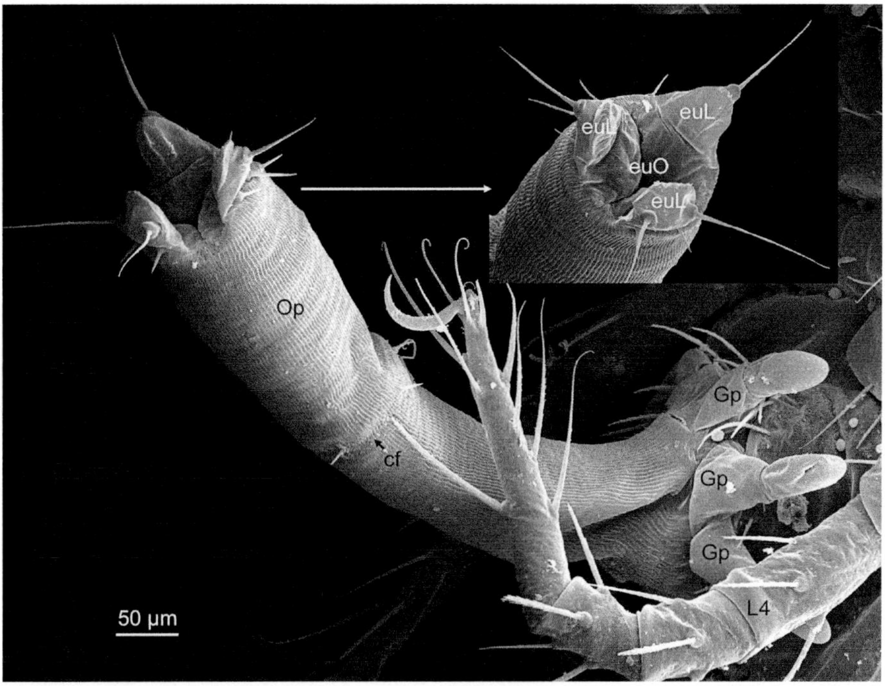

Fig. 12.3 Micrographs (SEM) of the extruded ovipositor and its appendages of *Archegozetes longisetosus* cf: circular fold, euL: eugenital lobes, euO: eugenital orifice, Gp: genital papillae, L4: walking leg, Op: ovipositor.

further suggested that muscles inserting on the pleated wall are responsible for oviposition, although he did not provide a mechanism.

12.2.3 Parthenogenesis

Parthenogenesis is widespread among oribatid mites, but the cytological mechanism involved is poorly understood. Available data suggest that automixis is the rule (Taberly 1987; Heethoff et al. 2006; see also Chapter 4). Taberly (1987) performed the first cytological study of meiosis in parthenogenetic oribatid mites. He observed that the parthenogenetic species *Platynothrus peltifer* (Koch) and *Trhypochthonius tectorum* (Berlese) restore diploidy by terminal fusion (fusion of the egg pronucleus with the second polar nucleus), a mechanism expected to produce homozygous offspring (Wrensch et al. 1994). This mechanism was also suggested for *Archegozetes longisetosus* (Heethoff et al. 2006; Laumann et al. 2008). However, using isozyme techniques, Palmer and Norton (1992) found fixed heterozygosity, absence of complete homozygosity and absence of recombination in nine parthenogenetic oribatid mite species. They proposed that apomixis or central fusion automixis may be

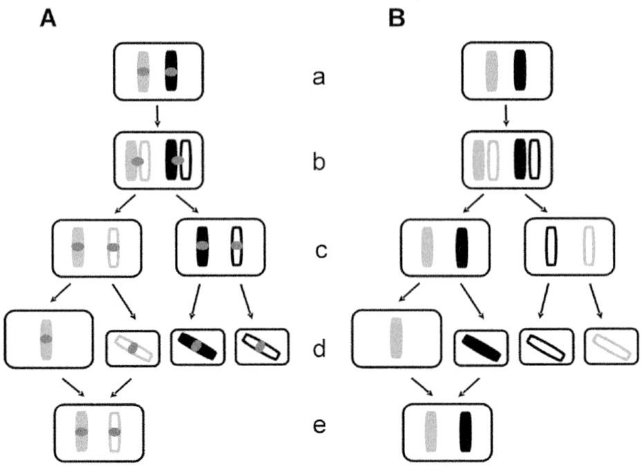

Fig. 12.4 Effects of inverted meiosis on the genetic reconstitution of the embryo under the aspect of terminal fusion automixis (exemplified by one homologous chromosome pair under the absence of recombination). **A** normal meiosis with monocentric chromosomes. **B** inverted meiosis with holokinetic chromosomes. *a*: diploid cell. *b*: chromosomes duplicated by replication prior to the initiation of meiosis. *c*: meiosis I: separation of homologous chromatids in normal meiosis (**A** reductional division) and sister chromatids in inverted meiosis (**B** equational division) leading to haploid cells (**A**) and diploid cells (**B**). *d*: meiosis II: equational division in normal meiosis (**A**), reductional division in inverted meiosis (**b**). *e*: the genetic constitution of the embryo as compared to the mother's genome dramatically differs in (**A**) and (**B**): **A** – embryo constitutes a diploid chromosome number containing sister chromatids of one of the initiating chromosomes of the mother, **B** – embryo has the same genetic constitution as the mother (and hence is in effect a clone)

common in parthenogenetic oribatid mites, even if neither mechanism had yet been discovered. However, Wrensch et al. (1994) suggested inverted meiosis as an alternative mechanism – made possible by holokinetic chromosomes of oribatid mites – in which the sequence of meiotic divisions (reductional and equational) is inverted compared to normal meiosis, leading to a reversal in effects of terminal and central fusion. Fixed heterozygosity and terminal fusion would thus be compatible observations (Wrensch et al. 1994; Heethoff et al. 2006; Laumann et al. 2008; Fig. 12.4).

We studied oogenesis of the parthenogenetic oribatid mite *Archegozetes longisetosus* Aoki, a widely used model organism for Chelicerata with mother-daughter genetic fidelity, i.e. no recombination of heterozygous isoenzyme genotypes (Palmer and Norton 1992). Oocytes grow inside the rhodoid of the ovary, vitellogenesis takes place in the meros of the ovary and afterwards, eggs enter the oviducts via an ovarial bulb (Bergmann et al. 2008). Meiotic divisions probably occur in the meros of the ovary, the first polar body seems to be expelled (Laumann et al. 2008). The first three to five embryonic cleavages are holoblastic and occur in the beginning/middle of the oviduct (M. Laumann, P. Bergmann and M. Heethoff, unpubl.).

Clearly, the chromosomal kinetics during meiotic divisions still has to be investigated in more detail since Schaefer et al. (2006) showed that the so-called "Meselson effect" (diverging allelic sequences in absence of recombination; Mark Welch and Meselson 2000), expected to occur in this scenario, does not occur in parthenogenetic oribatid mites even with the probable absence of genetic recombination. These authors used partial sequences of the genes of the elongation factor-1α (*ef-1α*) and the heat shock protein 82 (*hsp82*) to test the existence of the "Meselson effect" for putative ancient parthenogenetic oribatid mite lineages. Unexpectedly, the intra-individual divergence of parthenogenetic lineages was rather low and resembled that in sexual species. Additionally, strong selection pressure may keep both the *ef-1α* and the *hsp82* gene functioning. No evidence for recombination or gene conversion was found for sexual or parthenogenetic oribatid mite species in the *hsp 82* gene supporting the assumption that homogenizing mechanisms prevent the accumulation of sequence divergences, i.e. the "Meselson effect".

The sex determination mechanism in oribatid mites is unknown (Heethoff et al. 2006). In diplodiploid systems, sex determination is often accomplished by sex chromosomes and the sex ratio is assumed to be close to unity, i.e. 1:1 (Fisher 1930). Sex determination may also be influenced by external factors, such as temperature (Ewert et al. 1994), or by hormonal or pheromonal control (White 1973). Oribatid mites lack sex chromosomes despite their diplodiploidy (Sokolov 1954; Norton et al. 1993; Wrensch et al. 1994; Heethoff et al. 2006). Nevertheless, the sex ratio of sexual oribatid mite species is close to unity, whereas in parthenogenetic species, males are rare (spanandric) and sterile. Spermatophores are non-functional, as spermatogenesis is incomplete (Taberly 1988), and there is no evidence for recombination or incorporation of paternal genetic material into the offspring (Palmer and Norton 1992; see also Cianciolo and Norton 2006; Schaefer et al. 2006). A genetic mechanism based on inverted meiosis with terminal fusion automixis does not explain the occurrence of these spanandric males. Environmental sex determination is also unlikely because sexual and parthenogenetic oribatid mite species coexist under a wide variety of conditions, i.e. they are exposed to similar external stimuli. Further efforts are necessary to uncover the mechanism for the sporadic production of sterile males in parthenogenetic oribatid mite species (Heethoff et al. 2006).

12.2.4 Endosymbiotic Bacteria

In recent years, much has been written about the ability of endosymbiotic bacteria (*Wolbachia*, *Cardinium*) to control the reproductive biology of hosts, particularly arthropods, and the induction of parthenogenesis is one of several possible manifestations of infection (Stouthamer et al. 1999). The first use of DNA techniques to search for *Wolbachia* in parthenogenetic oribatid mites was negative for all species tested (Perrot-Minnot and Norton 1997), but *Wolbachia* was detected in an unidentified oribatid mite by Cordaux et al. (2001) and the parthenogenetic species *Oppiella nova* (Oudemans) can serve as host for both *Wolbachia* and *Cardinium* (Weeks et al. 2003). Ongoing research (A. Weeks, R. Stouthamer and R. A. Norton, unpubl.) suggests that both bacteria are present in a small range of oribatid mite species. While bacterial phenotypes remain unproven, the distribution of endosymbionts in relation to mite reproductive mode does not support inducement of parthenogenesis.

12.3 Phylogeny of Parthenogenetic Lineages

12.3.1 General Phylogeny

A morphology-based classification of oribatid mites that reflects "natural" phylogenetic groups was first established by Grandjean (1953, 1965, 1969). He grouped oribatid mites into six taxa, including (1) the basal Palaeosomata, with few species; (2) the Enarthronota, including e.g. the species-rich Brachychthonioidea and Hypochthonioidea; (3) the small group Parhyposomata; (4) the "Mixonomata", which includes the species-rich box mites (Phthiracaroidea and Euphthiracaroidea); (5) the "Desmonomata", with several species-rich groups; and (6) the diverse, species-rich Circumdehiscentiae (=Brachypylina). As indicated by quotes, two of these groups are probably paraphyletic (Haumann 1991; Norton et al. 1993; Weigmann 1996). Maraun et al. (2004) and Weigmann (2006) summarized a common view of oribatid mite phylogeny based on morphological characters. Phylogenetic studies using molecular markers (18S rDNA, 28S rDNA, *hsp82*, elongation factor 1 α) have largely confirmed this view (Maraun et al. 2004; I. Schaefer et al. unpubl.; Fig. 12.5), but efforts to refine phylogenetic hypotheses continue.

12.3.2 Radiation of Parthenogenetic Lineages

The existence of ancient parthenogenetic taxa questions the necessity of sexual reproduction for the evolution and diversification of lineages into discrete genetic and morphological entities (Barraclough et al. 2003; see also Lushai et al. 2003 and Chapter 10). A single parthenogenetic population may display a tree-like ancestry and give the appearance of discrete taxa, but the degree of diversification is determined by the time elapsed since the split of the last common ancestor. Diversification at this level could be achieved through isolation in different geographic areas or adaptive divergence in a heterogeneous environment (Barraclough et al. 2003; Laumann et al. 2007). Diversifying selection and geographic isolation drive speciation in sexual species and are expected to have similar effects in parthenogenetic species.

The radiation and diversification of ancient taxa of parthenogenetic species has empirically been little studied. It challenges theories on the evolution and maintenance of sex, and ancient parthenogenetic taxa therefore have been termed "evolutionary scandals" (Maynard Smith 1978; Judson and Normark 1996). It is generally assumed that parthenogenetic lineages are "dead ends" in evolution for various reasons. They are believed to be unable to keep up in the evolutionary arms race with pathogens and predators (cf. Red Queen hypothesis; van Valen 1973;

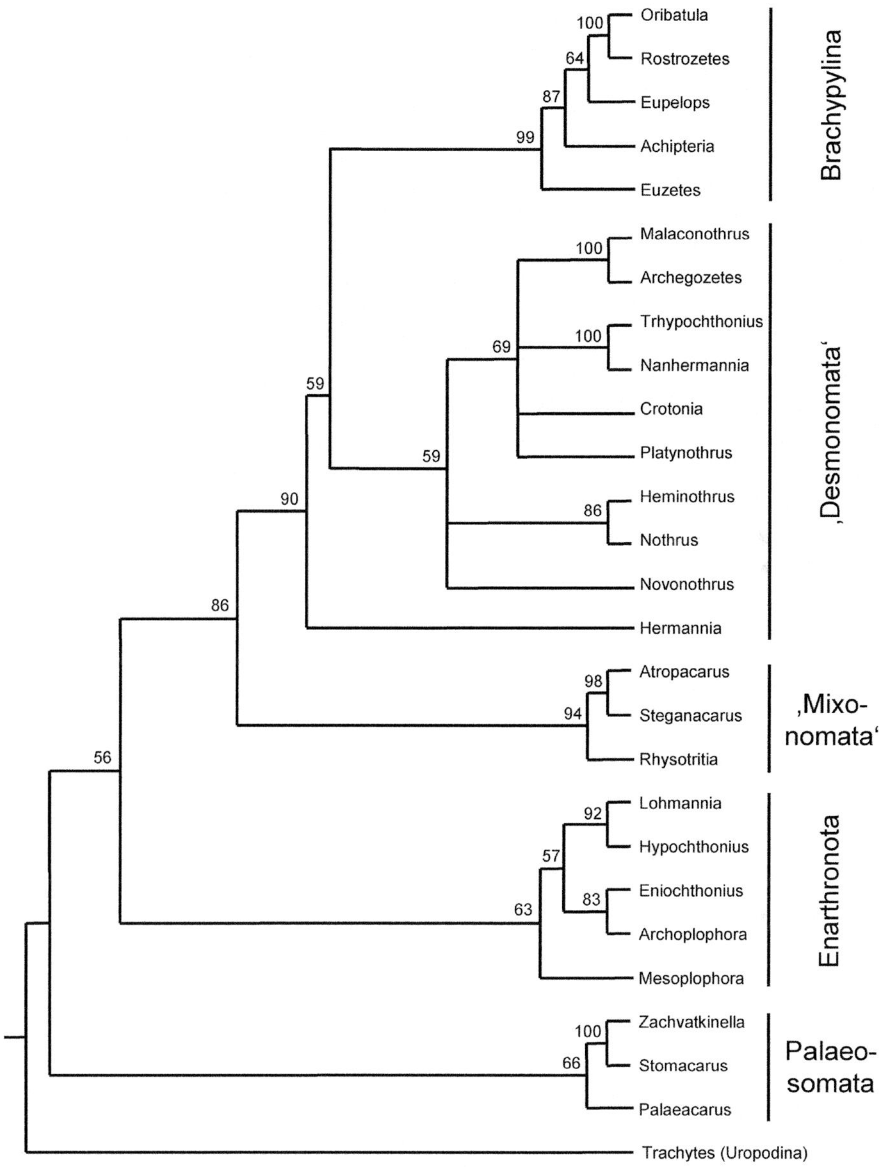

Fig. 12.5 Neighbour-joining tree of 26 genera of oribatid mites with *Trachytes* sp. (Mesostigmata, Uropodina) as outgroup taxon, constructed using the 18S rDNA sequences and a part of the heat shock protein (*hsp82*) region Numbers at nodes represent percentages of 1000 bootstrap replicates (only values above 50% are reported; I. Schaefer, K. Domes, S. Scheu and M. Maraun; unpubl. data).

Hamilton 1980; see also Chapter 7), they may be unable to adapt to different ecological niches (cf. Tangled bank hypothesis; Ghiselin 1974; Bell 1982), and they may accumulate deleterious mutations (cf. Muller's ratchet; Muller 1964; cf. Mutation load reduction theory; Kondrashov 1993; Crow 1994; see also Chapter 5).

Despite the theoretical arguments, there is increasing empirical evidence that oribatid mites comprise many clades of various size, extant members of which reproduce exclusively via parthenogenesis (Norton and Palmer 1991; Norton et al. 1993; Maraun et al. 2003b, 2004; Heethoff et al. 2007a; Laumann et al. 2007). These groups presumably radiated without sexual reproduction. The largest clades are within early to middle-derivative groups, such as Enarthronota (Brachychthoniidae; 102 spp.; Lohmanniidae; 156 spp.) and "Desmonomata" (Nanhermanniidae, 56 spp.; Malaconothridae, 104 spp.; Trhypochthoniidae, 68 spp.; Camisiidae, 92 spp.; and the genus *Nothrus* (Nothridae, 54 spp.)). Within the "Desmonomata", sexual reproduction reevolved in the family Crotoniidae. This evolutionary recovery of a complex trait (sexual reproduction) is a unique evolutionary scenario in the animal kingdom and was termed a "breaking of Dollo's law" (Domes et al. 2007). Violation of Dollo's law has been controversial (Goldberg and Igic 2008) but recent phylogenetic studies have demonstrated that the re-evolution of complex features occurs more often than previously assumed. If the underlying developmental pathway is retained after the loss of a certain character, it may be regained quickly when needed (Collin and Miglietta 2008).

Molecular evidence further suggests the existence of parthenogenetic speciation for several groups of oribatid mites. Maraun et al. (2003b) used comparisons of sequence divergences of the 28S rDNA D3-region of closely related parthenogenetic oribatid mite species to show that parthenogenesis is an ancient phenomenon for the genus *Tectocepheus* (Brachypylina) and lineages within the "Desmonomata". Phylogenetic analyses of the D3-region also suggest the existence and radiation of multiple parthenogenetic and diverse lineages, such as Nanhermanniidae, Malaconothridae, Camisiidae and the genus *Tectocepheus* (Maraun et al. 2004). Laumann et al. (2007) analysed four nuclear genes of three controversial species of *Tectocepheus* and showed that *T. velatus* (Michael), *T. sarekensis* (Träghård) and *T. minor* (Berlese), although similar in morphological aspects, have distinct genotypes and likely split into separate species while being parthenogenetic. Heethoff et al. (2007a) found seven distinct genetic lineages (which can be termed cryptic species) for the mitochondrial gene cytochrome oxidase 1 (*cox1*) of *Platynothrus peltifer* (Camisiidae), corresponding to their geographical origin (North America, Northern/Central Europe, Southern Europe, Northern Tyrolia, Eastern Europe and Japan).

The 4X rule is a quantitative measurement for speciation in parthenogentic species based on genetic distances and is used to identify species under the Evolutionary Genetic Species Concept (EGSC; W. Birky, pers. comm.; see also Chapter 10 in this book). The seven lineages of *Platynothrus peltifer* fulfill the conditions of the 4X rule and thus may be considered evolutionary genetic species. High genetic distances corresponded to divergence times up to 64 million years and it was concluded that the current distribution of lineages result from vicariance rather than dispersal. This supports the old theory that the general distribution of oribatid mites is largely related to patterns of continental drift (Hammer and Wallwork 1979).

Besides the phylogenetically clustered, ancient parthenogenetic lineages of oribatid mites, there are also recent and phylogenetically isolated parthenogenetic lineages with close sexual congeners (Cianciolo and Norton 2006). Of the proposed explanations for the ecological distribution of parthenogenetic oribatid mites, three were examined by Cianciolo and Norton (2006), namely a correlation with overall biotic uncertainty (as generated by competitors and predators); a lack of ecological correlation consistent with Muller's ratchet and the existence of general purpose genotypes (Lynch 1984; see also Chapter 6) among ancient parthenogenetic lineages and recent parthenogenetic lineages. They rejected biotic uncertainty as an explanatory mechanism for either ancient parthenogenetic lineages or recent parthenogenetic lineages; further, there was no evidence for the existence of general purpose genotypes in the two groups, although this clearly deserves further investigation. However, while Muller's ratchet (Muller 1964) seems unimportant for ancient parthenogenetic lineages, it may influence the distribution of recent parthenogenetic lineages (Cianciolo and Norton 2006). This is surprising since Muller's ratchet acts in the long-term and should therefore apply more strongly to ancient parthenogenetic lineages than to recent parthenogenetic lineages.

The large number of parthenogenetic lineages, the spectrum-like distribution of reproductive modes (ancient parthenogenetic lineages to recent parthenogenetic lineages) and the unusual mode of reproduction (terminal fusion automixis and possibly inverted meiosis) make oribatid mites a unique model system to analyse the ecological and evolutionary advantages and disadvantages of parthenogenetic reproduction. It is a challenge for future studies to elucidate the mechanisms that allow the presence and persistence of numerous parthenogenetic oribatid mite clades that vary widely in size and, presumably, in age.

12.4 Glossary

Hysterosoma: the hind body division of Acari consisting of the opisthosoma and the metapodosoma (segments bearing the two hind legs).
Meros: distal part of the ovary in Acari with vitellogenetic oocytes.
Proterosoma: the fore body division of Acari consisting of the gnathosoma (mouthparts) and the fore two segments with walking legs.
Rhodoid: central part of the ovary in Acari with previtellogenetic oocytes.
Vitellogenesis: process of yolk deposition and formation via nutrients being deposited in the oocyte.

References

Alberti G, Coons LB (1999) Acari: Mites. In: Harrison FW, Foelix RF (eds) Microscopic Anatomy of Invertebrates. Wiley-Liss Inc., New York, pp. 515–1215

Anderson JM (1975) The enigma of soil animal species diversity. Proc. 5th Int Coll Soil Zool, Prague, 1973, pp. 51–58

Badejo MA, Espindola JAA, Guerra JGM, De Aquino AM, Correa MEA (2002) Soil oribatid mite communities under three species of legumes in an ultisol in Brazil. Exp Appl Acarol 27: 283–296

Barraclough TG, Birky CW Jr, Burt A (2003) Diversification in sexual and asexual organisms. Evolution 57: 2166–2172

Bell G (1982) The Masterpiece of Nature. The Evolution and Genetics of Sexuality. University of California Press, Berkeley, CA

Behan VM, Hill SB (1978) Feeding-habits and spore dispersal of oribatid mites in the North-American arctic. Rev Ecol Biol Sol 15: 497–516

Behan-Pelletier VM (1999) Oribatid mite biodiversity in agroecosystems: role for bioindication. Agric Ecosys Environ 74: 411–423

Bergmann M, Laumann M, Cloetens P, Heethoff M (2008) Morphology of the internal reproductive organs of *Archegozetes longisetosus* Aoki (Acari, Oribatida). Soil Org 80: 171–195

Cianciolo JM, Norton RA (2006) The ecological distribution of reproductive mode in oribatid mites, as related to biological complexity. Exp Appl Acarol 40: 1–25

Collin R, Miglietta MP (2008) Reversing opinions on Dollo's law. Trends Ecol Evol 23: 602–609

Cordaux R, Michel-Salzat A, Bouchon D (2001) *Wolbachia* infection in crustaceans: novel hosts and potential routes for horizontal transmission. J Evol Biol 14: 237–243

Crow JF (1994) Advantages of sexual reproduction. Dev Gen 15: 205–213

Domes K, Norton RA, Maraun M, Scheu S (2007) Reevolution of sexuality breaks Dollo's law. Proc Natl Acad Sci USA 104: 7139–7144

Eisenhauer N, Partsch S, Parkinson D, Scheu S (2007) Invasion of a deciduous forest by earthworms: changes in soil chemistry, microflora, microarthropods and vegetation. Soil Biol Biochem 39: 1099–1110

Erdmann G, Floren A, Linsenmair KE, Scheu S, Maraun M (2006) Effect of forest age on oribatid mites from the bark of trees. Pedobiologia 50: 433–441

Erdmann G, Otte V, Langel R, Scheu S, Maraun M (2007) The trophic structure of bark-living oribatid mite communities analysed with stable isotopes (^{15}N; ^{13}C) indicates strong niche differentiation. Exp Appl Acarol 41: 1–10

Evans GO (1992) Principles of Acarology. CAB International, Wallingford

Ewert MA, Jackson DR, Nelson CE (1994) Patterns of the temperature-dependant sex determination in turtles. J Exp Zool 270: 3–15

Fisher RA (1930) The genetical theory of sexual selection. Clarendon Press

Ghiselin MT (1974) The Economy of Nature and the Evolution of Sex. University of California Press, Berkeley, CA

Goldberg EE, Igic B (2008) On phylogenetic tests of irreversible evolution. Evolution 62: 2727–2741

Grandjean F (1953) Essai de classification des Oribates (Acariens). Bull Soc Zool France 78: 421–446

Grandjean F (1956) Caracteres chitineux de l'ovipositeur, en structure normale chez les Oribates (Acariens). Arch Zool Exp Gen Paris 93: 96–106

Grandjean F (1965) Complément a mon travail de 1953 sur la classification des oribates. Acarologia 7: 713–734

Grandjean F (1969) Considérations sur le classement des oribates: leur division en 6 groupes majeurs. Acarologia 11: 127–153

Hamilton WD (1980) Sex versus non-sex versus parasites. Oikos 35: 282–290

Hammer M, Wallwork JA (1979) A review of the world distribution of oribatid mites (Acari: Cryptostigmata) in relation to continental drift. Biol Skr Dan Vid Selsk 22: 1–31

Hansen RA (1999) Red oak litter promotes a microarthropod functional group that accelerates its decomposition. Plant Soil 209: 37–45

Hansen RA (2000) Effects of habitat complexity and composition on a diverse litter microarthropod assemblage. Ecology 81: 1120–1132

Harding DJL, Stuttard RA (1974) Microarthropods. In: Dickinson CH, Pugh GJF (eds) Biology of Plant Litter Decomposition. Acad Press, London, pp. 489–532

Haumann G (1991) Zur Phylogenie primitiver Oribatiden (Acari: Oribatida). Verlag Technische Univ, Graz

Heethoff M, Bergmann P, Norton RA (2006) Karyology and sex determination of oribatid mites. Acarologia 46: 127–131

Heethoff M, Domes K, Laumann M, Maraun M, Norton RA, Scheu S (2007a) High genetic divergences indicate ancient separation of parthenogenetic lineages of the oribatid mite *Platynothrus peltifer* (Acari, Oribatida). J Evol Biol 20: 392–402

Heethoff M, Laumann M, Bergmann P (2007b) Adding to the reproductive biology of the parthenogenetic oribatid mite *Archegozetes longisetosus* (Acari, Oribatida, Trhypochthoniidae). Turk J Zool 31: 151–159

Heethoff M, Norton RA (2009) Role of musculature during defecation in a particle-feeding arachnid, *Archegozetes longisetosus* (Acari, Oribatida). J Morphol 270: 1–13

Judson PO, Normark BB (1996) Ancient asexual scandals. Trends Ecol Evol 11: 41–46

Kondrashov AS (1993) Classification of hypotheses on the advantage of amphimixis. J Hered 84: 372–387

Labandeira CC, Phillips TL, Norton RA (1997) Oribatid mites and the decomposition of plant tissues in Palaeozoic coal-swamp forests. Palaios 12: 319–353

Laumann M, Norton RA, Weigmann G, Scheu S, Maraun M, Heethoff M (2007) Speciation in the parthenogenetic oribatid mite genus *Tectocepheus* (Acari, Oribatida) as indicated by molecular phylogeny. Pedobiologia 51: 111–122

Laumann M, Bergmann P, Heethoff M (2008) Some remarks on the cytogenetics of oribatid mites. Soil Org 80: 223–232

Lindo Z, Visser S (2004) Forest floor microarthropod abundance and oribatid mite (Acari: Oribatida) composition following partial and clearcut harvesting in the mixed wooded boreal forest. Can J Forest Res 34: 998–1006

Lindo Z, Winchester NN (2006) A comparison of microarthropod assemblages with emphasis on oribatid mites in canopy suspended soils and forest floors associated with ancient western redcedar trees. Pedobiologia 50: 31–41

Lindquist EE (1984) Current theories on the evolution of major groups of Acari and on their relationships with other groups of Arachnida, with consequent implications for their classification. In: Griffith DA, Bowman CE (eds) Acarology VI. Ellis Horwood Publ, Chichester, pp. 28–62

Lions JC, Gourbiere F (1988) Adult and immature populations of *Adoristes ovatus* (Acari, Oribatida) in the litter needles of *Abies alba*. Rev Ecol Biol Sol 25: 343–352

Lushai G, Loxdale HD, Allen JA (2003) The dynamic clonal genome and its adaptive potential. Biol J Linn Soc 79: 193–208

Lynch M (1984) Destabilizing hybridization, general purpose genotypes and geographic parthenogenesis. Quart Rev Biol 59: 257–290

Maraun M, Migge S, Schaefer M, Scheu S (1998a) Selection of microfungal food by six oribatid mite species (Oribatida, Acari) from two different beech forests. Pedobiologia 42: 232–240

Maraun M, Visser S, Scheu S (1998b) Oribatid mites enhance the recovery of the microbial community after a strong disturbance. Appl Soil Ecol 9: 179–186

Maraun M, Alphei J, Bonkowski M, Buryn R, Migge S, Peter M, Schaefer M, Scheu S (1999) Middens of the earthworm *Lumbricus terrestris* (Lumbricidae): microhabitats for micro- and mesofauna in forest soil. Pedobiologia 43: 276–287

Maraun M, Scheu S (2000) The structure of oribatid mite communities (Acari, Oribatida): patterns, mechanisms and implications for future research. Ecography 23: 374–383

Maraun M, Alphei J, Beste P, Bonkowski M, Buryn R, Peter M, Migge S, Schaefer M, Scheu S (2001) Indirect effects of carbon and nutrient amendments on the soil meso- and microfauna of a beechwood. Biol Fertil Soil 34: 222–229

Maraun M, Martens H, Migge M, Theenhaus A, Scheu S (2003a) Adding to the 'enigma of soil animal diversity': fungal feeders and saprophagous soil invertebrates prefer similar food substrates. Eur J Soil Biol 39: 85–95

Maraun M., Heethoff M, Scheu S, Norton RA, Weigmann G, Thomas RH (2003b) Radiation in sexual and parthenogenetic oribatid mites (Oribatida, Acari) as indicated by genetic divergence of closely related species. Exp Appl Acarol 29: 265–277

Maraun M, Heethoff M, Schneider K, Scheu S, Weigmann G, Cianciolo J, Thomas RH, Norton RA (2004) Molecular phylogeny of oribatid mites (Oribatida, Acari): evidence for multiple radiations of parthenogenetic lineages. Exp Appl Acarol 33: 183–201

Maraun M, Illig J, Sandmann D, Krashevska V, Norton RA, Scheu S (2008) Soil Fauna. In: Beck E, Bendix J, Kottke I, Makeschin F, Mosandl R (eds) Gradients in a Tropical Mountain Ecosystem of Ecuador. Ecological Studies 198, Springer, Berlin, pp. 181–192.

Mark Welch D, Meselson M (2000) Evidence for the evolution of bdelloid rotifers without sexual reproduction or genetic exchange. Science 288: 1211–1215

Maynard Smith J (1978) The Evolution of Sex. Cambridge University Press, Cambridge, UK

Michael AD (1884) British Oribatidae, Vol. 1, Ray Society, London

Migge S, Maraun M, Scheu S, Schaefer M (1998) The oribatid mite community (Acarina) on pure and mixed stands of beech (*Fagus sylvatica*) and spruce (*Picea abies*) at different age. Appl Soil Ecol 9: 119–126

Migge-Kleian S, McLean MA, Maerz JC, Heneghan L (2006) The influence of invasive earthworms on indigenous fauna in ecosystems previously uninhabited by earthworms. Biol Invasions 8: 1275–1285

Minagawa M, Wada E (1984) Stepwise enrichment of ^{15}N along food chains: further evidence and the relationship between $\delta^{15}N$ and animal age. Geochim Cosmochim Acta 48: 1135–1140

Muller HJ (1964) The relation of recombination to mutational advance. Mutat Res 1: 2–9

Norton RA, Bonamo PM, Grierson JD, Shear WA (1988) Oribatid mite fossils from a terrestrial Devonian deposit near Gilboa, New York. J Paleontol 62: 421–499

Norton RA, Palmer SC (1991) The distribution, mechanisms and evolutionary significance of parthenogenesis in oribatid mites. In: Schuster R, Murphy W (eds) The Acari: Reproduction, Development and Life-History Strategies. Chapman and Hall, London, pp. 107–136

Norton RA, Kethley JB, Johnston DE, OConnor BM (1993) Phylogenetic perspectives on genetic systems and reproductive modes of mites. In: Wrensch DL, Ebbert MA (eds) Evolution and Diversity of Sex Ratios. Chapman and Hall, New York, pp. 8–99

Norton RA (1994) Evolutionary aspects of oribatid mite life histories and consequences for the origin of the Astigmata. In: Houck MA (ed) Mites: Ecological and Evolutionary Studies of Life-History Patterns. Chapmann and Hall, New York, pp. 99–135

Otto J (1997) Observations on prelarvae in Anystidae and Tenerifiidae. In: Mitchell R, Horn DJ, Needham GR, Welbourn WC (eds) Acarology IX. Proc Vol. 1. Ohio Biological Survey, Columbus, OH, pp. 343–354

Palmer SC, Norton RA (1992) Genetic diversity in thelytokous oribatid mites (Acari; Acariformes: Desmonomata). Biochem Syst Ecol 20: 219–231

Perrot-Minnot MJ, Norton RA (1997) Obligate thelytoky in oribatid mites: no evidence for *Wolbachia*-inducement. Can Entomol 129: 691–698

Peschel K, Norton RA, Scheu S, Maraun M (2006) Do oribatid mites live in enemy-free space? Evidence from feeding experiments with the predatory mite *Pergamasus septentrionalis*. Soil Biol Biochem 38: 2985–2989

Post DM (2002) Using stable isotopes to estimate trophic position: models, methods, and assumptions. Ecology 83: 703–718

Renker C, Otto P, Schneider K, Zimdars B, Maraun M, Buscot F (2005) Oribatid mites as potential vectors for soil microfungi: study of mite-associated fungal species. Microbiol Ecol 50: 518–528

Salamon JA, Alphei J, Ruf A, Schaefer M, Scheu S, Schneider K, Sührig A, Maraun M (2006) Transitory dynamic effects in the soil invertebrate community in a temperate deciduous forest: effects of resource quality. Soil Biol Biochem 38: 209–221

Sanders FH, Norton RA (2004) Anatomy and function of the ptychoid defensive mechanism in the mite *Euphthiracarus cooki* (Acari: Oribatida). J Morphol 259: 119–154

Schaefer I, Domes K, Heethoff M, Schneider K, Schön I, Norton RA, Scheu S, Maraun M (2006) No evidence for the 'Meselson effect' in parthenogenetic oribatid mites (Acari, Oribatida). J Evol Biol 19: 184–193

Schatz H (2002) Die Oribatidenliteratur und die beschriebenen Oribatidenarten (1758–2001) – eine Analyse. Abh Ber Naturkundemus Görlitz 72: 37–45

Scheu S, Setälä H (2002) Multitrophic interactions in decomposer food webs. In: Hawkins BA, Tscharntke T (eds) Multitrophic Level Interactions. Cambridge University Press, Cambridge, UK, pp. 223–264

Schmelzle S, Norton RA, Helfen L, Heethoff M (2008) The ptychoid defensive mechanism in Euphthiracaroidea (Acari: Oribatida): a comparison of exoskeletal elements. Soil Org 80: 233–247

Schneider K, Migge S, Norton RA, Scheu S, Langel R, Reineking A, Maraun M (2004) Trophic niche differentiation in oribatid mites (Oribatida, Acari): evidence from stable isotope ratios ($^{15}N/^{14}N$). Soil Biol Biochem 36: 1769–1774

Schneider K, Maraun M (2005) Feeding preferences among dark pigmented fungi (Dematiacea) indicate trophic niche differentiation of oribatid mites. Pedobiologia 49: 61–67

Schneider K, Renker C, Scheu S, Maraun M (2005) Oribatid mite feeding on ectomycorrhizal fungi. Mycorrhiza 16: 67–72

Shear WA, Bonamo M, Grierson JD, Rolfe WDI, Smith EL, Norton RA (1984) Early land animals on North America: evidence from Devonian age arthropods from Gilboa, New York. Science 224: 492–494

Sokolov II (1954) The chromosome complex of mites and its importance for systematics and phylogeny. Tr O Estestvoispyt 72: 124–159

Sovik G, Leinaas HP (2003) Long life cycle and high adult survival in an arctic population of the mite *Ameronothrus lineatus* (Acari, Oribatida) from Svalbard. Polar Biol 26: 500–508

Stouthamer R, Breeuwer JAJ, Hurst GDD (1999) *Wolbachia pipientis*: microbial manipulator of arthropod reproduction. Annu Rev Microbiol 53: 71–102

Subias LS (2004) Listado sistimatico, sininimico y biogeografico de los Acaros Oribatidos (Acariformes, Oribatida) del mundo (1748–2002). Graellsia 60: 3–305

Taberly G (1987) Recherches sur la parthéogenèse thélytoque de deux espèces d'acariens oribatides: *Trypochthonius tectorum* (Berlese) et *Platynothrus peltifer* (Koch). III. Etude anatomique, histologique et cytologique des femelles parthénogenétiques. Acarologia 28: 389–403

Taberly G (1988) Recherches sur la parthénogenèse thélytoque de deux espèces d'acariens oribatides: *Trypochthonius tectorum* (Berlese) et *Platynothrus peltifer* (Koch). IV. Observation sur les males ataviques. Acarologia 29: 95–107

Van Valen LM (1973) A new evolutionary law. Evol Theory 1: 1–30

Wallwork JA (1977) The structure of the ovipositor and the mechanisms of oviposition in the oribatid mite *Machadobelba symmetrica* Bal. (Acari: Cryptostigmata). Acarologia 19: 149–154

Walter DE, Proctor HC (1999) Mites. Ecology, Evolution and Behaviour. CAB International, Oxon

Weeks AR, Velten R, Stouthamer R (2003) Incidence of a new sex-ratio-distorting endosymbiotic bacterium among arthropods. Proc R Soc Lond B 270: 1857–1865

Weigmann G (1996) Hypostome morphology of Malaconothridae and phylogenetic conclusions on primitive Oribatida. In: Mitchell R, Horn DJ, Needham GJ, Welbourn WC (eds) Acarology IX, vol 1, Proc, Ohio Biological Survey, Columbus, OH, pp. 273–276

Weigmann G (2006) Hornmilben (Oribatida). Goecke and Evers, Keltern

White MJD (1973) Animal Cytology and Evolution. Cambridge University Press, Cambridge, UK

Wrensch DL, Kethley JB, Norton RA (1994) Cytogenetics of holokinetic chromosomes and inverted meiosis: keys to the evolutionary success of mites, with generalizations on eukaryotes. In: Houck MA (ed) Mites: Ecological and Evolutionary analyses of life-history pattern. Chapman and Hall, New York, pp. 282–343

Chapter 13
Bdelloid Rotifers: Progress in Understanding the Success of an Evolutionary Scandal

David B. Mark Welch, Claudia Ricci and Matthew Meselson

Abstract Despite considerable attention, what drives asexual populations to extinction and allows the maintenance of sexual reproduction remains a mystery. Bdelloid rotifers appear to be an exception to the general rule that multicellular eukaryotes that abandon sex are doomed to early extinction, and an understanding of their biology may reveal how they have avoided the fate of other asexual lineages. Here, we discuss aspects of the ecology, physiology, and genome evolution of bdelloid rotifers that may contribute to their evolutionary success. We propose that many of the unusual characteristics of bdelloids derive from their adaptation to desiccation-prone habitats but may also contribute to their long-term success in the absence of sexual reproduction. However, many unanswered questions about these characteristics and the basic biology of bdelloids remain. Population genetics and molecular ecology are promising approaches for the next generation of studies using this model system to provide answers to the questions of how and why sex is the dominant form of reproduction in plants and animals.

13.1 Introduction

As other chapters in this book make clear (see Chapters 1, 5, 6, and 7), although many hypotheses have been proposed, there is no general agreement in what maintains sexual reproduction and causes the early extinction of lineages that abandon it. A promising approach to the resolution of this fundamental problem in biology is the study of ancient asexuals (see also Chapters 11 and 12). Despite much observation of field and laboratory populations, neither meiosis nor males, hermaphrodites or vestigial male structures have ever been demonstrated in the highly unusual rotifers of the Class Bdelloidea; nevertheless, Bdelloidea has diversified and flourished for tens

D.B. Mark Welch (✉)
Josephine Bay Paul Center for Comparative Molecular Biology and Evolution, Marine Biological Laboratory, Woods Hole MA 02543, USA
e-mail: dmarkwelch@mbl.edu

I. Schön et al. (eds.), *Lost Sex*, DOI 10.1007/978-90-481-2770-2_13,

of millions of years. The evolutionary success of bdelloids in the apparent absence of sexual reproduction led John Maynard Smith to describe them as "something of an evolutionary scandal" (Maynard Smith 1986). The accumulated evidence that bdelloid rotifers have evolved asexually, together with the ease of their propagation in the laboratory, make them a uniquely promising experimental system for the investigation of the consequences of long-term evolution without meiosis and the elucidation of the factors responsible for the maintenance of sexual reproduction. Here, we review the evidence for obligate asexuality in bdelloids and discuss how their ecology, physiology, and genome evolution may have combined to contribute to their success.

First observed by Leeuwenhoek more than 300 years ago (Leeuwenhoek 1677, 1702), bdelloids have long been a favorite subject of amateur microscopists and can easily be observed with a low power microscope. Bdelloids are only 0.1–1.5 mm in length (adults of many species are 0.25–0.5 mm long) but contain about 1,000 nuclei and have ganglia; muscles; photo-, chemo-, and tactile sensory organs; structures for crawling, feeding, and swimming; digestive, secretory and excretory organs; and ovaries. Bdelloids propel themselves through water by rhythmic beating of their coronal cilia; when attached to a substrate by their foot, they may withdraw their corona and extend an apical structure called the rostrum. Bdelloids crawl by extending the rostrum to grab the substrate, then moving their foot to a position near their rostrum, then re-extending the rostrum; this inching motion is the source of their name (from the Greek ßδελλα, for their leech-like locomotion).

Bdelloidea is a class of the traditional phylum Rotifera (or sub-class of the Class Eurotatoria within Rotifera), a taxon of basal invertebrates belonging to the so-called Gnathifera, related to Gnathostomulida, Micrognathizoa, non-acoel flatworms, and perhaps Cycliophora. Bdelloidea is considered to be monophyletic (Melone and Ricci 1995; Wallace et al. 1996; Mark Welch 2000), while the relationship between Bdelloidea and other rotifer groups remains unresolved (Garey et al. 1996; Mark Welch 2000; Herlyn et al. 2003; Giribet et al. 2004; Funch et al. 2005; García-Varela and Nadler 2006; Sørensen and Giribet 2006). There are about 460 described bdelloid species, primarily European and reflecting the distribution of rotifer taxonomists (Segers 2007, 2008); more limited sampling in other parts of the world suggests that many additional species are undescribed. Molecular and morphometric evidence indicate that many described species are cryptic complexes; the question of what constitutes a bdelloid species has been under investigation and discussion for some time (Ricci 1988, 2001; Barraclough et al. 2003; Barraclough and Herniou 2003; Birky et al. 2005; Fontaneto et al. 2007b, 2008a).

Bdelloids can be found in almost any freshwater environment over a wide range of temperature and pH, and are often among the most common microinvertebrates, particularly in ephemerally aquatic environments such as rain gutters, birdbaths, moss, lichen, and temporary freshwater pools. The success of bdelloids is particularly surprising in light of the fact that they are believed to be obligately parthenogenetic and to have evolved from a common asexual ancestor.

13.2 Evidence of Long-Term Asexuality in Bdelloidea

Descriptions and drawings of bdelloid rotifers can be found in the letters and notebooks of Leeuwenhoek and other early microscopists, and bdelloids were avidly studied by 18th and 19th century naturalists. By the time Hudson and Gosse wrote their treatise on rotifers, they were able to observe with curiosity that although males of many species of monogonont rotifers were readily found and identified, there was no report of a male bdelloid (Hudson and Gosse 1886). In a 565-page monograph on rotifer reproduction, Wesenberg-Lund (1930) devoted just over one page to bdelloids, beginning, "As is well known, the *Bdelloida* are regarded as totally acyclic; males have never been found" (p. 185). He went on to provide the following account of rotifers he observed on the submerged roots of a chestnut tree, which occasionally is cited as evidence for male bdelloids:

> "With great hesitation I venture to remark, that twice I saw among thousands of Philodinidae (*Rotifer vulgaris* [=*Rotaria rotatoria,* a bdelloid]) a little creature, unquestionably a rotifer male, with a ciliary wreath resembling that on *Rhinops* [=*Rhinoglena frontalis*, a monogonont], but without any projecting rostrum, shaped like a *Rhinops* male and with two red eyespots rather a long distance from each other. The male was new to me, but both times I failed to get it isolated. It moved round and between the numerous females with extreme rapidity." (p. 186)

In his concluding remarks, Wesenberg-Lund assumes that bdelloids are asexual (p. 207). Rotifer treatises by Remane (1929), Donner (1965), and Voigt (1959) treat bdelloids as exclusively female.

Whether or not bdelloid females ever produce males, bdelloids do reproduce by parthenogenesis. A single egg will hatch into a female that can lay eggs that hatch into females. Bdelloid oogenesis was studied in detail by Hsu (1956a, b), who described the production of oocytes without chromosome pairing or reduction in chromosome number. His papers contain *camera lucida* drawings of iron-hematoxylin stained thin sections; a renewed study of bdelloid oogenesis with modern imaging technology is clearly warranted. Bdelloid mitotic chromosomes are relatively small, generally less than 5 μm. The number of chromosomes in described karyotypes ranges from 10 to 14 (Mark Welch and Meselson 1998). Karyotype evolution is another aspect of rotifer biology that is poorly explored.

If bdelloids are obligately asexual, it is much more likely that they evolved from a common asexual ancestor than that multiple bdelloid lineages lost sex independently and that all sexual forms have gone extinct or are unrecognized. However, the available observational and cytological evidence of bdelloid asexuality cannot exclude rare or unrecognized sexual reproduction or other forms of genetic exchange. Two unusual features of bdelloid genomes have been cited as additional support for the complete absence of sex in bdelloids: high sequence divergence between gene copies (Mark Welch and Meselson 2000), and the lack of high-copy number retrotransposable elements (Arkhipova and Meselson 2000), which will be considered here in turn.

That heterozygosity will increase in the absence of segregation and recombination has been recognized by geneticists since Muller's analysis of balancer chromosomes in *Drosophila* (Muller 1917, 1918); that divergence would accumulate between former haplotypes in an amictic lineage was predicted by Darlington (1931, 1939), Carson (1967) and White (1970, 1973). Expressed in terms of population genetics theory, heterozygosity at neutral loci in sexually reproducing populations is determined by the evolutionary forces of mutation and drift. Mutation increases heterozygosity and random drift of haplotypes produced by segregation drives alleles toward fixation or extinction, decreasing allelic diversity and individual heterozygosity. At equilibrium, these forces are balanced and neutral heterozygosity for a randomly mating population is approximately $4\ N_e\mu/(1\text{–}4\ N_e\mu)$, where N_e is the effective population size and μ the mutation rate (Kimura 1969). In the absence of segregation, there can be no independent drift of haplotypes and neutral mutations between formerly segregating alleles will continue to accumulate in the same nucleus (unless homogenized by gene conversion or mitotic crossing over). Neutral divergence, neglecting multiple mutations at the same site, will be $H_0+2\mu t$, where H_0 is the heterozygosity inherited from the sexual population and t is the time the lineage has been reproducing without segregation (if homogenizing processes continue to act on the genome, t is the time since the last homogenizing event in the region under consideration and H_0 will be zero).

The accumulation of heterozygosity in an asexual lineage has several important caveats. First, there is no prediction that heterozygosity will accumulate in a lineage that reproduces automictically, as segregation and recombination still occur. Second, there must be sufficient time since the loss of mixis for mutations to accumulate above the background level of heterozygosity predicted for a sexual population. For example, measured nucleotide level heterozygosity averaged over diverse regions in sexually reproducing invertebrate species is about 1% and the mutation rate in invertebrate nuclear genomes is about 1×10^{-8} per nucleotide per year (Lynch and Conery 2003; Lynch 2006). Thus, after a million years of amictic reproduction – a longer period of time than many models for the maintenance of sex can account for – divergence between former alleles would be $H_0+2\mu t \approx 1\% + 2 \times 10^{-8} \times 10^6 = 3\%$, not convincingly beyond the possible variance for sexually reproducing species. Third, nucleotide heterozygosity in some sexual species is considerably higher than 1%: for example, whole genome sequencing has revealed that heterozygosity in a single genome of the obligately sexual sea squirt *Ciona savignyi* is ~7% (the reason for this is not yet apparent; heterozygosity in *Ciona intestinalis* is closer to 1% (Vinson et al. 2005; Kim et al. 2007). Finally, forms of recombination not dependent on sexual reproduction, such as gene conversion and mitotic crossing-over, as well as chromosome loss and reduplication, can eliminate accumulated heterozygosity in an amictic lineage, so that the absence of high levels of heterozygosity does not prove that mixis must be occurring (Butlin 2002).

The expectation that sequence divergence between gene copies in a long-term amictic lineage would be greater than the allelic heterozygosity predicted for a sexual population was used to assess the ancient asexuality of bdelloids by Mark Welch and Meselson (2000). They amplified regions of several nuclear genes from DNA isolated from clonal populations of rotifer species by PCR and found that monogonont rotifers, which are facultatively sexual, had low levels of heterozygosity.

Bdelloids, in contrast, had much higher levels of divergence between copies of each gene examined. The possibility that each gene copy identified by PCR had a closely similar homolog indistinguishable by DNA sequence was ruled out by subsequent fluorescent *in situ* hybridization (FISH) of the cloned gene copies to embryo nuclei. The FISH results showed that each gene copy identified by PCR was on a different chromosome and that there were no additional copies in the genome, indicating that bdelloids are not highly homozygous sexual polyploids (Mark Welch 2001; Mark Welch et al. 2004b). While interpreted as supporting the asexuality of bdelloids, two aspects of the findings confounded a simple interpretation of long-term mitotic evolution of Bdelloidea from a common diploid ancestor: (1) for the two genes sampled from more than one species (*hsp82* and *tbp*), no unambiguous genealogy of the same two lineages was present in each species, as might be expected if all species evolved without conversion or other homogenizing process from a common diploid ancestor that abandoned sex (Mark Welch et al. 2004a). (2) PCR screens revealed the presence of more than two copies of the most thoroughly examined gene, *hsp82*, in 3 of the 4 bdelloid species examined (Mark Welch and Meselson 2000). Given that the other genes examined were found in only two copies, and that the presence of additional copies appeared to be sporadically distributed, along with the rareness of polyploidy in animals assumed at the time (White 1973), the available data suggested occasional segmental duplication and loss, which would not be an unexpected consequence of mitotic reproduction. However, recent genome surveys of *P. roseola* and *A. vaga*, discussed below, indicate that the additional copies do not result from independent small duplications and instead suggest that bdelloids are degenerate tetraploids that descended from a common tetraploid ancestor.

The second line of genomic evidence that appeared to support bdelloid asexuality comes from the expectation that for a lineage to survive without sexual reproduction over the long term, it should lack those deleterious transposable elements that do not engage in horizontal transfer (Arkhipova and Meselson 2005a; Dolgin and Charlesworth 2006). Once lost, these elements cannot be regained in an asexual lineage; in contrast, even if detrimental, they can be maintained in outcrossing sexual populations (Hickey 1982). To test this expectation, Arkhipova and Meselson conducted nested PCR screens for reverse transcriptase genes of two major superfamilies of retrotransposons: LINE-like, which are rarely horizontally transferred if ever, and gypsy-like, which are only occasionally transferred horizontally. Neither was found in any of five bdelloid species tested, representing three families. However, they were readily found in all (LINE-like) or nearly all (gypsy-like) of 39 sexually reproducing species, representing 23 animal phyla, including monogonont rotifers (Arkhipova and Meselson 2000). Also, except for certain retrovirus-like elements, retrotransposons are not present in any of several 40–50 kb clones from bdelloid heterochromatin-like sub-telomeric regions, where such elements are relatively abundant in *Drosophila* and other model eukaryotes (Bartolome et al. 2002). If present at all, non-viral retrotransposons must be at unusually low copy number in bdelloid genomes. However, as might be expected (Hickey 1992), bdelloid genomes do contain transposable elements of types that are particularly prone to move horizontally, including various DNA transposons and retrovirus-like retrotransposons. Even these, however, appear to be rare or absent in gene-rich proximal regions, being concentrated instead in sub-telomeric regions (Arkhipova and Meselson 2005b).

13.3 Environmental Adaptations in Bdelloidea

Bdelloids are found throughout the world in a wide variety of freshwater ecosystems. The majority of bdelloid species can be found in limno-terrestrial habitats, including freshwater sediments and terrestrial soils, where water, salt content, and temperature change quickly and unpredictably (Ricci and Balsamo 2000). The ability of bdelloids to occupy temporal habitats has been ascribed to the interaction between two peculiar bdelloid features: obligate parthenogenesis and anhydrobiosis (Ricci 1987). Parthenogenesis could allow bdelloids to rapidly re-occupy a habitat after it changed from unfavourable to favourable conditions or to rapidly recover from a very low population size following a change from inhospitable conditions (both predictions have clear consequences for population genetic structure that have yet to be tested). Anhydrobiosis, discussed in detail below, allows bdelloids to stably inhabit desiccation-prone habitats and also provides a means of dispersal: when desiccated, a bdelloid contracts into a small flat ellipsoid called a tun, which adheres firmly to substrates such as soil particles, mud, or moss fragments (Fig. 13.1). Tuns represent true propagules that may be passively transported over long distances; a single tun is sufficient to found a new population if it is deposited in a suitable habitat. Thus, although the faunistic knowledge of bdelloids "is so small that it borders on total ignorance" (Fontaneto et al. 2007a), bdelloid species have long been considered cosmopolitan. A survey of 302 bdelloid isolates from around the world, representing 61 species, confirms that most species are cosmopolitan, although biogeographical patterns can be discerned when genetic clusters are recognised within traditional species (Fontaneto et al. 2006, 2008a). Cosmopolitan distribution implies high rates of dispersal, which was confirmed for some species by Birky and coworkers in a study of the mitochondrial *cytochrome oxidase I* (*coxI*) gene from more than 100 bdelloid isolates representing 21 phylogenetically defined species (Birky et al. 2005). They found that the correlation between amino acid sequence divergence and geographic distance for isolates within a species implied that many species could have spread throughout the world since their origin. A list of bdelloid species from Europe (available at www.faunaeur.org), and studies from poorly sampled areas in Eastern Asia, Australia, and Antarctica also show that most species have a worldwide distribution, with only 10% of the species having a limited distribution and appearing to be endemic (De Ridder 1956; Dumont 1983; Ricci 1987, 2001; Fontaneto and Melone 2003; Ricci et al. 2003b; Fontaneto et al. 2008a). The fact that some bdelloid species seem to occur in specific areas only and to have restricted distribution may be ascribed to scarce faunistic knowledge of bdelloids, but also may suggest that these species are either poor propagules or that their habitat requirements are very selective (Ricci 2001; Fontaneto and Melone 2003; Ricci et al. 2003b; Segers and Shiel 2003, 2005; Fontaneto et al. 2006, 2007a).

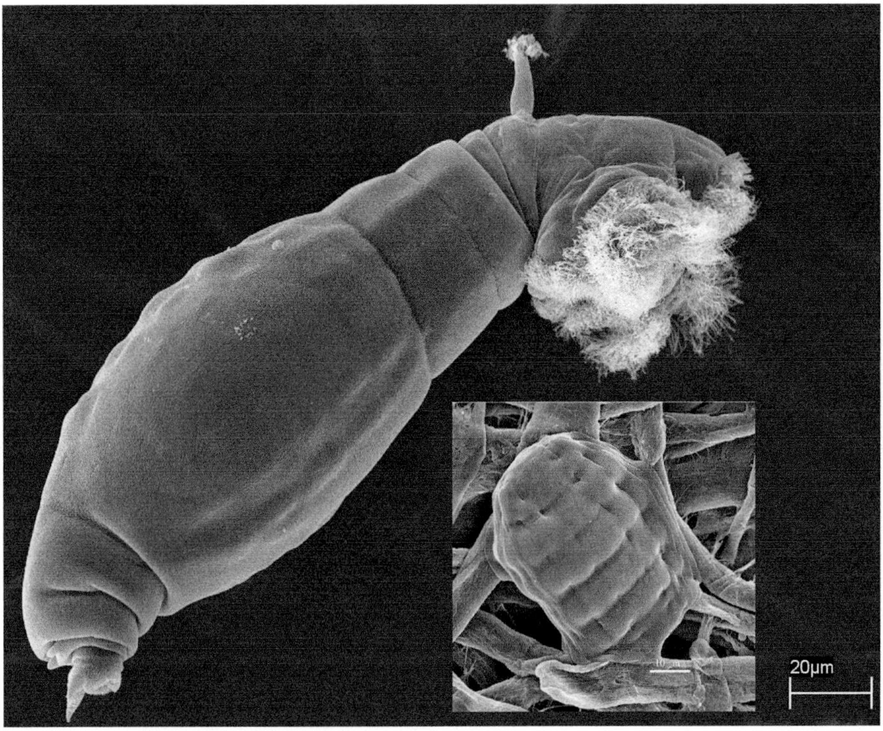

Fig. 13.1 SEM images of the bdelloid *Macrotrachela quadricornifera* hydrated, and desiccated on filter paper (inset). Courtesy of G. Melone.

If dispersal is common and if most species have opportunities to disperse, the local habitat represents an important factor shaping bdelloid communities. In other words, the chance for a bdelloid to successfully recover in a new location depends on abiotic characteristics of the new habitat. All other factors being equal, the presence of a bdelloid species in a given habitat can be predicted with a reasonable certainty on the basis of habitat characteristics and species preferences. This hypothesis is supported by two recent surveys of Alpine bdelloids, which show that although there is not a connection between geographic distance and species distribution, and that bdelloids do appear to disperse freely, individual bdelloid species predictably occupy different specific habitats (Fontaneto et al. 2006; Fontaneto and Ricci 2006). Thus, to a first approximation, bdelloids—like all microscopic animals– are everywhere, but, unlike protists, the environment selects at both the global and local levels. In addition to the implication that the dispersal patterns of bdelloids and protists are similar, this biogeographic hypothesis implies that individual bdelloid species have successfully adapted to different specific niches in the absence of sexual reproduction.

A bdelloid species in a given location often consists of several independent genetic lineages (whether clones or cryptic species is a matter of debate and definition; see also Chapter 9). For this coexistence to be dynamically stable, different clones must vary in ecological requirements so that reproductive advantage shifts from clone to clone with temporal and spatial micro-habitat heterogeneity. That this pattern indeed occurs in bdelloids is supported by isozyme variation in isolates of *Macrotrachela quadricornifera* collected from one moss patch throughout one year (Ricci et al. 1989), and *coxI* variation of isolates of *Philodina flaviceps* collected from 7 neighbouring moss patches in a single season (Fontaneto et al. 2008b). Also, an unpublished study by King and Ricci found that two clones of *M. quadricornifera* collected from the same piece of moss had very similar life history traits but that one consistently out-competed the other over a wide range of conditions, while the other had greater success recovering from desiccation, suggesting that their coexistence may be due to shifting fitness advantages in cyclically changing conditions. Furthermore, when five different clones of *M. quadricornifera* were assessed for fecundity and other fitness-related traits under different temperatures, they were found to differ in ways that could be related to adaptation to local conditions (Ricci 1991). Together, these results indicate that bdelloid species are able to diversify into distinct evolutionary entities, and that local coexistence of these independent lineages may be common.

13.4 Anhydrobiosis

When dried moss, detritus from rain gutters, or even tree bark is placed in water and observed under a dissecting microscope, adult bdelloid rotifers will often soon be seen actively swimming and feeding. Bdelloids thrive in such ephemerally aquatic environments because of their ability to survive desiccation through a form of quiescence called anhydrobiosis, or "latent life by drying" (Giard 1894). Anhydrobiosis in bdelloids differs from desiccation-resistant diapausing stages such as monogonont resting eggs or nematode dauer larvae, in that it can occur at any point during the life of an individual. Among animals, bdelloids, some tardigrades, and some nematodes (but not *Caenorhabditis elegans*) are capable of anhydrobiosis (reviewed in Càceres and Soluk 2002). Desiccation-resistant nematodes and tardigrades synthesize disaccharides such as trehalose as osmoprotectant molecules before entering anhydrobiosis (Crowe 1971; Westh and Ramlov 1991; Higa and Womersley 1993; Clegg 2001). In contrast, bdelloid rotifers lack trehalose and appear to lack other osmoprotectant disaccharides (Lapinski and Tunnacliffe 2003; Caprioli et al. 2004). Like other desiccation-resistant organisms, bdelloids synthesise proteins similar to the Late Embryogenesis Abundant (LEA) proteins known from plant seeds (Browne et al. 2002; Tunnacliffe et al. 2005). While the exact function of these proteins is unknown, they appear to undergo a conformational change during desiccation that may allow them to protect cellular components during anhydrobiosis (Goyal et al. 2003).

The most conspicuous modification of a desiccating bdelloid entering anhydrobiosis is a change in body shape and reduction in volume. An individual retracts its head and foot into the trunk and contracts longitudinal muscles to form a tun (Wright 2001; Ricci et al. 2003a). When drying, bdelloids lose virtually all unbound water and reduce their weight by 95% (Ricci et al. 2008). During this process, they modify their internal organization remarkably, condensing tissues, cells, and organelles and folding membranes into "myelin figures" (Dickson and Mercer 1967; Wharton and Lemmon 1998; Marotta et al. 2008). During dormancy, bdelloids appear to suspend respiration, metabolism, and aging. On recovery, bdelloids resume reproduction and other life history traits consistent with their age at the start of dormancy ignoring the time spent in anhydrobiosis; this has been termed the "Sleeping Beauty" strategy (Ricci et al. 1987; Ricci and Covino 2005). The ability of bdelloids to recover from anhydrobiosis is dependent on the amount of food in the individual's digestive system, although different authors have found contrasting effects (Lapinski and Tunnacliffe 2003; Ricci et al. 2004) and abiotic conditions such as the nature of substrate and rate of evaporation seem to play a role (Ricci et al. 2003a). Although rates of recovery change with the age of the individual (Orstan 1995; Orsenigo et al. 1998; Ricci 1998; Ricci and Caprioli 1998), repeated events of desiccation do not change the probability of recovery (Ricci and Caprioli 2005; Ricci and Covino 2005).

Rates of recovery from desiccation can be as high as 90% in many bdelloid species (Ricci and Caprioli 2005). Only a few species, scattered among the four bdelloid families, seem incapable of anhydrobiosis; these are found exclusively in stable aquatic systems (Fig. 13.2). On the basis of this observation, Ricci suggested that anhydrobiosis is an apomorphic trait that was present in the common ancestor

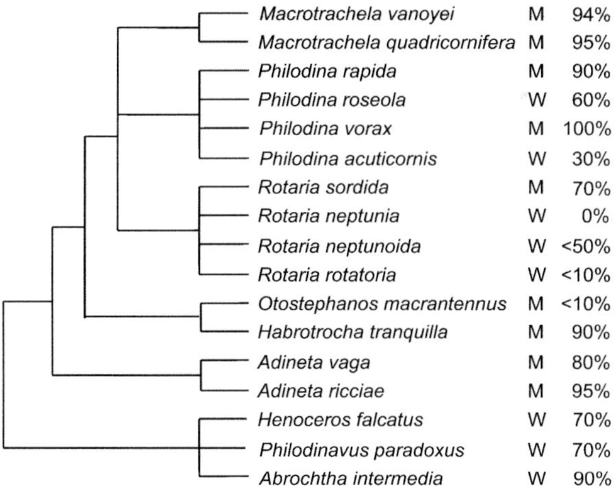

Fig. 13.2 Anhydrobiosis survival rates in different bdelloid species. The cladogram represents the systematic relationship of the bdelloid species examined; a species name is followed by the habitat type in which the species is generally associated: M= moss, soil, or other temporarily aquatic habitat prone to dehydration; W= permanently wet habitats. Following the habitat type is the percent of individuals recovering from anhydrobiosis; *Otostephanos macrantennus* has an anomalously low desiccation survival rate under the laboratory conditions used (Ricci 1998; Claudia Ricci, unpublished).

of bdelloids (Ricci 1998). The fact that anhydrobiosis can be lost in the absence of selection to maintain it implies that it is a physiologically and genetically complex phenomenon.

Experimental results indicate that anhydrobiosis does not result in a fitness cost to the animals that recover from desiccation. Rather, mothers who have been through desiccation produce daughters of increased fitness and longevity (Ricci and Covino 2005). This finding suggests the existence of repair processes associated with recovery from desiccation that may have a beneficial effect beyond desiccation tolerance. Further evidence of such repair processes comes from laboratory experiments showing a decline in fitness for bdelloid populations maintained in a hydrated state over many generations compared to populations that were cyclically desiccated (Ricci et al. 2007). This decline in fitness refers to the average fecundity of the population in the absence of selection, and some individuals keep reproduction at a high rate. These individuals are likely the ones that sustain the population, as large clonal populations of bdelloids have been kept in laboratories for 10–25 years (*Adineta vaga* and *Philodina roseola*on a diet of *Escherichia coli*, *Macrotrachela quadricornifera* on fish pellets or yeast; Ricci et al. 1987; Hur et al. 2009; David Mark Welch and Claudia Ricci, unpubl.). Some explanations for this result can be offered. The decline in reproductive capacity with continued hydration may result from epigenetic effects, as fitness can be restored after a single desiccation event (Claudia Ricci, unpubl.). But other explanations can not be not ruled out, including the possible presence of parasites that may be killed by desiccation. Epigenetic inheritance is well known in rotifers: mothers transmit age-dependent changes in life-history traits to offspring, which are reported to be cumulative and reversible, a phenomenon that has been called the "Lansing effect" (Lansing 1954). In bdelloids, the maternal age effect has also been found to operate on the biochemical phenotype (Ricci et al. 1999). A morphological basis for this cytoplasmic heredity may come from the process of bdelloid ovogenesis, where developing oocytes receive cytoplasm from the nurse gland (called the vitellarium) of the mother (Amsellem and Ricci 1982). As the mother ages, the cytoplasmic constituents she provides may change, perhaps succumbing to oxidative damage or simply decreasing in amount. Processes associated with recovery from desiccation (see below) may repair or increase the production of these constituents, resulting in more fit offspring.

13.5 Ionizing Radiation and Desiccation

In addition to their apparent lack of males and meiosis, their lack of high copy-number retrotransposons, and their ability to survive and resume reproduction after desiccation at any life stage, bdelloid rotifers are also extraordinary in their resistance to ionizing radiation. The ability of the bdelloids *Adineta vaga* and *Philodina roseola* to produce progeny is hardly affected by exposure to 200 Gray of gamma radiation, a dose that reduces fecundity in the monogonont rotifer *Euchlanis dilatata*, and all other animals for which there is applicable data, by at least 90%

(Gladyshev and Meselson 2008). Extraordinary resistance to ionizing radiation is likely to be characteristic of Bdelloidea generally, as *A. vaga* and *P. roseola* represent different families within Bdelloidea and because of the probable association between radiation resistance and stage-independent anhydrobiosis, as seen in the bacterium *Deinococcus radiodurans* (Mattimore and Battista 1996). Both bdelloid species incur DNA double-strand breaks (DSBs) at a rate of about 0.005 DSB per million base pairs per Gray, a rate comparable to that in species not resistant to ionizing radiation. A dose of 560 Gray creates hundreds of DSBs per *A. vaga* genome, most of which would certainly be lethal if not repaired, yet this dose reduces fecundity by only about 20%. Thus, the ability of bdelloids to remain fertile after massive DNA breakage must result not from protection against the occurrence of DSBs but from an extraordinary ability to repair them.

Like bdelloid rotifers, the bacterium *Deinococcus radiodurans* is resistant to both desiccation and radiation (Cox and Battista 2005). Radiation resistance in itself is unlikely to confer any selective benefit outside the laboratory and chemically-induced mutations that diminish radiation resistance in *D. radiodurans* have been found also to diminish its desiccation resistance. It therefore appears that the radiation resistance of *D. radiodurans* is a consequence of evolutionary adaptation to survive the desiccation frequently encountered in its characteristic habitats (Mattimore and Battista 1996). By analogy with *D. radiodurans*, the extraordinary radiation resistance of bdelloid rotifers is almost certainly an evolutionary adaptation to survive the desiccation they frequently encounter.

Even after several weeks of exposure of *Deinococcus radiodurans* to low humidity, during which hundreds of DNA double strand breaks accumulate per genome, recovery upon rehydration can be essentially complete and declines substantially only after even more protracted desiccation (Minton 1994). Although DNA integrity has not yet been examined in desiccated bdelloids, it is most likely that, as in *D. radiodurans*, desiccation breaks bdelloid DNA but that DNA breaks and other desiccation-induced damage can be repaired upon rehydration.

13.6 Genome Structure

The genome structure of bdelloids has been investigated most thoroughly in two distantly related species, *Philodina roseola* and *Adineta vaga*, by a combination of sequencing large-insert clones from genomic libraries and FISH of these clones to chromosomes of early embryos (Mark Welch 2001; Mark Welch et al. 2008; Hur et al. 2009). The available evidence now suggests that bdelloid genomes are composed of two ancient lineages (corresponding to the A and B lineages of Mark Welch & Meselson (2000) with two copies of each lineage (Mark Welch et al. 2008). Genes have been lost from each lineage and the lineages can only be reliably aligned within shared coding sequences, where divergence at synonymous positions (K_S) is 60–120%. Within a lineage, genes are preserved with co-linearity between the two copies, although there are numerous indels including differences in intron number; K_S ranges from zero to more than 20% between gene copies in these co-linear pairs.

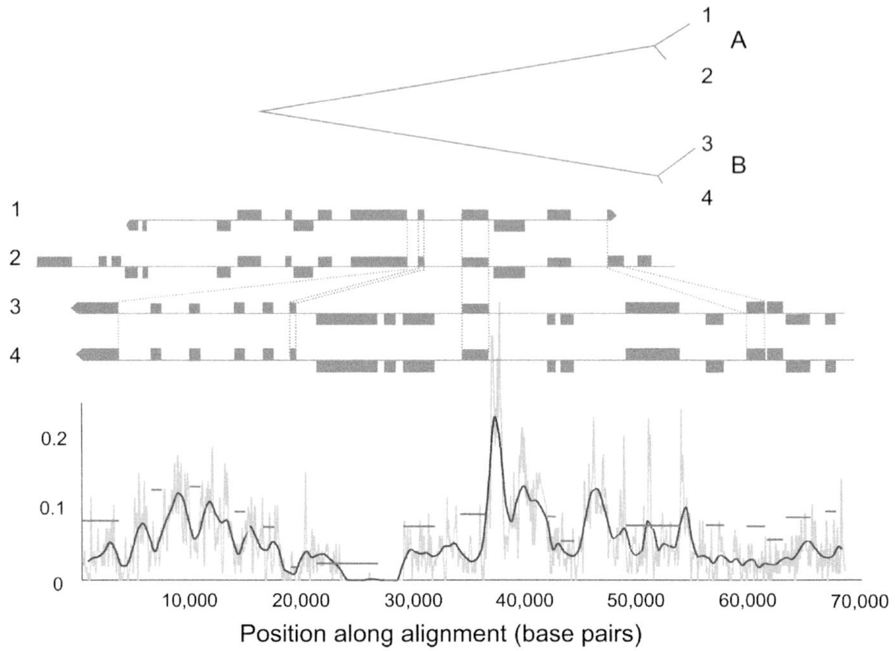

Fig. 13.3 The region around *hsp82* in *Philodina roseola*. *Top*: the neighbor-joining tree showing the relationship between *hsp82* regions, based on synonymous divergence of shared genes. *Middle*: cartoon of the four regions with genes shown as rectangles in the forward (above the *line*) or reverse (below the *line*) orientation, *dashed lines* indicate genes shared in all four copies; the cartoon is centered on *hsp82*. *Bottom*: divergence between the two B lineage copies of *hsp82* in *P. roseola* in a sliding window of 100 bases (*thin line*) or 1000 bases (*thick line*), with the synonymous divergence of each gene shown in the cartoon indicated as horizontal *thick lines*. Adapted from Mark Welch et al. (2008).

The four homologous regions of the *P. roseola* genome containing *hsp82* are shown in Fig. 13.3; FISH has demonstrated that each copy is on a separate chromosome (Mark Welch 2001; Mark Welch et al. 2004b).

Sequencing and FISH of regions containing two different Hox genes in *Philodina roseola*, and regions containing histone gene clusters in *P. roseola* and in *Adineta vaga* show this same basic structure of two ancient lineages each composed of two co-linear gene pairs (Hur et al. 2009; Van Doninck et al. 2009; J. Mark Welch, unpubl.). Earlier, Mark Welch and Meselson (2000) found three copies of *hsp82* in *A. vaga* and *Habrotrocha constricta* and two copies in *M. quadricornifera* by sequencing cloned PCR products. In the case of *A. vaga* and *H. constricta*, two of the three copies were closely related and a third was more distant. Subsequent genome-level analysis, FISH, and Southern hybridization has demonstrated that *A. vaga* in fact has four copies of *hsp82*, each on a separate chromosome, but that two copies are identical or nearly identical across the nearly 50 kb examined to date (Hur et al. 2009). The two non-identical but closely related copies have a co-linear gene structure, with some genes being conserved between this co-linear pair and the co-linear pair being composed of identical copies. Thus, the genome structure around *hsp82* in *A. vaga* is very similar to that in *P. roseola*, and it seems reasonable to assume that *H. constricta* may also harbor a fourth copy, identical over the surveyed region of *hsp82* to the copy that lacks an identified co-linear partner. *Mactotrachela quadricornifera* was not screened as thoroughly in the original PCR screens and has not yet been examined at the fosmid clone level or by FISH. The two copies of *hsp82* found in *M. quadricornifera* by PCR are clearly representatives of two ancient lineages, and are each closely related to copies of *hsp82* in *P. roseola*. This similarity contrasts sharply with much greater synonymous divergence values between *M. quadricornifera* and *P. roseola* for the mitochondrial gene *cox1* (Birky et al. 2005), which led Hillis to assert that the similarity between *hsp82* copies in *M. quadricornifera* and *P. roseola* must be due to genetic exchange between the two species (Hillis 2007). However, this assumes equal or nearly equal rates of change at *cox1* and *hsp82* when in fact *cox1* in bdelloids is nearly saturated at synonymous positions (due in part to extreme AT bias at third positions) while also having far fewer non-synonymous changes than *hsp82*. A greater understanding of the evolution of *cox1* in bdelloids is needed before the gene can be used as a standard by which to judge possible exchange of nuclear genes. That *hsp82* sequence divergence should be much less between *M. quadricornifera* and *P. roseola* than between either species and *A. vaga* or *H. constricta* is not surprising given their closer taxonomic affiliation (both are in the family Philodinidae, while *A. vaga* and *H. constricta* are each in separate families). Further, we note that in separate investigations, we (Matthew Meselson and David Mark Welch) have been unable to recover sequences of *hsp82* from *M. quadricornifera* that match those we reported previously. Clearly, the genome of *M. quadricornifera* bears further investigation.

Based on evidence from fosmid-scale genome sequencing and FISH, the basic structure of the bdelloid genome is that of a degenerate tetraploid. Under this model, the 12 chromosomes of *A. vaga* comprise three sets of four chromosomes, each consisting of two co-linear pairs that are homeologs. The karyotype of *Philodina roseola* is 13 chromosomes, including two dot chromosomes and a very long chromosome; FISH results are consistent with the long chromosome being an isochromosome (Mark Welch 2001). This would imply that *P. roseola* has the same basic structure as *Adineta vaga*, considering the two arms of the isochromosome as a colinear pair and ignoring the dot chromosomes as supernumerary or B chromosomes. The only other species for which there are published karyotypes do not yet have accompanying FISH analyses but are also consistent with the same basic genome structure, implying that the ancestral karyotype of bdelloids is $x=3$ (Fig. 13.4). The karyotypes of monogonont rotifers that have been examined suggest that the number of chromosomes in the haploid genome of most monogononts is larger, around 10 (reviewed in Mark Welch and Meselson 1998).

Analysis of several hundred thousand bases of aligned sequence of co-linear pairs reveals extensive variation in divergence along a pair. Particularly noticeable are tracts of near or complete identity; these tracts are not associated with the position of genes and they occur in different places in homologous regions of *P. roseola* and

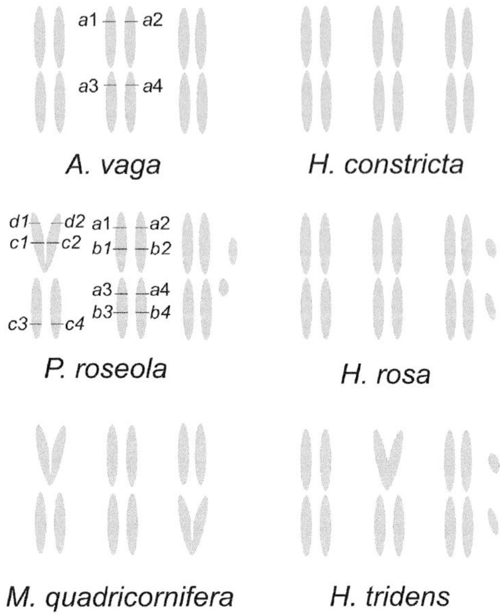

Fig. 13.4 Model of bdelloid karyotype organization. For each species, chromosomes are arranged in a karyogram with co-linear pairs in two rows representing the two ancient A and B lineages; the small supernumerary B chromosomes are considered to be transient and inconsequential. Potential isochrosomes are shown as V-shaped and should not necessarily be considered homologous. A limited number of homologous regions have been mapped to chromosomes by FISH: *a*: the region surrounding *hsp82* in *Philodina roseola* and *Adineta vaga* (Mark Welch et al. 2004b; Hur et al. 2009); *b*: the region surrounding a *Hox5*-like gene; *c*: the region surrounding a *Hox6*-like gene; *d:* an unsequenced YAC clone (J. Mark Welch, pers. comm.)

A. vaga. The simplest explanation for these tracts is occasional gene conversion. There is a statistically significant excess of blocks of identity greater than 80 bp in length; these blocks, and especially the regions of identity greater than 1 kb in length, suggest that gene conversion is occurring within pairs and implies that other discernable tracts of low divergence, such as the region containing *hsp82* in the A lineage pair (see Fig. 13.3), may be the result of earlier gene conversion events (Mark Welch et al. 2008). If co-linear pairs are subject to recurring homogenization, as by gene conversion or by mitotic crossing-over in G2, their age may be much greater than that suggested by their average K_S of ~7%. The maximum K_S observed to date between members of a colinear pair in *P. roseola* is ~26%, and even this value may represent the time since the last gene conversion event in that region, rather than the origin of the co-linear pair.

Gene conversion may also have shaped the pattern of divergence between the more ancient A and B lineages. For example, all phylogenetic analyses consistently show that the nucleotide sequences of all four copies of *hsp82* in *A. vaga* are more similar to the sequences of the A lineage in *P. roseola* than to the sequences of the B lineage (Mark Welch and Meselson 2000; Mark Welch et al. 2004a). However, examination of the 40–80 kb region surrounding each copy of *hsp82* in these two species shows that two of the four A-like copies in *A. vaga* are surrounded by genes found in the B lineage of *P. roseola* but not the A lineage (Hur et al. 2009). This suggests that the two *A. vaga* copies are on B lineage chromosomes and that gene conversion or other homogenizing events occurred between the *A. vaga* A and B lineages in the *hsp82* region before their divergence became too great. This also highlights the potential danger of attempting to reconstruct the phylogeny of Bdelloidea using nuclear DNA sequences.

If co-linear pairs have persisted without segregation for many tens of millions years, evolutionary theory predicts that, absent homogenizing events, they should diverge in function because maintenance of two non-segregating copies of a gene with identical function is not generally stable over this time span. The laboratory of Alan Tunnacliffe found two copies of a gene in *Adineta ricciae* that encode two different Late Embryogenesis Abundant proteins with different predicted physio-chemical properties (Pouchkina-Stantcheva et al. 2007), which would be consistent with this prediction. However, these copies may represent a gene duplication rather than belonging to a co-linear pair (Meselson and Mark Welch 2007). Therefore, we focus here on gene copies known to be members of co-linear pairs, several of which have patterns of sequence divergence consistent with a history of positive selection for divergent function. For example, in *P. roseola*, a locus for a phospho-glycerate mutase gene (*PGAM*) is shared by a co-linear pair of chromosomes; the two copies differ at 10% of synonymous positions and 2.5% of non-synonymous positions, indicating strong purifying selection along most of the gene. However, non-synonymous differences exceed synonymous ones in two localized areas, suggesting that there has been selection for differences in amino acid sequence between the copies. *Arabidopsis thaliana* also has two copies of *PGAM*, the result of a whole genome duplication event; the same two areas have nearly the same elevated ratio of non-synonymous to synonymous differences (Fig. 13.5). Of the 29 loci identified in the region around *hsp82* in *P. roseola*, 8 have such localized areas of accelerated fixation of non-synonomous changes between co-linear gene copies. As regions of elevated K_a/K_s are only one indicator of sub- or neo-functionalization, which can occur through much more subtle changes in amino acid sequence and, perhaps at least as often, through changes in transcriptional regulation, this may be an underestimate of the number of loci with gene copies that have diverged in function between co-linear pairs. This is quite different from what is observed in sexually reproducing diploids, where haplotype drift and the homozygous load imposed by segregation limit the opportunity for such differences to accumulate. However, while this observation is consistent with the absence of segregation between co-linear copies, it does not require that sex be absent in order to evolve.

13.7 Synthesis

The evidence for long-term asexuality of bdelloid rotifers remains circumstantial. The strongest evidence continues to be the consistent absence of males despite hundreds of years of scrutiny by numerous investigators. The anciently diverged genomic lineages originally observed by Mark Welch and Meselson may be due to the loss of sex in a diploid (possibly allodiploid) ancestor followed by a whole genome duplication, or perhaps more likely, to a whole genome duplication (or allotetraploidization) at or near the origin of bdelloids followed by extensive gene loss from one or the other homologous pair; either history could result in the observed degenerate tetraploid genome structure. The apparent absence of LINE-like and gypsy-like retrotransposons is consistent with long-term asexuality, but in the absence of a clear explanation for how these elements were lost, it remains possible that bdelloids simply have an efficient means of limiting the proliferation of horizontally transferred elements unconnected to their reproductive mode (Gladyshev and Meselson 2008). That bdelloid genomes appear to be composed of co-linear chromosome pairs raises the possibility that these pairs are in fact homologous chromosomes actively involved in segregation and syngamy. The large fraction of loci that appear to have evolved divergent function argues against the segregation of co-linear pairs, and the presence of potential gene conversion tracts between co-linear pairs and of the remarkable ability of bdelloids to recover from DSB damage suggests a different explanation for the preservation of co-linear pairs. Bdelloid germ-line cells in G1 lack sister chromatids, leaving co-linear chromosomes as the only templates available for the repair of DSBs. The sequence homology of such pairs could be maintained by homogenizing events, including those caused by DSB repair itself, and including gene conversion and perhaps mitotic crossing-over in G2. In addition to homogenizing events, the divergence of co-linear pairs would be limited by selection against clones in which it reaches levels that substantially reduce the efficiency or accuracy of repair. Whatever mechanism(s) operate in bdelloid DSB repair, it is reasonable to suppose that frequent exposure to a high level of DSBs and to other DNA damage caused by desiccation would, like exposure to ionizing radiation, be accompanied by a substantially elevated level of genomic lesions, including deletions. Such lesions will disrupt or remove portions of the genome that, if inessential, will decay or be eliminated from bdelloid lineages. Repeated cycles of desiccation may therefore have accelerated the loss or decay of deleterious retrotransposons and the loss of "missing" segments that may have been initially present but for which two copies in the genome are sufficient.

Thus, it may be that many of the unusual characteristics of bdelloid genomes derive from adaptation to a life style that exposes bdelloids to repeated bouts of desiccation and DNA damage repair. Indeed, the unusual structure and dynamics of the bdelloid genome may contribute to the long-term success of bdelloids. The lack of retrotransposons, perhaps in part due to DNA lesions resulting from desiccation, reduces the deleterious load from insertional elements (Gladyshev and Meselson 2008). If bdelloids are indeed asexual, the presence of co-linear pairs that do not reassort but do undergo occasional gene conversion can also provide a source of advantageous genetic variation through several mechanisms (Kirkpatrick and Jenkins 1989; Chasnov 2000; Gladyshev and Meselson 2008; Mark Welch et al. 2008). Much progress has recently been made in the study of Bdelloidea, particularly in characterizing their extraordinary resistance to desiccation and relating it to their biogeography, in determining the general structure of their genomes and in providing plausible hypotheses for their preservation of co-linear chromosome pairs, their lack of retrotransposons and high copy-number transposable elements generally, their highly unusual ectopic incorporation of foreign genes, and even their success as obligate asexuals. The stage now seems set for further investigations that will answer the question with which we began: what has allowed bdelloid rotifers to persist and thrive over millions of years when other taxa that abandon sexual reproduction suffer early extinction?

References

Amsellem J, Ricci C (1982) Fine structure of the female genital apparatus of *Philodina* (Rotifera, Bdelloidea). Zoomorphology 100: 89–105

Arkhipova IR, Meselson M (2000) Transposable elements in sexual and ancient asexual taxa. Proc Natl Acad Sci USA 97: 14473–14477

Arkhipova IR, Meselson M (2005a) Deleterious transposable elements and the extinction of asexuals. Bioessays 27: 76–85

Arkhipova IR, Meselson M (2005b) Diverse DNA transposons in rotifers of the class Bdelloidea. Proc Natl Acad Sci USA 102: 11781–11786

Barraclough TG, Birky CW, Burt A (2003) Diversification in sexual and asexual organisms. Evolution 57: 2166–2172

Barraclough TG, Herniou E (2003) Why do species exist? Insights from sexuals and asexuals. Zoology 106: 275–282

Bartolome C, Maside X, Charlesworth B (2002) On the abundance and distribution of transposable elements in the genome of *Drosophila melanogaster*. Mol Biol Evol 19: 926–937

Birky CW, Wolf C, Maughan H, Herbertson L, Henry E (2005) Speciation and selection without sex. Hydrobiologia 546: 29–45

Browne J, Tunnacliffe A, Burnell A (2002) Anhydrobiosis: Plant desiccation gene found in a nematode. Nature 416: 38

Butlin R (2002) The costs and benefits of sex: New insights from old asexual lineages. Nature Rev Genet 3: 311–317

Càceres CE, Soluk DA (2002) Blowing in the wind: A field test of overland dispersal and colonization by aquatic invertebrates. Oecologia 131: 402–408

Caprioli M, Katholm AK, Melone G, Ramlov H, Ricci C, Santo N (2004) Trehalose in desiccated rotifers: A comparison between a bdelloid and a monogonont species. Comp Biochem Physiol A Mol Integr Physiol 139: 527–532

Carson HL (1967) Permament heterozygosity. Evol Biol 1: 143–168

Chasnov JR (2000) Mutation-selection balance, dominance and the maintenance of sex. Genetics 156: 1419–1425

Clegg JS (2001) Cryptobiosis: A peculiar state of biological organization. Comp Biochem Physiol B Biochem Mol Biol 128: 613–624

Cox MM, Battista JR (2005) *Deinococcus radiodurans*, the consummate survivor. Nature Rev Microbiol 3: 882–892

Crowe JH (1971) Anhydrobiosis: An unsolved problem. Am Nat 105: 563–573

Darlington CD (1931) Cytological theory of inheritance in Oenothera. J Genet 24: 405–474

Darlington CD (1939) The Evolution of Genetic Systems. Cambridge University Press, Cambridge, UK

De Ridder M (1956) Enkele Beschouwingen over de Taxonomie der Raderdieren. Natuurwet Tijdschr 38: 160–166

Dickson MR, Mercer EH (1967) Fine structural changes accompanying desiccation in *Philodina roseola* (Rotifera). J Microsc 6: 331–348

Dolgin ES, Charlesworth B (2006) The fate of transposable elements in asexual populations. Genetics 174: 817–827

Donner J (1965) Ordnung Bdelloidea. Akademie-Verlag, Berlin

Dumont HJ (1983) Biogeography of rotifers. Hydrobiologia 104: 19–30

Fontaneto D, Barraclough TG, Chen K, Ricci C, Herniou E (2008a) Molecular evidence for broad-scale distributions in bdelloid rotifers: Everything is not everywhere but most things are very widespread. Mol Ecol 17: 3136–3146

Fontaneto D, Boschetti C, Ricci C (2008b) Cryptic diversification in ancient asexuals: Evidence from the bdelloid rotifer *Philodina flaviceps*. J Evol Biol 21: 580–587

Fontaneto D, Ficetola DF, Ambrosini R, Ricci C (2006) Patterns of diversity in microscopic animals: Are they comparable to those in protists or in larger animals? Global Ecol Biogeogr 15: 153–162

Fontaneto D, Herniou EA, Barraclough TG (2007a) On the global distribution of microscopic animals: New worldwide data on bdelloid rotifers. Zool Stud 46: 336–346

Fontaneto D, Herniou EA, Boschetti C, Caprioli M, Melone G, Ricci C, Barraclough TG (2007b) Independently evolving species in asexual bdelloid rotifers. PLoS Biology 5: e87

Fontaneto D, Melone G (2003) Redescription of *Pleuretra hystrix*, an endemic alpine bdelloid rotifer. Hydrobiologia 497: 153–160

Fontaneto D, Ricci C (2006) Spatial gradients in species diversity of microscopic animals: The case of bdelloid rotifers at high altitude. J Biogeogr 33: 1305–1313.

Funch P, Sorensen MV, Obst M (2005) On the phylogenetic position of Rotifera – have we come any further? Hydrobiologia 546: 11–28

García-Varela M, Nadler SA (2006) Phylogenetic relationships among Syndermata inferred from nuclear and mitochondrial gene sequences. Mol Phyl Evol 40: 61–72

Garey JR, Near TJ, Nonnemacher MR, Nadler SA (1996) Molecular evidence for Acanthocephala as a subtaxon of Rotifera. J Mol Evol 43: 287–292

Giard A (1894) L′anhydrobiose ou ralentissement des phenomones vitaux sous l′influence de la deshydration progressive. CR Acad Sci, Sci Vie 46: 497–500

Giribet G, Sorensen MV, Funch P, Kristensen RM, Sterrer W (2004) Investigations into the phylogenetic position of Micrognathozoa using four molecular loci. Cladistics 20: 1–13

Gladyshev E, Meselson M (2008) Extreme resistance of bdelloid rotifers to ionizing radiation. Proc Natl Acad Sci USA 105: 5139–5144

Goyal K, Tisi A, Basran A, Browne J, Burnell A, Zurdo J, Tunnacliffe A (2003) Transition from natively unfolded to folded state induced by desiccation in an anhydrobiotic nematode protein. J Biol Chem 278: 12977–12984

Herlyn H, Piskurek O, Schmitz J, Ehlers U, Zischler H (2003) The syndermatan phylogeny and the evolution of acanthocephalan endoparasitism as inferred from 18S rDNA sequences. Mol Phyl Evol 26: 155–164

Hickey DA (1982) Selfish DNA: A sexually-transmitted nuclear parasite. Genetics 101: 519–531.

Hickey DA (1992) Evolutionary dynamics of transposable elements in prokaryotes and eukaryotes. Genetica 86: 269–274

Higa LM, Womersley CZ (1993) New insight into the anhydrobiotic phenomenon: The effects of trehalose content and differential rates of adaptive evaporative water-loss on the survival of *Aphelenchus avena*. J Exp Zool 267: 12–129

Hillis DM (2007) Asexual evolution: Can species exist without sex? Curr Biol 17: R543

Hsu WS (1956a) Oogenesis in *Habrotrocha tridens* (Milne). Biol Bull 111: 364–374

Hsu WS (1956b) Oogenesis in the Bdelloidea rotifer *Philodina roseola* Ehrenberg. Cellule 57: 283–296

Hudson CT, Gosse PH (1886) The Rotifera or Wheel-Animalcules. Longmans, Green, London

Hur JH, Van Doninck K, Mandigo ML, Meselson M (2009) Degenerate tetraploidy was established before bdelloid rotifer families diverged. Mol Biol Evol 26: 375–383

Kim JH, Waterman MS, Li LM (2007) Diploid genome reconstruction of *Ciona intestinalis* and comparative analysis with *Ciona savignyi*. Genome Res 17: 1101–1110

Kimura M (1969) The number of heterozygous nucleotide sites maintained in a finite population due to steady flux of mutations. Genetics 61: 893–903

Kirkpatrick M, Jenkins CD (1989) Genetic segregation and the maintenance of sexual reproduction. Nature 339: 300–301

Lansing AI (1954) A nongenic factor in the longevity of rotifers. Ann New York Acad Sci 57: 455–464

Lapinski J, Tunnacliffe A (2003) Anhydrobiosis without trehalose in bdelloid rotifers. FEBS Lett 553: 387–390

Leeuwenhoek A (1677) Letter of October 9, 1676 to the Royal Society. Philos Trans R Soc Lond B 12: 821–831

Leeuwenhoek A (1702) Letter of December 25, 1702 to the Royal Society. Philos Trans R Soc Lond B 23: 1291–1311

Lynch M (2006) The origins of eukaryotic gene structure. Mol Biol Evol 23: 450–468

Lynch M, Conery JS (2003) The origins of genome complexity. Science 302: 1401–1404

Mark Welch DB (2000) Evidence from a protein-coding gene that acanthocephalans are rotifers. Invert Biol 119: 17

Mark Welch DB, Cummings MP, Hillis DM, Meselson M (2004a) Divergent gene copies in the asexual class Bdelloidea (Rotifera) separated before the bdelloid radiation or within bdelloid families. Proc Natl Acad Sci USA 101: 1622–1625

Mark Welch DB, Mark Welch JL, Meselson M (2008) Evidence for degenerate tetraploidy in bdelloid rotifers. Proc Natl Acad Sci USA 105: 5145–5149

Mark Welch DB, Meselson M (2000) Evidence for the evolution of bdelloid rotifers without sexual reproduction or genetic exchange. Science 288: 1211–1215

Mark Welch JL (2001) Cytological evidence for the absence of meiosis in Bdelloid rotifers. PhD, Departm Mol Cell Biol, Harvard University, Cambridge, MA

Mark Welch JL, Mark Welch DB, Meselson M (2004b) Cytogenetic evidence for asexual evolution of bdelloid rotifers. Proc Natl Acad Sci USA 101: 1618–1621

Mark Welch JL, Meselson M (1998) Karyotypes of bdelloid rotifers from three families. Hydrobiologia 387/388: 403–407

Marotta R, Ricci C, Melone G (2008) Morphological changes accompanying anhydrobiosis in the rotifer *Macrotrachela quadricornifera* (Rotifera Bdelloidea). J Morph 269: 1481

Mattimore V, Battista JR (1996) Radioresistance of *Deinococcus radiodurans*: Functions necessary to survive ionizing radiation are also necessary to survive prolonged desiccation. J Bact 178: 633–637

Maynard Smith J (1986) Contemplating life without sex. Nature 324: 300–301

Melone G, Ricci C (1995) Rotatory apparatus in bdelloids. Hydrobiologia 313/314: 91–98

Meselson M, Mark Welch D (2007) Evolution: Stable heterozygosity? Science 318: 202–203

Minton KW (1994) DNA repair in the extremely radio-resistant bacterium *Deinococcus radiodurans*. Mol Microbiol 13: 9–15

Muller HJ (1917) An Oenothera-like case in *Drosophila*. Proc Natl Acad Sci USA 3: 619–626

Muller HJ (1918) Genetic variability, twin hybrids and constant hybrids, in a case of balanced lethal factors. Genetics 3: 421–499

Orsenigo S, Ricci C, Caprioli M (1998) The paradox of bdelloid egg size. Hydrobiologia 387/388: 317–320

Orstan A (1995) Dessication survival of the eggs of the rotifer *Adineta vaga* (Davis, 1873). Hydrobiologia 313/314: 373–375

Pouchkina-Stantcheva NN, McGee BM, Boschetti C, Tolleter D, Chakrabortee S, Popova AV, Meersman F, Macherel D, Hincha DK, Tunnacliffe A (2007) Functional divergence of former alleles in an ancient asexual invertebrate. Science 318: 268–271

Remane A (1929) Rotatorien. Bronn's Klassen und Ordnungen des Tier-Reichs, Akademische Verlagsgesellschaft, Leipzig, pp. 1–576

Ricci C (1987) The ecology of bdelloids: How to be successful. Hydrobiologia 147: 117–127

Ricci C (1988) Il concetto di specie nei Rotiferi Bdelloidei. Prob Biol Storia Natura 1: 207–209

Ricci C (1991) Comparison of five strains of a parthenogenetic species, *Macrotrachela quadricornifera* (Rotifera, Bdelloidea). I. Life history traits. Hydrobiologia 211: 147–155
Ricci C (1998) Anhydrobiotic capabilities of bdelloid rotifers. Hydrobiologia 387/388: 321–326
Ricci C (2001) A reconsideration of the taxonomic status of *Macrotrachela quadricornifera* (Rotifera, Bdelloidea). J Zool 255: 273–277
Ricci C, Balsamo M (2000) The biology and ecology of lotic rotifers and gastrotrichs. Freshw Biol 44: 15–28
Ricci C, Caprioli M (1998) Stress during dormancy: Effect on recovery rates and life-history traits of anhydrobiotic animal. Aquat Ecol 32: 353–359
Ricci C, Caprioli M (2005) Anhydrobiosis in bdelloid species, populations, and individuals. Integr Comp Biol 45: 759–763
Ricci C, Caprioli M, Fontaneto D (2007) Stress and fitness in parthenogens: Is dormancy a key feature for bdelloid rotifers? BMC Evol Biol 7 (Suppl 2) S9
Ricci C, Caprioli M, Fontaneto D, Melone G (2008) Volume and morphology changes of a bdelloid roitfer species (*Macrotrachela quadricornifera*) during anhydrobiosis. J Morph 269: 233–239
Ricci C, Caprioli M, Santo N (2004) Feeding and anhydrobiosis in bdelloid rotifers: A preparatory study for an experiment aboard the International Space Station. Invert Biol 123: 283–288
Ricci C, Covino C (2005) Anhydrobiosis of *Adineta* sp 1: Costs and benefits. Hydrobiologia 546: 307–314
Ricci C, Melone G, Santo N, Caprioli M (2003a) Morphological response of a bdelloid rotifer to desiccation. J Morph 257: 246–253
Ricci C, Pagani M, Bolzern AM (1989) Temporal analysis of clonal structure in a moss bdelloid population. Hydrobiologia 186/187: 145–152
Ricci C, Santo N, Radaelli E, Bolzern AM (1999) Epigenetic inheritance systems in bdelloid rotifers. I. Maternal-age-related biochemical effects. Ital J Zool 66: 333–339
Ricci C, Shiel R, Fontaneto D, Melone G (2003b) Bdelloid rotifers recorded from Australia with description of *Philodinavus aussiensis n.sp*. Zool Anz 242: 241–248
Ricci C, Vaghi L, Manzini ML (1987) Desiccation of rotifers (*Macrotrachela quadricornifera*): Survival and reproduction. Ecology 68: 1488–1494
Segers H (2007) Annotated checklist of the rotifers (Phylum Rotifera), with notes on nomenclature, taxonomy and distribution. Zootaxa 1564: 1–104
Segers H (2008) Global diversity of rotifers (Rotifera) in freshwater. Hydrobiologia 595: 49–59
Segers H, Shiel RJ (2003) Microfaunal diversity in a biodiversity hotspot: New rotifers from southwestern Australia. Zool Stud 42: 516–521
Segers H, Shiel RJ (2005) Tale of a sleeping beauty: A new and easily cultured model organism for experimental studies on bdelloid rotifers. Hydrobiologia 546: 141–145
Sørensen MV, Giribet G (2006) A modern approach to rotiferan phylogeny: Combining morphological and molecular data. Mol Phyl Evol 40: 585–608
Tunnacliffe A, Lapinski J, McGee B (2005) A putative LEA protein, but no trehalose, is present in anhydrobiotic bdelloid rotifers. Hydrobiologia 546: 315–321
Van Doninck K, Mandigo M, Hur JH, Wang P, Guglielmini J, Milinkovitch MC, Lane WS, Meselson M (2009) Phylogenomics of unusual histone H2A variants in bdelloid rotifers. PLoS Genetics 5: e1000401
Vinson JP, Jaffe DB, O'Neill K, Karlsson EK, Stange-Thomann N, Anderson S, Mesirov JP, Satoh N, Satou Y, Nusbaum C, Birren B, Galagan JE, Lander ES (2005) Assembly of polymorphic genomes: Algorithms and application to *Ciona savignyi*. Genome Res 15: 1127–1135
Voigt M (1959) Rotatoria. Die Rädertiere Mitteleuropas. Gebr Borntraeger, Berlin-Nikolassee
Wallace RL, Ricci C, Melone G (1996) A cladistic analysis of pseudocoelomate (aschelminth) morphology. Invert Biol 115: 104–112
Wesenberger-Lund C (1930) Contributions to the biology of the rotifera, part II: The periodicity and sexual periods. Kgl Danske Vidensk Selsk Skrifter, Naturv Og Math Afd 9 II: 1–230
Westh P, Ramlov H (1991) Trehalose accumulation in the tardigrade *Andorybiotus coronifer* during anhydrobiosis. J Exp Zool 258: 303–311

Wharton DA, Lemmon J (1998) Ultrastructural changes during desiccation of the anhydrobiotic nematode *Ditlenchus dipsaci*. Tissue Cell 30: 312–323
White MJD (1970) Heterozygosity and genetic polymorphism in parthenogenetic animals. In: Hecht MK, Steere WC (eds) Essays in Evolution and Genetics in Honor of Theodosius Dobzhansky: A Supplement to Evolutionary Biology. Appleton-Century-Crofts, New York, pp. 237–262
White MJD (1973) Animal cytology and evolution. Cambridge University Press, Cambridge, UK
Wright JC (2001) Cryptobiosis: 300 years on from van Leuwenhoek: What have we learned about tardigrades? Zool Anz 240: 563–582

Chapter 14
Sex Loss in Monogonont Rotifers

Manuel Serra and Terry W. Snell

Abstract Monogonont rotifers are small, aquatic invertebrates capable of asexual and sexual reproduction. Sexual reproduction is required to produce diapausing eggs, which are able to survive adverse periods that typically occur every year. Their cyclically parthenogenetic life-cycle is believed to retain the advantages of recombination while minimizing the cost of sex. However, this life cycle is also thought to be unstable due to periodic loss of sexual reproduction by directional selection. Explaining the evolutionary dynamics of the monogonont rotifer life cycle is important for understanding how cyclical parthenogenesis is maintained, and for comparing monogononts with their close relatives, the bdelloid rotifers, which are ancient obligate asexuals. Our analysis clarifies that the cost of sex in monogononts is two-fold when compared to an obligate asexual lineage on an annual time-scale. However, when compared to an obligate sexual, cyclical parthenogens avoid the cost of sex in every parthenogenetic generation. In monogonont rotifers, where sexual reproduction is triggered by crowding, reproducible loss of sex has been reported in laboratory experiments. The mechanistic hypothesis is that some obligate asexual clones arise by spontaneous mutation, and they fail to respond to the sex triggering chemical signals produced by conspecifics. Hence, in these clones, asexual females never produce sexual daughters. Using a simple model, we show that as a result of this association of sex with dormancy, sex loss results in a huge short-term advantage, because sexual females only produce males or diapausing eggs, and do not contribute to current population growth. However, the requirement of sex for dormancy should result in a mid-term selection pressure to retain sex. It is this mid-term pressure that stabilizes cyclical parthenogenesis and allows it to persist. From this analysis, the periodic occurrence of obligate asexuals is predicted in monogonont rotifer populations, especially those with infrequent adverse periods.

M. Serra (✉)
Institute of Biodiversity and Evolutionary Biology, Universitat de València, A.O. 2085, E46061 Valencia, Spain
e-mail: Manuel.Serra@uv.es

I. Schön et al. (eds.), *Lost Sex*, DOI 10.1007/978-90-481-2770-2_14,

14.1 Introduction

Cyclical parthenogenesis is a life cycle combining asexual (parthenogenetic) and sexual reproduction. This life cycle is found in approximately 15,000 animal species (Hebert 1987) belonging to several taxa including aphids, cladocerans, and monogonont rotifers. It therefore has evolved independently several times. Cladocerans and rotifers are short-lived invertebrates commonly found in the zooplankton of ponds and lakes. In these habitats there is substantial temporal heterogeneity that is seasonal and/or unpredictable, conditions thought to favor cyclical parthenogenesis (Serra et al. 2003).

Rotifera is a phylum with a variety of life cycles regarding sex. The class Seisonidea reproduces by obligate sex, the class Bdelloidea reproduces by obligate parthenogenesis, and the class Monogononta – which is the taxon this chapter focuses on- reproduces by cyclical parthenogenesis (Wallace et al. 2006). Bdelloids are regarded as an "evolutionary scandal", because they are ancient obligate asexuals and a diversified group (Maynard Smith 1986). How this long-term survival and relative evolutionary success was possible in the absence of sex requires explanation because it is counter to evolutionary theory and is the subject of a large ongoing research effort (e.g., Mark Welch et al. 2004a, b; see Chapter 13 in this book). The most parsimonious explanation is that bdelloids and monogononts evolved from a cyclically parthenogenetic ancestor, but bdelloids have lost sexual reproduction from their life cycle (Normark et al. 2003). Therefore, knowledge of the monogonont life cycle, genomic organization, and selection pressures on sexual reproduction is crucial to understand the evolutionary success of obligate asexuality in bdelloids.

Another feature of monogonont rotifers that makes them useful for examining the evolution of sex is the phenomenon of sex loss. A common observation is that *Brachionus calyciflorus* newly isolated from natural populations has a high frequency of sexual reproduction (Boraas 1983). However, after 20–30 generations in chemostat culture, sexual reproduction of the cyclically parthenogenetic life cycle is often eliminated (Fussmann et al. 2003). This ability to experimentally manipulate sex loss in monogonont rotifers makes them an excellent model for exploring the adaptive dynamics of this process.

Our objectives in this chapter are to examine the cost of sex in cyclically parthenogenetic life cycles, to investigate the mechanisms of sex loss in monogononts and the selective processes involved, and to explore the constraints on monogonont rotifers that limit the loss of sex.

14.2 The Monogonont Life Cycle

Monogonont rotifers are an evolutionarily successful group. There are approximately 1450 named species (Wallace et al. 2006). This number is likely a substantial underestimation, since molecular techniques are revealing the existence of many

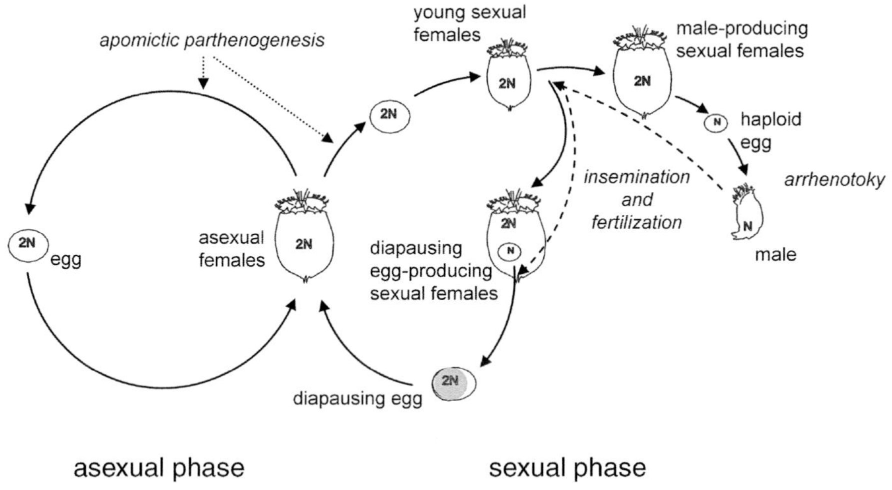

Fig. 14.1 Monogonont rotifer life cycle. Asexual reproduction continues indefinitely, until environmental cues trigger the production of sexual females. If a young sexual female is inseminated and her haploid eggs fertilized, she produces diapausing eggs. If unfertilized, she produces haploid males. Processes are shown in italics

cryptic species (Gómez et al. 2002; Suatoni et al. 2006). Monogononts have colonized a diversity of aquatic and moist habitats; they are found in the plankton of fresh and brackish waters, in the soil, in mosses, and other habitats that are wet for more than a few days (Wallace et al. 2006). However, only a few species inhabit in the open sea, presumably because their vulnerability to predation. Monogonont rotifers, despite their small size (<2000 μm), are quite efficient grazers, with an important role in many aquatic food webs (e.g. Armengol et al. 2001).

In monogonont populations, asexual (also called amictic) females produce asexual daughters by ameiotic (also called apomictic or amictic) parthenogenesis, so that the offspring is genetically identical to the mother in the absence of mutation (Fig. 14.1). This asexual reproduction can occur for an indefinite number of generations, causing clonal propagation. Episodically, in response to certain environmental cues such as population density and photoperiod (Gilbert 1963; Pourriot and Clément 1981), asexual females produce sexual daughters (also called mictic females). These sexual females produce meiotic eggs, which, if not fertilized, develop into haploid males (arrhenotoky). If these females are inseminated by a male, their fertilized eggs develop into cysts (diapausing or resting eggs) which undergo diapause. Diapausing eggs are resistant to adverse conditions such as drying and freezing, and can remain dormant in sediments for decades (Marcus et al. 1994; Kotani et al. 2001; Garcia-Roger et al. 2006). They are considered to be the main dispersal stage of monogononts. When diapausing eggs hatch, the asexual cycle is renewed as asexual females reproduce clonally until the next round of sexual reproduction. Some variation of this life cycle has been described in a few species, including eggs produced parthenogenetically that go into a short diapause (Gilbert 1995), females capable of producing both ameiotic and meiotic eggs (amphoteric females; e.g., King and Snell 1977), and sexual females hatched from diapausing eggs (Schröder et al. 2007). There are theoretical reasons to expect that half of sexual females will be male-producing, and half will be diapausing-egg producing. Fertilization rate of the sexual female eggs is controlled by the density of sexual females -which affects the density of males- and by the threshold age for fertilization. By developing the sex ratio theory for the monogonont life cycle, Aparici et al. (1998) showed that fertilization rate would evolve so that half of sexual females would not be inseminated before the threshold age, and so they will be male-producing, and half would be inseminated, and so they will produce diapausing eggs. There is some empirical evidence supporting these theoretical expectations (Aparici et al. 2002).

Most natural rotifer populations are temporary, inhabiting the water column of ponds and lakes for only a limited period of the year. Population growth typically begins when diapausing eggs in the sediment hatch. There is an initial phase of growth where the population is composed of exclusively asexual females. After a variable time period, sexual reproduction is triggered by environmental cues, and males and diapausing egg are produced, though asexual reproduction does not cease. The investment in sexual reproduction of a clone during a growing season can be described as a time series of the proportion of sexual daughters in the offspring of the asexual females (Serra et al. 2003). This time series has been called sexual reproduction pattern or mixis pattern. The pattern of sexual reproduction can be approximately described using two parameters: the timing of sexual reproduction initiation, and the sexual reproduction (mixis) ratio (i.e., the proportion of sexual daughters when sexual reproduction is initiated). In the genus *Brachionus*, where population density triggers sexual reproduction, the timing of sexual reproduction can be estimated by the threshold density for the sexual reproduction, a parameter that is easily quantified (Snell et al. 2006).

14.3 The Timing of Sex

Selection is expected to optimize sexual reproduction in rotifers from the trade-off in costs due to initiating sexual reproduction too early or too late (Snell 1987; Serra and Carmona 1993). If sex is early, population density would be low and diapausing egg production would be poor. The expected half-fertilization rate of sexual females might be unattainable because male-female encounters are unlikely. Moreover, investment in early sexual reproduction would waste the opportunity to achieve a larger population size by using up the available resources. Sex late in the season might also result in sex at low population size if growth rate has become negative due to environmental deterioration. This trade-off operating on the production of diapausing eggs has been well studied theoretically, and the optimal pattern of sexual reproduction has been related to different ecological scenarios (Serra et al.

2003). Either density-dependent growth or environmental uncertainty favor intermediate investment in sexual reproduction, while density-independent growth in predictable habitats favors a bang-bang pattern (early in the growing season all asexual reproduction, then all sexual reproduction).

The timing of sex affects not only the number of diapausing eggs produced, but also the amount of genetic diversity retained in the diapausing egg bank (Williams 1975; King, 1980). During the period of clonal propagation, natural selection acts on all the expressed genetic variance, either additive or non-additive, since the genome is inherited without recombination. Clonal selection typically leads to an erosion of clonal diversity, as it has been observed in a few studies (Gómez and Carvalho 2000; Ortells et al. 2006). As a result, populations with longer periods of parthenogenetic growth tend to contain lower genetic diversity (Ortells et al. 2006). Nevertheless, clonal selection on quantitative traits may cause selection of phenotypically similar clones, but having hidden genetic variance, which could be expressed after sexual recombination (Lynch and Deng 1994). If sexual reproduction pattern has an effect on genetic diversity and fitness, then this effect would be relevant to shaping the optimal timing of sex. However, the consequences of genetic diversity for the optimal timing of sex have yet to be explored, probably due to the difficulty of determining the optimal amount of genetic variance within diapausing eggs.

14.4 The Cost of Sex in Cyclically Parthenogenetic Life Cycles

Conventional wisdom on the monogonont life cycle states that it has evolved as an adaptation for fast population growth via parthenogenesis, in order to exploit ephemeral resource abundance without losing the advantages of sex. Cyclical parthenogenesis can be seen as combining the best of sexual and asexual reproduction, particularly because theoretical studies support the idea that a little sex is enough to fully provide all of the benefits of recombination (Peck and Waxman 2000). The paradox of sex -why sexual reproduction is so prevalent in the living world- is based on the assumption that sex has to compensate for its large cost when compared to asexual reproduction. The cost of sex in dioecious organisms is usually assumed to be two-fold (Maynard Smith 1978) because half of the offspring are typically males. Additionally, other costs are implied, such as those of maintaining the meiotic cytological mechanisms and the costs of mating activity. It is difficult to find a sufficiently large advantage for sex to compensate for such high costs. However, at least a small advantage is likely associated with recombination. Whichever the advantages of recombination, a little sex would secure these advantages while minimizing the cost of sex (Fig. 14.2). As a consequence, rather than a paradox of sex, the paradox is more accurately viewed as why cyclical parthenogenesis is so rare.

How much is the cost of sex in monogonont rotifers? A convenient way to address this question is to compare obligate asexuals, obligate sexuals and cyclical parthenogens, assuming that both obligate asexuals and obligate sexuals can produce non-diapausing and diapausing eggs, while in cyclical parthenogens sex is

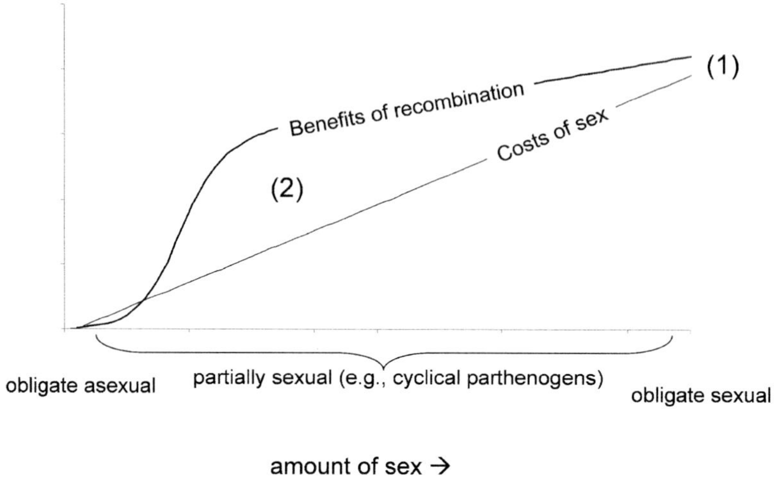

Fig. 14.2 Hypothetical relationship between the cost and benefits of sex and the amount (frequency) of sex. It is assumed that a little sex provides most of its benefits (Peck and Waxman 2000), and that obligate asexuals compensate for the two-fold cost of sex *(1)*. Cyclical parthenogenesis (partially sexual) maximizes the difference between benefits and costs *(2)*, and hence is optimal. Sigmoid curve for the benefits is assumed, so that a little sex is necessary and sufficient to compensate the cost of sex

necessary for dormancy. In order to neutralize the effect of dormancy on population growth, we will assume that the pattern of producing diapause stages is the same for all three life cycles.

As monogonont populations are usually seasonal, and need to be re-established yearly from diapausing eggs, the annual production of diapausing eggs can be assumed to determine the fitness of a rotifer clone, all the other things being equal (Serra and King 1999). Assuming that half of the sexual females are diapausing egg producers, the cost of sex is two-fold because two-fold more diapausing eggs could be produced if they could be produced asexually. Therefore, when compared to an obligatory asexual lineage, cyclical parthenogenetic rotifers incur the two-fold cost on an annual time-scale. However, when compared to an obligatory sexual lineage where the two-fold cost of sex is incurred every generation, cyclical parthenogens avoid this cost every parthenogenetic generation. There is no paradox in this different computation for the sex cost, as one is computed for every sexual generation (from diapausing egg to diapausing egg) of a cyclical parthenogen, and the other cost is the sex cost per generation, averaged over both asexual and sexual generations.

Another way to understand the cost of sex in monogononts is to view partial or complete sex loss as a way to minimize the cost of sex. Under this scenario, whenever selection favoring dormancy is relaxed (i.e. in chemostats), the loss of sex rapidly follows (Boraas 1983; Fussmann et al. 2003). If selection for recombination is also weak, this loss of sex would be evolutionarily stable. Consequently, if we can identify environments where sex is consistently lost, it will provide insight into the selective pressures maintaining sex in the cyclically parthenogenetic monogonont life cycle.

14.5 Mechanisms of Sex Loss in Monogononts

Obligate asexuality has been observed in some natural populations of cladocerans (Colbourne and Hebert 1996; see also Chapter 15). For monogonont rotifers, many limnological studies, which do not focus on sex loss, report rotifer populations where males have never been observed. However, this could be due to inappropriate sampling frequency (Snell 1989) or to the fact that rotifer males are dwarf and difficult to identify (Ricci and Melone 1998).

Loss of sex has been demonstrated in experimental populations of monogonont rotifers. Boraas (1983) found that newly established cultures of *Brachionus calyciflorus* collected from the field produced 40% mictic (sexual) females when induced. After 2–3 months in a chemostat, that percentage was reduced to 0 in similarly inducing environments. He argued that the loss of sex was permanent, due to selection against sexual reproduction. This work has been repeated by Bennet and Boraas (1989) and more recently by Fussmann et al. (2003). Bennet and Boraas (1989) founded their cultures with a single female, and hence genetic variation causing sex loss had to arise during experimental culture. In other experiments not initiated with one female, genetic variation for investment in sex might be present in the founder populations. Nevertheless, the observed evolutionary dynamics did result in selection for sex loss. It could be argued that the loss of sex in these chemostat cultures is due to the evolution of an increased density threshold for sexual reproduction, and thus is reversible. Loss of sex has been found to be stable in strains used in different laboratories for years (Stelzer 2008). On the other hand, absence of sex has been observed at densities much higher than those observed in natural rotifer populations, making loss of sex evolutionarily permanent.

These experiments show that rotifers are good models for investigating sex loss because the level of sex can be manipulated experimentally (Snell and Boyer 1988; Stelzer and Snell 2003; Snell et al. 2006). Also, experimental selection can reproducibly produce obligate parthenogens in laboratory chemostats in a few months (20–30 generations). It furthermore shows that in rotifers evolutionary and ecological time scales overlap (Fussmann et al. 2003; Yoshida et al. 2003).

In some cladocerans, such as *Daphnia pulex*, obligate asexuality is caused by a meiosis suppressor gene, which is expressed only in females (Innes and Hebert 1988; Crease et al. 1989). This gene spreads because males are produced by obligate asexual clones, and these males copulate with females belonging to cyclically parthenogenetic clones. Interestingly, while this "contagious" process is going on, asexuals are capturing genetic variance from their sexual conspecifics (Simon et al. 2003). Since in monogonont rotifers meiosis only occurs in females -males are haploids-, a sex-dependent meiosis suppressor gene is not possible, and obligate asexuality cannot spread as in cladocerans.

Brachionid rotifers obtain information about their current population density by sensing one or more chemicals produced by the rotifers themselves in a process analogous to quorum sensing in bacteria (Miller and Bassler 2001; Kubanek and Snell 2008). These compounds accumulate in the medium as population density increases to a threshold that triggers the switch from asexual to sexual reproduction; the threshold is about 70 rotifers per liter in *B. plicatilis* (Snell et al. 2006). There is extensive variation among clones from the same population in sexual response as well as among geographical isolates (Gilbert 2003; Stelzer and Snell 2006). Our current model of sex loss envisions mutations appearing somewhere in the mixis signal transduction pathway. This pathway probably includes genes involved in mixis signal production, signal reception, G-protein signaling, intracellular messengers, and transcription factors. None of these genes and proteins has been characterized from rotifers, but such pathways are known to be involved in the chemosensory mating behavior of *Drosophila* (Bray and Amrein 2003). Stelzer (2008) performed cross-induction experiments using both cyclically parthenogenetic strains and obligate asexual strains of *B. calyciflorus*. He found that conditioned medium from asexual strains was able to elicit sex in cyclically parthenogenetic strains, but not reciprocally in obligately asexual strains. This finding supports the hypothesis that asexual strains produce the mixis signal but do not respond to it.

Loss of sex by becoming unresponsive to the mixis signal has the added advantage of truncating the sex response at the first step, so that no energy is wasted on a failed sexual attempt. A single gene mutation could eliminate the ability of the mixis signal to bind to its receptor (Snell et al. 2006). It is known that some *Brachionus* species have delayed responsiveness to the mixis signal for several generations after diapause egg hatching (Gilbert 2002). A reversible physiological mechanism must exist capable of operating for several generations to block responsiveness to mixis signal. It is therefore easy to imagine a mutant capable of blocking mixis responsiveness indefinitely. There also is species-specificity in the response, so that signals from closely related *B. calyciflorus* species are sufficiently different so that there is little cross-reactivity (Gilbert 2003). It may be easy for rotifers to evolve a new reaction norm to mixis signals through changes in the receptor. However, Stelzer and Snell (2006) found that crowding by several putative species from the *B. plicatilis* species complex were capable of cross-inducing mixis. More work is needed on the selection pressures shaping the species-specificity of mixis signals among closely related species to understand when to expect cross-induction and when to expect differentiation.

14.6 Selection for Sex Loss

The rarity of cyclical parthenogenesis in nature could be due to the possible instability of such a life cycle (Simon et al. 2002). Following this idea, cyclical parthenogenesis is inherently unstable because the transition to obligate asexuality does not imply the acquisition of a new function, but only the loss of the sexual function. The rationale behind this argument can be better understood if the evolutionary dynamics of obligate sexual and obligate asexual counterparts is analyzed in a hypothetical species. If passing from obligate sexuality to obligate asexuality is difficult, but not impossible, asexuality could invade some populations due to its short-term advantage (i.e., the two-fold cost of sex). However, other populations would remain obligately sexual. These obligate sexual populations would have higher survival chances than obligate asexual populations, due to the higher capability of the former to adapt to environmental change and/or to purge deleterious mutations. As a result, most of the observed species will be sexual. In some sense, the unlikely transition from sex to asexuality makes similar the time scales of mutation and selection, both being long-term. This makes selection an efficient factor in maintaining sex. However, in cyclical parthenogens, the transition to asexuality would be much easier than in obligate sexuals. Therefore, they would become obligate asexuals due to the cost of sex, and consequently become more prone to extinction (but see the discussion below on the bdelloid case).

In order to clarify this issue, it is useful to analyze sex costs within an annual growing season of a cyclical parthenogen as compared to an obligate asexual when the latter is unable to produce diapausing eggs. This comparison is interesting because we imagine that this type of obligate asexual reproduction often arises in natural rotifer populations through selection for sex loss as described above. We will call this cost the within-growing-season cost of sex in cyclical parthenogens. We assume that when obligate asexuals appear, they produce only asexual daughters, with no recruitment from diapausing eggs. Under our assumptions, the within-growing-season cost of sex is the proportion of sexual daughters in the offspring of the asexual females belonging to the cyclical parthenogenetic lineage. Sexual daughters of cyclically parthenogenetic rotifers do not contribute to the current population growth, as they will only produce either males or diapausing eggs. By contrast, all of the daughters of the obligate asexual clone contribute to the current growth. Fig. 14.3 shows the dynamics for a simulated population composed of cyclical parthenogens and obligate asexuals, and for a simultated population of exclusively cyclical parthenogens. Notice that within-growing-season costs of sex in monogononts as stated here are due to the association between sex and dormancy. However, when a period of adverse conditions occurs, the production of resistant, diapausing eggs is required and it imposes a short-term (annual) selection pressure for the maintenance of sex, similar to what occurs in other cyclical parthenogens (Simon et al. 2002). Therefore, in an annual cycle of a seasonal population, we expect to observe fluctuating selection for sex. Early in the growing season, before sex is induced in cyclical parthenogens, both the cyclical parthenogens and the newly mutated obligate asexual lineages would have no cost of sex. After sex induction, cyclical parthenogens would be selected against due to its within-growing-season cost. In contrast, only dormant stages survive adverse environments, so obligate asexuals would have zero fitness when the water column becomes uninhabitable. Nevertheless, cyclical parthenogens might enjoy short-term benefits associated with recombination, which would compensate at least in part for the within-growing season cost of sex. For instance, genetic recombination might cause the production of some F1 clones with low mutational load, or decreased

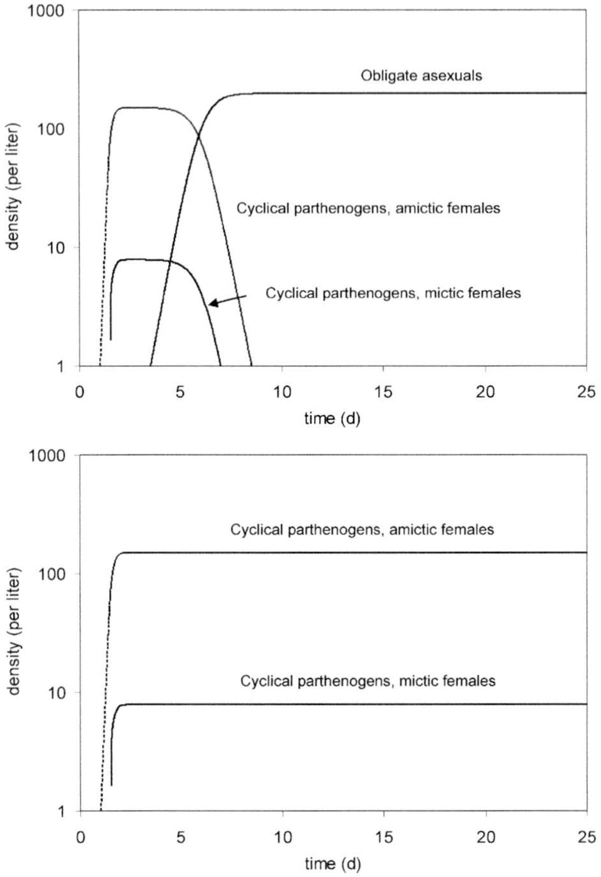

Fig. 14.3 Hypothetical dynamics of monogonont rotifer populations. *Upper panel*: Population composed of cyclical parthenogens (initial frequency: 99.99%) and an obligatorily asexual invader (initial frequency: 0.01%). *Lower panel*: Monomorphic cyclically parthenogenetic population. Curves were produced by simulating the following model (modified from Serra and King 1999): $dA/dt = b(N)(1 - m(N))A - qA$, $dM/dt = b(N)\,m(N)\;A - qM$, $dO/dt = b(N) - qO$, where A, M and O are respectively the densities of cyclical parthenogenetic asexual females, cyclical parthenogenetic sexual females, and obligatorily asexual females, q is the mortality rate (assumed to be 0.4 d^{-1}), N is the total density (i.e., A + M + O; initial N = 1 L^{-1}), b(N) is the birth rate, and m(N) is the proportion of asexual females, both assumed to be density-dependent. b was assumed to be $b_{max} - (b_{max} - q)(N/K)$, where b_{max} is the maximum birth rate (0.5 d^{-1}) and K is the carrying capacity (200 L^{-1}). m(N) was assumed to be 0 if $N < 70\ L^{-1}$) and 0.05 otherwise. Parameter values are conservative ones obtained from the literature (see Snell et al. 1998)

interclonal competition due to their diversity, as found in experimental populations of cladocerans (Tagg et al. 2005).

The pattern of selection observed in chemostats (Boraas 1983; Fussmann et al. 2003) may be duplicated in nature. Cyclical parthenogens probably dominate in seasonal environments (e.g. temperate lakes) where dormancy is required to survive the winter. However, obligate asexuals could be favored in regions where diapausing egg production is not necessary (tropics, permanent habitats). Unnoticed coexistence of cyclical parthenogens and obligate asexuals might be common in natural rotifer populations. More than 30% of aphid species are polymorphic mixes of cyclical parthenogens and obligate asexuals (Moran 1992). Rotifer populations commonly number in the billions of individuals, so most monogonont populations probably rapidly generate new obligate asexual mutants soon after diapausing egg hatching. Therefore, even in non-permanent populations, the frequency of clones with no or low investment in sex is expected to increase during the growing season due to clonal selection. The longer the period, in which cyclical parthenogens engage in sex, the higher the probability of observing asexual clones. Specific ecological conditions promoting long periods of sexual reproduction in cyclical parthenogens are (1) density-dependent population growth, and (2) a large variance in the length of the rotifer growing season (Serra and King 1999).

14.7 Dormancy and Sex

According to the fluctuating selection scheme for sex loss outlined above, the linkage between dormancy and sex promotes the maintenance of sex through short-term selection favoring dormancy (Simon et al. 2002). This linkage reduces the disadvantage of sex (the two-fold cost) by providing a correlated advantage (dormancy). The linkage does not need to be absolute, it only needs to be unlikely to be broken. Similarly to the obligate sexual-asexual dynamics described above, the long time scale for the dissociation between sex and dormancy could be equivalent to the time scale for the long-term advantages of recombination.

This argument prompts a crucial question: Why are dormancy and sex linked? The association could have arisen by chance, as a fortunate constraint allowing the evolution of a life cycle of cyclical parthenogenesis. More likely, it seems that when two types of reproduction are available (sexual and asexual), sexual reproduction tends to be associated with dispersal in time or space. This applies, for instance, to vegetative reproduction and seed production in plants. Williams (1975) stressed that natural selection would favor a correlation between parent-offspring genetic similarity and parent-offspring environmental similarity, as a cause to expect association between sex and dormancy in his rotifer-aphid model. In fact, the association between dormancy and sex makes it difficult to ignore a role for ecology in the evolution of sex.

Bdelloid rotifers have managed to uncouple dormancy and sex. Instead of producing diapausing eggs through fertilization like monogononts, bdelloids undergo anhydrobiosis where adults desiccate and can be revived after 20 or more years of dormancy (Caprioli and Ricci 2001; see also Chapter 13). An individual bdelloid rotifer can become anhydrobiotic in minutes without reproduction. Likewise, revival from an anhydrobiotic state is rapid and independent of reproduction. Uncoupling dormancy and sex has eliminated the correlated advantage of sex and it has been lost in bdelloids. In fact, bdelloid rotifers have done without sex for millions of years (Mark Welch and Meselson 2000, 2001) and they are arguably the most successful animal group of ancient obligate asexuals. The bdelloid case illustrates the delicate balance of selective forces maintaining sex in populations of cyclical parthenogens. Loss of a short-term correlated advantage like dormancy can lead to sex loss which can be stable over evolutionary time. The challenge is to explain the adaptations in the bdelloid genome that have allowed this group to continue to evolve at rates sufficient to avoid extinction in the absence of recombination.

Acknowledgements We thank Maria José Carmona for valuable comments that improved this paper. Two anonymous reviewers made useful comments that we incorporated into this chapter. This work was supported by the National Science Foundation grants BE/GenEn MCB-0412674 to TWS and by the Spanish Ministry of Education and Science grants CLG2006-27069-E/BOS and BOS2003-0075 to MS.

References

Aparici E, Carmona MJ, Serra M (1998) Sex allocation in haplodiploid cyclical parthenogens with density-dependent proportion of males. Am Nat 152: 652–657

Aparici E, Carmona MJ, Serra M (2002) Evidence for an even sex allocation in haplodiploid cyclical parthenogens. J Evol Biol 15:65–73

Armengol X, Boronat L, Camacho A, Wurtsbaugh W (2001) Grazing by a dominant rotifer *Conochilus unicornis* Rousselet in a mountain lake: In situ measurements with synthetic microspheres. Hydrobiologia 446/447: 107–114

Bennet WN, Boraas ME (1989) A demographic profile of the fastest growing metazoan – a strain of *Brachionus calyciflorus* (Rotifera). Oikos 55: 365–369

Boraas ME (1983) Population dynamics of food-limited rotifers in two-stage chemostat culture. Limnol Oceanogr 28: 546–563

Bray S, Amrein H (2003) A putative *Drosophila* pheromone receptor expressed in male-specific taste neurons is required for efficient courtship. Neuron 39: 1019–1029

Caprioli M, Ricci C (2001) Recipies for successful aanhydrobiosis in bdelloid rotifers. Hydrobiologia 446/447: 13–17

Colbourne JK, Hebert PDN (1996) The systematics of North American *Daphnia* (Crustacea: Anomopoda): a molecular phylogenetic approach. Philos Trans R Soc Lond 351: 349–360

Crease TJ, Stanton DJ, Hebert PDN (1989) Polyphyletic origins of asexuality in *Daphnia pulex* II. Mitochondrial-DNA variation. Evolution 43: 1016–1026

Fussmann GF, Ellner SP, Hairston NG Jr (2003) Evolution as a critical component of plankton dynamics. Proc R Soc Lond B 270: 1015–1022

Garcia-Roger EM, Carmona MJ, Serra M (2006) Patterns in rotifer diapausing egg banks: density and viability. J Exp Mar Biol Ecol 336: 198–210

Gilbert JJ (1963) Mictic female production in the rotifer *Brachionus calyciflorus*. J Exp Zool 153: 113–124

Gilbert JJ (1995) Structure, development and induction of a new diapause stage in rotifers. Freshw Biol 34: 263–270

Gilbert JJ (2002) Endogenous regulation of environmentally induced sexuality in a rotifer: a multigenerational parental effect induced by fertilisation. Freshw Biol 47: 1633–1641

Gilbert JJ (2003) Specificity of crowding response that induces sexuality in the rotifer *Brachionus*. Limnol Oceanogr 48: 1297–1303

Gómez A, Carvalho GR (2000) Sex, parthenogenesis and genetic structure of rotifers: microsatellite analysis of contemporary and resting egg bank populations. Mol Ecol 9: 203–214

Gómez A, Serra M, Carvalho GR, Lundt DH (2002) Speciation in ancient cryptic species complexes: evidence from the molecular phylogeny of *Brachionus plicatilis* (Rotifera). Evolution 56: 1431–1445

Hebert PDN (1987) Genotypic characteristics of cyclic parthenogens and their obligately asexual derivatives. In: Stearns SC (ed) The evolution of sex and its consequences. Birkhauser Verlag, Basel and Boston, pp. 175–195

Innes DJ, Hebert PDN (1988) The origin and genetic basis of obligate parthenogenesis in *Daphnia pulex*. Evolution 42: 1024–1035

King CE (1980) The genetic structure of zooplankton populations. In: Kerfoot WC (ed). Evolution and ecology of zooplankton communities. University Press of New England, Hanover, pp. 315–329

King CE, Snell TW (1977) Genetic basis of amphoteric reproduction in rotifers. Heredity 39: 361–364

Kotani T, Ozaki M, Matsuoka K, Snell TW, Hagiwara A (2001) Reproductive isolation among geographically and temporally isolated marine *Brachionus* strains. Hydrobiologia 446/447: 283–290

Kubanek J, Snell TW (2008) Quorum sensing in rotifers. In: Winans SC, Bassler BL (eds). Chemical communication among bacteria. ASM Press, Washington DC, pp. 453–461

Lynch M, Deng HW (1994) Genetic slippage in response to sex. Am Nat 144: 242–261

Marcus NH, Lutz R, Burnett W, Cable P (1994) Age, viability, and vertical distribution of zooplankton resting eggs from an anoxic basin: evidence of an egg bank. Limnol Oceanogr 39: 154–158

Mark Welch DB, Cummings MP, Hillis DM, Meselson M (2004a) When do gene copies in the asexual class Bdelloidea diverge? Proc Natl Acad Sci USA 101: 1622–1625

Mark Welch DB, Meselson M (2000) Evidence for the evolution of bdelloid rotifers without sexual reproduction or genetic exchange. Science 288: 1211–1215

Mark Welch DB, Meselson M (2001) Rates of nucleotide substitution in sexual and ancient asexual rotifers. Proc Natl Acad Sci USA 98: 6720–6724

Mark Welch, JL, Mark Welch DB, Meselson M (2004b) Cytogenetic evidence for asexual evolution of bdelloid rotifers. Proc Natl Acad Sci USA 101: 1618–1621

Maynard Smith J (1978) The evolution of sex. Cambridge University Press, Cambridge, UK

Maynard Smith J (1986) Contemplating life without sex. Nature 324: 300–301

Miller MB, Bassler BL (2001) Quorum sensing in bacteria. Annu Rev Microbiol 55: 165–199

Moran NA (1992) The evolution of aphid life cycles. Annu Rev Entomol 27: 321–348

Normark BB, Judson OP, Moran NA (2003) Genomic signatures of ancient asexual lineages. Biol J Linnean Soc 79: 69–84

Ortells R, Gomez A, Serra M (2006) Effects of duration of the planktonic phase on rotifer genetic diversity. Arch Hydrobiol 167: 203–216

Peck J, Waxman D (2000) What's wrong with a little sex? J Evol Biol 13: 63–69

Pourriot R, Clément P (1981) Action de facteurs externes sur la reproduction et le cycle reproducteur des Rotifers. Acta Oecologica Generale 2: 135–151

Ricci C, Melone G (1998) Dwarf males in monogonont rotifers. Aquatic Ecol 32: 361–365

Schröder T, Howard S, Arroyo ML, Walsh EJ (2007) Sexual reproduction and diapause of *Hexarthra* sp (Rotifera) in short-lived ponds in the Chihuahuan Desert. Freshw Biol 52: 1033–1042

Serra M, Carmona MJ (1993) Mixis strategies and resting egg production of rotifers living in temporally-varying habitats. Hydrobiologia 255–256: 117–126

Serra M, King CE (1999) Optimal rates of bisexual reproduction in cyclical parthenogens with density-dependent growth. J Evol Biol 12: 263–27.

Serra M, Snell TW, King CE (2003) The timing of sex in cyclically parthenogenetic rotifers. In: Moya A, Font E (eds) Evolution: from molecules to ecosystems. Oxford University Press, New York, pp. 135–146.

Simon JC, Delmotte F, Rispe C, Crease T (2003) Phylogenetic relationships between parthenogens and their sexual relatives: the possible routes to parthenogenesis in animals. Biol J Linn Soc 79: 151–163

Simon JC, Rispe C, Sunnucks P (2002) Ecology and evolution of sex in aphids. Trends Ecol Evol 17: 34–39

Snell TW (1987) Sex, population dynamics and resting egg production in rotifers. Hydrobiologia 144: 105–111

Snell TW (1989) Systematics, reproductive isolation and species boundaries in monogonont rotifers. Hydrobiologia 186/187: 299–310

Snell TW, Boyer EM (1988) Thresholds for mictic female production in the rotifer *Brachionus plicatilis* (Muller). J Exp Mar Biol Ecol 124: 73–85

Snell TW, Kubanek J, Carter J, Payne AB, Kim J, Hicks MK, Stelzer CP (2006) A protein signal triggers sexual reproduction in *Brachionus plicatilis* (Rotifera). Mar Biol 149: 763–773

Snell TW, Serra M, Carmona MJ (1998) Toxicity and sexual reproduction in rotifers: reduced resting egg production and heterozygosity loss. In: Forbes VE (ed) Genetics and Ecotoxicology. Taylor and Francis, London, pp. 169–185

Stelzer CP (2008) Obligate asexuality in a rotifer and the role of sexual signals. J Evol Biol 21: 287–293

Stelzer CP, Snell TW (2003) Induction of sexual reproduction in *Brachionus plicatilis* (Monogononta, Rotifera) by low density-dependent chemical cue. Limnol Oceanogr 48: 939–943

Stelzer CP, Snell TW (2006) Specificity of the crowding response in the *Brachionus plicatilis* species complex. Limnol Oceanogr 51: 125–130

Suatoni E, Vicario S, Rice S, Snell TW, Caccone A (2006) An analysis of species boundaries and biogeographic patterns in a cryptic species complex: the rotifer *Brachionus plicatilis*. Mol Phyl Evol 41: 86–98

Tagg N, Doncaster CP, Innes DJ (2005) Outcomes of reciprocal invasions between genetically diverse and genetically uniform populations of *Daphnia obtusa* (Kurz). Oecologia 143: 527–536

Wallace RL, Snell TW, Ricci C, Nogrady T (2006) Rotifera 1: Biology, Ecology and Systematics. Backhuys Publishers, Leiden

Williams GC (1975) Sex and Evolution. Princeton University Press, Princeton

Yoshida T, Jones LE, Ellner S, Fussmann GF, Hairston NG Jr (2003) Rapid evolution drives ecological dynamics in a predator-prey system. Nature 424: 303–306

Chapter 15
Cyclical Parthenogenesis in *Daphnia*: Sexual Versus Asexual Reproduction

Ellen Decaestecker, Luc De Meester and Joachim Mergeay

Abstract In the current chapter, we discuss the peculiar but successful reproduction mode of cyclical parthenogenesis, using the cladoceran genus *Daphnia* as a model. We first focus on the cyclically parthenogenetic life cycle of *Daphnia*, the phylogenetic backgrounds of this reproduction mode, and how cyclical parthenogenesis impacts the genetic structure of *Daphnia* populations. Further, we discuss the advantages of sex. Finally, we change perspective and discuss evolution from cyclical parthenogenesis to strict asexuality in this genus, contrasting the advantages and drawbacks of both strategies, starting from the selective environment of obligate asexuals.

15.1 Introduction

Most animal taxa use sexual reproduction to produce offspring, while a minority passes on their genes asexually. Both strategies have their advantages and weaknesses, and throughout evolution, a number of taxa have evolved independently a mixed strategy that seems to combine the best of both worlds: phases of asexual propagation are alternated with bouts of sexual reproduction, called cyclical parthenogenesis, holocycly or heterogony. Cladocerans, monogonont rotifers (see Chapter 14) and aphids (see Chapter 25) are the best-known cyclical parthenogens, but life cycles combining sexual and asexual reproduction are common in protists, cnidarians, bryozoans, and plants (De Meester et al. 2004). We here focus on recent studies and reviews of cyclical parthenogenesis in *Daphnia* so as to identify future avenues of research. Detailed reviews that also include older literature are given by Hebert (1978, 1987), Lynch (1984), Carvalho (1994) and De Meester (1996).

E. Decaestecker (✉)
Laboratory of Aquatic Ecology and Evolutionary Biology, K.U. Leuven, Ch. Debériotstraat 32, B-3000 Leuven, Belgium; Laboratory of Aquatic Biology, Interdisciplinary Research Center, K.U. Leuven Campus Kortrijk, E. Sabbelaan 53, B-8500 Kortrijk, Belgium
e-mail: Ellen.Decaestecker@kuleuven-kortrijk.be

I. Schön et al. (eds.), *Lost Sex*, DOI 10.1007/978-90-481-2770-2_15,

De Meester et al. (2004) focused on key ecological and evolutionary consequences of cyclical parthenogenesis, comparing cladocerans, monogonont rotifers and aphids.

15.2 Cyclical Parthenogenesis and Its Effect on the Genetic Structure of Daphnia Populations

Cyclical parthenogenesis arose within the Branchiopoda during the Permian (Taylor et al. 1999) when the Cladocera evolved as a taxon. Apart from a few strictly asexual derivates, all Cladocera are cyclical parthenogens. The success of this reproduction mode is reflected in the known 620 species that radiated within this order, this is more than half of the known Branchiopod species diversity and the estimated number of cladoceran species is even two to four times higher (Korovchinsky 1996; Adamowicz and Purvis 2005; Forró et al. 2008). Cladocera are ubiquitous components of inland aqueous habitats all around the world, but are rare in marine habitats. Within the Cladocera, the genus *Daphnia*, approximately 150 species rich, has been used as a key model to study ecological and evolutionary questions, including the consequences of cyclical parthenogenesis.

Daphnia (Crustacea, Branchiopoda, Cladocera, Daphniidae) is an important component of zooplankton in lakes and ponds. It has a short generation time (9–11 days at 20°C), but total life span is longer (> 60 days at 20°C, up to one year at colder temperatures; Gliwicz et al. 2001). Figure 15.1 shows the reproduction cycle of cyclically parthenogenetic *Daphnia*. Under favourable conditions, they reproduce by amictic parthenogenesis, producing genetically identical offspring that build up a population consisting of only females. This can be continued for several generations, resulting in an exponential growth of clonal lineages (Carvalho 1994). The

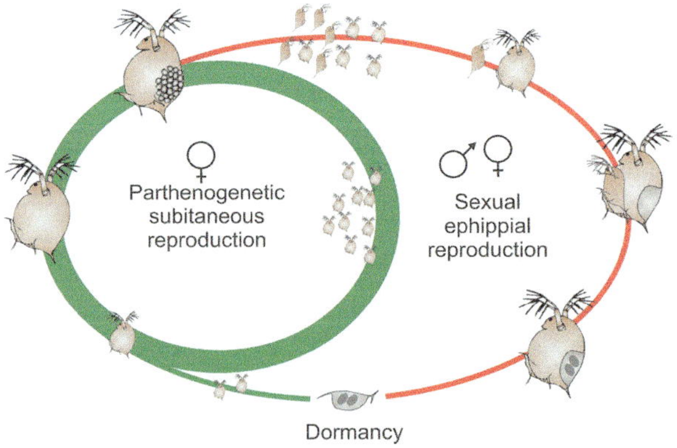

Fig. 15.1 The cyclically parthenogenetic life cycle of *Daphnia*. During favourable conditions, parthenogenetic reproduction takes place for one to several generations (*green*). Sexual reproduction (*red*) results in the production of long-lived dormant eggs, which can hatch once environmental conditions become favourable again. Some taxa have omitted males from the cycle and produce dormant eggs asexually

relative abundance of these clones reflects their relative success in the habitat. When unfavourable conditions arise (e.g., food shortage, overcrowding, presence of predators, change in day-length or temperature, Pijanowska and Stolpe 1996), the animals switch to sexual reproduction. Males are produced parthenogenetically, and females switch to the production of sexual eggs. A single female may first produce diploid amictic eggs and subsequently produce two meiotic haploid eggs that need to be fertilized (De Meester et al. 2004). After fertilization of the haploid eggs, they are encapsulated in an ephippium, a chitinuous membrane secreted around the brood pouch of the female carapace (Schultz 1977). Development of the eggs is arrested at the blastula stage, and the eggs go in diapause. At the next moult, the ephippium is shed with the old carapace (Zaffagnini 1987). The enclosed dormant eggs are able to withstand extreme conditions (drying, freezing, digestion, ...) and can remain viable for up to 150 years (Cáceres 1998; Brendonck and De Meester 2003).

Dormancy is therefore a strategy to bridge unfavourable periods (risk spreading over time), while also maximizing the chances of survival during passive dispersal by wind, waterfowl or other means (risk spreading in space, Cohen and Levin 1987). The dormant eggs may hatch the next season when favourable conditions are restored, but a significant portion will not hatch and remain in the sediment. As such, a series of overlapping generations gradually builds up year after year, in what is commonly called a persistent dormant egg bank (DeStasio 1989), analogous to seed banks in plants (Templeton and Levin 1979). Depending on prevailing selection forces, these dormant egg pools can accelerate or delay evolutionary responses to changing environments (Hairston and DeStasio 1988; Hedrick 1995; Hairston 1996). In addition to the evolutionary and ecological importance of the presence of a dormant egg bank in the stratified sediments of lakes and ponds, it also represents a unique archive of the history of the population over time (Hairston et al. 1999; Cousyn et al. 2001; Limburg and Weider 2002; Mergeay et al. 2006, 2007).

The genetic structure of cyclically parthenogenetic *Daphnia* populations is determined by the consequences of combining sexual and asexual reproduction in the same life cycle (Hebert 1987; Carvalho 1994; De Meester et al. 2006). At the start of the growing season, hatching of sexually produced dormant eggs introduces genetic variation in the population, with a one-to-one relationship between the number of hatchlings and the number of unique clones (De Meester 1996; De Meester et al. 2006). However, during the growing season, parthenogenetic reproduction results in the erosion of clonal diversity by natural selection and chance extinctions of clones, leading to lower genetic variation and deviations from Hardy-Weinberg equilibrium at the end of the growing season (clonal erosion, Tessier et al. 1992; De Meester 1996; Ortells et al. 2006; De Meester et al. 2006). The population may, however, hold considerable amounts of hidden genetic variation that is not genetically expressed as long as the animals reproduce clonally, but which may become expressed after sexual recombination (Deng and Lynch 1996; Pfrender and Lynch 2000). Natural selection acts differently during the two reproductive phases (Pfrender and Lynch 2000; King and Schonfeld 2001). During the asexual phase, all genes belong essentially to one linkage group, and selection thus acts upon the whole linked genome. As a result, clonal selection also acts on the interaction of genetic variation. In contrast, sexual reproduction breaks up the associations of these linked alleles.

De Meester et al. (2006) recently proposed a unifying conceptual framework to understand the genetic structure of cyclically parthenogenetic zooplankton populations. In this framework, the key factor that influences the genetic structure of cyclical parthenogens is the relative importance of sexual recombination and parthenogenetic reproduction. De Meester et al. (2006) list three main factors that determine the degree to which clonal erosion affects the genetic structure of cyclical parthenogens. A first factor is the population size as determined by the amount of hatchlings from the dormant egg bank. Populations that have larger dormant egg banks are expected to start the growing season with a higher number of hatchlings and thus a higher number of clones than populations with smaller dormant egg banks (Vanoverbeke and De Meester 1997). A second factor is the length of the growing season and the degree to which a population can persist in the habitat in the active phase, as this determines the number of parthenogenetic generations between sexual phases (Hebert 1987; Pfrender and Lynch 2000). Permanent populations and non-permanent populations that enjoy extended growing seasons show higher propensities of chance extinctions of clones and experience longer lasting selection that erodes clonal diversity. Furthermore, the impact of hatchlings in spring is likely to be lower in permanent than in intermittent populations, as the presence of a resident population often lowers the hatching response in *Daphnia* (Cáceres and Tessier 2003) and the hatchlings have to compete with the already established population (cf. priority effects; see De Meester et al. 2002). A third factor influencing the degree of clonal erosion is the strength of clonal selection, which may vary among populations and during the course of the growing season (De Meester et al. 2006).

15.3 Reasons to Maintain Sexual Reproduction in Daphnia

Sexual reproduction has costs relative to asexual reproduction (Lewis 1987), yet, sexual reproduction prevails in nature. A main reason for the maintenance of sexual reproduction is the improvement of fitness, despite the reduction in overall number of offspring ("two-fold" cost of sexual reproduction, Maynard Smith 1978). Many hypotheses focus on this issue, building on the notion that sexual recombination accelerates the creation of genetic variation in offspring by the induction of new gene combinations, the spread of advantageous mutations, and the removal of deleterious genes. As such, sexual reproduction leads to higher rates of adaptation and inhibits the accumulation of deleterious mutations (West et al. 1999; Pound et al. 2002; see also Chapter 5).

15.3.1 Local Genetic Adaptation

Cyclical parthenogenesis allows for rapid local genetic adaptation, as it combines effective selection on the whole genetic component of variation during the parthenogenetic phase with the release of hidden genetic variation during sexual recombination (Lynch and Gabriel 1983; De Meester 1996). Evolutionary potential is increased by the fact that sexual reproduction is coupled with the production of dormant eggs, as this results in the build-up of dormant egg banks integrating genetic variation that accumulates over different growing seasons (Hedrick 1995). One can thus predict that cyclical parthenogens are likely to adapt rapidly to local conditions through selection for genotypes with an adaptive combination of phenotypic plasticity responses (De Meester et al. 2004). Several studies have indeed provided evidence for efficient tracking of environmental changes over time in natural *Daphnia* populations (e.g., Hairston et al. 1999, 2001; Cousyn et al. 2001). Striking examples of adaptation through changes in phenotypic plasticity in *Daphnia* come from studies that show genetic differences in adaptive and inducible shifts in trait values in response to predators, including behavioural, morphological and life history traits (Spitze 1992; De Meester 1993a, 1996; Boersma et al. 1998; Tollrian and Harvell 1999; Cousyn et al. 2001). Adaptations to local conditions may, however, involve complex interactions between multiple antagonists, such as predators and parasites, leading to trade-offs between different response mechanisms (Decaestecker et al. 2002).

15.3.2 Red Queen Dynamics

When an environment changes, previously neutral or deleterious alleles can become favourable. If the environment changes sufficiently rapidly over time (i.e., between different generations), sexual reproduction reintroducing these alleles can be advantageous. This is especially so in systems in which parasites continuously track specific and common host genotypes, resulting in parasite driven time-lagged negative frequency dependent selection. The production of genetic variation among offspring provided by sexual recombination is thus important in confrontation with the fast and specific genetic adaptation of parasites to their hosts ("Red Queen" hypothesis, Van Valen 1973; Hamilton 1980; Stearns 1987; Maynard Smith 1989; Ebert and Hamilton 1996; Hurst and Peck 1996; see also Chapter 7).

In *Daphnia,* it has been shown that parasites are important selective forces. Many natural populations are infected by parasites, some of which induce severe virulence effects, resulting in fitness decline in the host (Green 1974; Stirnadel and Ebert 1997; Decaestecker et al. 2005; Ebert 2005; Johnson et al. 2006; Lass and Ebert 2006). Furthermore, there is evidence for local genetic adaptation of the parasite along spatial distance gradients (Ebert 1994) as well as for short-term parasite mediated selection in *Daphnia* (Capaul and Ebert 2003; Haag and Ebert 2004; Duncan et al. 2006; Zbinden et al. 2008), which affects host-parasite dynamics (Duffy and Sivars-Becker 2007). It has also been shown that parasite epidemics can select for resistance in *Daphnia* (Duncan and Little 2007).

Several lines of evidence suggest that *Daphnia*-parasite coevolution follows "Red Queen" dynamics with no directional increase in fitness of both antagonists over time (Hamilton et al. 1990; Woolhouse et al. 2002; Ebert 2008). It has been shown that there is ample genetic variation in *Daphnia* resistance against parasites (Little and Ebert 1999). Host clones within and between populations vary strongly in their resistance to parasites, but also parasites differ strongly in fitness components between host genotypes and populations (Carius et al. 2001; Decaestecker et al. 2003; Haag et al. 2003; Refardt and Ebert 2007; Ebert 2008). Host-parasite interactions are genotype-specific with no parasite isolates being able to infect all host genotypes, and no host genotypes that are able to resist all parasite genotypes (Carius et al. 2001). It has also been shown in the *D. galeata* x *D. hyalina* hybrid system that an under-infected taxon can become over-infected after an increase in frequency, and that this over-infection has a genetic basis (Wolinska et al. 2006).

The observation that both *Daphnia* and its microparasites can be "resurrected" from dormant propagule banks opened the possibility for a historical reconstruction of the coevolutionary dynamics of *Daphnia* and its parasite *Pasteuria ramosa* from layered sediment cores (Decaestecker et al. 2004). A time shift experiment, in which host clones from each sediment layer were exposed to parasite isolates from the same, the previous and the following sediment layer showed evidence for temporal adaptation with the contemporary parasites showing a higher infectivity than parasite isolates from the past and the future, resulting in temporal variation in parasite infectivity that changed little over time. This analysis revealed that the parasite fast tracks its host over a time period of only a few years, in line with specific antagonistic host-parasite coevolution based on negative frequency dependent selection (Decaestecker et al. 2007; Gandon et al. 2008; Ebert 2008).

Little and Ebert (2001) found in one of the studied host populations temporal adaptation to parasites showing higher infectivity in *Daphnia* from the same growing season than in hosts of a later growing season. However, this pattern could not be confirmed in other populations. The expected pattern of host-parasite coevolution depends, among other things, on the time lag between the host and parasite coevolutionary dynamics, on the time scale separating the various samples of the host populations and on the number of generations of hosts and parasites considered (Decaestecker et al. 2007; Gandon et al. 2008). Moreover, it has been suggested that parasite selection must be unrealistically severe to create rapid parasite mediated dynamics or that genotype-environment interactions impede detection of selection against environmental noise (Little and Ebert 2001; Duncan et al. 2006). Seasonal changes in temperature and predation alter the interaction between *Daphnia* and parasites such that differences in temperature and predation pressure will change parasite mediated selection, resulting in the maintenance of genetic variation of the traits involved (Mitchell et al. 2004, 2005; Duffy et al. 2005; Hall et al. 2006; Vale et al. 2008).

However, sex and immuno-competence do not necessarily go hand in hand, for example, when sexual reproduction is a byproduct of temporal parasite-avoidance (Duncan et al. 2006), in a similar way that sexual reproduction can be induced when *Daphnia* are confronted with predators (Sluzarczýk 1995). In such cases, the main goal of sexual reproduction is not genetic recombination in order to increase immuno-competence, as the sexual phase precedes parasitism (Duncan et al. 2006), but the formation of dormant stages to ensure persistence in the habitat.

15.3.3 Deleterious Mutations

Sexual reproduction can reduce the mutational load in offspring, as it enhances the removal of deleterious mutations. Sexual reproduction allows to reconstruct mutation-free individuals by recombination and by doing so inhibits the random loss of individuals without deleterious mutations. It was Muller (1932), who suggested that this stochastic process can lead to an inexorable decline in the fitness of clones in finite asexual populations ("Muller's ratchet"; Muller 1932; Felsenstein 1974; see also Chapter 5). In later studies, Muller's basic idea was extended by including cases in which mutation accumulation is decoupled from stochastic processes, such that an advantage to sex can accrue even in infinite populations ("Mutational Deterministic" hypothesis, Kondrashov 1982, 1988; Charlesworth 1990; Howard and Lively 1998). As selection against mutations of weak to intermediate deleterious effect is small, these mutations can accumulate to high frequencies (Whitlock et al. 2000; Ebert et al. 2002). There is evidence for high genetic loads in *Daphnia* (Innes 1989; De Meester 1993b; Lynch and Deng 1994). Further, Paland and Lynch (2006) showed that, because of permanent linkage of the whole genome, asexual *D. pulex* clones may be prone to the accumulation of deleterious mutations, leading to a higher ratio of the rate of amino acid to silent substitution (K_a/K_S) in mitochondrial protein coding genes in asexual lineages than in cyclically parthenogenetic ("sexual") lineages. This result suggests that sexual reproduction in the cyclically parthenogenetic *Daphnia* has the power to indeed reduce the accumulation of deleterious mutations, and obligately parthenogenetic *Daphnia* face a decline in fitness over time. Nevertheless, there are indictations that sufficient amounts of variation (relative to mutation rate) are generated in *Daphnia* by ameiotic recombination. Although this recombination is internal and does not allow genetic exchange across lineages as in outcrossing sex, it shows that asexual lineages do not only acquire variation through mutations (Omilian et al. 2006).

Evidence is growing that a pluralistic approach may be required to explain the maintenance of sexual reproduction (West et al. 1999). Support for this approach in *Daphnia* comes from an experimental study, in which fitness consequences of deleterious mutations were stronger when associated with parasite infection (Killick et al. 2006). As shown for parasite mediated selection in *Daphnia* (Mitchell et al. 2005; Vale et al. 2008), the fitness cost of deleterious mutations depends on environmental conditions as well (Killick et al. 2006).

Another explanation for the advantage of sexual reproduction in *Daphnia* may relate to the direct advantage caused by hybrid vigour or heterosis. Inbreeding depression has consequences at the *Daphnia* metapopulation level, as shown by Ebert et al. (2002) and Haag et al. (2002). In a *Daphnia* metapopulation inhabiting small coastal rockpools in Finland, crossing between immigrants and inbred residents leads to hybrid offspring that are superior competitors. This increases gene flow between populations and the spread of favourable alleles across the metapopulation (Ebert et al. 2002; Haag et al. 2002).

15.4 Evolution to Asexuality in Daphnia and Other Cladocerans

Although cyclical parthenogenesis is the rule in Cladocera (Taylor et al. 1999), a few taxa have modified this life cycle, and have shunted away from sexual reproduction, while retaining the possibility to produce resistant dormant stages. For instance, asexuality has been observed in allopolyploid lineages of *Sinobosmina* (Little et al. 1997), while an asexual or possibly pseudo-sexual life cycle was suggested for several *Holopedium* lineages (Korovchinsky 2005; Hebert et al. 2007). Within the genus *Daphnia,* evolution to obligate asexuality has evolved in at least four independent occasions by to three different mechanisms (Table 15.1).

In the animal kingdom, hybridization is probably one of the most used routes to asexuality, often in combination with genome duplication (allopolyploidy) (Kearney 2005). Obligate parthenogenesis of hybrid origin in *Daphnia* is known from the subgenera *Daphniopsis* (*D. truncata x pusilla*; Hebert and Wilson 2000), *Ctenodaphnia* (allopolyploid alpine populations of *D. thomsoni*; Hebert and Wilson 1994) and in various related species of the subgenus *Daphnia* (*D. pulex* complex; Hebert et al. 1989; Dufresne and Hebert 1994, 1997; Hebert 1995; Aguilera et al. 2007; Mergeay et al. 2008).

Apart from obligate parthenogenesis through hybridisation with or without polyploidy, asexuality has been acquired de novo in some populations of *D. cephalata* (Hebert 1981; Hebert and Wilson 1994), an Australian species of the *D. carinata* complex (Colbourne et al. 2006).

Thirdly, in certain lineages of panarctic *D.* "*pulex*" (a different species from the typical European *D. pulex*; see Mergeay et al. 2008, for an account on taxonomic issues in this group) females reproduce by obligate parthenogenesis, whereas the clonally propagated males produce functional haploid sperm that allows them to breed with sexual females of normal cyclically parthenogenetic lineages (contagious asexuality; Innes and Hebert 1988; Paland et al. 2005). On average half of the offspring will also consist of obligate parthenogens, although offspring viability is much reduced (Innes and Hebert 1988). Paradoxically, contagiously asexual lineages thus spread asexuality through sex. The genetic structure of contagiously asexual populations is undistinguishable from that of strictly asexual lineages. Although asexual lineages are bound to accumulate deleterious mutations (Paland and Lynch 2006) and face an evolutionary dead end, contagious asexuality allows the recurrent creation of genetically diverse asexual lineages with lower genetic loads and high micro-evolutionary potential (Innes and Hebert 1988; Simon et al. 2003). This interaction between asexual and sexual strains means that the maternal lineage actually partially escapes Muller's ratchet by creating new asexual lineages that carry half of its genes and have a reduced mutational load. As long as the cyclically parthenogenetic sister taxon is present, there is an opportunity to continuously rejuvenate part of the genome. Metapopulations of contagiously asexual lineages can thus be expected to be genetically diverse (Innes and Hebert 1988; Crease et al. 1989). Remarkably, however, there seems to be a certain degree of

Table 15.1 Overview of asexual *Daphnia* species and lineages

Taxon	Subgenus	Includes sexual lineages (Y/N)	Origin of asexuality	Ploidy	Habitat	Reference
Panarctic *D.* "*pulex*"	*Daphnia*	Y	Contagious	Diploid	Temperate ponds	Innes and Hebert (1988)
Polar *D. pulicaria*	*Daphnia*	N	Interspecific hybrid	Polyploid	Arctic lakes and ponds	Dufresne and Hebert (1997), Weider et al. (1999)
D. pulicaria	*Daphnia*	Y	Contagious	Diploid	Temperate lakes	Innes and Hebert (1988)
D."*pulex*" *x pulicaria*	*Daphnia*	N	Interspecific hybrid	Diploid	Temperate ponds and lakes	Dufresne and Hebert (1994), Hebert and Finston (2001)
D. pulicaria x "*pulex*"	*Daphnia*	N	Interspecific hybrid	Polyploid	Arctic lakes and ponds	Dufresne and Hebert (1994)
D. tenebrosa	*Daphnia*	Y	Intraspecific hybrid	Polyploid	Arctic lakes and ponds	Dufresne and Hebert (1995)
D. middendorffiana	*Daphnia*	N	Interspecific hybrid, multiple parent species	Polyploid	Arctic lakes and ponds	Dufresne and Hebert (1997)
South American *D.* "*pulicaria*" A	*Daphnia*	N	Interspecific hybrid	Polyploid	Alpine lakes, cold temperate lakes	Adamowicz et al. (2002), Mergeay et al. (2008)
South American *D.* "*pulicaria*" B	*Daphnia*	N	Interspecific hybrid	Polyploid	Alpine lakes	Mergeay et al. (2008)
South American *D.* "*pulicaria*" C	*Daphnia*	N	Interspecific hybrid	Polyploid	Alpine lakes	Mergeay et al. (2008)
D. cephalata	*Ctenodaphnia*	Y	Spontaneous	Diploid	Temperate ponds	Hebert (1981), Colbourne et al. (2006)
D. thomsoni	*Ctenodaphnia*	Y	Interspecific hybrid	Polyploid	Alpine lakes	Hebert and Wilson (1994)
D. truncata x pusilla	*Daphniopsis*	N	Interspecific hybrid	Not specified	Intermittent desert ponds	Hebert and Wilson (2000)

longitudinal geographic segregation between cyclically parthenogenetic and contagiously asexual lineages of *D. "pulex"* in North America, with few and relatively narrow contact zones (Hebert and Finston 2001). Most likely, asexuality in contagiously asexual lineages is determined by a sex-limited meiosis suppressor gene that is thought to have spread to at least some other species in the complex by hybridisation and introgression (Innes and Hebert 1988). Eight of eleven taxa of the *D. pulex* complex are known to include lineages that reproduce by obligate parthenogenesis (Table 15.1), but many of these are hybrids and even polyploids. It is therefore not clear whether the same meiosis suppressor mechanism is active in these lineages, or whether asexuality results directly from hybridization and meiotic incompatibility between the parents (for an overview, see Simon et al. 2003).

In addition to these three cases of asexuality, which are characterised by the disruption or suppression of meiosis at some stage ("fundamental asexuality"), there are *Daphnia* populations that are theoretically cyclically parthenogenetic, but whose genetic structure is undistinguishable from that of obligately parthenogenetic populations. These populations remain all year round in the lakes they inhabit, and seem to use solely the parthenogenetic phase of the life cycle (Fig. 15.1). An example is given by a *D. galeata* population studied by Gliwicz et al. (2001; identified as *D. longispina*) and several *D. pulicaria* populations studied by Cáceres and Tessier (2004a). Such populations can easily be mistaken for strictly asexual lineages, and care should be taken in the interpretation of their breeding modes. First, it is not certain whether these lineages have really lost the capacity to engage in sexual reproduction, and secondly, it is even not sure whether occasional sexual reproduction is not occurring in nature in these populations. Indeed, lack of observation of sexual stages does not preclude the very rare occurrence of sex. A special case of lineages that may totally rely on asexual reproduction is provided by hybrid offspring of members of the *D. longispina* complex that live in permanent habitats such as deep lakes. This complex includes species whose ancestors diverged more than eight million years ago, but which still hybridise readily, e.g., *D. galeata, D. cucullata, D. longispina.* (Schwenk and Spaak 1995). Although their hybrids have a strongly reduced fertility, they can survive and reproduce parthenogenetically for long periods and may in this sense behave as asexual strains that have lost the ability to produce viable dormant eggs (Schwenk and Spaak 1995).

15.5 Why Switch to Asexual Reproduction When You Can Be a Cyclical Parthenogen?

A large number of hypotheses have been proposed to explain the evolution of sex and later reversals to asexuality, accompanied by numerous empirical and theoretical studies (Maynard Smith 1971, 1978; Vrijenhoek 1979; Kondrashov 1988; Crow 1994; Peck 1994; Doncaster et al. 2000; Peck and Waxman 2000; Pound et al. 2002, 2004; Paland and Lynch 2006). Most of these hypotheses have focused on the cost of males, effects of sex on the mutational load or on the rate of evolutionary adaptation, or on ecological differences between sexual organisms and asexual derivates. To cut a long story short, it is the balance between the costs and the benefits of sex that determines whether sexual reproduction is more advantageous than asexual reproduction.

Cyclical parthenogens generally start the growing season from sexual offspring, and hence can start from a wide genetic array on which selection can act, just like in obligately sexual organisms with a r-strategy of reproduction. In comparison to obligately sexual organisms, however, this initial pool of sexually produced offspring will grow much faster as a result of consecutive bouts of parthenogenetic reproduction. Males are produced only during the much shorter sexual phase (Fig. 15.1), and will therefore represent an almost negligible cost, especially when seen in the light of the benefit of recombination. Moreover, clonal selection is efficient during the parthenogenetic phase as it acts on both the interactive and the additive component of genetic variation. Evolutionary adaptation can indeed proceed very rapidly in cyclical parthenogens, as witnessed by Hairston et al. (1999), Cousyn et al. (2001) and Decaestecker et al. (2007). In addition, cyclical parthenogens have a much higher mutational clearance than asexual relatives (Paland and Lynch 2006). Cyclical parthenogenesis thus seems to combine all the advantages of sex, while minimizing the drawbacks. In this light, the high prevalence and multiple origins of asexuality, as in the *D. pulex* complex, is intriguing. To gain more insights into the origin of obligate asexuality from cyclical parthenogenesis, we consider what the drawbacks of the cyclically parthenogenetic reproduction mode are. Below, we discuss existing hypotheses in the light of cyclical parthenogenesis, as well as integrating new concepts that might provide better insights into the evolution of asexuality in lineages of cyclical parthenogens. The first hypothesis is related to the timing of sexual reproduction, the second and third relate to constraints imposed by time stress during the growing season, and the fourth hypothesis is unrelated to time, but sees obligate parthenogenesis as a side-product of selection for polyploidy and/or hybrid vigour. Finally, we explore contagious asexuality in the light of selfish gene phenomena.

15.5.1 Clonal Erosion and Inbreeding

Probably one of the major assets of cyclical parthenogenesis is that it initially mimics an asexual life cycle for a number of generations before switching to a sexual mode. The timing of the switch is important, however. Clonal selection erodes the initial genetic diversity during the parthenogenetic phase, and in small and permanent habitats this may lead to the coexistence of only a very limited number of clones (Vanoverbeke and De Meester 1997; De Meester and Vanoverbeke 1999; De Meester et al. 2006). If only one to a few genotypes remain due to clonal erosion, sexual reproduction is disadvantageous as it results in inbreeding and concomitant fitness losses (De Meester 1993b; Lynch and Deng 1994; Ebert et al. 2002). Asexual genotypes do not suffer from inbreeding, and can be favoured over sexual lineages under such circumstances. In addition to time, habitat heterogeneity may also influence the degree of clonal erosion. The more clones can coexist in a given habitat, the less important inbreeding will be. Whereas the tangled bank hypothesis (Bell 1982) views sexual recombination as an adaptation to environmental diversity, we here suggest that environmental diversity may also result in reduced levels of inbreeding depression in cyclical parthenogenesiss as it allows more clones to coexist, reducing the scope for asexual lineages to dominate because of inbreeding depression in sexual offspring. The high prevalence of asexual *Daphnia* at high latitudes and altitudes (geographic parthenogenesis; see also Chapter 7), is associated with low habitat complexity, which is a typical feature of extreme environments like arctic ponds (disclimax habitats, Glesener and Tilman 1978).

15.5.2 Food Limitations and Time Stress

In the cyclical parthenogenetic life cycle, the first clutch usually consists of subitaneous parthenogenetic eggs, and it is important to note that most species can only produce dormant eggs from the second clutch, either sexually or ameiotically. (Only two species are known to produce ephippial eggs in a single cycle; Cáceres and Tessier 2004b; Aguilera et al. 2007). In extremely oligotrophic habitats, growth rates of *Daphnia* are so low that the whole growing season may be needed to reach maturity, which is the case in some alpine lakes (Gliwicz et al. 2001). *Daphnia* may suffer from the same stress in very ephemeral or other briefly suitable habitats. If time and food constraints are such that reproducing at least twice (once parthenogenetically, once sexually) becomes unlikely, immediate investment in dormant eggs may be a strategy to ensure persistence. So far, only one sexual species is known to bypass this limitation. *D. ephemeralis* is a cold stenotherm pond species that hatches earlier than any other species, but is rapidly outcompeted by *D. pulex* when temperatures rise in spring (Schwartz and Hebert 1987; Cáceres and Tessier 2004b). Hatchlings from dormant eggs of this species can consist of parthenogenetic females as well as sexual females and males (Schwartz and Hebert 1987). *D. ephemeralis* can thus reduce the cyclically parthenogenetic cycle to an entirely sexual cycle, in which the animals immediately invest in the production of sexual dormant eggs. However, this strategy of rapid investment in dormant eggs is easier to accomplish in obligately parthenogenetic species, where no males are needed to fertilize the eggs. Indeed, Spanish Pyrenees populations of *D. pulicaria* and Andean populations of an undescribed *D. pulex*-like species (Pérez-Martínez et al. 2007; Aguilera et al. 2007; Mergeay et al. 2008) are obligate asexuals that can produce ephippial dormant eggs from the first clutch. Although both strategies are similar and the reproductive output is low (only two eggs can be produced per ephippium), sexual species also suffer from the twofold cost of males (Maynard Smith 1978). In general, the cost of males in cyclically parthenogenetic *Daphnia* is highly diluted when many parthenogenetic generations precede the sexual generation. However, when the parthenogenetic phase is strongly reduced or even absent, cyclical parthenogenesis may become less advantageous than obligate parthenogenesis.

15.5.3 Genetic Slippage and Time Stress

Although sexual reproduction allows more rapid evolution than asexual reproduction, sexual reproduction can be disadvantageous as it disassembles previously successful gene combinations due to genetic recombination during meiosis (Allen and Lynch 2008). Especially in cyclical parthenogens, where clonal selection during the preceding phase of parthenogenetic reproduction also selects on genetic interaction effects, the result is that the initial average fitness of the sexual offspring (F1) is lower than that of the parents (F0), a phenomenon known as genetic slippage (Fig. 15.2a; Lynch and Deng 1994; Deng and Lynch 1996; Allen and Lynch 2008). During the growing season, however, clonal selection will lead to a re-increase of the average fitness up to or exceeding that of the previous generation (Fig. 15.3). Although in sexual lineages genetic slippage will be less pronounced the shorter the growing season is, the increase in average fitness will also be smaller due to a shorter period of clonal selection. In asexual populations, all else being equal, the fitness of F0 and F1 would remain the same (Fig. 15.2b). In very predictable, but time-stressed

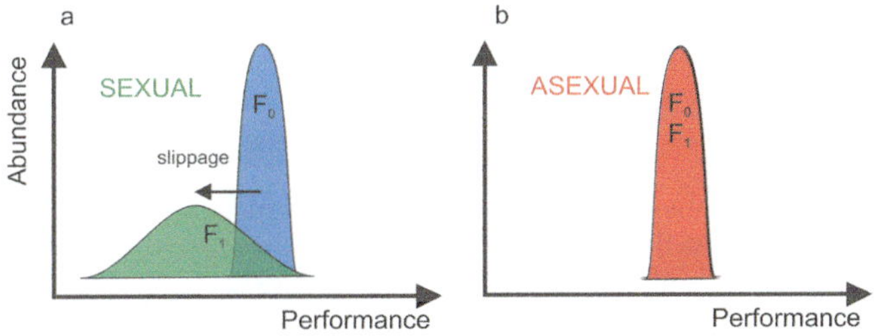

Fig. 15.2 Genetic slippage in sexual organisms as compared to amictic parthenogenetic organisms. F0: parental generation; F1: offspring generation

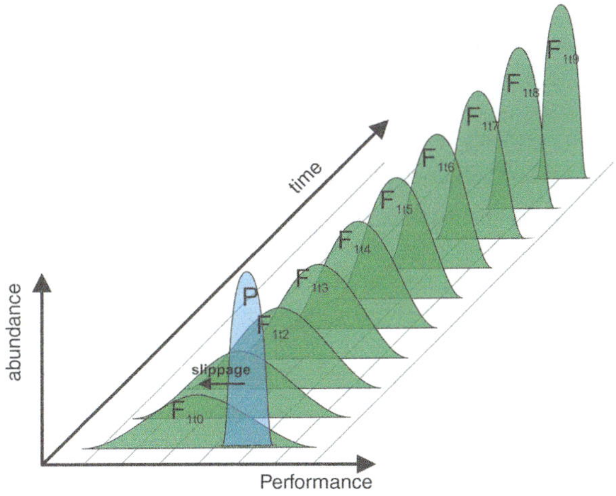

Fig. 15.3 The effect of clonal selection on average fitness in sexual populations after genetic slippage, shown at ten time intervals (t0–t9). F0 represents the parental generation (*blue*), while F1 represents the offspring (*green*). Under time stress, the final average fitness is lower than when time stress is relaxed

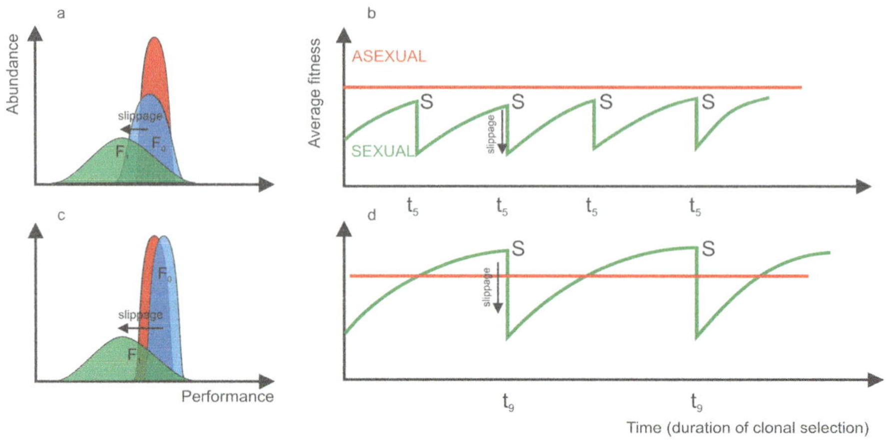

Fig. 15.4 The effect of time stress and genetic slippage on average fitness of a population at the time of sexual reproduction (S). **a + b**: Under time stress (S at t5), clonal selection may not be strong enough to purge the population from less fit genotypes, so that the average fitness remains well below the maximum achievable average fitness of a population that has abandoned sexual reproduction. **c + d**: When time stress is relaxed (S at t9) and clonal selection can act long enough, the average fitness of sexual populations can exceed that of asexuals. t5 and t9 refer to equal time intervals as in Fig. 15.3

habitats (habitats in which the ambient conditions for growth and development are short and which are thus characterized by a short growing season), asexuals may take advantage of such a temporal sex-related fitness reduction (Fig. 15.4a, b). Asexuals would be able to start the growing season with a population consisting entirely of equally well-adapted individuals, while sexual populations start with an on average lower fitness but also with greater variance in fitness over all individuals. Although the increased genetic variation on which subsequent clonal selection can act may compensate for this reduced fitness, this requires time for clonal selection to purge the population from the less fit genotypes (Fig. 15.4c, d). In combination with slow growth rates (e.g., in oligotrophic arctic or alpine systems), the advantage of asexuality could therefore be quite high (Fig. 15.4a, b), at least in habitats that show a similar selection regime over time (e.g., from year to year). Arctic or alpine ponds and lakes are harsh but relatively predictable habitats, in which co-adapted gene complexes are important to cope with the harsh conditions, but in which the uncertainties introduced by biotic interactions are relatively low.

Genetic surveys of asexual *Daphnia* in the arctic have shown that in several species a few common asexual genotypes have very wide geographic distributions, spanning over a thousand kilometres (Weider et al. 1999). These common clones may either be general-purpose genotypes (Lynch 1984; Weider et al. 1999) or their widespread distribution might merely reflect the presence of a common habitat type over large distances combined with an efficient screening of the best adapted genotype to that habitat (itself being a combination of high dispersal rates and local selection; see De Meester et al. 2002).

15.5.4 Polyploidy and Hybrid Vigour

Polyploidy, hybridization and asexuality are three phenomena that are firmly entwined with each other. Cause and effect are therefore often hard to distinguish (Simon et al. 2003; Kearney 2005). Most asexual *Daphnia* lineages are polyploids of hybrid origin (Table 15.1), as most other secondary asexual organisms (Kearney 2005). It is thought that polyploidy and the associated genetic redundancy represent an adaptation to increased mutagenic stress (Zakharov et al. 1970). It is thus conceivable that polyploidy and associated loss of sexual recombination have an adaptive value at high altitudes and latitudes. Asexuality in arctic or alpine regions may therefore to a certain degree be an epiphenomenon of selection for hybrid vigour (Kearney 2005) and/or polyploidy (Beaton and Hebert 1988). Circumstantial data support this hypothesis, by showing that polyploids occur more in UV radiation-stressed environments like alpine and boreal regions (Bierzychudek 1985). More specifically, within the *D. pulex* complex, only polyploid *Daphnia* are found in arctic settings (Beaton and Hebert 1988). The high incidence of cuticular melanisation in *Daphnia* from arctic or alpine regions (Hebert 1995; Černý and Hebert 1999) is indeed in line with the idea that UV radiation is an important stressor in these habitats. In addition, polyploidy seems to be adaptive at low temperatures. There is evidence that greater cell volumes, achieved through polyploidy, are favoured in arctic and alpine habitats (Otto and Whitton 2000), but diploid *Daphnia* can also achieve this through endopolyploidy (Gregory and Hebert 1999). Dufresne and Hebert (1998) showed that polyploid strains of *D. middendorffiana* performed better under cold conditions than diploid strains of the sister taxon *D. pulex*. Although their results were not phylogenetically independent and may just reflect species-specific differences not related to ploidy level, they fit the general trend of polyploidy as an adaptation to cold temperature stress. A similar example of geographic polyploidy accompanied by asexuality is found in alpine New Zealand populations of *D. thomsoni,* a species of the *D. carinata* complex (Hebert and Wilson 1994).

15.5.5 Contagious Asexuality: Selfish or Not?

One may argue that contagious asexuality is not an adaptive strategy per se, because it may merely be a selfish gene phenomenon, and that the question of the reason behind an asexual reproduction mode may thus be irrelevant. However, for a selfish gene, the meiosis suppressor allele may not be very efficient, as on average only one third of the offspring of contagiously asexual males and cyclically parthenogenetic females are viable (Innes and Hebert 1988). As a result, there seems to be an important net cost of male investment (see also Innes et al. 2000), as the fitness of males will on average be only one sixth of the fitness of obligately parthenogenetic females (one obligately parthenogenetic female that produces two dormant eggs has a maximal fitness of two, whereas three contagiously asexual males are needed to contribute to the equivalent of one viable egg). The maintenance of contagious asexuality is therefore more likely to be found in the advantages of occasional recombination (cf. supra) than that it represents a selfish gene phenomenon.

15.6 Conclusions

We provided an overview of possible scenarios for the evolution of asexuality in *Daphnia* that mostly involves either a contagious spread of a meiosis suppressor gene or hybridisation. Intriguingly, this genus shows a wide range of degrees of asexuality, from entirely sexual *D. ephemeralis* over normal cyclically parthenogenetic lineages that end every growing season with a bout of sexual recombination to effectively asexual populations of facultatively sexual species, contagiously obligate parthenogens that occasionally engage in sexual reproduction through their males, and strictly asexual lineages. Although asexuality is most abundant in *Daphnia* in the *D. pulex* species complex, the common incidence of this reproduction mode within this group and their wide geographic distribution reflects the success of this reproductive strategy under specific environmental conditions.

Acknowledgements We thank J. Vanoverbeke and D.J. Innes for comments on an earlier version of the manuscript. This project was supported by the K.U. Leuven Research Fund (projects OT/04/23, CREA/08/009 and STRT1/08/019) and by the ESF EURODIVERSITY project BIOPOOL.

References

Adamowicz SJ, Gregory TR, Marinone MC, Hebert PDN (2002) New insights into the distribution of polyploid *Daphnia*: the Holarctic revisited and Argentina explored. Mol Ecol 11: 1209–1217

Adamowicz SJ, Purvis A (2005) How many branchiopod crustacean species are there? Quantifying the components of underestimation. Global Ecol Biogeogr 14: 455–478

Aguilera X, Mergeay J, Wollebrants A, Declerck S, De Meester L (2007) Asexuality and polyploidy in *Daphnia* from the tropical Andes. Limnol Oceanogr 52: 2079–2088

Allen D, Lynch M (2008) Both costs and benefits of sex correlate with relative frequency of asexual reproduction in cyclically parthenogenetic *Daphnia pulicaria* populations. Genetics 179: 1497–1502

Beaton MJ, Hebert PDN (1988) Geographical parthenogenesis and polypoidy in *Daphnia pulex*. Am Nat 132: 837–845

Bell, G (1982) The masterpiece of nature: the evolution and genetics of sexuality. University of California Press, Berkeley

Bierzychudek, P (1985) Patterns in plant parthenogenesis. Experientia 41: 1255–1264

Boersma M, Spaak P, De Meester L (1998) Predator-mediated plasticity in morphology, life history, and behavior of *Daphnia*: the uncoupling of responses. Am Nat 152: 237–248

Brendonck L, De Meester L (2003) Egg banks in freshwater zooplankton: evolutionary and ecological archives in the sediment. Hydrobiologia 491: 65–84

Cáceres CE (1998) Interspecific variation in the abundance, production, and emergence of *Daphnia* diapausing eggs. Ecology 79: 1699–1710

Cáceres CE, Tessier AJ (2003) How long to rest: the ecology of optimal dormancy and environmental constraint. Ecology 84: 1189–1198

Cáceres CE, Tessier AJ (2004a) Incidence of diapause varies among populations of *Daphnia pulicaria*. Oecologia 141: 425–431

Cáceres CE, Tessier AJ (2004b) To sink or swim: variable diapause strategies among *Daphnia* species. Limnol Oceanogr 49: 1333–1340

Capaul M, Ebert D (2003) Parasite-mediated selection in experimental *Daphnia magna* populations. Evolution 57: 249–260

Carius HJ, Little TJ, Ebert D (2001) Genetic variation in a host-parasite association. Evolution 55: 1146–1152

Carvalho GR (1994) Genetics of aquatic clonal organisms. In: Beaumont A (ed) Genetics and evolution of aquatic organisms. Chapman and Hall, London, pp. 291–323

Černý M, Hebert PDN (1999) Intercontinental allozyme differentiation among four holarctic *Daphnia* species. Limnol Oceanogr 44: 1381–1387

Cohen D, Levin SA (1987) The interaction between dispersal and dormancy strategies in varying and heterogeneous environments. Lect Notes Biomath 71: 110–122

Colbourne JK, Wilson CC, Hebert PDN (2006) The systematics of Australian *Daphnia* and *Daphniopsis* (Crustacea: Cladocera): a shared phylogenetic history transformed by habitat-specific rates of evolution. Biol J Linn Soc 89: 469–488

Cousyn C, De Meester L, Colbourne JK, Brendonck L, Verschuren D, Volckaert F (2001) Rapid, local adaptation of zooplankton behavior to changes in predation pressure in the absence of neutral genetic changes. Proc Natl Acad Sci USA 98: 6256–6260

Crease TJ, Stanton DJ, Hebert PDN (1989) Polyphyletic origins of asexuality in *Daphnia pulex*. II mitochondrial-DNA variation. Evolution 43: 1016–1026

Crow JF (1994) Advantages of sexual reproduction. Dev Genet 15: 205–213

Carius HJ, Little TJ, Ebert D (2001) Genetic variation in a host-parasite association: potential for coevolution and frequency-dependent selection. Evolution 55: 1146–1152.

Charlesworth, B (1990) Mutation-selection balance and the evolutionary advantage of sex and recombination. Genet Res 55: 199–221

De Meester L (1993a) Genotype, fish-mediated chemicals, and phototaxis in *Daphnia magna*. Ecology 74: 1467–1474

De Meester L (1993b) Inbreeding and outbreeding depression in *Daphnia*. Oecologia 96: 80–84

De Meester L (1996) Local genetic differentiation and adaptation in freshwater zooplankton populations: patterns and processes. Ecoscience 3: 385–399

De Meester L, Gómez A, Okamura B, Schwenk K (2002) The monopolization hypothesis and the dispersal-gene flow paradox in aquatic organisms. Acta Oecol-Intern J Ecol 23: 121–135

De Meester L, Gómez A, Simon JC (2004) Evolutionary and ecological genetics of cyclical parthenogens. In Moya A, Font E (eds) Evolution from molecules to ecosystems. Oxford University Press, Oxford, pp. 122–134

De Meester L, Vanoverbeke J (1999) An uncoupling of male and sexual egg production leads to reduced inbreeding in the cyclical parthenogen *Daphnia*. Proc Roy Soc Lond B 266: 2471–2477

De Meester L, Vanoverbeke J, De Gelas K, Ortells R, Spaak P (2006) Genetic structure of cyclic parthenogenetic zooplankton populations – a conceptual framework. Arch Hydrobiol 167: 217–244

DeStasio, BT (1989) The seed bank of a freshwater crustacean – copepodology for the plant ecologist. Ecology 70: 1377–1389

Decaestecker E, De Meester L, Ebert D (2002) In deep trouble: habitat selection constrained by multiple enemies. Proc Natl Acad Sci USA 91: 5481–5484

Decaestecker E, Declerck S, De Meester L, Ebert D (2005) Ecological implications of parasites in natural *Daphnia magna* populations. Oecologia 144: 382–390

Decaestecker E, Lefever C, De Meester L, Ebert D (2004) Haunted by the past: evidence for dormant stage banks and epibionts of *Daphnia*. Limnol Oceanogr 49: 1355–1364

Decaestecker E, Vergote A, Ebert D, De Meester L (2003) Evidence for strong host clone-parasite species interactions in the *Daphnia*-microparasite system. Evolution 57: 784–792

Decaestecker E, Gaba S, Raeymaekers J, Stoks R, Van Kerckhoven L, Ebert D, De Meester L (2007) Host-parasite red queen dynamics archived in pond sediment. Nature 450: 870–873
Deng HW, Lynch M (1996) Change of genetic architecture in response to sex. Genetics 143: 203–212
Doncaster CP, Pound GE, Cox SJ (2000) The ecological cost of sex. Nature 404: 281–285
Duffy MA, Hall SR, Tessier AJ, Huebner M (2005) Selective predators and their parasitized prey: Are epidemics in zooplankton under top down control? Limnol Oceanogr 50: 412–420
Duffy MA, Sivars-Becker L (2007) Rapid evolution and ecological host-parasite dynamics. Ecol Lett 10: 44–53
Dufresne F, Hebert PDN (1994) Hybridization and origins of polyploidy. Proc Roy Soc Lond B 258: 141–146
Dufresne F, Hebert PDN (1995) Polyploidy and clonal diversity in an arctic cladoceran. Heredity 75: 45–53
Dufresne F, Hebert PDN (1997) Pleistocene glaciations and polyphyletic origins of polyploidy in an arctic cladoceran. Proc Roy Soc Lond B 264: 201–206
Dufresne F, Hebert PDN (1998) Temperature-related differences in life history characteristics between diploid and polyploid clones of the *Daphnia pulex* complex. Ecoscience 5: 433–437
Duncan A, Mitchell SE, Little TJ (2006) Parasite-mediated selection and the role of sex and diapause in *Daphnia*. J Evol Biol 19: 1183–1189
Duncan A, Little TJ (2007) Parasite-driven genetic change in a natural population of *Daphnia*. Evolution 61: 796–803
Ebert D (1994) Virulence and local adaptation of a horizontally transmitted parasite. Science 265: 1084–1086
Ebert D (2005) Ecology, Epidemiology, and Evolution of Parasitism in *Daphnia* [Internet]. Bethesda (MD): National Library of Medicine (USA), National Center for Biotechnology Information
Ebert D (2008) Host-parasite coevolution: insights from the *Daphnia*-parasite model system. Curr Opin Microbiol 11: 290–301
Ebert D, Haag C, Kirkpatrick M, Riek M, Hottinger JW, Pajunen VI (2002) A selective advantage to immigrant genes in a *Daphnia* metapopulation. Science 295: 485–488
Ebert D, Hamilton WD (1996) Sex against virulence: the coevolution of parasitic diseases. Trends Ecol Evol 11: 79–82
Felsenstein J (1974) The evolutionary advantage of recombination. Genetics 78: 737–756
Gandon S, Buckling A, Decaestecker E, Day T (2008) Host-parasite coevolution and patterns of adaptation across time and space. J Evol Biol 21: 1861–1866
Forró L, Korovchinsky NM, Kotov AA, Petrusek A (2008) Global diversity of cladocerans (Cladocera, Crustacea) in freshwater. Hydrobiologia 595: 177–184
Glesener RR, Tilman D (1978) Sexuality and the components of environmental uncertainty: clues from geographic parthenogenesis in terrestrial animals. Am Nat 112: 659–673
Gliwicz ZM, Slusarczyk A, Slusarczyk M (2001) Life history synchronization in a long-lifespan single-cohort *Daphnia* population in a fishless alpine lake. Oecologia 128: 368–378
Green J (1974) Parasites and epibionts of cladocera. Trans Zool Soc Lond 32: 417–515
Gregory TR, Hebert PDN (1999) The modulation of DNA content: proximate causes and ultimate consequences. Genome Res 9: 317–324
Haag CR, Ebert D (2004) Parasite-mediated selection in experimental metapopulations of *Daphnia magna*. Proc Roy Soc Lond B 271: 2149–2155
Haag CR, Hottinger JW, Riek M, Ebert D (2002) Strong inbreeding depression in a *Daphnia* metapopulation. Evolution 56: 518–526
Haag CR, Sakwinska O, Ebert D (2003) A test of a synergistic interaction between infection and inbreeding in *Daphnia magna*. Evolution 57: 777–783
Hairston NG (1996) Zooplankton eggs as biotic reservoirs in changing environments. Limnol Oceanogr 41: 1087–1092

Hairston, NG, DeStasio BT (1988) Rate of evolution slowed by a dormant propagule pool. Nature 336: 239–242

Hairston NG, Lampert W, Cáceres CE, Holtmeier CL, Weider LJ, Gaedke U, Fischer JM, Fox JA, Post DM (1999) Rapid evolution revealed by dormant eggs. Nature 401:446

Hairston NG, Holtmeier CL, Lampert W, Weider LJ, Post DM, Fischer JM, Caceres CE, Fox JA, Gaedke U (2001) Natural selection for grazer resistance to toxic Cyanobacteria: evolution of phenotypic plasticity? Evolution 55: 2203–2214

Hall SR, Tessier AJ, Duffy MA, Huebner M, Cáceres CE (2006) Warmer does not have to mean sicker: temperature and predators can jointly drive timing of epidemics. Ecology 87: 1684–1695

Hamilton WD (1980) Sex versus non-sex versus parasites. Oikos 35: 282–290

Hamilton WD, Axelrod R, Tanese R (1990) Sexual reproduction as an adaptation to resist parasites (A review). Proc Natl Acad Sci USA 87: 3566–3573

Hebert PDN (1978) Population biology of *Daphnia* (Crustacea, Daphnidae). Biol Rev Cambridge Phil Soc 53: 387–426

Hebert PDN (1981) Obligate asexuality in *Daphnia*. Am Nat 117: 784–789

Hebert PDN (1987) Genetics of *Daphnia*. In: Peters RH, De Bernardi R (eds) *Daphnia*. Mem Inst Ital Idrobiol 45: 439–460

Hebert PDN (1995) The *Daphnia* of North America: an illustrated fauna (on CD-rom). Cyber Natural Software, Guelph, Ontario

Hebert PDN, Beaton MJ, Schwartz SS, Stanton DJ (1989) Polyphyletic origins of asexuality in *Daphnia pulex*. I. Breeding-system variation and levels of clonal diversity. Evolution 43: 1004–1015

Hebert PDN, Finston TL (2001) Macrogeographic patterns of breeding system diversity in the *Daphnia pulex* group from the United States and Mexico. Heredity 87: 153–161

Hebert PDN, Rowe CL, Adamowicz SJ (2007) Life at low temperatures: A novel breeding-system adjustment in a polar cladoceran. Limnol Oceanogr 52: 2507–2518

Hebert PDN, Wilson CC (1994) Provincialism in plankton – endemism and allopatric speciation in Australian *Daphnia*. Evolution 48: 1333–1349

Hebert PDN, Wilson CC (2000) Diversity of the genus *Daphniopsis* in the saline waters of Australia. Can J Zool 78: 794–808

Hedrick PW (1995) Genetic polymorphism in a temporally varying environment: effects of delayed diapause. Heredity 75: 164–170

Howard RS, Lively CM (1998) The maintenance of sex by parasitism and mutation accumulation under epistatic fitness functions. Evolution 52: 604–610

Hurst LD, Peck JR (1996) Recent advances in understanding of the evolution and maintenance of sex. Trends Ecol Evol 11: 46–52

Innes DJ (1989) Genetics of *Daphnia obtusa:* genetic load and linkage analysis in a cyclical parthenogen. J Hered 80: 6–10

Innes DJ, Fox CJ, Winsor GL (2000) Avoiding the cost of males in obligately asexual *Daphnia pulex* (Leidig). Proc Roy Soc Lond B 267: 991–997

Innes DJ, Hebert PDN (1988) The origin and genetic basis of obligate parthenogenesis in *Daphnia pulex*. Evolution 42: 1024–1035

Johnson PTJ, Longcore JE, Stanton DE, Carnegie RB, Shields JD, Preu ER (2006) Chytrid infections of *Daphnia pulicaria*: development, ecology, pathology and phylogeny of *Polycarium leave*. Freshw Biol 51: 634–648

Kearney M (2005) Hybridization, glaciation and geographical parthenogenesis. Trends Ecol Evol 10: 495–502

Killick SC, Carlsson AM, West SA, Little TJ (2006) Testing the pluralist approach to sex: the influence of environment on synergistic interactions between mutation load and parasitism in *Daphnia magna*. J Evol Biol 19: 1603–1611

King CE, Schonfeld J (2001) The approach to equilibrium of multilocus genotype diversity under clonal selection and cyclical parthenogenesis. Hydrobiologia 446: 323–331

Kondrashov AS (1982) Selection against harmful mutations in large sexual and asexual populations. Genet Res 40: 325–332
Kondrashov AS (1988) Deleterious mutations and the evolution of sexual reproduction. Nature 336: 435–440
Korovchinsky NM (1996). How many species of Cladocera are there? Hydrobiologia 321: 191–204
Korovchinsky NM (2005). New species of *Holopedium* Zaddach, 1855 (Crustacea: Cladocera: Ctenopoda) from Greenland. J Limnol 64: 103–112
Lass S, Ebert D (2006) Apparent seasonality in parasite dynamics: analysis of cyclical prevalence patterns. Proc Roy Soc Lond B 273: 109–206
Lewis WM (1987) The cost of sex. In Stearns SC (ed), The evolution of sex and its consequences. Birkhäuser Verlag, Basel, p. 33–57
Limburg PA, Weider LJ (2002) 'Ancient' DNA in the resting egg bank of a microcrustacean can serve as a palaeolimnological database. Proc Roy Soc Lond B 269: 281–287
Little TJ, Demelo R, Taylor DJ, Hebert PDN (1997) Genetic characterization of an arctic zooplankter: insights into geographic polyploidy. Proc Roy Soc Lond B 264: 1363–1370
Little TJ, Ebert D (1999). Associations between parasitism and host genotype in natural populations of *Daphnia* (Crustacea: Cladocera). J Anim Ecol 67: 134–149
Little TJ, Ebert D (2001) Temporal patterns of genetic variation for resistance and infectivity in a *Daphnia*-microparasite system. Evolution 55: 1146–1152
Lynch M (1984) Destabilizing hybridization, general purpose genotypes and geographic parthenogenesis. Quart Rev Biol 59: 257–290
Lynch M, Deng HW (1994) Genetic slippage in response to sex. Am Nat 144: 242–261
Lynch M, Gabriel W (1983) Phenotypic evolution and parthenogenesis. Am Nat 122: 745–764
Maynard Smith J (1971) The origin and maintenance of sex. In Williams GC (ed) Group selection. Aldine-Atherton, Chicago
Maynard Smith, J (1978) The evolution of sex. Cambridge University Press, Cambridge, UK
Maynard Smith J (1989) Evolutionary genetics. Oxford University Press, Oxford
Mergeay J, Verschuren D, De Meester L (2006) Invasion of an asexual American water flea clone throughout Africa, and rapid displacement of a native sibling species. Proc Roy Soc Lond B 273: 2839–2944
Mergeay J, Vanoverbeke J, Verschuren D, De Meester L (2007) Extinction, recolonisation and dispersal through time in a planktonic crustacean. Ecology 88: 3032–3043
Mergeay J, Aguilera X, Declerck S, Petrusek A, Huyse T, De Meester L (2008) The genetic legacy of polyploid Bolivian *Daphnia*: the tropical Andes as a source for the North and South American *D. pulicaria* complex. Mol Ecol 17: 1789–1800
Mitchell SE, Read AF, Little TJ (2004) The effect of a pathogen epidemic on the genetic structure and reproductive strategy of the crustacean *Daphnia magna*. Ecol Lett 7**:** 848–858
Mitchell SE, Rogers ES, Little TJ, Read AF (2005) Host-parasite and genotype-by-environment interactions: temperature modifies potential for selection by a sterilizing pathogen. Evolution 59: 70–80
Muller HJ (1932) Some genetic aspects of sex. Am Nat 66: 118–138
Omilian AR., Cristescu MEA, Dudycha JL, Lynch M (2006) Ameiotic recombination in asexual lineages of *Daphnia*. Proc Natl Acad Sci USA 103: 18638–18643
Ortells R, Gomez A, Serra M (2006) Effects of duration of the planktonic phase on rotifer genetic diversity. Arch Hydrobiol 167: 203–216
Otto SP, Whitton J (2000) Polyploid incidence and evolution. Annu Rev Genet 34: 401–437
Paland S, Colbourne J, Lynch M (2005) Evolutionary history of contagious asexuality in *Daphnia pulex*. Evolution 59: 800–813
Paland S, Lynch M (2006) Transitions to asexuality result in excess amino acid substitutions. Science 311: 990–992
Peck JR (1994) A ruby in the rubbish – beneficial mutations, deleterious mutations and the evolution of sex. Genetics 137: 597–606
Peck JR, Waxman D (2000) Mutation and sex in a competitive world. Nature 406: 399–404

Pérez-Martínez C, Barea-Arco J, Conde-Porcuna JM, Morales-Baquero R (2007) Reproduction strategies of *Daphnia pulicaria* population in a high mountain lake of Southern Spain. Hydrobiologia 594: 75–82

Pfrender M, Lynch M (2000) Quantitative genetic variation in *Daphnia*: temporal changes in genetic architecture. Evolution 54: 1502–1509

Pijanowska J, Stolpe G (1996) Summer diapause in *Daphnia* as a reaction to the presence of fish. J Plankton Res 18: 1407–1412

Pound GE, Doncaster CP, Cox SJ (2002) A Lotka-Volterra model of coexistence between a sexual population and multiple asexual clones. J Theor Biol 217: 535–545

Pound GE, Cox SJ, Doncaster CP (2004) The accumulation of deleterious mutations within the frozen niche variation hypothesis. J Evol Biol 17: 651–662

Refardt D, Ebert D (2007) Inference of parasite local adaptation using two different fitness components. J Evol Biol 20: 921–929

Schultz TW (1977) Fine structure of the ephippium of *Daphnia pulex* (Crustacea: Cladocera). Trans Am Microsc Soc 96: 313–321

Schwartz SS, Hebert PDN (1987) Breeding system of *Daphniopsis ephemeralis*: adaptations to a transient environment. Hydrobiologia 145: 195–200

Schwenk K, Spaak P (1995) Evolutionary and ecological consequences of interspecific hybridization in cladocerans. Cell Mol Life Sci 51: 465–481

Simon JC, Delmotte F, Rispe C, Crease T (2003) Phylogenetic relationships between parthenogens and their sexual relatives: the possible routes to parthenogenesis in animals. Biol J Linn Soc 79: 151–163

Sluzarczyk M (1995) Predator-induced diapause in *Daphnia*. Ecology 76: 1008–1013

Spitze K (1992) Predator mediated plasticity of prey life history and morphology: *Chaoborus americanus* predation on *Daphnia pulex*. Am Nat 139: 229–247

Stearns SC (1987) The evolution of sex and its consequences. Birkhäuser Verlag, Basel

Stirnadel HA, Ebert D (1997) Prevalence, host specificity and impact on host fecundity of microparasites and epibionts in three sympatric *Daphnia* species. J Anim Ecol 66: 212–222

Taylor DJ, Crease TJ, Brown WM (1999) Phylogenetic evidence for a single long-lived clade of crustacean cyclic parthenogens and its implications for the evolution of sex. Proc Roy Soc Lond B 266: 791–797

Templeton AR, Levin DA (1979) Evolutionary consequences of seed pools. Am Nat 114: 232–249

Tessier AJ, Young A, Leibold M (1992) Population dynamics and body-size selection in *Daphnia*. Limnol Oceanogr 37: 1–13

Tollrian R, Harvell CD (1999) The ecology and evolution of inducible defenses. Princeton University Press, Princeton

Vale PF, Stjernman M, Little TJ (2008) Temperature dependent costs of parasitism and the maintenance of polymorphism under genotype by environment interactions. J Evol Biol 21: 1418–1427

Vanoverbeke J, De Meester L (1997) Among-populational genetic differentiation in the cyclical parthenogen *Daphnia magna* (Crustacea, Anomopoda) and its relation to geographic distance and clonal diversity. Hydrobiologia 360: 135–142

Van Valen L (1973) A new evolutionary law. Evol Theory 1: 1–30

Vrijenhoek RC (1979) Factors affecting clonal diversity and coexistence. Am Zool 19: 787–797

Weider LJ, Hobaek A, Hebert PDN, Crease TJ (1999) Holarctic phylogeography of an asexual species complex – II. Allozymic variation and clonal structure in Arctic *Daphnia*. Mol Ecol 8: 1–13

West SA, Lively CM, Read AF (1999) A pluralist approach to sex and recombination. J Evol Biol 12: 1003–1012

Whitlock, MC, Ingvarsson PK, Hatfield T (2000) Local drift load and the heterosis of interconnected populations. Heredity 84: 452–457

Wolinska J, Bittner K, Ebert D, Spaak P (2006) The coexistence of hybrid and parental *Daphnia*: the role of parasites. Proc Roy Soc Lond B 273: 1977–1983

Woolhouse MEJ, Webster JP, Domingo E, Charlesworth B, Levin BR (2002) Biological and biomedical implications of the co-evolution of pathogens and their hosts. Nature Genet 32: 569–577

Zaffagnini F (1987) Reproduction in *Daphnia*. In Peters RH, De Bernardi R (eds) *Daphnia*. Mem Inst Ital Idrobiol 45: 245–284

Zakharov, IA, Suslova NG, Fedorova IV (1970) Polyploidy and radioresistance in yeast. I. The effects of mutation on radioresistance of polyploid strains of *Saccharomyces cerevisiae* (in Russian). Genetika 6: 100–106

Zbinden M, Haag CR, Ebert D (2008) Experimental evolution of *Daphnia magna* populations in response to parasite treatment. J Evol Biol 21: 1068–1078

Chapter 16
Metasexual Stick Insects: Model Pathways to Losing Sex and Bringing It Back

Valerio Scali

Abstract This chapter illustrates the variety of parthenogenetic mechanisms within three genera of phasmids (stick insects) occurring around the Mediterranean and the close ties between sexuals and asexuals. The link between parthenogenesis and polyploidy appears to provide an escape from cytogenetic disruption. When compared to diploids, both auto-polyploids and allo-polyploids can take advantage of higher mutation rates and the resulting increase of heterozygosity; they can also escape the accumulation of mutations. Detailed investigations of the karyotype structure, reproduction modes and phylogenetic relationships reveal a wide variety of reproductive modes including repeated backcrosses to parental species. Diploid hybrids can also reproduce hemiclonally, with one parental genome being eliminated and the hybrid being produced anew in each generation. However, also the reversal route from hybrids to the fathering species occurs through androgenesis, which resembles sexual reproduction with most of its genetic traits. All of these reproductive modes can be considered as patterns of "reticulate" evolution. Automicts, apomicts and hybrids with different amounts of recombination and levels of polyploidy (from diploids to tetraploids) have been found. The genetic diversity in these asexuals is kept high in mitochondrial DNA, rDNA cistrons, satellite DNA and coding genes. Other addressed issues, clearly linked to phasmid parthenogenesis are the peculiarities of spindle structure and the microtubule and γ-tubulin patterns for the egg development: these might be one of the reasons why phasmids are so exceptional in their variety of reproductive modes.

16.1 Introduction

16.1.1 Asexuals and Sexuals

Some genetic systems, although clearly derived from sexuals, are not conform with classic sexual reproduction involving recombination and equal chromosome

V. Scali (✉)
Department of Biologia Evoluzionistica Sperimentale, University of Bologna, Via Selmi 3, 40126, Bologna, Italy
e-mail: valerio.scali@unibo.it

I. Schön et al. (eds.), *Lost Sex*, DOI 10.1007/978-90-481-2770-2_16,

assortment. These "asymmetrics", as they have been called, typically show thelytoky, haplodiploidy and/or parent-specific gene expression (Normark 2006). It appears a coarse definition to identify asexuality only by the lack of recombination and/or fertilization (see, for example, Bell 1982 or Baxevanis et al. 2006). Also, the assumption that asexuals lack both and are therefore barred from any genetic variance and evolutionary potential (Baxevanis et al. 2006) is not correct. Some all-female taxa are neither invariant nor faithful genetic replicas of their mothers (McKinnel 1985; Hughes 1989; Mittwoch 2002). On the contrary, most of the investigated thelytokous animals have shown a variety of genetic diversifying mechanisms at work that produce new clones. This is realized either by real recombination, as in automicts and even some apomicts, or by mutations, incorporation of other genomes and polyploidy (see below). Thus, asexual organisms can maintain significant levels of genetic diversity (White 1970; Turner et al. 1983; Good and Wright 1984; Densmore et al. 1989; Hedges et al. 1992; Quattro et al. 1992; Haddal et al. 1994; Tinti and Scali 1996; Johnson et al. 1999; Schön et al. 2000; Halkett et al. 2005).

16.1.2 Parthenogenesis and Polyploidy

Parthenogenetic animals, particularly apomicts, are often polyploid (see Otto and Whitton 2000 for a data base; Pongratz et al. 2003; Stenberg et al. 2003; Halkett et al. 2005; Gomez-Zurita et al. 2006). The link between parthenogenesis and polyploidy provides an avenue for escaping cytogenetic disruption due to sex-chromosomes and their dosage compensation in polyploid germ lines and embryo cells (Orr 1990), particularly in cases of uneven ploidy. As a consequence, polyploidy is more often found in animals with genetic sex-determination than with defined hetero-chromosomes (Dufresne and Hebert 1994; Evans et al. 2004; Holloway et al. 2006).

When compared to diploids, both auto-polyploids and allo-polyploids can take advantage of a higher mutation rate and the resulting increase of heterozygosity. A general advantage of polyploids is their better buffering ability against the accumulation of non-functional mutations in long term parthenogenes, thus maintaining genetic functionality (Lokki 1976a, b). However, the largest short-term advantage is gained by those allopolyploids with a considerable number of heterozygous loci, which are immediately generated through hybridisation. This refers to relatively recent polyploidization events with increasing numbers of chromosome sets and disregards ancient polyploidizations as became obvious as gene duplications in deep phylogenetic analyses of animal genomes (Gibson and Spring 2000; Otto and Whitton 2000; Maere et al. 2005; Friedman and Hughes 2007).

16.2 Sexual and Asexual Stick Insects

Phasmida (Otte and Brock 2005; = Phasmatodea or Cheleutoptera of previous literature) are an orthopteroid insect order, which holds about three thousand species. A few years ago, Zompro (2005) erected the Timematodea as a new order, hereby separating them from Phasmida, within which they had been hitherto embodied. Among the Phasmida, about 10% of taxa, scattered in different families, are parthenogenetic. The same applies to the Timematodea (Sandoval et al. 1998; Law and Crespi 2002). Phasmid parthenogenesis is invariably thelytokous (i.e., producing all-female offspring), with just one, still unexplained exception for the deuterotokous *Ctenomorphodes tessulatus* (Hadlington 1961). Detailed investigations of the karyotype structure, reproduction and phylogenetic relationships of the holomediterranean genus *Bacillus* reveal a wide variety of reproductive modes. These include facultative and obligate parthenogenesis, hybridogenesis and androgenesis and interracial and interspecific hybridizations, leading to both diploid and polyploid taxa. Owing to the occurrence of recombination in parthenogens and to the use of non-classic reproductive modes of stick insects, the term "metasexual" has been proposed for them (Scali et al. 2003). More recent analyses of the Iberian *Leptynia* and *Pijnackeria* suggest a similar array of reproductive modes and species interactions (Passamonti et al. 2004; Ghiselli et al. 2007; Scali 2009).

These study cases are here analysed in detail, since they represent good model groups with general significance for evolutionary research.

16.2.1 **Bacillus** *(Latreille)*

The genus includes the following taxa:

1) The Western-Mediterranean facultative parthenogen *Bacillus rossius* (Rossi) (2n = 35/36, X0/XX). North-African races are strictly sexual, while the two Italian ones consist of either sexual or all-female populations (see Mantovani et al. 1999). In Italy, only the most Northern areas of the distribution range show geographic parthenogenesis, whereas in Southern Tyrrhenian and the Adriatic and Ionian regions, a patchy distribution of sexual and parthenogenetic populations is found. Westwards, along the coast of the Mediterranean Sea, sexual populations have been reported from some areas of Southern France (Bullini 1966) and Spanish Catalaunia (Tinti 1993). Elsewhere in France and Spain, so far only females appear to be present (Scali, unpublished). Also, the Eastern area of the species' range (Slovenia, Croatia, Albania and Greece) is inhabited by all-female populations. Males of any site readily mate with parthenogenetic females to produce fertile, sexual offspring (Scali 1968). On the other hand, sudden switches from sexual to parthenogenetic reproduction have been observed in two populations (at Otranto and Erice, Italy; Scali 1996a). Investigations of these and additional all-female populations failed to reveal any obvious trigger of parthenogenesis, including the micro-organism *Wolbachia* (Scali and Parker, unpublished).
2) The sexual *Bacillus grandii* Nascetti & Bullini (2n = 33/34, X0/XX) is endemic to small areas of Sicily and formally split into three subspecies, namely *B. grandii grandii*; *B. grandii benazzii* and *B. grandii maretimi*.

3) The Central/Eastern Mediterranean species *Bacillus atticus* (Brunner), at present an all-female obligate parthenogenetic complex is differentiated into three karyological and allozymic races (*B. atticus atticus*, 2n = 34, XX; *B. atticus cyprius*, 2n = 32, XX; *B. atticus carius*, few populations with 2n = 34, XX, and 3n = 48–51, XXX).

Bacillus rossius redtenbacheri, *B. grandii grandii* and *B. atticus atticus* are the ancestors of two Sicilian hybrids, namey the diploid *Bacillus whitei* Nascetti & Bullini (= *rossius redtenbacheri/ grandii grandii*, 2n = 35, XX) and the triploid *Bacillus lynceorum* Bullini, Nascetti & Bianchi Bullini (= *rossius redtenbacheri/ grandii grandii/ atticus atticus*, 3n = 51, XXX). In addition, females of *Bacillus rossius* and *B. grandii* males hybridise to produce two different hybridogenetic strains (sensu Schultz 1961): *B. rossius redtenbacheri-grandii grandii* and *B. rossius redtenbacheri-grandii benazzii*. These two strains will pass an invariant, maternal *rossius redtenbacheri* haploset on to their offspring, while the paternal genome (either *grandii grandii*, for the southern hybridogen, or *grandii benazzii*, for the northern one) will be renewed each generation (hemiclonal reproduction). Through hybridogenesis, offspring of both sexes are produced but males are mostly lethal and always sterile (Mantovani and Scali 1992; Mantovani et al. 1999). Mitochondrial DNA sequencing analyses (see below) show that *Bacillus* species are hybrids and originated through asymmetrical crosses, always with *B. rossius* as the maternal parent. On the whole, *Bacillus* taxa experience a wide array of reproductive modes as is illustrated by a complex net of reproductive and phylogenetic interactions (Fig. 16.1; see also Bullini 1994; Mantovani et al. 1999).

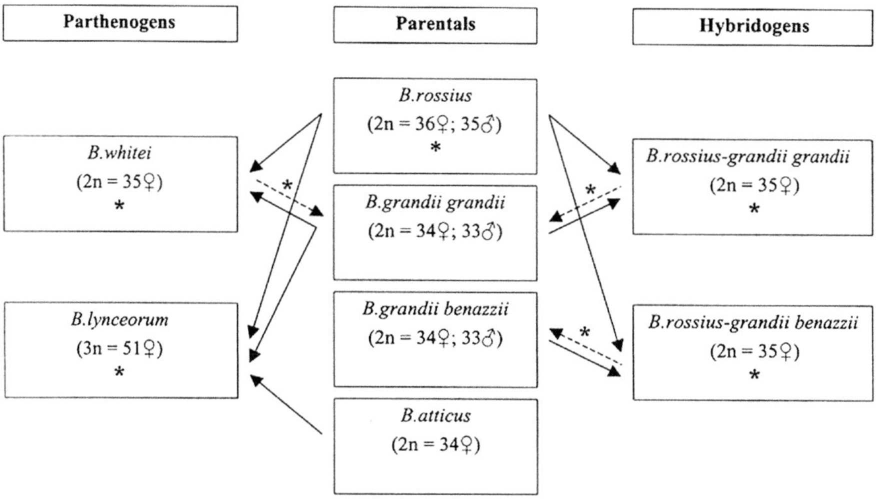

Fig. 16.1 Reproductive relationships among *Bacillus* taxa. Stars mark taxa with *rossius* mitochondrial genome, while sketched arrows indicate reversions from hybrid to non-hybrid paternal ancestors. From Scali et al. (2003)

The two hybridogenetic strains *Bacillus rossius-grandii grandii* and *B. rossius-grandii benazzii* are organisms that violate the equal transmission of genomes to the progeny by eliminating the *grandii* paternal genome and just passing the haploid *rossius* chromosome set on to their offspring. Keeping a pure *rossius* genome in the female germ line allows hybrids to overcome the meiotic constraints of heterospecific chromosome pairing. Pairing chromosomes are in fact sisters, which is why their recombination and/or assortment cannot produce any genetic diversification. Consequently, the only source of novel genetic variability is the incoming haploid set of the *grandii* spermatozoon (Fig. 16.2). Moreover, allozyme analyses do not reveal any polymorphisms in the maternal genome, so that the *rossius* hemi-clone is identical in all hybrid and hybridogenetic specimens. This finding is also mirrored in the genetic homozygosity observed in present-day, sympatric asexual populations (Tinti and Scali 1995).

Interestingly, hybrid females can escape their mode of reproduction, revert to sexuality and produce progeny with the nuclear genetic structure of the paternal species in just one generation if they shift to androgenesis during their egg development (Figs. 16.1 and 16.2).

Phasmid eggs are physiologically polyspermic; when – apparently through the expression of androgenesis alleles (McKone and Halpern 2003) – syngamy with the egg pronucleus fails, two male pronuclei may fuse instead and produce a progeny of both sexes with pure *Bacillus grandii* nuclear gene composition. These offspring no longer behave as hemiclonally reproducing hybrids, but as sexually

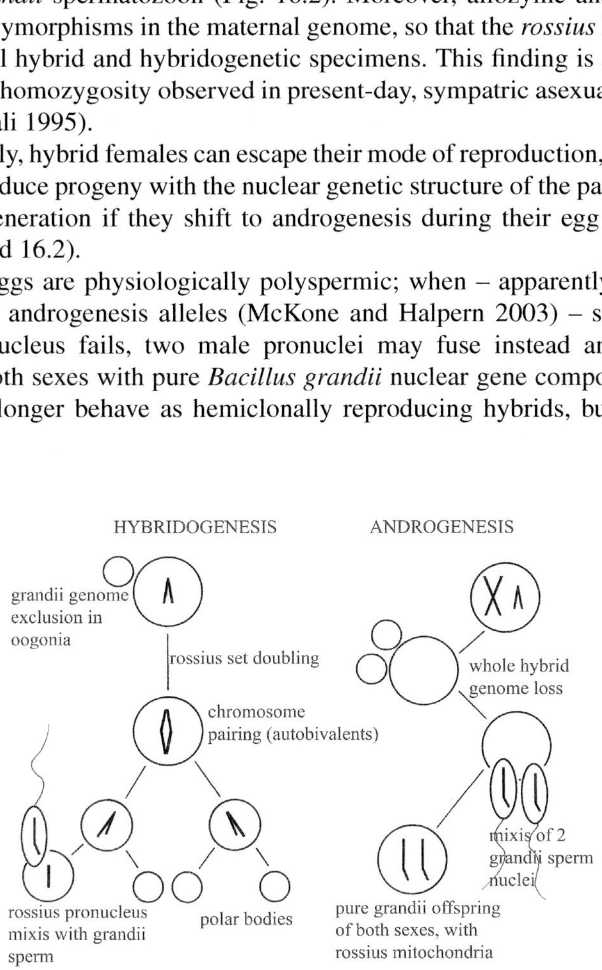

Fig. 16.2 Hybridogenesis and androgenesis in *Bacillus* taxa.(see Fig. 16.1). In the hybridogenetic female germ line, the elimination of the whole *grandii* haploset leads to the doubling of the *rossius* chromosomes, which undergo a mechanically normal meiosis with autobivalents; fertilization of the reduced egg by a new *grandii* sperm follows. In the androgenetic egg, no mixis of *grandii* sperm with the *rossius* pronucleus takes place. Instead, a fusion of two sperm heads occurs because of physiological polyspermy

reproducing specimens. The production of androgens actually frees the hybridogenetic strains from being sexual parasites of the paternal *Bacillus grandii* species. Their hybridogenetic origin is clear, however, since they possess *grandii* nuclear alleles (often in a heterozygous condition), together with a *rossius* mitochondrial genome. It has been this peculiar genetic feature that has allowed a sound assessment of androgenetic specimens occurring in the field (Tinti and Scali 1996; Mantovani et al. 2001). Surprisingly, instances of reversion from hybridisation to the parental species have been also observed in *Bacillus whitei* offspring, supporting sperm access to parthenogenetic eggs and the availability of androgenesis to parthenogenetic species.

Thelytokous *Bacillus* taxa produce progenies with different amounts of genetic variability. For a deeper understanding of those differences, we need to analyse the processes of egg maturation in each taxon. The egg maturation of the facultative parthenogen *Bacillus rossius* is accomplished through a normal meiosis followed by the embryo development as a haploid germ (A+X). Only after thousands of blastula nuclei have been produced, germ diploidization starts in a few cells by means of anaphase restitution, which immediately restores homozygosity at all loci. Only about 50 cells beneath the micropylar area promote further development of the diploid female embryo (AA+XX), whereas in about ten days, the vast majority of haploid nuclei degenerate (Scali 1969, 1972). If mated, these females will again produce sons and daughters. An identical situation with a similar embryonic developmental pattern has been demonstrated in *Medaura extradentata* (=*Clitumnus extradentatus*) by Bergerard (1958, 1962). In *Bacillus rossius*, shifts from sexual to all-female populations must have occurred many times since in both Tyrrhenian and Adriatic locations, the genetic features of parthenogenetic females are related to those of the geographically closest sexual populations (Gasperi et al. 1983; Tinti 1993). When such a shift occurs, different alleles at a given locus segregate into different haploid eggs, which, after diploidization (gamete duplication), develop into genetically different embryos, thus producing polyclonal offspring, which are homozygous at all loci. If parthenogenetic reproduction is maintained, these clones will become fixed and a polyclonal population established. This is the picture that is always observed in all-female populations of *Bacillus rossius* when compared to the closest sexual populations (Mantovani and Scali 1991; Tinti et al. 1992).

The di/triploid obligate parthenogen *Bacillus atticus* has an automictic maturation mechanism differing from the one of *B. rossius*. In diploid females, after a normal first meiotic division allowing chromosome pairing and recombination, the two haploid nuclei soon show an interphase swollen appearance and immediately fuse to restore an unreduced egg nucleus (Fig. 16.3A). A second division follows, clearly mimicking meiosis II, to yield a pycnotic polar body and a pronucleus, which starts embryonic development. During the first meiotic division in triploid females, also univalent and multivalent chromosome associations are formed in addition to bivalents, leading to chromosomally unbalanced sister nuclei. However, the unbalanced segregation of chromosomes is without harmful consequences, since the unbalanced products of the 1st division soon fuse back, similarly to the mechanism in diploids. The reconstituted triploid nucleus undergoes a second division to

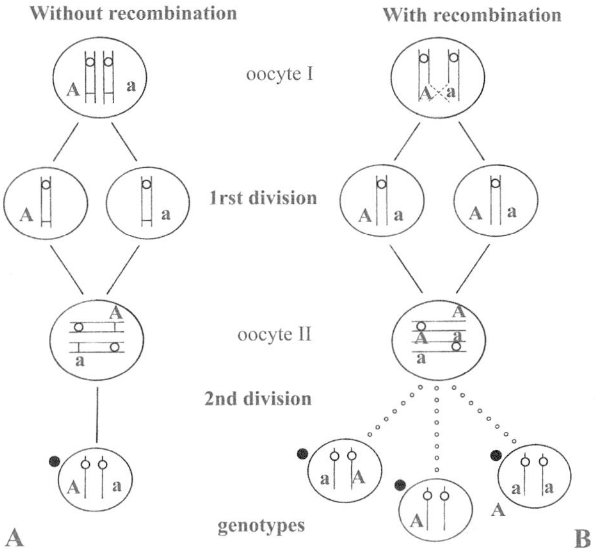

Fig. 16.3 Meiotic restitution of parthenogenetic eggs of *Bacillus atticus* during 2nd prophase. Without recombination, the maternal heterozygosity at a given locus is kept in the progeny, (A) whereas homozygous offspring are also produced if crossing-over occurs at certain chromosomal loci (B) (modified from Marescalchi and Scali 2003)

produce a degenerating polar body and an unreduced and yet chromosomally and genetically balanced, triploid pronucleus (Marescalchi and Scali 2003). Besides the triploid constitution of the egg, the meiotic mechanism of triploids is the same as in diploids, also including recombination. The described gametogenetic process can account for both the clonal maintenance of chromosomal rearrangements – causing some wobbling in their number (48/51) – and the transmission of partially fixed heterozygosities at some loci. Also, the production of homozygous offspring genotypes from the heterozygous mothers may be envisaged as a by-product of this uncommon egg maturation process when an appropriate cross-over occurs during the first division of egg maturation (Fig. 16.3B). These cytological features are in line with the complex clonal population structure of the species, as illustrated in Table 16.1. The clonal richness makes the obligate parthenogen *Bacillus atticus* genetically diversified and allows it to persist in a variety of habitats along the Mediterranean basin, from Sardinia and Sicily to Israel, the Italian mainland, the former Yugoslavia countries, Greece and Turkey. It is relevant that interspecific F1 diploid hybrids are produced in the laboratory and the field through crosses of diploid *B. atticus* females with *Bacillus rossius* or *B. grandii* males. These can obviously only be obtained if the normal meiotic process is restored in inseminated eggs, suppressing the parthenogenetic process of the virgin eggs undergoing the nuclear fusion after the 1st meiosis (Marescalchi and Scali 2003).

The obligate parthenogens *Bacillus whitei* and *B. lynceorum* share the same apomictic mechanism that mechanically mimicks a normal meiosis. Their maturation process is significantly different, however, since the synapsis is interrupted during prophase I, the somatic number of double stranded chromosomes is re-established and an additional synthesis of DNA takes place to produce pachytene chromosomes with four chromatids each (Marescalchi et al. 1991; Fig. 16.4). Two

Table 16.1 Zymotype diffusion and frequency in the automictic parthenogen *Bacillus atticus*

	Populations	Specimens	Total zymotypes	Specific zymotype/ population∗	Number of different zymotypes/ population∗∗
B. atticus (2n)	15	156	52 {27}	3/7	8/1
				21/6	7/1
				20/4	5/1
				1/2	4/3
				2/1	3/5
				4–19/1	2/1
				22–27/1	1/3
B. atticus (3n)	9	145	44 {21}	11/4	10/1
				18/3	9/1
				10/2	5/1
				13/2	4/4
				17/2	3/1
				1–9/1	1/1
				12/1	
				14–16/1	
				19–21/1	

In parentheses, samples of *2n* and *3n* zymotypes are analysed in detail. From Scali et al. (2003).
∗The first number refers to the ordinal number of the zymotype, the second to the number of populations in which each zymotype has been found.
∗∗ Frequency of different zymotypes per population.

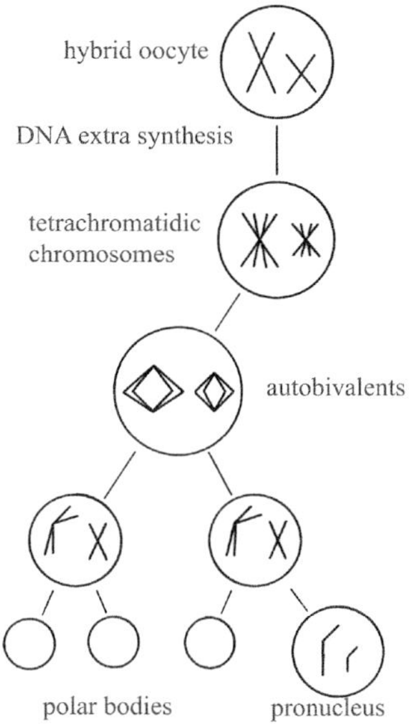

Fig. 16.4 Apomixis in *Bacillus whitei* and *B. lynceorum*. The apomictic parthenogenesis in *Bacillus whitei* and *B. lynceorum* undergoes a transient chromosome pairing, allowing a low rate of recombination before the extra-doubling of DNA takes place at pachytene; eggs with somatic chromosome number are produced in both parthenogens, namely *B. whitei*, 2n = 35 and *B. lynceorum*, 3n = 52 (from Marescalchi et al. 1991)

divisions follow and an unreduced pronucleus plus three degenerating polar bodies are produced. The same cytological traits have been described from the polyploid stick insects *Carausius morosus* and *Sipyloidea sipylus* (Pijnacker 1966; Pijnacker and Ferwerda 1978, 1986).

Due to the described maturation process, *Bacillus whitei* and *B. lynceorum* have almost invariably maintained the genetic structure of their parental ancestors, with many fixed heterozygous loci (about 65%) because of the sharp genetic differentiation of their ancestors, *B. rossius* and *B. atticus/B. grandii*. The transmission of the maternal genetic structure to the thelytokous progeny has some variation, however, since the short pairing phase of chromosomes during early prophase I leaves room for recombination to occur at a low rate. This could also explain why different clones are found at the population level (Mantovani et al. 1992; Scali et al. 1995). A polyphyletic origin of these hybrids might have further contributed to their polyclonal structure (Mantovani 1998; Scali and Tinti 1999; Scali et al. 2003), but among the 6 clones (zymotypes) characterised in 12 populations of *Bacillus whitei* and among the 15 clones found in 13 populations of *B. lynceorum* (Table 16.2), a few must have derived from such recombination events.

In *Bacillus whitei*, it has been noticed that some loci that should be heterozygous are in fact homozygous and lack the allele of one parental species. Similarly,

Table 16.2 Zymotype diffusion and frequency in the apomictic parthenogens *Bacillus whitei* and *B. lynceorum*

	Populations	Specimens	Total zymotypes	Specific zymotype/ population*	Number of different zymotypes/ population**
B. whitei					
(2n)	12	128	6	1/12	4/1
				4/2	2/3
				2/1	1/8
				3/1	
				5/1	
				6/1	
B. lynceorum					
(3n)	13	64	15	5/5	5/2
				7/4	3/1
				1/3	2/2
				10/2	1/8
				2,3/1	
				4,6/1	
				8,9/1	
				11–15/1	

From Scali et al. (2003).
* The first number refers to the ordinal number of the zymotype, the second to the number of populations in which each zymotype has been found.
** Frequency of different zymotypes per population.

but at a higher rate, in the trihybrid *Bacillus lynceorum*, the alleles of one or even two parental species have been reciprocally substituted, thus producing unusual homozygous loci and, as a consequence, new genotypes (Mantovani et al. 1992). Experimental crosses have furthermore demonstrated that recombination and whole genome incorporation can take place in these apomicts (Tinti and Scali 1996). Although it is clear that these apomicts are much less diversified than the automictic *Bacillus atticus*, their egg maturation process does not completely suppress recombination, so that a strictly clonal inheritance is not maintained.

An often-overlooked source of variability in hybrids is the pattern of chromosomal localization and utilisation of rDNA cistrons. If we summarize the findings in the trihybrid *Bacillus lynceorum* when using the AgNO3 marking method, all populations with one exception showed an individual marking pattern: in each specimen active NORs ranged from 1 to 3 and eight populations out of 20 expressed only the maternal *rossius* rDNA (with either one or two active clusters), seven both *rossius* and *grandii* (5) or *rossius* and *atticus* (2), four *atticus* and *grandii*, and one the three different genomes, *rossius, atticus, grandii* (Manaresi et al. 1993). Their differential activity could well be of adaptive significance.

An additional source of genetic variability in both parental and derived hybrid species is the peculiar pattern of pericentromeric satellite DNA. In *Bacillus rossius,* a complete "library" of the Bag320 family variants of the genus is present in low copy numbers beside private sequences. The *B. rossius* monomers became differentially amplified and characterized along the evolutionary lineages leading to the present-day taxa, namely to *B. grandii grandii* and *B. atticus* on one hand and to *B. grandii benazzii* and *B. grandii maretimi* on the other. These relationships are in full agreement with mitochondrial phylogenies (Cesari et al. 2003; Luchetti et al. 2003). In *Bacillus grandii* and *B. atticus*, only specific monomer subsets of the Bag320 library appear. In addition, these show specific different clustering features: absolute variability of monomer sequences is lower in *Bacillus grandii* than in *B. atticus* (Table 16.3), but there is clearly a higher homogeneity within each subspecies than

Table 16.3 Mean p-distance (pD) ± standard error (SE) of the Bag320 family repeats for the bisexual *Bacillus grandii* species and subspecies and the unisexuals *B. atticus*, *B. whitei* and *B. lynceorum*

Taxon	Mean pD ± S.E.
B. grandii	0.121 ± 0.008
B. grandii grandii	0.073 ± 0.005
B. grandii benazzii	0.086 ± 0.005
B. grandii maretimi	0.093 ± 0.006
B. atticus	0.146 ± 0.007
B. atticus atticus	0.146 ± 0.007
B. atticus cyprius	0.146 ± 0.007
B. whitei	0.071 ± 0.005
B. lynceorum	0.129 ± 0.008

From Mantovani (1998).

between subspecies. It seems that absolute variability of sequences is shaped by concerted evolution in both sexual and asexual *Bacillus* but their fixation only occurs in sexually reproducing species (Dover 1982, 1986; Mantovani 1998; Luchetti et al. 2003). In other words, each *Bacillus grandii* subspecies has a well-defined homogeneous cluster, whereas *B. atticus* shows more differentiated Bag320 monomers (mean p-distance = 0.146$\pm$ 0.007 *vs* 0.073$\pm$0.005 of *B. grandii grandii*) but these do not form sub-clusters. In each single *Bacillus atticus* female, the Bag320 sequences show the same range of variability as in different females of the same or of different populations, or even of different subspecies (Luchetti et al. 2003). In addition to the *rossius* private monomers, which are clearly derived from the maternal ancestor, *grandii grandii*-like sequences have been recovered from *Bacillus whitei* and *B. lynceorum*. In the latter species, because of its specific triple hybrid structure, also *atticus*-like sequences are found (Mantovani 1998; Luchetti et al. 2003).

On the whole, specimen complexity for satellite DNA monomers is high in *Bacillus* parthenogenes, in which they are shaped by the reproductive mode and the hybrid/polyploid structure. Even if most repeated satellite DNA sequences are neutral, some satellite DNA is known to be involved in different molecular functions such as centromere structure and dynamics, karyotype evolution and sex/tissue-specific transcripts (Tautz 1993; Elder and Turner 1995; Renault et al. 1999; Henikoff et al. 2001; Schueler et al. 2001; Slamovits et al. 2001). No specific role has yet been assigned to the Bag320 family, but it seems wise to take satellite DNA monomer sequences and their copy number into account when evaluating possible specific cytological properties of *Bacillus* parthenogenes and their adaptive significance, regardless of their mode of origin and ploidy.

16.2.2 **Leptynia** *Pantel and* **Pijnackeria** *Scali*

The two species complexes of the Iberian *Leptynia* stick insects comprise the nominal species *L. attenuatal, L. montana* and *L. caprai* (Scali 1996b), and 1 additional, undescribed species, being genetically and phylogenetically distinct (Passamonti et al. 1999, 2004) (see Box 16.1and Fig. 16.5).

Box 9.1: Taxonomic confusion in Iberian stick insects

Pantel (1890) split the Iberian *Leptynia* stick insects into *L. attenuata* and *L. hispanica* according to a conspicuous pointed ending of the anal segment in the females that was missing in *L. attenuata*. Also the presence of a wholly membranous subanal vomer, quite distinct from that of *B. rossius*, was described for males collected with *L. hispanica*. Comparative analyses of recently collected samples of Iberian stick insects reveal that *Leptynia hispanica* is in fact an all-female taxon and the syntopic males belong to *L. attenuata*. Furthermore, a subanal vomer is not found in males of truly sexual *Leptynia hispanica*. Afterwards, within the two nominal species, the occurrence of several distinct species was ascertained, so that it seemed more appropriate to refer to the two species as species – complexes (Bullini and Nascetti 1987; Scali 1996b; Passamonti et al. 1999, 2004). Eventually, the presence/absence status of the subanal vomer, which has been even proposed as a "guide character" at higher systematic levels (Key 1970), together with karyotype structure and other taxonomic distinctive traits of bodies and eggs, strongly suggested the splitting of *Leptynia* into two different genera. The splitting has actually been done by maintaining the name *Leptynia* for the taxa with the subanal vomer in the males and giving the new name *Pijnackeria* to the former *hispanica* – complex; the parthenogenetic *P. hispanica* became the type species of the new genus (Scali 2009).

Those studies reveal that the *attenuata* complex only contains sexual species, whose chromosomal repatterning had a major role in speciation and even affected

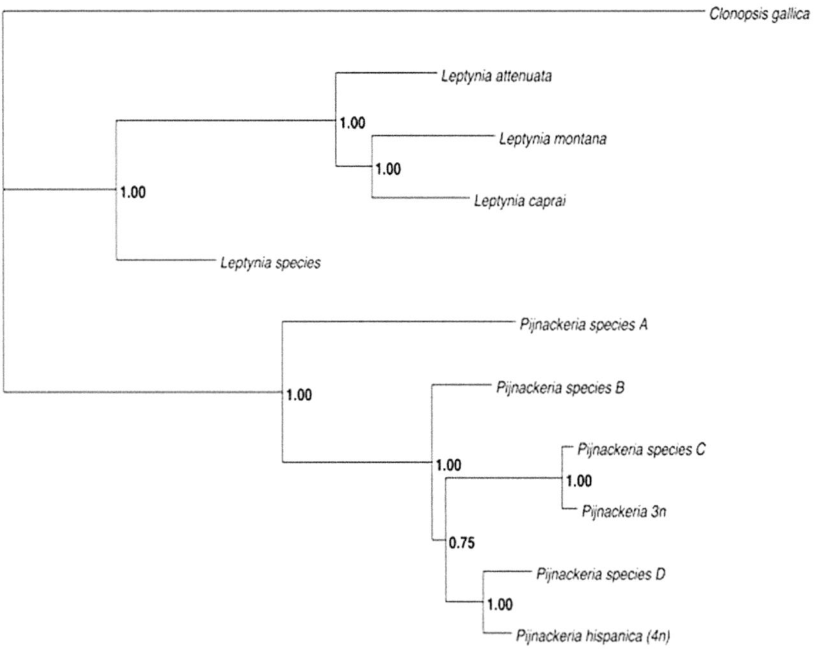

Fig. 16.5 Phylogenetic relationships within the genera *Leptynia* and *Pijnackeria*. Phylogenetic tree, obtained from *cox2* mitochondrial gene sequences (Mr Bayes 3.1; 10,000,000 generations), giving an updated systematic status of Iberian taxa. *Clonopsis gallica* has been used as outgroup. Bootstrap values (100 replicates) are given above branches (from Scali 2009)

the sex-chromosome determination, producing neo-Y species, and this way, generating a parapatric distribution of the species ranges. The whole picture perfectly matches what has been described from the Australian *Didymuria violescens* complex and other Australian phasmids (Craddock 1970, 1972, 1975; White 1976)

Parallel data for *Pinackeria* confirmed the existence of distinct sexual diploid taxa, all with 37/38 chromosomes owing to an XO-male/XX-female sex-chromosome determination. Also, the occurrence of two all-female taxa is confirmed: a triploid species (3n = 57, XXX) and a tetraploid one, the actual *P. hispanica* sensu stricto (4n = 76, XXXX; Fig. 16.5) (Nascetti et al. 1983; Bianchi 1992; Bianchi and Meliado 1998; Lelong 1992; Brock 1993). The cytology of the egg maturation is not known for those parthenogenes, mainly due to the very low number of eggs laid by each female and to the high fragility of their chorion, which makes egg handling rather difficult to study maturation divisions. Allozyme analyses could not be used either to establish the phylogenetic relationships of polyploids, since electrophoretic patterns are not as simple as in Nascetti et al. (1983) but rather too complex for unambiguous interpretation (Scali et al. unpublished). Therefore, DNA sequencing has been applied instead to disentangle the relationships within the *Pijnackeria* genus. Mitochondrial *cox2* sequence data from recently collected samples (2003–2006) distinguish 6 distinct groups, also reflecting geographical distribution (Ghiselli et al. 2007). Four groups of diploid taxa (A–D, not *sensu* Nascetti et al. 1983) are found, one group of triploids plus one group of tetraploids (Fig. 16.5). It is furthermore apparent that sequences derived from triploids are identical to the diploid sexuals from Alcocebre (*Pinackeria sp. C*), which are now considered to be the maternal ancestor of triploids (*Pijnackeria sp. 3n*). Also, the link of tetraploid haplotypes to diploid sexuals from Tiscar (Sierra de Cazorla) is obvious, although some sequence variability is observed. Therefore, *Pinackeria sp. D* can be considered as the maternal ancestor of tetraploids. The lower variability of the triploid haplotypes as compared to tetraploids may be due to the older age of tetraploids, which is estimated as 2.86 $\pm$ 1.01 Myr. The higher ploidy level might furthermore have allowed more mutational changes to accumulate (Ghiselli et al. 2007).

The distribution of *Pinackeria* parthenogens clearly follows a geographic pattern (Fig. 16.6). The situation of the maternal ancestors (*Pinackeria sp. C and Pinackeria sp. D*) of *P. sp. 3n* and of *P. hispanica*, respectively, is similar to the relic *Bacillus grandii* distribution when compared to the distribution ranges of *Bacillus whitei* and *B. lynceorum* in Sicily. The parental species is apparently displaced by hybrid descendants, with the paternal ancestor being very likely outcompeted and the maternal ancestor showing a constrained distribution. A variety of hypotheses has been proposed to account for geographical parthenogenesis (see also Chapter 8). Because of the wide distribution range of polyploid asexual *Leptynia* and *Bacillus*, it seems likely that hybridisation and uniparental reproduction may provide physiological robustness and a strong colonizing ability, allowing the parthenogens to displace their ancestors and to exploit remarkably different niches.

In order to reveal the putative paternal ancestor(s) of the two parthenogenetic taxa, the intra-individual variability of the nuclear *elongation factor-1α* gene has

Fig. 16.6 Distribution ranges of *Pijnackeria* species. The two diploid species close to the 3n and 4n parthenogens (first and fourth in the list to the right) are their maternal ancestors

been estimated for 5 diploid specimens of each sexual *Pijnackeria species*, 5 triploids and 2 tetraploids, respectively. The results are rather unexpected as there is more than one copy of the elongation *factor- 1α* gene per haploset in all taxa, namely 3, 8, 11 in diploids, triploids and tetraploids, respectively; this seems to represent the first instance of such a result for a hemimetabolus insect. Phylogenetic trees constructed with three different methods (Maximum Likelyhood, Minimum Evolution, Maximum Parsimony) gave the same topologies. Their main clusters were four sets of haplotypes related to diploid sexuals species (*Pijnackeria spp. A, B, C, D*, respectively); two sets of sequences obtained from *Pijnackeria sp. 3n* – one related to *Pijnackeria sp. C*, its maternal ancestor, and the other to a presumed, unknown fathering species, and one rather complex set derived from the *P. hispanica* (4n). In the latter, no sequence was related to the maternal *Pijnackeria sp. D* nor to any known taxon. While triploid *Pijnackeria sp 3n* can be explained as a species hybrid and likely originated via a diploid hybrid parthenogen backcrossing to the fathering taxon, the tetraploid *P. hispanica* parthenogens, although certainly hybrids, require a different explanation, since no nuclear *elongation factor-1α* alleles appear to be derived from the mitochondrially identified maternal ancestor (Ghiselli et al. 2007).

It is possible that a diploid androgen, escaping a former hybrid structure, has been an intermediate step to *L. hispanica* tetraploids. It could actually be envisaged that such an androgen hybridized again with a closely related species to produce an all-female parthenogen through unreduced, tetraploid eggs. Although other routes to

L. hispanica tetraploid origin are possible, the presented one fits best with the reproductive features of phasmids. A similar origin has been suggested for some hybrid parthenogens of *Daphnia pulex* (Dufresne and Hebert 1994), since genetic analysis has not provided any evidence for nuclear alleles to be derived from maternal ancestors as would have been expected from their mitochondrial constitution.

16.3 Centrosome Dynamics and γ-Tubulin(s) in Stick Insects

From a cytological point of view, a parthenogenetic egg has to accomplish two major tasks in order to start development: achieving/maintaining the right chromosome number and building the first embryonic spindle without sperm contribution. The great variety of mechanisms, by which the first goal can be reached in *Bacillus* has been already analysed; the second aspect, most often disregarded, requires further explanation.

Spindle microtubules (MTs) play a key role in cellular organization and, more specifically, in chromosome segregation during mitosis and meiosis. In animal cells, the main MT organizing centre (MTOC) is the centrosome; it generally consists of two orthogonally arranged centrioles encircled by a mix of several proteins, collectively known as pericentriolar material, from which MTs originate. In the most commonly recognized model of spindle cell-cycle, the centrosome duplicates during the interphase and when the cell is entering the prophase, each duplicated centrosome moves apart and emanates orderly arranged MTs (aster and hemispindle). The joining of the two hemi-spindles builds up the whole MT apparatus, which brings about the correct chromosome segregation (Compton 1998). For the structural and biochemical characterization of the centrosome, much attention has been given to the role of γ-tubulin, a ubiquitous, highly conserved protein of the tubulin superfamily, mainly found in the pericentriolar material of the MTOCs, where it starts MTs nucleation (see Marescalchi et al. 2002a and references therein). During gametogenesis of both sexes, the MTOC partially disassembles and acquires specific features of morphological and functional differentiation (Gonzales et al. 1998). In the male germ line, the centrosome possesses a standard composition, but at spermiogenesis, its pericentriolar material, including γ-tubulin, is stripped off, while the centrioles, still capable of duplication, are maintained. Conversely, at the onset of meiosis in the female germ line, the centrosome loses centrioles, but its pericentriolar material is kept, including γ-tubulin; however, it becomes dispersed throughout the egg cytoplasm (Fig. 16.7). These remnants of the centrosome are not capable of reorganization and duplication.

In insect spermatogenesis, a notable exception to the common pattern of centrosome duplication occurs: during the 1st meiotic division, the pericentriolar material encircles a centriole doublet at each pole. However, no centriole duplication takes place before the second division, so that during the 2nd division, dividing spermatocytes only have a single centriole at each spindle pole. Afterwards, only one centriole is found in the neck region of differentiating spermatozoa (Gonzales et al. 1998; Callaini et al. 1999; Krioutchkova and Onishchenko 1999).

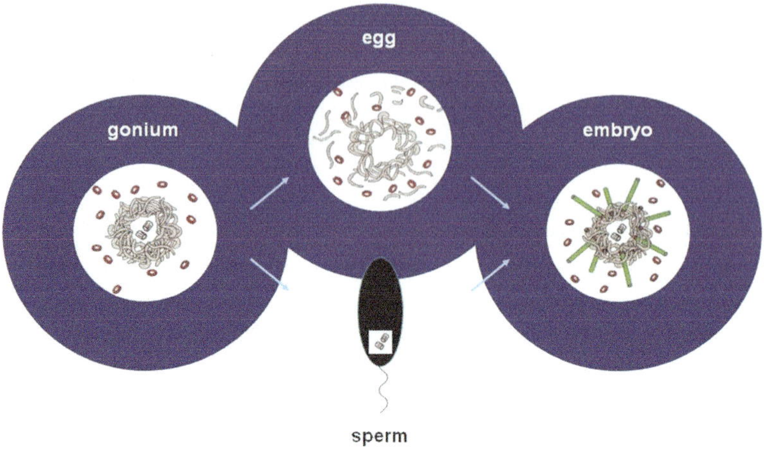

Fig. 16.7 Centrosome dynamics in male and female germ cells. The sperm contributes centioles and the egg the pericentriolar material to the reconstituted zygotic centrosome, which is then able to nucleate the aster first and the spindle microtubules later. Rings represent γ-tubulin complexes, twisted segments the other, numerous centrosome proteins and straight segments indicate some newly polymerized aster microtubules

In sexually reproducing animals, the centrosome is typically reassembled soon after sperm penetration into the egg by the recruitment of the pericentriolar material from the cytoplasm, including γ-tubulin and additional factors needed for future embryo polarization around the single, sperm-derived centriole (Schatten 1994; Riparbelli et al. 1997; Callaini et al. 1999; Tassin and Bornens 1999; Wu and Palazzo 1999; Cowan and Hyman 2006). In fertilized eggs, the maturation of the new centrosome coincides with the nucleation of the spermaster MTs and is followed by centriole/centrosome duplication, which apparently plays a major role in the reciprocal rejoining and fusion of male and female pronuclei. Therefore, in general terms, the sperm-associated centrosome has a key function in syngamy and onset of embryo development. Sperm contribution of additional paternal factors is also important in insect fertilization and zygote viability, but these aspects will not be considered here as they are beyond the task of this chapter.

The relative contribution to the zygote centrosome by each gamete is a variable process, the most clear and common example being parthenogenetic development (Simon et al. 2003), which has been particularly well analysed in insects (Suomalainen et al. 1987; Normark 2003).

As all other phasmids, *Bacillus* females possess panoistic ovarioles showing a clear distinction between germarium and vitellarium (Fig. 16.8). The germarium contains very small, flat follicular (somatic) cells, mingled with oocytes that are blocked in a resting pachy-diplotenic phase since early development of the first instar, whose oogonia stop dividing (Taddei et al. 1993). When resting oocytes resume growth and enter the vitellarium, follicular cells actively multiply and keep the centrosomic γ-tubulin throughout vitellogenesis. In growing oocytes, γ-tubulin

Fig. 16.8 Confocal picture of an ovariolar tip of *Bacillus rossius*. Note its panoistic organization with a gemarium crowded with round previtellogenetic oocytes, neatly separated from the vitellarium, built up by a string of growing oocytes, delimited by follicular cells with wedge-shaped nuclei. Flat external nuclei belong to cells of the ovariole delimiting membrane. Chromatin is labelled with green, α tubulin with red. (From Maurizii et al., unpublished). For terminology explanation, see glossary

disperses and the centrosome, together with centrioles, is lost as is the general rule in animals. Attempts at scoring γ-tubulin at spindle poles fail all throughout female meiosis. At the end of meiosis, the degenerating polocytes are engulfed in a thick tangle of MTs, whereas the pronucleus is free of them; however, it starts to nucleate MTs just before syngamy or parthenogenetic development (Marescalchi et al. 2002a).

In dividing spermatogonia and spermatocytes of *Bacillus*, MTs form slender and tapered anastral spindles (Fig. 16.9A), whose poles clearly reveal centrioles and spots of γ-tubulin in the centrosome area (Fig. 16.9B). In spermatids, however, γ-tubulin is gradually reduced and then lost altogether in mature spermatozoa. At the same time, the centriole is no longer detectable in the neck region of spermatozoa (Fig. 16.9C). Therefore, at syngamy, neither centriole nor MT-aster contributes to the zygotic spindle (Marescalchi et al. 2002a).

The *Bacillus* fertilization process is quite peculiar in having abolished the spermatic centriole contribution to the egg. On the other hand, the egg's ability to derive the first embryonic spindle wholly from MTs being nucleated around the egg chromatin is of significance, since it makes up for the lack of sperm contribution to the MTOC (Heald et al. 1996; Hyman and Karsenti 1996; Karsenti et al. 1996; Marescalchi et al. 2002a).

In both fertilized and parthenogenetic germ "anlagen", γ-tubulin foci can only be immunolabelled when the embryo has gone through the first third of its development (Fig. 16.10). When their different ploidy level and mode of origin are not considered, bisexual and parthenogenetic embryos of *Bacillus* do not show any remarkable differences at the onset of development: a functional first spindle can be formed either

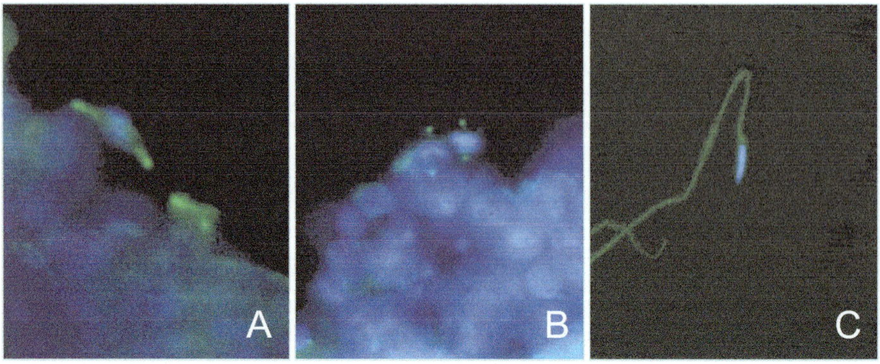

Fig. 16.9 Immunolabelling of male germ cells of *Bacillus rossius*. **A**: tapered anastral spindle of a dividing spermatocyte marked with DAPI (*blue*) and immunolabelled against β tubulin (*green*). **B**: dividing spermatocytes marked with DAPI (*blue*) and immunolabelled against γ tubulin (*green*), demonstrating the presence of a normal centrosome. **C**: spermatozoon marked with DAPI and immunolabelled against both β and γ tubulins: while DAPI reacts with the sperm head and β tubulin antibody marks the tail axoneme, no γ tubulin is found in the neck region (Modified from Marescalchi et al. 2002a). For terminology, see glossary

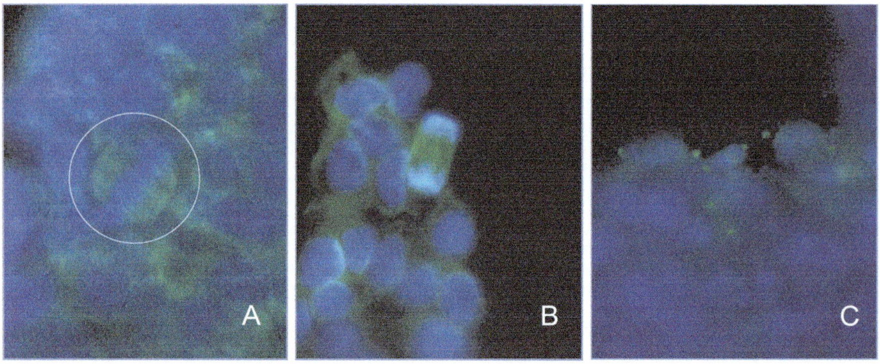

Fig. 16.10 Spindles of embryonic cells of two *Bacillus* species at increasing times of development. **A**: spherical anastral spindle of a very early bisexual embryo of *B. rossius*. (DAPI + anti-β-tubulin-immunolabelling). **B**: anastral barrel shaped spindle of a parthenogenetic embryo of *B. atticus* (DAPI + anti-β-tubulin immunolabelling). **C**: γ tubulin spots at the spindle poles of dividing cells in a *B. rossius* embryo at an intermediate stage of development (DAPI +. anti-γ-tubulin immunolabelling). At the beginning and during the early stages of embryo development, no standard centrosomes appear to be formed, whereas typically organized centrosomes are found later on. Asters are never seen (modified from Marescalchi et al. 2002a)

by the fusion of male and female pronuclear MTs or just by MTs nucleated from the female pronucleus alone. The flexibility in the building of the first spindle appears to be a major pre-adaptation of phasmid species to parthenogenesis.

The lack of sperm contribution to the MTOC during fertilization is not the only unusual feature of the phasmid mitotic apparatus since, as already mentioned, all mitotic divisions fail to form astral MTs (see Figs. 16.9A, 16.10A, B). This uncommon cellular trait suggests that γ-tubulin could be involved since asters are the most direct descents of γ-tubulin MT-nucleating activity (Schatten 1994; Schatten et al. 1986; Riparbelli et al. 1997; Callaini et al. 1999; Tassin and Bornens 1999; Cowan and Hyman 2006).

Investigations on the γ-tubulin expression and molecular characteristics in *Bacillus rossius* with immunoblots reveal that in germ line and somatic tissues (eggs, testes, midgut, muscles), the monoclonal anti-γ-tubulin antibody recognises one protein with a molecular weight (MW) of about 50 kDa without cross reactivity to other proteins. This is lighter than in *Drosophila* and other animals (Megraw and Kaufman 2000). Furthermore, the γ-tubulin of different *Bacillus* taxa has different MWs, even at the intrageneric level: *B. rossius* and *B. grandii* show γ-tubulins of different MWs, which is why two separate bands are expressed in their hybrids (Marescalchi et al. 2002b). Western blots using the same tissues in distantly related species of different subfamilies, such as Tropidoderinae, Lonchodinae, Phasmatinae and Bacillinae, have been used to investigate expression of different tubulin isoforms. All tested species appear to possess a lighter γ-tubulin than that of *D. melanogaster;* among the subfamilies, several different mobility patterns are recognized (Marescalchi et al. 2002b). This finding of γ-tubulin variability between species of the same genus and within the same class is quite surprising, especially in view of the fact that γ-tubulin is a very conserved protein, and similar from yeast to man (Joshi et al. 1992). Sequencing of the phasmid γ-tubulin(s) DNA is required to gain deeper knowledge of the partenogenetic development of stick insects.

16.4 Conclusions – Are Stick Insects True Asexuals?

Investigations of asexual animals, particularly parthenogens, are appealing because of the results they provide on sexual, reproductive, ecological and evolutionary issues. Parthenogenesis can reduce genetic variability but does not wholly suppress it. If all organisms would behave as bdelloid rotifers, parthenogenetic reproduction would be the same as loosing sex (*sensu* Bell 1982). But the complex interactions between phasmids demonstrate another point of view as these open unusual and additional pathways to sexual and asexual reproduction.

Every approach to studying parthenogenesis is valid but to avoid flaws in interpreting results, the widest array of methods should be used for case studies of parthenogenesis. Genetic analyses and cytological studies are difficult and time-consuming, but egg maturation processes have turned out to be central to our understanding of asexual stick insects. In *Bacillus*, such joint investigations reveal insights into the genetic features of asexual taxa *per se* and as compared to their bisexual relatives. As illustrated above, it has been possible to demonstrate that in these phasmids, parthenogenesis of different kinds occurs: in *B. rossius*, automictic thycoparthenogenesis arose many times at different geographic locations, while the recent, automictic *B. atticus* can rather be regarded as diploid or triploid interracials than interspecific hybrids (Bullini 1994; Marescalchi and Scali 1997, 2003). In parthenogenetic eggs of phasmids, a meiotic program is at work. This is either complete, as in automicts, or interrupted, as in apomicts, and allows an additional synthesis of DNA, leading to an unreduced number of chromosomes with four chromatids. Although chromosome pairing is very short and the somatic number soon restored, the egg maturation still keeps the meiotic program with two-divisions and the formation of synaptonemal complexes (Pijnacker 1966; Koch et al. 1972; Pijnacker and Ferwerda 1982, 1986); all these features are clear remnants of sexual reproduction.

The inclusion of conspecific or heterospecific paternal genomes into diploid hybrid asexuals appears to be a common pattern in the route to polyploids, as it has for example also been observed in the fresh water planarian complex *Schmidtea polychroa* (Pongratz et al. 2003; see also Chapter 18); the curculionid weevils (Stenberg et al. 2003) and *Calligrapha* beetles (Gomez-Zurita et al. 2006); the fish complexes *Cobitis* (Vasil'ev et al. 1989), *Poecilia, Poeciliopsis* and *Squalius alburnoides* (see also Chapter 19; Vrijenhoek 1978, 1998; Vrijenhoek and Lerman 1982; Bulger and Schultz 1979, 1982; Turner et al. 1980, 1983, 1990; Pala and Coelho 2005), *Rutilus* (= *Tropidophoxinellus*) *alburnoides* (Collares-Pereira 1989; Alves et al. 1999), *Phoxinus eos-neogaeus* (Dawley and Bogart 1989; Goddard et al. 1989); the urodele *Ambystoma* complex (Bogart et al. 1987; Kraus 1989), the African clawed frogs *Xenopus* (Evans et al. 2004) and the teiid lizards (Dessauer and Cole 1989; Moritz et al.1989; Parker et al. 1989; Peccinini-Seale 1989; see also Chapter 21).

On the other hand, apomictic parthenogenesis with some recombination has been observed in the species hybrids *Bacillus withei* and *B. lynceorum*. The different traits of parthenogenetic mechanisms in *Bacillus* taxa are reflected in their patterns of inheritance for single copy and repetitive DNA as in the chronology of their cladogenesis (Mantovani et al. 1992; Scali et al. 1995; Mantovani 1998; Luchetti et al. 2003). The low but existing rate of recombination, together with hybrid and polyploid structures, helps to understand the relatively long persistence of *Bacillus* parthenogens and suggests that the same reasoning might apply to other ancient, differentiated genera of parthenogenetic stick insects such as the Australian *Sipyloidea* (John et al. 1987), *Acanthoxyla* from New Zealand (Morgan-Richard and Trewick 2005) and the related North American *Timema* (Sandoval et al. 1998).

Owing to methodological constraints, genetic investigations on phasmids detect only part of the diversity in coding genes, leaving variability and polymorphisms of structural proteins completely unexplored. Other sources of genetic diversification in diploid or polyploid hybrid parthenogenes such as rDNA cistrons (Manaresi et al. 1993; Marescalchi and Scali 1997) or non-coding DNA (e.g., satDNAs or microsatellites (Andersen et al. 2005)), should not be disregarded either when considering possible adaptive responses at the organismal or cellular level.

In *Bacillus* stick insects, comparative allozymic, chromosomal and gametogenetic investigations have made it possible to reveal the occurrence of hybridogenetic and androgenetic reproductive modes, which otherwise would have gone unnoticed. Hybridogenetic strains of phasmids represent up to now the only instance of hemiclonal land animals. The other known examples belong to fish and amphibian complexes (reviewed in Dawley and Bogart 1989; Carmona et al. 1997). *Bacillus* hybridogens compensate the invariant genetic structure of the *rossius* haploset (hemiclonality) by acquiring a genetically different *grandii* component in each generation, ensuring a high level of unfixed heterozygosity, which is much higher (for single-copy and repetitive nuclear genes) than that of *B. rossius* females. The most interesting genetic novelty is produced when a switch from hybridogenesis to androgenesis occurs: hybrid offspring then display the nuclear constitutions of a pure species. Their sexual reproduction does not only represent a reversion from hybridogenesis, but generates at the same time novel genetic combinations with enhanced evolutionary perspectives, since heterologous mitochondrial and nuclear genomes will suddenly have to cooperate in the same organism. Although more rarely, the same escape from hybridogenesis has been demonstrated in the obligate parthenogen *B. whitei*: the finding further reinforces the network of reproductive interactions between asexuals and sexual relatives, hereby bridging their supposedly sharp and definite distinction and making it difficult to regard stick insects as true asexuals. Phasmids are not unique in having genetic connections between sexual and asexual taxa. Similar tangled interactions have also been observed in salamander hybrids (Bogart et al. 1987; Bogart 1989) and in a complex of *Hyla versicolor* (Holloway et al. 2006).

Another reason for the difficulty of assigning the asymmetrically reproducing systems of stick insects to asexuals is the presence of sex differentiating chromosomes, which continue expressing themselves in a taxon-specific pattern of dosage compensation as they did in the sexual ancestors (Orr 1995, 1999, 2000; Larsson and Meller 2006). This is quite clearly observed in polyploid parthenogenetic phasmids, where all deviations from the 1:1 X/A ratio produce intersexes (Pijnacker and Ferwerda 1980, 1982; Tinti and Scali 1995). The dosage compensation of sex-linked genes finds a corresponding situation in parent-specific gene expression because genome imprinting appears to operate the origin and maintainance of many thelytokous systems (Normark 2006).

In stick insects, it should be emphasized that the ability of MTs nucleation around the egg chromatin is, together with the regulation of chromosome numbers, the most outstanding trait of virgin egg development. In phasmids, this trait could somehow be related to the unusual γ-tubulin molecule and the lack of aster formation throughout the whole life cycle (Marescalchi et al. 2002a, b). Because other parthenogenetic insects with asters build up the first embryo spindle rather differently – *Muscidifurax oviraptor, Nasonia vitripennis* and *Drosophila mercatorum*, the best analyzed insect parthenogens, all make use of two asters, among the many autonomously formed from the egg cytoplasm (Riparbelli et al. 1998; Callaini et al. 1999; Tram and Sullivan 2000) – phasmids are apparently not able to form asters. The alternative routes to the building of the first spindle could be related to different ovariole types, being panoistic in the stick insects and meroistic in wasps and flies. Despite the development without asters, *B. rossius* embryos are able to synthetize *de novo* centrioles later on, thus reinforcing the observation that centrioles are not strictly required for the correct spindle formation and function (Callaini et al. 1999).

In this context, it is important to recall that in general terms, the alternative occurrence of either parthenogenesis or gynogenesis within animal groups – such as in rotifers, crustaceans, several insect orders and reptiles on one hand and in turbellarians, nematodes, annelids, fish and amphibia on the other – might have to do with the ability of the egg to quickly build spindles at the beginning of embryonic development (with or without centrioles to start with) in the first group and the inability of the second group to do so without sperm. This theory needs to be further investigated and could hereby open exciting new avenues of research.

Acknowledgements I wish to thank Dr Federico Plazzi for his kind help in editing the figures. I also want to acknowledge the extra help given me by the editor Isa Schön in shaping the text and in allowing me to publish more figures than most other chapters.

Glossary

Androgenesis: In this mode of reproduction, the eggs, often from a hybrid female, are exploited for their cytoplasmic components, while their nucleus is ignored, by unreduced sperm or by fusing haploid spermatozoa of conspecific or heterospecific males. A progeny with complete paternal derivation is obtained and it may consist of either only males or of both sexes.
Aster: The mitotic spindle originating from centers (centrosomes) and being made of radiating microtubules (asters = stars) of α and β tubulins. In animals, all cells but the eggs have "astral spindles". The aster organized by the sperm centriole after entering the egg is called spermaster.
Bag320 family: A series of tandemly repeated similar sequences (repetitive DNA; satellite DNA) discovered in the stick insect species *Bacillus atticus* and *B. grandii*; they are mainly localized near the centromere of several chromosome pairs.
Germarium and **vitellarium**: Each ovariole shows a fore part containing young, small oocytes (germarium) and a hind section, where eggs grow by uptaking large amounts of nutritious stuff (vitellum) to be utilized by the future embryo; this ovariole section is called vitellarium.
Hybridogenesis: It is a mode of reproduction realized by a hybrid taxon whose females discard one complete genome of either parental species, commonly the paternal one. Their eggs are fertilized by males of the species of the discarded genome to produce anew, each generation, the hybrid condition in the progeny. Hybridogenetic females realize a hemiclonal transmission of genes since they pass their invariant set on to the offspring, while the genetic contribution of the fathering species is variable.
Microtubule (MT) nucleation: The process by which new tubulin molecules are added to the growing end of the MT is called MT nucleation. The γ tubulin present in the centrosome is the major center of MT nucleation.
NOR: The acronym for Nucleolar Organizing Region(s), corresponding to the locations of the tandemly repeated ribosomal genes on specific chromosomes.

Pachytene: The phase of the first meiotic division when the homologous chromosomes are paired and can recombine their four chromatids (tetrachromatidic pairs, two from each homolog). In hybrids with partially homologous chromosomes, chromosome pairing may be incomplete and recombination reduced or completely lacking. In phasmid hybrids, the partially paired pachytenic chromosomes separate again and allow additional synthesis of DNA, leading to single chromosomes with four chromatids. Afterwards, a mechanically normal sequence of two divisions, mimicking meiosis, takes place although the obtained egg is unreduced.
Panoistic ovarioles: Ovaries of most invertebrates are made of subunits called ovarioles. In insects, there are two main types of ovarioles: those in which all germ cells become eggs (panoistic) and those in which part of the germ cells help in egg ripening and become nurse cells (meroistic). Panoistic ovarioles are generally found in basal insects, such as Phasmida, while meroistic ones occur in more evolved groups such as Diptera.
Pycnotic polar body: A polar body is one of the 3 degenerating cells deriving from a meiotic process in the female germ cell line. When actually degenerating, a polar body becomes deeply stained (pycnosis).
Spermiogenesis: The term indicates the post-meiotic part of the process (spermatogenesis) leading to sperm formation. During spermiogenesis, the round spermatids deriving from the meiotic divisions of germ cells (spermatocytes) acquire the specific structure and morphology of mature male gametes (spermatozoa).
Thycoparthenogenesis: If an egg of a sexual species accidentally self-activates to produce a spontaneous parthenogenetic offspring, this kind of reproduction is known as thycoparthenogenesis and it is opposed to the parthenogenesis triggered by hybridization.
Zymotype: Indicates an intraspecific taxon on the basis of its allozymatic characterization. Specific enzymatic proteins under direct genetic control can be analysed through the electrophoretic technique owing to their different mobility; such enzymes controlled by different alleles of the same gene are called allozymes.

References

Alves MJ, Coelho MM, Prospero MI, Collares Pereira MJ (1999) Production of fertile unreduced sperm by hybrid males of *Rutilus alburnoides* complex (Teleostei, Cyprinidae): an alternative route to genome tetraploidization in unisexuals. Genetics 151: 177–183

Andersen DH, Pertoldi C, Loeschcke V, Scali V (2005) Primer note: characterization of microsatellite loci in the stick insects *Bacillus rossius rossius*, *Bacillus rossius redtenbacheri* and *Bacillus whitei* (Insecta: Phasmida). Mol Ecol Notes 5: 576–578

Baxevanis AD, Kappas I, Abatzopoulos TJ (2006) Molecular phylogenetics and asexuality in the brine shrimp *Artemia*. Mol Phyl Evol 40: 724–738

Bell G (1982) The masterpiece of nature. The evolution of genetics of sexuality. Croom Helm Ltd.

Bergerard J (1958) Etude de la parthénogenèse facultative de *Clitumnus extradentatus* Br. (Phasmidae). Bull Biol France Belg 92: 87–182

Bergerard J (1962) Parthenogenesis in the Phasmidae. Edeavour 2: 137–143

Bianchi AP (1992) Karyological studies of Mediterranean stick-insects belonging to the genera *Clonopsis* and *Leptynia* (Insecta Phasmatodea). Caryologia 45: 1–19

Bianchi AP, Meliado P (1998) Analysis of the karyotypes of four species of the *Leptynia attenuata* complex (Insecta Phasmatodea). Caryologia 51: 207–219
Bogart JP (1989) A mechanism for interspecific gene exchange via all-female salamander hybrids. In: Dawley RM, Bogart JP (eds) Evolution and ecology of unisexual vertebrates, Bulletin 466. New York State Museum, Albany, New York, pp.170–179
Bogart JP, Lowcock LA, Zeyl CW, Mable BK (1987) Genome constitution and reproductive biology of the *Ambystoma* hybrid salamanders on Kelleys Island in Lake Erie. Can J Zool 65: 2188–2201
Brock PD (1993) Studies on stick-insects of the genus *Leptynia* in Spain. Amateur Entomol Soc Bull 52: 165–172
Bulger AJ, Schultz RJ (1979) Heterosis and interclonal variation in thermal tolerance in unisexual fishes. Evolution 33: 848–859
Bulger AJ, Schultz RJ (1982) Origin of thermal adaptations in northern versus southern populations of a unisexual hybrid fish. Evolution 36: 1041–1050
Bullini L (1966) Spanandria e partenogenesi geografica in *Bacillus rossius* (Rossi). Atti Acc Naz Lincei, Rend Sci Fis Mat Nat 45: 926–932
Bullini L (1994) Origin and evolution of animal hybrid species. Trends Ecol Evol 9: 422–426
Bullini L, Nascetti G (1987) Speciation by hybridization in Othopteroid insects. In: Baccetti B (ed) Evolutionary biology of orthopteroid insects. Ellis Horwood, Chichester, pp. 134–147
Callaini G, Riparbelli MG, Dallai R (1999) Centrosome inheritance in insects: fertilization and parthenogenesis. Biol Cell 91: 355–366
Carmona JA, Sanjur OI, Doadrio I, Machordom A, Vrijenhoeck RC (1997) Hybridogenetic reproduction and maternal ancestry of polyploid Iberian fish: the *Tropidophoxinellus alburnoides* complex. Genetics 146: 983–993
Cesari M, Luchetti A, Passamonti M, Scali V, Mantovani B (2003) Polymerase chain reaction amplification of the *Bag*320 satellite family reveals the ancestral library and past gene conversion events in *Bacillus rossius* (Insecta, Phasmatodea). Gene 312: 298–295
Collares-Pereira MJ (1989) Hybridization in European cyprinids: evolutionary potential of unisexual populations. In: Dawley RM, Bogart JP (eds) Evolution and ecology of unisexual vertebrates, Bulletin 466. New York State Museum, Albany, New York, pp. 281–288
Compton DA (1998) Focusing on spindle poles. J Cell Sci 111: 1477–1481
Cowan CR, Hyman AA (2006) Cyclin E-Cdk2 temporally regulates centrosome assembly and establishment of polarity in *Caenorhabditis elegans* embryos. Nature Cell Biol 12: 1441–1447
Craddock EM (1970) Chromosome number variation in a stick insect *Didymuria violescens* (Leach). Science 167: 1380–1882
Craddock EM (1972) Chromosomal diversity in the Australian Phasmatodea. Austral J Zool 20: 445–462
Craddock EM (1975) Intraspecific karyotype differentiation in the Australian phasmatid *Didymuria violescens* (Leach). Chromosoma 53: 1–24
Dawley RM, Bogart JP (1989) Evolution and ecology of unisexual vertebrates. New York State Mus Bull 466. Albany, New York
Densmore LD, Moritz CC, Wright JW, Brown WM (1989) Mitochondrial DNA analyses and the origin and relative age of parthenogenetic lizards (genus *Cnemidophorus*). Nine *sexlineatus* – group parthenogens. Evolution 43: 969–983
Dessauer HC, Cole CJ (1989) Diversity between and within nominal forms of unisexual teiid lizards. In: Dawley RM, Bogart JP (eds) Evolution and ecology of unisexual vertebrates, Bulletin 466. New York State Museum, Albany, New York, pp. 49–71
Dover GA (1982) Molecular drive: a cohesive mode of species evolution. Nature 299: 111–117
Dover GA (1986) Molecular drive in multigene families: how biological novelties arise, spread and are assimilated. Trends Genet 2: 159–165
Dufresne F, Hebert PDN (1994) Hybridization and origins of polyploidy. Proc R Soc Lond B 258: 141–146
Elder JF, Turner BJ (1995) Concerted evolution of repetitive DNA sequences in eukaryotes. Quart Rev Biol 70: 297–320

Evans BJ, Kelley DB, Tinsley RC, Melnick DJ, Cannatella DC (2004) A mitochondrial DNA phylogeny of African clawed frogs: phylogeography and implications for polyploid evolution. Mol Phyl Evol 33: 197–213

Friedman R, Hughes AL (2007) Pattern and timing of gene duplication in animal genomes. Genome Res 11: 1842–1847

Gasperi G, Malacrida A, Scali V (1983) Variabilità enzimatica in popolazioni italiane e nord-africane di *Bacillus rossius* (Insecta, Phasmatodea, Bacillidae). Atti XIII Congresso Nazionale Italiano Entomologia. Sestriere, Torino, Italy: 647–649

Gibson TJ, Spring J (2000) Evidence in favour of ancient octaploidy in the vertebrate genome. Biochem Soc Trans 28: 259–264

Ghiselli F, Milani L, Scali V, Passamonti M (2007) The *Leptynia hispanica* species complex (Insecta Phasmida): polyploidy, parthenogenesis, hybridization and more. Mol Ecol 16: 4256–4268

Goddard KA, Dawley RM, Dowling TE (1989) Origin and genetic relationships of diploid, triploid and diploid-triploid mosaic biotypes in the *Phoxinus eos-neogaeus* unisexual complex. In: Dawley RM, Bogart JP (eds) Evolution and ecology of unisexual vertebrates, Bulletin 466. New York State Museum, Albany, New York, pp. 268–280

Gonzales C, Tavosanis G, Mollinari G (1998) Centrosome and microtubule organization during *Drosophila* development. J Cell Sci 111: 2697–2706

Good DA, Wright JW (1984) Allozymes and hybrid origin of the parthenogenetic lizard *Cnemidophorus exanguis*. Experientia 40: 1012–1014

Gomez-Zurita J, Funk DJ, Vogler A (2006) The evolution of unisexuality in *Calligrapha* leaf beetles: molecular and ecological insights on multiple origins via interspecific hybridization. Evolution 60: 328–347

Haddal CFB, Pombal JP, Batistic RF (1994) Natural hybridization between diploid and tetraploid species of leaf frog, genus *Phyllomedusa* (Amphibia). J Herpetol 28: 425–430

Hadlington P (1961) Diapause and parthenogenesis in the eggs of three species of Phasmatodea. Proc Linn Soc New South Wales 86: 268–279

Halkett F, Plantegenest M, Prunier-Leterme N, Mieuzet L, Delmotte F, Simon JC (2005) Admixed sexual and facultatively asexual aphid lineages at mating sides. Mol Ecol 14: 325–336

Heald R, Tournebize R, Blank T, Sandaltzopoulos R, Becker P, Hyman A, Karsenti E (1996) Self organization of microtubules into bipolar spindles around artificial cromosomes in *Xenopus* egg extracts. Nature 382: 420–425

Hedges SB, Bogart JP, Maxson LR (1992) Ancestry of unisexual salamanders. Nature 356: 708–710

Henikoff S, Ahmad K, Malik HS (2001) The centromere paradox: stable inheritance with rapidly evolving DNA. Science 293: 1098–1202

Hyman AA, Karsenti E (1996) Morphogenetic properties of microtubules and mitotic spindle assembly. Cell 84: 401–410

Holloway AK, Cannatella DC, Gerhardt HC, Hillis MD (2006) Polyploids with different origin and ancestors form a single sexual polyploid species. Am Nat 167: 88–101

Hughes RN (1989) A functional biology of clonal animals. London, Chapman and Hall

John B, Rentz DCF, Contreras N (1987) Extensive chromosome variation in the stick insect genus *Sipyloidea* Brunner von Watenwyl (Phylliidae: Necrosciinae) within Australia, and description of three new species. Invertebr Taxonomy 1: 603–630

Johnson SG, Hopkins R, Goddard K (1999) Constraints on elevated ploidy in hybrid and nonhybrid parthenogenetic snails. J Hered 90: 659–662

Joshi H, Palacios MJ, McNamara L, Cleveland DW (1992) Gamma tubulin is a centrosomal protein required for cell cycle-dependent microtubule nucleation. Nature 356: 80–83

Karsenti E, Boleti H, Vernos I (1996) The role of microtubule dependent motors in centrosome movements and spindle pole organization during mitosis. Stem Cell Dev Biol 7: 367–378

Key KHL (1970) Phasmatodea (Stick-insects). In: Insects of Australia. CSIRO Ed Melbourne, Melbourne University Press, Melbourne, pp. 348–359

Koch P, Pijnacker LP, Kreke J (1972) DNA reduplication during meiotic prophase in the oocytes of *Carausius morosus* Br. (Insecta, Cheleutoptera). Chromosoma 36: 313–321

Kraus F (1989) Constraints and evolutionary history of unisexual salamanders of the *Ambystoma texanum-laterale* complex as revealed by mitochondrial DNA analysis. In: Dawley RM, Bogart JP (eds) Evolution and ecology of unisexual vertebrates, Bulletin 466. New York State Museum, Albany, New York, pp. 218–217

Krioutchkova MM, Onishchenko GE (1999) Structural and functional characteristics of the centrosome in gametogenesis and early embryogenesis of animals. Int Rev Cytol 185: 107–186

Larsson J, Meller VH (2006) Dosage compensation, the origin and the afterlife of sex chromosomes. Chromosome Res 14: 417–431

Law JH, Crespi BJ (2002) The evolution of geographic parthenogenesis in *Timema* walking-sticks. Mol Ecol 11: 1471–1489

Lelong P (1992) Le genre *Leptynia* dans la peninsule Iberique. Le Monde des Phasmes 17: 10–15

Lokki J (1976a) Genetic polymorphism and evolution in parthenogenetic animals VII. The heterozygosity in diploid populations. Hereditas 83: 57–74

Lokki J (1976b) Genetic polymorphism and evolution in parthenogenetic animals VIII. Heterozygosity in relation to polyploidy. Hereditas 83: 15–56

Luchetti A, Cesari M, Carrara G, Cavicchi S, Passamonti M, Scali V (2003) Unisexuality and molecular drive: Bag320 sequence diversity in *Bacillus* taxa (Insecta, Phasmatodea). J Mol Evol 56: 587–596

Maere S, De Bodt S, Raes J, Casneuf T, Van Montagu M, Kuiper M, Van de Peer Y (2005) Modelling gene and genome duplications in eukaryotes. Proc Natl Acad Sci USA 102: 5454–5459

Manaresi S, Marescalchi O, Scali V (1993) The trihybrid genome constitution of *Bacillus lynceorum* (Insecta, Phasmatodea) and its karyotypic variations. Genome 36: 317–326

Mantovani B (1998) Satellite sequence turnover in parthenogenetic systems: the apomictic triploid hybrid *Bacillus lynceorum* (Insecta, Phasmatodea). Mol Biol Evol 15: 1288–1297

Mantovani B, Passamonti M, Scali V (1999) Genomic evolution in parental and hybrid taxa of the genus *Bacillus* (Insecta, Phasmatodea). Ital J Zool 66: 265–272

Mantovani B, Passamonti M, Scali V (2001) The mitochondrial cytochrome oxidase II gene in *Bacillus* stick insects: ancestry of hybrids, androgenesis and phylogenetic relationships. Mol Phyl Evol 19: 157–163

Mantovani B, Scali V (1991) Allozymic characterization of Sardinian *Bacillus rossius* (Rossi) and *B. atticus* Brunner (Insecta, Phasmatodea). Genetica 83: 275–287

Mantovani B, Scali V (1992) Hybridogenesis and androgenesis in the stick insect *Bacillus rossius-grandii benazzii*. Evolution 46: 783–796

Mantovani B, Scali V, Tinti F (1992) New morphological and allozymic characterization of *Bacillus whitei* and *B. lynceorum* hybrid complexes (Insecta, Phasmatodea). Biol Zen Bl 111: 87–103

Marescalchi O, Galassi A, Scali V (2002b) Peculiar immunocytochemical features of γ tubulin in gametes and embryos of stick insects. Anim Biol 11: 13

Marescalchi O, Pijnacker L.P., Scali V (1991) Cytology of parthenogenesis in *Bacillus whitei* and *Bacillus lynceorum* (Insecta, Phasmatodea). Invertebr Reprod Dev 20: 75–81

Marescalchi O, Scali V (1997) Chromosomal and NOR patterns in the polyclonal stick insect *Bacillus atticus atticus* (Insecta, Phasmatodea). Genome 40: 261–270

Marescalchi O, Scali V (2003) Automictic parthenogenesis in the diploid-triploid stick insect *Bacillus atticus* and its flexibility leading to heterospecific diploid hybrids. Invertebr Reprod Dev 43: 163–173

Marescalchi O, Zauli C, Scali V (2002a) Centrosome dynamics and inheritance in related sexual and parthenogenetic *Bacillus* (Insecta, Phasmatodea). Mol Reprod Dev 63: 89–95

McKinnel RG. 1985. Cloning of frogs, mice, and other animals cloning: A biologist reports. Revised edition. University of Minnesota Press, Minneapolis

McKone MJ, Halpern SL (2003) The evolution of androgenesis. Am Nat 161: 641–656

Megraw TL, Kaufman TC (2000) The centrosome in *Drosophila* oocyte development. Curr Top Dev Biol 49: 385–407

Mittwoch U (2002) "Clone": the history of a euphonious scientific term. Med Hist 46: 381–402

Morgan-Richards M, Trewick SA (2005) Hybrid origin of a parthenogenetic genus? Mol Ecol 14: 2133–2142

Moritz C, Brown WM, Densmore LD, Wright JW, Vyas D, Donnellan S, Adams M, Baverstock P (1989) Genetic diversity and the dynamics of hybrid parthenogenesis in *Cnemidophorus* (Teiidae) and *Heteronotia* (Gekkonidae). In: Dawley RM, Bogart JP (eds) Evolution and ecology of unisexual vertebrates, Bulletin 466. New York State Museum, Albany, New York, pp. 87–112

Nascetti G, Bianchi Bullini AP, Bullini L (1983) Speciazione per ibridazione nei fasmidi del bacino mediterraneo (Cheleutoptera: Bacillidae). Atti XIII Congresso Nazionale Italiano Entomologia, Sestriere, Torino, pp. 475–478

Normark BB (2003) The evolution of alternative genetic systems in insects. Annu Rev Entomol 48: 397–423

Normark BB (2006) Perspective: maternal kin groups and the origin of asymmetric genetic systems-genomic imprinting, haplodiploidy, and parthenogenesis. Evolution 60: 631–642

Orr HA (1990) "Why polyploidy is rarer in animals than in plants" revisited. Am Nat 136: 759–770

Orr HA (1995) The population genetics of speciation: the evolution of hybrid incompatibilities. Genetics 139: 1805–1813

Orr HA (1999) Does hybrid lethality depend on sex or genotype? Genetics 152: 1767–1769

Orr HA (2000) The rate of adaptation in asexuals. Genetics 155: 961–968

Otte D, Brock PD (2005) Phasmida species file. Catalogue of the stick and leaf insects of the world. 2nd ed. The insect diversity association at the Acad Nat Sci, Philadelphia, 1–414

Otto S, Whitton J (2000) Polyploid incidence and evolution. Annu Rev Genet 34: 401–437

Pala J, Coelho MM (2005) Contrasting views over a hybrid complex: between speciation and evolutionary "dead end". Gene 347: 283–294

Pantel PJ (1890). Notes Orthoptérologiques. II Les Phasmides d'Europe et des pays limitrophes. An Soc Esp Hist Nat 19: 371–404

Parker ED, Walker JM, Paulissen MA (1989) Clonal diversity in *Cnemidophorus*: ecological and morphological consequences. In: Dawley RM, Bogart JP (eds) Evolution and ecology of unisexual vertebrates, Bulletin 466. New York State Museum, Albany, New York, pp. 72–86

Passamonti M, Mantovani B, Scali V (1999) Karyotype and allozyme characterization of the Iberian *Leptynia attenuata* species complex (Insecta Phasmatodea). Zool Sci 16: 675–684

Passamonti M, Mantovani B, Scali V (2004) Phylogeny and karyotype evolution of the Iberian *Leptynia attenuata* species complex (Insecta Phasmatodea). Mol Phyl Evol 30: 87–96

Peccinini-Seale DM (1989) Genetic studies on bisexual and unisexual populations of Amazonian *Cnemidophorus*. In: Dawley RM, Bogart JP (eds) Evolution and ecology of unisexual vertebrates, Bulletin 466. New York State Museum, Albany, New York, pp. 241–251

Pijnacker LP (1966) The maturation divisions of the parthenogenetic stick insect *Carausius morosus* Br. (Orthoptera, Phasmidae). Chromosoma 19: 99–112

Pijnacker LP, Ferwerda MA (1978) Additional chromosome duplication in female meiotic prophase of *Sipyloidea sipylus* Westwood (Insecta, Phasmida) and its absence in male meiosis. Experientia 34: 1558–1559

Pijnacker LP, Ferwerda MA (1980) Sex chromosomes and origin of males and sex mosaics of the parthenogenetic stick insect *Carausius morosus* Br. Chromosoma 79: 105–114

Pijnacker LP, Ferwerda MA (1982) Normal and BrdC-substituted chromosomes during spermatogenesis with an endomeiotic chromosome reduplication in *Carausius morosus* Br. Chromosoma 85: 67–82

Pijnacker LP, Ferwerda MA (1986) Development of the synaptonemal complex of two types of pachytene in oocytes and spermatocytes of *Carausius morosus* Br. (Phasmatodea). Chromosoma 93: 281–290

Pongratz N, Stohas M, Carranza S, Michiels NK (2003) Phylogeography of competing sexual and parthenogenetic forms of a freshwater flatworm: patterns and explanations. BMC Evol Biol 3: 23–37

Quattro JM, Avise JC, Vrijenhoek RC (1992) Mode of origin and sources of genotypic diversity in triploid fish clones (Poeciliopsis: Poecilidae). Genetics 130: 621–628

Renault S, Roulex-Bonnin F, Periquet G, Bigot Y (1999) Satellite DNA transcription in *Diabromus pulchellus* (Hymenoptera). Insect Biochem Mol Biol 29: 103–111

Riparbelli MG, Whitfield WGF, Dallai R, Callaini G (1997) Assembly of the zygotic centrosome in fertilized *Drosophila* eggs. Mech Dev 65: 135–144

Riparbelli MG, Stouthamer R, Dallai R, Callaini G (1998) Microtubule organization during the early development of the parthenogenetic egg of the hymenopteran *Muscidifurax uniraptor*. Dev Biol 195: 89–99

Sandoval C, Carmean DA, Crespi BJ (1998) Molecular phylogenetics of sexual and parthenogenetic *Timema* walking-sticks. Proc R Soc Lond B 265: 589–595

Scali V (1968) Biologia riproduttiva del *Bacillus rossius* (Rossi) nei dintorni di Pisa con particolare riferimento all'influenza del fotoperiodo. Atti Soc Tosc Sci Nat, Mem Serie B 75: 108–139

Scali V (1969) Osservazioni citologiche sullo sviluppo embrionale di *Bacillus rossius*. Atti Acc Naz Lincei, Rend Sci Fis Mat Nat 46: 110–116

Scali V (1972) La citologia della partenogenesi di *Bacillus rossius*. Boll Zool 39: 567–573

Scali V (1996a) Distribution, genetic diversification and reproductive mechanisms of unisexual *Bacillus* stick insects. 5th Intern Congress of System Evol Biol Pars Ldt Budapest, p. 165

Scali V (1996b) Descrizione di due specie incipienti di insetti stecco (Phasmatodea) del complesso *Leptynia attenuata* Pantel: *L. montana* n. sp. e *L. caprai* n. sp. Redia 79: 123–136

Scali V (2009) Revision of the Iberian stick insect genus *Leptynia* Pantel and description of the new genus *Pijnackeria* It J Zool (in press)

Scali V, Passamonti M, Marescalchi O, Mantovani B (2003) Linkage between sexual and asexual lineages: genome evolution in *Bacillus* stick insects. Biol J Linn Soc 79: 137–150

Scali V, Tinti F (1999) Satellite DNA variation in parental and derived unisexual hybrids of *Bacillus* stick insects (Phasmatodea). Insect Mol Biol 8: 557–564

Scali V, Tinti F, Mantovani B, Marescalchi O (1995) Mate recognition and gamete cytology features allow hybrid species production and evolution in *Bacillus* stick insects. Boll Zool 62: 59–70

Schatten G (1994) The centrosome and its mode of inheritance: the reduction of the centrosome during gametogenesis and its restoration during fertilization. Dev Biol 165: 299–335

Schatten H, Schatten G, Mazia D, Balczon R, Simerly C (1986) Behavior of centrosomes during fertilization and cell division in mouse oocytes and in sea urchin eggs. Proc Natl Acad Sci USA 83: 105–109

Schön I, Gandolfi A, Di Masso E, Rossi V, Griffiths HI, Martens K, Butlin RK (2000) Persistence of asexuality through mixed reproduction in *Eucypris virens* (Crustacea, Ostracoda). Heredity 84: 161–169

Schueler MG, Higgins AW, Rudd MK, Gustashaw K, Willard HF (2001) Genomic and genetic definition of a fuctional human centromere. Science 294: 109–115

Schultz RJ (1961) Reproductive mechanism of unisexual and bisexual strains of the viviparous fish *Poeciliopsis*. Evolution 15: 302–305

Simon J-C, Delmotte F, Rispe C, Crease T (2003) Phylogenetic relationships between parthenogens and their sexual relatives: the possible routes to parthenogenesis in animals. Biol J Linn Soc 79: 151–163

Slamovits CH, Cook JA, Kessa EP, Rossi MS (2001) Recurrent amplification and deletions of satellite DNA accompanied chromosomal diversification in South American tuco-tucos (genus *Ctenomys*, Rodentia: Octodontidae): A phylogenetic approach. Mol Biol Evol 18: 1708–1719

Stenberg P, Lundmark M, Knutelski S, Saura A (2003) Evolution of clonality and polyploidy in a weevil system. Mol Biol Evol 20: 1626–1632

Suomalainen E, Saura A, Lokki J (1987) Cytology and evolution in parthenogenesis. CRC Press, Boca Raton, Florida

Taddei C, Maurizii MG, Chicca M, Scali V (1993) The germarium of the panoistic ovarioles of *Bacillus rossius* (Insecta Phasmatodea): structure and function during imaginal life. Invertebr Reprod Dev 23: 203–210

Tassin AM, Bornens M (1999) Centrosome structureand microtubule nucleation in animal cells. Biol Cell 91: 343–354

Tautz D (1993) Notes on the definition and nomenclature of tandemly repetitive DNA sequences. In: Pena SDJ, Chakraborty R, Epplen JT, Jeffreys AJ (eds) DNA fingerprinting: State of the science 67, Birkhäuser Verlag, Basel, pp. 21–28

Tinti F (1993) Morphological, genetic and chromosomal characterization of Corsican and Spanish *Bacillus rossius* (Insecta, Phasmatodea). Vie Milieu 43: 109–117

Tinti F, Mantovani B, Scali V (1992) Caratterizzazione allozimatica di popolazioni *di Bacillus rossius* dell'Italia centro- meridionale e della Sicilia. Boll Soc Entomol Ital 123: 184–194

Tinti F, Scali V (1995) Allozymic and cytological evidence for hemiclonal, all-paternal and mosaic offspring of the hybridogenetic stick insect *Bacillus rossius-grandii grandii*. J Exp Zool 273: 149–159

Tinti F, Scali V (1996) Androgenetics and triploids from an interacting parthenogenetic hybrid and its ancestors in stick insects. Evolution 50: 1251–1258

Tram U, Sullivan W (2000) Reciprocal inheritance of centrosome in the parthenogenetic hymenopteran *Nasonia vitripennis*. Curr Biol 10: 1413–1419

Turner BJ, Brett BL, Miller RR (1980) Interspecific hybridization and the evolutionary origin of a gynogenetic fish, *Poecilia formosa*. Evolution 34: 917–922

Turner BJ, Balsano JS, Monaco PJ, Rash EM (1983) Clonal diversity and evolutionary dynamics in a diploid-triploid breeding complex of unisexual fishes (*Poecilia*). Evolution 37: 798–809

Turner BJ, Elder Jr. JF, Laughlin TF, Davis WP (1990) Genetic variation in clonal vertebrates detected by simple sequence DNA fingerprinting. Proc Natl Acad Sci USA 87: 5653–5657

Vasil'ev PV, Vasil'eva KD, Osinov AG (1989) Evolution of a diploid-triploid-tetraploid complex in fishes of the genus *Cobitis* (Pisces, Cobitidae). In: Dawley RM, Bogart JP (eds) Evolution and ecology of unisexual vertebrates, Bulletin 466. New York State Museum, Albany, New York, pp. 153–169

Vrijenhoek RC (1978) Coexistence of clones in a heterogeneous environment. Science 199: 549–552

Vrijenhoek RC (1998) Animal clones and diversity. BioScience 48: 617–628

Vrijenhoek RC, Lerman S (1982) Heterozygosity and environmental stability under sexual and asexual breeding systems. Evolution 36: 768–776

White MJD (1970) Heterozygosity and genetic polymorphism in parthenogenetic animals. In: Hecht MK, Steere WC (eds) Essays in evolution and genetics in honor of Theodosius Dobzhansky. Appleton-Century-Crofts, New York

White MDJ (1976) Insect 2. In: John B (ed) Animal cytogenetics. Gebrüder Bornträger, Berlin, pp. 1–75

Wu X, Palazzo RE (1999) Differential regulation of maternal *versus* paternal centrosomes. Proc Natl Acad Sci USA 96: 1397–1402

Zompro O (2005) Inter- and intra-ordinal relationships of the Mantophasmatodea, with comments on the phylogeny of polyneopteran ordes (Insecta: Polyneoptera). Mitt Geol-Paläontol Instit Univ Hamburg 89: 85–116

Chapter 17
Thelytoky in Hymenoptera with *Venturia canescens* and *Leptopilina clavipes* as Case Studies

Irene Mateo Leach, Bart A. Pannebakker, Maria Victoria Schneider, Gerard Driessen, Louis van de Zande and Leo W. Beukeboom

Abstract The insect order of Hymenoptera comprises around 200.000 described species of ants, bees, wasps and sawflies, many of which serve important ecological and economic functions. All Hymenoptera have a haplodiploid mode of reproduction. Males always develop from unfertilized eggs and are haploid. Females are always diploid and can develop from both fertilized and unfertilized eggs. Within haplodiploidy, arrhenotoky is the most common mode of reproduction: unfertilized eggs develop into males that are haploid and 100% related to their mother, whereas fertilized eggs yield diploid females with a haploid complement of both parents. Thelytoky is a less common mode of reproduction. Thelytokous females develop parthenogenetically from unfertilized eggs after restoration of diploidy and are 100% related to their mother. Two distinctive classes of thelytoky can be distinguished based upon the causal mechanism: thelytoky can be induced by nuclear genes or be based on cytoplasmic genes including microorganisms. Most thelytokous hymenopterans reproduce by some form of automixis: both terminal fusion and central fusion have been found, while most cases of microbe-induced thelytoky are a form of gamete duplication. These different mechanisms can have a number of important implications for the genetic make-up of individuals and the amount and structure of genetic variation in populations. We discuss these implications and their evolutionary consequences, with a special focus on the ichneumonid parasitoid wasp *Venturia canescens*, in which thelytoky has a genetic basis, and the figitid parasitoid wasp *Leptopilina clavipes*, which has *Wolbachia*-induced thelytoky.

17.1 Thelytoky in Hymenoptera

17.1.1 Introduction

The insect order of Hymenoptera comprises around 200.000 described species of ants, bees, wasps and sawflies. The actual number of existing species may be well

I. Mateo Leach (✉)
Evolutionary Genetics, Centre for Ecological and Evolutionary Studies, University of Groningen, P.O. Box 14, NL-9750 AA Haren, The Netherlands
e-mail: I.Mateo-Leach@rug.nl

I. Schön et al. (eds.), *Lost Sex*, DOI 10.1007/978-90-481-2770-2_17,

over one million (Grimaldi and Engel 2005 and references therein), thereby making it one of the largest groups among the insects. The major groups of Hymenoptera are the Symphyta (primitive sawflies) and the Apocrita ("wasp-waisted" Hymenoptera), the latter being subdivided into the Parsitica (parasitoid wasps) and the Aculeata (ants, bees and wasps) (Gauld and Bolton 1996).

Many hymenoptera serve important ecological functions, such as flower pollination (e.g., bees, bumblebees) and controlling other insect populations (e.g., ants, parasitoid wasps) (Gauld and Bolton 1996). This is also reflected in their economic importance. Bumblebees play a crucial role as pollinators of a variety of crops and the honeybee is domesticated for honey production. Parasitoid wasps are successfully used as biocontrol agents to regulate and control pests. Only few hymenopterans are harmful in the sense that they destroy crops (e.g., phytophagous sawfly larvae), infest our homes or directly injure man (e.g., the hornet).

17.1.2 Reproductive Modes

All species within the Hymenopteran order have a haplodiploid mode of reproduction. Males always develop from unfertilized eggs and are haploid. Females are always diploid and can develop from both fertilized and unfertilized eggs. Within haplodiploidy, arrhenotoky is the most common mode of reproduction: unfertilized eggs develop into males that are haploid and 100% related to their mother, whereas fertilized eggs yield diploid females with a haploid complement of both parents. Hence, virgin arrhenotokous females can reproduce, but produce all-male progenies. Mated females typically store sperm in the spermatheca and can control the sex of their offspring by allowing a sperm to fertilize the egg upon oviposition.

Thelytoky is a less common mode of reproduction. Thelytokous females develop parthenogenetically from unfertilized eggs after restoration of diploidy and are 100% related to their mother. Males do not occur. Arrhenotoky is thought to be the ancestral mode of reproduction and thelytoky has evolved from it in several groups independently (Cook 1993; Godfray 1994). Note that the terms arrhenotoky and thelytoky refer to sexual and parthenogenetic female production under haplodiploidy, respectively. Males are always produced parthenogenetically in arrhenotokous species only. Hence, arrhenotoky can be considered as a mixed mode of sexual and parthenogenetic reproduction, whereas thelytoky is strictly parthenogenetic.

17.1.3 Types and Incidence of Thelytoky in Hymenoptera

Thelytoky occurs in all major groups of Hymenoptera, but is especially present among sawflies (Symphyta) and some parasitoid families such as the Chalcidoidea and Cynipoidea (Cook 1993; van Wilgenburg et al. 2006). Two distinctive classes of thelytoky can be distinguished based upon the causal mechanism: thelytoky can be induced by nuclear genes or be based on cytoplasmic genes including microorganisms. Most cases of thelytoky concern species that are infected with parthenogenesis-inducing microorganisms (Stouthamer 1997; Braig et al. 2002; van Wilgenburg et al. 2006). A genetic basis for thelytoky has been shown for few studied species only (see below).

A genetic basis for thelytokous parthenogenesis has been shown for *Trichogramma cacoeciae* (Stouthamer et al. 1990b; Vavre et al. 2004), several species of the genus *Lysiphlebus* (Belshaw et al. 1999), the ant species *Plathythyrea puncata* (Schilder et al. 1999), the ichneumonid *Venturia canescens* (Beukeboom and Pijnacker 2000) and the cape honeybee *Apis mellifera capensis* (Tucker 1958; Lattorff et al. 2005). Virtually nothing is known about the underlying genetics of thelytokous parthenogenesis in these hymenopterans. An exception are the egg laying workers of the cape honeybee. From crosses between arrhenotokous and thelytokous subspecies, Ruttner (1988) concluded that thelytokous parthenogenesis is under the control of a single gene. Lattorff et al. (2005) expanded on this study by introgressing sexual genes into an asexual background by means of backcrosses between thelytokous queens and arrhenotokous males. The segregation pattern of the reproductive mode in the offspring of the second generation of these backcrosses indicated that thelytokous parthenogenesis is a qualitative character determined by a single major recessive gene, called thelytoky (*th*). There is a large need for genetic studies of thelytoky in Hymenoptera. Given the availability of the total honeybee genome sequence, identification of the first thelytokous gene in Hymenoptera may be expected to come from further studies on the cape honeybee.

A different type of thelytoky is the one caused through infection by intracellular bacteria. So far, thelytoky-inducing bacteria have only been found among the Gram-negative bacteria, in the genera *Wolbachia*, *Cardinium* and *Rickettsia* (O'Neill et al. 1997; Stouthamer et al. 1999; Zchori-Fein and Perlman 2004; Perlman et al. 2006). They predominantly occur in the insect order Hymenoptera (Huigens and Stouthamer 2003; Zchori-Fein et al. 2001, 2004; Hagimori et al. 2006), but sporadically in other groups as well, such as Coleoptera (Werren et al. 1995), Thysanoptera (Arakaki et al. 2001) and mites (Weeks and Breeuwer 2001; Groot and Breeuwer 2006). In general, bacterial induction of thelytoky is restricted to host species with haplodiploid sex determination (see Section 17.1.3), although Weeks et al. (2001) found it in an entirely haploid mite species.

Besides inducing thelytoky, these bacteria are involved in a wider array of reproductive manipulations, including cytoplasmic incompatibility, feminization and male-killing (O'Neill et al. 1997; Werren 1997; Perlman et al. 2006) *Wolbachia* is the most widespread and best known of these reproductive parasites, the manipulative phenotypes of *Cardinium* and *Rickettsia* have only recently started to become evident (Zchori-Fein and Perlman 2004; Perlman et al. 2006). All these bacteria live within the reproductive and other tissues of their host and are maternally inherited through the egg cytoplasm. Sperm cells do not contain enough cytoplasm and males are therefore considered an evolutionary dead end for these bacteria. The existence of these symbionts can therefore most easily be explained by their selective advantage of increasing the production of infected female offspring, and hence their transmission to the next generation. Their most extreme phenotype is the induction of all-female offspring by inducing thelytoky.

17.1.4 Cytology and Genetic Consequences of Thelytoky

Several cytological mechanisms are responsible for parthenogenesis in Hymenoptera, each with different genetic consequences for the genetic make-up of the offspring (Fig. 17.1). Although any classification of parthenogenetic forms may be disputable (Suomalainen et al. 1987), a first distinction can be made based upon whether the maternal ploidy level is maintained without fusion of nuclei (apomixis) or restored after fusion of two division products from a single cell (automixis). In apomictic thelytoky, offspring are mitotically produced, whereas automictic forms of thelytoky involve meiosis and recombination (see also Chapter 4). Apomictic reproduction (Fig. 17.1A) retains heterozygosity and results in offspring that are genetically identical to their mother because sister chromatids pair during meiosis (Suomalainen et al. 1987; Beukeboom and Zwaan 2005). Although apomixis is the most common form of parthenogenesis in insects (Suomalainen et al. 1987), it occurs only sporadically in Hymenoptera. It is found in the sawfly *Strongylogaster maculata* (Peacock and Sanderson 1939), the spring generation of the gall wasp *Neuroterus bacarum* (Doncaster 1916; Dodds 1939), the egg laying workers of queenless groups of the weaver ant *Oecophylla longinoda* (Ledoux 1954), the parasitoid wasp *Trichogramma cacoeciae* (Vavre et al. 2004) and the little fire ant *Wasmannia auropunctata* (Fournier et al. 2005). Vavre et al. (2004) found high levels of heterozygosity in field populations and no segregation of heterozygous genetic markers, which is consistent with apomictic reproduction.

Most thelytokous hymenopterans reproduce by some form of automixis. Typically, early stages of meiosis in automixis (meiotic oogenesis) are normal: pairing of chromosomes, crossing over, bivalent formation and chromosome reduction to form a haploid ovum. Diploidy is restored by fusion of two haploid nuclei from a single dividing oogonium. The restoration of diploidy may occur in different ways, each with different consequences for the genetic variation among offspring (Lamb and Willey 1987; Suomalainen et al. 1987; Beukeboom and Zwaan 2005). Terminal fusion (Fig. 17.1B) is the process in which two haploid daughter cells, the second polar body nucleus and the egg nucleus, fuse to form a diploid egg. This leads to an increase in homozygosity at loci proximal of cross-overs (Suomalainen et al. 1987). The saw flies *Pristiphora rufipes* (Comrie 1938) and *Diprion polytonim* (Smith 1941) and the parasitoid wasp *Aphytis mytilaspides* (Rössler and DeBach 1973) produce their eggs through terminal fusion.

In central fusion (Fig. 17.1C) the second polar body fuses with a descendant of the first polar body to form a diploid egg. These two nuclei are non-sister ootids (originate from different secondary oocytes) and although a certain level of heterozygosity will be maintained, homozygosity increases each generation. Loci close to the centromere have a higher chance to remain heterozygous because less recombination occurs in this region of the chromosome. This mechanism is known from the cape honeybee *Apis mellifera capensis* (Tucker 1958; Verma and Ruttner 1983), the parasitoid wasp *Venturia canescens* (Speicher 1937; Beukeboom and Pijnacker 2000, see Section 17.2.2), and the ant *Cataglyphis cursor* (Pearcy et al. 2004).

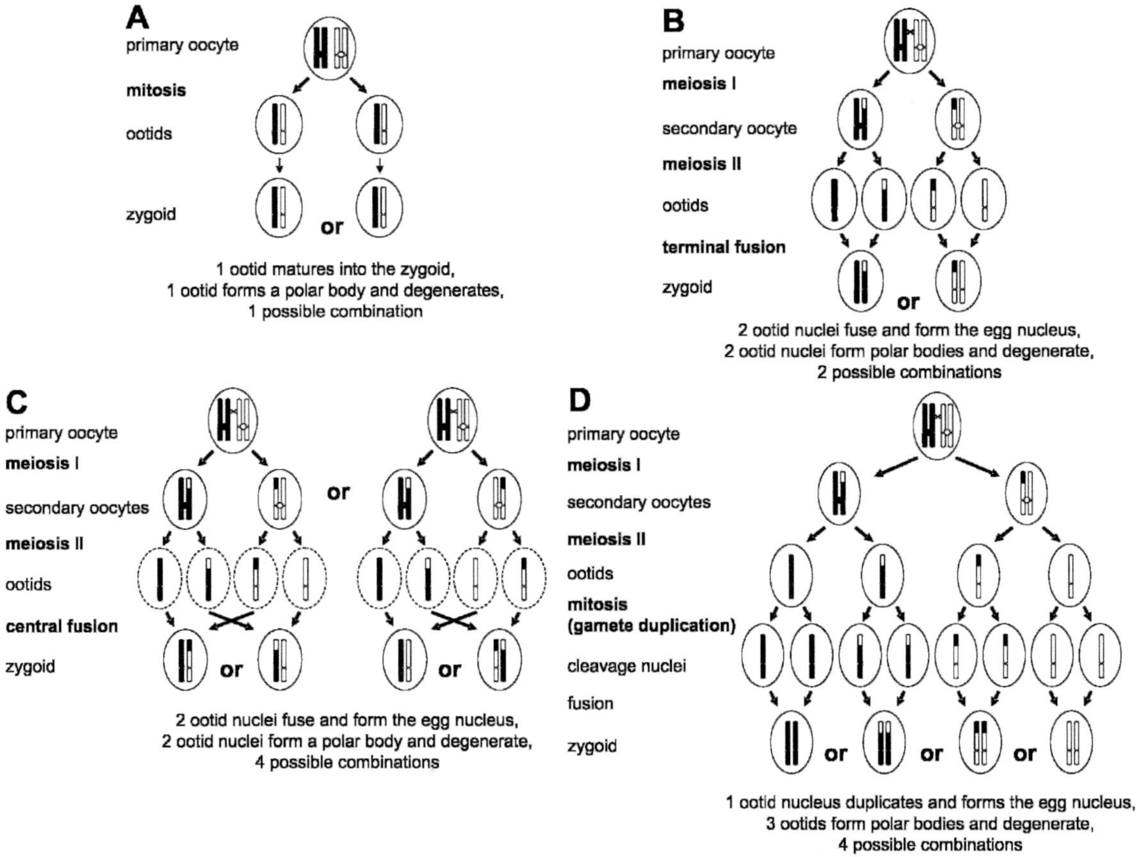

Fig. 17.1 Cytological mechanisms of thelytoky in Hymenoptera. (**A**) apomixis, (**B–D**) automixis, (**B**) terminal fusion, (**C**) central fusion and (**D**) gamete duplication. Diagrams show all possible combinations of nuclei, but each primary oocyte always yields only one zygoid. Alternative segregations occur depending on orientation on the metaphase plate in secondary oocytes. Adapted from Suomalainen et al. (1987)

Gamete duplication (Fig. 17.1D) is a post-meiotic process in which chromosome number doubles during the first cleavage division of the egg (fusion of cleavage nuclei) through endomitosis. To date all studied cases of thelytoky induced by *Wolbachia* involve gamete duplication, resulting in complete homozygosity of the offspring. This has been shown for the gall wasps *Diplolepis rosae* (Stille and Dävring 1980) and *Diplolepis spinosissimae* (Plantard et al. 1998), and the parasitic wasps *Muscidifurax uniraptor* (Legner 1985; Gottlieb et al. 2002), *Leptopilina clavipes* (Pannebakker et al. 2004b) and some species of *Trichogramma* (Stouthamer and Kazmer 1994). Although no cytological study was done, genetic analysis of thelytoky in the *Wolbachia*-infected mite *Bryobia praetiosa* showed the production of heterozygous progeny, suggesting a different mechanism of thelytoky than gamete duplication (Weeks and Breeuwer 2001). The cytological mechanisms involved in *Cardinium* and *Rickettsia*-induced thelytoky have not yet been clarified. Other mechanisms of diploidy restoration induced by *Wolbachia* are likely to be found when more systems are cytologically investigated.

17.1.5 Thelytoky and Sex Determination

Different sex determination mechanisms exist in Hymenoptera, but little is known about the underlying genetics (Beukeboom 1995). Under haplodiploidy, sex determination does not depend on heteromorphic sex chromosomes, but on the number of chromosome sets (haploid males and diploid females). Under single locus complementary sex determination (sl-CSD), first described by Whiting (1943), sex depends on the allelic composition at a single *csd* locus: hemizygous haploids develop into males, heterozygous diploids into females and *csd*-homozygous diploids into diploid males. Diploid males typically have low viability or fertility, or produce diploid sperm that will yield sterile triploid offspring (Agoze et al. 1994). Hence, CSD is considered disadvantageous under inbreeding conditions because more homozygous diploid males are produced. They pose a genetic load to the population and selection will therefore favour rare *csd* alleles in the population (Cook and Crozier 1995). There is some recent evidence for multiple *csd* loci (multilocus CSD) in the parasitoid wasp *Cotesia vestalis* (De Boer et al. 2007), which may be an alternative way to reduce diploid male frequencies and lessen the effects of inbreeding. Van Wilgenburg et al. (2006) have recently reviewed the incidence of CSD and found it to be present in over 60 species of Hymenoptera occurring in each major subgroup, including the social Hymenoptera. This suggests that it is the ancestral mode of sex determination although the basal groups have been poorly studied (Cook and Crozier 1995). The *csd* gene has recently been cloned and sequenced in the honey bee but the molecular regulation of CSD remains unknown (Beye et al. 2003).

The fact that thelytokous species can have a form of complementary sex determination may look contradictory because most forms of thelytoky lead to increased homozygosity, which would result in a high proportion of diploid males. The most extreme case is gamete duplication mediated by microorganisms which leads to complete homozygosity within a single generation. Therefore, it is believed that CSD may prevent the evolution of thelytoky induced by *Wolbachia*. There is some phylogenetic information that supports this idea: *Wolbachia* induced thelytoky appears mostly absent in hymenopteran groups for which CSD has been described, such as the sawflies (Tenthredinoidea) and bees and wasps (Wenseleers and Billen 2000), and is particularly abundant in the non-CSD prasitoid groups Chalcidoidea and Cynipidae (van Wilgenburg et al. 2006). However, not all cases of thelytoky are incompatible with complementary sex determination. Heterozygosity may be completely (apomixis) or partially preserved (central and terminal fusion). An example is the parasitoid wasp *Venturia canescens* which combines sl-CSD with thelytokous reproduction (see Section 17.2.2).

Alternative models to CSD have been proposed to explain sex determination in groups of Hymenoptera where homozygous diploids develop into females (reviewed in Cook 1993). Most information to date comes from the parasitoid wasp *Nasonia vitripennis* and points towards a balance between a maternally produced sex determiner and the number of chromosome sets present in the developing egg (Beukeboom et al. 2007).

17.1.6 Evolutionary Consequences of Thelytoky

The evolutionary consequences of thelytoky are manifold and have been considered by many authors (see also Chapter 5). Advantages of thelytoky are that the cost of sex (Maynard Smith 1978) and meiosis (Williams 1975) are avoided. Apomictic thelytoky fixes heterotic combinations of genes and reduces gene loss (Slobodchikoff and Daly 1971). On the other hand, arrhenotokous reproduction allows for a reduction of genetic load because recessive deleterious alleles are exposed to selection in haploid males (Crozier 1985). Haccou and Schneider (2004) showed theoretically that segregation and recombination do reduce the accumulation of deleterious alleles under several forms of automictic thelytoky, although not to the same degree as under arrhenotoky.

In strictly thelytokous populations, genes involved in sexual reproduction are no longer maintained by selection and the traits determined by these genes are expected to become reduced or disappear completely (Fong et al. 1995). This process occurs in thelytokous populations regardless of the underlying mechanism, but it is easiest studied for cases of microbe-induced thelytoky because these can be reverted to the sexual mode of reproduction by curing infected females from their infection using high temperature or antibiotic treatments (Stouthamer et al. 1990a). Cured females produce males from unfertilized eggs and this allows for testing deterioration of traits involved in sexual reproduction by determining the functionality of male and female sexual traits, such as sperm production and egg fertilization ability.

So far, curing experiments have been done on 19 Hymenoptera species infected with *Wolbachia* (for reviews see Stouthamer 1997; Huigens and Stouthamer 2003), two infected with *Cardinium* (Zchori-Fein et al. 2001, 2004) and one infected with *Rickettsia* (Hagimori et al. 2006). Of these, only several *Wolbachia*-infected

Trichogramma species that co-occur with uninfected individuals in mixed populations showed full sexual functionality in both males and females (Stouthamer et al. 1990a, b). In *Wolbachia*-infected species where completely infected and uninfected populations occur in allopatry (e.g., *Apoanagyrus diversicornis* (Pijls et al. 1996), *Telonomus nawai* (Arakaki et al. 2000) and *Leptopilina clavipes* (Pannebakker et al. 2004c, see Section 17.3.3)), antibiotic curing induced males from the infected population proved to be (partially) functional when mated with females from uninfected populations. However, they did not sire offspring with females from the infected population. In species in which thelytoky-inducing *Wolbachia* have infected all individuals within populations (e.g., *Encarsia formosa* (Zchori-Fein et al. 1992), *Eretmocercus mundus* (De Barro and Hart 2001) and *Muscidifurax uniraptor* (Gottlieb and Zchori-Fein 2001), females did not fertilize their eggs when mated to restored males. Hence, with the exception of the mixed populations of *Trichogramma* species, females from parthenogenesis-inducing *Wolbachia* infected populations have lost the ability to reproduce sexually, whereas males induced from these populations are, at least partially, functional. The recent discovery of mixed populations of the parasitoid *Tetrastichus coeruleus* (S. Wielaard, B.A. Pannebakker and J.J.M. van Alphen, unpublished results) in the Netherlands offers excellent opportunities to further test this link between population infection status and sexual functionality. In the two *Encarsia* species being infected with thelytoky-inducing *Cardinium*, to which antibiotic curing was applied, *E. hispida* did produce males, but the functionality of these males was not tested (Zchori-Fein et al. 2004). Cured *E. pergandiella* females produced almost no offspring and no males, but did change their oviposition behavior (Zchori-Fein et al. 2001). In the only known case of thelytoky-inducing *Rickettsia*, antibiotic curing resulted in male offspring, although the functionality of these males was not further tested (Hagimori et al. 2006).

Several hypotheses have been put forward to explain the sexual degradation associated with thelytoky-inducing *Wolbachia*, and these are equally valid for thelytoky-inducing *Cardinium* and *Rickettsia*. The first hypothesis is the neutral accumulation of mutations in genes coding for sexual function that are no longer under selection in parthenogenetic populations (Muller 1949; Carson et al. 1982). In addition, if these mutations are not neutral and improve the parthenogenetic performance of females, they may even be selected for (Pijls et al. 1996; Werren 1998). The second hypothesis is the virginity mutation hypothesis proposed by Huigens and Stouthamer (2003). They stated that mutations disrupting female sexual function are likely to be strongly selected for during the initial stages of a thelytoky-inducing *Wolbachia* infection in a population. Their argument is based on sex ratio selection and is as follows: if half of the females in a population are infected during these initial stages, uninfected females that produce only male offspring will have a large selective advantage over both infected and uninfected females because their sons can mate with the surplus of females present in such a population. Remember that in haplodiploid species, all offspring are male (haploid) when a female does not mate or does not fertilise her eggs. Thus, a mutation that induces all-male offspring (virginity mutation) will have a large fitness advantage in the female-biased population and will spread rapidly when the mutation-carrying-males mate with the infected females.

Because *Wolbachia* transmission to the offspring is rarely fully efficient (i.e., more than 90%), infected females produce rare males carrying the mutation, and such males can only mate with the remaining wild type females. This causes the mutation to also spread to the uninfected part of the population and as a consequence the number of uninfected females decreases. This results in complete fixation of parthenogenesis-inducing *Wolbachia* in a population, even with a vertical transmission efficiency that is lower than 100% (R. Stouthamer, pers. comm.). Consequentially, the population will consist entirely of mutant females that are no longer able to mate or fertilise their eggs (Huigens and Stouthamer 2003). Evidence for the genetic basis of such mutations has recently been found in females of the egg parasitoid *Telonomus nawai* (Jeong and Stouthamer 2005).

We will now describe two hymenopteran species, in which genetic diversity and evolutionary consequences of their thelytokous reproduction have been investigated in some detail. The first is the ichneumonid parasitoid wasp *Venturia canescens* in which thelytoky has a genetic basis. The second is the figitid parasitoid wasp *Leptopilina clavipes*, which has *Wolbachia*-induced thelytoky.

17.2 Case Study I: Venturia canescens

17.2.1 Introduction

Venturia canescens (Gravenhorst) (Fig. 17.2) is a solitary endoparasitoid wasp of lepidopteran larvae (Beling 1932; Salt 1976) that has been widely used as a biological model in behavioural, population dynamical, genetic and physiological studies (references in Beukeboom and Pijnacker 2000; Thiel et al. 2006). It has both sexual (arrhenotokous) and parthenogenetic (thelytokous) reproduction and females of either mode occur sympatrically in Southern Europe (Schneider et al. 2002). Thelytokous females have an extended distribution range because they also inhabit man-made environments, such as bakeries and granaries. These environments are considered relatively constant, which would allow thelytokous strains to outcompete arrhenotokous ones due to the reproductive advantage of parthenogenesis over sexuality (Schneider et al. 2002).

17.2.2 Cytology and Genetics of Thelytoky

Thelytokous females produce haploid eggs meiotically that subsequently undergo diploidy restoration and develop into females (Speicher 1937; Beukeboom and Pijnacker 2000). Beukeboom and Pijnacker (2000) showed that thelytoky in *Venturia canescens* is not due to infection with *Wolbachia* and supposed a genetic basis for thelytoky. Mateo Leach unpublished data refined the study by investigating the possibility that endosymbionts other than *Wolbachia* may cause thelytoky

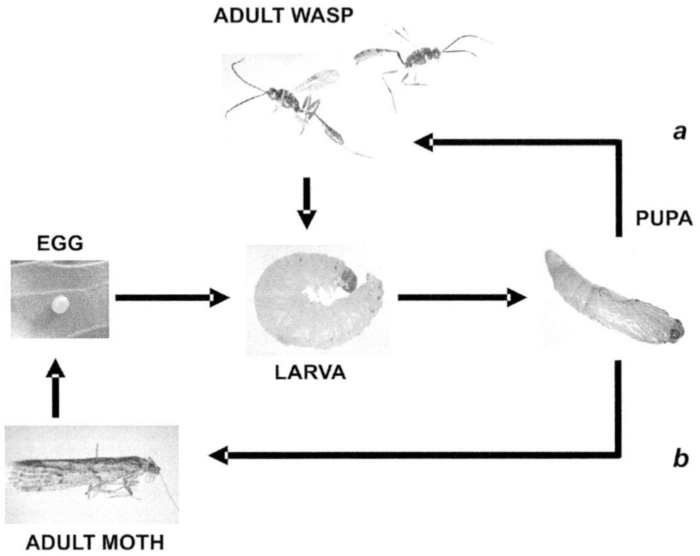

Fig. 17.2 Life cycle of the parasitoid *Venturia canescens* (**a**) and its host *Ephestia kuehniella* (**b**). The *Venturia* adult and pupa and the adult *Ephestia* pictures were taken at the Evolutionary Genetics laboratory in Groningen. The *Ephestia* egg and *Venturia* larva pictures were taken from the internet without a reference source

in *V. canescens*. Antibiotic treatments and the combined prokaryote and *Wolbachia* diagnostic PCR-based assays confirmed the absence of bacteria in the ovaries of both arrhenotokous and thelytokous individuals. These results make any prokaryotic endosymbiont infection as the cause for parthenogenesis in these wasps very unlikely, and strongly indicate that parthenogenesis in *V. canescens* indeed has a genetic basis.

Diploidy restoration in the thelytokous strains of *V. canescens* is described as a form of central fusion automictic parthenogenesis (Beukeboom and Pijnacker 2000) (Fig. 17.3). Speicher (1937) already described the cytological mechanism to consist of an aberrant first meiotic division. Later genetic studies (Speicher et al. 1965) were ambivalent about the nature of the second division, it either being reductional or equational. Beukeboom and Pijnacker (2000) showed that the first meiotic division is followed by an equational division of the restituted number of diploid chromosomes. The chromatid of one sister chromosome can segregate with either one of the other sister chromosome, resulting in two possible segregation combinations (Fig. 17.3). Subsequently, one of the two nuclei becomes a polar body and degenerates; whereas the other develops into the diploid embryo. This cytological mechanism of diploidy restoration does not instantly lead to complete homozygosity. It enables heterozygosity to be maintained for loci close to the centromere, but distal loci will become homozygous over generations in half of the segregation combinations. The fact that thelytokous females can be genetically heterozygous (Mateo Leach, unpublished results) is consistent with this.

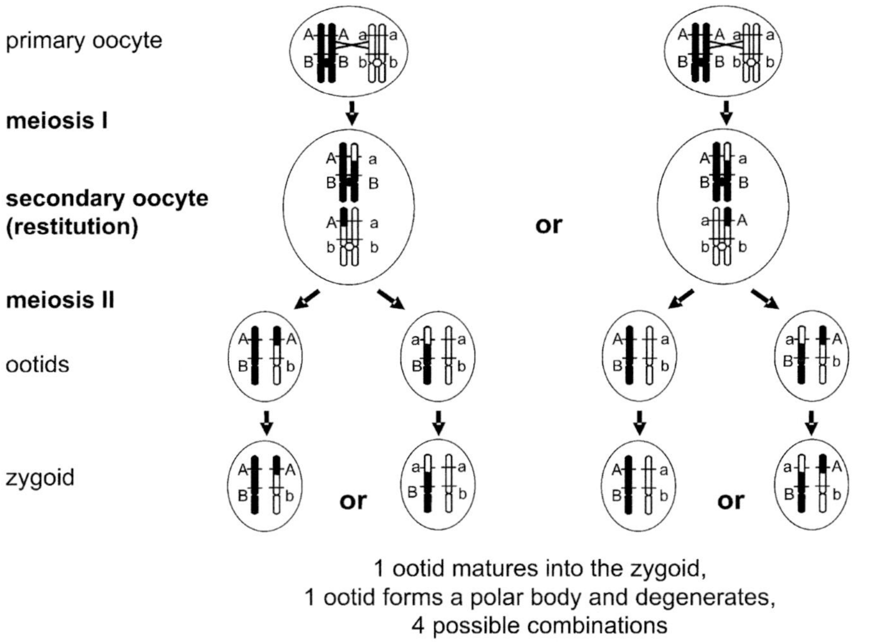

Fig. 17.3 The cytological mechanism of thelytoky in *Venturia canescens*. The first meiotic division results in a diploid restitution metaphase, which divides and of which one nucleus becomes a polar body and one nucleus develops into the zygoid. Alternative segregations occur depending on orientation on the metaphase plate in secondary oocytes. Loci distal of crossing-over (locus A) have a 50% chance of becoming homozygous (*left diagram*) or remain heterozygous (*right diagram*) depending on the segregation combination), loci proximal of crossing-over (locus B) always remain heterozygous. For details, see Beukeboom and Pijnacker (2000)

The sex determining mechanism in *Venturia canescens* is a single locus complementary sex determination (sl-CSD), which was originally discovered by comparing sex ratios between crosses of related and unrelated individuals using an arrhenotokous strain (Beukeboom 2001). Typically, sl-CSD is tested by brother-sister matings in which, due to haplodiploidy, 50% of males are expected to share a *csd* allele with their sisters. Such matched crosses will yield 50% homozygous and 50% heterozygous *csd* offspring which develop into diploid males and females respectively. Crosses between unrelated males and females are used to compare progeny sizes, because diploid males in some species are unviable (Agoze et al. 1994). These inbreeding experiments resulted in higher sex ratios at frequencies that were consistent with the presence of viable diploid males as predicted under sl-CSD (for details, see Beukeboom 2001). Recently, the generation of diploid males under inbreeding was confirmed by flow cytometry (Mateo Leach, unpublished data). How can the mechanism of diploidy restoration in *V. canescens* be compatible with sl-CSD? Both can operate as long as the *csd* locus is located in a chromosomal region where heterozygosity is maintained, e.g., close to a centromere on one of the eleven autosomes or in a region in which recombination is prevented by inversions.

Interestingly, occasionally males are found in thelytokous laboratory populations (Mateo Leach, unpublished data). The most straightforward explanation for their origin is a rare recombinational event leading to homozygosity of the *csd* locus. Such males would be expected to be diploid. However, flow cytometric analyses proved them to be haploid. Hence, these rare males in thelytokous populations are likely the result of a failure in the mechanism of diploidy restoration, for example due to unequal chromosome segregation after restitution. We tested four of these asexual males for reproductive ability. None of them showed courtship behaviour for any of the arrhenotokous or thelytokous females they were offered. This supports the theory that such males do not have an active role in the thelytokous population.

Very little is known about the genetic basis of parthenogenesis in animals and in hymenopterans in particular. In the cape honeybee, parthenogenesis in the egg-laying worker is a heritable trait probably determined by a single recessive locus (Latorff et al. 2005, Section 17.1.4). In *Venturia canescens*, we do not know whether a single or multiple genes are responsible for thelytoky. We also cannot exclude the possibility that thelytoky is encoded by mitochondrial genes that are maternally inherited. Schneider et al. (2003) performed crosses between arrhenotokous males and thelytokous females and discovered paternal genetic markers in the hybrid female offspring. Such introgression of arrhenotokous genes would allow for studying the genetic basis of thelytoky in *V. canescens*. Assuming a nuclear basis of thelytoky and recombination, segregation of arrhenotokous and thelytokous reproduction among offspring may be expected when a sufficient proportion of arrhenotokous genes were introduced into a thelytokous background. We set out to do just this, but, unfortunately, were not able to repeat the results of Schneider et al. (2003) even though several laboratory and field strains were used under different conditions. Although thelytokous females mated with arrhenotokous males and males transferred sperm, as evidenced from dissection of spermatheca of mated females, females did not lay eggs that contained the paternal genotype (Mateo Leach, unpublished data). Unfortunately, the particular laboratory strain (collected from Golfe, southern France) used by Schneider et al. (2003) was lost. Hence, the most likely explanation is that gene exchange between arrhenotokous males and thelytokous females is very rare.

17.2.3 Thelytoky and Genetic Diversity

Central fusion automixis has important consequences for genetic variation in thelytokous populations. Due to cross-overs in the primary oocytes and the subsequent fusion of nuclei, recombination leads to an increase in homozygosity rather than a decrease as may intuitively be expected. Each generation, loci distal of a chiasma have a 50% chance of becoming permanently homozygous in an individual (Fig. 17.3). Hence, depending on the number and location of chiasmata in the primary oocytes of the parental generation, offspring will become more and more homozygous. Moreover, at the level of individual genomes, an increasing homozygosity gradient is expected to become established from the centromeres to the telomeres. In theory, this process of "genome homozygozation" can be followed using polymorphic genetic markers and the degree of homozygosity may be used as a "genomic clock" to determine the age of thelytokous lineages. At this moment genetic linkage maps of *Venturia canescens* have insufficient detail to perform such an analysis. However, in the cape honeybee, *Apis mellifera capensis*, such a study has been done (Baudry et al. 2004). The authors indeed observed gradients of homozygosity and were able to map centromere positions for most of the linkage groups. However, they also found that the recombination rate was reduced by more than tenfold during meiosis in thelytokous workers. This points towards the existence of genetic mechanisms for reducing homozygosity in thelytokous Hymenoptera which will have important evolutionary consequences (see Section 17.1.6). More studies are needed on the effects of particular cytological mechanisms of thelytoky on genetic variation in populations. Such studies should investigate how recombination events change individual genomes over successive generations and how this affects the structure of genetic variation in natural populations.

The simultaneous occurrence of arrhenotokous and thelytokous reproduction poses an evolutionary problem because thelytokous populations are expected to rapidly outcompete arrhenotokous ones due to the twofold cost of sex (Maynard Smith 1978). Schneider et al. (2002) studied the geographical distribution and genetic diversity of arrhenotokous and thelytokous populations of *V. canescens* in southern Europe. Arrhenotokous wasps were more abundant than thelytokous ones, but simultaneous occurrence of both reproductive modes in the same localities was found frequently. Analysis of the genetic structure of the populations at the Côte d'Azur revealed that there was one widespread thelytokous lineage and a few rare ones (Fig. 17.4B). The authors base this conclusion on the significant clustering of thelytokous individuals in a genetic similarity tree constructed from multi-locus genotypes using Amplified Fragment Length Polymorphisms (AFLP). They propose two explanations for the occurrence of few thelytokous individuals with high genetic similarity to arrhenotokous ones: apart from the widespread lineage, new lineages might recently have arisen from local arrhenotokous populations by a loss of sex, and/or (2) introgression of arrhenotokous genes into thelytokous lineages had taken place through occasional sex between males and thelytokous females. Schneider et al. (2003) showed that thelytokous females can mate with arrhenotokous males and produce biparental offspring. However, as mentioned before, all our attempts to repeat such crosses under laboratory conditions were unsuccessful (Mateo Leach, unpublished).

In a later study, Schneider (2003) also performed a mitochondrial DNA restriction fragment length analysis on the original Cote d'Azur individuals and on wasps collected at 22 sites along a 500 km transect along the Rhône Valley in southern France (Fig. 17.4A). Of a total of 234 female wasps that emerged from baits, 12% were thelytokous and found at 3 sites, whereas 88% were arrhenotokous and occurred at 19 sites. A tree was constructed based on genetic similarity data using nuclear markers (104 AFLP fragments) of 34 arrhenotokous and 9 thelytokous individuals. Significant clustering of all thelytokous wasps was shown with a bootstrapping method (Schneider et al. 2002), indicating that thelytokous wasps from the

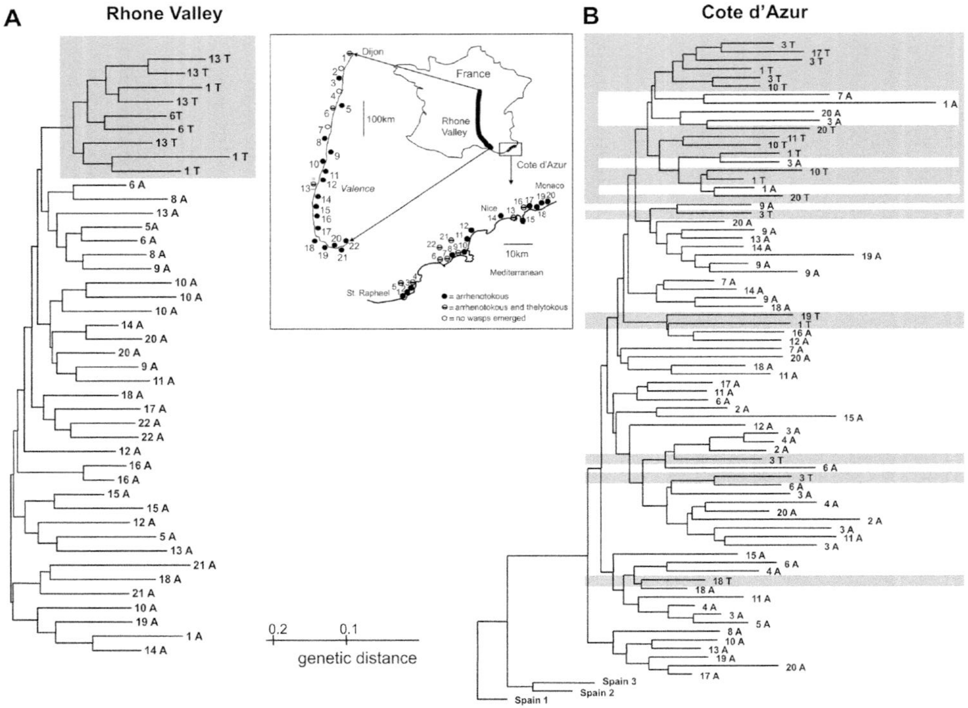

Fig. 17.4 Genetic distance Neighbour Joining tree of arrhenotokous and thelytokous *Venturia canescens* collected along two transects in south-east France, (**A**) Cote d'Azur en (**B**) Rhone Valley. Thelytokous individuals are shown in shaded boxes, arrhenotokous in white. In the Cote d'Azur, some arrhenotokous individuals closely resemble the major thelytokous lineage shown in the top of the tree, and some thelytokous individuals cluster with arrhenotokous ones in the whole tree. Localities of the Rhône transect: 1 =Dijon, 2 = Ladoix Serrigny, 3 = Forêt de Beauregard, 4 = St Ambreuil , 5 = St Oyen, 6 = Creches , 7 = St George de Reneis, 8 = Limonest, 9 = Brignais, 10 = St Cyr sur Rhône, 11 = Serrieres, 12 = Arras sur Rhône, 13 = Valence, 14 = Payre, 15 = Ciments Lafarge, 16 = Lapalud, 17 = Orange, 18 = St Saturin les Avignon, 19 = La Lempe, 20 = La Montauronne, 21= Les Bannettes, 22 = St Maximin la Ste Baume. Localities of the Cote d'Azur transect : 1 = Cap du Dramont, 2 = Agay, 3 = Anthéor, 4 = La Trajas, 5 = Saint Guitte, 6 = La Napoule, 7 = La Croix de Gardes, 8 = Vallauris, 9 = Juan le Pins, 10 = Cap d'Antibes, 11 = La Brague, 12 = Biot, 13 = Nice, 14 = Panoramèr, 15 = Mont Boron, 16 = Mont Vinagrier, 17 = Mont Gros, 18 = Villefranche, 19 = Eze, 20 = St Laurent, 21 = Valbonne, 22 = Vallée Verte. Spain = Spanish individuals used as outgroup. For details see Schneider et al. (2002) and Schneider (2003). Modified after Schneider et al. (2002) and Schneider (2003)

three localities (95–248 km apart) were genetically more similar to each other than to the co-occurring arrhenotokous wasps. Hence, a widespread thelytokous lineage was again found but at this larger geographical scale, no thelytokous wasps were detected with close resemblance to arrhenotokous ones, as was the case in the Cote d'Azur. The mitochondrial DNA analysis was subsequently used to compare maternal haplotypes between both reproductive modes. In the Rhône Valley transect, all nine thelytokous wasps had the same mitochondrial type (haplotype I), which was shared with only 3 of 34 arrhenotokous individuals. All other arrhenotokous wasps had haplotype II (Fig. 17.4A). In the Cote d'Azur, all arrhenotokous wasps (95 individuals from 22 sites) carried haplotype II, and except for one thelytokous wasp with haplotype II (at Eze), all thelytokous wasps (31 from 7 sites) had haplotype I. These individuals were not the same as the ones used for the nuclear marker analysis and therefore not shown in Fig. 17.4B. Schneider (2003) concluded that mitochondrial haplotypes are largely divergent between both reproductive modes.

What do the nuclear and mitochondrial marker data reveal about the genetic structure of *V. canescens* populations? As mentioned in Schneider (2003), both transects show the occurrence of one wide-spread thelytokous lineage. When new thelytokous lineages arise regularly from local arrhenotokous populations, thelytokous wasps with the arrhenotokous haplotype II should have been found. Since only one such case was found, it can be concluded that recurrent arousal of new thelytokous lineages may occur but probably plays a minor role, if any, at the level of the local dynamics of both reproductive modes. If this particular case was the result of a recent mutational loss of sex, a high nuclear similarity of this wasp would be expected with its sympatric arrhenotokous conspecifics. This was not the case (see Fig. 17.4) and therefore this wasp might be a member of another thelytokous lineage. In addition to thelytokous females resembling arrhenotokous ones, Schneider et al. (2002) also found some arrhenotokous females that shared many nuclear markers with females of the widespread thelytokous lineage. This indicates occasional introgression of thelytokous genes into the arrhenotokous cytotypes. One possible explanation is that thelytokous females occasionally produce males that mate with arrhenotokous females. Rare males were observed in thelytokous laboratory populations (Schneider et al. 2003 and Mateo Leach, unpublished) and were haploid. However, when tested (N=4), such males were never reproductively active under laboratory conditions (Mateo Leach, unpublished).

Many questions about the dynamics of thelytokous and arrhenotokous reproduction in *Venturia canescens* remain unanswered: What are the genetic mechanisms involved, i.e., how can sex be lost and is a return to sex of thelytokous wasps possible? More detailed population genetic analyses of co-occurring thelytokous and arrhenotokous populations as well as laboratory crosses are needed to reveal which mechanisms can cause gene flow between both reproductive modes in nature. The persistence at the same time and place of both reproductive modes within this species remains an intriguing paradox which might be enlarged even more by the presence of gene flow between the modes (Schneider 2003). Studies on the ecological mechanisms behind their coexistence are highly relevant for gaining a better understanding of the problem of the "maintenance of sex".

17.2.4 Coexistence of Arrhenotokous and Thelytokous Wasps

The vast majority of publications on *Venturia canescens* during the passed 80 years dealt with wasps from thelytokous populations that were collected from granaries, mills and other food storages infested with phycitid moths. It was only recently discovered that arrhenotokous populations are widespread under outdoor conditions (Beukeboom et al. 1999; Schneider et al. 2002). Occasionally, wasps of both kinds have been found at the same site or even caught in the same tree. Arrhenotokous wasps have never been reported from indoor samplings: a year round monitoring of sticky traps in a granary in South of France yielded thousands of female wasps but not one male (L. Lapchin, pers. comm.). From a theoretical point of view, all else being equal, we would expect that the sexual form would be outcompeted by the asexual one due to the demographic costs of sex. However, the above observations suggest that the reproductive modes occupy different niches and that the coexistence could be facilitated by ecological (Schneider 2003).

The conditions for wasps in a stored product environment differ fundamentally from the natural situation. Depending on sanitary management, the host abundance inside storage buildings can vary greatly from small spots with a few host larvae to large patches with hundreds of larvae covered over the stored grain during periods of population outbreaks. Outdoors *Venturia* attack pyralid moth larvae that live in dried fruits of fig and medlar, almond husks and carob pods (Driessen and Bernstein 1999). Individual trees can carry hundreds of fruits, but typically only one host lives in a fruit, occasionally two or three. Natural conditions therefore seem to be more uniform and stable. Furthermore, food for adult wasps is practically absent inside granaries and mills, while for wasps living under natural conditions food is easily accessible (Casas et al. 2003).

A number of studies have now compared behaviour and life-history traits of arrhenotokous and thelytokous *Venturia canescens*. When foraging on single patches both kind of wasps have similar patch residence times and ovipositions (Lafortune and Driessen, unpublished; Amat et al. 2006; Thiel et al. 2006). However, when offered a sequence of patches, arrhenotokous wasps are less sensitive to patch encounter rate and keep to an oviposition strategy that results in a spreading of offspring across the habitat. This can be interpreted as an adaptation to CSD. Thelytokous wasps, on the other hand, adjust their oviposition much more in relation to patch encounter rate and follow a strategy that maximizes their foraging efficiency in a variable environment (Thiel et al. 2006).

Arrhenotokous wasps are more sensitive to temperature changes compared to thelytokous wasps: they increase patch time and the number of ovipositions in response to a sudden drop in temperature (Amat et al. 2006). This can be seen as an adaptive response to living outdoors, since sudden temperature drops may indicate unfavourable weather conditions in the immediate future and increase mortality risks. Luchetta et al. (2007) investigated the response to host and food availability of wasps under different feeding regimes. Their results suggest that thelytokous wasps are more sensitive to the presence of food in the environment when they are exploiting a host patch. Food sources are scarce to absent in indoor conditions and hence they are much more valuable for thelytokous wasps.

Pelosse et al. (2007) studied differences in energy allocation between arrhenotokous and thelytokous wasps. Arrhenotokous wasps emerge with more energy reserves, glycogen in particular, which can be seen as an adaptation to the outdoor life-style, where wasps have to search for host patches and food more actively. Thelytokous wasps, on the other hand, put more energy in eggs, which allows them to take advantage of occasional extreme host densities in granaries and mills (Ellers et al. 2000). Higher egg loads at emergence in thelytokous wasps were also reported by Schneider (2003) and Barke et al. (2005) (Fig. 17.5). The difference appeared to be due a morphological difference: thelytokous wasps have a higher number

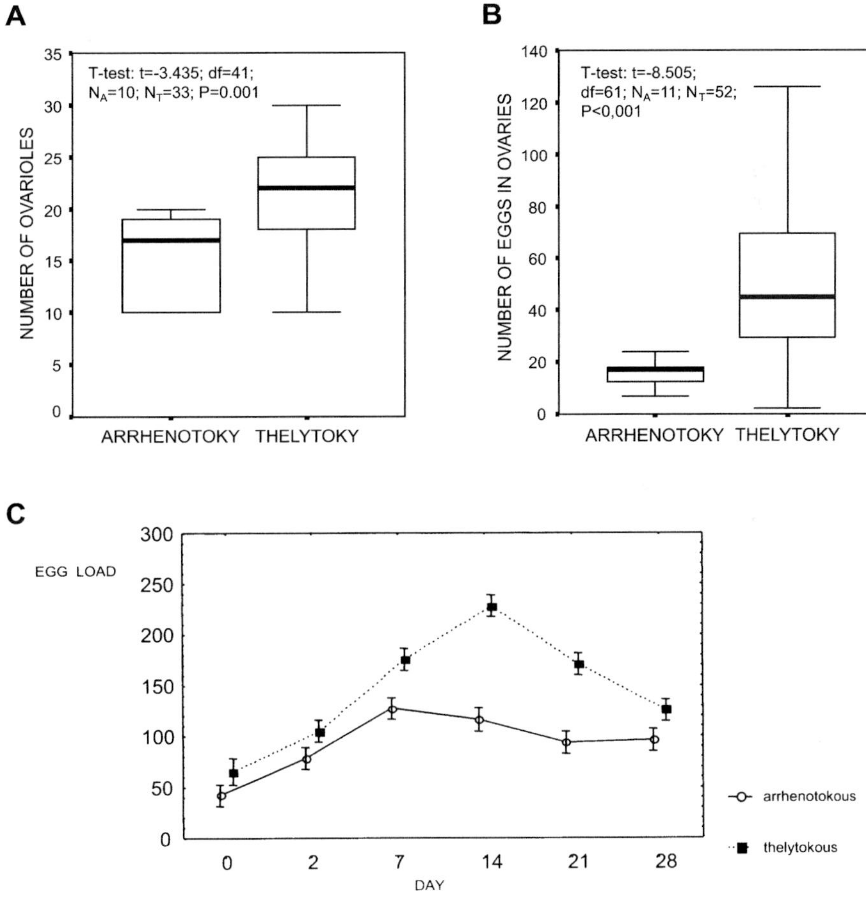

Fig. 17.5 Number of (**A**) ovarioles and (**B, C**) eggs in the ovaries of arrhenotokous and thelytokous *Venturia canescens*. Ovariole number and egg load in (**B**) was determined in one-day old females. The box plot shows the median (*bold line*), the interquartile range (*height of box*), the 1.5 interquartile range (*vertical line*) and extreme data points. Egg load (average ± S.E.) in (**C**) was measured in samples of 40 individuals at emergence and different ages up to 4 weeks. Egg load was significantly higher in arrhenotokous females at all ages (One-way ANOVA, P<0.001). Data are from Barke et al. (2005) for **A** and **B**, and Schneider (2003) for **C**

of ovarioles than arrhenotokous wasps (Fig. 17.5A, Barke et al. 2005). Therefore, thelytokous wasps have a higher egg production throughout their lives resulting in higher life-time fecundity (Fig. 17.5C, Schneider 2003). All in all, these studies suggest that each reproductive form has developed specific behavioural and life-history traits that make it the better competitor in its "own" habitat and that niche differentiation is indeed an explanation for the coexistence of arrhenotokous and thelytokous forms of *V. canescens*.

17.3 Case Study II: Leptopilina clavipes

17.3.1 Introduction

Our second case study comprises *Leptopilina clavipes*. *L. clavipes* (Figitidae) is a solitary endoparasitoid wasp of *Drosophila* larvae (Fig. 17.6) that occurs throughout Europe (Nordlander 1980). *L. clavipes* is found in forested areas where it parasitizes *Drosophila* larvae living in fungal fruit bodies (Vet 1983; Driessen et al. 1990). In northern Europe, its main host is *D. phalerata*, but larvae of *D. kuntzei*, *D. transversa* and *D. subobscura* are also parasitized (Driessen et al. 1990). Southern European *L. clavipes* parasitize *D. melanogaster* larvae as well, which mainly breed in fermenting fruits (Pannebakker et al. 2004c).

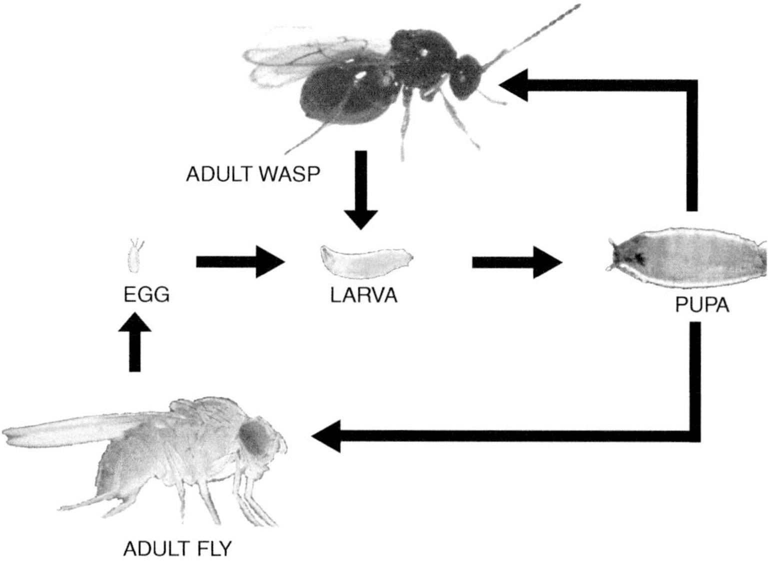

Fig. 17.6 Life cycle of the parasitoid *Leptopilina clavipes* (**a**) and its host *Drosophila phalerata* (**b**). Adapted from Pannebakker (2004)

17.3.2 Cytology and Genetics of Thelytoky

All known northwestern European populations (The Netherlands, England, Germany, Denmark and Sweden) reproduce thelytokously (Nordlander 1980; Vet 1983). In contrast, southern European populations (Spain) reproduce arrhenotokously (Pannebakker et al. 2004c). Thelytoky in *Leptopilina clavipes* is induced by infection with *Wolbachia* bacteria (Werren et al. 1995; Schidlo et al. 2002). The southern European arrhenotokous populations are uninfected. Thelytoky in *L. clavipes* involves diploidization of the haploid eggs through anaphase restitution during the first somatic mitosis (Pannebakker et al. 2004b). This mechanism is a form of gamete duplication that results in the generation of completely homozygous progeny (Suomalainen et al. 1987). If *L. clavipes* were to have complementary sex determination, all offspring would develop into diploid males. The fact that this is not the case implies that sex determination has a different molecular regulation. Although *Wolbachia* bacteria are subject to much research, the molecular and biochemical processes being responsible for inducing gamete duplication have not yet been elucidated.

17.3.3 Thelytoky and Genetic Diversity

The production of homozygous offspring by *Wolbachia*-infected females does not necessarily result in a reduction of genetic variation in infected compared to uninfected populations (Simon et al. 2003). In a population genetic analysis, Pannebakker et al. (2004c) determined the effect of thelytoky-inducing *Wolbachia* on the level of genetic variation in *Leptopilina clavipes* by comparing the genetic diversity of infected and uninfected populations throughout Western Europe. The two forms of *L. clavipes* are clearly genetically differentiated (Fig. 17.7), suggesting a barrier to gene flow between them. Besides possible ecological barriers, such as phenology differences or disjoint distribution, the presence of thelytoky-inducing *Wolbachia* can create a barrier to gene flow by itself when infected *L. clavipes* females stop fertilizing their eggs or mating at all (see Section 17.1.6).

Besides its different effects on the genetic differentiation of thelytokous and arrhenotokous populations, *Wolbachia* has a profound effect on the genetic diversity within and among the infected populations. *Wolbachia*-induced thelytoky and the associated production of fully homozygous offspring (see Section 17.1.4) could be expected to result in low genetic differentiation between the infected populations. In *Leptopilina clavipes*, however, the mean genetic distance between infected populations is not different from that between uninfected populations. Within the sampled populations, at least two dominant clonal lineages were present. Because the same *Wolbachia* strain was found in all infected populations, different *L. clavipes* lineages probably originated through horizontal transmission of the *Wolbachia* bacteria from infected to various uninfected populations.

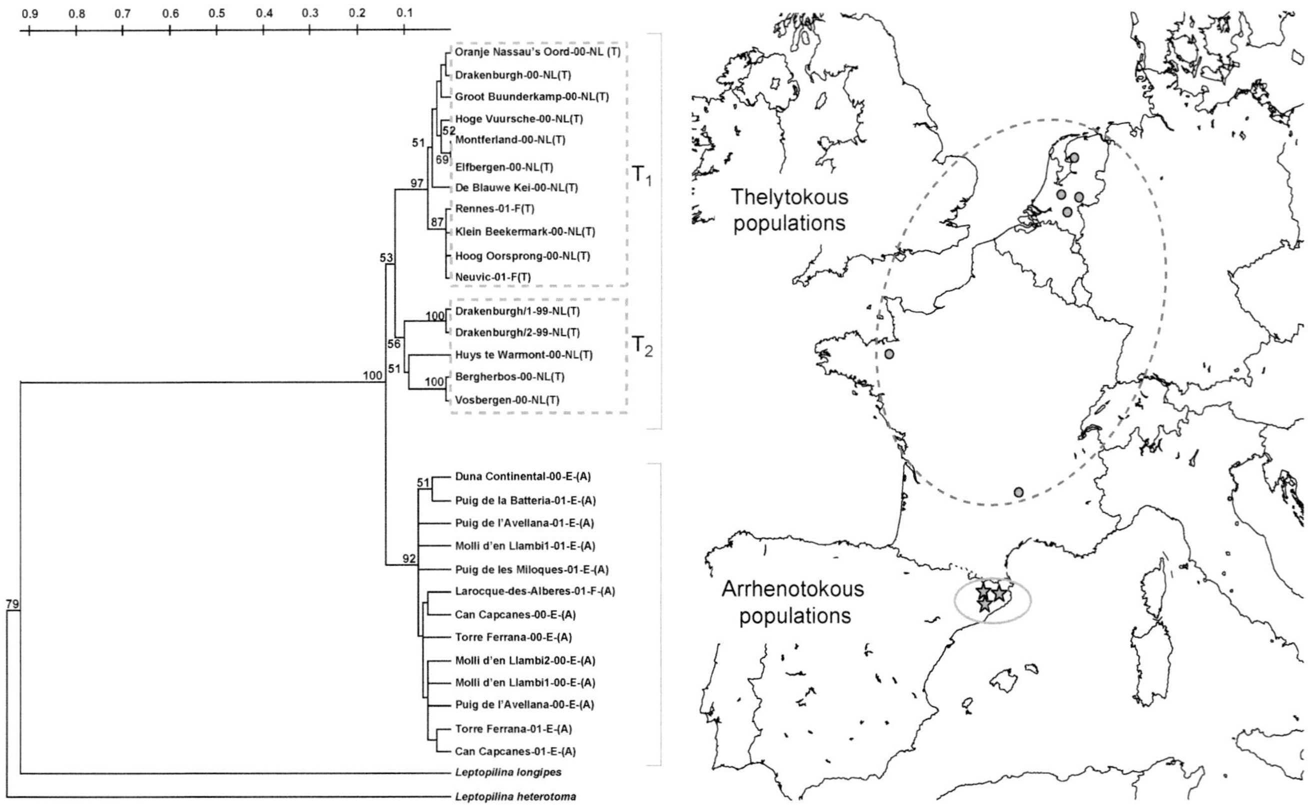

Fig. 17.7 UPGMA tree for genetic distance between arrhenotokous and thelytokous populations of European *Leptopilina clavipes*. T1 and T2 represent different clonal genotypes. Scale bar indicates genetic distance (Nei and Li 1979). Numbers at nodes represent bootstrap values over 50% (1000 replicates). Adapted from Pannebakker et al. (2004c) with permission from Blackwell Publishing

Various mechanisms can be involved in the maintenance of the observed clonal diversity in *L. clavipes*. Because both clonal lineages are widespread and were collected in similar habitats using the same array of baits, it can be assumed that they are not differentiated in niche use. Without niche differentiation, clonal coexistence is likely to be temporary, because one clonal lineage is expected to eventually replace the other through competitive exclusion or clonal drift (Jaenike et al. 1980). However, due to the metapopulation structure typical of species restricted to western European forest areas (Hanski 1999), competitive exclusion may proceed very slowly reducing the probability for complete extinction of one clonal lineage in *L. clavipes*. Another factor influencing diversity in clonal species is the accumulation of deleterious mutations due to Muller's Ratchet (Muller 1964; see also Chapter 5), which can eventually result in the extinction of clonal lineages (mutational meltdown, Gabriel et al. 1993). However, in species where thelytoky is *Wolbachia*-induced, Muller's Ratchet will be much less effective due to the mechanism of gamete duplication (Werren 1998), resulting in complete homozygosity of the offspring. Any deleterious mutations are immediately expressed homozygously and are therefore more likely to be purged by selection before drifting to fixation (Werren 1998; Haccou and Schneider 2004). Therefore, clonal extinction due to Muller's Ratchet is likely to play but a minor role in determining the clonal diversity of *Wolbachia* infected *L. clavipes* populations.

17.3.4 Evolutionary Consequences of Thelytoky

Thelytokous reproduction reduces selection on traits involved in sexual reproduction (see Section 17.1.6). Genes coding for these traits are no longer maintained by selection, and mutations in these genes can accumulate freely. On the other hand, certain genes may also be favoured by selection if they improve the performance of the infected females (Pijls et al. 1996; Werren 1998). In *Leptopilina clavipes*, male and female sexual functions were determined in populations infected by *Wolbachia* (Pannebakker et al. 2005). Males were obtained by antibiotic curing of females from thelytokous populations of different geographic origins. Most of these antibiotic curing induced males performed complete courtship behaviour with arrhenotokous females, resulting in successful copulation and sperm transfer. Interestingly, for all the thelytokous lines tested, the sex ratio (proportion of males) of the offspring resulting from these crosses, was significantly higher than that from crosses between arrhenotokous individuals (in both intra- and interpopulation crosses) Because of haplodiploid sex determination, and because *L. clavipes* does not have complementary sex determination, a higher sex ratio implies a lower fertilization success for the thelytokous males. Infected females were willing to mate with arrhenotokous males but they did not use the received sperm. Hence, restored males from thelytokous populations are sexually only partially functional, and females from thelytokous populations apparently lost their sexual functionality (Pannebakker et al. 2005). These results only partially parallel those of *Venturia canescens* in which thelytokous females were also found to not use stored sperm in most cases, although with the difference that they were inseminated by arrhenotokous males. Moreover, the genetic rather than infectious basis of thelytoky in *V. canescens* does not allow easy generation of males from thelytokous females by antibiotic curing. Rather, spontaneous aberrations of diploidy restoration appear responsible for occasional males in thelytokous *V. canescens* populations. Because in *L. clavipes* the thelytokous and arrhenotokous populations occur allopatrically, the observed reduced fertility in mixed crosses could have been the result of incompatibilities that had arisen after the separation of the two modes of reproduction. This alternative hypothesis was rejected after paternal and maternal molecular markers showed equal recovery in hybrid offspring. This suggests that genomic incompatibilities between the reproductive modes are absent and implies further that the reduced male fertility and non-fertilization of infected females are truly evolutionary consequences of asexual reproduction induced by *Wolbachia* bacteria.

If curing of thelytokous lines results in (partially) functional males that can produce viable offspring with uninfected arrhenotokous females, crossing experiments between these lines can help to determine the genetic basis of traits involved in sexual reproduction. This approach was used in *Leptopilina clavipes*, where the variation in male fertility between thelytokous and arrhenotokous lines was used to determine the genetic basis of this trait (Pannebakker et al. 2004a). Restored males from a thelythokous line (collected at Klein Beekermark, s'Heerenbergh, The Netherlands) were crossed to females from an arrhenotokous line (collected at Duna Continental, L'Estartit, Spain). Using amplified fragment length polymorphism (AFLP) markers, a genetic linkage map was generated spanning a total distance of 219.9 cM and consisting of five linkage groups, likely representing the five chromosomes of *L. clavipes*. The two parental lines differed in sex ratio, whereas the recombinant F_2 males produced an intermediate sex ratio (Fig. 17.8). In a small scale quantitative trait locus (QTL) study, a single genetic factor of large effect (explaining 46.5% of the phenotypic variance) was identified for male fertility which was called *male fertility factor* (*mff*). Although its effect size is likely to be overestimated due to small sample size, these results do indicate the presence of a major gene or a few tightly linked genes affecting male fertility in *L. clavipes*.

Different mechanisms could account for the phenotypic effect of the male fertility factor. However, visual inspection of the reproductive tracts of males of arrhenotokous and thelytokous lines failed to reveal differences between the two modes and showed the presence of viable sperm in the seminal vesicles and testes of males of both lines. This rules out a reduction in sperm production in the restored males as a possible mechanism explaining the effect of the male fertility factor, which is more likely to act later on in the fertilization process. Although the gene(s) responsible for the fertility reduction have not yet been identified, this study shows that curing of hymenopterans with *Wolbachia*-induced thelytoky can be very useful for the study of the genetic basis of character loss in the absence of selection.

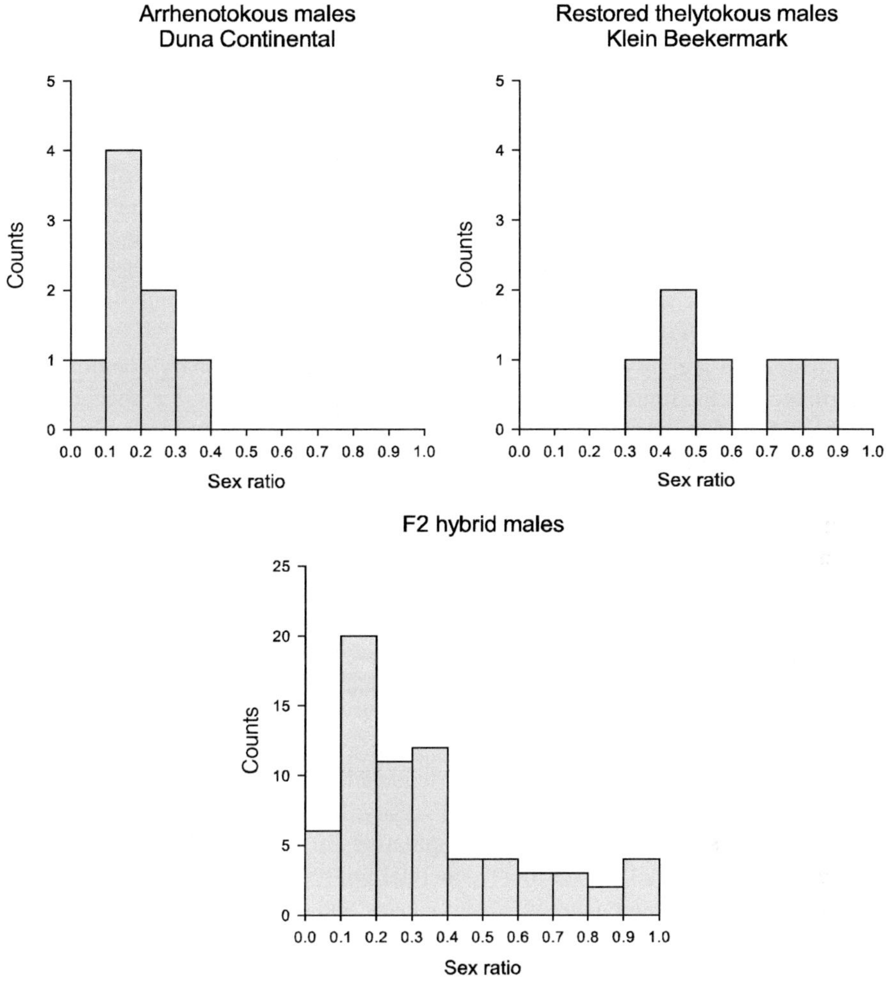

Fig. 17.8 Sex ratios of parental and F2 hybrid *Leptopilina clavipes* males. Adapted from Pannebakker et al. (2004a) with permission from the Genetics Society of America

17.4 Outlook

Egg development without fertilization occurs in all Hymenoptera. Most species reproduce by arrhenotoky (haploid males, diploid females), in which males develop parthenogenetically from unfertilized haploid eggs. Females develop from unfertilized eggs under thelytokous reproduction which require cytological mechanisms that maintain or restore diploidy. Most thelytokous hymenopterans have a form of automictic parthenogenesis involving fusion of meiotic products. Virtually nothing is known about the genetic regulation of these processes. Such information is required for a better understanding of the evolutionary dynamics of parthenogenesis and the twofold cost of sex paradox.

Thelytokous reproduction in Hymenoptera can have a number of important implications for the genetic make-up of individuals and the amount and structure of genetic variation in populations. Although several different cytological mechanisms are involved, all automictic forms are believed to increase the level of homozygosity within individuals and populations. However, we are only at the beginning of understanding the interplay between the genetic consequences of the cytological mechanism of thelytoky and the population level processes that shape genetic variation in thelytokous populations. This requires more detailed population genetic studies with a larger array of thelytokous species.

Hymenoptera are very suitable for investigating the evolutionary consequences of parthenogenetic reproduction. This is particularly the case for species with microbe-induced thelytoky, in which we have the unique ability to change thelytokous towards arrhenotokous reproduction by antibiotic treatment. Restored males can be used to test theories about mutation accumulation in the absence of selection for sexual function. The next step to be made is to take such studies to the genome level and identify the molecular processes responsible for trait loss and degeneration.

Studies on *Venturia canescens* have uncovered a number of basic differences in life-history traits between arrhenotokous and thelytokous individuals, which may facilitate their co-occurrence. However, ecological aspects of thelytokous reproduction have hardly been considered for other Hymenoptera. For example, does infection with parthenogenesis-inducing *Wolbachia* alter host life histories? Another little studied aspect of thelytoky is its consequences at the community level. Does arrhenotoky allow for more ecological diversity, such as a broader spectrum of hosts in parasitoids? Are arrhenotokous populations more able to adapt to changing environments? All in all, much more genetical and ecological research is needed for obtaining a full understanding of the causes and consequences of lost sex in Hymenoptera.

References

Agoze M, Drezen JM, Renault S, Periquet G (1994) Analysis of the reproductive potential of diploid males in the wasp *Diadromus pulchellus* (Hymentoptera: Ichneumonidae). Bull Entomol Res 84: 213–218

Amat I, Castelo M, Desouhant E, Bernstein C (2006) The influence of temperature and host availability on the host exploitation strategies of sexual and asexual parasitic wasps of the same species. Oecologia 148: 153–161

Arakaki N, Noda H, Yamagishi K (2000) *Wolbachia*-induced parthenogenesis in the egg parasitoid *Telenomus nawai*. Entomol Exp Appl 96: 177–184

Arakaki N, Miyosi T, Noda H (2001) *Wolbachia*-mediated parthenogenesis in the predatory thrips *Franklintothrips vespiformis* (Tysanoptera: Insecta). Proc R Soc Lond B 268: 1011–1016

Barke J, Mateo Leach, I, Beukeboom LW (2005) Fitness of arrhenotokous and thelytokous *Venturia canescens*. Proc Exp Appl Entomol Neth NEV Amsterdam 16: 27–35

Baudry E, Kryger P, Allsopp M, Koeniger N, Vautrin D, Mougel F, Cornuet JM, Solignac M (2004) Whole-genome scan in thelytokous-laying workers of the Cape honeybee (*Apis mellifera capensis*): central fusion, reduced recombination rates and centromere mapping using half-tetrad analysis. Genetics 167: 243–252

Beling I (1932) Zur Biologie von *Nemeritis canescens* Grav (Hymen Ophion). Z Angew Entomol 19: 223–249

Belshaw R, Quicke DLJ, Volkl W, Godfray HCJ (1999) Molecular markers indicate rare sex in a predominantly asexual parasitoid wasp. Evolution 53: 1189–1199

Beukeboom LW (1995) Sex determination in Hymenoptera: a need for genetic and molecular studies. BioEssays 17: 813–817

Beukeboom LW (2001) Single-locus complementary sex determination in *Venturia canescens* (Hymenoptera: Ichneumonidae). Neth J Zool 51: 1–15

Beukeboom LW, Pijnacker LP (2000) Automictic parthenogenesis in the parasitoid *Venturia canescens* (Hymenoptera: Ichneumonidae) revisited. Genome 43: 939–944

Beukeboom LW, Zwaan BJ (2005) Genetics. In: Jervis MA (ed) Insect natural enemies. Springer, Dordrecht, pp. 167–218

Beukeboom LW, Driessen G, Luckerhoff L, Bernstein C, Lapchin L, van Alphen JJM (1999), Distribution and relatedness of sexual and asexual *Venturia canescens* (Hymenoptera). Proc Exp Appl Entomol Neth NEV Amsterdam 10: 23–28

Beukeboom LW, Kamping A, van de Zande L (2007) Sex determination in the haplodiploid wasp *Nasonia vitripennis* (Hymenoptera: Chalcidoidea): a critical consideration of models and evidence. Semin Cell Dev Biol 18: 371–378

Beye MM, Hasselmann M, Fondrk MK, Page RE, Omholt SW (2003) The gene csd is the primary signal for sexual development in the honeybee and encodes an SR-type protein. Cell 114: 419–429

Braig HR, Turner BD, Normark BB, Stouthamer R (2002) Microorganism-induced parthenogenesis. In: Hughes RN (ed) Reproductive biology of invertebrates. Vol XI. Progress in asexual reproduction. Wiley and Sons, Chicester, pp. 1–62

Casas J, Driessen G, Mandon N, Wielaard S, Desouhant E, Van Alphen J, Lapchin L, Rivero A, Christides JP, Bernstein C (2003) Energy dynamics in a parasitoid foraging in the wild. J Anim Ecol 72: 691–697

Carson HL, Chang LS, Lyttle TW (1982) Decay of female sexual behavior under parthenogenesis. Science 281: 68–70

Comrie LC (1938) Biological and cytological observations on Tenthredinid parthenogenesis. Nature 142: 877–878

Cook JM (1993) Sex determination in the Hymenoptera: a review of models and evidence. Heredity 71: 421–435

Cook JM, Crozier RH (1995) Sex determination and population biology in the Hymenoptera. Trends Ecol Evol 10: 281–286

Crozier RH (1985) Adaptive consequences of male halpoidy. In: SM Helle W (ed) Spider mites, their biology, natural enemies, and control. Elsevier, Amsterdam, pp. 201–222

De Barro PJ, Hart PJ (2001) Antibiotic curing of parthenogenesis in *Eretmocerus mundus* (Australian parthenogenetic form). Entomol Exp Appl 99: 225–230

De Boer JG, Ode PJ, Vet LEM, Whitfield J, Heimpel GE (2007) Complementary sex determination in the parasitoid wasp *Cotesia vestalis* (*C. plutellae*). J Evol Biol 20: 340–348

Dodds KS (1939) Oogenesis in *Neuroterus baccarum* L. Genetica 21: 177–190

Doncaster L (1916) Gametogenesis and sex determination of the gall fly *Neuroterus lenticularis* (*Spathegaster baccarum*) III. Proc R Soc Lond B 89: 183–200

Driessen G, Bernstein C (1999) Patch departure mechanisms and optimal host exploitation in an insect parasitoid. J Anim Ecol 68: 445–459

Driessen G, Hemerik L, van Alphen JJM (1990) *Drosophila* species, breeding in the stinkhorn (*Phallus impudicus* Pers.) and their larval parasitoids. Neth J Zool 40: 409–427

Ellers J, Sevenster JG, Driessen G (2000) Egg load evolution in parasitoids. Am Nat 156: 650–665

Fong DW, Kane TC, Culver DC (1995) Vestigialization and loss of nonfunctional characters. Annu Rev Ecol Syst 26: 249–268

Fournier D, Estoup A, Orivel J, Foucaud J, Jourdan H, Le Breton J, Keller L (2005) Clonal reproduction by males and females in the little fire ant. Nature 435: 1230–1234

Gabriel W, Lynch M, Burger R (1993) Muller's ratchet and mutational meltdowns. Evolution 47: 1744–1757

Gauld ID, Bolton B (1996) The Hymenoptera. Oxford University Press, Oxford

Godfray HCJ (1994) Parasitoids: behavioral and evolutionary ecology. Princeton University Press, Princeton, NJ

Gottlieb Y, Zchori-Fein E (2001) Irreversible thelytokous reproduction in *Muscidifurax uniraptor*. Entomol Exp Appl 100: 271–278

Gottlieb, Y, Zchori-Fein, E, Werren, JH, Karr, TL (2002) Diploidy restoration in Wolbachia-infected *Muscidifurax uniraptor* (Hymenoptera: Pteromalidae). J Invertebr Pathol 81: 166–174

Grimaldi D, Engel MS (2005) Evolution of the insects. Cambridge University Press, Cambridge, UK

Groot TVM, Breeuwer JAJ (2006) Cardinium symbionts induce haploid thelytoky in most clones of three closely related *Brevipalpus* species. Exp Appl Acarol 39: 257–271

Haccou P, Schneider MV (2004) Modes of reproduction and the accumulation of deleterious mutations with multiplicative fitness effects. Genetics 166: 1093–1104

Hagimori T, Abe Y, Date S, Miura K (2006) The first finding of a Rickettsia bacterium associated with parthenogenesis induction among insects. Curr Microbiol 52: 97–101

Hanski I, (1999) Metapopulation ecology. Oxford University Press, Oxford

Huigens ME, Stouthamer R (2003) Parthenogenesis associated with *Wolbachia*. In: Bourtzis K, Miller TA (eds) Insect symbiosis. CRC Press Boca Raton, Florida, pp. 247–266

Jaenike J, Parker ED, Selander RK (1980) Clonal niche structure in the parthenogenetic earthworm *Octolasion tyrtaeum*. Am Nat 166: 196–205

Jeong G, Stouthamer R (2005) Genetics of female functional virginity in the parthenogenesis-Wolbachia infected parasitoid wasp *Telenomus nawai* (Hymenoptera: Scelionidae) Heredity 94: 402–407

Lamb RY, Willey RB (1987) Cytological mechanisms of thelytokous parthenogenesis in insects. Genome 29: 367–369

Lattorff HMG, Moritz RFA, Fuchs S (2005) A single locus determines thelytokous parthenogenesis of the laying honeybee workers (*Apis mellifera capensis*). Heredity 94: 533–237

Ledoux A (1954) Recherches sur le cycle chromosomique de la fourmi fileuse *Oecophylla longinoda* Latr. Insect Soc 1: 149–175

Legner EF (1985) Effects of scheduled high temperature on male production in thelytokous *Muscidifurax uniraptor* (Hymenoptera: Pteromalidae). Can Entomol 117: 383–389

Luchetta P, Desouhant E, Wajnberg E, Bernstein C (2007). Small but smart: the interaction between environmental cues and internal state modulates host-patch exploitation in a parasitic wasp. Behav Ecol Sociobiol 61: 1409–1418

Maynard-Smith J (1978) The evolution of sex. Cambridge University Press, Cambridge

Muller HJ (1949) The Darwinian and modern conceptions of natural selection. Proc Am Philos Soc 93: 459–470

Muller HJ (1964) The relation of recombination to mutational advance. Mutat Res 1: 2–9

Nei M, Li WH (1979) Mathematical model for studying genetic variation in terms of restriction endonucleases. Proc Natl Acad Sci USA 76: 5269–5273

Nordlander G (1980) Revision of the genus *Leptopilina* Forster, 1869, with notes on the status of some other genera (Hymenoptera, Cynipoidea: Eucoilidae). Entomol Scand 11: 428–453

O'Neill SL, Hoffmann AA, Werren JH (1997) Influential passengers: Microbes and invertebrate reproduction. Oxford University Press, Oxford

Pannebakker BA (2004) Evolutionary consequences of *Wolbachia*-induced parthenogeneses in the parasitoid *Leptopilina clavipes*. PhD Thesis Leiden University, Leiden

Pannebakker BA, Beukeboom LW, van Alphen JJM, Brakefield PM Zwaan BJ. (2004a) The genetic basis of male fertility in relation to haplodiploid reproduction in *Leptopilina clavipes*. Genetics 168: 341–349

Pannebakker BA, Pijnacker LP, Zwaan BJ, Beukeboom LW (2004b) Cytology of *Wolbachia*-induced parthenogenesis in *Leptopilina clavipes* (Hymenoptera: Figitidae). Genome 47: 299–303

Pannebakker BA, Zwaan BJ, Beukeboom LW, van Alphen JJM (2004c). Genetic diversity and *Wolbachia* infection of the *Drosophila* parasitoid *Leptopilina clavipes* in western Europe. Mol Ecol 13: 1119–1128

Pannebakker BA, Schidlo NS, Boskamp GJ, Dekker L, van Dooren TJ, Beukeboom LW, Zwaan BJ, Brakefield PM, van Alphen JJ (2005) Sexual functionality of *Leptopilina clavipes* (Hymenoptera: Figitidae) after reversing *Wolbachia*-induced parthenogenesis. J Evol Biol 18: 1019–1028

Peacock AD, Sanderson AR (1939) The cytology of the thelytokous sawfly *Thrinax maculata*. Trans R Soc Edin 59: 647

Pearcy M, Aron S, Doums C, Keller L (2004) Conditional use of sex and parthenogenesis for worker and queen production in ants. Science 306: 1780–1783

Pelosse P, Bernstein C, Desouant E (2007) Differential energy allocation as an adaptation to different habitats in the parasitic wasp *Venturia canescens*. Evol Ecol 21: 669–685

Perlman SJ, Hunter MS, Zchori-Fein E (2006) The emerging diversity of Rickettsia. Proc R Soc Lond B 273: 2097–2106

Pijls JWAM, van Steenbergen HJ, Alphen JJM (1996) Asexuality cured: the relations and differences between sexual and asexual *Apoanagyrus diversicornis*. Heredity 76: 506–513

Plantard O, Rasplus J-Y, Mondor G, Le Clainche I, Solignac M (1998) *Wolbachia*-induced thelytoky in the rose gall wasp *Diplolepis spinosissimae* (Giraud) (Hymenoptera: Cynipidae), and its consequences on the genetic structure of its host. Proc R Soc Lond B 265: 1075–1080

Rössler Y, DeBach P (1973) Genetic variability in a thelytokous form of *Aphytis mytilaspidis* (Le Baron) (Hymenoptera: Aphelinidae). Hilgardia 42: 149–175

Ruttner F (1988) Taxonomy and biogeography of honeybees. Springer, Berlin

Salt S (1976) The host of *Nemeritis canescens*: a problem in the host specificity of insect parasitoids. Ecol Entomol 1: 63–67

Schidlo NS, Pannebakker BA, Zwaan BJ, Beukeboom LW, Van Alphen JJM (2002) Curing thelytoky in the *Drosophila* parasitoid *Leptopilina clavipes* (Hymenoptera: Figitidae). Proc Exp Appl Entomol Neth NEV Amsterdam 13: 93–96

Schilder K, Heinze J, Gross R, Hölldobler B (1999) Microsatellites reveal clonal structure of populations of the thelytokous ant *Platythyrea punctata* (F. Smith) (Hymenoptera; Formicidae). Mol Ecol 8: 1497–1507

Schneider MV (2003) Modes of reproduction in the solitary parasitoid wasp *Venturia canescens*. PhD thesis, University of Leiden, Leiden

Schneider MV, Beukeboom LW, Driessen G, Lapchin L, Bernstein C, van Alphen JJM (2002) Geographical distribution and genetic relatedness of sympatrical thelytokous and arrhenotokous populations of the parasitoid *Venturia canescens* (Hymenoptera). J Evol Biol 15: 191–200

Schneider MV, Driessen G, Beukeboom LW, Boll R, van Eunen K, Selzner A, Talsma J, Lapchin L (2003) Gene flow between arrhenotokous and thelytokous populations of *Venturia canescens* (Hymenoptera). Heredity 90: 260–267

Simon JC, Delmotte F, Rispe C, Crease T (2003) Phylogenetic relationships between parthenogens and their sexual relatives: the possible routes to parthenogenesis in animals. Biol J Linn Soc 79: 151–163

Slobodchikoff CN, Daly HV (1971) Systematic and evolutionary implications of parthenogenesis in the hymenoptera. Am Zool 11: 273–282

Smith SG (1941) A new form of the spruce sawfly identified by means of its cytology and parthenogenesis. Sci Agr 21: 245–305

Speicher BR (1937) Oogenesis in a thelytokous wasp *Nemeritis canescens* (Grav). J Morphol 61: 453–471

Speicher BR, Speicher KG, Roberts FL (1965) Genetic segregation in the unisexual wasp *Devorgilla*. Genetics 52: 1035–1041

Stille B, Dävring L (1980) Meiosis and reproductive strategy in the parthenogenetic gall wasp *Diplolepis rosae*. Genetica 67: 145–151

Stouthamer R (1997) *Wolbachia*-induced parthenogenesis. In: O'Neill SL, Hoffman AA, Werren, JH (eds) Influential passengers: inherited microorganisms and arthropod reproduction. Oxford University Press, Oxford, pp. 102–124

Stouthamer R, Kazmer DJ (1994) Cytogenetics of microbe-associated parthenogenesis and its consequences for gene flow in *Trichogramma* wasps. Heredity 73: 317–327

Stouthamer R, Luck R, Hamilton WD (1990a) Antibiotics cause parthenogenetic *Trichogramma* (Hymenoptera/Trichogrammatidae) to revert to sex. Proc Natl Acad Sci USA 87: 2424–2427

Stouthamer R, Pinto JD, Platner GR, Luck R (1990b) Taxonomic status of the thelytokous forms of *Trichogramma* (Hymenoptera, Trichogrammatidae). Ann Entomol Soc Am 83: 475–481

Stouthamer R, Breeuwer JAJ, Hurst GDD (1999) *Wolbachia pipientis*: Microbial manipulator of arthropod reproduction. Annu Rev Microbiol 53: 71–102

Suomalainen E, Saura A, Lokki J (1987) Cytology and evolution in parthenogenesis. CRC Press Boca Raton Inc, Florida

Thiel A, Driessen G, Hoffmeister TS (2006) Different habitats, different habits? Response to foraging information in the parasitic wasp *Venturia canescens*. Behav Ecol Sociobiol 59: 614–623

Tucker KW (1958) Automictic parthenogenesis in the honey bee. Genetics 43: 299–316

Van Wilgenburg E, Driessen G, Beukeboom LW (2006) Single locus complementary sex determination in Hymenoptera: an "unintelligent" design? Frontiers Zool 3: 1

Vavre F, de Jong JH, Stouthamer R (2004) Cytogenetic mechanism and genetic consequences of thelytoky in the wasp *Trichogramma cacoeciae*. Heredity 93: 592–596

Verma S, Ruttner F (1983) Cytological analysis of the thelytokous parthenogenesis in the Cape honeybee (*Apis mellifera capensis* Escholtz). Apidologie 14: 41–57

Vet LEM (1983) Host-habitat location through olfactory cues by *Leptopilina clavipes* (Hartig) (Hymenoptera: Eucoilidae), a parasitoid of fungivorous *Drosophila*: the influence of conditioning. Neth J Zool 33: 225–248

Weeks AR, Breeuwer JA. (2001) *Wolbachia*-induced parthenogenesis in a genus of phytophagous mites. Proc R Soc Lond B 268: 2245–2251

Weeks AR, Marec F, Breeuwer JAJ (2001) A mite species that consists entirely of haploid females. Science 292: 2479–2482

Wenseleers T, Billen J (2000) No evidence for *Wolbachia*-induced parthenogenesis in the social Hymenoptera. J Evol Biol 13: 277–280

Werren JH (1997) Biology of *Wolbachia*. Annu Rev Entomol 42: 587–609

Werren JH (1998) *Wolbachia* and speciation. In: Howard DJ, Berlocher SH (eds) Endless forms: species and speciation. Oxford University Press, New York, pp. 245–260

Werren JH, Zhang W, Guo LR (1995) Evolution and phylogeny of *Wolbachia*: reproductive parasites of arthropods. Proc R Soc Lond B 261: 55–63

Whiting PW (1943) Multiple alleles in complementary sex determination in *Habrobracon*. Genetics 28: 365–382

Williams GC (1975) Sex and evolution. Monographs in population biology. Princeton University Press, Princeton

Zchori-Fein E, Perlman SJ (2004) Distribution of the bacterial symbiont *Cardinium* in Arthropods. Mol Ecol 13: 2009–2016

Zchori-Fein E, Roush RT, Hunter MS (1992) Male production induced by antibiotic treatment in *Encarsia formosa* (Hymenoptera: Aphelinidae), an asexual species. Experientia 48: 102–105

Zchori-Fein E, Gottlieb Y, Kelly SE, Brown JK, Wilson JM, Karr TL, Hunter MS (2001) A newly discovered bacterium associated with parthenogenesis and a change in host selection behaviour in parasitoid wasps. Proc Natl Acad Sci USA 98: 12555–12560
Zchori-Fein E, Perlman J, Kelly SE, Katzir N, Hunter MS (2004) Characterization of a 'Bacteroidetes' symbiont in *Encarsia* wasps (Hymenoptera: Aphelinidae): proposal of '*Candidatus Cardinium hertegii*'. Int J Syst Evol Micr 54: 961–968

Chapter 18
Sex in Parthenogenetic Planarians: Phylogenetic Relic or Evolutionary Resurrection?

Thomas G. D'Souza and Nico K. Michiels

Abstract Besides their remarkable capability of regeneration, planarian flatworms are well known for their wide range of reproductive modes. In this chapter, we elucidate the evolutionary significance of sexual and parthenogenetic reproduction in the freshwater planarian *Schmidtea polychroa*. In accordance with the major theories of sex, parthenogenetic *S. polychroa* seem to suffer from both a higher mutation load and parasite load. Nevertheless, parthenogenesis is still maintained in *S. polychroa* and is more prevalent across Europe than sexual reproduction. The success of parthenogenesis can be explained by occasional sexual events in predominantly parthenogenetic types. We stress the evolutionary consequences of such rare sexual processes for the success and maintenance of parthenogenesis. At the end of the chapter, we speculate on what future perspectives this system still has to offer.

18.1 The Uniqueness of Planarians

Freshwater planarian flatworms have long been known for their remarkable diversity in reproductive modes, both within and between species, including fissioning, parthenogenesis and sexual outcrossing, usually correlated with various ploidy levels. Gamete-producing forms are always hermaphroditic, with very rare exceptions. Although planarians offer many advantages for the study of reproductive modes, very few research groups have actually used them to address questions concerning the costs and benefits of sex. Much of the research on planarian reproduction has been initiated by the extensive work on flatworm cytogenetics by Mario Benazzi and his wife Guiseppina Benazzi-Lentati, who, for many decades, described the various spectacular and diverse reproductive mechanisms present in these animals. Current interest in planarian reproduction is rising rapidly because of their basal position

T.G. D'Souza (✉)
Faculty of Biology, Institute for Evolution and Ecology, Animal Evolutionary Ecology, University of Tuebingen, Auf der Morgenstelle 28, D-72076, Tuebingen, Germany
e-mail: thomas.dsouza@uni-tuebingen.de

I. Schön et al. (eds.), *Lost Sex*, DOI 10.1007/978-90-481-2770-2_18,

within the bilateral metazoans. Initiated by work in the group of Jaume Baguña in Barcelona in the 1980s, several large groups worldwide are now using planarians as a model system for regeneration and stem cell research, yielding many molecular tools for use in evolutionary biology and population genetics.

In this chapter, we first give some background on the life history and ecology of flatworms in general and *Schmidtea polychroa*, our house model, in particular. We put special emphasis on the advantages of *S. polychroa* for comparisons between obligate outcrossing sexual and parthenogenetic forms. The latter are unusual in that their female germ line is essentially asexual, but requires sperm from conspecifics to activate embryogenesis. This process results in occasional "leakage" of paternal chromosomes into the zygote. The male germ line of (hermaphroditic!) parthenogens is capable of fertilising eggs of both sexual and parthenogenetic conspecifics. These features make this system unique among animals.

18.1.1 Flatworms in General

Planarians are free-living flatworms (Platyhelminthes). They are widely distributed in marine, terrestrial and freshwater habitats. Of these, the freshwater planarians (Order Seriata, Suborder Tricladida, Infraorder Paludicola or Continenticola) are best studied. As all other flatworms, they possess a basic, bilateral body plan and a simple centralised nervous system. Their primitive position shows in the lack of circulatory, respiratory and skeletal structures, as well as the presence of a blind-ending gut. Rather than having a body cavity or coelom, as all higher animals do, a flatworms' body is filled with loose tissue or parenchyma (Benazzi and Benazzi Lentati 1976).

18.1.2 Freshwater Planarians

What is considered a "typical" planarian is a 1–2 cm long, flattened, mucous covered worm that glides with a cilia-driven motion without signs of peristaltic contractions. The common *Dugesia*-type is usually dark-brown with two characteristic cross-eyed-looking, simple cup-eyes (Figs. 18.1A, 18.2). They possess a remarkable regenerative power (Salo 2006; Birnbaum and Sánchez Alvarado 2008). Responsible for regeneration are neoblasts, multi-potent, proliferating stem cells which can differentiate into any functional cell type. Regeneration allows for fissioning, a simple form of vegetative reproduction. Mostly, planarians first divide and then regenerate the missing parts (archytomy). The reverse strategy to grow and differentiate a full new individual before fission (paratomy) is less common (Benazzi and Benazzi Lentati 1993). Other members of this group reproduce parthenogenetically. In most cases, parthenogens are sperm-dependent (pseudogamy) and require sperm from a partner to trigger embryogenesis (Benazzi and Benazzi Lentati 1993; Beukeboom and Vrijenhoek 1998), despite the fact that – being

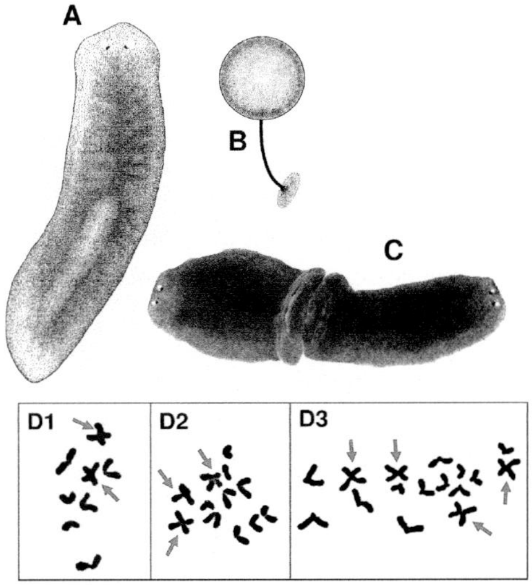

Fig. 18.1 **A**: Adult specimen of *Schmidtea polychroa* (length approximately 15 mm), **B**: Spherical stalked cocoon; **C**: Mating pair, **D1–D3**: Karyotypes. *Arrows* indicate large metacentric chromosomes. **D1**: Diploid sexual (2x = 8), **D2**: triploid parthenogen (3x = 12), **D3**: tetraploid parthenogen (4x = 16)

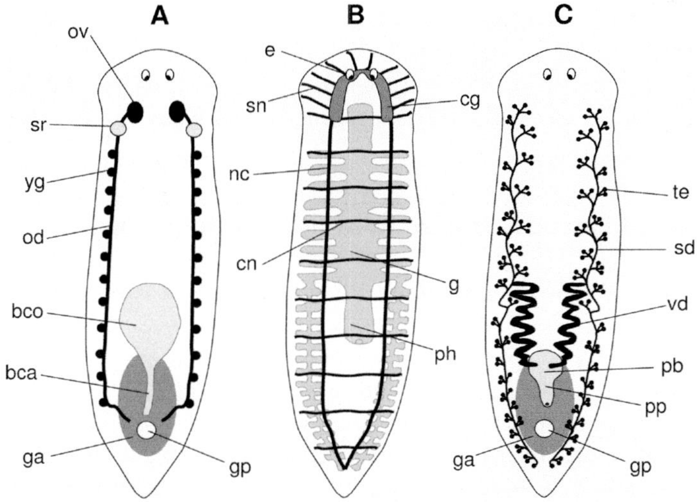

Fig. 18.2 Schematic representation of the internal anatomy of *Schmidtea polychroa*. Structures are not drawn to scale. **A**: Female reproductive organs; **B**: Gut and nervous system; **C**: Male reproductive organs. Abbreviations: bca, bursal canal; bco, bursa copulatrix; cg, cephalic ganglion; cn, commissural neurons; e, eye; g, gut system; ga, genital atrium; gp, gonopore; nc, nerve cord; od, ovovitelline duct; ov, ovary; pb, penis bulb; ph, pharynx; pp, penis papilla; sd, sperm duct; sn, sensory neurons; sr, seminal receptacle; te, testes; vd, vas deferens; yg, yolk glands

hermaphroditic – they produce sperm themselves. In many planarian species parthenogenesis is associated with polyploidy whereby ploidy levels can go up from triploidy (most common) to decaploidy (very rare) (Benazzi and Benazzi Lentati 1976; Benazzi 1982). In sexually (diploid) reproducing planarians, cross-fertilisation is the rule and selfing is virtually absent. Different reproductive modes can be present within one species or even within a single individual, but all individuals are simultaneous hermaphrodites, with a tendency to slight protandry in some species (Benazzi and Benazzi Lentati 1976). The only planarians with pure males and females are marine species of the genus *Sabussowia* (Tekaya et al. 1999). Adding to this bewildering diversity is the bizarre capacity of some fissioning planarians to switch to sexual reproduction after ingesting sexual forms of other planarian species. This indicates that species-unspecific, sexualising substances may trigger sexual reproduction (Kobayashi et al. 1999a, 1999b, 2002; Hoshi et al. 2003). The high incidence of mixed sexual/vegetative reproduction or sexual/parthenogenetic reproduction suggests that the archetype of Plathyhelminthes combined sexual and asexual reproduction (Gremigni et al. 1982; Rieger 1986).

Freshwater planarians of the Family Dugesiidae in general and *Schmidtea mediterranea* in particular have become an important model system for studying the cytological and molecular basis of development and regeneration (Newmark and Sánchez Alvarado 2002; Sánchez Alvarado et al. 2002; Sánchez Alvarado 2003, 2004; Orii et al. 2005; Salo 2006; Sato et al. 2006). This has led to the development of many molecular and cytogenetic tools for *S. mediterranea* (Robb et al. 2008), such as ESTs (Sánchez Alvarado et al. 2002; Zayas et al. 2005) and RNAi techniques (Sánchez Alvarado and Newmark 1999; Reddien et al. 2005). The whole genome of *S. mediterranea* is currently being sequenced (http://genome.wustl.edu/), which will complete the already existing genetic data bases for this species. This chapter focuses on a sister species of *S. mediterranea*, *S. polychroa*.

18.2 General Introduction to Schmidtea polychroa

18.2.1 General Characteristics

Schmidtea polychroa is a generalist that can be found in meso- to eutrophic freshwater habitats such as lowland rivers, streams, ditches, and lakes. It can be collected from the underside of stones in shallow water. In habitats lacking solid substrates individuals may be observed directly on the sediment or on plant surfaces. Like other freshwater planarians, *S. polychroa* mainly feeds on small gastropods, oligochaets and isopods (Reynoldson and Davies 1970). Feeding is done with the help of an eversible pharynx, a muscular tube which is protruded through the mouth and may penetrate the prey by peristaltic movement and secretion of digestive fluids. The semi-fluid interior of the prey is then taken up into the gut diverticles. Because the gut is blind-ending, indigestible material is discharged through the pharynx (Ball and Reynoldson 1981; Fig. 18.2B).

18.2.2 Reproductive Organs, Life Cycle and Development

Both male and female gonadal ducts and the bursal canal from the copulatory sac (*bursa copulatrix*) open into a common genital atrium, which connects to the outside via the ventral gonopore (see Fig. 18.2A, C). Numerous, small testes extend dorsally over most of the body and drain into two large sperm ducts proceeding posteriorly. Before entering the penis bulb, the sperm ducts are widened and serve as a storage organ for autosperm. The male copulatory organ consists of a muscular penis bulb (wide base) and papilla (slender tip). During copulation the penis is reciprocally inserted into the partner's genital atrium and a single sperm mass is transferred into the partner's *bursa* (Peters et al. 1996). During the hours following sperm receipt, allosperm migrate out of the *bursa* and along the paired ovovitelline ducts towards the sperm receptacles in the "neck" region, adjacent to the ovaries. Although 99.7% of the received sperm is resorbed within the *bursa* and along the ovovitelline duct soon after receipt, the rest can be stored and used for months (Sluys 1989; Streng 1999; Pongratz and Michiels 2003). Immediately after leaving the ovaries, the small eggs (25–30 μm) are fertilised. While moving towards the genital atrium, they are joined by a large quantity of yolk cells from yolk glands bordering the ovovitelline ducts. Once they have reached the genital atrium, eggs and yolk cells are enclosed in a leathery, orange-brown, almost spherical cocoon shell (average ø 2 mm), which is attached to the substrate with a stalk, and turns into a deep chestnut black during the following hours (Ball and Reynoldson 1981, Fig. 18.1B).

Schmidtea polychroa is perennial and can breed repeatedly for several years under natural conditions (Reynoldson 1966). Lifespan in the laboratory has exceeded 30 years in rare cases (Benazzi 1992). The main reproductive season is during spring and early summer in Britain (Reynoldson 1977), but may start earlier and last longer, with a summer break, in warmer parts of Europe (own observations). Mating individuals lift their "tail" tips and press the ventral gonopores against their partner's (Fig. 18.1C). Pairs stay in copula from 5 min to 2.5 h (Peters et al. 1996). In *S. polychroa*, sperm exchange typically occurs after having been *in copula* for more than 30 min. Insemination is fast and happens virtually synchronously. After insemination, partners remain *in copula* for about another hour before separating.

Several ectolecithal (yolk-poor) eggs are encapsulated within one cocoon, which is for the most part filled with yolk cells. Mature individuals produce up to one cocoon per day in the laboratory, from which about 4–5, but up to ten offspring will hatch after three weeks at 14°C. Embryo development takes place within the cocoon in a very peculiar and unique fashion ("blastomere anarchy") (Cardona et al. 2005, 2006). During the early developmental stages, embryos form a transient, embryonic pharynx to feed on common yolk cells. In a later stage, this pharynx is absorbed and replaced by a definite pharynx. Most probably, the embryonic pharynx has evolved as a consequence of strong resource competition among embryos. Except for their reproductive system, hatchlings are fully developed and may reach reproductive maturity after six to eight weeks.

18.2.3 Reproductive Types

The characterisation of reproductive types as well as the phylogenetic relationships within the genus *Schmidtea* (formerly *Dugesia*) has been unravelled by Benazzi and co-workers and was mainly based on karyology. Within the *polychroa-lugubris* group Benazzi distinguished seven biotypes (A–G) that can be assigned to four reproductively isolated species (Benazzi et al. 1970; Benazzi 1982). Based on comparative morphology of the genitalia and chromosomes, biotypes A, B, C and D were found to correspond to *S. polychroa*, whereas the remaining three form chromosomally differentiated and reproductively isolated species: Biotype E *S. lugubris*, biotype F *S. nova nomen nudum*, biotype G *S. mediterranea* (Benazzi et al. 1975; Benazzi 1982).

In the following, we only refer to biotypes A–D, whose reproductive cytogenetics are described in detail in Benazzi Lentati (1970). All four biotypes are functional simultaneous hermaphrodites and share the same basic haploid chromosome number (see Fig. 18.1D). Biotype A is diploid ($2x = 8$) and amphimictic with normal meiosis in both germ lines (Fig. 18.3A). The remaining three biotypes are (auto) polyploid parthenogens, probably derived from sexual, diploid ancestors (Benazzi

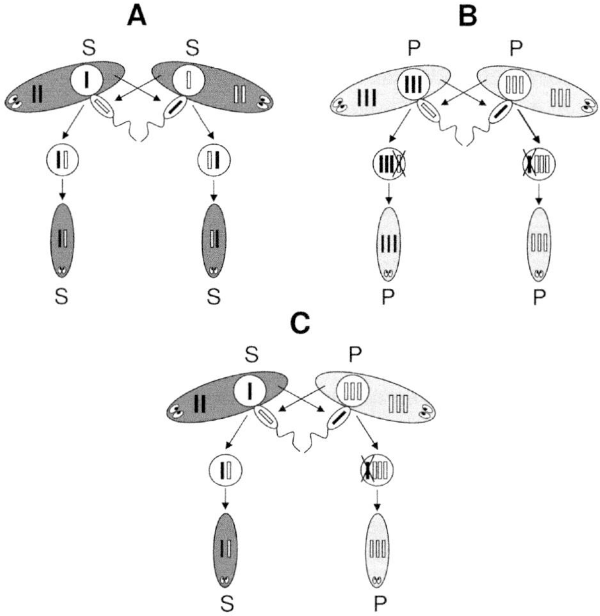

Fig. 18.3 Reproductive types of *Schmidtea polychroa* and their possible interactions. Sexual individuals (S) are marked in dark grey, parthenogens (P) in light grey. The number of bars indicates the ploidy level. **A**: Sexual reproduction: Diploid sexual individuals reciprocally donate haploid sperm which is used to fertilise the partner's haploid egg. **B**: Sperm-dependent parthenogenesis: Triploid parthenogens produce triploid eggs which require haploid sperm from the partner to trigger embryo development. The paternal chromosome set is discarded from the zygote with the consequence that parthenogenetic offspring and their "mother" are genetically identical. **C**: Mixed mating: In mixed crosses, sexual individuals receive haploid sperm from the polyploid parthenogens. Via this process diploid, sexual offspring arise which carry parthenogenetic chromosomes (*open bars*). On the other hand, haploid sperm from sexuals only induces development of triploid eggs from the parthenogenetic partner. In this case, the whole parthenogenetic genome (*open bars*) is transmitted to the next generation without any paternal contribution. Reproduction of tetraploid parthenogens proceeds in a similar fashion but with tetraploid eggs instead of triploids

1963, 1982). Both biotypes B and C are somatically triploid ($3x = 12$) and produce triploid eggs. But the process by which this is achieved is fundamentally different. Oogenesis in biotype B involves premeiotic doubling of the triploid chromosome set via endomitosis leading to hexaploid oogonia. Subsequently, reduction proceeds by regular two-step meiosis (automixis). Thus, the triploid condition is restored. As synapsis (chromosome pairing) occurs between previously doubled chromosomes, crossing over does not lead to effective recombination (Benazzi Lentati 1970). Contrary to B, biotype C neither shows premeiotic doubling nor meiosis during oogenesis and eggs are produced mitotically (apomixis). Biotype D is somatically tetraploid ($4x = 16$) and produces eggs mitotically, as described for biotype C. Additionally, pentaploids ($5x = 20$) and triploid-tetraploid mixoploids occur, whose exact mode of parthenogenesis is at present undescribed (Beukeboom et al. 1996; D'Souza et al. 2004).

Parthenogens are sperm-dependent, i.e., in order to proliferate, parthenogenetic eggs require penetration by a spermatozoon (sperm-dependent parthenogenesis or pseudogamy, Beukeboom and Vrijenhoek 1998). However, the sperm nucleus does not fuse with the female pronucleus. Instead, it is expelled from the zygote with one of the polar bodies (biotype B) or degenerates within the zygote (biotypes C and D) shortly after gamete fusion (Benazzi Lentati 1970; Fig. 18.3B). As selfing does not occur in this species, sperm dependence implicates the necessity of copulation in parthenogenetic as well as sexual biotypes.

What makes *Schmidtea polychroa* truly unique is that all parthenogenetic forms retain their functional male organs and produce haploid and functional sperm, despite the fact that they are polyploid (Benazzi Lentati 1970). At an early stage during spermatogenesis, one (in triploids) or two (in tetraploids) chromosome sets are excluded from the male germ line. This results in diploid spermatocytes, which are subsequently reduced by regular meiosis. Thus, rather than being reproductively isolated due to their clonal mode of reproduction, parthenogenetic biotypes B, C and D may interbreed with their sexual conspecifics of type A and sire viable diploid progeny (Benazzi 1963; Benazzi Lentati 1966, 1970; Fig. 18.3C). This is not only significant for population genetic processes in mixed populations. It also provides a valuable experimental tool to investigate the genetic basis (Benazzi Lentati 1970) and the consequences of clonal reproduction (Storhas et al. 2000).

18.2.4 Reproductive Behaviour

Although belonging to the more primitive metazoans, *Schmidtea polychroa* shows a relatively complex reproductive and mating behaviour. Unlike other planarians (Vreys and Michiels 1997; Vreys et al. 1997), *S. polychroa* shows no clear indication for pre-copulatory partner assessment (Peters et al. 1996). No evidence was found for avoidance of inbreeding with close kin (Peters and Michiels 1996a), for mate choice with respect to body size (Peters and Michiels 1996b) or to reproductive mode (Storhas 2001). However, individuals are more likely to donate sperm to partners that reciprocate insemination, which indicates *in copula* mate assessment

(Michiels and Streng 1998). Although unilateral sperm donations are possible, reciprocal exchange is more common than expected. At the other end of the scale, up to a third of all matings are terminated without sperm transfer (Michiels and Streng 1998). This pattern is indicative of conditional sperm donation, meaning that individuals are more likely to donate sperm, when their partner shows signs that it will reciprocate (Michiels and Streng 1998; Michiels and Bakovski 2000). This strategy of "sperm trading" is also present in parthenogens (Michiels and Kuhl 2003). Hence, parthenogens may require sperm production and donation to ensure full maternal fertility – despite the fact that in purely parthenogenetic populations paternal fitness is close to zero (Michiels and Kuhl 2003).

As suggested by the lack of pre-copulatory mate choice, sexual forms of *Schmidtea polychroa* are highly promiscuous and multiple paternity is the rule within cocoons. More than 80% of all cocoons produced during 4 weeks in groups of 10 individuals had two to five fathers for only three to five offspring per cocoon. Last-male sperm precedence is low: the immediate paternity a sperm donor can expect is about 25% (Pongratz and Michiels 2003). It was also shown that male reproductive success increases with the number of mates. There was, however, no sign that sperm trading led to a corresponding level of reciprocal fertilisation.

18.2.5 Phylogeographic Distribution and Population Genetics

The geographic distribution of reproductive types of *Schmidtea polychroa* in Europe is characterised by the absence of sexuals from most of Europe north of the Pyrenees and Alps (Beukeboom et al. 1996; Pongratz et al. 1998, 2003). Exceptions are sexual populations found in Swedish lakes by Melander (1963). In southern Europe, especially Italy, sexuals are common and often coexist in sympatry with parthenogens. Among parthenogens, triploids are particularly common (especially biotype B triploids), whereas higher ploidy levels (tetraploids and pentaploids) are rare (Reynoldson 1966; Beukeboom et al. 1996).

Strong intraspecific divergence among mtDNA lineages within and between populations suggests that major lineages split long before the quaternary ice ages. These may have survived the ice ages in ice-free zones in Central, Eastern and Western Europe. Recolonisation then ensued most probably from persisting nearby populations rather than from remote refugia (Pongratz et al. 2003). Based on allozymes and mitochondrial markers, parthenogenesis originated repeatedly from sexual conspecifics (Pongratz et al. 1998, 2003). In mixed populations, evidence for recent origins of parthenogenetic lineages supports the fact that most parthenogenetic lineages may well be less than 500,000 years old (Pongratz et al. 2003).

Even on a local scale *Schmidtea polychroa* shows extremely high genetic differentiation. Neighbouring samples just 13 m apart already revealed significant differentiation (Pongratz et al. 2002). This high sub-structuring is typical for sexuals as well as for parthenogens and indicates restricted gene flow due to low mobility and dispersal rates (Pongratz et al. 2002; D'Souza and Michiels 2006).

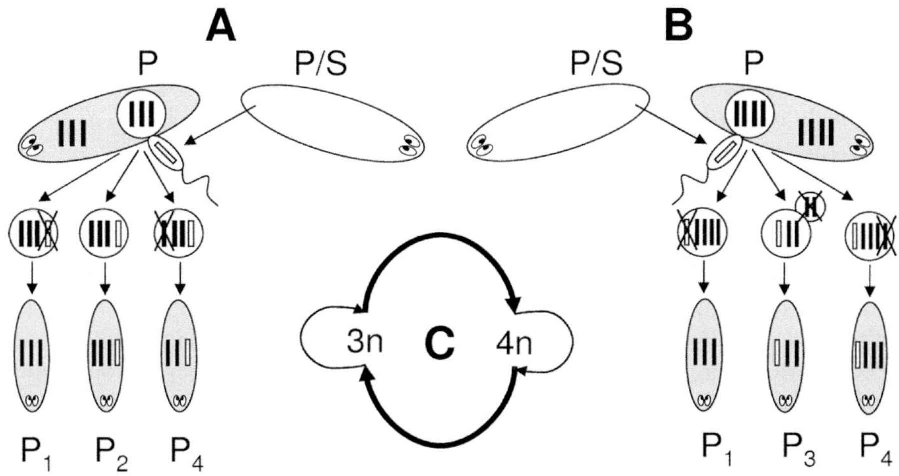

Fig. 18.4 Occasional sex in sperm-dependent parthenogenetic triploids (**A**) and tetraploids (**B**). The number of bars indicates the ploidy level. Sperm receiver is parthenogenetic (P), while sperm donor may be sexual (S) or parthenogen (P). Four different outcomes of sperm-egg-interaction are shown: 1. Pseudogamy: Sperm is used to trigger zygote development and resulting offspring are clonally produced. This applies both to triploids and tetraploids (P_1). 2. Chromosome addition: Sperm pronucleus fuses with triploid egg and fails to be expelled. Resulting offspring is tetraploid (P_2). 3. Chromosome loss: Tetraploid individual produces diploid, reduced egg, which fuses with haploid sperm nucleus. The resulting zygote then develops into triploid offspring (P_3). 4. Chromosome displacement: Instead of paternal chromosomes, one maternal chromosome set is expelled from the zygote which results in offspring with a mixture of paternal and maternal alleles but unaltered ploidy level (P_4). **C** illustrates the unique two-step cycle of occasional sex including ploidy alterations from triploidy to tetraploidy and back to triploidy. Alternatively, sex may occur without ploidy changes. Modified from D'Souza et al. (2008)

18.3 Indications for a Cost of Sex

Applying the cost of sex concept to hermaphrodites is not as simple as it seems at first sight (Charlesworth 1980; Joshi and Moody 1995, 1998; Mogie 1996). The cost of sex in hermaphrodites strongly depends on the sex allocation of coexisting parthenogenetic and sexual forms (Charlesworth 1980), and the possibility and type of reproduction by the male function of parthenogens (Joshi and Moody 1995, 1998). Most theoretical studies predict that the cost of sex in hermaphrodites is generally less than two-fold and may reach a maximum of 1.5-fold under "optimal" conditions (Charlesworth 1980; Joshi and Moody 1995, but see Mogie 1996). The general prediction is that an asexual (parthenogenetic) hermaphrodite with a simple genetic system should benefit from quickly abolishing the male function.

In *Schmidtea polychroa*, the picture is complicated because parthenogens are on the one hand sperm-dependent, but on the other hand they can sire offspring through their male function. Sperm may not only fertilise sexual eggs (Storhas et al. 2000; Fig. 18.3C), but sometimes also evade expulsion in parthenogenetic eggs, leading to alternative sexual processes (D'Souza et al. 2004, 2006; see also below and Fig. 18.4). Therefore, investment into male function of parthenogens is not only a proximate, mechanistic requirement demanded by pseudogamy, it also has an evolutionary function by offering a way out of the "asexual dead-end" faced by the female germ line. Hence, at least "some" sperm should be produced. In fact, male function in *S. polychroa* parthenogens is strongly reduced – as expected for parthenogenetic hermaphrodites (Weinzierl et al. 1998), but rarely completely abolished – as expected for sperm-dependent parthenogens with occasional sex. This reduction in maleness coincides with increased female fecundity (Weinzierl et al. 1998, 1999b). Parthenogens produce 1.43 times more cocoons than sexuals, illustrating that the cost of sex in hermaphrodites may stay well below two-fold.

Since both sexuals and parthenogens need to find mates and copulate, there are no further exclusive disadvantages for the sexual form that would increase the cost of sex. Furthermore, there are no direct indications for ecological separation of both reproductive types (Weinzierl et al. 1999a).

18.4 Indications for a Benefit of Sex

The cost of sex in *Schmidtea polychroa*, expressed as high male-relative-to-female investment, should offer parthenogens a competitive advantage over sexuals. Indeed, parthenogens are more common and widespread across Europe, while sexuals are largely restricted to southern and eastern Europe (Beukeboom et al. 1996; Pongratz et al. 1998, 2003). In those regions, sexuals have been coexisting with parthenogens for at least several decades. As no obvious niche separation is present (Weinzierl et al. 1999a), sexual reproduction has to offer immediate benefits when directly competing with more prolific parthenogens. Indications for the validity of the mutation hypothesis and the parasite hypothesis (Red Queen) for sex have been addressed in *S. polychroa* and are summarised in the next sections.

18.4.1 Accumulation of Deleterious Mutations

One of the main hypotheses to explain the success of sexuality is that it purges deleterious mutations more efficiently from the population, leading to a lower mutation load relative to asexuals at equilibrium (Deterministic mutation hypothesis (e.g., Kondrashov 1988), Muller's ratchet (e.g., Muller 1964), see also Chapter 5). Work on *Schmidtea polychroa* in this context has focused on embryo development because irregularities and mortality is relatively common at this stage and easy to quantify. It is also likely that accumulation of deleterious mutations directly affects embryonic development. The reason why developmental instabilities may be tolerated more in planarians such as *S. polychroa* is because mortality of some embryos, simply leads to an increase in size of the surviving hatchlings. Hence, the loss of some offspring is compensated by increased survival of others. This effect results from the fact that planarians put several ectolecithal eggs with a large mass of common yolk in a single cocoon (Greeff et al. 1999). The ensuing fierce competition for yolk ascertains that the investment only goes to the fitter ones (Cardona et al. 2005, 2006).

Initial tests showed that embryo mortality is higher in parthenogenetic than in sexual forms, as expected (Storhas et al. 2000). Whether this is due to the presence of more genetic defects in parthenogens was tested using the ability of parthenogens to fertilise eggs from sexuals (see Fig. 18.3C). Since parthenogens transmit a haploid complement to the haploid (sexual) egg, any mutations and their effect on development in the sexual offspring can be studied without the confounding effect of ploidy differences between sexuals and parthenogens. By comparing such crosses with those within sexuals and parthenogens (Fig. 18.3), physiological and genetic degradation of sperm can be discriminated. Storhas et al. (2000) found that in purely parthenogenetic and mixed sexual/parthenogenetic crosses, parthenogens fathered fewer offspring, which may result from low sperm quantity or quality. On the other hand, cocoons from parthenogens showed the highest embryo mortality independent of the father's reproductive mode. However, when sperm from parthenogens is used to fertilise sexual eggs, the proportion of undeveloped embryos significantly increases compared to sperm from sexuals. In a second study, genetically distinct parthenogenetic lineages were tested for both male and female fertility (Storhas 2001). The results showed that they differed notable in their capability to have offspring with sexual partners. Male and female fertility correlated significantly among parthenogenetic lineages, which is strong evidence in favour of a genetic component to fertility, making it very likely that reduced fertility is caused by increased embryo mortality and an accumulation of deleterious mutations in parthenogens (Storhas 2001).

18.4.2 Parasites (Red Queen)

The Red Queen hypothesis predicts that sexuality is favoured when virulent parasites adapt quickly to host genotypes (e.g., Hamilton 1980; see also Chapter 7). In *Schmidtea polychroa*, several protozoan symbionts have been characterised. Among them are amoeba-like forms, ectoparasitic ciliates and gregarines (Michiels et al. 2001; Bruvo 2005; Bruvo et al. 2007). Especially for the amoeba-like symbionts, some of the basic predictions from the Red Queen hypothesis could be confirmed (Michiels et al. 2001). In locations where sexuals and parthenogens coexist, parthenogens have a consistently higher likelihood of being infected than sexuals. The number of amoeba in infected individuals is also much higher in parthenogens (Michiels et al. 2001).

Genetic analysis using allozymes as genetic markers revealed genotype-specific infection in parthenogens. Parthenogenetic lineages differed in number of infected individuals as well as in the number of parasites per infected individuals. Such an effect was missing from sexuals – which is expected for recombining forms whenever the susceptibility locus and allozyme locus are unlinked, as is very likely.

In contrast to other systems (e.g., *Potamopyrgus antipodarum*, Lively (1996)), time-lagged genotype-specific parasite-host dynamics were never studied directly in *Schmidtea polychroa*. However, common parthenogenetic genotypes tended to have more parasites, which may indicate negative frequency-dependent selection leading to cyclic patterns of infection across *S. polychroa* genotypes (Michiels et al. 2001; Bruvo et al. 2007). What are the fitness consequences of these infections? Fitness effects of amoeban parasites were weak and could only be shown as a small reduction in fertility with increasing parasite load in parthenogens (Michiels et al. 2001). However, it is not clear whether the parasite number caused fertility reduction or whether less fit individuals are more parasitised because of decreased immunocompetence. No indications for reduced growth or fecundity were found in infected individuals. Despite low virulence, such "symbionts" may still be important in parthenogens with a weakened immune system caused by genome degradation, as we shall discuss next.

18.4.3 Pluralism

Theoretically, the Red Queen and the mutation hypotheses should amplify their effects whenever they act together (West et al. 1999). Increased genome deterioration caused by accumulated mutations in parthenogens may weaken the immune response and hence strengthen the effect caused by parasites. On the other hand, increased parasite infection of parthenogens may reduce group size, which in turn accelerates the accumulation of deleterious mutations by Muller's ratchet.

Indications of a synergistic effect were revealed by analysing the prevalence of different parasite species and embryo mortality, the latter again indicating accumulated deleterious mutations (Bruvo et al. 2007). Strongly parasitised parthenogenetic clusters of related genotypes showed higher embryo mortality compared to less parasitised clusters. The significant positive correlation between both parameters indicates a link between parasite and mutational models for sex. It is unlikely that developmental problems are directly caused by parasites, as parasites have never been found to be linked to the reproductive system. Therefore, the results suggest that the deterioration of the genome caused by the accumulation of deleterious mutations decreases immunocompetence and hence enhances susceptibility of parasites.

18.5 Maintenance of Parthenogenesis

The data thus far suggest on the one hand that parthenogens indeed suffer from both the accumulation of deleterious mutations and elevated susceptibility to parasites, with synergistic effects possibly amplifying both. On the other hand, however, parthenogens remain geographically more widespread and dominant than their sexual conspecifics. This discrepancy complicates general predictions about failure or success of parthenogens as the cost-benefit balance appears to be quite dynamic and at least population-specific. Having this variability makes *Schmidtea polychroa* interesting in itself, as it allows looking into invasion and displacement dynamics from the perspective of both sexuals and asexuals, using field data.

One important reason for the question why the fate of parthenogenetic populations is hard to predict is their enormous karyotypic, genetic and phenotypic diversity, which is likely to buffer some disadvantages of parthenogenesis. Clonal diversity allows a parthenogenetic population to respond more dynamically to changing environments. Polyploidy offers additional benefits as multiple gene copies may buffer deleterious mutations and slow down their accumulation (Otto and Whitton 2000). Moreover, the genome duplication resulting from polyploidisation may lead to sub-functionalisation of gene copies and can eventually lead to increased functional gene diversity (Otto and Whitton 2000). As for *Schmidtea polychroa*, however, triploids generally perform worse than diploids (see above), and tetraploids do worse than triploids (D'Souza et al. 2005). If our conclusions formulated above hold, at least some of this reduced performance in polyploids can be attributed to the accumulation of deleterious mutations in parthenogens.

Microsatellite diversity is high in parthenogens and may reach values typical for sexual populations (Storhas 2001; D'Souza and Michiels 2006). Significant interclonal variation has been described for several traits, such as male allocation and fertility (Storhas 2001) and offspring number, fecundity, body size and hatching time (D'Souza et al. 2005). This indicates that either clones are of different age and hence with accumulated different amounts of deleterious mutations, or that clones follow different ecological strategies and occupy different micro-niches. The latter supports the idea of the Frozen Niche Variation hypothesis to explain the maintenance of clonal diversity (Vrijenhoek 1998; Jokela et al. 2003; see also Chapter 6).

In what follows, we want to address some of the peculiar consequences of sperm-dependent parthenogenesis in *S. polychroa* and search for reasons why parthenogens appear so resilient to invasion and displacement by sexuals. Although being sperm-dependent appears to be a complicated way of having no sex, we shall see that these parthenogens do not necessarily waste resources when investing in a (small) male function.

18.5.1 Costs of Sperm-Dependent Parthenogenesis

Requiring sperm for offspring production is problematic in purely parthenogenetic populations as sexual sperm donors are missing. A mutant allocating all resources to the female function will gain a clear fitness advantage over the residents and increase in frequency. Hence, parthenogenetic populations may go extinct due to a lack of sperm donors. Maintaining an albeit small, residual male function is thus essential for long-term survival of parthenogenetic populations. One possible solution to this problem may be sperm trading (Michiels and Bakovski 2000; Michiels and Kuhl 2003): By only giving sperm when receiving sperm, cheaters that accept sperm without ever reciprocating, cannot spread. Parthenogens do indeed strongly insist on reciprocity: Unilateral sperm transfer is rare relative to bilateral or no transfer (Michiels and Kuhl 2003). If one partner is reluctant to give sperm, copulation is aborted without exchange in any direction in 18% of the copulations. Sperm trading may offer an explanation for why a seemingly "useless" male function is maintained. It cannot, however, prevent gradual degradation of sperm quality: Parthenogens seem to reduce the proportion of sperm and transfer more accessory fluid, which may represent another form of cheating (A. Streng, personal communication).

18.5.2 Benefits of Sperm-Dependent Parthenogenesis

The key benefit of sperm-dependency in parthenogens is the option of obtaining paternal fitness by fertilising eggs as a male. Here, we describe how sperm from parthenogens can mediate gene flow from parthenogens to sexuals as well as between parthenogens and how this allows a combination of the benefits of asexual (female) and sexual (male) reproduction.

18.5.2.1 Gene Flow Between Sexuals and Parthenogens

Since parthenogens produce haploid, fertile sperm, they are able to fertilise eggs from sexual partners (e.g., Benazzi Lentati 1970; Storhas et al. 2000). Hence, parthenogenetic genes can be transmitted via sperm to sexual biotypes (see Fig. 18.3C). Offspring from such mixed-matings have been frequently obtained in the laboratory. They are usually like normal sexuals, but sometimes produce unreduced ($2x$) eggs (Benazzi Lentati 1970). After fertilisation, these can lead to triploid offspring, which reproduce parthenogenetically. Field-collected sexuals produced cocoons with 1–2% triploid hatchlings, indicating that this pathway indeed exists in nature (Weinzierl et al. 1999b). This process can explain the high clonal diversity and recurrent origin of parthenogens in mixed populations (Pongratz et al. 1998, 2003).

Repeated origin of new lineages enhances the sustainability of a parthenogenetic subpopulation as their gene pool is "refreshed" with genetic material from sexuals. This compensates for extinction of lineages caused by mutational meltdown or increased susceptibility to disease. Sexuals, however, loose when engaging in mixed mating as they donate large amounts of high-quality sperm, which boosts female fertility in the parthenogenetic partner (Storhas et al. 2000; Storhas 2001), but offers no paternal fitness. Reversely, poor-quality parthenogenetic sperm causes reduced fertility and "contaminates" sexual offspring with harmful mutations (Storhas et al. 2000). Hence, mixed matings between parthenogens and sexuals offer advantages for parthenogens, not for sexuals. This leads one to predict that sexuals would avoid such matings, but they obviously lack the ability to do so (Storhas et al. 2000).

18.5.2.2 Gene Flow Among Parthenogens

In addition to the gene flow from parthenogens to sexuals and (indirectly) back to parthenogens in future generations, some form of sex is also present in purely parthenogenetic populations (Fig. 18.4). These occasional sexual processes were shown with paternity analysis for parthenogens mated in the field (D'Souza et al. 2004) and for controlled crosses in the laboratory (D'Souza et al. 2006). Sex among parthenogens can take different forms that may or may not involve ploidy alternations. In all types, haploid sperm fuses with a parthenogenetic egg, but without subsequent elimination. This process allows paternal genes to "leak" into the next parthenogenetic generation. A ploidy change follows when haploid sperm is incorporated into a triploid egg giving rise to a tetraploid offspring (Fig. 18.4A). Tetraploids in turn can produce triploid offspring. This requires the production of reduced, diploid eggs which are fertilised by haploid sperm (Fig. 18.4B). Through the two-step cycle from the common triploid forms to the rare tetraploids and back, triploid genomes effectively recombine through the production of tetraploids (D'Souza et al. 2004, Fig. 18.4C). Hence, although rare, tetraploids represent a crucial intermediate for occasionally sex in parthenogenetic *Schmidtea polychroa*. This is stressed by the positive correlation between tetraploid frequency and the genetic diversity of triploids. Tetraploid frequency can therefore be used as a predictor for the amount of occasional sex (D'Souza and Michiels 2006, 2008). Since tetraploids have a clearly lower fitness than triploids, they would disappear from any population if occasional sex did not occur (D'Souza et al. 2005). Hence, tetraploidy emerges and is maintained as a result of occasional sex.

Other known forms of occasional sex do not involve a change in ploidy, but are based on a one-step mechanism in which a maternal chromosome set is substituted by the incorporated paternal set. The result is that triploid mothers sometimes produce triploid offspring with two maternal and one paternal chromosome set (Fig. 18.4A). The same process can be observed in tetraploids (Fig. 18.4B). We estimated that 5–12% of all offspring produced in a purely parthenogenetic population arise via one of these mechanisms (D'Souza et al. 2004, 2006).

Occasional sex has several important consequences for the stability and composition of parthenogenetic populations. The stable presence of tetraploids, for instance, is a direct consequence of occasional sex since tetraploids are less fit than triploids. Hence, tetraploids would go extinct if not constantly generated de novo. This "dead-end" can be seen as a cost of occasional sex. Combined, occasional sex and low tetraploid fitness maintain relatively stable proportions of both ploidy types (D'Souza et al. 2005). Occasional sex also leaves telltale population genetic signatures in genotypic diversity and evenness that coincide with different amounts of occasional sex (D'Souza and Michiels 2006). Such variation is unique and useful as it makes it possible to test the consequences of different (but low!) levels of sex at the population level among otherwise absolutely identical biotypes. For instance, by comparing a more clonal population with a nearby population that shows occasional sex, we could show that sperm are smaller in clonal *Schmidtea polychroa*. Crosses within and between these populations showed that the ability to fertilise eggs was lowest for sperm donors taken from the clonal population. This suggests that occasional sex selectively favours better ejaculates through sperm quality and/or quantity (D'Souza et al. 2008). The crucial bottom line is that occasional sex eventually increases population's fitness: Proportion of tetraploids, used as a measurement of the degree of sex in a population, correlated positively with fitness variance across six subpopulations. Due to increased variability, selection may act more effectively, which explains that subpopulations with a higher degree of sex also had highest mean fitness (D'Souza and Michiels 2008).

Summarised, although costly, maintaining a (small) male function keeps the option of rare sex open, and may contribute to the long-term survival of parthenogenetic populations as it increases population fitness.

18.6 Evaluation: Fate of Parthenogenetic and Sexual Populations

The unusual aspects of parthenogenesis in *Schmidtea polychroa* discussed above have many implications for the interactions and ultimate fate of parthenogenetic and sexual populations.

Although purely sexual populations may enjoy all the advantages of sexual recombination, they may be susceptible to invasion by parthenogens. Once parthenogens appear, a new parthenogenetic lineage should spread because of its increased female fecundity and presumed (initial) low level of mutation accumulation and susceptibility to disease. Furthermore, such a strain benefits from mixed matings as they boost maternal fitness (through high-quality sperm from sexuals) and allows for paternal fitness and the generation of additional new lineages. These effects are particularly strong as long as parthenogens are rare. A theoretical model predicts that when these conditions are applied to an unstructured environment, parthenogens will always outcompete sexuals (Bruvo 2005). Complete displacement of the sexuals must not automatically ensue as some of the benefits for parthenogens are negatively frequency-dependent, keeping them at intermediate frequencies.

Established, purely parthenogenetic populations face very different scenarios. Here, the future is primarily determined by a lack of recombination. Under purely clonal reproduction, parthenogens pay for the "classic" costs of clonality that are the focus of this book. Sperm-dependency and occasional sex, however, might be just enough to hold off the worst-case scenario. Our results show that occasional sex indeed increases populations' fitness and therefore might facilitate long-term persistence (D'Souza and Michiels 2008). However, it is still not clear whether rare sex itself is evolutionary stable. On the long-term, rare sex may lead to either obligate sex or obligate asex (Hurst and Peck 1996; Peck and Waxman 2000). Under heterozygote advantage, a mixed reproductive mode may evolve to obligate asex when sexual processes become too rare and hence the benefits of sex vanish (Peck and Waxman 2000). Moreover, the fate of partial sexuality is often condition-dependent (Peck et al. 1997; Rispe and Pierre 1998). Population size, number of loci under selection, and stability of the environment strongly determine the benefits of partial sexuality compared to obligate reproductive modes (Peck 1996; Peck et al. 1997)

Even if sexual reproduction may eventually be more advantageous than parthenogenesis in *Schmidtea polychroa*, the invasion of a parthenogenetic population by sexuals is difficult as sexuals would obtain virtually no paternity, but instead improve the maternal success of their parthenogenetic partners by giving them a high-quality ejaculate. Moreover, sexuals would allow parthenogens to obtain more paternal fitness and increase the likelihood that new parthenogenetic lineages will arise, which may strengthen the parthenogenetic resident population even more. If a little bit of sex can eliminate most of the disadvantages associated with asex, at a fraction of the real cost of "regular" recombination, parthenogens may effectively be invincible. In a recent model, complete displacement of parthenogens by sexuals was limited to a small set of rare conditions (Bruvo 2005).

Given all these scenarios, how can we explain that parthenogens and sexuals live in relatively stable coexistence in some locations in, e.g., Italy (Pongratz et al. 1998, 2003; Storhas 2001)? A theoretical model revealed that the likelihood of coexistence is clearly increased in structured populations with niches of different quality (Bruvo 2005). Moreover, partial or total separation of biotypes into distinct habitat patches due to subtle environmental preferences greatly facilitates coexistence as it decreases mixed matings and their consequences (Bruvo 2005). Although ecological niche differentiation seems weak (Weinzierl et al. 1999a), the same study also found that that local proportions of sexuals and parthenogens vary widely. Given the poor dispersal capacity that is characteristic of planarians (Pongratz et al. 2002), the apparent overall (e.g., a lake) stable coexistence may actually be a mosaic of many, slow temporal invasion and extinction events at local sites. Different parasite loads (Bruvo 2005) or differences in reproductive timing may further contribute to ecologically separate the two biotypes and facilitate long-term coexistence.

18.7 Outlook

Summarised, the unique feature of the planarian *Schmidtea polychroa* is that parthenogens have retained the ability to produce fertile sperm and can gain paternal fitness. The fact that rates of occasional sex vary from 0 to 12% in otherwise cytologically, reproductively and ecologically identical field populations allows the study of the costs and benefits of sex versus asex on a truly "all else being equal" basis. This feature is truly unique and will help to understand many aspects in the evolutionary significance of sex and parthenogenesis in future projects. Other advantages are accessibility in the field and methods to assess developmental stability and infection rates from field-sampled individuals.

"Infecting" sexuals with parthenogenetic genes/chromosomes is a useful tool to examine the consequences of genome deterioration in the absence of recombination. As diploids carrying parthenogenetic genes can be compared with diploids carrying only genes from sexuals, indications for mutational meltdown can be demonstrated without the confounding effect of different ploidy levels. Especially genes responsible for the male function, such as spermatogenesis, which are predicted to be degraded in parthenogens, can be studied under these conditions.

A close look at the genetics of diploid offspring from mixed sexual x parthenogenetic crosses and subsequent generations may clarify the mechanisms and causes of parthenogenesis in *Schmidtea polychroa*, as new parthenogenetic lineages arise from such mixed matings. Such analyses may also allow identifying genes and gene combinations responsible for parthenogenesis. Additionally, we may find answers for why parthenogenesis is associated with polyploidy and why diploid parthenogens have never been observed.

The combination of sexual and asexual reproduction in parthenogenetic *Schmidtea polychroa* opens another door for extensive future research. For instance, the presence of rare sex enables to investigate the effects of sex without having to compare biotypes of different ploidy levels. It is important to also explore why parthenogenetic lineages and populations differ in the amount of sex. Perhaps parthenogens are able to adjust the frequency of sex as a response to environmental stress, caused, e.g., by parasites and/or to the amount of accumulated mutations.

With the help of modern molecular techniques developed for *Schmidtea mediterranea* and improved culture conditions, we may answer one of the key questions in the context of the adaptive value of sex in parthenogens: Is sex in parthenogens just a phylogenetic relic or a rediscovered tool to prevent parthenogens from going extinct?

Acknowledgements We must say "thank you very much" to dozens of people who worked on planarians in our previous groups (MPI Seewiesen, University of Muenster) in the past and now at the University of Tuebingen. Those who have most strongly contributed to the development of *Schmidtea polychroa* as a model system are Leo W. Beukeboom, Jaco M. Greeff, Anne Peters, Rolf P. Weinzierl, Andrea Streng, Martin G. Storhas, Norbert Pongratz and Ružica Bruvo. Earlier drafts have been improved by comments from two anonymous referees. This work has been supported by several grants from the German Science Foundation (DFG).

References

Ball IR, Reynoldson TB (1981) British planarians. Cambridge University Press, Cambridge

Benazzi Lentati G (1966) Amphimixis and pseudogamy in fresh-water Triclads: experimental reconstitution of polyploid pseudogamic biotyps. Chromosoma 20: 1–14

Benazzi Lentati G (1970) Gametogenesis and egg fertilization in planarians. Int Rev Cytol 27: 101–179

Benazzi M (1963) Genetics of reproductive mechanisms and chromosome behaviour in some freshwater triclads. In: Daugherty EC, Brown ZN (eds) The lower Metazoa. University of California Press, Berkeley, pp. 405–422

Benazzi M, Puccinelli I, Del Papa R (1970) The planarians of the *Dugesia lugubris-polychroa* group: taxonomic inference based on cytogenetic and morphologic data. Lincei – Rend Sc fis mat e nat 48: 369–376

Benazzi M, Baguna J, Ballester R, Puccinelli I, Del Papa R (1975) Further contributions to the taxonomy of the '*Dugesia lugubris-polychroa* group' with description of *Dugesia mediterranea* n. sp. (Tricladida, Paludicola). Bull Zool 42: 81–89

Benazzi M, Benazzi Lentati G (1976) Platyhelminthes. Gebrüder Bornträger, Berlin, Stuttgart
Benazzi M (1982) Speciation events evidenced in turbellaria. Mechanisms of speciation. Alan R Liss, New York, pp. 307–344
Benazzi M (1992) Platyhelminthes – Turbellaria. In: Adiyodi KG, Adiyodi RG (eds) Reproductive biology of invertebrates. V. Sexual differentiation and behaviour. Oxford and IBH Publishing Co. PVT Ltd, New Delhi, pp. 53–74
Benazzi M, Benazzi Lentati G (1993) Platyhelminthes – Turbellaria. In: Adiyodi KG, Adiyodi RG (eds) Reproductive biology of invertebrates. VI. Asexual propagation and reproductive strategies. Oxford & IBH Publishing Co. PVT. Ltd., New Delhi, Bombay, pp. 107–139
Beukeboom LW, Weinzierl RP, Reed KM, Michiels NK (1996) Distribution and origin of chromosomal races in the freshwater planarian *Dugesia polychroa* (Turbellaria: Tricladida). Hereditas 124: 7–15
Beukeboom LW, Vrijenhoek RC (1998) Evolutionary genetics and ecology of sperm-dependent parthenogenesis. J Evol Biol 11: 755–782
Birnbaum KD, Sánchez Alvarado A (2008) Sclicing across kingdoms: regeneration in plants and animals. Cell 132: 697–710
Bruvo R (2005) Importance of parasites and deleterious mutations for the maintenance of sex in the freshwater planarian flatworm *Schmidtea polychroa*. PhD thesis. Westfälische Wilhelms-University, Münster
Bruvo R, Schulenburg H, Storhas M, Michiels NK (2007) Synergism between mutational meltdown and Red Queen in parthenogenetic biotypes of the freshwater planarian *Schmidtea polychroa*. Oikos 116: 313–323
Cardona A, Hartenstein V, Romero R (2005) The embryonic development of the triclad *Schmidtea polychroa*. Dev Genes Evol 215: 109–131
Cardona A, Hartenstein V, Romero R (2006) Early embryogenesis of planaria: A cryptic larva feeding on maternal resources. Dev Genes Evol 216: 667–681
Charlesworth B (1980) The cost of sex in relation to mating system. J Theor Biol 84: 655
D'Souza TG, Storhas M, Schulenburg H, Beukeboom LW, Michiels NK (2004) Occasional sex in an 'asexual' polyploid hermaphrodite. Proc R Soc Lond B 271: 1001–1007
D'Souza TG, Storhas M, Michiels NK (2005) The effect of ploidy level on fitness in parthenogenetic flatworms. Biol J Linn Soc 85: 191–198
D'Souza TG, Michiels NK (2006) Genetic signatures of occasional sex in parthenogenetic subpopulations of the freshwater planarian *Schmidtea polychroa*. Freshw Biol 51: 1890–1900
D'Souza TG, Schulte RD, Schulenburg H, Michiels NK (2006) Paternal inheritance in parthenogenetic forms of the planarian *Schmidtea polychroa*. Heredity 97: 97–101
D'Souza TG, Bellenhaus V, Wesselmann R, Michiels NK (2008) Sperm length and quality in sperm-dependent parthenogens. Biol J Linn Soc 93: 81–87
D'Souza TG, Michiels NK (2008) Correlation between sex rate estimates and fitness across predominantly parthenogenetic flatworm subpopulations. J Evol Biol 21: 276–286
Greeff JM, Storhas MG, Michiels NK (1999) Reducing losses to offspring mortality by redistributing resources. Funct Ecol 13: 786–792
Gremigni V, Nigro M, Puccinelli I (1982) Evidence of male germ cell redifferentiation into female germ cells in planarian regeneration. J Exp Morphol 70: 29
Hamilton WD (1980) Sex versus non-sex versus parasite. Oikos 35: 282–290
Hoshi M, Kobayashi K, Arioka S, Hase S, Matsumoto M (2003) Switch from asexual to sexual reproduction in the planarian *Dugesia ryukyuensis*. Integr Comp Biol 43: 242–246
Hurst LD, Peck JR (1996) Recent advances in understanding of the evolution and maintenance of sex. Trends Ecol Evol 11: 46–52
Jokela J, Lively CM, Dybdahl MF, Fox JA (2003) Genetic variation in sexual and clonal lineages of a freshwater snail. Biol J Linn Soc 79: 165–181
Joshi A, Moody ME (1995) Male gamete output of asexuals and the dynamics of populations polymorphic for reproductive mode. J Theor Biol 174: 189–197
Joshi A, Moody ME (1998) The cost of sex revisited: Effects of male gamete output of hermaphrodites that are asexsual in their female capacity. J Theor Biol 195: 533–542

Kobayashi K, Koyanagi R, Matsumoto M, Cabrera JP, Hoshi M (1999a) Switching from asexual to sexual reproduction in the planarian *Dugesia ryukyuensis*: Bioassay system and basic description of sexualizing process. Zool Sci 16: 291–298

Kobayashi K, Koyanagi R, Matsumoto M, Hoshi M (1999b) Switching from asexual to sexual reproduction in the planarian *Dugesia ryukyuensis*. Invertebr Reprod Dev 36: 153–158

Kobayashi K, Arioka S, Hase S, Hoshi M (2002) Signification of the sexualizing substance produced by the sexualized planarians. Zool Sci 19: 667–672

Kondrashov AS (1988) Deleterious mutations and the evolution of sexual reproduction. Nature 336: 435–440

Lively CM (1996) Host-parasite coevolution and sex – do interactions between biological enemies maintain genetic variation and cross-fertilization? Bioscience 46: 107–114

Melander Y (1963) Cytogenetic aspects of embryogenesis in Paludicola Tricladida. Hereditas 49: 119–166

Michiels NK, Streng A (1998) Sperm exchange in a simultaneous hermaphrodite. Behav Ecol Sociobiol 42: 171–178

Michiels NK, Bakovski B (2000) Sperm trading in a hermaphroditic flatworm: reluctant fathers and sexy mothers. Anim Behav 59: 319–325

Michiels NK, Beukeboom LW, Pongratz N, Zeitlinger J (2001) Parthenogenetic flatworms have more symbionts than their coexisting, sexual conspecific, but does this support the Red Queen? J Evol Biol 14: 110–119

Michiels NK, Kuhl A (2003) Altruistic sperm donation in a sperm-dependent parthenogenetic hermaphrodite is stabilized by reciprocal sperm exchange. J Zool 259: 77–82

Mogie M (1996) Is there a cost of sex in hermaphrodites? Naturwissenschaften 83: 225

Muller HJ (1964) The relation of recombination to mutational advantage. Mutat Res 1: 2–9

Newmark PA, Sánchez Alvarado A (2002) Not your father's planarian: A classic model enters the era of functional genomics. Nat Rev Genet 3: 210–219

Orii H, Sakurai T, Watanabe K (2005) Distribution of the stem cells (neoblasts) in the planarian *Dugesia japonica*. Dev Genes Evol 215: 143–157

Otto SP, Whitton J (2000) Polyploid incidence and evolution. Annu Rev Genet 34: 401–437

Peck JR (1996) Limited dispersal, deleterious mutations and the evolution of sex. Genetics 142: 1053–1060

Peck JR, Barreau G, Heath SC (1997) Imperfect genes, fisherian mutation and the evolution of sex. Genetics 145: 1171–1199

Peck JR, Waxman D (2000) What's wrong with a little sex? J Evol Biol 13: 63–69

Peters A, Michiels NK (1996a) Evidence for lack of inbreeding avoidance by selective mating in a simultaneous hermaphrodite. Invertebr Biol 115: 99–103

Peters A, Michiels NK (1996b) Do simultaneous hermaphrodites choose their mates? Effects of body size in a planarian flatworm. Freshw Biol 36: 623–630

Peters A, Streng A, Michiels NK (1996) Mating behaviour in a hermaphroditic flatworm with reciprocal insemination: Do they assess their mates during copulation? Ethology 102: 236–251

Pongratz N, Sharbel TF, Beukeboom LW, Michiels NK (1998) Allozyme variability in sexual and parthenogenetic freshwater planarians: evidence for polyphyletic origin of parthenogenetic lineages through hybridization with coexisting sexuals. Heredity 81: 38–47

Pongratz N, Gerace L, Michiels NK (2002) Genetic differentiation within and between populations of a hermaphroditic freshwater planarian. Heredity 89: 64–69

Pongratz N, Michiels NK (2003) High multiple paternity and low last-male sperm precedence in a hermaphroditic planarian flatworm: consequences for reciprocity patterns. Mol Ecol 12: 1425–1433

Pongratz N, Storhas M, Carranza S, Michiels NK (2003) Phylogeography of competing sexual and parthenogenetic forms of a freshwater flatworm: patterns and explanations. BMC Evol Biol 3: 23

Reddien PW, Oviedo NJ, Jennings JR, Jenkin JC, Sánchez Alvarado A (2005) Developmental biology: SMEDWI-2 is a PIWI-like protein that regulates planarian stem cells. Science 310: 1327–1330

Reynoldson TB (1966) The distribution and abundance of lake-dwelling triclads – towards a hypothesis. Adv Ecol Res 3: 1–71

Reynoldson TB, Davies RW (1970) Food niche and co-exsistence in lake-dwelling Tricalds. J Anim Ecol 39: 599

Reynoldson TB (1977) The population dynamics of *Dugesia polychroa* (Schmidt) (Turbellaria, Tricladida) in a recently-constructed Anglesey pond. J Anim Ecol 46: 63–77

Rieger RM (1986) Asexual reproduction and the turbellarian archetype. Hydrobiologia 132: 35–45

Rispe C, Pierre JS (1998) Coexistence between cyclical parthenogens, obligate parthenogens, and intermediates in a fluctuating environment. J Theor Biol 195: 97–110

Robb SMC, Ross E, Alvarado AS (2008) SmedGD: the *Schmidtea mediterranea* genome database. Nucleic Acids Res 36: D599–D606

Salo E (2006) The power of regeneration and the stem-cell kingdom: freshwater planarians (Platyhelminthes) Bioessays 28: 546–559

Sánchez Alvarado A, Newmark PA (1999) Double-stranded RNA specifically disrupts gene expression during planarian regeneration. Proc Natl Acad Sci USA 96: 5049–5054

Sánchez Alvarado A, Newmark PA, Robb SMC, Juste R (2002) The *Schmidtea mediterranea* database as a molecular resource for studying platyhelminthes, stem cells and regeneration. Development 129: 5659–5665

Sánchez Alvarado A (2003) The freshwater planarian *Schmidtea mediterranea*: Embryogenesis, stem cells and regeneration. Curr Opin Genet Dev 13: 438–444

Sánchez Alvarado A (2004) Planarians. Curr Biol 14: R737–R738

Sato K, Shibata N, Orii H, Amikura R, Sakurai T, Agata K, Kobayashi S, Watanabe K (2006) Identification and origin of the germline stem cells as revealed by the expression of nanos-related gene in planarians. Dev Growth Differ 48: 615–628

Sluys R (1989) Sperm resorption in triclads (Platyhelminthes, Tricladida). Invertebr Reprod Dev 15: 89–95

Storhas M, Weinzierl RP, Michiels NK (2000) Paternal sex in parthenogenetic planarians: a tool to investigate the accumulation of deleterious mutations. J Evol Biol 13: 1–8

Storhas M (2001) Sex versus asex in a hermaphroditic flatworm. Schueling, Muenster

Streng A (1999) Sperm exchange in simultaneous hermaphroditic freshwater flatworms. PhD thesis. Ludwig-Maximilians University of München

Tekaya S, Sluys R, Zghal F (1999) Cocoon production, deposition, hatching, and embryonic development in the marine planarian *Sabussowia dioica* (Platyhelminthes, Tricladida, Maricola). Invertebr Reprod Dev 35: 215–223

Vreys C, Michiels NK (1997) Flatworms flatten to size up each other. Proc R Soc Lond B 264: 1559–1564

Vreys C, Schockaert ER, Michiels NK (1997) Unusual pre-copulatory behaviour in the hermaphroditic planarian flatworm, *Dugesia gonocephala* (Tricladida, Paludicola). Ethology 103: 208–221

Vrijenhoek RC (1998) Animal clones and diversity. Bioscience 48: 617–628

Weinzierl RP, Berthold K, Beukeboom LW, Michiels NK (1998) Reduced male allocation in the parthenogenetic hermaphrodite *Dugesia polychroa*. Evolution 52: 109–115

Weinzierl RP, Beukeboom LW, Gerace L, Michiels NK (1999a) Spatial and ecological overlap between coexisting sexual and parthenogenetic *Schmidtea polychroa* (Tricladida; Platyhelminthes). Hydrobiologia 392: 179–185

Weinzierl RP, Schmidt P, Michiels NK (1999b) High fecundity and low fertility in parthenogenetic planarians. Invertebr Biol 118: 87–94

West SA, Lively CM, Read AF (1999) A pluralist approach to sex and recombination. J Evol Biol 12: 1003–1012

Zayas RM, Hernandez A, Habermann B, Wang Y, Stary JM, Newmark PA (2005) The planarian *Schmidtea mediterranea* as a model for epigenetic germ cell specification: Analysis of ESTs from the hermaphroditic strain. Proc Natl Acad Sci USA 102: 18491–18496

Chapter 19
Sperm-Dependent Parthenogenesis and Hybridogenesis in Teleost Fishes

Dunja K. Lamatsch and Matthias Stöck

... there is always some danger in using sperm: even if you do not want, you run the risk of being fertilized... (Dubois 1990)

Abstract In so-called unisexual teleost fishes, a broad spectrum of evolutionary stages with varying amounts of sexual elements has evolved. These range from pure sperm-dependent parthenogenesis (gynogenesis) without or with different amounts of paternal leakage to hybridogenesis with hemiclonal diploid gametogenesis or genome elimination followed by meiosis (meiotic hybridogenesis). All of these phenomena are of hybrid origin.

Many of these fish form complexes which involve the coexistence of one or more sexually reproducing species with derived all-female forms that have various ploidy levels and reproductive modes, including gynogenesis, (meiotic) hybridogenesis and sexual reproduction. In teleosts, parthenogenetic reproduction is strictly dependent on sperm to initiate embryonic development. As opposed to true parthenogenesis, sperm-dependent parthenogenetic teleost lineages must primarily coexist with their "sperm donor", usually males from a parental sexual lineage or from a related sexual species. In some systems, gynogens were able to escape from their initial sperm donors ("host switch") and therefore, to enlarge their ranges and ecological niches. Sperm donors normally do not contribute genetically to the next generation. However, paternal leakage is observed in many systems contributing differing amounts of genetic material (from microchromosomes to entire chromosome sets) allowing interaction between genomes of different origin. Hybridogenesis is similar to gynogenesis in depending upon coexistence with sexual species but incorporates recombined genetic material by true fertilization. While hybridogens usually form clonal gametes, some triploids are capable of genome elimination followed by a normal diploid meiosis. Sperm-dependent parthenogenesis and hybridogenesis

D.K. Lamatsch (✉)
Institute for Limnology, Austrian Academy of Sciences, Mondseestrasse 9, A – 5310 Mondsee, Austria
e-mail: dunja.lamatsch@oeaw.ac.at

I. Schön et al. (eds.), *Lost Sex*, DOI 10.1007/978-90-481-2770-2_19,

combine disadvantages and advantages from both sexuality and asexuality. Here, we give an overview of sperm-dependent breeding complexes in fishes, discuss the evolutionary consequences of paternal leakage, and speculate about the evolutionary significance of intergenomic (re)combination.

19.1 Introduction

Fifteen years ago, the topic "unisexual fish" and their evolution have been reviewed to a great completeness (Vrijenhoek 1994). While the total number of newly discovered teleost complexes with unisexual bias has been limited, an enormous number of papers examining the different systems with new techniques have been published. The present chapter attempts to introduce to the various teleost complexes with some key papers and major recent findings.

19.1.1 Unisexual Reproduction

The need for at least some recombination seems to govern most eukaryotic life. This general rule may be also responsible for the fact that unisexual or all-female reproduction is very rare among vertebrates, comprising just <0.1% of all vertebrate species (Dawley 1989). Unisexuals seem exclusively to arise as a consequence of hybridization between sexually reproducing progenitors. It has been hypothesized that a particular combination of genomes shifts the sex ratio in an interspecific hybrid towards all-female and alters meiosis in the hybrids so they produce eggs without reduction in ploidy and supposedly without recombination. This was articulated by Wetherington et al. (1987) and called the "balance hypothesis" by Moritz et al. (1989). Genetic divergence of parental genomes has to be sufficiently large to cause a high proportion of unreduced gametes, but not too large to significantly decrease the viability or fertility of hybrids (Moritz et al. 1989). Hybridization is particularly common among fish (Scribner et al. 2000), in which it occurs more frequently than in other vertebrate groups. Fishes might hybridize more frequently because of their (usually) external fertilization, unequal abundance of parental species, competition for limited spawning habitats and susceptibility to secondary contact (Campton 2008). In fishes, unisexual reproduction occurs exclusively in the form of sperm-dependent parthenogenesis (i.e., gynogenesis) and hybridogenesis, forcing them into close ecological associations with their progenitor sexual species (Beukeboom and Vrijenhoek 1998).

19.1.2 Gynogenesis

In gynogenetic systems, unreduced eggs are normally produced by an all-female species, but egg development must be triggered by allospecific sperm from males of a related species (Fig. 19.1). Normally, the sperm does not contribute any genetic material to the offspring. Due to this exploitation of the host, gynogenesis has also been called "sperm parasitism" (Hubbs 1964). Although not a common mode of reproduction, sperm-dependent parthenogenesis has evolved multiple times within seven phyla (Beukeboom and Vrijenhoek 1998). However, regular gynogenesis is absent in some major groups of vertebrates, like birds and mammals. It is assumed that genomic imprinting plays a role in the absence of natural parthenogenesis in mammals (Georgiades et al. 2001; Scott and Spielman 2006), in which it leads to death during embryogenesis (Rougier and Werb 2001; see also Chapter 26).

19.1.3 Paternal Leakage

Gynogens are interesting models because they seem to combine some disadvantageous traits from both sexuality (e.g., finding mating partners, exposure to predation during mating, risk of diseases) and asexuality ("Muller's ratchet", Muller 1964; see also Chapter 5) including "mutational meltdown" (Lynch et al. 1993; for more details, see Schlupp 2005). Occasional leakage of genes from a paternal host into sperm-dependent clones may however provide a source of adaptive variation to circumvent the disadvantages of asexuality (Fig. 19.1). This additional genetic material may also cause the formation of a small proportion of males (Lamatsch et al. 2000). Expression of paternal genes may provide a local adaptive advantage in physiological or phenotypic sexual mimicry traits (Beukeboom and Vrijenhoek 1998). It has also been argued that paternal leakage leading to the expression of paternal genes plays a pivotal role to stop Muller's ratchet (Schartl et al. 1995a; Schlupp 2005; Loewe and Lamatsch 2008). However, the observation of paternal leakage should not be confused with true recombination, a reason for discussions of the ratchet-stopping potential of paternal leakage (Beukeboom et al. 1995; Beukeboom and Batenburg 1999). It has been speculated that the paternal genome might be used as a template for DNA repair, but its precise role remains unclear (Beukeboom and Vrijenhoek 1998). Despite its obscure role, paternal leakage enables interactions of a "frozen" (unrecombined) genome with a recently recombined one.

19.1.4 Hybridogenesis

Hybridogenesis is a hemiclonal form of reproduction with features of both, sperm-dependent parthenogenesis and sexuality (Fig. 19.1) (Schultz 1969; Vrijenhoek et al. 1977; see also Chapters 4, 16 and 18). Diploid hybridogenetic females (e.g., AB) transmit a haploid, non-recombinant, maternal genome (i.e., hemiclone A) to their ova. Diploidy is restored by true fertilization with sperm from males of species B. The hemiclonal A genome is combined with a new recombined B genome in each generation; therefore, only maternal genes and chromosomes are perpetuated across generations of the unisexual biotype. Although variation from species B is phenotypically expressed by a hybridogenetic lineage, it is substituted in each generation and is not heritable.

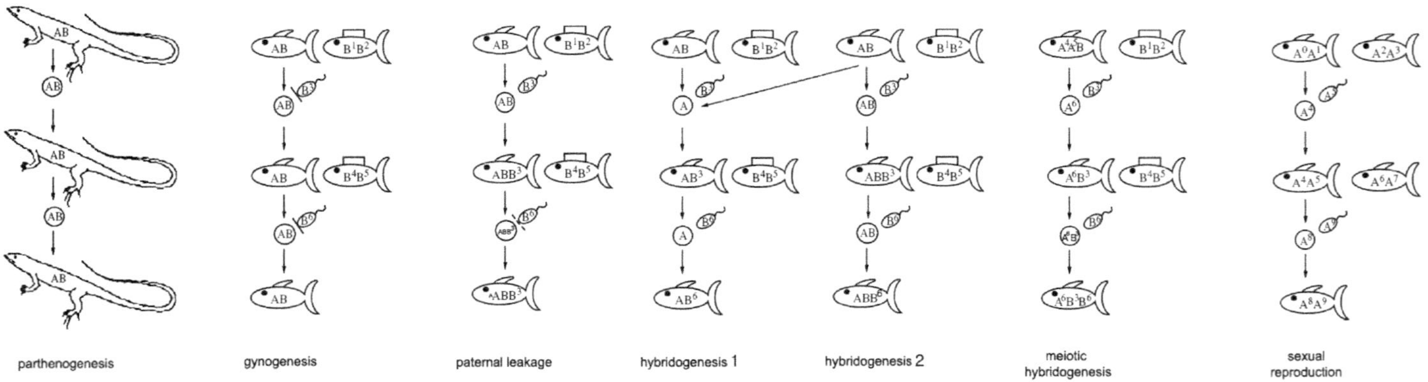

Fig. 19.1 Principles of major reproductive modes in teleost fishes with increasing quantities of sexual elements (expanded from Lampert and Schartl 2008). Capitals illustrate genomes, different letter codes indicate different species. Numbers (0–9) differentiate individual genomes derived from recombination. *Parthenogenesis* (restricted to reptiles): hybrid females produce unreduced ova (AB) that develop into all-female offspring, genetically identical to their mother (AB); *Gynogenesis* (sperm-dependent parthenogenesis): hybrid females (AB) produce unreduced ova (AB) that develop into all-female offspring (AB); sperm are, however, needed from a closely related sexual species to trigger embryonic development without contributing to the offspring's genotypes; reproduction in gynogenesis is truly clonal; *Paternal leakage*: Instead of only triggering the embryo's development, the sperm and egg pronucleus fuse and the offspring show paternal genetic contribution in form of an additional chromosome set resulting in triploids (1), or in form of microchromosomes (2); *Hybridogenesis 1*: the hybrid female's genome (A) is passed on clonally to the offspring while the other genome is substituted every generation (B, B^3, B^6) (hemiclonal); *Hybridogenesis 2*: the hybrid female's genome (AB) is passed on clonally to the offspring while the sexual species' genome (B^3) first elevates the ploidy level and then is substituted every generation (B^6); *Meiotic hybridogenesis*: After pairing of homologous chromosomes (A^4A^5), the third set of unmatched chromosomes (B) is eliminated, and the remaining ones form bivalents and undergo normal meiosis. The recombined haploid eggs (A^6) are fertilized by haploid sperm of the host resulting in diploid females which produce unreduced diploid eggs (A^6B^3). The allotriploidy is restored by fertilization ($A^6B^3B^6$); *Sexual reproduction*: both mating partners produce haploid germ cells that are unique due to recombination, and result in highly variable individual offspring

It has been a longstanding enigma why so few organisms exist that combine parthenogenetic and sexual cycles of reproduction (Green and Noakes 1995). A number of recent studies have shown that parthenogens can have cryptic sex (e.g., D′Souza et al. 2006 and Chapter 18; Omilian et al. 2006) and suggest that rare sexual processes may be more common than previously thought (Beukeboom 2007). In the following sections, we will provide an overview on gynogenetic and hybridogenetic teleost fishes that apparently show amazingly complex reproductive modes (see also Table 19.1). However, deeper insight suggests they may well be efficiently exploiting both, the benefits of sexual and asexual reproduction.

19.2 Poeciliidae (Livebearing Toothcarps)

Members of this family have internal fertilization using an insemination apparatus (gonopodium), with females giving birth to broods of live young. More than 200 species in approximately 30 genera inhabit freshwater and brackish environments of North and South America. Two genera (*Poecilia* and *Poeciliopsis*) contain unisexual representatives; in each case they include both diploid and triploid unisexual forms.

19.2.1 Poecilia formosa

The Amazon Molly is a diploid gynogen, native to freshwater habitats in northeastern Mexico and southeastern Texas (Kallman 1962b; Turner et al. 1983; Schlupp et al. 2002;). When discovered by Hubbs and Hubbs (1932), *Poecilia formosa* was the first vertebrate conclusively demonstrated to be clonal using tissue transplantation experiments (Kallman 1962a). Molecular genetic data revealed that this all-female species arose via hybridization approximately 81,000–280,000 years ago (Schartl et al. 1995b; Lampert and Schartl 2008; Loewe and Lamatsch 2008) between two sexual species: the shortfin molly, *Poecilia mexicana,* as the ancestral female parent and an unidentified ancestor of the sailfin molly, *Poecilia latipinna,* as the male parent (Avise et al. 1991; Schartl et al. 1995b) (Fig. 19.2). For reproduction, the gynogens depend on either of these parental species, or on the broadspotted molly, *P. latipunctata*, as their sexual hosts (Niemeitz et al. 2002; Schlupp et al. 2002). Two forms of paternal leakage have been shown so far: polyploidy results if the whole sperm genome remains in the ovum, and supernumerary microchromosomes can be observed if only small quantities of the sperm's chromosome set fail to be eliminated (as usual in gynogenesis) and remain in the ovum (for reviews, see Lampert and Schartl 2008; Schlupp et al. 1998). Polyploids occur in natural habitats as triploid biotypes, mostly where diploids are sympatric with a subspecies of *P. mexicana* (mlm), but only rarely where diploids occur sympatrically with *P. latipinna* (mll) (Schultz and Kallman 1968; Rasch and Balsano 1989; Lampert

Table 19.1 List of the known, so-called "unisexual" teleost fishes

Family *Genus/biotype*	Ploidy	Genome composition (∗original maternal genome if known)	Karyotype	Reprod. mode	Host switch (if known)	References
Poeciliidae *Poecilia*						
P. formosa (mexicana-latipinna)	2n	*m∗l*	2n = 46	G	*P. latipunctata*	Hubbs and Hubbs (1932); Lampert and Schartl (2008); Loewe and Lamatsch (2008); Schlupp et al. (2002)
mexicana-2latipinna	3n	*m∗ll*	3n = 69	G		Schultz and Kallman (1968)
2mexicana-latipinna	3n	*m∗lm*	3n = 69	G		Lampert et al. (2005); Schories et al. (2007)
Poeciliopsis						
monacha-lucida	2n	*M∗L*	2n = 48	H		Angers and Schlosser (2007); Schultz (1969, 1977)
monacha-latidens	2n	*M∗Lat*	2n = 48	H		Schultz (1977)
monacha-occidentalis	2n	*M∗O*	2n = 48	H		Schultz (1977)
monacha/ viriosa x *lucida*	2n	*(M∗xV)L*	2n = 48	H		Schultz (1977); Vrijenhoek and Schultz (1974); Mateos and Vrijenhoek (2002)
2monacha-lucida	3n	*M∗ML*	3n = 72	G	*Poeciliopsis ssp.*	Schultz (1967, 1969, 1977)
monacha-2lucida	3n	*M∗LL*	3n = 72	G	*Poeciliopsis ssp.*	Schultz (1967, 1969, 1977)
monacha-lucida-viriosa	3n	*M∗LV*	3n = 72	G	*Poeciliopsis ssp.*	Schultz (1967, 1977); Mateos and Vrijenhoek (2005)
Cyprinodontidae fundulus						
diaphanus-heteroclitus	2n	*d∗h*	2n = 48	G		Fritz and Garside (1974); Dawley (1992)
2diaphanus-heteroclitus	3n	*d∗hd*	3n = 72	?		Dawley (1992)

Table 19.1 (continued)

Family *Genus/biotype*	Ploidy	Genome composition (*original maternal genome if known)	Karyotype	Reprod. mode	Host switch (if known)	References
Atherninidae						
Menidia clarkhubbsi (beryllina x peninsulae-"like"*)*	2n	*B*P*	?	G		Echelle and Moisier (1982); Echelle and Echelle (1997)
2beryllina-peninsulae	3n	*B*BP*	?	?		Echelle et al. (1988, 1989)
beryllina-2peninsuae	3n	*B*PB*	?	?		Echelle et al. (1988, 1989)
Cyprinidae						
Carassius						
gibelio	3n	*Carassius auratus* x *Cyprinus carpio*	3n = 156–162	G		Wei et al. (2003)
gibelio	4n	3n *gibelio* x *C. carpio*	4n = 212	G		Zhu and Gui (2007)
langsdorfii	3n	*langsdorfii* 2n x *C.a. auratus*	3n = 150–156	G		Murayama et al. (1986); Murakami and Fujitani (1997)
langsdorfii	4n	?	4n = 206	G		Murayama et al. (1986)
Phoxinus						
eos-neogaeus	2n	*e*n*	2n = 50	G		Goddard et al. (1989); Goddard and Dawley (1990)
2eos-neogaeus	3n	*e*ne*	3n = 75	"triploid hybridogenesis"		Goddard et al. (1989); Goddard and Dawley (1990)
eos- 2neogaeus	3n	*e*nn*	3n = 75	G		Goddard et al. (1989); Goddard and Dawley (1990)
eos-neogaeus / 2eos-neogaeus	2n/3n mosaic	*e*n/e*en*	2n = 50 / 3n = 75	G		Goddard et al. (1989); Goddard and Dawley (1990)
eos-neogaeus /eos-2neogaeus	2n/3n mosaic	*e*n/e*ne*	2n = 50/3n=75	G		Goddard et al. (1989)
Squalius						
alburnoides (*pyrenaicus*-unknown ancestor)	2n	P*A	2n = 50	clonal eggs+fertilization	*S. carolitertii*	Alves et al. (2001)

Table 19.1 (continued)

Family *Genus/biotype*	Ploidy	Genome composition (*original maternal genome if known)	Karyotype	Reprod. mode	Host switch (if known)	References
carolitertii-unknown ancestor	2n	C*A	2n = 50	H		Alves et al. (2001)
all-male lineage	2n	AA	2n = 50	Meiotic		Alves et al. (2001)
triploid hybrids	3n	P*AA	3n = 75	Meiotic hybridogenesis		Alves et al. (2001)
	3n	P*PA	3n = 75	Meiotic hybridogenesis		Alves et al. (2001)
	3n	C*AA	3n = 75	Meiotic hybridogenesis ?		Alves et al. (2001) ;
	3n	C*CA	3n = 75	Meiotic hybridogenesis		Alves et al. (2001)
Cobitidae						
Misgurnus anguillicaudatus	2n	?	2n = 50	G		Morishima et al. (2002, 2008a)
Misgurnus anguillicaudatus	3n	?	3n = 75	Meiotic hybridogenesis		Zhang and Arai (1999); Morishima et al. (2002, 2008a,b); Itono et al. (2007); Oshima et al. (2005); Arai (2003)
Misgurnus anguillicaudatus	4n	?	4n = 100	?		Arai et al. (1991, 1993)
Misgurnus anguillicaudatus	2n/3n mosaic	?	2n = 50/ 3n = 75	Meiotic hybridogenesis, clonal eggs+fertilization		Morishima et al. (2004)
Cobitis taenia complex						
elongatoides-taenia	2n	*ET*	2n = 49	G		Janko et al. (2007)
elongatoides-taenia / 2 elongatoides-taenia	2n/3n	ET/EET mosaic	2n = 49/ 3n = 74	G		Janko et al. (2007)
2 *elongatoides-taenia*	3n	*EET*	3n = 74	G		Janko et al. (2007)
elongatoides-2taenia	3n	*ETT*	3n = 73	G		Janko et al. (2007)
3*elongatoides-taenia*	4n	*EEET*	4n = 99	G?		Janko et al. (2007)
2*elongatoides-2taenia*	4n	*EETT*	4n = 98	G?		Janko et al. (2007)

Table 19.1 (continued)

Family *Genus/biotype*	Ploidy	Genome composition (∗original maternal genome if known)	Karyotype	Reprod. mode	Host switch (if known)	References
elongatoides-3taenia	4n	*ETTT*	4n = 97	G?		Janko et al. (2007)
elongatoides-tanaitica	2n	*ETa*	2n = 50	G?		Janko et al. (2007)
2elongatoides-tanaitica	3n	*ETaE*	3n = 75	G?		Janko et al. (2007)
elongatoides-2tanaitica	3n	*ETaTa*	3n = 75	G?		Janko et al. (2007)
elongatoides-tanaitica-taenia	3n	*ETaT*	3n = 74	G?		Janko et al. (2007)
3elongatoides-tanaitica	4n	*EEETa*	4n = 100	G?		Janko et al. (2007)
Triploid biotype	3n	C. *elongatoides-taurica-taenia*	3n = 75	G?		Janko et al. (2007)
Diploid biotype	2n	E-*albicolaris(=strumicae)*	2n = 50	G?		Janko et al. (2007)
Triploid biotype	3n	C. *elongatoides-tanaitica-albicolaris(=strumicae)*	3n = 75	G?		Janko et al. (2007)
Tetraploid biotype	4n	C. 2*elongatoides-taenia-melanoleuca*	4n = 98?	G?		Janko et al. (2007); Vasil'ev et al. (2007)
Tetraploid form (3n + 24)	4n	*taenia* (2) x unknown sp.[a] (2)	4n = 98	G?		Vasil'ev et al. (1989)
Tetraploid form (3n + 25)	4n	*taenia* x unknown sp.[a] (2) x *granoei*	4n = 99	G?		Vasil'ev et al. (1989)
Tetraploid form (3n + 24)	4n	*taenia* (2) x unknown sp.[a] (2)	4n = 98	G?		Vasil'ev et al. (1989)
Tetraploid form (3n + 25)	4n	*taenia* x unknown sp.[a] (2) x "*granoei*"(=*melanoleuca)*	4n = 99	G?		Vasil'ev et al. (1989)
C. 2hankugensis-longicorpus	3n	*Hlh*	3n = 73	Meiotic hybridogenesis		Kim and Lee (1990, 2000)
C. hankugensis-2longicorpus	3n	*hll*	3n = 74	Meiotic hybridogenesis?		Kim and Lee (1990, 2000)

[a] Unknown sp. of Vasil'ev et al. (1989) is probably *C. elongatoides* (Janko, personal communication).

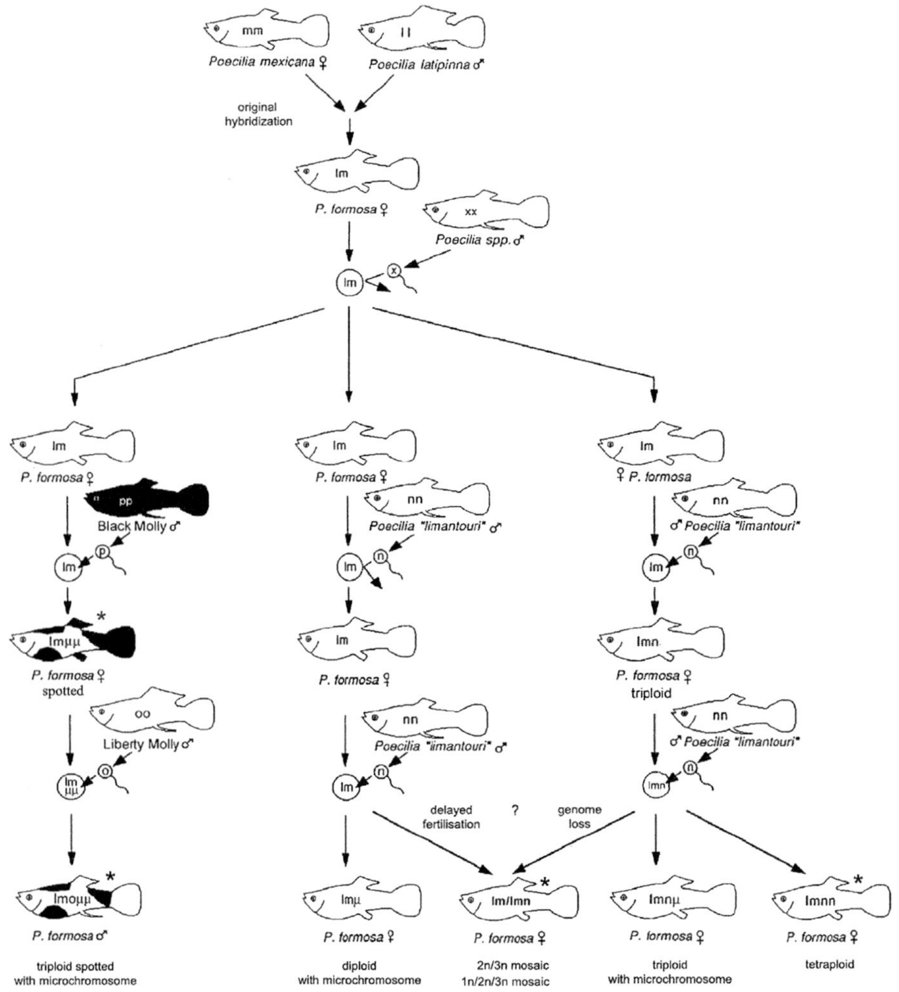

Fig. 19.2 Graphic summary of the breeding complex of *Poecilia formosa*, updated after Dawley (1989). * Only observed under laboratory conditions

et al. 2005). Under laboratory conditions, tetraploids may also occur (Lampert et al. 2008) and even somatic mosaics (Lamatsch et al. 2002; Lampert et al. 2007a), in which the reproductive mode is also gynogenetic. Microchromosomes are found in nature (Sola et al. 1993; Lamatsch et al. 2004) as well as under laboratory conditions (Schartl et al. 1995a; Nanda et al. 2007) and show that microchromosomes are not necessarily as genetically inert as widely assumed (Camacho et al. 2000). A combination of these two types of introgression events in the laboratory has been found to result in unusual triploid males (Lamatsch et al. 2000; and Lamatsch, unpublished data). Additional genetic material derived from paternal introgression events might possibly ensure the reproductive success and evolutionary longevity of *Poecilia formosa* (Loewe and Lamatsch 2008).

19.2.2 Poeciliopsis

Unisexual species of *Poeciliopsis* are native to desert arroyos in northwestern Mexico. Diploid forms originate through crosses between the sexual Headwater livebearer (*Poeciliopsis monacha*; *M*, as the female parent) and the Clearfin livebearer (*Poeciliopsis lucida*; *L*) (Schultz 1973), Gila topminnow (*P. occidentalis*; *O*) or the Lowland livebearer (*P. latidens*; *lat*), as the male parent, respectively (Fig. 19.3; for historical biogeography of the genus, see Mateos et al. 2002). These

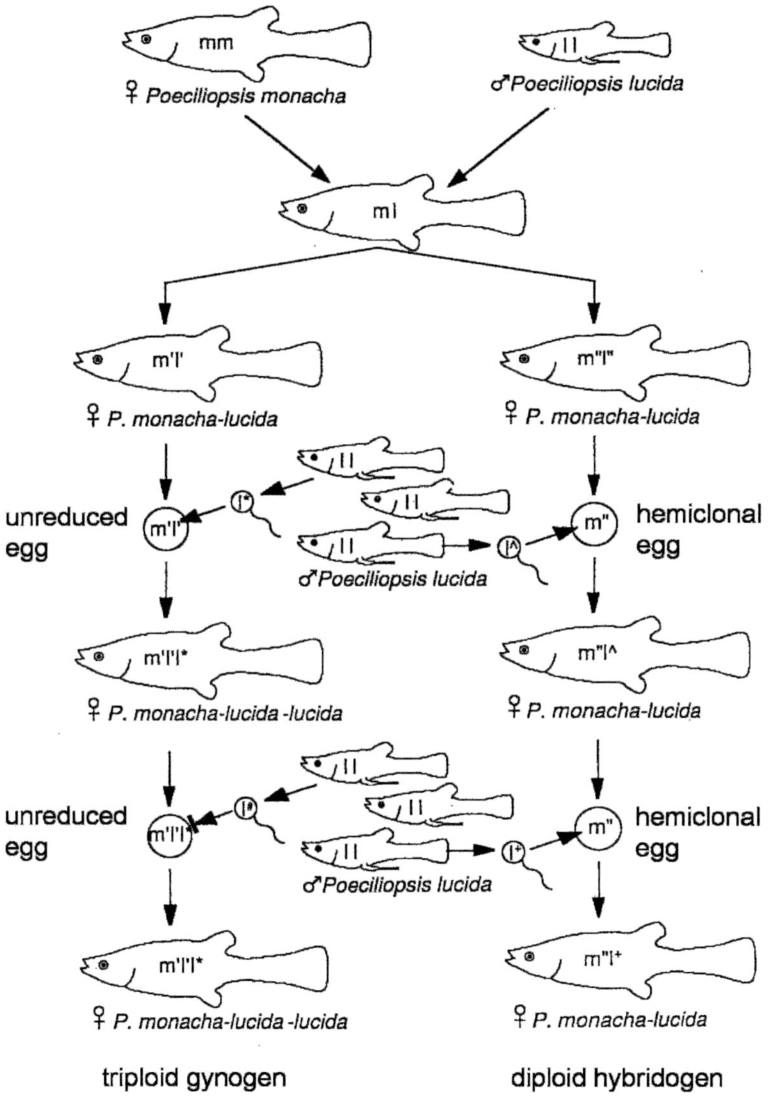

Fig. 19.3 Example of the breeding complex of *Poeciliopsis*, after Dawley (1989) that shows *P. lucida* as sperm-donor but see Table 19.1 for all possible combinations

diploid hybrids reproduce via hybridogenesis, in which the haploid *M* genome is transmitted clonally to eggs, whereas the paternal genome is excluded and replaced in each generation by insemination from the respective *Poeciliopsis* males (Schultz 1967; Mateos and Vrijenhoek 2005). Triploid forms, however, reproduce gynogenetically (Schultz 1967). Three different triploid biotypes are known: *P. 2monacha-lucida* (*MML*), *P monacha-2lucida* (*MLL*) and *P. monacha-lucida-viriosa* (*MLV*, a "trihybrid") that originated by the fertilization of an unreduced *ML*-egg by a *V*-sperm. Although in nature *MML*, *MLL* and *MLV* appear to use sperm from sympatric *P. monacha, P. lucida* and *P. viriosa*, respectively, these triploids can alternatively use sperm from different *Poeciliopsis* species to activate development of the clonal egg (Schultz 1967). In contrast to allodiploidy, allotriploidy has arisen only a few times in *Poeciliopsis* and is of rather recent origin (Mateos and Vrijenhoek 2005; Quattro et al. 1991, 1992). Further ploidy elevation (e.g., tetraploidy) has not been found so far.

19.3 Cyprinodontidae (Pupfishes)

The family Cyprinodontidae contains more than 100 species distributed mostly in fresh and brackish waters throughout the Americas, Africa and Eurasia. Like all killifishes, pupfishes have external fertilization, are egg-layers and mostly reproduce bisexually.

19.3.1 Fundulus diaphanus-heteroclitus

One of the few departures from bisexual reproduction involves an all-female clonal biotype within the otherwise sexual species of *Fundulus*. This biotype, known from two sites in Nova Scotia, Canada, is probably gynogenetic (Dawley 1992) and arose through hybridization between sexual *F. diaphanus* (banded killifish) and *F. heteroclitus* (mummichog) (Dawley et al. 1999, 2000; Hernandez Chavez and Tuergeon 2007). Interspecies hybridization between *F. heteroclitus* and *F. diaphanus* occurs over a wide geographic range (Hernandez Chavez and Tuergeon 2007), but gynogens were only observed at two sites, indicating that only specific crosses between parental species result in clonal reproduction. With all clones bearing the same *F. diaphanus* mtDNA-haplotype, its origin probably goes back to a few rather recent hybridization events (Dawley 1992; Hernandez Chavez and Tuergeon 2007) while microsatellite data raise the possibility of several independent origins of asexuality (Hernandez Chavez and Tuergeon 2007). The unisexual hybrids are mostly diploid and only rarely triploid. Triploids show DNA contents close to what would be expected for a hybrid with a double dose of the *F. diaphanus* genome and a single dose of the *F. heteroclitus* genome (*DDH*) (Dawley 1992), indicating paternal leakage as source of triploidy. However, the potential role of paternal introgression, sexual reproduction, occasional recombination and mutational events has not been conclusively addressed to date.

19.4 Atherinopsidae (Neotropical Silversides)

This family contains approximately 104 species in 13 genera that are distributed throughout the tropical and temperate waters of the Americas, including both marine and freshwater habitats. So far, unisexuality has only been discovered in one species.

19.4.1 Menidia clarkhubbsi

Species of *Menidia* are small, silvery-sided, planktivorous fishes that form dense, highly mobile foraging schools along the Texas coast and eastward at sites on the northern Gulf of Mexico. They are egg-layers with external fertilization and have no striking sexual dimorphism. The unisexual-bisexual complex comprises the bisexual Inland silverside (*M. beryllina*) and the Tidewater silverside (*M. peninsulae*), F_1 hybrids between the two species and several different all-female clones (*M. clarkhubbsi*, Texas silverside) that arose through multiple hybridizations between males of *M. beryllina* and females of an extinct or as-yet undetected species genetically similar to *M. peninsula* (Echelle and Moisier 1982; Echelle et al. 1983; Echelle and Echelle 1997). The complex consists mainly of diploid females and there is no compelling evidence of persistent polyploid clones, although rare occurrence of wild-caught triploids has been stated (Echelle et al. 1988, 1989). It is not yet resolved whether these triploids occur by paternal leakage involving a diploid egg from *M. clarkhubbsi* (*BP*) and a haploid sperm of *M. peninsulae* (*P*) (Echelle et al. 1989) or by back-crosses of F_1 hybrids, producing unreduced gametes, to the parental species (*BPP*, *BBP*) (Echelle et al. 1988), and how they reproduce. Low abundance seems characteristic of this unisexual complex: Despite intense research, no environmental situations were found where unisexual *Menidia* are predictably more abundant than their bisexual relatives (Echelle et al. 1989; Echelle and Echelle 1997). This may reflect competition with the diversity of other forms of *Menidia* (two bisexual species, their hybrids and backcross progeny) as well as the lack of opportunity for origins of new unisexual species due to the absence of one of its bisexual progenitors (the missing *M. peninsulae*-like form). Thus, the existing *M. clarkhubbsi* species complex may be a relict of a once more diverse, and therefore more abundant assemblage of clones.

19.5 Cyprinidae (Minnows and Allies)

Members of this huge taxonomic assemblage are native to North America, Eurasia and Africa. With more than 1,600 species in nearly 300 genera, this is the most species-rich family of fishes. However, only a few clonal or hemiclonal biotypes are known.

The Japanese crucian carp (*Carassius auratus*) was morphologically classified into several subspecies: kinbuna (*C. a. ssp.*), nagabuna (*C. a. burgeri*), nigorobuna (*C. a. grandoculis*), gengorobuna (*C. a. cuvieri*), ginbuna (*C. a. langsdorfii*) and Prussian carp (*C. a. gibelio*). Recent research based on mitochondrial *Cytochrome b* sequences (Kalous et al. 2007), however, has characterized *C. a. cuvieri*, *langsdorfii* and *gibelio* as distinct species, leaving only *grandoculis*, *burgeri* and the kinbuna as subspecies of *C. auratus*.

19.5.1 Carassius gibelio

The gynogenetic Prussian carp was originally described in a population from the Shuangfeng Reservoir in northern China and now invasively spreads in freshwater streams, ponds and lakes over a wide geographic range from northern Europe to Asia (e.g., Veetema et al. 2005; Flajshans et al. 2007; Verreycken et al. 2007; Leonardos et al. 2008). The triploid gynogen originated from an ancient hybridization event, with *Carassius auratus* being the maternal and *Cyprinus carpio* being the paternal ancestor (Chun et al. 2001). Seven different clones have been identified, differing significantly in body shape, growth rate, spawning time, serum protein phenotype, karyotype etc. (Zhou et al. 2000; Yi et al. 2003). This species shows two unusual characteristics for unisexual breeding complexes: (1) populations contain up to 20% males in natural habitats (Abramenko et al. 1998); (2) two reproductive modes exist: gynogenesis and gonochoristic reproduction; besides its role in activating the egg, sperm contribute to the progeny in a high percentage of cases. Cytological observations have revealed two different patterns of sperm development: if eggs are inseminated with heterologous sperm (i.e., from other species), the entered sperm does not decondense and is then eliminated from the zygote. This is the normal process of gynogenesis and gives rise to all-female progeny. However, when the eggs are inseminated with homologous sperm of silver crucian carp males, the sperm undergoes normal decondensation and pronucleus formation and fuses with the female pronucleus. The fused nucleus of the zygote undergoes recombination and extra chromosomes (about half of the maternal chromosomes) are eliminated from the egg. In this case, genetically diverse offspring are produced (including males), similarly as in gonochoristic reproduction. This form of gynogenesis has been referred to as "allogynogenesis" (Jiang et al. 1983). As a consequence, the complex consists of diploid individuals (males and some of the females), mostly triploid individuals (almost exclusively females) and rare tetraploid females (Fan and Liu 1990; Zhu and Gui 2007). In one clone, paternal leakage of subgenomic amounts of genetic material has been detected in the offspring. Phenotype similarity with the sperm-donor implied that these microchromosomes might carry genes that are expressed in the foreign genetic background (Yi et al. 2003). Although this is an unusual situation since supernumerary chromosomes are often found to be genetically inert (Camacho et al. 2000), it has also been demonstrated in microchromosomes of *Poecilia formosa* expressing the macromelanophore locus (Schartl et al. 1997).

19.5.2 *Carassius langsdorfii*

The Japanese silver crucian carp, ginbuna, is widely distributed in Japan. There are three forms of females in this complex: a bisexual diploid form as well as a gynogenetic triploid and tetraploid form (Kobayashi et al. 1970, 1977). Although the hybrid origin of polyploid ginbuna has been revealed with nuclear markers (Murakami and Fujitani 1997), its ancestral parents have not yet been identified. According to Murakami et al. (2001), triploid ginbuna have been derived from two different maternal lineages approximately 70,000–160,000 years ago and it seems likely that the goldfish, *C. a. auratus,* contributed to the ploidy elevation from diploid hybrid ginbuna (Murakami et al. 2002). In mating experiments of triploid ginbuna with male goldfish (*C. a. auratus*) diploid-triploid and diploid-triploid-tetraploid mosaic offspring were obtained, most of which turned out to be males (Murayama et al. 1986). Since no paternal contribution could be detected, the exact mechanism remains unclear. Other authors have discovered rare triploid (Muramoto 1975) and tetraploid males (Murakami and Fujitani 1997) from natural populations. Despite warnings based on experiences with other species of *Carassius* invading native fish communities (Crivelli 1995; Fraser and Adams 1997), individuals of *C. langsdorfii* have lately been found in Europe in the River Elbe basin (Czech Republic) (Kalous et al. 2007), probably accidentally being introduced along with imports of commercially important fishes like Koi carps (*Cyprinus carpio*). Their impact on endemic fish fauna needs to be assessed urgently.

19.5.3 *Phoxinus eos-neogaeus*

The *Phoxinus eos-neogaeus* complex is widely distributed in north-eastern America and occupies very heterogeneous habitats (Angers and Schlosser 2007). It originated by multiple hybridization between males of the northern redbelly dace (*P. eos*) and females of the finescale dace (*P. neogaeus*) (Dawley et al. 1987; Goddard et al. 1989; Angers and Schlosser 2007). Altough these diploid hybrids reproduce by sperm-dependent parthenogenesis (Goddard et al. 1998), the exclusion mechanism, which normally clears the egg from the sperm in gynogenesis, often fails in this hybrid complex, leading to an unusually high level of sperm incorporation (Fig. 19.4). As a consequence, five different hybrid biotypes are found in the complex: (1) the strictly clonal, all-female diploid *P. eos-neogaeus* lineage (*en*), (2) triploid *P. 2eos-neogaeus (ene)*, (3) triploid *P. eos-2neogaeus (enn)*, (4) mosaic *P. eos-neogaeus / 2eos-neogaeus (en/ene)* and (5) mosaic *P. eos- neogaeus / eos-2neogaeus* (*en/enn*) (Goddard et al. 1989; Goddard and Dawley 1990). The first four of these hybrid biotypes have been found in natural populations (Doeringsfeld et al. 2004), whereas the latter mosaic biotype with an additional *neogaeus* genome has only been reported from a laboratory mating (Goddard et al. 1989). Where the diploid clonal hybrid occurs sympatrically with only one parental species (typically *P. eos*), the complex is comprised of three biotypes: the diploid clone augmented with triploids and mosaics carrying an additional genome from that species (Dawley

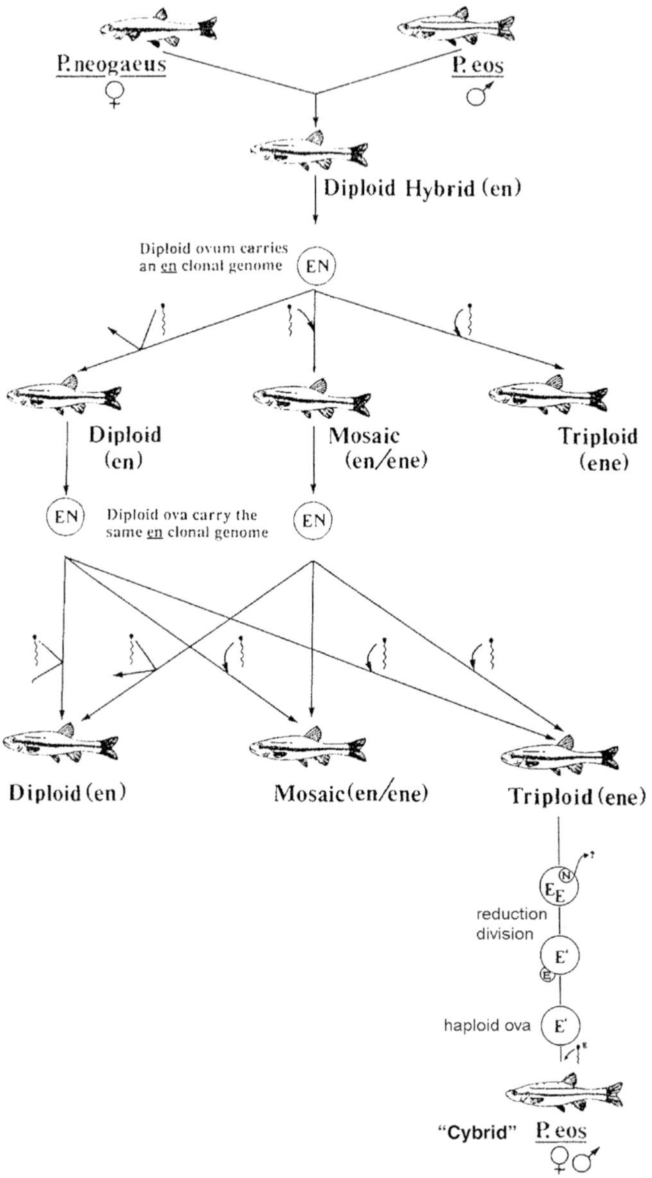

Fig. 19.4 Graphic example of the breeding complex of *P. eos-neogaeus* showing diploid, triploid and diploid-triploid mosaic hybrids when mating with *P. eos* (modified from Goddard and Dawley (1990) and Goddard and Schultz (1993)

et al. 1987). Where it occurs in sympatry with both parental species, all five biotypes may be present (Goddard et al. 1989; Doeringsfeld et al. 2004). Mosaic individuals (*en/een*) produce diploid eggs (*en*), which may develop gynogenetically or incorporate sperm. Triploids (*een*), however, exclude the *neogaeus* genome before they undergo reduction division, resulting in haploid (*e*) ova, a process comparable to meiotic hybridogenesis (see *Squalius, Misgurnus*). When fertilized by *P. eos* sperm, the resultant offspring are indistinguishable from males and females of *P. eos* (Goddard and Schultz 1993) although they carry the mitochondrial DNA of *P. neogaeus*. These are called "cybrids" in contrast to nuclear hybrids (gynogens, triploids and mosaics). Doeringsfeld et al. (2004) conclude that the distribution and ecological success of the hybrid complex is a function of both, an apparently broadly adapted clonal lineage and additional genetic and phenotypic variation expressed by various polyploid biotypes.

19.5.4 Squalius alburnoides

The Iberian fishes of the *Squalius* (previously *Leuciscus*, *Tropidophoxinellus* or *Rutilus*) *alburnoides* complex (for nomenclature, see Kottelat 1997 and Collares-Pereira et al. 1999) contain mixed reproductive systems of diploid, triploid and tetraploid forms with highly female biased sex ratios. The complex arose through interspecific crosses between *S. pyrenaicus* (*P*-genome) and males of an apparently extinct species (*A*-genome) (Alves et al. 2001). The most common form of the complex includes hybrid females and males with diploid (*PA*), triploid (*PAA* and *PPA*) and tetraploid (*PPAA*, *PAAA* and *PPPA*) genomes (Gromicho et al. 2006) (see Fig. 19.5). The second form of the complex comprises a diploid nuclear non-hybrid but all-male lineage (*AA*) with *pyrenaicus*-mtDNA that is hypothesized to have been reconstituted within the complex by triploid *PAA*-females (Alves et al. 2002) (Fig. 19.6). Recent research revealed that the paternal ancestor of the complex was an *Anaecypris hispanica*-like species according to evidence from cytogenetics (Gromicho et al. 2006), microsatellites (Crespo-Lopez et al. 2007) and nuclear sequence data (Robalo et al. 2006). While this ancestral species seems to be extinct in most or all relevant river basins, its nuclear genome is preserved in the all-male form (with *pyrenaicus*-mtDNA). The oogenesis of triploid *PAA*-females is mostly achieved by "meiotic hybridogenesis" (Alves et al. 1998), which involves elimination of the *P* (*Squalius*) genome, followed by random segregation and recombination between the two remaining genomes, generating *A*-ova. This mechanism has been first assumed to be operating in certain triploid hybrid *Rana esculenta* frogs (Günther et al. 1979) and seems to be similar to males of all-triploid Batura-toads (Stöck et al. 2002). When these recombined *A*-ova are fertilized with recombined haploid *A*-sperm of *AA*-males, new *AA*-all-male (with *pyrenaicus*-mtDNA) progeny is restored. Alves et al. (2004), however, reported a triploid female that generated both large triploid and small haploid eggs resulting in all-female progeny. In diploid hybrid *PA*-females, few eggs (< 3%) develop by gynogenesis. The majority of *PA*-females transmits the complete hybrid genome to the egg and fertilization results in triploid progeny (Alves et al. 1998). In all other forms of the complex, reproductive modes include syngamy (Alves et al. 2001). Males in the *S. alburnoides* complex are fertile and play a role in the dynamics of the complex: diploid hybrids (*PA*) produce unreduced sperm, while others designated as diploid "nuclear non-hybrid males" (*AA*) produce reduced sperm (Alves et al. 1999).

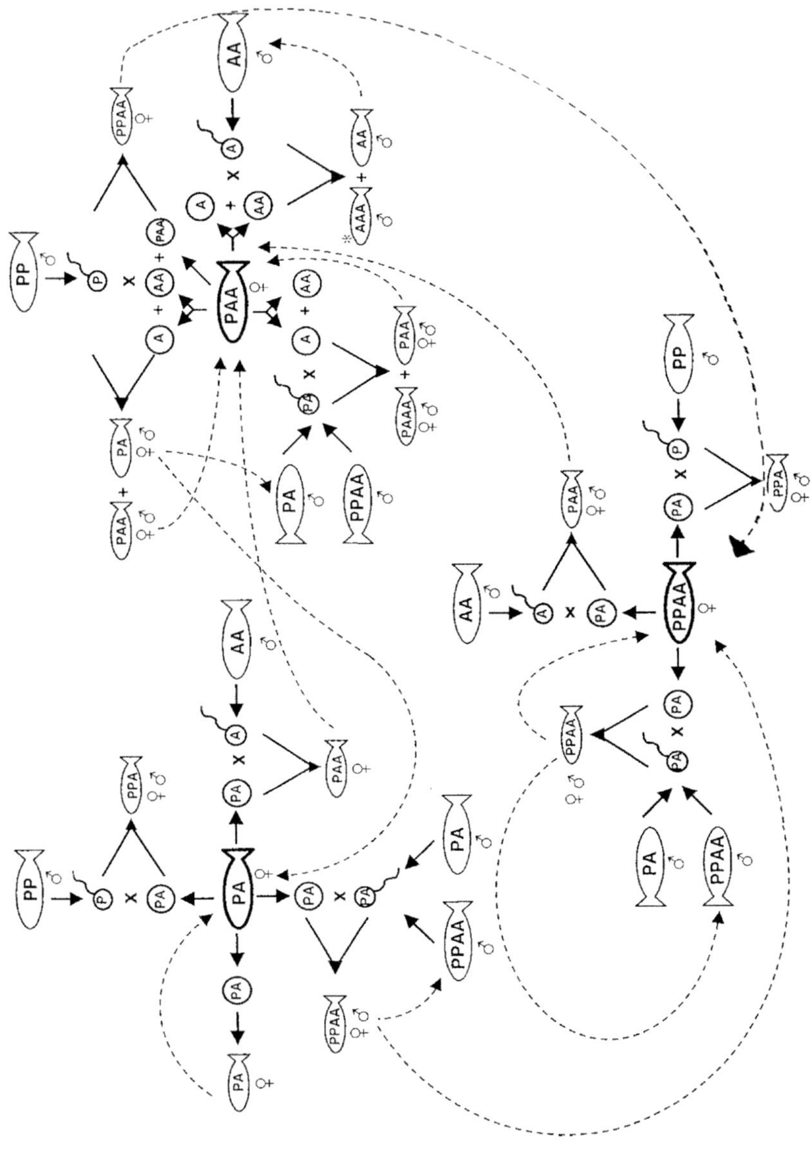

Fig. 19.5 Putative relationships between the various forms of the *S. alburnoides* complex modified from Alves et al. (2001). * Never observed in natural populations, but in breeding experiments

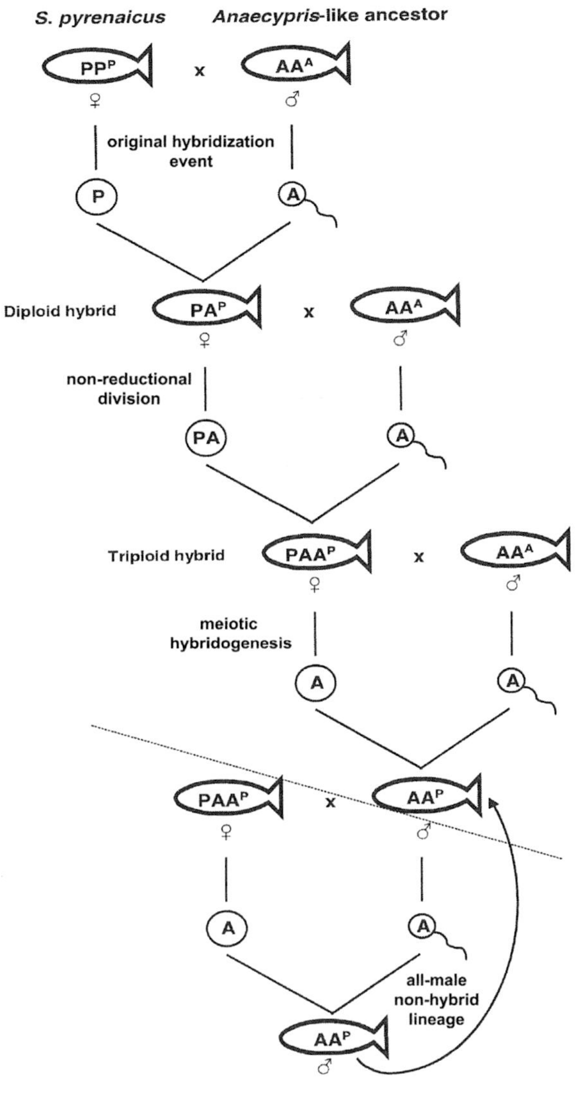

Fig. 19.6 Hypothetical evolutionary trajectory of the all-male non-hybrid lineage within the hybrid *S. alburnoides* complex. Above dashed line, mechanism of the origin of the hybrid complex; below dashed line, perpetuating mechanism of the all-male non-hybrid lineage (modified from Gromicho et al. 2006)

In the absence of *S. pyrenaicus* (e.g., N-Portugal), the complex seems to be maintained by crosses with males of *S. carolitertii* (*CC*) and by diploid hybrid males (*CA*), although the mtDNA found in *S. alburnoides* is *S. pyrenaicus*-like (Cunha et al. 2004; Pala and Coelho 2005; Sousa Santos et al. 2006). Recently, also extensive mtDNA-introgression from a related species (*S. aradensis*) into the *S. alburnoides* has been detected (Sousa Santos et al. 2006). A spectacular novel finding has been added to the knowledge about the evolutionary dynamics of the complex: In two populations from the NW-Iberian Douro drainage, tetraploid individuals represent 85.6–97.5% of the population, with no observed sex ratio bias.

Using flow cytometry of blood and sperm, microsatellite data and experimental crosses, Cunha et al. (2008) describe two gonochristic allotetraploid populations (*CCAA*) with normal meiosis. This illustrates how the evolutionary dynamics of a hybrid complex may contribute to polyploid speciation. Such processes have been predicted for polyploid fishes at least as early as 30 years ago by Schultz (1979).

Studying gene expression, Pala et al. (2008) found in some triploid forms of the *S. alburnoides* complex "that a compensation mechanism exists, reducing transcript levels to the diploid state". Their data suggest a silencing of one of the three alleles, although unexpectedly, it is not a whole haplome that is inactivated. The allelic expression patterns differ between genes and between different tissues for one and the same gene.

19.6 Cobitidae (Loach Fishes)

This family comprises about 150 extant freshwater species that inhabit Eurasia, with species diversity being the highest in southern Asia. These fish are typically bottom-dwellers with downward facing mouths and "wormlike" or fusiform (spindle-shaped) bodies. A recent molecular phylogeny was proposed by Slechtova et al. (2008).

19.6.1 Cobitis

Spined loaches form a monophyletic group within the cypriniforms, which contains more than 42 species of freshwater fishes (Sawada 1982). Apparently, nothing is known about the occurrence of hybrid complexes among the Iberian species. In Central and Eastern Europe, at least seven diploid gonochoristic spined loach species are found (Vasil'ev et al. 2007). In addition, since the beginning of the 1980's, diploid, triploid and tetraploid all-female spined loach forms were discovered (Vasil'ev and Vasil'eva 1982; Vasil'ev et al. 1989; Bohlen 2000; Rab et al. 2000; Bohlen and Rab 2001; Boron et al. 2003; Janko et al. 2003) (Fig. 19.7). The

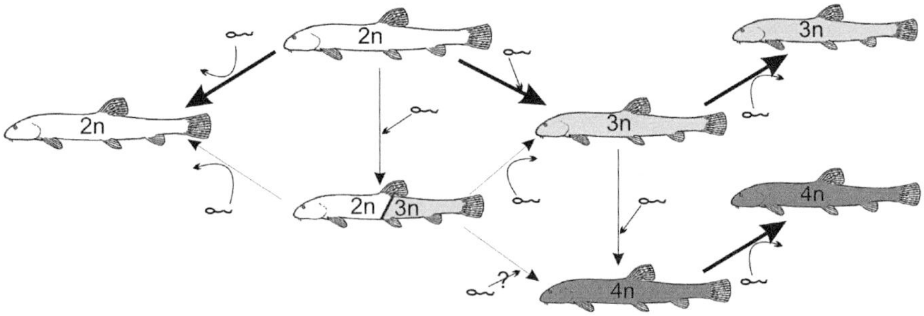

Fig. 19.7 Representation of the European *Cobitis* breeding complex (kindly provided by K. Janko; with modifications)

C. taenia complex sensu Janko comprises six parapatric species known to hybridize (*C. elongatoides, C. taenia, Cobitis tanaitica, Cobitis taurica, Cobitis strumicae and Cobitis melanoleuca*) (Choleva et al. 2008).

These asexual lineages arose by hybridization between three species: *Cobitis taenia*, *C. elongatoides* and *C. tanaitica.* Diploid, triploid and tetraploid all-female hybrids between *C. elongatoides* and *C. taenia* are found, as well as triploid and tetraploid all-female hybrids between *C. elongatoides* and *C. tanaitica*, co-occurring over a large range with their parental species (Slechtova et al. 2000; Bohlen and Rab 2001). Triploids predominate in most populations, but diploid hybrids are common at some localities in the Oder and Elbe river basins (Bohlen et al. 2002; Slechtova et al. 2000). For the *C. elongatoides-C.taenia* hybrid complex (Vasil'ev et al. 1989; Saat 1991), gynogenetic reproduction has been reported. This is also presumed for the *C. elongatoides-tanaitica* hybrids as they are always associated with the parental species (Bohlen and Rab 2001). The latter form all-female populations with mostly triploids, but without any evidence for hybridogenesis. MtDNA analyses of both hybrid complexes showed that hybridization resulting in *C. elongatoides-taenia* asexuals was reciprocal, and that the asexual lineages are of recent polyphyletic origin. Taking advantage of previous knowledge of the genome composition of polyploids, Janko et al. (2003) further concluded that polyploidy has been achieved by backcrosses of diploid F_1-hybrids to both parental species. Mezhzherin and Chudakorova (2002) analysed the Dnieper river *C. taenia* hybrid complex using allozymes and found it to consist of 87% polyploid females triploids, tetraploids and "possibly a few pentaploids". The tetraploid hybrids in the Moscow River comprise some males without normal spermatozoa. However, experimental crosses between gynogenetic triploid females and tetraploid males revealed that these males could trigger gynogenesis of clonal forms in a few cases (Vasil'ev et al. 2003). Using multilocus fingerprinting, Vasil'ev et al. (2007) found tetraploids in the Don Basin to exhibit clonal inheritance, while female tetraploids in the Moscow River may also have arisen de novo, i.e., by fertilization of clonal triploid eggs. Some tetraploids are supposedly trihybrids (Vasil'ev et al. 2007), as also described by Choleva et al. (2008).

A comprehensive review of diversity and systematics of the Central and Eastern European *Cobitis taenia* complex has been published by Janko et al. (2007). Janko et al. (2005) and Culling et al. (2006) reconstructed the Quaternary biogeography of the sexual parental and clonal hybrid lineages of European *Cobitis* from two separate refuges. The authors found multiple Prae-Würmian and Holocene origins of asexuality, irrespective of the parental populations involved and similar dispersal potential of diploid and triploid lineages.

Another hybridogenetic complex of *Cobitis* occurs in East Asia. In the Korean *C. hankugensis* (previously *sinensis*: *S*) – *C. longicorpus* (*O*) complex, diploid *SO*-hybrids produce diploid ova, whose fertilization leads to triploids, which produce haploid ova (Kim and Lee 1990, 2000; Saitoh et al. 2004). According to Kim and Lee (2000), *SSO*-triploids eliminate the *O*-genome and perform meiosis with the remaining *S*-genomes (i.e., meiotic hybridogenesis). When these haploid *S*-eggs are fertilized by normal *C. hankugensis* males, diploid *C. hankugensis* are regenerated but carry foreign *C. longicorpus* mtDNA (nucleo-cytoplasmic hybrids; Saitoh et al. 2004). Kim and Lee (2000) present a hypothetical scheme on the interactions in the breeding complex (not shown).

19.6.2 Misgurnus anguillicaudatus

The oriental weather loach, *Misgurnus anguillicaudatus*, is a common freshwater fish that inhabits shallow ponds and paddy fields all over Japan, Korea, China, Vietnam and other Asian regions (Arai 2003). Although bisexually reproducing, diploid individuals are most common in the wild populations of Japan, a relatively small number of triploid individuals have also been discovered (Morishima et al. 2002; Arai 2003; Oshima et al. 2005). In wild populations of Japan, no natural tetraploids have been found despite intense screening (Arai, personal communication). Ojima and Takai (1979) and Arai et al. (1991) found tetraploids; however, all of these were of "commercial" origin; possibly from the Yang-tze (= Chiangjiang) River basin of China, where natural tetraploids and diploids occur sympatrically (Li et al. 2008). In 2002, Morishima and colleagues discovered a diploid clonal lineage within the wild population of the northern area of Hokkaido, although its origin remains unclear (Morishima et al. 2008a) (Fig. 19.8). These females generate unreduced diploid eggs by premeiotic endomitosis (Itono et al. 2006), which are activated by sperm from bisexually reproducing diploid loaches and develop gynogenetically into clonal offspring (Itono et al. 2007). Paternal leakage into unreduced eggs may lead to triploid individuals (Morishima et al. 2002; Oshima et al. 2005) or diploid-triploid mosaics (Morishima et al. 2004). Triploid females have been reported to produce mainly haploid eggs by meiotic hybridogenesis (Morishima et al. 2008b; see above): After pairing of homologous chromosomes, the third set of unmatched chromosomes is eliminated and the remaining bivalents undergo normal meiosis, resulting in haploid eggs. Therefore, diploid gynogenetic progeny can be produced from haploid eggs of these triploid loaches after normal fertilization with the sperm of bisexual diploids. Small numbers of diploid, triploid and aneuploid eggs have also been reported from triploid loaches (Oshima et al. 2005). Triploid males, however, appear to be sterile (Itono et al. 2006). In contrast, diploid-triploid mosaic males produce fertile diploid sperm (Morishima et al. 2004), whereas diploid-triploid mosaic females have been found to lay haploid, diploid and triploid eggs simultaneously (Yoshikawa et al. 2007). The occurrence of meiotic hybridogenesis in triploids derived from clonal gynogenetic *M. anguillicaudatus* suggests the presence of two distinct genomes in the clone. Arias-Rodriguez et al. (in press) found that inter-populational hybrid loaches produced unreduced or other unusual eggs. Taken together, these data suggest a hybrid origin of *M. anguillicaudatus*, but its exact origin and ancestors are still not clarified.

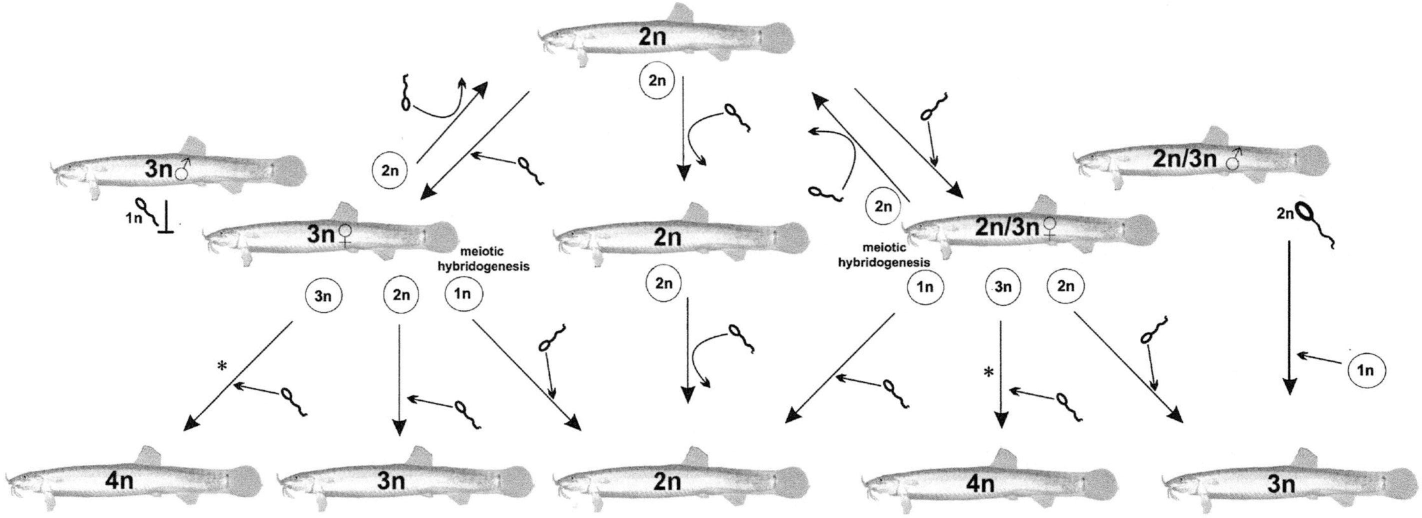

Fig. 19.8 Putative representation of the cryptic clonal lineages in the loach *Misgurnus anguillicaudatus* from Japan. ∗ Never observed in natural populations, tetraploid offspring only observed in breeding experiments!

19.7 Conclusions

As we have outlined above, teleost fishes have not completely "lost sex". In contrast, the entire range of imaginable stages between sperm-dependent parthenogenesis without or with different amounts of paternal leakage (e.g., *Poecilia, Carassius gibelio*) to hybridogenesis with hemiclonal diploid gametogenesis (e.g., *Poeciliopsis*) or genome elimination, followed by a normal meiosis (e.g., *Squalius, Misgurnus, Phoxinus*), has evolved in this group (Fig. 19.1).

Moreover, some systems include a combination of these phenomena and form species complexes that are governed by the coexistence of several reproductive modes in addition to normal sexual reproduction (e.g., *Squalius, Cobitis, Phoxinus*) and/or may even lead back to the evolution of sexuality in allotetraploids via triploids (e.g., *Squalius*).

True parthenogenesis (i.e., sperm-independent reproduction) has not been found in fishes; in vertebrates, it seems restricted to reptiles (see also Chapter 21). All described teleost complexes are of hybrid origin (even the so far mysterious *Misgurnus anguillicaudatus* seems to be of hybrid origin, see Morishima et al. 2008b) and the resulting clonal or hemiclonal reproduction can be viewed as a serendipitous effect of miscegenation between two species, which are sufficiently closely related to form viable hybrids and too distantly related for their genomes to execute normal meiosis (Moritz et al. 1989). However, between sperm-dependent parthenogenesis and true sexual reproduction, all kinds of intermediate states have evolved in teleosts. Nevertheless, all of these are sperm-dependent. This dependence may be explained by the need for a mechanical and/or chemical trigger to initiate embryogenesis in teleost ova (cf. Dawley et al. 1987; Pandian and Koteeswaran 1998). Once gynogenesis is established in a system as insemination without paternal genetic input, rare failure of the normal sperm-exclusion might lead to paternal leakage: a tiny genetic contribution of these "pseudo-fathers" that may or may not be phenotypically expressed. A higher amount of paternal leakage may comprise the complete incorporation of the foreign sperm nucleus and, as a consequence, ploidy elevation in a usually still all-female situation.

As opposed to the relatively rare paternal leakage, regular fertilization and biparternal gene expression in the offspring has evolved in hybridogenesis, but one genome is pre-meiotically eliminated. Here, we find either hemiclonal transmission of one genome without recombination (as in diploid hybridogens) and replacement of the other genome from a normal sexually recombining species. Alternatively, after the elimination of one complete chromosome set, some triploids may produce clonal diploid eggs, while others even enter a normal meiosis and produce recombined gametes (meiotic hybridogenesis).

Regarding the observed stages of complete or partial asexuality, two important aspects should be addressed by future research: (I) a hybrid may have a given reproductive mechanism immediately after it has formed (e.g., caused by the alteration of normal meiosis to automixis; Lampert et al. 2007b), just as a result of the genomic distance between the parental forms (e.g., for true gynogens, the "balance hypothesis" according to Moritz et al. 1989). (II) Ongoing interactions between sexual progenitors and hybrids (genome shuffling, addition of a foreign genome, intergenomic recombination and exchange, introgression) enabled by paternal leakage, genome addition, and various forms of hybridogenesis may, over time, lead to a higher compatibility and co-evolution of the hybridizing genomes with the result that these interspecies interactions may experience a true evolutionary transition from one reproductive mode to another. Indications of such "evolution in action" have been observed in *Squalius alburnoides*, for example (Alves et al. 2001, Cunha et al. 2008). We encounter these complexes in a certain stage but we need to understand if and how ongoing interspecies and intergenomic interactions may change in evolutionary time.

Interestingly, the speculative evolutionary (re-) transition from asexuality to sexuality (gynogenesis → paternal leakage → hybridogenesis → meiotic hybridogenesis → meiotic allotetraploids) in teleosts seems to be accompanied by an increasing number of males (e.g., *Squalius*, *Cobitis*) shifting the all-female (e.g., *Poecilia*) to a female-biased situation or perhaps again to a balanced situation as in bisexual allotetraploids (*Squalius*). This effect, however, seems rather a bi-product of introgression of genetic material (microchromosomes, entire chromosome sets) than a directed evolutionary tendency.

Another question that should be addressed by future research is the reason for the occurrence or absence of "meiotic hybridogenesis" in some triploids. Similarly to triploid *Rana esculenta* frogs (Günther et al. 1979) and triploid *Bufo baturae* toads (Stöck et al. 2002), triploids in *Squalius, Cobitis*, *Misgurnus* and *Phoxinus* have developed mechanisms that can exclude one entire chromosome set (the one that is "minority" in such 2n + 1n triploids) before they enter an apparently normal diploid meiosis. In contrast, hybrids in *Poeciliopis* (which are hybridogenetic in their diploid forms) become gynogenetic as triploids. We speculate if the reason could be some kind of trihybridity (the involvement of a third, slightly different genome) in the latter form.

Sperm-dependent unisexuals (gynogens and hybridogens) are normally primarily restricted to the range(s) of the one (or more) bisexual species on which they depend, the parental forms in most cases. It has been documented, however, that asexual lineages may rarely use sperm from a non-parental species or even switch their host (Choleva et al. 2008). This latter phenomenon has been discussed by Choleva et al. (2008) who found it in four genera of (partly) "asexual" lineages of fish and two amphibians.

Recent research (e.g., Ogielska et al. 2004 in hybridogenetic frogs; and Bi and Bogart 2006 and Bi et al. 2007 in gynogenetic salamanders) shows that once the genetic material of two species is present in the same nucleus (as in gynogens or hybridogens), a variety of intergenomic exchange events ("intergenomic (re)combination") can be expected to occur (Mable 2007). These phenomena remain to be investigated in several of the described teleost complexes with adequate molecular-cytogenetic methods. Such techniques might reveal that the intergenomic barriers between the genomes of the parental species are likely much more "porous" than assumed in the "classical" concepts of gynogenesis and hybridogenesis. One of the challenges for future research, using the technological advances that provide access to genomic information, is to address how frequently inter-species interactions result in intergenomic gene transfer and recombination. These phenomena may be the key points to understand the nature of these outstanding hybrid breeding-complexes. By broadening the view of Moritz et al. (1989), we regard them as the result of interactions between lineages reflecting a wide spectrum of interspecies genomic exchange: Thus these viable, often female-biased hybrids with ameiotic or partly meiotic gametogenesis may be viewed as "genome-shuttles" that shuffle genetic material between parental lineages over long evolutionary periods in a framework that can be considered to be a "mobile hybrid zone". "Homologous recombination, as in sex, is important for population genetics – shuffling of minor variants, but relatively insignificant for large-scale evolution. Evolutionary innovations depend much more on illegitimate recombination, which makes novel genes by gene duplication and by gene chimaerism – essentially mutational forces" (Cavallier-Smith 2002).

Acknowledgements We thank Robert B. Vrijenhoek, Karel Janko, Kathryn Goddard-Doms, Katsutoshi Arai, Robert M. Dawley, Maria Collares-Pereira and Manuela Coelho for valuable comments on earlier versions of the manuscript as well as for the permission to develop some of our figures based on their previous versions, and Manfred Schartl for discussion on some aspects. The English has kindly been improved by Duncan Cameron.

Glossary (for details: see also Fig. 19.1)

Gynogenesis: sperm-dependent parthenogenesis. Hybrid females produce unreduced ova that develop into all-female offspring; sperm are needed from a closely related sexual species to trigger embryonic development without genetic contribution to the offspring.

Hemiclonal reproduction, Hybridogenesis: the hybrid female's genome is passed on clonally to the offspring while the other genome is substituted every generation (hemiclonal). It is also possible that the genome of the sexual species first elevates the ploidy level before it gets substituted.

Parthenogenesis: Hybrid females produce unreduced ova that develop into all-female offspring being genetically identical to their mother.

Paternal leakage: Instead of only triggering embryogenesis (Gynogenesis), the offspring show paternal genetic contribution in form of an additional chromosome set (triploids) or microchromosomes.

References

Abramenko MI, Poltavseva TG, Vasetskii SG (1998) Discovery of triploid males in lower Don populations of the crucian carp *Carassius auratus gibelio* (Bloch). Doklady Akademii Nauk 363: 415–418

Alves MJ, Coelho MM, Collares-Pereira MJ (1998) Diversity in the reproductive modes of females of the *Rutilus alburnoides* complex (Teleostei, Cyprinidae): a way to avoid the genetic constraints of uniparentalism. Mol Biol Evol 15: 1233–1242

Alves MJ, Coelho MM, Collares-Pereira MJ (2001) Evolution in action through hybridisation and polyploidy in an Iberian freshwater fish: a genetic review. Genetica 111: 375–385

Alves MJ, Coelho MM, Prospero MI, Collares-Pereira MJ (1999) Production of fertile unreduced sperm by hybrid males of the *Rutilus alburnoides* complex (Teleostei, cyprinidae). An alternative route to genome tetraploidization in unisexuals. Genetics 151: 277–283

Alves MJ, Collares-Pereira MJ, Dowling TE, Coelho MM (2002) The genetics of maintenance of an all-male lineage in the *Squalius alburnoides* complex. J Fish Biol 60: 649–662

Alves MJ, Gromicho M, Collares-Pereira MJ, Crespo-Lopez ME, Coelho MM (2004) Simultaneous production of triploid and haploid eggs by triploid *Squalius alburnoides* (Teleostei: Cyprinidae). J Exp Zool 301A: 552–558

Angers B, Schlosser IJ (2007) The origin of *Phoxinus eos-neogaeus* unisexual hybrids. Mol Ecol 16: 4562–4571

Arai K (2003) Genetics of the loach, *Misgurnus anguillicaudatus*: recent progress and perspective. Folia Biol (Krakow) 51: 107–117

Arai K, Matsubara K, Suzuki R (1991) Chromosomes and developmental potential of progeny of spontaneous tetraploid loach *Misgurnus anguillicaudatus*. Nippon Suisan Gakkaishi 57: 2173–2178

Arai K, Matsubara K, Suzuki R (1993) Production of polyploids and viable gynogens using spontaneously occurring tetraploid loach, *Misgurnus anguillicaudatus*. Aquaculture 117: 227–235

Arias-Rodriguez L, Yasui GS, Katsutoshi Arai (in press) Disruption of normal meiosis in artificial inter-populational hybrid females of *Misgurnus* loach. Genetica 136: 49–52

Avise JC, Trexler JC, Travis J, Nelson WS (1991) *Poecilia mexicana* is the recent female parent of the unisexual fish *P. formosa*. Evolution 45: 1530–1533

Beukeboom LW (2007) Sex to some degree. Heredity 98: 123–124

Beukeboom LW, Batenburg FHD (1999) The effect of paternal leakage on the rate of Muller's ratchet. 7th Conference of ESEB, Barcelona, Spain 1999, poster presentation

Beukeboom LW, Vrijenhoek RC (1998) Evolutionary genetics and ecology of sperm-dependent parthenogenesis. J Evol Biol 11: 755–782

Beukeboom LW, Weinzierl RP, Michiels NK (1995) Amazon molly and Muller's ratchet. Nature 375: 111–112

Bi K, Bogart JP (2006) Identification of intergenomic recombinations in unisexual salamanders of the genus *Ambystoma* by genomic in situ hybridization (GISH). Cytogenet Genome Res 112: 307–312

Bi K, Bogart JP, Fu J (2007) Intergenomic translocations in unisexual salamanders of the genus *Ambystoma* (Amphibia, Caudata). Cytogenet Genome Res 116: 289–297

Bohlen J (2000) Similarities and differences in the reproductive biology of loaches (*Cobitis* and *Sabanejewia*) under laboratory conditions. Folia Zoologica 49: 179–186

Bohlen J, Rab P (2001) Species and hybrid richness in spined loaches of the genus *Cobitis* (Teleostei: Cobitidae), with a checklist of European forms and suggestions for conservation. J Fish Biol 59: 75–89

Bohlen, Rab P, Slechtova V, Rabova M, Ritterbusch D, Freyhof J (2002) Hybridogenous biotypes in spined loaches (genus *Cobitis*) in Germany with implications for conservation. In: Collares-Pereira MJ, Coelho MM, Cowx IG (eds) Conservation of freshwater fishes: options for the future. Blackwell Science, Oxford, pp. 311–321

Boron A, Culling M, Pulym A (2003) Cytogenetic characteristics of the fish genus *Cobitis* from England. Folia Biol (Krakow) 51 Suppl: 13–16

Camacho JP, Sharbel TF, Beukeboom LW (2000) B-chromosome evolution. Philos Trans R Soc Lond B 355: 163–178

Campton (2008) Natural hybridization and introgression in fishes: methods of detection and genetic interpretations. In: Ryman N, Utter F (eds) Population genetics and fishery management. University of Washington Press, Seattle, pp. 161–192

Cavallier-Smith T (2002) Origins of the machinery of recombination and sex. Heredity 88: 125–141

Choleva L, Apostolou A, Rab P, Janko K (2008) Making it on their own: sperm-dependent hybrid fishes (*Cobitis*) switch the sexual hosts and expand beyond the ranges of their original sperm donors. Philos Trans R Soc B 363: 2911–2919

Chun FL, Yang ST, Gui JF (2001) Differential screening and characterization analysis of the egg envelope glycoprotein ZP3 cDNAs between gynogenetic and gonochoristic crucian carp. Cell Res 11: 17–27

Collares-Pereira MJ, Alves MJ, Coelho MM (1999) Reassessment of the generic position of the Iberian cyprinid *alburnoides* complex: its return to the genus *Leuciscus*. J Fish Biol 54: 465–468

Crespo-Lopez ME, Pala I, Duarte TE, Dowling TE, Coelho MM (2007) Genetic structure of the diploid–polyploid fish *Squalius alburnoides* in southern Iberian basins Tejo and Guadiana, based on microsatellites. J Fish Biol 71: 423–436

Crivelli AJ (1995) Are fish introductions a threat to endemic freshwater fishes in the northern Mediterranean region? Biol Conserv 72: 311–319

Culling MA, Janko K, Boron A, Vasil'ev VP, Cote IM, Hewitt GM (2006) European colonization by the spined loach (*Cobitis taenia*) from Ponto-Caspian refugia based on mitochondrial DNA variation. Mol Ecol 15: 173–190

Cunha C, Coelho MM, Carmona JA, Doadrio I (2004) Phylogeographical insights into the origins of the *Squalius alburnoides* complex via multiple hybridization events. Mol Ecol 13: 2807–2817

Cunha C, Doadrio I, Coelho MM (2008) Speciation towards tetraploidization after intermediate processes of non-sexual reproduction. Philos Trans R Soc B 363: 2921–2929

Dawley (1989) An introduction to unisexual vertebrates. In: Dawley RM, Bogart JP (eds) Evolution and ecology of unisexual vertebrates, Bulletin 466. New York State Museum, Albany, New York, pp. 1–18

Dawley RM (1992) Clonal hybrids of the common laboratory fish *Fundulus heteroclitus*. Proc Natl Acad Sci USA 89: 2485–2488

Dawley RM, Chrzanowski D, Phiel KL, Beaulieu K, Goddard KA (1999) Allozyme analysis of clonal diversity among unisexual hybrids of the killifishes *Fundulus diaphanus* and *F. heteroclitus*. J Hered 90: 607–612

Dawley RM, Schultz RJ, Goddard KA (1987) Clonal reproduction and polyploidy in unisexual hybrids of *Phoxinus eos* and *Phoxinus neogaeus* (Pisces:Cyprinidae). Copeia 1987: 275–283

Dawley RM, Yeakel AM, Beaulieu KA, Phiel KL (2000) Histocompatibility analysis of clonal diversity in unisexual hybrids of the killifishes *Fundulus heteroclitus* and *Fundulus diaphanus*. Can J Zool 78: 923–930

Doeringsfeld MR, Schlosser IJ, Elder JF, Evenson DP (2004) Phenotypic consequences of genetic variation in a gynogenetic complex of *Phoxiuns eos-neogaeus* clonal fish (Pisces: Cyprinidae) inhabiting a heterogeneous environment. Evolution 58: 1261–1273

D'Souza TG, Schulte RD, Schulenburg H, Michiels NK (2006) Paternal inheritance in parthenogenetic forms of the planarian *Schmidtea polychroa*. Heredity 97: 97–101

Dubois A (1990) Nomenclature of parthenogenetic, gynogenetic and "hybridogenetic" vertebrate taxons: new proposals. Alytes (Paris) 8: 61–74

Echelle AA, Echelle AF (1997) Patterns of abundance and distribution among members of a unisexual-bisexual complex of fishes (Atherinidae: Menidia). Copeia 1997: 249–259

Echelle AA, Echelle AF, Crozier CD (1983) Evolution of an all-female fish, *Menidia clarkhubbsi* (Atherinidae). Evolution 37: 772–784

Echelle AA, Echelle AF, DeBault LE, Durham DW (1988) Ploidy levels in silverside fishes (Atherinidae, Menidia) on the Texas coast: flow-cytometric analysis of the occurrence of allotriploidy. J Fish Biol 32: 835–844

Echelle, Echelle AF, Middaugh DP (1989) Evolutionary biology of the *Menidia clarkhubbsi* complex of unisexual fishes (Atherinidae): Origins clonal diversity, and mode of reproduction. In: Dawley RM, Bogart JP (eds) Evolution and ecology of unisexual vertebrates, Bulletin 466. New York State Museum, Albany, New York, pp. 144–152

Echelle AA, Moisier DT (1982) *Menidia clarkhubbsi*, n. sp. (Pisces: Atherinidae), an all-female species. Copeia 1982: 533–540

Fan Z, Liu G (1990) The ploidy and reproductive mechanism of crucian carp, *Carassius auratus gibelio*. J Fish Biol 36: 415–419

Flajshans M, Rodina M, Halacka K, Vetesnik L, Papousek I, Gela D, Luskova V, Lusk S (2007) Ploidy levels and reproductive patterns of *Carassius auratus gibelio* as revealed by means of artificial hybridization and flow cytometry. Aquaculture 272: S256

Fraser D, Adams CE (1997) A crucian carp *Carassius carassius* (L.) in Loch Rannoch, Scotland: further evidence of the threat posed to unique fish communities by introduction of alien fish species. Aquatic Conserv 7: 323–326

Fritz ES, Garside ET (1974) Identification and description of hybrids of *Fundulus heteroclitus* and *F. diaphanus* (Pisces: Cyprinodontidae) from Porters Lake, Nova Scotia, with evidence for absence of backcrossing. Can J Zool 52: 1433–1442

Georgiades P, Watkins M, Burton GJ, Ferguson-Smith AC (2001) Roles for genomic imprinting and the zygotic genome in placental development. Proc Natl Acad Sci USA 98: 4522–4527

Goddard KA, Dawley RM (1990) Clonal inheritance of a diploid nuclear genome by a hybrid freshwater minnow (*Phoxinus eos-neogaeus*, Pisces: Cyprinidae). Evolution 44: 1052–1065

Goddard, Dawley RM, Dowling TE (1989) Origin and genetic relationships of diploid, triploid and diploid-triploid mosaic biotypes in the *Phoxinus eos-neogaeus* unisexual complex. In: Dawley RM, Bogart JP (eds) Evolution and ecology of unisexual vertebrates, Bulletin 466. New York State Museum, Albany, New York, pp. 268–280

Goddard KA, Megwinoff O, Wessner LL, Giaimo F (1998) Confirmation of gynogenesis in *Phoxinus eos-neogaeus* (Pisces: Cyprinidae). J Hered 89: 151–157

Goddard KA, Schultz RJ (1993) Aclonal reproduction by polyploid members of the clonal hybrid species *Phoxinus eos-neogaeus* (Cyprinidae). Copeia 3: 650–660

Green RF, Noakes DLG (1995) Is a little bit of sex as good as a lot? J Theor Biol 174: 87–96

Gromicho M, Coelho MM, Alves MJ, Collares-Pereira MJ (2006) Cytogenetic analysis of *Anaecypris hispanica* and its relationship with the paternal ancestor of the diploid-polyploid *Squalius alburnoides* complex. Genome 49: 1621–1627

Günther R, Uzzel T, Berger L (1979) Inheritance patterns in triploid *Rana* "*esculenta*" (Amphibia, Salientia). Mitt Zool Mus Berlin 55: 35–37

Hernandez Chavez C, Tuergeon J (2007) Asexual and sexual hybrids between *Fundulus diaphanus* and *F. heteroclitus* in the Canadian Atlantic region. Mol Ecol 16: 1467–1480

Hubbs C (1964) Interactions between a bisexual fish species and its gynogenetic sexual parasite. Bull Texas Memorial Mus 8: 1–72

Hubbs CL, Hubbs LC (1932) Apparent parthenogenesis in nature, in a form of fish of hybrid origin. Science 76: 628–630

Itono M, Morishima K, Fujimoto T, Bando E, Yamaha E, Arai K (2006) Premeiotic endomitosis produces diploid eggs in the natural clone loach, *Misgurnus anguillicaudatus* (Teleostei: Cobitidae). J Exp Zool 305A: 513–523

Itono M, Okabayashi N, Morishima K, Fujimoto T, Yoshikawa H, Yamaha E, Arai K (2007) Cytological mechanisms of gynogenesis and sperm incorporation in unreduced diploid eggs of the clonal loach, *Misgurnus anguillicaudatus* (Teleostei: Cobitidae). J Exp Zool 307A: 35–50

Janko K, Flajshhans M, Choleva L, Bohlen J, Slechtova V, Rabova M, Lajbner Z, Slechta V, Ivanovak P, Dobrovolok I, Culling M, Persat H, Kotusz J and Rab P (2007) Diversity of European spined loaches (genus *Cobitis* L.): an update of the geographic distribution of the *Cobitis taenia* hybrid complex with a description of new molecular tools for species and hybrid determination. J. Fish Biol. 71 Suppl C: 387–408

Janko K, Culling MA, Rab P, Kotlik P (2005) Ice age cloning – comparison of the Quaternary evolutionary histories of sexual and clonal forms of spiny loaches (*Cobitis*: Teleostei) using the analysis of mitochondrial DNA variation. Mol Ecol 14: 2991–3004

Janko K, Kotlik P, Rab P (2003) Evolutionary history of asexual hybrid loaches (*Cobitis*: Teleostei) inferred from phylogenetic analysis of mitochondrial DNA variation. J Evol Biol 16: 1280–1287

Jiang JG, Yu HX, Chen BD, Liang SC, Shan SX, Yang DL, Lin SE, Shen GQ (1983) Biological effect of heterologous sperm on gynogenetic offspring in *Carassius auratus gibelio*. Acta Hydrobiol Sin 8: 1–13

Kallman KD (1962a) Gynogenesis in the teleost, *M. formosa*, with discussion of the detection of parthenogenesis in vertebrates by tissue transplantation. J Genet 58: 7–21

Kallman KD (1962b) Population genetics of the gynogenetic teleost *Mollienesia formosa* (Girard). Evolution 16: 497–504

Kalous L, Slechtova V, Bohlen J, Petrtyl M, Svatora M (2007) First European record of *Carassius langsdorfii* from the Elbe basin. J Fish Biol 70 Suppl A: 132–138

Kim IS, Lee EH (2000) Hybridization experiment of diploid-triploid cobitid fishes, *Cobitis sinensis-longicorpus* complex (Pisces: Cobitidae). Folia Zool 49: 17–22

Kim IS, Lee JH (1990) Diploid-triploid hybrid complex of the Spined loach *Cobitis sinensis* and *C. longicorpus* (Pisces, Cobitidae). Korean J Ichthyol 2: 203–210

Kobayashi H, Kawashima Y, Takeuchi N (1970) Comparative chromosome studies in the genus *Carassius*, especially with a finding of polyploidy in the ginbuna (*C. auratus langsdorfii*). Japanese J Ichthyol 17: 153–160

Kobayashi H, Nakano K, Nakamura M (1977) On the hybrids, 4n ginbuna (*Carassius auratus langsdorfii*) x kinbuna (*Carassius auratus* subsp.), and their chromosomes. Bull Jap Soc Sci Fisheries 46: 413–418

Kottelat M (1997) Environmental biology of fishes. An heuristic checklist of the freshwater fishes of Europe (exclusive of former USSR), with an introduction for non-systematists and comments on nomenclature and conservation. Biologia (Bratislava) Sect Zool 52 Suppl: 1–271

Lamatsch DK, Nanda I, Epplen JT, Schmid M, Schartl M (2000) Unusual triploid males in a microchromosome-carrying clone of the Amazon molly, *Poecilia formosa*. Cytogenet Cell Genet 91: 148–156

Lamatsch DK, Nanda I, Schlupp I, Epplen JT, Schmid M, Schartl M (2004) Distribution and stability of supernumerary microchromosomes in natural populations of the Amazon molly, *Poecilia formosa*. Cytogenet Genome Res 106: 189–194

Lamatsch DK, Schmid M, Schartl M (2002) A somatic mosaic of the gynogenetic Amazon molly. J Fish Biol 60: 1417–1422

Lampert KP, Lamatsch DK, Epplen JT, Schartl M (2005) Evidence for a monophyletic origin of triploid clones of the Amazon molly, *Poecilia formosa*. Evolution 59: 881–889

Lampert KP, Lamatsch DK, Fischer P, Schartl M (2008) A tetraploid Amazon molly, *Poecilia formosa*. J Hered 99: 223–226

Lampert KP, Schartl M (2008) The origin and evolution of a unisexual hybrid: *Poecilia formosa*. Philos Trans R Soc Lond B 363: 2901–2909

Lampert KP, Steinlein C, Schmid M, Fischer P, Schartl M (2007a) A haploid-diploid-triploid mosaic of the Amazon molly, *Poecilia formosa*. Cytogenet Genome Res 119: 131–134

Lampert KP, Lamatsch DK, Fischer P, Epplen JT, Nanda I, Schmid M, Schartl M. (2007b) Automictic reproduction in interspecific hybrids of poeciliid fish. Curr Biol 17: 1948–1953

Leonardos ID, Tsikliras AC, Eleftheriou V, Cladas Y, Kagalou I, Chortatou R, Papigioti O (2008) Life history characteristics of an invasive cyprinid fish (*Carassius gibelio*) in Chimaditis Lake (northern Greece). J Appl Ichthyol 24: 213–217

Li Y, Yin J, Wang J, Yuan X, Wei J, Sun X, Arai K (2008) A study on the distribution of polyploid loaches in China. Nippon Suisan Gakkaishi 74: 177–182

Loewe L, Lamatsch DK (2008) Quantifying the threat of extinction from Muller's ratchet in the diploid Amazon molly (*Poecilia formosa*). BMC Evol Biol 8: 88–108

Lynch M, Burger R, Butcher D, Gabriel W (1993) The mutational meltdown in asexual populations. J Hered 84: 339–344

Mable BK (2007) Sex in the postgenomic era. Trends Ecol Evol 22: 559–561

Mateos M, Vrijenhoek RC (2002) Ancient versus reticulate origin of a hemiclonal lineage. Evolution 56: 985–992

Mateos M, Vrijenhoek RC (2005) Independent origins of allotriploidy in the fish genus *Poeciliopsis*. J Hered 96: 32–39

Mateos M, Sanyur OI, Vrijenhoek RC (2002) Historical biogeography of the livebearing fish genus *Poeciliopsis* (Poeciliidae: Cyprinodontiformes). Evolution 56: 972–984

Mezhzherin SV, Chudakorova TY (2002) The genetic structure of the diploid-polyploid complex of the spined loach *Cobitis taenia* (Cypriniformes: Cobitidae) from the middle Dnieper basin. Russ J Genet 38: 70–75

Morishima K, Nakamura-Shiokawa Y, Bando E, Li YJ, Boron A, Khan MMR, Arai K (2008a) Cryptic clonal lineages and genetic diversity in the loach *Misgurnus anguillicaudatus* (Teleostei: Cobitidae) inferred from nuclear and mitochondrial DNA analyses. Genetica 132: 159–171

Morishima K, Yoshikawa H, Arai K (2008b) Meiotic hybridogenesis in triploid *Misgurnus* loach derived from a clonal lineage. Heredity 100: 581–586

Morishima K, Oshima K, Horie S, Fujimoto T, Yamaha E, Arai K (2004) Clonal diploid sperm of the diploid-triploid mosaic loach, *Misgurnus anguillicaudatus* (Teleostei:Cobitidae). J Exp Zool 301A: 502–511

Morishima K, Horie SYE, Arai K (2002) A cryptic clonal line of the loach *Misgurnus anguillicaudatus* (Teleostei: Cobitidae) evidenced by induced gynogenesis, interspecific hybridization, microsatellite genotyping and multilocus DNA fingerprinting. Zool Sci 19: 565–575

Moritz C, Brown WM, Densmore LD, Wright JW, Vyas D, Donnellan S, Adams M,Braverstock, P (1989). Genetic diversity and the dynamics of hybrid parthenogenesis in *Cnemidophorus* (Teiidae) and *Heteronotia* (Gekkonidae). In: RM Dawley and JB Bogart (eds) Evolution and ecology of unisexual vertebrates. Bulletin 466, New York State Museum, Albany, New York, pp. 87–112

Muller HJ (1964) The relation of recombination to mutational advance. Mutat Res 1: 2–9

Murakami M, Fujitani H (1997) Polyploid-specific repetitive DNA sequences from triploid ginbuna (Japanese silver crucian carp, *Carassius auratus langsdorfi*). Genes Genet Syst 72: 107–113

Murakami M, Matsuba C, Fujitani H (2001) The maternal origins of the triploid ginbuna (*Carassius auratus langsdorfi*): phylogenetic relationships within the *C. auratus* taxa by partial mitochondrial D-loop sequencing. Genes Genet Syst 76: 25–32

Murakami M, Matsuba C, Fujitani H (2002) Characterization of DNA markers isolated from the gynogenetic triploid ginbuna (*Carassius auratus langsdorfi*) by representational difference analysis. Aquaculture 208: 59–68

Muramoto J (1975) A note on triploidy of the *Funa* (Cyprinidae, Pisces). Proc Japan Acad 51: 583–587

Murayama Y, Hijikata M, Kojima M, Noda M, Kajishima T (1986) The appearance of diploid-triploid and diploid-triploid-tetraploid mosaic individuals in polyploid fish, ginbuna (*Carassius auratus langsdorfii*). Experientia 42: 187–188

Nanda I, Schlupp I, Lamatsch DK, Lampert KP, Schmid M, Schartl M (2007) Stable inheritance of host species-derived microchromosomes in the gynogenetic fish, *Poecilia formosa*. Genetics 177: 917–926

Niemeitz A, Kreutzfeld R, Schartl M, Parzefall J, Schlupp I (2002) Male mating behaviour of a molly, *Poecilia latipunctata*: a third host for the sperm-dependent Amazon molly, *Poecilia formosa*. Acta Ethologica 5: 45–49

Ogielska M, Kierzkowski P, Rybacki M (2004) DNA content and genome composition of diploid and triploid water frogs belonging to the *Rana esculenta* complex (Amphibia, Anura). Can J Zool 82: 1894–1901

Omilian AR, Cristescu MEA, Dudycha JL, Lynch M (2006) Ameiotic recombination in asexual lineages of *Daphnia*. Proc Natl Acad Sci USA 103: 18638–18643

Ojima Y, Takai A (1979) The occurrence of spontaneous polyploid in the Japanese common loach, *Misgurnus anguillicaudatus*. Proc Japan Acad 55: 487–491

Oshima K, Morishima K, Yamaha E, Arai K (2005) Reproductive capacity of triploid loaches obtained from Hokkaido Island, Japan. Ichthyol Res 52: 1–8

Pala I, Coelho MM (2005) Contrasting views over a hybrid complex: between speciation and evolutionary "dead-end". Gene 347: 283–294
Pala I, Coelho MM, Schartl M. (2008) Dosage compensation by gene-copy silencing in a triploid hybrid fish. Curr Biol 18: 1344–1348
Pandian TJ, Koteeswaran R (1998) Ploidy induction and sex control in fish. Hydrobiologia 384: 167–243
Quattro JM, Avise JC, Vrijenhoek RC (1991) Molecular evidence for multiple origins of hybridogenetic fish clones (Poeciliidae: *Poeciliopsis*). Genetics 127: 391–398
Quattro JM, Avise JC, Vrijenhoek RC (1992) Mode of origin and sources of genotypic diversity in triploid gynogenetic fish clones (*Poeciliopsis*: Poeciliidae). Genetics 130: 621–628
Rab P, Rabova M, Bohlen J, Lusk S (2000) Genetic differentiation of the two hybrid diploid-polyploid complexes of loaches, genus *Cobitis* (Cobitidae), involving *C. taenia*, *C. elongatiodes* and *C. spp.* in the Czech Republic. Folia Zoologica 49 Suppl1: 55–66
Rasch, Balsano JS (1989) Trihybrids related to the unisexual molly fish, *Poecilia formosa*. In: Dawley RM, Bogart JP (eds) Evolution and ecology of unisexual vertebrates, Bulletin 466. New York State Museum, Albany, New York, pp. 252–267
Robalo JI, Sousa Santos C, Levy A, Almada VC (2006) Molecular insights on the taxonomic position of the paternal ancestor of the *Squalius alburnoides* hybridogenetic complex. Mol Phyl Evol 39: 276–281
Rougier N, Werb Z (2001) Minireview: parthenogenesis in mammals. Mol Reprod Dev 59: 468–474
Saat TV (1991) Reproduction of the diploid and polyploid spinous loach (*Cobitis*, Teleostei). Oocyte maturation and fertilization in the triploid form. Soviet J Dev Biol 22: 332–338
Saitoh K, Kim IS, Lee EH (2004) Mitochondrial gene introgression between spined loaches via hybridogenesis. Zool Sci 21: 795–798
Sawada Y (1982) Phylogeny and zoogeography of the superfamily Cobitoidea (Cyprinoidei, Cypriniformes). Mem Fac Fisher Hokkaido Univ 28: 65–223
Schartl A, Hornung U, Nanda I, Wacker R, Muller-Hermelink HK, Schlupp I, Parzefall J, Schmid M, Schartl M (1997) Susceptibility to the development of pigment cell tumors in a clone of the Amazon molly, *Poecilia formosa*, introduced through a microchromosome. Cancer Res 57: 2993–3000
Schartl M, Nanda I, Schlupp I, Wilde B, Epplen JT, Schmid M, Parzefall J (1995a) Incorporation of subgenomic amounts of DNA as compensation for mutational load in a gynogenetic fish. Nature 373: 68–71
Schartl M, Wilde B, Schlupp I, Parzefall J (1995b) Evolutionary origin of a parthenoform, the Amazon molly *Poecilia formosa*, on the basis of a molecular genealogy. Evolution 49: 827–835
Schlupp I (2005) The evolutionary ecology of gynogenesis. Annu Rev Ecol Evol Syst 36: 399–417
Schlupp I, Nanda I, Dobler M, Lamatsch DK, Epplen JT, Parzefall J, Schmid M, Schartl M (1998) Dispensable and indispensable genes in an ameiotic fish, the Amazon molly *Poecilia formosa*. Cytogenet Cell Genet 80: 193–198
Schlupp I, Parzefall J, Schartl M (2002) Biogeography of the Amazon molly, *Poecilia formosa*. J Biogeogr 29: 1–6
Schories S, Lampert KP, Lamatsch DK, Garcia de Leon FJ, Schartl M (2007) Analysis of a possible independent origin of triploid *P. formosa* outside of the Río Purificación river system. Frontiers Zool 4:13
Schultz RJ (1967) Gynogenesis and triploidy in the viviparous fish *Poeciliopsis*. Science 157: 1564–1567
Schultz RJ (1969) Hybridization, unisexuality, and polyploidy in the teleost *Poeciliopsis* (Poeciliidae) and other vertebrates. Am Nat 103: 605–661
Schultz RJ (1973) Unisexual fish: laboratory synthesis of a "species". Science 179: 180–181
Schultz RJ (1977) Evolution and ecology of unisexual fishes. Evol Biol 10: 277–331
Schultz RJ, Kallman KD (1968) Triploid hybrids between the all-female teleost *Poecilia formosa* and *Poecilia sphenops*. Nature 219: 280–282

Schultz RJ (1979) Role of polyploidy in the evolution of fishes. Basic Life Sci 13: 313–340

Scott RJ, Spielman M (2006) Genomic imprinting in plants and mammals: how life history constrains convergence. Cytogenet Genome Res 113: 53–67

Scribner KT, Page KS, Bartron ML (2000) Hybridization in freshwater fishes: a review of case studies and cytonuclear methods of biological inference. Rev Fish Biol Fisheries 10: 293–323

Slechtova V, Bohlen J, Perdices A (2008) Molecular phylogeny of the freshwater fish family Cobitidae (Cypriniformes : Teleostei): Delimitation of genera, mitochondrial introgression and evolution of sexual dimorphism. Mol Phyl Evol 47: 812–831

Slechtova V, Luskova V, Slechta V, Lusk S, Halacka K, Bohlen J (2000) Genetic differentiation of two diploid-polyploid complexes of spined leach, genus *Cobitis* (Cobitidae), in the Czech Republic, involving *C. taenia*, *C. elongatoides,* and *C.* spp.: Allozyme interpopulation and interspecific differences. Folia Zoologica 49: 67–78

Sola L, Rossi AR, Bressanello S, Rasch EM, Monaco PJ (1993) Cytogenetics of bisexual/unisexual species of *Poecilia*. V. Unisexual poeciliids with anomalous karyotypes from northeastern Mexico. Cytogenet Cell Genet 63: 189–191

Sousa Santos C, Collares-Pereira MJ, Almada VC (2006) Evidence of extensive mitochondrial introgression with nearly complete substitution of the typical *Squalius pyrenaicus*-like mtDNA of the *Squalius alburnoides* complex (Cyprinidae) in an independent Iberian drainage. J Fish Biol 68 Suppl B: 292–301

Stöck M, Lamatsch DK, Steinlein C, Epplen JT, Grosse WR, Hock R, Klapperstuck T, Lampert KP, Scheer U, Schmid M, Schartl M (2002) A bisexually reproducing all-triploid vertebrate. Nat Genet 30: 325–328

Turner BJ, Balsano JS, Monaco PJ, Rasch EM (1983) Clonal diversity and evolutionary dynamics in a diploid-triploid breeding complex of unisexual fishes (*Poecilia*). Evolution 37: 798–809

Vasil'ev VP, Akimova NV, Emel′yanova NG, Pavlov DA, Vasil'eva ED (2003) Reproductive capacities in the polyploid males of spined loaches from the unisexual-bisexual complex, occurred in the Moscow River. Folia Biol (Krakow) 51 Suppl: 67–73

Vasil'ev VP, Vasil'eva KD (1982) A new diploid-polyploid complex in fishes. Dokl Akad Nauk SSSR 266: 250–252

Vasil'ev VP, Vasil'eva KD, Levenkova ES (2007) Trihybrid origin of an all-female tetraploid form of fish from the genus *Cobitis* (Pisces, Cobitidae). Doklady Biol Sci 413: 107–110

Vasil'ev, Vasil'eva KD, Osinov AG (1989) Evolution of a diploid-triploid-tetraploid complex in fishes of the genus *Cobitis* (Pisces, Cobitidae). In: Dawley RM, Bogart JP (eds) Evolution and ecology of unisexual vertebrates, Bulletin 466. New York State Museum, Albany, New York, pp. 153–169

Veetema M, Eschbaum R, Albert A, Saat T (2005) Distribution, sex ratio and growth of *Carassius gibelio* (Bloch) in coastal and inland waters of Estonia (north-eastern Baltic Sea). J Appl Ichthyol 21: 287–291

Verreycken H, Anseeuw D, Van Thuyne G, Quataert P, Belpaire C (2007) The non-indigenous freshwater fishes of Flanders (Belgium): review, status and trends over the last decade. J Fish Biol 71: 160–172

Vrijenhoek RC (1994) Unisexual fish: model systems for studying ecology and evolution. Annu Rev Ecol Syst 25: 71–96

Vrijenhoek RC, Angus RA, Schultz RJ (1977) Variation and heterozygosity in sexually vs. clonally reproducing populations of *Poeciliopsis*. Evolution 31: 767–781

Vrijenhoek RC, Schultz RJ (1974) Evolution of a trihybrid unisexual fish (*Poeciliopsis*, Poeciliidae). Evolution 28: 306–319

Wei WH, Zhang J, Zhang YB, Zhou L, Gui JF (2003) Genetic heterogeneity and ploidy level analysis among different gynogenetic clones of the polyploid gibel carp. Cytometry Part A 56A: 46–52

Wetherington JD, Kotora KE, Vrijenhoek RC (1987) A test of the spontaneous heterosis hypothesis for unisexual vertebrates. Evolution 41: 721–723

Yi MS, Li YQ, Liu JD, Zhou L, Yu QX, Gui JF (2003) Molecular cytogenetic detection of paternal chromosome fragments in allogynogenetic gibel carp, *Carassius auratus gibelio* Bloch. Chromosome Res 11: 665–671

Yoshikawa H, Morishima K, Fujimoto T, Yamaha E, Arai K (2007) Simultaneous formation of haploid, diploid and triploid eggs in diploid–triploid mosaic loaches. J Fish Biol 71: 250–263

Zhang Q, Arai K (1999) Aberrant meioses and viable aneuploid progeny of induced triploid loach (*Misgurnus anguillicaudatus*) when crossed to natural tetraploids. Aquaculture 175: 63–76

Zhou L, Wang Y, Gui JF (2000) Analysis of genetic heterogeneity among five gynogenetic clones of silver crucian carp, *Carassius auratus gibelio* Bloch, based on detection of RAPD molecular markers. Cytogenet Cell Genet 88: 133–139

Zhu HP, Gui JF (2007) Identification of genome organization in the unusual allotetraploid form of *Carassius auratus gibelio*. Aquaculture 265: 109–117

Chapter 20
Masked Damage: Mutational Load in Hemiclonal Water Frogs

Christoph Vorburger, Dirk S. Schmeller, Hansjürg Hotz, Gaston-Denis Guex and Heinz-Ulrich Reyer

Abstract Hemiclonal hybrids of Western Palearctic water frogs of the *Rana esculenta* complex transmit only one parental genome to their offspring without recombination (hybridogenesis). Such genomes are thus prone to accumulate deleterious mutations. The frog complex is unique among hybridogens in that hemiclonal hybrids occur in both sexes. This provides the opportunity of using experimental crosses to produce offspring possessing two clonal genomes of various origins and thereby study their homozygous and heterozygous effects on fitness. Here we review work that made use of this possibility to assess the evolutionary consequences of clonal inheritance in water frogs. Overall, these studies indicate that clonally transmitted genomes bear a substantial load of fixed deleterious mutations, yet these mutations appear to have minor effects on fitness in the heterozygous state. We also point out potential mechanisms for episodic recombination by which otherwise clonal genomes may be purged of deleterious alleles, and we present evidence for such episodic recombination to occur in natural populations of hybridogenetic frogs. Finally, we provide an outlook on work in progress that exploits the peculiarities of this system to obtain relevant estimates of the frequency of segregating lethal mutations in sexual populations of water frogs.

20.1 Hybridogenesis

The term hybridogenesis was introduced by Schultz (1969) to describe a remarkable type of non-Mendelian inheritance that was detected in interspecific hybrids of fishes from the genus *Poeciliopsis* in Mexico (see also Chapter 19). These all-female hybrids form distinct taxa that are somatically intermediate between the parental species, but they routinely exclude the entire paternal chromosome set

C. Vorburger (✉)
Institute of Zoology, University of Zürich, Winterthurerstrasse 190, 8057 Zürich, Switzerland
e-mail: christoph.vorburger@zool.uzh.ch

I. Schön et al. (eds.), *Lost Sex*, DOI 10.1007/978-90-481-2770-2_20,

from the germline to produce haploid eggs only containing the unrecombined maternal genome. For their reproduction, hybrids depend entirely on backcrossing with males of their paternal species, with which they are thus always sympatric. Because hybridogens always possess one sexually and one clonally transmitted genome, individuals sharing the same clonal genome can be referred to as a hemiclone (Vrijenhoek et al. 1977).

Since Schultz' original description of the phenomenon, hybridogenesis has been found in other fishes, Iberian minnows of the *Squalius alburnoides* complex (Carmona et al. 1997) and in Korean fish of the genus *Cobitis* (Saitoh et al. 2004), stick insects of the genus *Bacillus* (Mantovani and Scali 1992; Tinti et al. 1995), and – the topic of this review – Western Palearctic water frogs of the genus *Rana*[1] (Tunner 1974). However, this unusual mode of reproduction may well have remained undetected in other species complexes with frequent hybridization. Hybridogenesis can arise as a spontaneous consequence of interspecific hybridization, as evidenced by the successful "de novo" formation of hybridogens in *Poeciliopsis* and *Rana* (Schultz 1973; Hotz et al. 1985; Vorburger and Reyer 2003). The molecular and cytological mechanisms by which genome exclusion is achieved in hybridogens are still poorly understood and may well differ among the different taxa. What is important for this review is the genetic consequence: Hybridogenesis leads to the clonal transmission of a haploid genome.

20.2 The Special Case of Water Frogs

It was an extensive series of crossing experiments performed by Leszek Berger in Poland that uncovered the fact that *Rana esculenta*, described by Linnaeus and until then considered a proper species, was in fact a hybrid between *R. ridibunda* Pallas and *R. lessonae* Camerano (Berger 1967, 1968, 1970). *Rana lessonae* is the smallest of the Western Palearctic water frogs (females up to 80 mm snout-to-vent length); it preferentially breeds in shallow, small and vegetated ponds and overwinters terrestrially. *Rana ridibunda*, on the other hand, is Europe's largest amphibian (females up to 140 mm snout-to-vent length) and occurs in larger, deeper, more open water bodies, in which it also overwinters (Plötner 2005). The hybrid *R. esculenta* is intermediate in size as well as in traits relevant for mate choice like advertisment calls and coloration. Its habitat preferences are also intermediate yet very broad, almost encompassing both parental niches. This may at least in part explain its wide distribution and high abundance in Europe.

The striking non-Mendelian inheritance of morphological traits and of serological markers led Tunner (1973) to realize that only one parental chromosome set

[1] Please note that according to the latest revision of amphibian classification (Frost et al. 2006), the Western Palearctic water frogs are now contained in the genus *Pelophylax*, formerly considered a subgenus of *Rana* (Dubois 1992). The new classification is likely to take hold (Vences 2007). For consistency with the literature we review, however, we adhere to the old classification in this chapter.

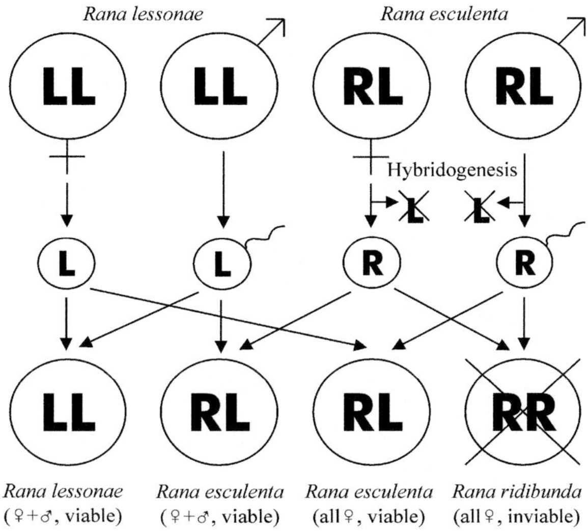

Fig. 20.1 Gamete production, possible mating combinations and resulting offspring in the L-E system, i.e., mixed populations of *Rana lessonae* (LL) and *R. esculenta* (RL), a hemiclonal hybrid between *R. ridibunda* (RR) and *R. lessonae.* Reproduced with permission from Vorburger (2001c)

is transmitted by hybrids, and he consequentially applied the term hybridogenesis to *R. esculenta*'s mode of reproduction (Tunner 1974). Unlike all other known hybridogens, *R. esculenta* is bisexual, and both sexes reproduce by hybridogenesis, offering unique research opportunities that will be discussed later on. Over much of central Europe, *R. esculenta* occurs outside the range of *R. ridibunda* and forms mixed populations with only one of its parental species, *R. lessonae*. This is referred to as the L-E system (Uzzell and Berger 1975) and is illustrated in Fig. 20.1. In such populations, *R. esculenta* excludes the *lessonae* genome from the germline and clonally transmits the *ridibunda* genome. Hybrids persist by backcrossing with *R. lessonae* every generation anew to produce *R. esculenta* offspring that again exclude the *lessonae* genome from the germline. *Rana esculenta* is thus a sexual parasite of *R. lessonae*. Matings among hybrids also occur in the L-E system. Such matings would produce *R. ridibunda* offspring, but these are typically inviable and die at an early larval stage (Fig. 20.1). Another interesting characteristic of the L-E system is that hybrid males – with a few exceptions – only produce daughters. The reason for this is that these frogs exhibit X–Y sex determination (Berger et al. 1988), and that for behavioural reasons, primary hybridizations tend to occur between females of *R. ridibunda* and males of *R. lessonae*. As a consequence, the clonally transmitted *ridibunda* genome in the hybrid *R. esculenta* contains an X chromosome and male hybrids thus produce X-bearing sperm.

The sensational discovery of their hemiclonal reproduction raised interest in the water frogs and subsequent field surveys revealed a very complex picture of different population systems across Europe (reviewed in Graf and Polls Pelaz 1989; Günther 1990; Plötner 2005). A second hybridogen, *R. grafi*, occurs in southern France and north-eastern Spain. It is a hybrid between *R. ridibunda* and *R. perezi*, with which it coexists to form the so called P-G system. A third hybridogen, a little confusingly referred to as *R. hispanica*, occurs in central and southern Italy. This taxon is a hybrid between *R. ridibunda* and *R. bergeri*, a water frog of the Italian peninsula with a somewhat uncertain taxonomic status. The mirror image of the L-E system, the R-E system, occurs in parts of north-eastern Europe (Uzzell and Berger 1975; Uzzell et al. 1977). There, *R. esculenta* males form mixed populations with *R. ridibunda* and often exclude the *ridibunda* rather than the *lessonae* genome from the germline. Finally and most remarkably, all-hybrid populations exist in northern Europe (Ebendal 1979; Günther et al. 1979). These populations consist of diploid and triploid hybrids with different combinations of parental genomes and different types of genome exclusion. Yet by far the most widespread and best studied is the L-E system, to which we will restrict ourselves for the rest of this chapter.

20.3 Hemiclonality – A Predisposition to Mutation Accumulation

Hybridogenesis obviously serves the selfish interests of the clonally transmitted genome very well, but it also seems like the perfect recipe for its rapid mutational degradation. Clonal populations of finite size suffer from the accumulation of deleterious mutations through Muller's ratchet (Muller 1964). In the haploid case, it has been shown that the rate of fixation of deleterious alleles equals the rate of accumulation, such that over time, a large part of the mutations in the population are shared by all the individuals of the population (Higgs and Woodcock 1995; Charlesworth and Charlesworth 1997). The rate at which this fixation occurs depends on the deleterious mutation rate U, on the average selection coefficient against deleterious alleles s, and on the population size N. We have no reason to assume that U would be higher for clonally transmitted genomes than for sexually transmitted genomes in the L-E system. In fact, U may even be lower as explained later on. Selection against deleterious mutations, on the other hand, will be weak. This is because clonal *ridibunda* genomes are constantly sheltered by sexual *lessonae* genomes in *R. esculenta*, preventing their occurrence in the homozygous state and reducing selection against deleterious alleles by their dominance coefficient h. Thus, deleterious mutations with very low h are invisible to selection and can accumulate roughly at the rate they occur. What about population size? Water frogs are common at permanent water bodies and tend to occur in large populations. Yet with respect to mutation accumulation, the hybrids' population history may be of particular relevance. The immigration of *R. esculenta* from areas of sympatry of the parental species into Central and Western Europe may have been accompanied by many colonization events, each possibly representing a bottleneck of only few founders. Founder events promote genetic drift and thus the fixation of deleterious mutations (e.g., Chao 1990). A second source of bottlenecks on a more regular basis is the fact the *R. esculenta* relies on matings with *R. lessonae* for its persistence. Depending on the ecological characteristics of the breeding habitat, water frog populations vary widely in their composition (Rybacki and Berger 1994, 2001; Holenweg Peter et al. 2002). Commonly, they are dominated by *R. esculenta* and in some cases contain less than 5% of *R. lessonae* (e.g., Blankenhorn et al. 1973; Rybacki and Berger 1994, 2001). At such sites, the number of successfully breeding hybrids may only be a fraction of the total population.

Finally, a particularly intriguing possibility is that mutation accumulation on clonally transmitted genomes may be promoted by sexual selection. While male *R. lessonae* do not appear to discriminate (and may not even benefit from doing so, see Schmeller et al. 2005a), female *R. lessonae* have been shown to exhibit a preference for males of their own species (Abt and Reyer 1993). This preference is crucial for the stability of the L-E system (Guex et al. 1993; Hellriegel and Reyer 2000; Som et al. 2000). If a loss-of-function mutation in a gene affecting a male trait used in female choice occurs on a clonally transmitted *ridibunda* genome, it may cause a more "*lessonae*-like" phenotype in *R. esculenta* males bearing it and thus increase their chances of mating with females of the sexual host. Although undoubtedly deleterious in a homospecific background, such mutations could be beneficial to hybrids and thus even be under positive selection within an L-E system.

20.4 Mutational Load in *Rana esculenta*

Based on these considerations, the expectation is that *ridibunda* genomes in natural L-E systems suffer from a high mutational load. The first and most manifest evidence that this may indeed be so comes from the very fact that defines an L-E system – the lack of adult *R. ridibunda* in the population. It cannot be explained by the lack of *R. esculenta* × *R. esculenta* matings that give rise to *R. ridibunda* offspring. Although female *R. esculenta* prefer male *R. lessonae* over hybrid males in two-choice situations (Abt and Reyer 1993; Roesli and Reyer 2000; Engeler and Reyer 2001), this preference is not always realized in more complex situations (Bergen et al. 1997). Abt Tietje and Reyer (2004) further showed that a substantial fraction of clutches found in natural ponds are the result of matings between two hybrids. Thus, *R. ridibunda* are produced in L-E systems but are apparently inviable. This was confirmed by numerous crossing experiments during early investigations of the system (e.g., Berger 1967; Blankenhorn et al. 1971; Heusser and Blankenhorn 1973). These findings were consistent with a high mutational load on clonal *ridibunda* genomes, but could not reveal whether the inviability is caused (i) by the homozygosity of recessive deleterious alleles at particular loci or (ii) by the cumulative load from the general deterioration of the non-recombining *ridibunda* genomes, independent of homozygosity. Although not mutually exclusive, these hypotheses create different predictions. Under the first hypothesis, matings between different hemiclones, i.e., between *R. esculenta* possessing evolutionarily independent *ridibunda* genomes from different primary hybridizations, should produce viable offspring, because it is very unlikely that different clonal lineages are just by chance fixed for the same mutations. Under the second hypothesis, all hybrid × hybrid matings should produce inviable progeny, with a possible correlation between the age of a clonal lineage and the severity of observed genetic defects (although that might be impossible to quantify, because a tadpole cannot be more than dead). In the following two paragraphs, we summarize studies that explicitly addressed these predictions and provided support for the first hypothesis, although with some very informative inconsistencies.

Vorburger (2001a) performed a diallele crossing experiment with *R. esculenta* parents from three different populations in Switzerland. The two northern populations (Elliker Auen and Alpnach) were separated from the southern population (Seseglio) by the Alps, which represent an insurmountable dispersal barrier for water frogs. Allozyme analysis revealed that the parents from Seseglio and Alpnach each belonged to a single but different hemiclone, while parents from Elliker Auen comprised three hemiclones, one of which was indistinguishable from the one in Alpnach. The clearest result was provided by the crosses within and between populations Alpnach and Seseglio: All within-population crosses produced inviable progeny, whereas all crosses between populations – using the same parents – produced viable tadpoles that successfully completed metamorphosis. Importantly, tadpoles of within-population crosses from Alpnach suffered from different developmental abnormalities than tadpoles from Seseglio (Fig. 20.2). This pattern clearly supports the hypothesis that the two clonal *ridibunda* genomes from the two localities are fixed for different recessive deleterious mutations that are responsible for the observed inviability of hybrid × hybrid crosses in the natural populations. Crosses within population Elliker Auen and between Elliker Auen and Alpnach produced less consistent results. It was not surprising that many of the crosses between the two populations were inviable, because one hemiclone occurred at both sites. It was also not too surprising that some within-population crosses from Elliker Auen survived, because the parents belonged to three different hemiclones. However, there were also two viable crosses between parents belonging to the same hemiclone, and five inviable crosses between parents belonging to different hemiclones. While the former could be explained by the limited resolution of the allozyme markers used, the latter seems to contradict the hypothesis that offspring inviability is caused by homozygosity for recessive deleterious alleles, unless the assumption is unjustified that different clonal *ridibunda* genomes found in the Elliker Auen population are

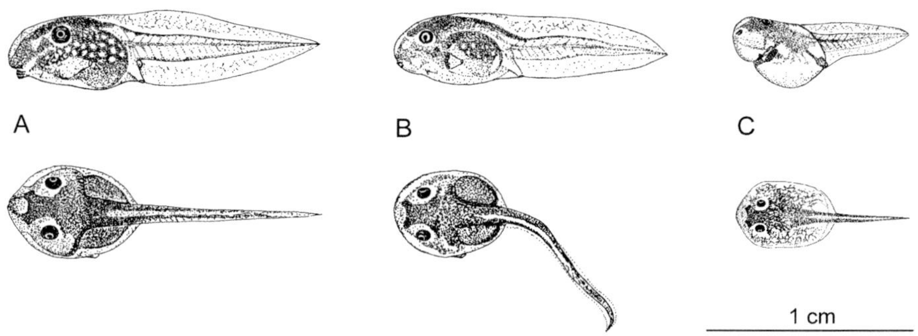

Fig. 20.2 Lateral and dorsal view of representative tadpoles from *Rana esculenta* × *R. esculenta* crosses between Alpnach and Seseglio (**A**), within population Alpnach (**B**), and within population Seseglio (**C**). Drawings were produced from photographs taken approximately three weeks after hatching. Reproduced with permission from Vorburger (2001a)

evolutionarily independent (see below). All Swiss hemiclones were also backcrossed with sexual *R. ridibunda* to produce offspring possessing one clonal and one sexual *ridibunda* genome. These crosses were generally viable and larval life-history traits were comparable to those of normal, sexually produced *R. ridibunda* tadpoles, suggesting that the heterozygous fitness effects of fixed mutations on the clonally transmitted *ridibunda* genomes are minor.

In a second study by Guex et al. (2002), crosses were performed within and among two different hemiclones from a Swiss population (Gütighausen) and a single hemiclone from Sicily. The inclusion of the Sicilian hemiclone added a temporal dimension; while *ridibunda* genomes found in Swiss hemiclones may plausibly have persisted without recombination for about 5,000 or fewer generations (Guex et al. 2002), paleogeography indicates that the Sicilian *ridibunda* genome must have experienced a minimum of about 20,000 generations of clonal transmission (Santucci et al. 1996). As expected, all crosses within the Sicilian hemiclone were inviable. Yet despite the age of the Sicilian hemiclone, it produced viable offspring when crossed with the two Swiss hemiclones, also indicating that homozygosity for particular recessive deleterious alleles rather than the cumulative mutational load is responsible for the observed inviabilities in natural populations. All crosses among Swiss *R. esculenta*, on the other hand, were inviable, even those between the two different hemiclones occurring at Gütighausen. Again, this finding is inconsistent with the homozygosity hypothesis, unless the two *ridibunda* genomes from Gütighausen are not evolutionarily independent. Backcrosses of Swiss and Sicilian hemiclones with *R. ridibunda* were viable, with at most minor reductions in larval performance compared to offspring with two *R. ridibunda* parents.

Both of the above studies thus suggest that when in the heterozygous state, the negative effects of fixed deleterious alleles on non-recombining *ridibunda* genomes are very small. This seems to be supported by the high fitness and competitive ability of *R. esculenta* reported in many studies (e.g., Semlitsch 1993; Rist et al. 1997). However, this evidence is weak because comparisons with sexual *R. ridibunda* reported above are based on only a small number of crosses, and because the high fitness of *R. esculenta* may be a consequence of heterosis observed in these hybrids (Hotz et al. 1999), compensating for the *ridibunda* genome's mutational load. A valid test must therefore uncouple clonal inheritance from hybridity. Vorburger (2001b) thus used artificial fertilizations to cross each of several *R. ridibunda* females with six males, three *R. ridibunda* males from three different populations, and three *R. esculenta* males from three populations. The resulting offspring were raised under benign (high food) and stressful (low food) conditions. In both treatments, tadpoles with *R. esculenta* fathers performed just as well and for one trait (size at metamorphosis) even better than tadpoles with *R. ridibunda* fathers, confirming that clonal *ridibunda* genomes in the heterozygous state have little, if any, negative effects on fitness. Because the evolutionary fate of the hybrid taxon *R. esculenta* depends primarily on these heterozygous effects, it may not be as gloomy as suggested by Milinski (1994).

In the following sections, we discuss mechanisms by which clonal *ridibunda* genomes may be able to purge deleterious alleles through episodic recombination and thus maintain a bearable load. However, it is also worth mentioning that their deleterious mutation rate may be reduced from the outset. The reason is that clonal *ridibunda* genomes in L-E systems contain an X chromosome and are therefore passed on through females more often than through males (Som and Reyer 2006). As mutation rates are generally higher in male than in female germlines (Redfield 1994; Hurst and Ellegren 1998; Ellegren and Fridolfsson 2003), clonal autosomes may overall experience a lower per-generation input of deleterious mutations than sexually transmitted autosomes, which on average spend equal amounts of time in both sexes. This effect may even be stronger in hybridogenetic systems other than water frogs, because there clonal genomes occur exclusively in females (Som et al. 2007).

20.5 Lost Load: Occasional Recombination Between Hemiclones

The studies by Vorburger (2001a) and Guex et al. (2002) both found that self-incompatible hemiclones can produce viable *R. ridibunda* offspring when crossed with different hemiclones, thus supporting the homozygosity hypothesis. Interestingly, both studies also shared the same inconsistency, namely that this outcome was entirely predictable for hemiclones that were widely separated geographically, but not for hemiclones occurring within the same population. We believe there is a common reason for this similarity: the potential for episodic recombination among clonal *ridibunda* genomes, resulting from the presence of both sexes in hybridogens.

Hotz et al. (1992) found a subpopulation of *R. ridibunda* females within an L-E system at Trubeschloo, Switzerland, and used mtDNA RFLPs to demonstrate conclusively that these females were produced by matings among *R. esculenta*. Apparently, compatible hemiclones came into contact at this site. Similarly, Vorburger (2001c) detected numerous *R. ridibunda* metamorphs in another Swiss L-E population (Elliker Auen – the multiclonal population also used in one of the crossing experiments described above). The allozyme genotypes of these froglets were consistent with combinations of clonal genomes present at the site and they all developed into females, also proving that they were formed by matings among *R. esculenta*. Such successful hybrid × hybrid matings cannot found independently reproducing populations of *R. ridibunda*, because all offspring are female. However, these females are expected to exhibit normal Mendelian meiosis and thus recombine the two clonal genomes they inherited from their hybrid parents. Subsequent matings with syntopic *R. lessonae* males may then found new hemiclones possessing recombined *ridibunda* genomes. This potential for occasional recombination, illustrated in Fig. 20.3, led Schmidt (1993) to compare hybridogenetic water frogs with cyclical parthenogens. As a consequence of this process, coexisting hemiclones would no longer be evolutionarily independent. Deleterious mutations present in the original *ridibunda* genomes may either be purged (Som and Reyer 2007) or else become linked to new combinations or marker alleles, making it impossible to predict offspring viability from the combination of parental hemiclones. We believe that

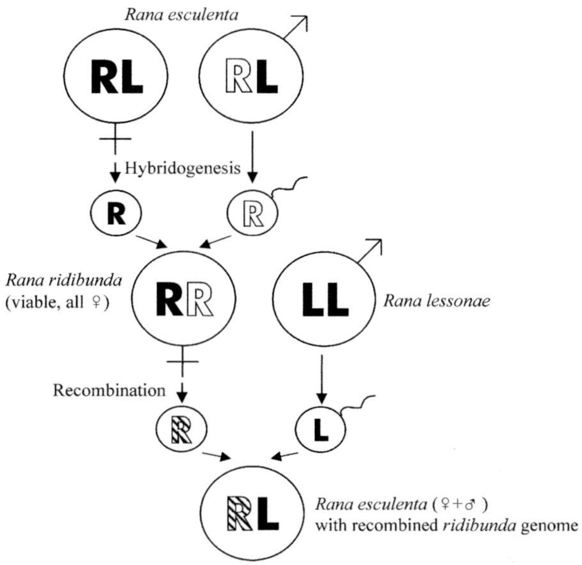

Fig. 20.3 Proposed mechanism for occasional recombination between otherwise clonally transmitted *ridibunda* genomes in L-E systems, mediated by viable *R. ridibunda* females produced by matings among two different, genetically compatible hemiclones of *R. esculenta*. Reproduced with permission from Vorburger (2001c)

this is the reason for what we called "informative inconsistencies" of the crossing experiments described above. These findings also urge some caution in the use of *R. esculenta* as a model to study the long-term consequences of the lack of recombination, because many *ridibunda* genomes found in L-E systems today may not have an uninterrupted history of clonal inheritance.

20.6 Lost Load Continued: Hybridogens as Vehicles for Gene Transfer

An alternative mechanism of genetic exchange for clonal genomes in hybridogenetic water frogs is provided when hybridogenesis is "leaky", i.e., when occasional recombination between parental genomes is possible in hybrids (Uzzell et al. 1977). Such exchange has far-reaching consequences because hybridogens can then act as vehicles for gene transfer among the parental species (introgression). This is common in hybrid zones with fertile hybrids (e.g., Szymura and Barton 1991), but typically precluded in hybridogens by the complete elimination of one parental genome prior to meiosis. However, deviations from this standard model of hybridogenesis are known to occur (reviewed by Schmeller 2004), particularly in populations where both parental species are present (i.e., outside the L-E system) or in "disturbed" L-E systems into which *R. ridibunda* has been introduced by humans, often from multiple origins (Pagano et al. 2003). This may be due to the fact that genomes of *R. ridibunda* vary geographically in their ability to induce hybridogenesis (Hotz et al. 1985). The evidence for recombination among parental genomes in hybridogenetic water frogs is mainly inferred from observed introgression of species-specific alleles in population studies employing allozyme electrophoresis (Schmeller 2004). Direct evidence for recombination from comparisons of parent and offspring genotypes is still virtually lacking. This is not surprising given that the genetic signature of a presumably rare event may be detected more easily in large population samples than the event itself in a limited number of breeding experiments. Vorburger and Reyer (2003) found just a single *R. esculenta* female exhibiting recombination in the germline among a large number of frogs from a "disturbed" L-E system (i.e., one also containing introduced *R. ridibunda*) used in experimental crosses. But even if rare, this process should not be ignored (Pagano and Schmeller 1999). It may provide a way for expanding hemiclones to acquire locally adapted alleles (Schmeller et al. 2005b), and it may allow for purging of deleterious mutations from clonally transmitted genomes. This, again, urges caution in the use of hemiclonal water frogs to study the long-term consequences of clonal inheritance.

Maybe not of direct relevance for purging, but worth mentioning here, is another form of introgression in which hybrids act as vehicles for gene transfer: the introgression of *lessonae* mtDNA into *R. ridibunda.* As explained earlier, primary hybridizations occur between females of *R. ridibunda* and males of *R. lessonae*, thus producing *R. esculenta* lineages that possess *ridibunda* mtDNA. In L-E systems, this would remain so as long as female *R. esculenta* mated with male *R. lessonae*, which is the more common combination. However, the reciprocal mating combination between females of *R. lessonae* and males of *R. esculenta* also occurs (Fig. 20.1), and such matings generate *R. esculenta* progeny possessing *lessonae* mtDNA. In the absence of syntopic *R. ridibunda*, this transfer of *lessonae* mtDNA into a hemiclonal hybrid lineage is irreversible, which explains why *R. esculenta* in natural L-E systems are indeed found to possess the mtDNA of their local host rather than *ridibunda* mtDNA (Spolsky and Uzzell 1986; Hotz et al. 1992). Within the range of *R. ridibunda*, such hemiclonal hybrid lineages can act as a vehicle for directional introgression of *lessonae* mtDNA into *R. ridibunda*, because matings between *R. esculenta* females bearing *lessonae* mtDNA and male *R. ridibunda* produce *R. ridibunda* progeny with mtDNA derived from *R. lessonae* (Spolsky and Uzzell 1984). Indeed, 42% of *R. ridibunda* in central Poland have been found to carry such introgressed *lessonae* mtDNA (Spolsky and Uzzell 1984).

20.7 Open Questions and Outlook: Spontaneous Mutational Load in Natural Populations of *Rana ridibunda*

The crossing experiments reported above provide evidence that non-recombining *ridibunda* genomes occurring in L-E systems bear a substantial mutational load, but they cannot distinguish between mutations that have accumulated over generations of clonal transmission and mutations that were already present in the original population of *R. ridibunda* and became "frozen" in clonal genomes through hybridization. In fact, the experiments only prove that clonal *ridibunda* genomes in

L-E systems contain at least one lethal equivalent, which may well be the case for any *ridibunda* genome even in a sexual population. To assess the relative importance of the two processes requires a quantitative estimate in terms of lethal equivalents of the spontaneous mutational load in sexual populations of *R. ridibunda*. Such an estimate is currently lacking, but the amphisexuality of *R. esculenta* lineages together with the relative ease to generate hemiclonally-reproducing F1 hybrids provides a unique opportunity to obtain this information. This work is in progress and includes four steps:

(1) Crosses between *R. ridibunda* from particular regions such as central Poland and *R. lessonae* spontaneously generate hemiclonal offspring (Hotz et al. 1985). Such crosses are used to obtain F1 individuals containing a "frozen" *ridibunda* haplotype.
(2) Each of these haplotypes is perpetuated in many copies by crossing such F1 hybrids with *R. lessonae*, producing a set of hemiclonal lineages that all contain the same, clonally transmitted *ridibunda* genome.
(3) The hybrids so obtained are crossed back with sexual *R. ridibunda* to produce *R. ridibunda* offspring that inherit a "frozen" *ridibunda* haplotype from their hybrid parent and a recombined *ridibunda* genome from their *R. ridibunda* parent. Such offspring exhibit normal Mendelian recombination.
(4) Finally, these backcrossed *R. ridibunda* are crossed with hybrids that possess the same *ridibunda* haplotype as their hybrid parent, generating offspring that are on average homozygote for 50% of their genome. Their viability compared to control crosses provides an estimate of the number of lethal equivalents on this *ridibunda* haplotype. Survival will be reduced to 0.5 for one lethal equivalent and to 0.5^n for n lethal equivalents.

Such estimates of the spontaneous mutational load in sexual populations of *R. ridibunda* will provide a plausible null hypothesis for testing the operation and magnitude of Muller's ratchet in natural hemiclones. The estimates will be even more important in their own right. Spontaneous deleterious mutations are an issue of central focus in evolutionary genetics because they critically influence a large array of biological phenomena such as extinction risk of small populations, ploidy level, Y chromosome degeneration, senescence, inbreeding avoidance and mate choice, as well as the maintenance of sexual reproduction (reviewed by Charlesworth and Charlesworth 1998). Yet relevant estimates of such loads are still scarce and such data are therefore badly needed (McCune et al. 2002; Halligan and Keightley 2003).

References

Abt G, Reyer H-U (1993) Mate choice and fitness in a hybrid frog: *Rana esculenta* females prefer *Rana lessonae* males over their own. Behav Ecol Sociobiol 32: 221–228
Abt Tietje G, Reyer H-U (2004) Larval development and recruitment of juveniles in a natural population of *Rana lessonae* and *Rana esculenta*. Copeia 3: 638–646

Bergen K, Semlitsch RD, Reyer H-U (1997) Hybrid female matings are directly related to the availability of *Rana lessonae* and *Rana esculenta* males in experimental populations. Copeia 2: 275–283

Berger L (1967) Embryonal and larval development of F1-generation of green frogs different combinations. Acta Zool Cracov 12: 123–160

Berger L (1968) Morphology of F1-generation of various crosses within *Rana esculenta*-complex. Acta Zool Cracov 13: 301–324

Berger L (1970) Some characteristics of the crosses within *Rana esculenta* complex in postlarval development. Ann Zool 27: 373–416

Berger L, Uzzell T, Hotz H (1988) Sex determination and sex ratios in wstern Palearctic water frogs: XX and XY female hybrids in the Pannonian basin? Proc Acad Nat Sci Phila 140: 220–239

Blankenhorn HJ, Heusser H, Notter P (1973) Zur Verbreitung von *Rana esculenta* Linnaeus und *Rana lessonae* Camerano im Zürcher Oberland. Rev Suisse Zool 80: 662–666

Blankenhorn HJ, Heusser H, Vogel P (1971) Drei Phänotypen von Grünfröschen aus dem *Rana esculenta*-Komplex in der Schweiz. Rev Suisse Zool 78: 1242–1247

Carmona JA, Sanjur OI, Doadrio I, Machordom A, Vrijenhoek RC (1997) Hybridogenetic reproduction and maternal ancestry of polyploid Iberian fish: the *Tropidophoxinellus alburnoides* complex. Genetics 146: 983–993

Chao L (1990) Fitness of RNA virus decreased by Muller's ratchet. Nature 348: 454–455

Charlesworth B, Charlesworth D (1997) Rapid fixation of deleterious alleles can be caused by Muller's ratchet. Genet Res 70: 63–73

Charlesworth B, Charlesworth D (1998) Some evolutionary consequences of deleterious mutations. Genetica 102/103: 3–19

Dubois A (1992) Notes sur la classification des Ranidae (amphibiens anoures). Bull Mens Soc Linn Lyon 61: 305–352

Ebendal T (1979) Distribution, morphology and taxonomy of the Swedisch green frogs (*Rana esculenta* complex). Mitt Zool Mus Berl 55: 143–152

Ellegren H, Fridolfsson AK (2003) Sex-specific mutation rates in salmonid fish. J Mol Evol 56: 458–463

Engeler B, Reyer HU (2001) Choosy females and indiscriminate males: mate choice in mixed populations of sexual and hybridogenetic water frogs (*Rana lessonae*, *Rana esculenta*). Behav Ecol 12: 600–606

Frost DR, Grant T, Faivovich J, Bain RH, Haas A, Haddad DFB, de Sá RO, Channing A, Wilkinson M, Donnellan SC, Raxworthy CJ, Campbell JA, Blotto BL, Moler P, Drewes RC, Nussbaum RA, Lynch JD, Green DM, Wheeler WC (2006) The amphibian tree of life. Bull Am Mus Nat Hist 297: 1–370

Graf J-D, Polls Pelaz M (1989) Evolutionary genetics of the *Rana esculenta* complex. In: Dawley RM, Bogart JP (eds) Evolution and ecology of unisexual vertebrates. Bulletin 466, New York State Museum, Albany, New York, pp. 289–301

Guex G-D, Beerli P, Barbour AD, Hotz H (1993) A dynamic model to describe equilibrium conditions in mixed populations of a hemiclonal hybrid and its sexual host in European water frogs. In: Catzeflis FM, Gautier M (eds) Evolution 93. 4th congress of the European society for evolutionary biology. Université Montpellier II, Montpellier, pp. 158

Guex GD, Hotz H, Semlitsch RD (2002) Deleterious alleles and differential viability in progeny of natural hemiclonal frogs. Evolution 56: 1036–1044

Günther R (1990) Die Wasserfrösche Europas (Anura – Froschlurche). A. Ziemsen Verlag, Wittenberg Lutherstadt

Günther R, Uzzell T, Berger L (1979) Inheritance patterns in triploid *Rana "esculenta"* (Amphibia, Salientia). Mitt Zool Mus Berl 55: 35–57

Halligan DL, Keightley PD (2003) How many lethal alleles? Trends Genet 19: 57–59

Hellriegel B, Reyer H-U (2000) Factors influencing the composition of mixed populations of a hemiclonal hybrid and its sexual host. J Evol Biol 13: 906–918

Heusser H, Blankenhorn HJ (1973) Crowding-Experimente mit Kaulquappen aus homo- und heterotypischen Kreuzungen der Phänotypen *esculenta*, *lessonae* und *ridibunda* (*Rana esculenta*-Komplex, Anura, Amphibia). Rev Suisse Zool 80: 543–569

Higgs PG, Woodcock G (1995) The accumulation of mutations in asexual populations and the structure of genealogical trees in the presence of selection. J Math Biol 33: 677–702

Holenweg Peter AK, Reyer HU, Tietje GA (2002) Species and sex ratio differences in mixed populations of hybridogenetic water frogs: The influence of pond features. Ecoscience 9: 1–11

Hotz H, Beerli P, Spolsky C (1992) Mitochondrial DNA revelas formation of nonhybrid frogs by natural matings between hemiclonal hybrids. Mol Biol Evol 9: 610–620

Hotz H, Mancino G, Bucci-Innocenti S, Ragghianti M, Berger L, Uzzell T (1985) *Rana ridibunda* varies geographically in inducing clonal gametogenesis in interspecies hybrids. J Exp Zool 236: 199–210

Hotz H, Semlitsch RD, Gutmann E, Guex G-D, Beerli P (1999) Spontaneous heterosis in larval life-history traits of hemiclonal frog hybrids. Proc Natl Acad Sci USA 96: 2171–2176

Hurst LD, Ellegren H (1998) Sex biases in the mutation rate. Trends Genet 14: 446–452

Mantovani B, Scali V (1992) Hybridogenesis and androgenesis in the stick insect *Bacillus rossius-grandii benazzii* (Insecta, Phasmatodea). Evolution 46: 783–796

McCune AR, Fuller RC, Aquilina AA, Dawley RM, Fadool JM, Houle D, Travis J, Kondrashov AS (2002) A low genomic number of recessive lethals in natural populations of bluefin killifish and zebrafish. Science 296: 2398–2401

Milinski M (1994) Hybridogenetic frogs on an evolutionary dead end road. Trends Ecol Evol 9:62

Muller HJ (1964) The relation of recombination to mutational advance. Mutat Res 1: 2–9

Pagano A, Dubois A, Lesbarrères D, Lodé T (2003) Frog alien species: a way for genetic invasion? C R Biol 326: S85–S92

Pagano A, Schmeller D (1999) Is recombination less negligible than previously described in hybridogenetic water frogs? In: Miaud C, Guyétant R (eds) 9th OG meeting – current studies in Herpetology. SEH, Le Bourget du Lac, France, pp. 351–356

Plötner J (2005) Die westpaläarktischen Wasserfrösche. Laurenti-Verlag, Bielefeld

Redfield RJ (1994) Male mutation rates and the cost of sex for females. Nature 369: 145–147

Rist L, Semlitsch RD, Hotz H, Reyer H-U (1997) Feeding behaviour, food consumption, and growth efficiency of hemiclonal and parental tadpoles of the *Rana esculenta* complex. Funct Ecol 11: 735–742

Roesli M, Reyer H-U (2000) Male vocalization and female choice in the hybridogenetic *Rana lessonae*/*Rana esculenta* complex. Anim Behav 60: 745–755

Rybacki M, Berger L (1994) Distribution and ecology of water frogs in Poland. Zoologica Poloniae 39: 293–303

Rybacki M, Berger L (2001) Types of water frog populations (*Rana esculenta* complex) in Poland. Mitt Mus Naturkd Berlin, Zool Reihe 77: 51–57

Saitoh K, Kim IS, Lee EH (2004) Mitochondrial gene introgression between spined loaches via hybridogenesis. Zool Sci 21: 795–798

Santucci F, Nascetti G, Bullini L (1996) Hybrid zones between two genetically differentiated forms of the pond frog *Rana lessonae* in southern Italy. J Evol Biol 9: 429–450

Schmeller DS (2004) Tying ecology and genetics of hemiclonally reproducing waterfrogs (*Rana*, Anura). Ann Zool Fenn 41: 681–687

Schmeller DS, O'Hara R, Kokko H (2005a) Male adaptive stupidity: male mating pattern in hybridogenetic frogs. Evol Ecol Res 7: 1039–1050

Schmeller DS, Seitz A, Crivelli A, Veith M (2005b) Crossing species' range borders: interspecies gene exchange mediated by hybridogenesis. Proc R Soc Lond B 272: 1625–1631

Schmidt BR (1993) Are hybridogenetic frogs cyclical parthenogens? Trends Ecol Evol 8: 271–273

Schultz RJ (1969) Hybridization, unisexuality, and polyploidy in the teleost *Poeciliopsis* (Poeciliidae) and other vertebrates. Am Nat 103:605–619

Schultz RJ (1973) Unisexual fish – laboratory synthesis of a species. Science 179: 180–181

Semlitsch RD (1993) Asymmetric competition in mixed populations of tadpoles of the hybridogenetic *Rana esculenta* complex. Evolution 47: 510–519

Som C, Anholt BR, Reyer H-U (2000) The effect of assortative mating on the coexistence of a hybridogenetic waterfrog and its sexual host. Am Nat 156: 34–46

Som C, Bagheri HC, Reyer HU (2007) Mutation accumulation and fitness effects in hybridogenetic populations: a comparison to sexual and asexual systems. BMC Evol Biol 7

Som C, Reyer HU (2006) Variation in sex ratio and evolutionary rate of hemiclonal *Rana esculenta* populations. Evol Ecol 20: 159–172

Som C, Reyer HU (2007) Hemiclonal reproduction slows down the speed of Muller's ratchet in the hybridogenetic frog *Rana esculenta*. J Evol Biol 20: 650–660

Spolsky C, Uzzell T (1984) Natural interspecies transfer of mitochondrial DNA in amphibians. Proc Natl Acad Sci USA 81: 5802–5805

Spolsky C, Uzzell T (1986) Evolutionary history of the hybridogenetic hybrid frog *Rana esculenta* as deduced from mtDNA analyses. Mol Biol Evol 3: 44–56

Szymura JM, Barton NH (1991) The genetic structure of the hybrid zone between the fire-bellied toads *Bombina bombina* and *B. variegata* – comparisons between transects and between loci. Evolution 45: 237–261

Tinti F, Mantovani B, Scali V (1995) Reproductive features of homospecific hybridogenetically-derived stick insects suggest how unisexuals can evolve. J Evol Biol 8: 81–92

Tunner HG (1973) Das Albumin und andere Bluteiweisse bei *Rana ridibunda* Pallas, *Rana lessonae* Camerano, *Rana esculenta* Linné und deren Hybriden. Z Zool Syst Evol 11: 219–233

Tunner HG (1974) Die klonale Struktur einer Wasserfroschpopulation. Z Zool Syst Evol 12: 309–314

Uzzell T, Berger L (1975) Electrophoretic phenotypes of *Rana ridibunda*, *Rana lessonae*, and their hybridogenetic associate, *Rana esculenta*. Proc Acad Nat Sci Phila 127: 13–24

Uzzell T, Günther R, Berger L (1977) *Rana ridibunda* and *Rana esculenta*– a leaky hybridogenetic system (Amphibia-Salientia). Proc Acad Nat Sci Phila 128: 147–171

Vences M (2007) The amphibian tree of life: Ideologie, Chaos oder biologische Realität? Z Feldherpetol 14: 153–162

Vorburger C (2001a) Fixation of deleterious mutations in clonal lineages: Evidence from hybridogenetic frogs. Evolution 55: 2319–2332

Vorburger C (2001b) Heterozygous fitness effects of clonally transmitted genomes in waterfrogs. J Evol Biol 14: 602–610

Vorburger C (2001c) Non-hybrid offspring from matings between hemiclonal hybrid waterfrogs suggest occasional recombination between clonal genomes. Ecol Lett 4: 628–636

Vorburger C, Reyer HU (2003) A genetic mechanism of species replacement in European waterfrogs? Conserv Genet 4: 141–155

Vrijenhoek RC, Angus RA, Schultz RJ (1977) Variation and heterozygosity in sexually vs. clonally reproducing populations of *Poeciliopsis*. Evolution 31: 767–781

Chapter 21
Lost Sex in the Reptiles: Constraints and Correlations

Michael Kearney, Matthew K. Fujita and Jessica Ridenour

Abstract Reptiles are the only truly parthenogenetic vertebrates, making it especially fascinating to understand how and why some reptilian taxa have broken free of sexual reproduction. In this review we consider the evolutionary and genomic constraints, consequences, and ecological correlations of reptile parthenogenesis, and how this informs us more generally about the loss of sex in organisms. In reviewing the taxonomic distribution of parthenogenesis we find that some lineages are particularly likely to evolve parthenogenesis (e.g., teiid lizards) and others biased strongly against parthenogenesis (e.g., colubrid snakes). Moreover, all but one of the natural cases also involves hybridization. The geographical and ecological tendencies of parthenogenetic reptiles suggest a bias toward "open" environments, yet there are often surprisingly high levels of coexistence and niche overlap between parthenogenetic lineages and their related sexual forms. To the extent that these fascinating patterns can be deciphered we will learn much about the constraints and selective forces acting on the evolution of parthenogenesis in nature.

21.1 Introduction

Two of the most important questions about asexual reproduction are (1) "When is the transition from sexuality to asexuality evolutionarily constrained?", and (2) "When is asexuality favoured by natural selection?" The latter question is of course crucial to a complete understanding of the evolutionary and ecological significance of sexual versus asexual reproduction. Yet, it cannot be fully answered without also understanding the extent to which the genomic processes and developmental pathways of organisms limit their ability to evolve parthenogenetic reproduction.

In this chapter, we will consider how reptiles as a group enable us to answer these questions. Reptiles are the only living chordates to have evolved true parthenogenesis and thus are of special interest in comparative studies of constraints on asexual

M. Kearney (✉)
Department of Zoology, The University of Melbourne, Victoria 3010, Australia
e-mail: mrke@unimelb.edu.au

I. Schön et al. (eds.), *Lost Sex*, DOI 10.1007/978-90-481-2770-2_21,

reproduction. Parthenogenesis has also evolved in taxonomically and ecologically diverse groups within the reptiles, providing the opportunity to look for generalities in the forces likely to be favouring parthenogenesis.

The ecology and evolution of parthenogenesis in reptiles was last summarized by Darevsky et al. (1985), with prior reviews by Maslin (1971) and Cole (1975). Since then, a number of new cases of reptilian parthenogenesis have been discovered and our knowledge of the ecology and evolution of many of these species has widened considerably. In the present review, we first summarise the phylogenetic distribution of parthenogenesis in reptiles. In doing so, we indicate what is known of the mechanisms by which parthenogenesis has evolved in these groups and note whether there are other genetic phenomena associated with the transition, such as hybridization and polyploidy. We then survey the geographical and ecological correlates of parthenogenesis within the reptiles. After reviewing this material, we assess what it tells us about the likely constraints operating on the evolution of parthenogenesis in reptiles, the possible evolutionary opportunities associated with the transition to parthenogenesis, and the likely selective forces that maintain parthenogenesis in nature.

21.2 The Phylogenetic Distribution and Genetic Correlates of Parthenogenesis in Reptiles

21.2.1 Phylogenetic Distribution

Parthenogenesis has never been recorded in crocodiles or turtles, or the relictual tuatara from New Zealand. We do not consider these groups further in this review but rather focus on the squamates, which include the amphisbaenians, snakes and lizards. Previous reviews of parthenogenesis in reptiles describe 29 cases of squamate parthenogenesis within 8 families (Darevsky et al. 1985; Vrijenhoek et al. 1989). Since these reviews, some additional cases have been discovered and our knowledge of their origins has increased dramatically (Table 21.1). Detailed studies and revisions of some groups have revealed additional species, although the concept of a parthenogenetic "species" remains problematic (Dickinson 1999). The most notable addition, however, is the first case of parthenogenesis in the Scincidae, *Menetia greyii* from Australia (Adams et al. 2003b). We have discounted the only known case of parthenogenesis within the Chamaeleonidae, *Rhampholeon boulengeri* (formerly *Brookesia spectrum affinis*), as the original evidence for parthenogenesis was weak (sex ratio bias, Hall 1970) and no subsequent data has come to light since the original proposal. Thus, there are currently 40 known cases of obligate parthenogenesis in the reptiles within 8 families (Table 21.1).

In addition to these obligate cases of parthenogenesis, there have been numerous anecdotal reports suggesting occasional parthenogenesis in captive reptiles kept in isolation, particularly in snakes (Dubach et al. 1997; Magnusson 1979; Scalka and Vozenilek 1986; Schuett et al. 1997). Proof that these were not the result of sperm

Table 21.1 A summary of known cases of parthenogenesis in reptiles, updated from the table presented in Darevsky et al. (1985)

Family	Taxon	Ploidy	Hybrid origin	Clonal diversity	References
Gekkonidae	*Nactus arnouxii* Formerly:*Cyrtodactylus pelagicus*	2n	Yes	Low	Moritz (1987)
	Hemidactylus garnotii-vietnamensis complex:			Unknown	Ota and Hikida (1989), Ota et al. (1989, 1996)
	H. vietnamensis	3n	Yes		
	H. garnotti	3n	Yes		
	H. stejnegeri	3n	Yes		
	H. sp.	2n	Yes		
	Heteronotia species	3n	Yes	High	Moritz (1983), Moritz et al. (1989b)
	Lepidodactylus lugubris	2n	Yes	Medium	Boissinot et al. (1997), Radtkey et al. (1995), Volobouev et al. (1993)
Agamidae	*Leiolepis*		Yes	Medium	Darevsky and Kupriyanova (1993), Malysheva et al. (2006), Schmitz et al. (2001)
	L. boehemi	2n			
	L. guentherpetersi	3n			
	L. triploida	3n			
Xantusiidae	*Lepidophyma flavimaculatum* (southern populations)	2n	No?	Unknown	Bezy (1989), Bezy and Carmarillo (2002), Bezy and Sites (1987), Sinclair et al. (2006)
	L. reticulatum	2n			
Lacertidae	*Darevskia* Formerly: *Lacerta*			Low-medium	Fu et al. (2000), Moritz et al. (1992a), Murphy et al. (1997)
	D. armeniaca	2n	Yes		
	D. rostombekovi	2n	Yes		
	D. dahli	2n	Yes		
	D. unisexualis	2n	Yes		
	D. uzzelli	2n	Yes		
	D. sapphirina	2n	Yes		
	D. bendimahiensis	2n	Yes		
Teiidae	*Cnemidophorous*			Low-medium	Cole and Dessauer (1993), Dessauer and Cole (1989), Reeder et al. (2002), Sites et al. (1990)
	C. cryptus (formerly *C. lemniscatus*)	2n	Yes		
	C. pseudolemniscatus (formerly *C. lemniscatus*)	3n	Yes		

Table 21.1 (continued)

Family	Taxon	Ploidy	Hybrid origin	Clonal diversity	References
	Aspidoscelis Formerly:*Cnemidophorus tesselata* complex:			Low-High	Densmore et al. (1989a, b), Dessauer and Cole (1989), Hernández-Gallegos et al. (2003), Manriquez et al. (2000), Moritz et al. (1989c, d, 1992c), Parker et al. (1989), Reeder et al. (2002), Taylor et al. (2003)
	A. tesselata	2n	Yes		
	A. dixoni	2n	Yes		
	A. neotesselata	3n**	Yes		
	A. flagellicauda complex	3n	Yes		
	A. sonorae complex	3n	Yes		
	A. neomexicana	2n	Yes		
	A. laredonsis complex	2n	Yes		
	A. exsanguis	3n**	Yes		
	A. opate complex	3n	Yes		
	A. uniparens complex	3n	Yes		
	A. velox complex	3n	Yes		
	cozumela species group:				
	A. cozumela	2n	Yes		
	A. rodecki	2n	Yes		
	A. maslini	2n	Yes		
	Kentropyx borckianus	2n	Yes	Low	Cole et al. (1995), Reeder et al. (2002)
	Teius suquiensis	2n	Yes	Unknown	Avila and Martori (1991), Cabrera and Monguillot (2007)
Gymnopthalmidae	*Gymnophthalmus underwoodi*	2n	Yes	Low	Reeder et al. (2002), Cole et al. (1990, 1993), Kizirian and Cole (1999)
	Leposoma percarinatum	3n	Yes	Unknown	Pellegrino et al. (2003)
Scincidae	*Menetia greyii*	3n	Yes	High	Adams et al. (2003b)
Typhlopidae	*Ramphotyphlops braminus*	3n	Yes	Unknown	Wynn et al. (1987)
Boidae	*Python molurus bivittatus* *	2n	No	Not clonal	Groot et al. (2003)
Varanidae	*Varanus panoptes* *	2n	No	Not clonal	Lenk et al. (2005)
	Varanus komodoensis *	2n	No	Not clonal	Watts et al. (2006)

Ploidy, hybrid state and the level of clonal diversity are indicated where known. Levels of clonal diversity are qualitative approximations based on the number of clones identified given the number of individuals and populations sampled. Species identified with an asterisk are captive cases of facultative parthenogenesis; double asterisks under "Hybrid Origin" indicate triploid lineages whose ancestry involves three sexual species. For brevity, we only provide references published subsequent to Darevsky et al. (1985).

storage or unnoticed matings has been lacking, but recently three genetically verified cases have come to light involving an additional snake species, the Burmese python (*Python molurus bivittatus*, Groot et al. 2003), as well as two varanid lizards: the yellow spotted monitor (*Varanus panoptes*, Lenk et al. 2005) and the Komodo dragon (*Varanus komodoensis*, Watts et al. 2006). Most genetically analysed cases of occasional parthenogenesis in captive reptiles demonstrate an automictic restitution process leading to increased homozygosity and, in the case of species with ZW sex-determining mechanisms, such as the Komodo dragon, all male offspring (Watts et al. 2006). Tantalizingly, however, the case of the Burmese python appears to involve a truly clonal mechanism (Groot et al. 2003).

Even including the facultative cases from captive animals, only 0.6% of the roughly 7000 squamates can reproduce parthenogenetically. An important question to ask of the phylogenetic distribution of parthenogenesis in the squamates is whether the pattern is non-random, or whether parthenogenesis is biased towards or against certain lineages. Figure 21.1 shows the phylogeny for the squamates with the independent origins of parthenogenesis mapped onto it. Visual inspection suggests a non-random distribution with apparently high frequencies of parthenogenesis in some groups. We have statistically tested whether the frequencies of parthenogenesis in each squamate group are greater or less than expected by chance. We did this by randomizing the distribution of parthenogenesis across the squamate families 10,000 times to determine the expected frequencies, taking into account the number of species in each family. We then compared the observed and expected frequencies using G-tests (Sokal and Rohlf 1995), the results of which are presented in Table 21.2 and Fig. 21.1. This analysis shows that only Lacertidae, Teiidae and Varanidae show a frequency of parthenogenesis higher than expected by chance, while the Colubridae, Iguanidae and Scincidae have frequencies lower than expected by chance (Table 21.2). When reanalysed using the number of parthenogenetic genera, rather than species, only the Teiidae have a higher than expected frequency of parthenogenesis, and only the Coloubridae have a lower than expected frequency.

21.2.2 Hybridization, Polyploidy and Genetic Diversity

A fascinating genetic pattern in natural cases of parthenogenesis in the squamates, and indeed the vertebrates generally, is the close association with hybridization. Hybridization can be defined in a variety of ways; while it typically refers to the interbreeding of two different species, crosses between two populations within a species that differ for one or more heritable characteristics are also considered cases of hybridization (Arnold 1997). The vagueness of the definition of hybridization reflects the quasi-continuum between population and species. Irrespective of these shades of grey, almost all the well studied cases of parthenogenetic reptiles are unambiguously considered to be hybrids. While some cases were initially doubted (e.g., Cuellar and Kluge 1972), the application of increasingly sophisticated genetic

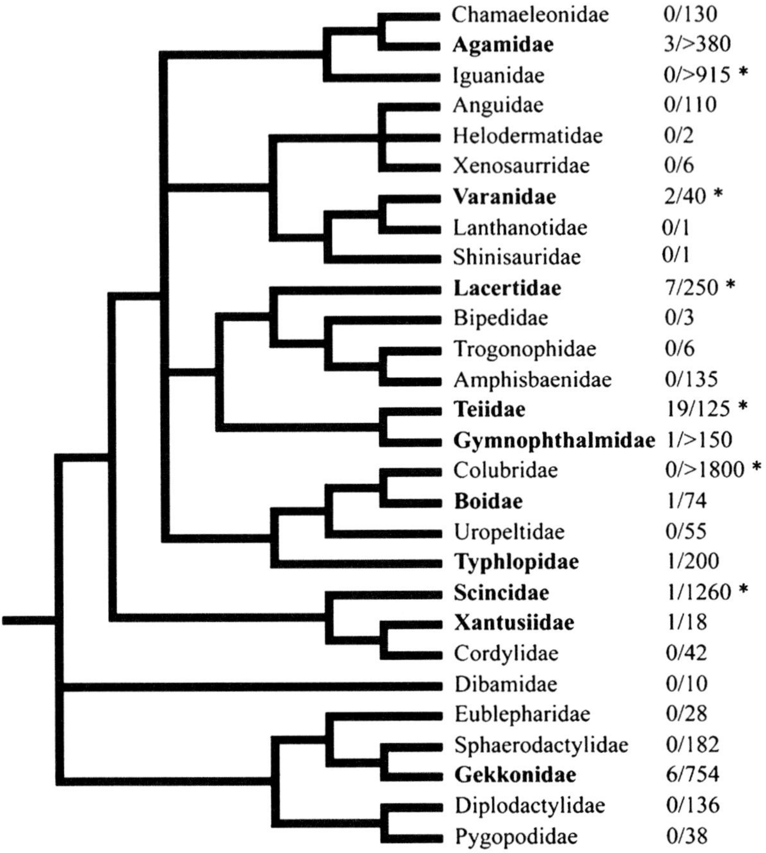

Fig. 21.1 Phylogenetic distribution of parthenogenesis in lizards at the family level. The topology is based on recent molecular phylogenetic analyses (Slowinski and Lawson 2002; Townsend et al. 2004). Families with parthenogenetic species are indicated in bold. Numbers next to branches indicate the number of parthenogenetic species relative to the total number of species included in the family. The number of species from each family are from Pough et al. (2004) and from Tony Gamble (for geckos, Gamble et al. (2008) and Gamble, "personal communication"). Asterisks indicate significant departures from expected number of occurrences of parthenogenesis as summarized in Table 21.2

markers to the problem has provided resolution. There is only one fascinating exception to this rule – xantusiid lizards of the genus *Lepidophyma*. Karyotypic, morphological, and allozymic analysis have shown no evidence for a hybrid origin (Bezy 1989; Bezy and Carmarillo 2002). Wider taxonomic sampling may demonstrate otherwise, as may the application of finer resolution markers to the problem (Sinclair et al. 2006), as recently demonstrated in a case of aphid parthenogenesis (Delmotte et al. 2003).

Another very strong genetic pattern in reptile parthenogenesis is the association with polyploidy. Around 40% of cases of reptile parthenogenesis are polyploid and

Table 21.2 Observed and expected frequencies of parthenogenesis in squamate families based on 10,000 randomizations of the distribution of parthenogenesis across the groups, taking into account the number of species per group

Family	Observed	Expected	G	P	Direction
Agamidae	3.0	2.4	0.142	0.706	+
Amphisbaenidae	0.0	0.9	1.547	0.214	–
Anguidae	0.0	0.7	1.259	0.262	–
Bipedidae	0.0	0.0	0.031	0.860	–
Boidae	1.0	0.5	0.402	0.526	+
Chamaeleonidae	0.0	0.8	1.495	0.221	–
Colubridae	**0.0**	**11.2**	**21.390**	**< 0.0001**	**–**
Cordylidae	0.0	0.3	0.483	0.487	–
Dibamidae	0.0	0.1	0.111	0.739	–
Diplodactylidae	0.0	0.8	1.532	0.216	–
Eublepharidae	0.0	0.2	0.321	0.571	–
Gekkonidae	7.0	4.7	0.900	0.343	+
Gymnophthalmidae	1.0	1.0	0.002	0.969	+
Helodermatidae	0.0	0.0	0.024	0.876	–
Iguanidae	**0.0**	**5.8**	**10.661**	**0.001**	**–**
Lacertidae	**7.0**	**1.6**	**9.195**	**0.002**	**+**
Lanthanotidae	0.0	0.0	0.011	0.917	–
Pygopodidae	0.0	0.2	0.427	0.513	–
Scincidae	**1.0**	**7.9**	**9.147**	**0.002**	**–**
Shinisauridae	0.0	0.0	0.010	0.922	–
Sphaerodactylidae	0.0	1.1	2.048	0.152	–
Teiidae	**19.0**	**0.8**	**79.041**	**< 0.0001**	**+**
Trogonophidae	0.0	0.0	0.064	0.800	–
Typhlopidae	1.0	1.3	0.050	0.823	–
Uropeltidae	0.0	0.4	0.631	0.427	–
Varanidae	**2.0**	**0.3**	**4.225**	**0.040**	**+**
Xantusiidae	1.0	0.1	2.201	0.138	+
Xenosauridae	0.0	0.0	0.074	0.786	–

G statistics and their associated *P* values (with William's correction) are presented and the direction of the difference from expectation is also indicated. Statistically significant comparisons are indicated in boldface.

indeed this was often the first strong evidence for parthenogenesis. Polyploidy is generally the norm in parthenogenetic species, thus the reptiles provide rare opportunities to disentangle the relative influences of polyploidy and hybridization on the phenotypes of parthenogenetic species. This is especially the case in *Aspidoscelis*, where 50% of parthenogenetic lineages are diploid. While some parthenogenetic insects have increased their genome to hexaploidy (Takenouchi 1976), polyploid parthenogenetic reptiles are only ever viable as triploids. Polyploidy in reptiles has most probably occurred through matings between the original diploid hybrid lineage and males of a sexual lineage. This is most apparent in triploid *Aspidoscelis neotesselata* which contains the nuclear genomes of three different species (Parker and Selander 1976).

The combination of hybrid origins and polyploidy results in extremely high within-individual diversity (heterozygosity) in parthenogenetic reptiles. Indeed, the highest observed heterozygosity within reptiles is generally observed in the parthenogens. In addition to their within-individual diversity, many parthenogenetic reptiles have very high between-individual, or clonal, diversity (Table 21.1).

21.2.3 Dynamics of Hybrid Origins

Given the ubiquity of hybrid origins of natural parthenogenesis within the squamates, another important question we can ask about the phylogenetic distribution of parthenogenesis in reptiles is whether there is any pattern in the genetic distance between the hybridizing parental species. This question can be best examined with the whiptail lizards of the Americas, of the genera *Aspidoscelis* and *Cnemidophorus*, and Caucasian rock lizards of the genus *Darevskia* (Moritz et al. 1992b; Murphy et al. 2000). Both groups include numerous independent origins of parthenogenesis from different sexual parental species, and good phylogenies of the sexual species are available. This allows tests for whether parthenogens were more likely to arise from hybridizations between close or distant congeners. In the case of the whiptail lizards, there is no consistent pattern regarding the maternal parental species and, with one exception, the hybridizing parental lineages were always from different clades (Moritz et al. 1992b). Similarly, only between-clade hybridizations have produced parthenogenetic lineages in *Darevskia*, despite numerous records of within-clade hybridizations occurring in nature (Murphy et al. 2000). However, the maternal parents in the *Darevskia* crosses always come from the same clade. Parthenogenetic *Heteronotia* are also of interest in this regard; while only two sexual lineages were involved in the hybridizations leading to the parthenogens, there have been two independent origins, and the maternal parent differed in each case (Moritz 1993).

21.3 The Geography and Ecology of Parthenogenesis in Reptiles

21.3.1 Geographical Tendencies

The geographical tendencies of parthenogenetic organisms have been of considerable importance in developing and testing ideas about the maintenance of sexual reproduction (Vandel 1928; Glesener and Tilman 1978; Bell 1982; Lynch 1984;Bierzychudek 1985; Cuellar 1994; Kearney 2005; Hörandl 2006). These studies show that, when compared with their nearest sexual relatives, parthenogenetic lineages are found to be biased toward disturbed or ecotonal habitats, higher latitudes and altitudes, drier habitats, and islands (see also Chapter 8).

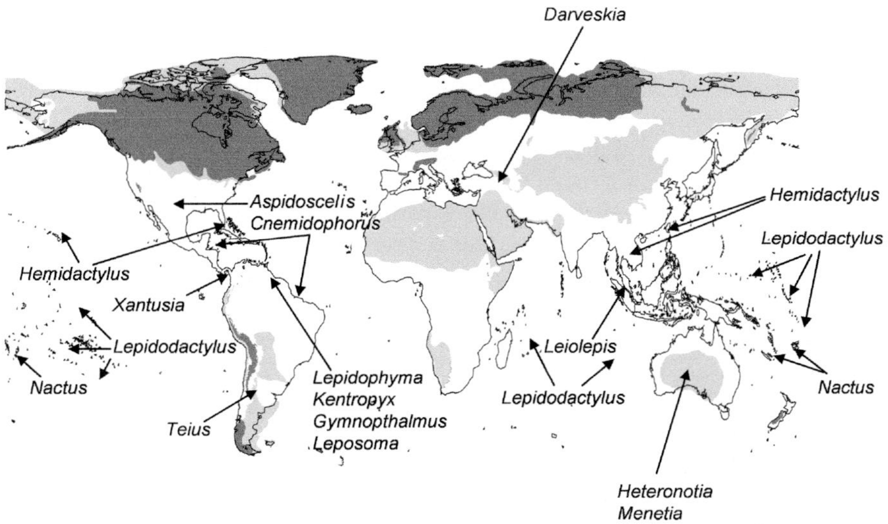

Fig. 21.2 The distribution of natural cases of parthenogenetic reptiles. The snake *Ramphotyphlops braminus* is cosmopolitan and is not indicated on the map. The dark grey shading represents areas that were glaciated during the Last Glacial Maximum, while the light grey areas represent hyper arid regions (less than 2% vegetation), based on Ray and Adams (2001)

Reptiles have played an important role in the development of these generalizations, most prominently in regard to the weed hypothesis of Wright and Lowe (1968). These authors drew a parallel between the agamic plant complexes, many of which are weeds, and the parthenogenetic whiptail lizards. Their analysis was perhaps most perceptive in noting remarkable similarities in the habitats occupied by plants and animals exhibiting the same genomic curiosities of hybridization, polyploidy and parthenogenetic (or, in the botanical terminology, apomictic) reproduction. Yet their study became best known for the analogy of the whiptail lizards with weeds, i.e., fugitive, colonizing species that do best in disturbed or ecotonal habitats, and eventually disappear from climax communities.

Perhaps the most stereotypic manifestation of geographical parthenogenesis involves biases to very high altitudes or latitudes, very often including glaciated habitats. These patterns are not as dramatic in reptilian cases as they are in other taxa (e.g., weevils and moths of northern Europe; Suomalainen et al. 1987) due to the dependence of reptiles on relatively warm habitats. Indeed, most parthenogenetic reptiles occur in the tropics (Fig. 21.2). Nonetheless, parthenogenetic lacertid lizards of the genus *Darevskia* are thought to be biased towards environments that were previously glaciated (Darevsky 1962; Darevsky et al. 1985), and parthenogenetic *Heteronotia* have a high-latitude bias to their distribution in comparison to their sexual ancestors (Kearney et al. 2003).

Parthenogenetic *Darevskia*, *Heteronotia*, *Menetia* and *Aspidoscelis* are found in arid environments (Wright and Lowe 1968; Darevsky et al. 1985; Adams et al. 2003b; Kearney et al. 2003), another general pattern under the umbrella of geographical parthenogenesis. The aridity of the environments occupied by parthenogenetic lineages of *Heteronotia* has been quantified relative to the sexual races (Kearney et al. 2003). While the two sexual progenitor lineages of parthenogenetic *Heteronotia* also occur within the arid zone of Australia, the parthenogenetic forms inhabit the driest regions and their distributions are very tightly associated with rainfall contours (Kearney et al. 2003; Strasburg et al. 2007). Moreover, the most recently discovered case of natural parthenogenesis in reptiles, the scincid lizard *Menetia greyii*, also occurs in this region (Adams et al. 2003b), along with a diverse array of other parthenogenetic taxa (Kearney 2003).

A number of tropical reptiles have also evolved parthenogenesis, but such cases appear mostly restricted to islands, coastal habitats, disturbed areas or human habitation. For instance, gekkonid geckos of three genera, *Nactus*, *Hemidactylus* and *Lepidodactylus*, include parthenogenetic lineages widespread throughout tropical islands in the Pacific. The sexual ancestors of parthenogenetic *Lepidodactylus* also occur on islands (Radtkey et al. 1995). In contrast, the sexual relatives of parthenogenetic *Nactus* can be found on both continental landmasses and islands, while the parthenogens are only found on islands (Moritz 1987). Parthenogenetic lineages of *Hemidactylus* occur in continental (Vietnam) and island areas (Taiwan, Hawaii and numerous smaller islands) but the direct sexual descendents of these geckos are unknown and perhaps extinct (Ota and Hikida 1989; Ota et al. 1989, 1996; Moritz et al. 1993).

Two lineages of parthenogenetic whiptail lizards occur in tropical regions of Central and South America, and retain the original genus name *Cnemidophorus*. The Central American lineage, *C. cozumela*, occurs on the Yucatan Peninsula where it inhabits sandy coastal areas and islands (Wright and Lowe 1968). The South American lineages *C. cryptus* and *C. pseudolemniscatus* (formerly *C. lemniscatus*), occur in the Amazon riparian habitats where they are closely associated with human habitation and disturbed areas (Vanzolini 1970, 1978). Another, tropically distributed parthenogenetic species, the Brahminy snake *Ramphotyphlops braminus*, is also associated with humans. The latter represents an unusual case of a cosmopolitan parthenogenetic species that has been transported throughout the tropical environments of the world in the novel environment of flowerpots (Wynn et al. 1987), and is the only snake known to be naturally parthenogenetic. There are no obvious candidates as yet for the sexual progenitors of this species, making it difficult to interpret its geographical distribution, but presumably the ancestors are highly restricted geographically, or perhaps extinct.

The only obvious exception to the stereotypic patterns of geographical parthenogenesis involves the xantusiid genus *Lepidophyma*. In this case, the unisexual populations of *L. flavimaculatum* are restricted to low elevation rainforest habitats of Costa Rica and Panama. The entirely parthenogenetic *L. reticulatum* is also confined to low-elevation sites in western Costa Rica (Bezy and Carmarillo 2002).

21.3.2 Patterns of Exclusion and Coexistence

While the broad-scale comparisons of geographical patterns in reptilian parthenogenesis reveal some interesting patterns, an understanding of the process driving them requires knowledge of ecological interactions at the local scale. In particular, we must understand the processes leading to coexistence or otherwise between sexual and parthenogenetic lineages, and between clones within parthenogenetic lineages.

At present, most of what we know about these topics within the reptiles comes from descriptive, anecdotal studies or uncontrolled experiments, but they do provide important insights. First of all, it seems clear that there are a number of instances where little or no geographic overlap occurs between related sexual and parthenogenetic lineages. For instance, in the whiptail lizards, the sexual *A. tigris* does not occur sympatrically with the parthenogenetic lineage *A. uniparens*, and in the case of *Heteronotia*, the sexual parental lineage SM6 abuts but never extends into the range of the parthenogenetic lineages (Kearney et al. 2003). The two major mechanisms proposed to account for cases where sexual and parthenogenetic lineages have non-overlapping distributions are competitive interactions and destabilizing hybridization (Lynch 1984). The latter process involves the formation of sterile hybrids in a zone of overlap, resulting in a "no-man's land" (so-to-speak) of "mutual self-destruction" (White and Contreras 1979). In this zone, parthenogens rob the sperm of sexual males, which is detrimental to the sexual lineage, and mated parthenogens produce offspring of deleteriously high ploidy.

There are no replicated experimental field-studies testing the causes of parapatrically distributed parthenogenetic and sexual reptiles. The only field data on the topic comes from Cuellar (1979, 1993) who observed the response of the sexual whiptail *Aspidoscelis tigris* upon removing the parthenogenetic *A. uniparens* from an area of sandy riparian floodplain over the course of nine years. It is unclear to what extent he was successful in removing *A. uniparens* from the site. Whatever the case, the abundance of *A. uniparens* recovered quickly after the manipulation, although throughout this recovery low numbers of *A. tigris* were observed at the study site. Thus, at most there was only a minor response to the manipulation.

More often than not, however, parthenogenetic and sexual lineages of reptile are found to coexist to some degree. Cuellar (1979) and Case (1990) describe numerous instances of coexistence between sexual and parthenogenetic whiptails, with Case (1990) demonstrating that the five unisexual and seven sexual species he considered occurred together as often as expected by chance. A detailed study of patterns of overlap in two species of *Aspidoscelis* by Paulissen et al. (1992) found coexistence at 62% of the sites where the parthenogen is found, with no consistent differences in abundance between sexual and parthenogenetic forms. Coexistence between sexual and parthenogenetic lineages occurs frequently in many other reptilian taxa including *Darevskia* (Darevsky et al. 1985), *Lepidodactylus* (Radtkey et al. 1995), *Heteronotia* (Kearney et al. 2003), and *Menetia* (Adams et al. 2003b).

Reptiles, particularly lizards, have been important model organisms in the development of theories about ecological coexistence generally, and patterns of niche partitioning are well documented in this group (Pianka 1973; Roughgarden 1995). Yet, attempts to understand the coexistence of sexual and parthenogenetic lizards by comparing their niches have found surprisingly high degrees of overlap. For instance, Schall (1993) found high overlap in diet, activity times and habitat use among five coexisting lineages of whiptail lizard (two parthenogenetic and three sexual), although the parthenogens tended to occur in a wider variety of micro- and macrohabitats. Similarly, Price et al. (1993) compared the niches of a sympatric pair of sexual and parthenogenetic whiptail and found high overlap in microhabitat use, while strong overlap in diet, microhabitat use and activity time has been reported in yet another pair of sympatric sexual and parthenogenetic whiptail species (Paulissen et al. 1992; Paulissen 2001). Moreover, Price et al. (1993) found little evidence for short term responses in population size in the sexual and parthenogenetic species when they reciprocally altered the abundance of each species. Dietary comparisons do indicate, however, that there is higher variation between individuals in sexual populations of whiptail than there is between individuals in parthenogenetic populations (Case 1990). Sympatric clones of whiptail lizard also show differences in diet (Paulissen et al. 1988). The significance of interclonal variation in niche characteristics with respect to the coexistence of sexual and parthenogenetic lineages is further discussed below.

21.3.3 Phenotypic Comparisons

The patterns in geographical distribution and coexistence of parthenogenetic and sexual reptiles may well be driven by the demographic consequences of parthenogenesis itself, including rapid population growth, resilience to small population size and superior colonizing abilities. Yet, the widespread occurrence of hybridization and polyploidy in parthenogenetic reptiles clearly shows all else is not equal between sexual and parthenogenetic lineages. Knowledge of the phenotypic consequences of a hybrid and/or polyploid transition from sex to parthenogenesis is thus critical for a complete understanding of these patterns.

The most detailed phenotypic comparisons of parthenogenetic and sexual reptiles are for the lizards *Aspidoscelis* and *Heteronotia*. Such comparisons are most informative when they involve the parthenogenetic lineage and its direct sexual ancestor, or ancestors in the case of a hybrid. While there have been numerous comparative studies of morphology, physiology and behaviour in *Aspidoscelis*, the majority have involved a parthenogenetic lineage and only one of its parental species (e.g., Congdon et al. 1978; Schall 1978; Parker 1979b; Bowker and Johnson 1980; Leuck 1985; Sievert and Paulissen 1996). These studies have produced mixed results. For instance, fecundity of parthenogenetic lineages has been reported to be both higher than (Congdon et al. 1978) and equal to (Schall 1978) that of related sexual lineages. Thermoregulatory precision of parthenogens has also been found to be greater than (Bowker and Johnson 1980) or similar to (Sievert and Paulissen 1996) that of related sexuals.

A very comprehensive analysis was presented by Cullum (1997), who conducted a phylogenetically independent analysis of a series of physiological traits (burst speed, endurance, maximum exertion, standard metabolic rate and evaporative water loss rate) among field-caught individuals of six sexual and six parthenogenetic lineages of *Aspidoscelis*. The only significant pattern observed in this study was a lower endurance in parthenogens relative to sexual lineages. Cullum (2000) also found reduced variance in parthenogens relative to sexuals for traits relating to locomotion, but not for metabolic or water loss rates.

A wide array of traits has also been compared among parthenogenetic *Heteronotia* and their two sexual parents. Fecundity of field-caught parthenogenetic *Heteronotia* is 30% lower than in the parental sexual species (Kearney and Shine 2005) but development of parthenogenetic embryos appears less sensitive to thermal fluctuations (Kearney and Shine 2004a). In direct contrast to the case in *Aspidoscelis*, field-caught adult parthenogens had dramatically higher endurance than their sexual progenitors (Kearney et al. 2005). Field-caught parthenogens also had higher water loss rates, lower heat tolerance and lower thermal preferences than sexuals, and some of these patterns persisted into the second generation (Kearney and Shine 2004b). The higher water-loss rates of parthenogenetic *Heteronotia* may be at least partially explained by higher infestations of parasitic mites (Kearney and Shine 2004b).

Clearly then, these phenotypic analyses demonstrate that parthenogenetic reptiles often differ from their sexual relatives, but not in a systematic manner. The few studies that have examined variation among clones within parthenogenetic reptiles have also found significant differences. For instance, two well-studied clones of *Aspidoscelis laredoensis* differ in thermal behaviour (Paulissen 1988; Sievert and Paulissen 1996), as did clones of *Lepidodactylus* (Bolger and Case 1994). In parthenogenetic *Heteronotia*, all clones are triploid and often represent the traits of the sexual parent for which they have a double nuclear dosage (Kearney and Shine 2004b). These dosage-related patterns are also robustly mirrored in their broad climatic associations (Kearney et al. 2003). Microsatellite-identified clones also differ for morphological and physiological traits in this species (Kearney and Strasburg, unpublished data).

21.3.4 Resistance to Parasites

Studies on reptiles have provided some confusing insights into the relevance of the Red Queen hypothesis for the maintenance of sex (see Chapter 7). This hypothesis predicts that recombination will provide an advantage in time-lagged, coevolutionary arms races between pathogen and host by allowing the host to construct a continually changing immune system "code" (Jaenike 1978; Bell 1982; Hamilton et al. 1990).

The Red Queen hypothesis predicts that parthenogens will have higher parasite loads than their sexual progenitors, a prediction clearly borne out in the case of desert-dwelling *Heteronotia*. Field-collected parthenogenetic individuals are 150 times more likely to suffer infestations of parasitic mites than are their sympatric sexual progenitors (Moritz et al. 1991). This pattern persists after many months in captive conditions (Kearney and Shine 2004b). Yet entirely the opposite pattern occurs in *Lepidodactylus* geckos of the tropical Pacific (Hanley et al. 1995), where sexual lineages had naturally higher prevalence, abundance and intensity of mites than their parthenogenetic offshoots, and were more likely to become infested under experimental conditions. This is despite parthenogenetic lineages having much lower diversity for immune-system genes than sexual forms (Radtkey et al. 1996). While similar comparisons remain to be done for other parthenogenetic reptiles and for other kinds of parasites, knowledge of the kinds of parasites harboured by these reptiles is continuing to grow (e.g., McAllister et al. 2003).

21.4 Genetic Constraints and Opportunities in the Evolution of Reptile Parthenogenesis

21.4.1 General Constraints

Two genetic constraints must be overcome in the transition from sex to parthenogenesis. First, it must be possible for egg development to initiate and proceed independent of sperm. Second, meiosis must be modified or circumvented such that ploidy is maintained. Among the chordates, only the reptiles have been able to overcome both of these constraints. The anamniotes, fish and amphibians, are apparently unable to break their dependence on sperm, perhaps because of their reliance on external fertilization (Uzzell 1970). Mammals have the constraint of genomic imprinting, where epigenetic modification of approximately 100 genes occurs differentially in either the sperm or the egg in a parent-of-origin manner (reviewed by Morison et al. 2005; Platonov 2005; Hore et al. 2007). The constraints preventing birds from reproducing parthenogenetically are unknown, but they are apparently strong with only one known case (Olsen 1966). To our knowledge, the presence or absence of genomic imprinting in non-avian reptiles has not yet been examined, yet is an important avenue of research that will help identify why and how squamates can bypass these constraints to parthenogenesis.

How did some squamates break these constraints and rid themselves of sexual reproduction? We consider the sperm-egg interaction constraint first. In animals, egg development proceeds through meiosis until metaphase II, when it maintains an arrested state by a complex activity called the cytostatic factor that inhibits anaphase onset (CSF, reviewed in Tunquist and Maller 2003). A key interaction in this pathway is the inhibition of the anaphase promoting complex (APC) by Emi2 (Schmidt et al. 2005). Recent work has shown that a calcium signal triggered by fertilization tags Emi2 for degradation, subsequently releasing the APC from inhibition and allowing the egg to complete meiosis (Lui et al. 2004; Rauh et al. 2005) (Fig. 21.3). Disruption of this pathway, perhaps as a result of reduced Emi2 regulation of the APC, could allow the calcium signal to proceed in the absence of sperm. Reduced

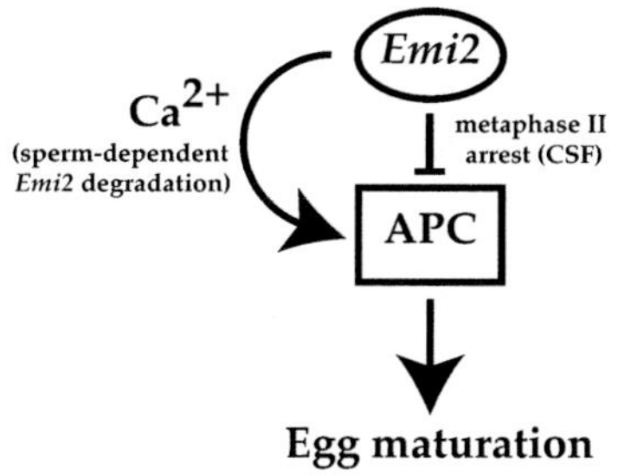

Fig. 21.3 A simplified pathway of Emi2 inhibition and sperm-dependent activation of the APC. Upon fertilization, an increase in Ca^{2+} concentration causes the degradation of Emi2, releasing inhibition of the APC, allowing the progression of egg maturation past the metaphase II arrest. Bypassing the Emi2 inhibition may be one mechanism to produce mature eggs without fertilization by sperm

regulation may occur in parthenogenetic lizards because of their hybrid genomes, in which their divergent genomic complements may interact at a sub-optimal level, including interactions between components of this pathway. While further research is required to identify the exact mechanisms allowing parthenogenetic reptiles to bypass the sperm-egg interaction, the disruption of regulatory activities, such as the CSF, by hybridization may provide a mechanism unifying the incidences of parthenogenesis across the squamate phylogeny.

The second major constraint in the transition from sex to parthenogenesis is the origin of a mechanism for the formation of eggs without a reduction in ploidy level. There are two basic ways to accomplish this: by producing eggs without the reducing affects of meiosis (e.g., by mitosis, also called apomixis), or by restitution of ploidy (see also Chapter 4). The most detailed study investigating the mechanism for ploidy maintenance in parthenogenetic lizards was done by Cuellar (1971), who documented the chromosomal count throughout meiosis in the whiptail lizard *Aspidoscelis uniparens*. Because the number of bivalents present at metaphase I generally matched the number of chromosomes in a somatic cell, Cuellar concluded that there must have been a doubling of chromosomes prior to the onset of meiosis, producing hexaploid germ cells; maturation of the oocyte via meiosis then restored the normal triploid genome (Cuellar 1971). This mechanism, called premeiotic endomitosis, is fairly common in vertebrates, and is seen in fish (see also Chapter 20; e.g., *Poeciliopsis* (Cimino 1972), medaka hybrids (Shimizu et al. 2000) and loach (Itono et al. 2006; Zhang et al. 1998)) and *Xenopus* hybrids (Kobel and DuPasquier 1975). Darevsy and Kulikova (1961) did not observe premeiotic chromosomal doubling in the rock lizard *Darevskia armeniaca*, concluding that ploidy is maintained by fusion of haploid nuclei or by defective meiosis II. Cuellar (1971) questioned these claims, arguing that the data were not sufficient to dismiss premeiotic endomitosis. Nevertheless, a meiotic mechanism without premeiotic chromosomal doubling is not without precedence, and occurs in the carp *Carassius auratus* as a result of defective meiosis I (Yamashita et al. 1993). More studies are required for other parthenogenetic lizards to determine the variations of ploidy maintenance during egg development, and whether any have adopted ameiotic mechanisms (apomixis) as an alternative to premeiotic endomitosis.

We have shown above that parthenogenesis is non-randomly distributed within the reptiles, occurring more often than expected by chance in certain squamate lineages. This fact, combined with the multiple independent origins of parthenogenesis within the squamates, provides the opportunity to look for common features of gametogenesis and development that may predispose certain lineages to evolving parthenogenesis. This is yet to be done systematically, but comparative analyses targeting the traits and pathways mentioned above for groups with and without parthenogenesis may provide further insights into how the constraints on parthenogenesis have been broken.

21.4.2 Consequences and Opportunities

The origin of parthenogenesis in lizards invariably coincides with several dramatic genomic events, which most certainly have important consequences and may provide evolutionary opportunities. First, nearly all natural parthenogenetic lizards have genomes that exhibit high heterozygosity as a result of their hybrid origin. Numerous allozyme studies have discovered extensive variation in parthenogens compared to their sexual ancestors (Table 21.1). Second, many lineages of parthenogenetic lizards are triploid, resulting in a genome with one genomic complement from one sexual ancestor, and two complements from the other sexual ancestor (Table 21.1). Third, parthenogenetic lizards often exhibit aberrant mitochondrial genome structures that contain large, tandem duplications (Moritz 1991; Moritz and Brown 1987; Zevering et al. 1991). Collectively, these genomic anomalies clearly demonstrate that there are genomic consequences of a unisexual life history in vertebrates.

Through recombination and sex, most organisms have the ability to optimize the functionality of their genomes by keeping the parts that work and ridding the parts that do not work. Reptile parthenogens not only lack this ability, but further complications may arise because their genomes are hybrid and often polyploid. At some level, the different genomic constituents must interact with each other, either in forming protein complexes, participating in enzyme pathways, or in metabolic or housekeeping functions. Depending on the divergence between the parental species, these inter-genomic interactions may not be optimal. Despite the clonal reproductive strategy of parthenogenetic lizards, the hybrid genomes are dynamic and exhibit mechanisms that can alleviate sub-optimal inter-genomic interactions. For example, Hillis et al. (1990) demonstrated that biased gene conversion consistently changes ribosomal DNA (rDNA) arrays to one of the two parental forms in parthenogenetic lineages of the Bynoe′s gecko, *Heteronotia binoei*. By biasing the genome to a composition that more closely reflects one parental form, gene conversion can maintain proper interactions. It is conceivable that this can occur at a variety of other loci as well, though allozyme studies have demonstrated that both parental copies are expressed at most loci (Moritz *et al.*, 1989b). More research is required to determine the extent of gene conversion within parthenogenetic genomes, and how taxonomically widespread it is among the unisexual squamate taxa.

Though they may be genetically identical, hybrid parthenogenetic genomes can differ from each other in their expression profiles. Epigenetic regulation provides another mechanism to preserve proper genetic interactions by silencing gene copies that may cause inter-genomic conflict. Furthermore turning on or turning off different parental gene copies by epigenetic mechanisms (e.g., DNA methylation) can cause individuals within a clonal population to vary phenotypically. Tissues can also have different expression profiles. For example, in a hybrid tetraploid cotton, Adams et al. (2003a) demonstrated that expression levels of the two parental forms at several loci varied dramatically in several tissues. Importantly, they found that such differential expression initiates almost immediately upon hybridization (Adams et al. 2003a). Whether differential gene expression was caused by the hybridization, the polyploidy, or both, remains uncertain (Otto 2003). Differential expression profiles of hybrid genomes between tissues and between individuals, combined with genome dosage effects as well as biased gene conversion, provide important mechanisms for generating potentially large amounts of phenotypic variation within a clonal population. This is an area of research that has not been explored with parthenogenetic lizards, but promises to produce exciting discoveries once evolutionary and functional genomics becomes more readily accessible to these systems.

A curious genomic anomaly correlated with the transition from sex to parthenogenesis in squamates is the presence of large, tandem duplications in their mitochondrial genomes (e.g., Moritz and Brown 1987; Moritz 1991; Zevering et al. 1991). These tandem duplications typically span several kilobases in length (up to 10.4 kb in the gecko *Heteronotia binoei*, Moritz 1991), and contain protein coding, rRNA, and tRNA genes. However, not all duplications from different parthenogenetic taxa share the same characteristics. For example, in several parthenogenetic *Aspidoscelis*, these tandem duplications abut the control region but do not contain it (Moritz and Brown 1987), while the tandem duplications in *Heteronotia* do include the control region (Moritz 1991). Differences such as these imply that the evolutionary dynamics of the duplications can be different in diverse taxa (Fujita et al. 2007). Nevertheless, the widespread presence of the anomalous mitochondrial genomes in parthenogenetic squamates suggests that the transition from sexual reproduction to unisexuality plays an important role in generating the tandem duplications (Zevering et al. 1991). Given the importance of the nuclear genome in mitochondrial DNA replication, and the hybrid nature of parthenogenetic squamate nuclear genomes, it is possible that sub-optimal inter-genomic interactions with DNA replication machinery play central roles in generating the tandem duplications (Moritz 1991; Zevering et al. 1991).

21.4.3 The Role of Hybridization

A critical question in any discussion of the evolution of parthenogenesis in vertebrates, is "what is the role of hybridization?" (Darevsky et al. 1985). There are two very different ways to view the link between hybridization and parthenogenesis; either parthenogenesis provides a means to the end of an advantageous hybrid state, or the hybrid state itself provides a means to become parthenogenetic (Darevsky et al. 1985; Cuellar 1987; Vrijenhoek 1989; Kearney 2005). Under the hybrid advantage hypothesis, one might expect that it is relatively easy to evolve parthenogenesis, but it is rarely favoured by selection. To paraphrase the late evolutionary biologist, Stephen Jay Gould, parthenogenesis is often proposed, but almost as often disposed in favour of the maintenance of sex. It is only in the context of a high-fitness hybrid genetic background that parthenogenesis is favoured, because of strong selection for that hybrid state; a state than cannot be maintained in any other way than a clonal mode of reproduction. In contrast, a causal link between parthenogenesis and hybridization would suggest that parthenogenesis is very rarely proposed because there are only a handful of pathways to this genetic system. But when avenues to parthenogenesis do open up, in this case through the destabilizing effects of hybridization, then selection strongly *imposes*.

The role of hybridization in the evolution of parthenogenesis has been very rigorously debated by students of reptile parthenogenesis (e.g., Cuellar, 1974, 1977, 1978; Wright 1978; Moritz et al. 1989a). There have been two major issues within this general debate that have not always been clearly separated: (1) are parthenogenetic reptiles hybrids? and (2) does hybridization actually cause parthenogenesis? In their review, Darevsky et al. (1985) cited the former question as the most urgent and, as we have shown, it is now largely settled; virtually all well studied cases of parthenogenetic reptiles are in fact hybrids, and were hybrids at the time of their origin. The latter point is important because, in the case of polyploid parthenogens, it is possible that parthenogenesis arose spontaneously within a lineage to produce a diploid which subsequently hybridized to form a polyploid (Cuellar 1974; Darevsky et al. 1985). Yet, the second issue regarding the causal role of hybridization remains murky. Can our current understanding of the phylogenetic distribution of parthenogenesis within the reptiles shed any light on this ongoing debate?

One causal explanation for the link between hybridization and parthenogenesis is that meiosis or gametogenesis is somehow disrupted in hybrids such that ploidy is maintained. Moritz et al. (1989a) proposed the "balance hypothesis" whereby the genetic divergence between hybridizing taxa should not be so high that the hybrids are unviable, but sufficiently great that meiosis is appropriately disrupted. The phylogenetic analyses of the hybridizing taxa within *Aspidoscelis* and *Darevskia*, reviewed earlier in this chapter, provide some support for this idea, since the parental species almost always come from different clades (Moritz et al. 1989a; Murphy et al. 2000). Whether such a process would arise from a general divergence across multiple genes involved in gametogenesis, or from a smaller number of lineage-specific factors, is still debatable (Darevsky et al. 1985; Moritz et al. 1989a; Murphy et al. 2000).

A difficulty with the hybrid dysgenesis hypothesis is that it would require hybridization to affect the conditions leading to meiosis in a very similar manner across diverse taxa and genomic states – e.g., premeitoic endomitosis in grasshoppers, lizards, fish and worms (Cuellar 1971). Moreover, the link between hybridization and asexuality in plants is also strong, extending the need for a general effect even further. While it is intuitively hard to imagine how such an indiscriminate force as hybridization could have such a general effect on a single process, one simple mechanism would be to prevent pairing of chromosomes in normal diploid germline cells because of hybrid mismatch. Then, if there is a background level of endomitosis that is normally not expressed during germline proliferation, these endomitotic cells will be all that make it through meiosis, thereby leading to parthenogenesis (Shimizu et al. 2000). A non-causal role for hybridization is also problematic since it requires that the newly formed hybrid also independently evolves parthenogenesis – two unlikely events.

A pluralistic view is also possible, whereby hybridization simultaneously induces parthenogenesis as well as directly conferring traits on the parthenogenetic lineage that aid it its establishment and maintenance. In this case, hybridization is not the only pathway to parthenogenesis, but it is the pathway most likely to succeed because of the additional bonuses of high heterozyosity within clones, potentially novel phenotypes (in addition to parthenogenesis) as well as the potential for multiple origins to generate a diverse set of clones. Many of these issues would come a long way towards being resolved if it proves possible to observe the process of hybrid origins of parthenogenesis in action within the laboratory.

21.5 Selective Pressure on Parthenogenesis in Reptiles

If there is a common theme among the geographical tendencies of parthenogenetic reptiles, it might best be characterized as a disposition toward "open" environments (with the exception of *Lepidophyma*, as noted above). That is, toward ecological vacuums left in the wake of major disturbances such as climatic extremes (cold or dry), retreating glaciers and dune fields, human alterations to landscapes, or inherently vacant habitats like islands. The simple explanation for the bias of parthenogenesis toward such open environments would be that the parthenogens are better able to colonize these habitats due to their all-female, clonal reproduction, which allows them to initiate populations from single females (Cuellar 1994).

All-female reproduction, however, is not the only possible reason for the colonizing success of asexuals. In her classic analysis of the evolution of niche width, Roughgarden (1972) showed (using a lizard as an example!) that an asexual population is better able to expand into an open and novel environment than a sexual population because it can more rapidly evolve a niche that matches the resource distribution in the new habitat. This is essentially because the blending effects of sexual reproduction tend to cause over crowding of phenotypes at the central part of a resource distribution, and under-exploitation at the tails. In contrast, a parthenogenetic lineage with sufficient clonal (or "between-phenotype") diversity can more rapidly attain the optimum phenotypic distribution to exploit the new habitat. Vrijenhoek (1979, 1984) extended these ideas to explain the coexistence of sexual and parthenogenetic lineages, coining the "Frozen Niche Variation Model" (see also Chapter 6).

Thus, we have two potential mechanisms that could explain the colonizing success of asexuals. Quoting Roughgarden (1972, p. 171):

> The relative importance of the two reasons for the good colonizing ability of asexual populations should vary according to the probability of dispersion into the new environment. If the new environment is easy to reach and asexual populations are successful there, then their potency for between-phenotype release should account for the success, while if the environment is hard to reach, then their reproductive ability in the absence of mates should account for any success. The genetic variability of the asexual population in the environment provides a conservative estimate of the number of colonists and hence could be used as an index to apportion accountability to the two explanations for the colonizing prowess of asexual populations.

Assessments of genetic, ecological, and phenotypic diversity in lizards provide information in this regard. For instance, some parthenogenetic geckos from islands show very low clonal diversity, suggesting all-female reproduction per se may provide the colonizing advantage. However, many parthenogenetic reptiles from continental areas show very high clonal diversity, particularly *Heteronotia* (Moritz et al. 1989b) and *Aspidoscelis* (Parker and Selander 1976; Parker 1979a; Parker et al. 1989) (Table 21.1), suggesting niche-width related explanations. Indeed, the ecological amplitude and geographic range of *Aspidoscelis* parthenogens appears to be positively associated with clonal diversity (Parker et al. 1989). While broad comparisons of niche variables between parthenogenetic and sexual lineages of lizards have found few major differences, phenotypic comparisons between parthenogens and sexuals demonstrate that they are indeed different in many ways. Most importantly, the few studies to compare the niches and phenotypes among clonal lineages have found differences without too much difficulty (Paulissen et al. 1988; Bolger and Case 1994; Sievert and Paulissen 1996). Perhaps explanations for divergence in geographical distributions, and patterns of coexistence, of sexual and parthenogenetic lizard lineages will be more forthcoming if the parthenogenetic lineages are more often studied at the resolution of individual clones (Parker et al. 1989).

Many have noted, however, that the patterns of geographical parthenogenesis also appear in polyploid lineages generally, both with and without asexuality (Stebbins 1971, 1984; Ehrendorfer 1980; Bierzychudek 1985; Little et al. 1997; Stenberg et al. 2003; Lundmark and Saura 2006). Kearney (2005) emphasized the potential role of the diversity generated by hybridization generally in driving these patterns, since most parthenogens and polyploids exhibiting the geographical patterns in question are also hybrids. Hybridization, parthenogenesis and polyploidy are all, in fact, means of rapid speciation and diversification (Coyne and Orr 2004). Thus the process described by Roughgarden and Vrijenhoek in the context of parthenogenesis could be seen as part of a more general tendency for mechanisms allowing rapid speciation being favoured in rapidly appearing environments; environments where the gradual, Darwinian diversification processes are unable to exploit the new ecological opportunities.

Finally, it is important to consider the forces selecting against parthenogenesis. This is perhaps the topic most relevant to the question of sex and its maintenance and yet it is one of the least studied within parthenogenetic reptiles. In theory, the most likely problems to be faced by parthenogenetic lineages are the accumulation of both parasites and deleterious mutations (West et al. 1999). As we have shown, the two most detailed comparisons of parasite loads in sexual and parthenogenetic lizards show opposite results. Experimental work on this topic will be a fruitful avenue of future research. We know nothing at all about the importance of mutation accumulation in parthenogenetic reptiles.

21.6 Conclusions

Much progress has occurred in our understanding of reptile parthenogenesis since the field was last reviewed by Darevsky et al. (1985). To what extent can our current understanding provide answers to the two questions we posed in the introduction to this review? First, when is the evolution of parthenogenesis constrained in reptiles? The phylogenetic distribution of parthenogenesis across the squamates suggests that the constraints to parthenogenesis in reptiles are particularly relaxed in some groups such as the teiid and lacertid lizards, but they are tight in other groups such as the colubrid snakes and the scincid and iguanid lizards. The origins of many parthenogenetic reptiles have now been resolved through the application of modern molecular techniques, and all but one of the natural cases have now been demonstrated to be of hybrid origin. Recent facultative cases of parthenogenesis in captive reptiles clearly show that reptiles have the capacity to evolve parthenogenesis by non-hybrid means, yet non-hybrid pathways to parthenogenesis apparently do not succeed. This pattern remains as one of the most mysterious in the ecology and evolution of parthenogenetic vertebrates generally. We now need detailed comparative analyses of the mechanistic nature of the constraints on parthenogenesis in reptiles, and why they are sometimes relaxed.

When is asexuality in reptiles favoured by natural selection? Consideration of the geographical distribution and ecological tendencies of parthenogenetic reptiles suggests that they do best in "open" environments. Yet despite the clear biases in the distributions of parthenogenetic reptiles relative to their sexual progenitors, some degree of coexistence is often observed. Moreover, niche analyses often find surprisingly high levels of overlap between parthenogenetic and sexual lineages. The processes allowing the coexistence of sexual and parthenogenetic reptiles represent another significant mystery that requires resolution. We suggest that detailed comparisons of the niches of individual clones relative to their sexual parental lineages may provide insights into this question of coexistence, and more generally into the question of the selective pressures favouring parthenogenesis in nature. Experimental studies of the dynamics of host-parasite interactions in sexual and parthenogenetic reptiles, as well as the importance of mutation accumulation, are also urgently needed to properly answer these questions.

Acknowledgments We thank Craig Moritz for lively discussions regarding many of the topics addressed in this review. J.W. Sites provided insights that improved the quality and content of this manuscript.

References

Adams KL, Cronn R, Percifield R, Wendel J (2003a) Genes duplicated by polyploidy show unequal contributions to the transcriptome and organ-specific reciprocal silencing. Proc Natl Acad Sci 100: 4649–4654

Adams M, Foster R, Hutchinson MN, Hutchinson RG, Donnellan SC (2003b) The Australian scincid lizard *Menetia greyii*: a new instance of widespread vertebrate parthenogenesis. Evolution 57: 2619–2627

Arnold MJ (1997) Natural Hybridization and Evolution. Oxford Univ Press, New York

Avila LJ, Martori RA (1991) A unisexual species of *Teius* Merrem 1820 (Sauria Teiidae) from central Argentina. Trop Zool 4: 193–201

Bell G (1982) The Masterpiece of Nature. University of California Press, Berkeley, California

Bezy RL (1989) Morphological differentiation in unisexual and bisexual Xantusiid lizards of the genus *Lepidophyma* in Central America. Herpetol Monogr 3: 61–80

Bezy RL, Carmarillo JL (2002) Systematics of xantusiid lizards of the genus *Lepidophyma*. Contrib Sci 493: 1–41

Bezy RL, Sites JWJ (1987) A preliminary study of allozyme evolution in the lizard family Xantusiidae. Herpetologica 43: 280–292

Bierzychudek P (1985) Patterns in plant parthenogenesis. Experentia 41: 1255–1264

Boissinot S, Ineich I, Thaler L, Guillaume C-P (1997) Hybrid origin and clonal diversity in the parthenogentic gecko, *Lepidodactylus lugubris* in French Polynesia. J Herpetol 31: 295–298

Bolger DT, Case TJ (1994) Divergent ecology of sympatric clones of the asexual gecko, *Lepidodactylus lugubris*. Oecologia 100: 397–405

Bowker RG, Johnson OW (1980) Thermoregulatory precision in three species of whiptail lizards (Lacertilia: Teiidae). Physiol Zool 53: 176–185

Cabrera MR, Monguillot JC (2007) Reptilia, Squamata, Teiidae, *Teius suquiensis*: new evidence of recent expansion of this parthenogenetic lizard? Check List 3: 180–184

Case TJ (1990) Patterns of coexistence in sexual and asexual species of *Cnemidophorus* lizards. Oecologia 83: 220–227

Cimino MC (1972) Meiosis in triploid all-female fish (*Poeciliopsis*, Poecilidae). Science 175: 1484–1486

Cole CJ (1975) Evolution of parthenogenetic species of reptile. In: Reinboth R (ed) Intersexuality in the animal kingdom. Springer-Verlag, New York, pp. 340–355

Cole CJ, Dessauer HC (1993) Unisexual and bisexual whiptail lizards of the *Cnemidophorus lemniscatus* complex (Squamata: Teiidae) of the Guiana Region, South America, with descriptions of a new species. Am Mus Novit 3018: 1–30

Cole CJ, Dessauer HC, Markezich AL (1993) Missing link found: the second ancestor of *Gymnophthalmus underwoodi* (Squamata: Teiidae), a South American unisexual lizard of hybrid origin. Am Mus Novit 3055: 1–13

Cole CJ, Dessauer HC, Townsend CR, Arnold MG (1990) Unisexual lizards of the genus *Gymnophthalmus*(Reptilia: Teiidae) in the neotropics: genetics, orgin, and systematics. Am Mus Novit 2994: 2–29

Cole CJ, Dessauer HC, Townsend CR, Arnold MG (1995) *Kentropyx borckiana* (Squamata: Teiidae): A Unisexual lizard of hybrid origin in the Guiana region, South America. Am Mus Novit 2671: 1–5

Congdon JD, Vitt LJ, Hadley NF (1978) Parental investment: comparative reproductive energetics in bisexual and unisexual lizards, genus *Cnemidophorus*. Am Nat 112: 509–521

Coyne JA, Orr HA (2004) Speciation. Sinauer Ass, Sunderland

Cuellar O (1971) Reproduction and mechanisms of meiotic restitution in the parthenogenetic lizard *Cnemidophorus uniparens*. J Morphol 133: 139–165

Cuellar O (1974) On the origin of parthenogenesis in vertebrates: the cytogenetic factors. Am Nat 108: 625–648

Cuellar O (1977) Animal parthenogenesis: A new evolutionary-ecological model is needed. Science 197: 837–843

Cuellar O (1978) Parthenogenetic lizards. Science 201:1155

Cuellar O (1979) On the ecology of coexistence in parthenogenetic and bisexual lizards of the genus *Cnemidophorus*. Am Zool 19: 773–786

Cuellar O (1987) The evolution of parthenogenesis: a historical perspective. In: Moens PB (ed) Meiosis. Acad. Press, Orlando, pp. 43–104

Cuellar O (1993) Further observations on competition and natural history of coexisting parthenogenetic and bisexual whiptail lizards. In: Wright JW, Vitt LJ (eds) Biology of Whiptail Lizards (Genus *Cnemidophorus*). Oklahoma Mus Nat Hist, Norman Oklahoma, pp. 345–370

Cuellar O (1994) Biogeography of parthenogenetic animals. Biogeographica 70: 1–13

Cuellar O, Kluge AG (1972) Natural parthenogenesis in the gekkonid lizard *Lepidodactylus lugubris*. J Genet 61: 14–26

Cullum A (1997) Comparisons of physiological performance in sexual and asexual whiptail lizards (genus *Cnemidophorus*): implications for the role of heterozygosity. Am Nat 150: 24–47

Cullum A (2000) Phenotypic variability of physiological traits in populations of sexual and asexual whiptail lizards (genus *Cnemidophorus*). Evol Ecol Res 2: 841–855

Darevsky IS (1962) On the origin and biological role of natural parthenogenesis in a polymorphic group of Caucasian rock lizards, *Lacerta saxicola* Eversmann. Zool Zh USSR 41: 397–408

Darevsky IS, Kulikova VN (1961) Natürliche Parthenogenese in der polymorphen Gruppe der Kaukasischen Felseidechse (*Lacerta saxicola* Eversmann). Zool Jahrb Syst 89: 119–176

Darevsky IS, Kupriyanova LA (1993) Two new all-female lizard species of genus *Leiolepis* CUVIER 1829 from Thailand and Vietnam (Squamata: Sauria: Uromastycinae). Herpetozoa 6: 3–20

Darevsky IS, Kupriyanova LA, Uzzell TM (1985) Parthenogenesis in reptiles. In: Gans C, Billett F (eds) Biology of the Reptilia. John Wiley and Sons Inc, New York, pp. 412–526

Delmotte F, Sabater-Munoz B, Prunier-Leterme N, Latorre A, Sunnucks P, Rispe C, Simon J-C (2003) Phylogenetic evidence for hybrid origins of asexual lineages in an aphid species. Evolution 57: 1291–1303

Densmore LDI, Moritz C, Wright JW, Brown WM (1989a) Mitochondrial DNA analyses and the origin and relative age of parthenogenetic lizards (Genus: *Cnemidophorus*). IV. Nine *sexlineatus*-group unisexuals. Evolution 43: 969–983

Densmore LDI, Wright JW, Brown WM (1989b) Mitochondrial DNA analyses and the origin and relative age of parthenogenetic lizards (Genus: *Cnemidophorus*). II. *C. neomexicanus* and the *C. tesselatus* complex. Evolution 43: 943–957

Dessauer HC, Cole CJ (1989) Diversity between and within nominal forms of unisexual Teiid lizards. In: Dawley RM, Bogart JP (eds) Evolution and Ecology of Unisexual Vertebrates. University of the State of New York, Albany, New York, pp. 49–71

Dickinson TA (1999) Species concepts in agamic complexes. In: van Raamsdonk LWD, den Nijs JCM (eds) Plant Evolution in Man-Made Habitats. Hugo de Vries Laboratory, Amsterdam, pp. 319–339

Dubach J, Sajewicz A, Pawley R (1997) Parthenogenesis in the Arafuran file snake (*Acrochordus arafurae*). Herpetol Nat Hist 5: 11–18

Ehrendorfer F (1980) Polyploidy and distribution. In: Lewis WH (eds) Polyploidy, biological relevance. Plenum Press, New York and London, pp. 45–60

Fu J, MacCulloch RD, Murphy DJ, Darevsky IS (2000) Clonal variation in the Caucasian rock lizard *Lacerta armeniaca* and its origin. Amphib-reptil 21: 83–89

Fujita MK, Boore JL, Moritz C (2007) Multiple origins and rapid evolution of duplicated mitochondrial genes in parthenogenetic geckos (*Heteronotia binoei*; Squamat, Gekkonidae). Mol Biol Evol 24: 2775–2786

Gamble T, Bauer AM, Greenbaum E, Jackman TR (2008) Evidence of Gondwanan vicariance in an ancient clade of the geckos. J Biogeogr 35: 88–104

Glesener RR, Tilman D (1978) Sexuality and the components of environmental uncertainty: clues from geographic parthenogenesis in terrestrial animals. Am Nat 112: 659–673

Groot TVM, Bruins E, Breeuwer JAJ (2003) Molecular genetic evidence for parthenogenesis in the Burmese python, *Python molurus bivittatus*. Heredity 90: 130–135

Hall WP (1970) Three probable cases of parthenogenesis in lizards (Agamidae, Chamaeleontidae, Gekkonidae). Experientia 26: 1271–1273

Hamilton WD, Axelrod R, Tanese R (1990) Sexual reproduction as an adaptation to resist parasites (a review). Proc Natl Acad Sci 87: 3566–3573

Hanley KA, Fisher RN, Case TJ (1995) Lower mite infestations in an asexual gecko compared with its sexual ancestors. Evolution 49: 418–426

Hernández-Gallegos O, Méndez F, Villagrán-Santa Cruz M, Cuellar O (2003) Genetic homogeneity between populations of *Aspidoscelis rodecki*, a parthenogenetic lizard from the Yucatan Peninsula. J Herpetol 37: 527–532

Hillis DM, Moritz C, Porter CA, Baker RJ (1990) Evidence for biased gene conversion in concerted evolution of ribosomal DNA. Science 251: 308–310

Hörandl E (2006) The complex causality of geographical parthenogenesis. New Phytol 171: 525–538

Hore TA, Rapkins RW, Graves JAM (2007) Construction and evolution of imprinted loci in mammals. Trends Genet 23: 440–448

Itono M, Morishima K, Fujimoto T, Bando E, Yamaha E, Arai K (2006) Premeiotic endomitosis produces diploid eggs in the natural clone loach, *Misgurnus anguillicaudatus* (Teleostei: Cobitidae). J Exp Zool 305A: 513–523

Jaenike J (1978) An hypothesis to account for the maintenance of sex within populations. Evol Theor 3: 191–194

Kearney M (2003) Why is sex so unpopular in the Australian desert? Trends Ecol Evol 18: 605–607

Kearney M (2005) Hybridization, glaciation and geographical parthenogenesis. Trends Ecol Evol 29: 495–502

Kearney M, Moussalli A, Strasburg J, Lindenmayer D, Moritz C (2003) Geographic parthenogenesis in the Australian arid zone: I. A climatic anaysis of *Heteronotia binoei* complex (Gekkonidae). Evol Ecol Res 5: 953–976

Kearney M, Shine R (2004a) Developmental success, stability and plasticity in closely related parthenogenetic and sexual lizards (*Heteronotia, Gekkonidae)*. Evolution 58: 1560–1572

Kearney M, Shine R (2004b) Morphological and physiological correlates of hybrid parthenogenesis. Am Nat 164: 803–813

Kearney M, Shine R (2005) Lower fecundity in parthenogenetic geckos than in sexual relatives in the Australian arid zone. J Evol Biol 18: 609–618

Kearney M, Wahl R, Autumn K (2005) Increased capacity for sustained locomotion at low temperature in parthenogenetic geckos of hybrid origin. Physiol Biochem Zool 78: 316–324

Kizirian DA, Cole CJ (1999) Origin of the unisexual lizard *Gymnophthalmus underwoodi* (Gymnophthalmidae) inferred from mitochondrial DNA nucleotide sequences. Mol Phyl Evol 11: 394–400

Kobel HL, DuPasquier L (1975) Production of large clones of histocompatible, fully identical clawed toads (*Xenopus*). Immunogenetics 2: 87–91

Lenk P, Eidenmueller B, Staudter H, Wicker R, Wink M (2005) A parthenogenetic *Varanus*. Amphib-reptil 26: 507–514

Leuck BE (1985) Comparative social behavior of bisexual and unisexual whiptail lizars (*Cnemidophorus*). J Herpetol 19: 492–506

Little TJ, Demelo R, Taylor DJ, Herbet PDN (1997) Genetic characterization of an actic zooplankter: insights into geographic polyploidy. Proc R Soc Lond B 264: 1363–1370

Lui S, Sun Y, Zhang C, Luo K, Lui Y (2004) Production of gynogenetic progeny from allotetraploid hybrids red crucian carp X common carp. Aquaculture 236: 193–200

Lundmark M, Saura A (2006) Asexuality alone does not explain the success of clonal forms in insects with geographical parthenogenesis. Hereditas 113: 23–32

Lynch M (1984) Destabilizing hybridization, general-purpose genotypes and geographic parthenogenesis. Quart Rev Biol 59: 257–290

Magnusson WE (1979) Production of an embryo by an *Acrochordus javanicus* isolated for years. Copeia 1878: 744–745

Malysheva DN, Darevsky IS, Tokarskaya ON, Petrosyan V, Martirosyan IA, Ryskov AP (2006) Analysis of genetic variation in unisexual and bisexual lizard species of the genus *Leiolepis* from Southeast Asia. Genetika 42: 581–586

Manriquez NL, Villagran-Santa Cruz M, Mendez-De La Cruz F (2000) Origin and evolution of the parthenogenetic lizards, *Cnemidophorus maslini* and *C. cozumela*. J Herpetol 34: 634–637

Maslin TP (1971) Parthenogenesis in reptiles. Am Zool 11: 361–380

McAllister CT, Cordes JE, Walker JM (2003) Helminth parasites of unisexual and bisexual whiptail lizards (teiidae) in North America. X. The western marbled whiptail (*Cnemidophorus tigris marmoratus*). Texas J Sci 55: 307–314

Morison IM, Ramsay JP, Spencer HG (2005) A census of mammalian imprinting. Trends Genet 21: 457–465

Moritz C (1983) Parthenogenesis in the endemic Australian lizard *Heteronotia binoei* (Gekkonidae). Science 220: 735–736

Moritz C (1987) Parthenogenesis in the tropical gekkonid lizards, *Nactus arnouxii* (Sauria: Gekkonidae). Evolution 41: 1252–1266

Moritz C (1991) Evolutionary dynamics of mitochondrial DNA duplications in parthenogenetic geckos, *Heteronotia binoei*. Genetics 129: 221–230

Moritz C (1993) The origin and evolution of parthenogenesis in the *Heteronotia binoei* complex: synthesis. Genetica 90: 269–280

Moritz C, Brown WM (1987) Tandem duplication in animal mitochondrial DNAs: variation in incidence and gene content among lizards. Proc Natl Acad Sci 84: 7183–7187

Moritz C, Brown WM, Densmore LD, Wright JW, Vyas D, Donnellan S, Adams M, Baverstock P (1989a) Genetic diversity and the dynamics of hybrid parthenogenesis in *Cnemidophorus* (Teiidae) and *Heteronotia* (Gekkonidae). In: Dawley RM, Bogart JP (eds) Evolution and Ecology of Unisexual Vertebrates. University of the State of New York, Albany, New York, pp. 87–112

Moritz C, Case TJ, Bolger DT, Donnellan S (1993) Genetic diversity and the history of pacific island house geckos (*Hemidactylus* and *Lepidodactylus*). Biol Linn Soc 48: 113–133

Moritz C, Donnellan S, Adams M, Baverstock PR (1989b) The origin and evolution of parthenogenesis in *Heteronotia binoei* (Gekkonidae): extensive genotypic diversity among parthenogens. Evolution 43: 994–1003

Moritz C, McCallum H, Donnellan S, Roberts JD (1991) Parasite loads in parthenogenetic and sexual lizards (*Heteronotia binoei*): support for the Red Queen hypothesis. Proc R Soc Lond B 244: 145–149

Moritz C, Uzzell TM, Spolsky C, Hotz H, Darevsky IS, Kupriyanova L, Danielyan F (1992a) The maternal ancestry and approximate age of parthenogenetic species of Caucasian rock lizards (*Lacerta*: Lacertidae). Genetica 87: 53–62

Moritz C, Wright JW, Brown WM (1989c) Mitochondrial DNA analyses and the origin and relative age of parthenogenetic Cnemidophorus: *C. velox* and *C. exsanguis*. Evolution 43: 958–968

Moritz C, Wright JW, Brown WM (1989d) Mitochondrial DNA analyses and the origin and relative age of parthenogenetic lizards (Genus: *Cnemidophorus*). III. *C. velox* and *C. exsanguis*. Evolution 43: 958–968

Moritz C, Wright JW, Brown WM (1992b) Mitochondrial DNA analyses and the origin and relative age of parthenogenetic *Cnemidophorus*: Phylogenetic constraints on hybrid origins. Evolution 46: 184–192

Moritz C, Wright JW, Singh RS, Brown WM (1992c) Mitochondrial DNA analyses and the origin and relative age of parthenogenetic *Cnemidophorus*. V. The *cozumela* species group. Herpetologica 48: 417–424

Murphy RW, Darevsky IS, MacCulloch RD, Jinzhong F, Kupriyanova LA, Upton DE, Danielyan F (1997) Old age, multiple formations or genetic plasticity? Clonal diversity in the uniparental Caucasian rock lizard, *Lacerta dahli*. Genetica 101: 125–130

Murphy RW, Fu J, MacCulloch RD, Darevsky IS, Kupriyanova LA (2000) A fine line between sex and unisexuality: the phylogenetic constraints on parthenogenesis in lacertid lizards. Zool J Linn Soc 130: 527–549

Olsen MW (1966) Segregation and replication of chromosomes in turkey parthenogens. Nature 212: 435–436

Ota H, Hikida TM (1989) A new triploid*Hemidactylus* (Gekkonidae: Sauria) from Taiwan, with comments on morphological and karyological variation in the *H. garnotii-vietnamensis* complex. J Herpetol 23: 50–60

Ota H, Hikida TM, Lue K-Y (1989) Polyclony in a triploid gecko, *Hemidactylus stejnegeri*, from Taiwan, with notes on its bearing on the chromosomal diversity of the *H. garnotii-vietnamensis* complex (Sauria: Gekkonidae). Genetica 79: 183–189

Ota H, Hikida TM, Matsui M, Chan-Ard T, Nabhitabhata J (1996) Discovery of a diploid population of the *Hemidactylus garnotii-vietnamensis* complex (Reptilia: Gekkonidae). Genetica 97: 81–85

Otto SP (2003) In polyploids, one plus one does not equal two. Trends Ecol Evol 18: 431–433

Parker ED Jr (1979a) Phenotypic consequences of parthenogenesis in *Cnemidophorus* lizards. I. Variability in parthenogenetic and sexual populations. Evolution 33: 1150–1166

Parker ED Jr (1979b) Phenotypic consequences of parthenogenesis in *Cnemidophorus* lizards. II. Similarity of *C. tessleatus* to its sexual parental species. Evolution 33: 1167–1179

Parker ED Jr., Selander RK (1976) The organisation of genetic diversity in the parthenogenetic lizard *Cnemidophorus tesselatus*. Genetics 84: 791–805

Parker ED Jr, Walker JM, Paulissen MA (1989) Clonal diversity in *Cnemidophorus*: ecological and morphological consequences. In: Dawley RM, Bogart JP (eds) Evolution and ecology of unisexual vertebrates. Univ. of the State of New York, Albany, New York, pp. 72–86

Paulissen MA (1988) Ontogenetic and seasonal shifts in microhabitat use by the lizard *Cnemidophorus sexlineatus*. Copeia 1988: 1021–1029

Paulissen MA (2001) Ecology and behavior of lizards of the parthenogenetic *Cnemidophorus laredoensis* complex and their gonochoristic relative *Cnemidophorus gularis*: Implications for coexistence. J Herpetol 35: 282–292

Paulissen MA, Walker JM, Cordes JE (1988) Ecology of syntopic clones of the parthenogenetic whiptail lizard, *Cnemidophorus* 'laredoensis'. J Herpetol 22: 331–342

Paulissen MA, Walker JM, Cordes JE (1992) Can parthenogenetic *Cnemidophorus laredoensis* (Teiidae) coexist with its bisexual congeners? J Herpetol 26: 153–158

Pellegrino KCM, Rodrigues MT, Yonenaga-Yassuda Y (2003) Triploid karyotype of *Leposoma percarinatum* (Squamata, Gymnophthalmidae) J Herpetol 37: 197–199

Pianka ER (1973) The structure of lizard communities. Annu Rev Ecol Syst 4: 53–74

Platonov ES (2005) Genomic imprinting and the problem of parthenogenesis in mammals. Russ J Dev Biol 36: 247–255

Pough FH, Andrews RM, Cadle JE, Savitzky AH, Wells KD (2004) Herpetology, 3rd ed. Pearson Prentice Hall, Upper Saddle River, NJ

Price AH, Lapointe JL, Atmar JW (1993) The ecology and evolutionary implications of competition and parthenogenesis in *Cnemidophorus*. In: Wright JW, Vitt LJ (eds) Biology of whiptail lizards (Genus *Cnemidophorus*). Oklahoma Mus. Nat. Hist, Norman Oklahoma, pp. 371–410

Radtkey RR, Becker B, Miller RD, Riblet R, Case TJ (1996) Variation and evolution of Class I Mhc in sexual and parthenogenetic geckos. Proc R Soc Lond B 263: 1023–1032

Radtkey RR, Donnellan SC, Fisher RN, Moritz C, Hanley KA, Case TJ (1995) When species collide: the origin and spread of an asexual species of gecko. Proc R Soc Lond B 259: 145–152

Rauh NR, Schmidt A, Bormann J, Nigg EA, Mayer TU (2005) Calcium triggers exit from meiosis II by targeting the APC/C inhibitor Xerp1 for degradation. Nature 437: 1048–1052

Ray N, Adams JM (2001) A GIS-based vegetation map of the world at the last glacial maximum (25,000–15,000 BP). Internet Archaeol 11 (http://intarch.ac.uk/journal/issue11/rayadams_toc.html)

Reeder TW, Cole CJ, Dessauer HC (2002) Phylogenetic relationships of whiptail lizards of the Genus *Cnemidophorus* (Squamata: Teiidae): A test of monophyly, reevaluation of karyotypic evolution, and review of hybrid origins. Am Mus Novit 3365: 1–61

Roughgarden J (1972) The evolution of niche width. Am Nat 106: 683–718

Roughgarden J (1995) Anolis Lizards of the Caribbean: Ecology, Evolution, and Plate Tectonics. Oxford Univ Press, London

Scalka P, Vozenilek P (1986) Case of parthenogenesis in water snakes, *Neroidia sipedon*. Fauna Bohemiae 11: 81–82

Schall JJ (1978) Reproductive strategies in sympatric whiptail lizards (*Cnemidophorus*): Two parthenogenetic and three bisexual species. Copeia 1978: 108–116

Schall JJ (1993) Community ecology of *Cnemidophorus* lizards in southwestern Texas: a test of the weed hypothesis. In: Wright JW, Vitt LJ (eds) Biology of whiptail lizards (Genus *Cnemidophorus*). Oklahoma Mus. Nat. Hist, Norman Oklahoma, pp. 319–343

Schmidt A, Duncan PI, Rauh NR Sauer G, Fry AM, Nigg EA, Mayer TU (2005) *Xenopus* polo-like kinase Plx1 regulates Xerp1, a novel inhibitor of APC/C activity. Genes Dev 19: 502–513

Schmitz A, Vences M, Weitkus S, Ziegler T, Böhme W (2001) Recent maternal divergence of the parthenogenetic lizard *Leiolepis guentherpetersi* from *L. guttata*: molecular evidence (Reptilia: Squamata: Agamidae). Zool Abh Mus Tierkde 51: 355–360

Schuett GW, Fernandez PJ, Gergits WF, Casna NJ, Chizar D, Smith HM, Mitton JB, Mackessy SP, Odum RA, Demlong. MJ (1997) Production of offspring in the absence of males: evidence for facultative parthenogenesis in bisexual snakes. Herpetol Nat Hist 5: 1–10

Shimizu Y, Shibata N, Sakaizumi M, Yamashita M (2000) Production of diploid eggs through premeiotic endomitosis in the hybrid medaka between *Oryzias latipes* and *O. curvinotus*. Zool Sci 17: 951–958

Sievert LM, Paulissen MA (1996) Temperature selection and thermoregulatory precision of bisexual and parthenogenetic *Cnemidophorus* lizards from southern Texas, U.S.A. J Therm Biol 21: 15–20

Sinclair EA, Scholl R, Bezy RL, Crandall KA, Sites JW (2006) Isolation and characterization of di- and tetranucleotide microsatellite loci in the yellow-spotted night lizard *Lepidophyma flavimaculatum* (Squamata: Xantusiidae). Mol Ecol Notes 6: 233–236

Sites JW, Peccinini-Seale DM, Moritz C, Wright JW, Brown GP (1990) The evolutionary history of parthenogenetic *Cnemidophorus lemniscatus* (Sauria, Teiidae). I. Evidence for a hybrid origin. Evolution 44: 906–921

Slowinski JB, Lawson R (2002) Snake phylogeny: evidence from nuclear and mitochondrial genes. Mol Phyl Evol 24: 194–202

Sokal RR, Rohlf FJ (1995) Biometry, 3rd ed. Freeman and Company, San Francisco

Stebbins GL (1971) Chromosome Evolution in Higher Plants. Edward Arnold, London

Stebbins GL (1984) Polyploidy and the distribution of arctic-alpine flora: new evidence and a new approach. Bot Helv 94: 1–13

Stenberg P, Lundmark M, Knutelski S, Saura A (2003) Evolution of clonality and polyploidy in a weevil system. Mol Biol Evol 20: 1626–1632

Strasburg J, Kearney M, Moritz C, Templeton AR (2007) Integrating phylogeography with distribution modeling: multiple pleistocene range expansions in a parthenogenetic gecko. PLoS One 2: e760

Suomalainen E, Saura A, Lokki J (1987) Cytology and Evolution in Parthenogenesis. CRC Press, Florida

Takenouchi Y (1976) A study of polyploidy in races of Japanese weevils (Coleoptera: *Curculionidae*). Genetica 46: 327–334

Taylor AC, Cole CJ, Dessauer HC, Parker ED, Jr. (2003) Congurent patterns of genetic and morphological variation in the parthenogenetic lizard *Aspidoscelis tesselata* (Squamata: Teiidae) and the origins of colour pattern classes and genotypic clones in eastern New Mexico. Am Mus Novit 3424: 1–40

Townsend TM, Larson A, Louis E, Macey JR (2004) Molecular phylogenetics of squamata: the position of snakes, amphisbaenians, and dibamids, and the root of the squamate tree. Syst Biol 53: 735–757

Tunquist BJ, Maller JL (2003) Under arrest: cytostatic factor (CSF)-mediated metaphase arrest in vertebrate eggs. Genes Dev 17: 683–710

Uzzell TM (1970) Meiotic mechanisms of naturally occurring unisexual vertebrates. Am Nat 104: 433–445

Vandel A (1928) La parthénogénèse geographique. Contribution à l'étude biologique et cytologique de la parthénogénèse naturelle. Bull Biol France Belg 62: 164–281

Vanzolini DE (1978) Parthenogenetic lizards. Science 201: 1152

Vanzolini PE (1970) Unisexual *Cnemidophorus lemniscatus* in the Amazonas valley: a preliminary note (Sauria: Teiidae). Papéis Avulsos Mus Zool São Paulo 23: 63–49

Volobouev V, Pasteur G, Ineich I, B. D (1993) Chromosomal evidence for a hybrid origin of diploid parthenogenetic females from the unisexual-bisexual *Lepidodactylus lugubris* complex (Reptilia, Gekkonidae). Cytogenet Cell Genet 63: 194–199

Vrijenhoek RC (1979) Factors affecting clonal diversity and coexistence. Am Zool 19: 787–797

Vrijenhoek RC (1984) Ecological differentiation among clones: the frozen niche variation model. In: Wöhrmann K, Loeschcke V (eds) Population Biology and Evolution. Springer-Verlag, Heidelberg, pp. 217–231

Vrijenhoek RC (1989) Genetic and ecological constraints on the origins and establishment of unisexual vertebrates. In: Dawley RM, Bogart JP (eds) Evolution and Ecology of Unisexual Vertebrates. University of the State of New York, Albany, New York, pp. 24–31

Vrijenhoek RC, Dawley RM, Cole CJ, Bogart JP (1989) A list of the known unisexual vertebrates. In: Dawley RM, Bogart JP (eds) Evolution and Ecology of Unisexual Vertebrates. University of the State of New York, Albany, New York, pp. 19–23

Watts PC, Buley KR, Sanderson S, Boardman W, Ciofi C, Gibson R (2006) Parthenogenesis in Komodo dragons. Nature 444: 1021–1022

West SA, Lively CM, Read AF (1999) A pluralist approach to sex and recombination. J Evol Biol 12: 1003–1012

White MJD, Contreras N (1979) Cytogenetics of the parthenogenetic grasshopper *Warramaba* (formerly *Moraba*) *virgo* and its bisexual relatives. V. Interaction of *W. virgo* and a bisexual species in geographic contact. Evolution 33: 85–94

Wright JW (1978) Parthenogenetic lizards. Science 201: 1152–1154

Wright JW, Lowe CH (1968) Weeds, polyploids, parthenogenesis, and the geographical and ecological distribution of all-female species of *Cnemidophorus*. Copeia 1968: 128–138

Wynn AH, Cole CJ, Gardner AL (1987) Apparent triploidy in the unisexual Brahminy blind snake, *Ramphotyphlops braminus*. Am Mus Novit 2868: 1–7

Yamashita M, Jiang J, Onozato H, Nakanishi T, Nagahama Y (1993) A tripolar spindle formed at meiosis I assures the retention of the original ploidy in the gynogenetic triploid crucia carp, Ginbuna *Carassius auratus langsdorfii*. Dev Growth Diff 35: 631–636

Zevering CE, Moritz C, Heideman A, Sturm RA (1991) Parallel origins of duplications and the formation of pseudogenes in mitochondrial DNA from parthenogenetic lizards (*Heteronotia binoei*; Gekkonidae). J Mol Evol 33: 431–441

Zhang Q, Arai K, Yamashita M (1998) Cytogenetic mechanisms for triploid and haploid egg formation in the triploid loach *Misgurnus anguillicaudatus*. J Exp Zool 281: 608–619

Chapter 22
An Apomixis-Gene's View on Dandelions

Peter Van Dijk, Hans de Jong, Kitty Vijverberg and Arjen Biere

Abstract In asexual organisms, the clone constitutes a level above the individual. Most dandelions (*Taraxacum officinale s.l.*) reproduce asexually through apomixis, asexual reproduction through seeds. A clone can be seen as a superorganism that is born, that growths, degenerates and eventually dies. Apomixis in dandelions is controlled by a few dominant genes, the so called apomixis-genes. This implies that there should be three hierarchical levels in a field of dandelions: 1. the individual plant, 2. the clone and 3. the apomixis gene. Using co-dominant genetic markers that are linked to a dominant apomixis gene, we provide evidence that this hierarchical structure indeed exists in apomictic dandelion populations. The apomixis gene view implies that whereas individual clones may go extinct due to deleterious mutation accumulation or the lack of adaptive potential, apomixis genes can prevail much longer periods of evolutionary time in a succession of clones. We provide evidence that an apomixis-gene in *Taraxacum* is not transmitted to diploid offspring, which could explain the absence of apomixis in diploid dandelions. Haploid non-transmission may be caused by a mutation load that is linked to the apomixis genes as a consequence of the deep asexual reproduction history of these genes residing in many clones in the past.

22.1 Introduction

In the last chapter of his book *The Extended Phenotype* (1989), Richard Dawkins writes about different ways to view a field of dandelions (*Taraxacum officinale s.l.*): "A certain type of ecologist may gain illumination from comparing a field full of dandelions with a single tree. But for other purposes it is important. . .to see the single dandelion ramet [plant] as analogous to the tree". Dawkins referred to an article written by Daniel Janzen in 1977 in *The American Naturalist* with the intriguing title: "What are dandelions and aphids?" (see also Chapter 25). Most dandelions

---

P. Van Dijk (✉)
Keygene N.V., Agro Business Park 90, 6708 PW Wageningen, The Netherlands
e-mail: peter.van-dijk@keygene.com

I. Schön et al. (eds.), *Lost Sex*, DOI 10.1007/978-90-481-2770-2_22,

reproduce by apomixis, i.e., asexual reproduction by seed. Janzen considered an apomictic dandelion clone as a kind of superorganism and argued that the individual dandelion plants, the clone mates, were comparable to the leaves of a tree. Clone mates and tree leaves are both genetically identical units, only differing in whether or not they are connected. Janzen argued that the whole dandelion clone and not the individual apomictic plant is the evolutionary individual, just like we consider the tree and not the individual leaf as the evolutionary individual. In this chapter, we will argue that there is even a higher level of biological hierarchy above the clone (see Chapter 9), namely that of the genes that encode for apomixis (shortly the apomixis genes). Through the pollen, the male sexual function of hermaphroditic flowering plants, these apomixis genes can enter the sexual gene pool and generate new clones. This way the apomixis genes can survive clones, which may have a limited evolutionary life span, and persist for much longer periods of evolutionary time than the individual clones. Here, we will provide new evidence that this apomixis gene level is a reality and that the apomixis genes can explain why apomixis is restricted to polyploids in *Taraxacum* and why diploid apomicts have not been found in nature. For an introduction to the biology of apomixis in plants, we refer to Chapter 3.

22.2 The Genetics of Apomixis in *Taraxacum*

Dandelions are hermaphrodictic plants, which produce both egg cells and pollen grains (Van Dijk 2003). Most dandelions in Northern Europe and North America are triploid ($3x = 24$) and produce seeds through apomixis: parthenogenetic development of an unreduced egg cell into an embryo and autonomous development of the central cell into the endosperm. Although most apomicts produce pollen, fertilization is not required for the production of seeds. Hence, from the point of view of the individual plant, pollen has no function.

In Central Europe, diploid sexuals ($2x = 16$) are found, co-occurring with triploid apomicts (higher ploidy levels are rare). Diploid sexuals depend on insect pollination and cross fertilization for seed set. Mixed populations of sexual and apomictic dandelions are rather common in Central Europe. Sexual dandelions reach their Northern distribution limit in Europe in The Netherlands; at higher latitudes only polyploid apomicts are found (Van Dijk 2003).

Under experimental conditions, it is rather easy to demonstrate apomixis. Sexuals do not set seeds in the absence of pollinators, whereas apomicts produce abundant seeds. When the top of a flower bud is decapitated, apomicts will still produce many viable seeds, despite the fact that the styles and the anthers of the florets are removed. Of course, sexuals fail to produce seeds after decapitation. In triploids, high seed set is an indicator for apomixis, because sexual triploids would even after cross pollination hardly produce any seeds due to aneuploidy caused by an unbalanced meiosis. Triploidy is therefore a good indicator for apomixis in the field. Triploidy can be determined in the field microscopically by the irregularly sized pollen grains or in the laboratory by DNA flow cytometry of field collected leaf samples. Apomictic reproduction can of course also be proven by genetically identical heterozygous offspring, but this is more time consuming.

Apomicts that produce functional pollen can act as pollen donors in crosses with diploid sexuals, enabling the study of the genetics of apomixis. Two unlinked dominant apomixis loci have been identified, controlling: 1. The avoidance of meiotic reduction (*DIPLOSPOROUS-Dip*) and 2. Parthenogenesis of the embryo (*PARTHENOGENESIS-Par*). In *Dip*-plants female meiosis I is restitutional, resulting in unreduced $2n$ egg cells. Van Dijk and Bakx-Schotman (2004) investigated the transmission genetics of the *Dip*-allele using two linked co-dominant Simple Sequence Repeat (SSR) markers (in coupling phase, both at 7 cM from the *Dip* allele). This led to the conclusion that the *Dip*-allele is dominant and that triploid apomicts can be genotypically represented as *Ddd*. The expression of the *Dip*-allele is sex-specific and does not affect pollen meiosis. Egg cells are therefore always *Ddd*, whereas pollen can be *Dd*, *dd*, *d* or *D* (discarding aneuploid pollen, which is inviable in *Taraxacum*). Genotypically triploid apomicts are *Ppp* for the *Par*-locus. There are indications that a third major locus is involved in apomixis, which may be related to the autonomous endosperm development, but this needs further investigation. A single dominant allele is sufficient in triploids to suppress the sexual reproductive pathway. Interestingly, this implies that loss of an apomixis allele, for example by non-disjunction, could lead to a reversal from apomixis to sexuality. This has indeed been observed by Sørensen and Gudjonsson (1946) and Sørensen (1958) who showed that rare 3x-1 offspring (aberrant *tenuis*) of a triploid apomictic lineage could be pollinated by other triploid apomicts, resulting in some diploid sexually reproducing offspring. *Tenuis* aberrants had lost one of the three Nucleolus Organizer Region chromosomes and the frequency of occurrence was estimated as 5×10^{-4} (3 out of 6000 clone mates). Since the *Dip*-locus was shown to be located on the NOR-chromosome (Van Dijk and Bakx-Schotman 2004), the modern interpretation is that a NOR chromosome carrying the D-allele was lost in the *tenuis* plants. According to Sørensen (1958), not all *tenuis* plants reverted to sexuality, which is consistent with the *Ddd* genotypic constitution. Most research so far has focused on the *Dip*-locus and therefore, the remainder of this chapter will deal with this locus.

22.3 Clones as Superorganisms

As mentioned before, mixed sexual and apomictic populations are not uncommon in Central Europe. Through crosses of diploid sexuals with polyploid apomicts (mainly triploid) new clones can be formed. In terms of a superorganism this is the "birth" of a new clone. Verduijn et al. (2004) found that in mixed populations with diploid sexuals and triploid apomicts in the Netherlands, ~2% of the natural progeny of diploid sexual plants were triploids. It was estimated that in a rather small peripheral sexual population of 2500 reproducing sexual plants, each year some 10,000 triploid seeds would be produced by diploids. If three apomixis genes would segregate independently, ~ 29% of these seeds could reproduce by apomixs. In reality most *neo*apomicts are partly seed sterile, probably depending on the action of additional modifier genes (Van Dijk 2007). Since established apomictic clones have full apomictic seed set, there will be strong fertility selection between neoclones. Even neoclones with high fitness run a high risk of extinction by demographic stochasticity shortly after formation when the numbers of clone mates are still low. Nevertheless, it is clear that many new clones are generated in mixed populations. Some of these clones will be well adapted and will proliferate, or grow in terms of the clone as a superorganism.

It is widely accepted that clones are prone to early extinction because of the lack of adaptability and the accumulation of slightly deleterious mutations. According to the Red Queen Hypothesis (Bell 1982; see also Chapter 7), rapidly evolving pathogens could cause the elimination of clones. The frequency distributions of clones in apomictic *Taraxacum* populations are typically L-shaped, with a few common clones and many rare ones (Van der Hulst et al. 2000; Van Dijk 2003). Frequency-dependent rust fungus infection has been demonstrated in apomictic populations of *Condrilla juncea*, a close relative of *Taraxacum* (Chaboudez and Burdon 1995).

Theoretically, asexual lineages will accumulate slightly deleterious mutations due to stochastic loss of the least loaded class in small populations (Muller's Ratchet, Muller 1964; see also Chapter 5), genetic hitchhiking with beneficial mutations (Rice 1987) and inefficient purging in large populations (Kondrashov 1982). Frequency-dependent selection may accelerate the accumulation of slightly deleterious mutations (West et al. 1999). Active transposable elements can cause slightly deleterious mutations and theoretical studies indicate that it is likely that asexual lineages become extinct because of transposon accumulation (Dolgin and Charlesworth 2006). Docking et al. (2006) analyzed transposable elements in a number of Canadian apomictic lineages of *Taraxacum*. Sequence comparisons indicated purifying selection acting on these transposable elements since substitution rates at non-synonymous sites were much lower than at synonymous sites. Transposons, such as Ty1-*copia*, Ty2-*gypsy* and LINE-like retroelements, were probably still functional in the *Taraxacum* apomicts and could reduce the fitness of clone mates. The decline of the number of clone mates and, eventually, the extinction of the apomictic clone causes the degeneration and the death of the superorganism (Fig. 22.1).

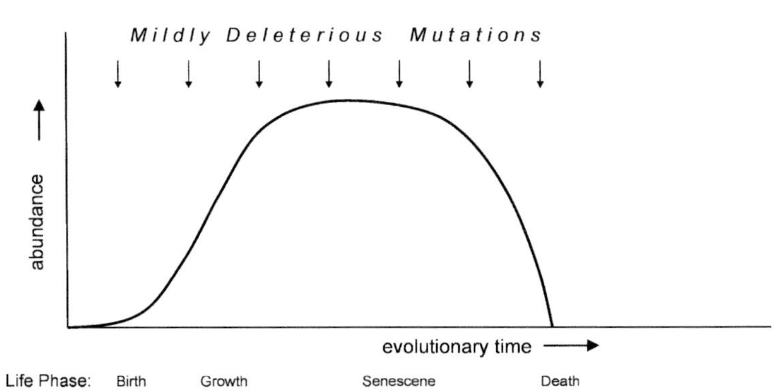

Fig. 22.1 The life phases of a clone viewed as a superorganism. Abundance refers to the number of clone mates. During its life time a clone will accumulate slightly deleterious mutations

22.4 The Superstructure of Asexual Populations

Taraxacum officinale apomicts are considered to be obligate apomicts. In other *Taraxacum* species, rare facultative apomixis has been reported, but we have never observed residual sexuality in apomictic *T. officinale*, despite searches for it. Notwithstanding the lack of sexuality, even in regions far north of the sexual distribution area dandelions exhibit a great deal of morphological variation. For a long time, morphologically distinct clones have been described as microspecies by certain taxonomists. Many different microspecies can be found growing together in grasslands. Recently, molecular studies have discovered even more genetic variation. Van der Hulst et al. (2000, 2003) investigated the clonal structure of a triploid dandelion population in a park in Viborg, Denmark (Jutland), more than 700 km north of the nearest known sexual diploid population (in the Netherlands). This study revealed an interesting hierarchical population structure and will be described here in some more detail. Allozymes, cpDNA, AFLPs and SSR markers were used to analyse a sample of 65 plants. AFLP fingerprinting generated on average about 100 fragments per plant. Thirty three different AFLP fingerprints were found, of which 14 occurred more than one time and 19 only once. The redundancy of these 14 fingerprints clearly indicated clonal reproduction by apomixis in this population. Two AFLP fingerprints were identical to AFLP fingerprints found in a Dutch apomictic population, more than 600 km to the South. These two clones had thus likely spread over a distance of 600 km or more. This finding suggests that it should be possible that clones are generated in the area of sympatry between sexual and apomictic dandelions and spread over hundreds of kilometres and over evolutionary time frames. Dandelions are effectively dispersed by wind and clones may have a long lifespan. This provides an explanation for the high clonal diversity found in the asexual regions. Alternatively the diversity of clones may have a local origin possibly caused by a rare local reversal to sex caused by loss of one or more apomixis genes, as discussed above.

Analysis of the Viborg sample for three SSR-loci resulted in 32 groups, closely resembling the AFLP groupings. Figure 22.2 shows schematically the variation at the three SSR-loci, after removal of the clone mate redundancy. Some somatic mutations within clones were detected at the *MSTA72* locus. However, more interesting was the genetic variation at *MSTA78*. All clones shared one allele, 164 bp, at this locus. The locus however, was not fixed; there was a lot of variation for the other two alleles. *MSTA78* is one of the SSR-loci that was shown to be linked to the *Dip*-locus by Van Dijk and Bakx-Schotman (2004). The other two SSR loci, *MSTA72* and *MSTA64*, are not linked to *Dip* (nor to *Par*). The fixation of the *MSTA78-164* allele can thus be explained by strong linkage disequilibrium between this allele and an allele controlling diplospory, an essential element of apomixis. Van Dijk and Bakx-Schotman (2004) originally estimated the size of the linked-allele of *MSTA78* to be 160 bp, but later, single gel analysis showed that this allele is of the same size as the fixed 164 bp allele in Viborg.

A more extensive survey furthermore indicated that the *MSTA78-164* allele was almost completely fixed in apomictic dandelion populations in the Netherlands,

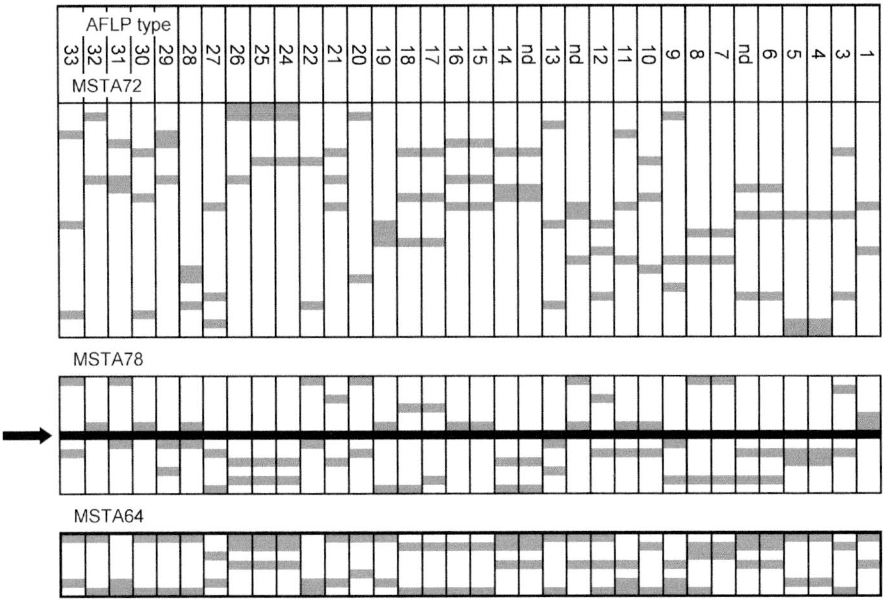

Fig. 22.2 A virtual SSR gels of the different apomictic clones in the Viborg population (Van der Hulst et al. 2003), based on real data. At the top, the different AFLP fingerprint types are indicated. All clones share the 164 bp allele at locus MSTA78 (between the arrows, black fragments). This allele is most likely in linkage disequilibrium with the *Dip* allele, which controls diplospory, an element of apomixis in dandelions

Northern Germany and Denmark. In contrast, in France, the 164 bp allele was not fixed in apomicts and was also found in diploid sexuals (Van Dijk, Van Culemborg, Vijverberg and Van der Hulst, unpublished results).

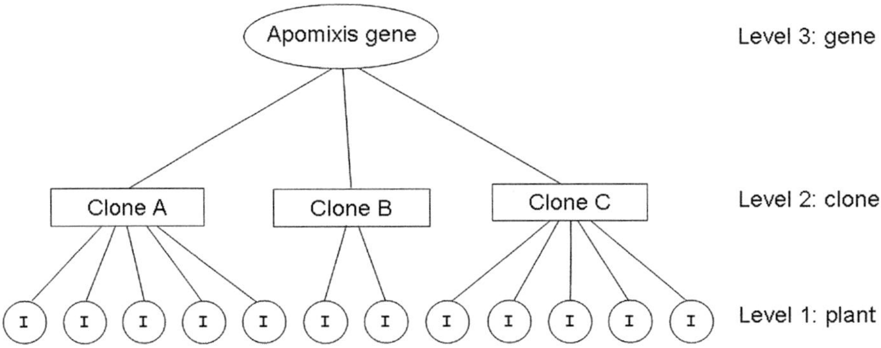

Fig. 22.3 Apomixis at three hierarchical levels. The individual plants form the lowest level (*level 1*). Plants can be clone mates sharing the same genotype, belonging to the same clone (*level 2*). Different clones may share the same apomixis gene(s) (*level 3*) because of common ancestry. *Clones A* and *C* are larger clones than *clone B*, because they have more clone mates

The structure of the Viborg population clearly shows that there are three hierarchical levels within apomictic dandelions (Fig. 22.3). From the lowest to the highest level these are: 1. the individual plant or clone mate, 2. the clone, or the evolutionary individual, and 3. the apomixis gene. In the remainder of the chapter we will have a closer look at the evolutionary implications of the highest level, the apomixis genes.

22.5 Apomixis Genes are Older than Clones

In mixed sexual-apomict populations, a so called sex-asex cycle will operate (Fig. 22.4). Through the pollen, the male sexual function, genes can cycle between the sexual gene pool and the asexual clone pool. Most of the pollen formed by triploid apomicts will be chromosomically unbalanced, but a low percentage will possess the full haploid or diploid genome. Through haploid pollen genes from the clones can flow into the sexual gene pool and through diploid pollen new clones will be formed. As we will show below, apomixis genes cannot be transmitted through haploid pollen and cannot enter the diploid gene pool. For the moment the most important aspect of the cycle is that new clones are being formed and that apomixis genes can survive within successive generations of clones. While non-apomixis genes can cycle between the sexual and the asexual complement, apomixis genes can only cycle within the asexual complement.

In the sexual gene pool, slightly deleterious mutations will be purged. Hence, the haploid genome of sexual egg cells will be cleansed from deleterious mutations. If such a haploid egg cell becomes fertilized by a diploid pollen grain carrying a dominant apomixis gene, a new apomictic clone will be formed, which will have a

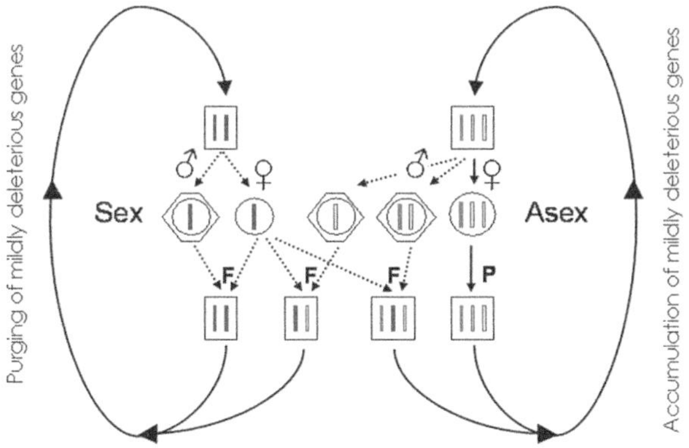

Fig. 22.4 The sex-asex cycle in dandelions. Sexual diploid dandelions and hermaphroditic apomictic triploid dandelions are linked by functional pollen produced by the apomicts. Only one chromosome set is shown. Pollen grains are shown with thick cell walls, egg cells with thin cell walls. F = fertilization; P = parthenogenesis. The arrows indicate directions of possible gene flow. The chromosomes from the sexual gene pool are in dark blue, the chromosome carrying a dominant apomixis gene is in red. The non-apomixis gene chromosomes in the triploid apomicts are indicated in light blue

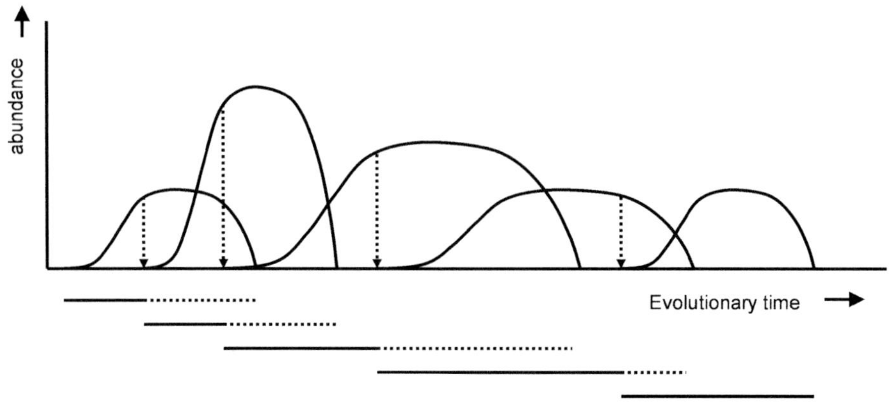

Fig. 22.5 This figure illustrates the idea that apomixis genes can predate apomictic clones. Abundance means the number of clone mates. Each time a new clone is formed (*dashed arrows*), the apomixis genes "jump" into a new clone. Apomixis genes can survive much longer in evolutionary time than individual clones and will have a far deeper asexual reproduction history. Consequently, the chromosomal regions surrounding the apomixis genes will have strong signatures of a long asexual reproduction history, especially when recombination during pollen meiosis in these regions is suppressed

reduced mutation load compared to its father clone. Apomixis genes become associated with a new genetic background that is partly cleansed. If a clone degenerates by an increasing mutation load, the apomixis gene may escape extinction by backcrossing to sexuals. The idea that apomixis genes can be much older than the genes that they reside in, is illustrated in Fig. 22.5.

Looking backwards in time, the three *Dip*-alleles in any triploid dandelion clone will differ in their asexual reproduction history. One recessive allele, *d1*, descends from the sexual mother and has only experienced asexual reproduction since the formation of the present clone. The other recessive allele, *d2*, descends from the triploid apomictic father clone and there is an equal chance (0.5) that it descends from the apomictic grandfather or the sexual grandmother. However, the chance that the *d2*-allele has resided in more than six previous clone generations is less than five percent ($(0.5)^5 = 0.031$). In contrast, the *D*-allele has cycled through the asexual part of the sex–apomixis cycle in numerous successive clones since its origin by mutation.

22.6 A Mutation Load Linked to Apomixis-Genes

During pollen meiosis, the *D* chromosomes can recombine and exchange flanking regions with the *d1* and *d2* chromosomes which were recently derived from the sexual gene pool. Through this process, large parts of the *D*-linked mutation load can be removed; however, uncoupling will be infrequent for regions close to the apomixis loci. The regions close to the apomixis loci are therefore expected to carry a higher

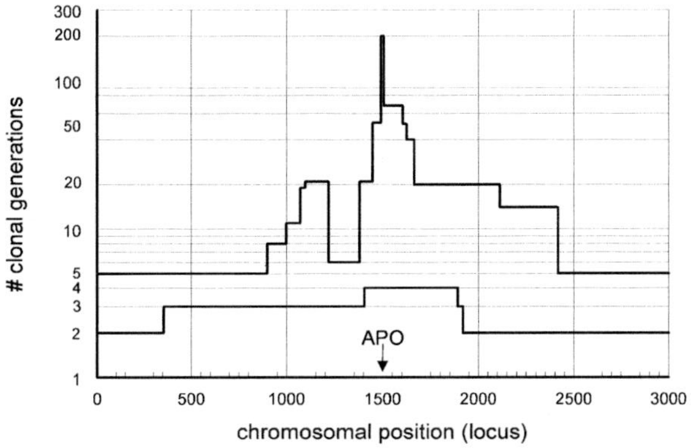

Fig. 22.6 The depth of the asexual reproduction history along a chromosome carrying a dominant single dose apomixis gene, after 200 clonal generations cycling in the sex-asex cycle. The three simulated chromosomes contain 3000 loci with one apomixis locus in the middle. One chromosome carries the dominant apomixis allele *D*, the other two the recessive sexual reproduction allele *d*. Bivalent chromosome pairing occurs at random (autotriploidy) and one randomly positioned cross-over occurs on each flank of the apomixis locus

genetic load than regions elsewhere in the genome. We have simulated this for a triploid with a single chromosome of 3000 loci carrying an apomixis locus in the middle (Van Dijk, unpublished results). Crossovers were randomly positioned on each flank of the apomixis locus. Figure 22.6 shows the results of a typical simulation after 200 clonal generations as the depth of the asexual reproduction history -the number of clonal generations that the chromosomal region resided in a clone- across the three chromosomes. The *d1*-chromosome of the sexual mother has only resided in the present clone. Parts of the *d2*-chromosome have resided in three ancestral clones. Almost one third of the *D*-chromosome has resided in more than 19 clonal generations. A small region surrounding the *D*-allele has resided in clones since the origin of the D-allele itself. The mutation load across the chromosome would have a similar pattern, thus increasing towards the apomixis locus. A chromosome walk towards the apomixis gene will be a walk into deeper asexual history and potentially offers a way to study the long term effects of asexual reproduction. This will also apply to other apomixis loci. The genome of an apomictic clone is thus a mosaic of regions differing in asexual reproduction history and asexual mutation load. Neoclones originate from a single cell, a zygote. Therefore, the new clone will be fixed for all the ancestral deleterious mutations in the fertilizing diploid pollen grain. Single cell descent is a powerful mechanism for slightly deleterious mutation fixation in the asexual lineage. This is illustrated in Fig. 22.7.

Over evolutionary time, recombination between apomixis chromosomal regions and their non-apomixis homologs will further decrease. Since the chromosomal regions surrounding apomixis genes do not undergo exchanges with their homologs in the sexual gene pool, they will diverge from these homologs. Sequence divergence will further decrease recombination (Opperman et al. 2004; Li et al. 2005), which will further increase sequence divergence, as a self-reinforcing process.

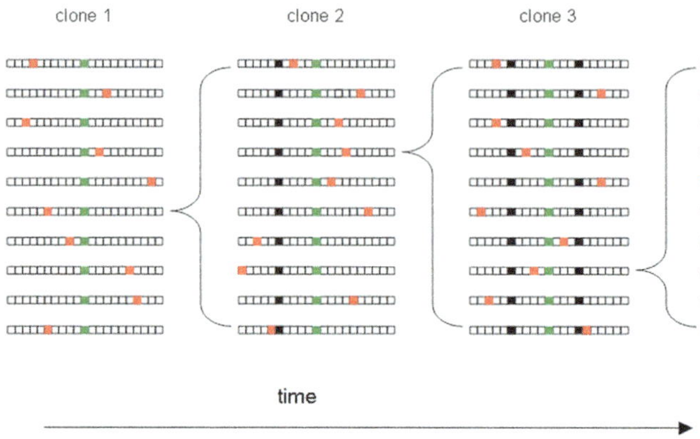

Fig. 22.7 The sampling effect in pollen grains giving rise to new apomictic clones. Each successive clone is represented by 10 clone mates. Only the chromosome carrying the dominant apomixis allele (*green square*) and 21 other loci (*white square*) is shown. For clearity, recombination is not included. Slightly deleterious mutations that newly occurred during the life of a clone are indicated in red. The mutations inherited from the previous clonal generation are indicated in black. The result is a fixation of slightly deleterious mutations in the vicinity of the apomixis locus

Moreover, apomixis chromosomal regions will accumulate chromosomal rearrangements (inversions, translocations), because there is no meiotic sterility sieve against chromosomal rearrangements in apomictic clones, as female meiosis is circumvented in seed formation. Mutants in apomictic clones with chromosomal rearrangements will not have decreased in seed fertility. Chromosomal rearrangements will suppress recombination, again leading to increased sequence divergence. Thus, the chromosomal region with a deep asexual reproduction history surrounding the apomixis gene will expand as will the mutational footprint of asexual reproduction.

22.7 Recombination and Structure of Apomixis Chromosomal Regions

Studies with genetic markers have indicated that recombination around apomixis loci is suppressed in many species, as predicted above. Strong suppression of recombination in an Apomixis Specific Chromosome Region (ASCR) was first reported in the apomictic grass *Pennisetum squamulatum* (Ozias-Akins et al. 1998). A large number of genetic markers were co-segregating with the apomixis trait. Subsequently, suppression of recombination has been described in many other apomict species (reviewed in Ozias-Akins and Van Dijk 2007). Interestingly, especially in grass species, apomixis is inherited as a single dominant factor, whereas in the Asteraceae family, apomeiosis and parthenogenesis are inherited as separate genetic factors, e.g., *Taraxacum* (Tas and Van Dijk 1999), *Hieracium* (Catanach et al. 2006) and *Erigeron* (Noyes et al. 2007). This suggests that the single apomixis locus in some apomicts may in fact contain several tightly linked genes with different function. Given that apomeiosis and parthenogenesis are different processes occurring in different cells, control by different genes at different points in time is to be expected.

Analysis of the structure of ASCRs is also consistent with the idea of a deep asexual reproduction history. Fluorescent in situ hybridisation (FISH) has revealed hemizygosity of the ASCR in tetraploid apomictic *Pennisetum squamulatum* (Goel et al. 2003). The size of the non-recombining ASCR in *Pennisetum squamulatum* has been estimated to be about 50 Mb. Recently, the first sequences of ASCRs have been published. Shot gun sequencing of Bacterial Artificial Chromosomes of the ASCR in *Pennisetum* indicated many duplications and insertions of transposable elements and only few genes (Conner et al. 2008). Similar results were obtained in another apomictic grass species, *Paspalum simplex* (Calderini et al. 2006). Transposon insertions, small deletions and point mutations caused a loss of coding capacity in the ASCR compared to homologous rice chromosomal regions. Chromosomal FISH showed that in this apomict the ASCR in *Paspalum* is also hemizygous.

In *Taraxacum*, there is no indication for suppression of recombination around the *Dip*-locus (Vijverberg et al. 2004). In contrast, recombination around the *Par*-locus is strongly suppressed (Van Dijk, unpublished results). Although there is no evidence of suppression of recombination around the *DIP* locus, below we will provide evidence for segregation distortion of *Dip*-alleles in haploid pollen grains.

22.8 Why are Apomicts Not Diploid?

Above, we referred to the study by Van Dijk and Bakx-Schotman (2004) who investigated the inheritance op the *Dip*-locus in a cross between a sexual diploid and a diplosporous tetraploid plant. This *Dd1d2d3* tetraploid had balanced pollen meiosis and produced highly fertile diploid pollen (*Dd1*, *Dd2*, *Dd3*, *d1d2*, *d1d3* and *d2d3*). A *Dd1d2* triploid apomictic pollen donor would produce both haploid and diploid balanced pollen (*D*, *d1*, *d2* and *Dd1*, *Dd2* and *d1d2*, respectively), albeit at a low rate, because most pollen grains would be aneuploid. A recessive mutation load coupled to the *D*-allele, in a triploid apomict would result in a segregation bias against the *D*-allele in haploid pollen, but not in diploid pollen where the recessive lethals would be largely masked by the wild type alleles linked to *d1* or *d2*. To test this hypothesis of segregation distortion in haploid pollen grains, we made a diploid sexual X triploid apomict cross (TJX3-20 X A68; see Appendix). It was previously noticed that although diploid sexuals have a sporophytic self incompatibility system, high selfing rates occur in diploid sexual X triploid apomict crosses. Since no selfing occurs in diploid sexual X diploid sexual crosses, it seems that pollen from triploids has a mentor pollen effect on selfing (Tas and Van Dijk 1999). To avoid the confusing effects of selfing, a male-sterile diploid sexual plant TJX3-20 was used. Although pollen fertility of the triploid clone A68 was very low, by crossing more than 62 diploid inflorescences 192 hybrid offspring could be raised in total. Although pollen grains of triploid A68 were highly irregularly sized - indicating many aneuploid pollen grains-, DNA flow cytometry revealed that only viable euploid offspring plants were produced: 96 diploids, 95 triploids and 1 tetraploid. This suggests that only pollen grains with balanced genomes are capable of fertilization, or that only genomically balanced zygotes are viable. Since the diploid mother plant only produces haploid eggs, the ploidy level of the fertilizing pollen grains can be deduced. In the case of diploid offspring, the pollen grain from A68 was haploid, in triploid offspring the pollen grain from A68 was diploid and in the single tetraploid plant, the pollen grain from A68 must have been triploid (unreduced). Carrying the full paternal apomictic genome of A68, the tetraploid F1 hybrid was as expected apomictic. The 92 triploid hybrids segregated for apomixis, with approximately one third being apomictic (30.4%) and two thirds non-apomictic (69.6%). This was as expected, because apomixis in dandelions is controlled by several unlinked loci, which can recombine during pollen meiosis. Most remarkable, however, none of the 97 diploid hybrids reproduced apomictically.

To investigate whether the lack of apomixis in the diploids was due to non-expression or non-transmission of the apomixis genes, the transmission of two SSR markers that were linked to the *Dip*-allele was analyzed. As mentioned before, *MSTA78* and *MSTA53* had been mapped at 7 cM distance on the same side of the *Dip*-locus in a diploid x tetraploid cross (Van Dijk and Bakx-Schotman 2004 and Van Dijk, unpublished results; Fig. 22.8A). High allelic diversity between the crossed plants allowed the unambiguous deduction of all egg cell and pollen grain genotypes. Approximately two third of the triploids carried the *MSTA78-164* allele and the *MSTA53-202* allele. All apomicts carried these two alleles, however, not all

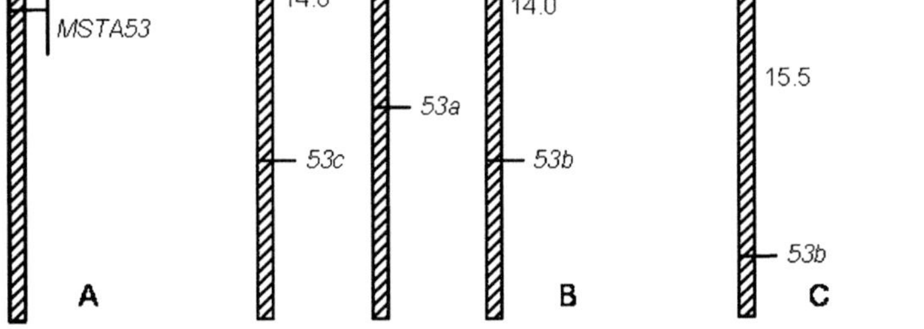

Fig. 22.8 Three genetic maps of the *Dip* chromosomal region in *Taraxacum officinale* based on: A. a *dd* X *Dddd* cross (Van Dijk and Bakx-Schotman 2004), B. Diploid pollen grains in a *dd* X *Ddd* cross and C. haploid pollen grains in the same *dd* X *Ddd* cross, assuming a recessive pollen lethal factor completely linked to the D-allele. The *MSTA78a* allele is the 164 bp allele, the *MST53b* allele is the 202 bp allele. For further explanation see the appendix

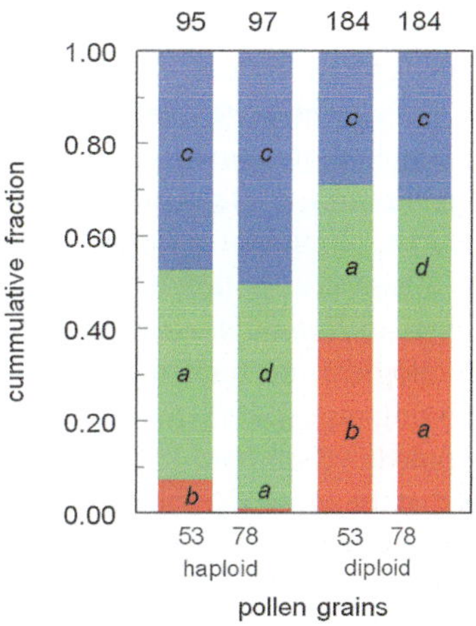

Fig. 22.9 Segregation of the *MSTA78* and the *MSTA53* SSR loci in a *dd* X *Ddd* cross, for haploid and diploid pollen grains. The number of alleles analyzed is indicated above the bars. The *MSTA78a* allele in the graph is the 164 bp allele, the *MST53b* allele is the 202 bp allele

carriers of these alleles were apomictic, which can be explained by the lack of other elements of apomixis, due to segregation.

Figure 22.9 shows the segregation of the SSR alleles in the haploid and the diploid pollen grains of the apomictic pollen donor A68. In the diploid pollen grains, the segregation ratios of the three paternal alleles were not significantly different from 1:1:1 Mendelian equality (*MSTA78*: χ^2 =1.96; d.f. = 2; $P = 0.375$; *MSTA53*: $\chi^2 = 2.36$; d.f. = 2; $P = 0.307$). In haploid pollen however, the segregation ratios were highly distorted (*MSTA78*: $\chi^2 = 45.61$; d. f. = 2; $P < 0.001$; *MSTA53*: $\chi^2 = 28.89$; d.f. = 2; $P < 0.001$). Haploid segregation distortion was caused by the two SSR-alleles that were linked to the *D*-allele. The *MSTA78-164* allele was found only in one out of 97 diploid plants, the *MSTA53-202* only in seven out of 95 diploid plants.

Figure 22.8B shows the genetic map of the *Dip*-chromosomal region, as constructed from the diploid pollen grains. The *MSTA78-164* and the *MSTA53-202* alleles were linked to the *D*-allele.

The allele-specific segregation distortions in haploid pollen grains cannot be explained by preferential chromosome pairing during pollen meiosis, because the frequencies of di-allelic genotypes in diploid pollen did not differ significantly from random assortment (*MSTA78*: $\chi^2 = 3.35$; d. f. = 2; $P = 0.187$; *MSTA53*: $\chi^2 = 4.19$; d. f. = 2; $P = 0.123$). A plausible explanation is that the *D*-allele was not transmitted via haploid pollen because of recessive pollen lethality due to a *D*-specific mutation load, as predicted above. A pseudotest-cross indicated that the single diploid F_1 plant carrying the paternal *164 bp*-allele for *MSTA78* lacked the dominant *D*-allele. A crossover probably uncoupled this marker from the *D*-allele and its linked mutational load. The seven haploid pollen grains transmitting the *MSTA53b* allele can also be explained this way. The genetic map, based on the haploid pollen grains and assuming recessive pollen lethality is shown in Fig. 22.8C. There were no segregation distortions between the other alleles at the SSR loci, suggesting that *d2* had not accumulated a significant number of linked deleterious mutations in the previous clonal generation (s) compared to *d1*, which has only resided in clone A68.

An alternative explanation is that the *D*-specific segregation distortion is not due to a linked mutation load, but that the *D*-allele itself has, besides the female meiosis I restitution, a pleiotropic recessive pollen lethal effect. Nogler (1984) suggested that the dominant apospory (*A*) factor in apomictic *Ranunculus auricomus* acted as a recessive pollen lethal. Both the *Dip*-gene in *Taraxacum officinale* and the *A*-factor in *Ranunculus auricomus* avoid meiotic reduction, but via entirely different cytological mechanisms, namely meiotic restitution (diplospory) and meiotic by-passing (apospory), respectively (see Chapter 3). It is unlikely that dissimilar genes would have similar pleiotropic effects. In contrast, what these genes have in common is that they reside permanently in asexual lineages and we consider it therefore more likely that recessive lethality is a consequence of a linked mutation load.

Apomixis has been described in $\sim$ 400 plant taxa (Bicknell and Koltunow 2004). Nearly all investigated gametophytic apomicts (apomixis *s.s.*) are polyploids and reports of natural diploid apomicts are rare. The well-known case of diploid apomixis in *Potentilla argentea* was shown to be selfing by the use of genetic markers (Holm et al. 1997). At present, *Boechera holboellii* is the only robust case of a natural diploid apomict (Naumova et al. 2001; see also Chapter 23). Non-transmission of apomixis genes in the haploid state caused by a linked recessive mutation load may be a general explanation for the strong association between gametophytic apomixis and polyploidy. A similar hypothesis was developed by Richards (1996), however without explaining how the load would become specifically linked to the dominant apomixis allele, nor providing empirical evidence for its existence.

22.9 Conclusions

If asexual lineages go extinct because of the accumulation of deleterious mutations or a lack of adaptability, the apomixis genes may escape extinction via crossing with sexuals. Apomixis genes become incorporated into new clones, with a genetic background that is partly drawn from the sexual gene pool, that is freed from deleterious mutations and that is potentially adaptive.

There are at least two conditions for this system to persist. First, it is essential that the sexuals do not go extinct. This is a problem, because theoretically, a dominant gene for apomixis will rapidly go to fixation in a sexual outcrossing population (Marshall and Brown 1981; Van Dijk 2007). The haploid non-transmission of the *D*-gene described here protects diploid sexual dandelion populations from being taken over by apomixis. Since the haploid non-transmission of apomixis genes due to a linked genetic load takes time to build up, the diploid sexual gene pool must have initially been isolated from the pollen producing apomicts, e.g., by habitat differentiation or geographic isolation. Another possibility would be a reversal to sexuality by loss of apomixis genes after the linked genetic load was established.

A second condition is that the clone must continue to produce at least some functional pollen; otherwise, the apomixis genes will become trapped in a clone and will go extinct with it. Because pollen is not needed for seed production, pollen function will degenerate over evolutionary time. There may even be selection against pollen production in clones if resources for pollen production can be reallocated to increase fitness (e.g., more or bigger seeds). This creates a conflict of interest between apomixis genes and apomictic clones. Indeed, dandelion clones that lack pollen are not uncommon. It would be interesting to theoretically explore the parameter values under which sexuals and pollen producing apomicts can coexist over long periods of time.

The apomixis gene transmission system depends on the male function in hermaphroditic apomictic organisms. Hermaphroditic apomixis is not restricted to plants; certain animals like earth worms or flatworms also reproduce by hermaphroditic parthenogenesis. Moreover, the apomixis/parthenogenesis gene transmission system is not limited to asexual hermaphrodites but also applies to asexual organisms, in which males can be induced by environmental triggers and in which there is a genetic basis for asexual reproduction, as has been demonstrated in aphids (Delmotte et al. 2001; see also Chapter 25) and in *Daphnia* (Innes and Hebert 1988; Paland et al. 2005; Lynch et al. 2008; see also Chapter 15). The three-level superstructure of asexual populations may therefore be more common.

John Maynard Smith wrote in his book *The Evolution of Sex*: "Asexual clones of the dandelion *Taraxacum officinale* continue to produce functionless yellow petals and many produce functionless pollen. It is difficult to suggest any explanation of these facts, other than that these clones may be relatively recent in origin, and that evolutionary adaptation in asexual populations is slow, so that the maladapted features are retained" (p 41, second edition, 1978). The apomixis-gene view provides an additional explanation, namely that without a pollen transmission system for apomixis-genes, apomictic dandelions would likely have become extinct long ago. Clearly, biologists may gain illumination from an apomixis-gene's view on dandelions.

Acknowledgments We are grateful to Ron Van der Hulst for donating the male sterile diploid plant TJX3-20 and to Marcel Van Culemborg and Tanja Bakx-Schotman for technical assistance. We thank Rolf Hoekstra, Bill Rice and Deborah and Brian Charlesworth for useful suggestions on earlier drafts of the manuscript.

Appendix

A pollen-sterile sexual diploid dandelion (TJX3-20) was crossed with a pollen-fertile apomictic triploid (A68) (Fig. 22.10). TJX3-20 originated from Langres (France). A68 originated from Heteren (The Netherlands). Viable seed set in the TJX3-20 X A68 cross was low (on average 2.1%), reflecting the high frequency of inviable aneuploid pollen grains produced by the unbalanced pollen meiosis of triploid A68. Sixty-two crossed capitula contained only 192 viable seeds in total. Ninety-six F_1 plants were diploid (50%), 95 triploid (49.5%) and one (0.5%) was

Fig. 22.10 Spontaneous seed set of bagged flowers (excluding cross pollen) in the diploid sexual male sterile TJX3-20 seed parent (*left*) and the triploid apomictic pollen parent A68 (*right*). Large seed heads indicate apomictic seed set, small seed heads indicate the absence of spontaneous seed development

tetraploid. These plants were the products of the fertilization of a haploid egg cell by a haploid, diploid and triploid pollen grain, respectively.

To induce flowering, eight week old F_1 plants were vernalized for 9 weeks in a cold room at 4°C. One hundred eighty two F_1 plants (94.7%) were tested for the ability to form apomictic seeds (six seedlings died and four adult plants did not flower). In order to prevent contamination by cross-pollination, the flowers were covered with small paper bags before opening. All F_1 plants were male sterile, like TJX3-20, hence seed set due to selfing can be excluded. The development of a large seed head is an indication for apomictic seed set (see Fig. 22.10). To determine the degree of apomictic seed set, for each F1 plant two batches of 50 randomly chosen seeds were germinated and the number of seedlings germinating was counted. Most of the apomicticly reproducing triploid F_1 plants had a high penetrance of apomixis (> 90% seed set), some however had a much lower penetrance.

The segregation of two microsatellite loci, *MSTA53* and *MSTA78*, which were known to be linked to the *DIP*-locus were analysed, was investigated using the methods described in Falque et al. (1998) and Van Dijk and Bakx-Schotman (2004). The *MSTA53* and *MSTA78* genotypes of TJX3-20 were respectively 202/202 and 162/166 (in base pairs). For convenience these genotypes are renamed as *b*/*b* and *a*/*b*. The *MSTA53* and *MSTA78* genotypes of A68 were respectively 198/202/222 and 164/170/174. For convenience these genotypes are renamed as *a*/*b*/*c* and *a*/*c*/*d*.

All 28 F_1 triploids that reproduced apomictically carried the paternal *MSTA78-164* allele ($\chi^2 = 12.65$; d. f. = 1; $P = 0.0004$), supporting the previously reported tight linkage between *MSTA-78* and the *Dip*-locus. Twenty four of the 28 F_1 triploids that reproduced apomictically carried the paternal *MSTA53-202*-allele ($\chi^2 = 2.05$; d.f. = 1; $P = 0.15$). This implies that the *MSTA78-164* allele is closer to *D* than the *MSTA53-202* allele – in the 2x X 4x cross described by Van Dijk and Bakx-Schotman (2004) no recombinants between *MSTA53* and *MSTA78* were found. The other paternal and maternal alleles of *MSTA78* and *MSTA53* were not significantly associated with apomixis in the triploid offspring, supporting the *Dip*-genotype constitution *Ddd*.

The fact that the microsatellites are codominant and that *D* occurs in a single dose, allowed genetic mapping of all three homologs in the diploid pollen grains of A68 (Wu et al 1992; Van Dijk and Schotman 2004). Because the genotypes in diploid pollen grains derived from a triploid segregate in a 2:1 and not in a 1:1 ratio, we balanced the data set by constructing a complementary haploid counterpart of each diploid pollen grain. The modified data set was analyzed with Joinmap® 3.0 (Van Ooijen and Voorrips 2001) using the BC1 module and the Kosambi mapping function. Figure 22.8B shows the genetic map of the *D*-chromosomal region, based on the diploid pollen grains.

For the linkage between the haploid pollen lethal and the *MSTA78-164* allele we assumed that the number of 164-pollen grains formed was equal to the number of *c* and *d*-pollen grains, on average 48. In only one of these *MSTA78-164*-pollen grains there was a cross-over between the *164*-allele and the recessive lethals, resulting in a Kosambi distance of 2.1 cM. Similarly a Kosambi-distance between the *MSTA53-202* allele and the recessive pollen lethal was estimated as 23.3 cM. Figure 22.8C shows the genetic map, based on haploid pollen grains, assuming recessive pollen lethality completely linked to the *Dip*-allele.

References

Bell G (1982) The masterpiece of nature: the evolution and genetics of sexuality. University of California Press, Berkeley

Bicknell RA, Koltunow AM (2004) Understanding apomixis: recent advances and remaining conundrums. Plant Cell 16: S228–S245

Calderini O, Chang SB, de Jong H, Bustil A, Paolocci F, Arcioni S, de Vries SC, Abma-Henkens MHC, Klein Lankhorst RH, Donnison IS, Pupilli F (2006) Molecular cytogenetics and DNA sequence analysis of an apomixis-linked BAC in *Paspalum simplex* reveal a non pericentromere location and partial microcolinearity with rice. Theor Appl Genet 112: 1179–1191

Catanach AS, Erasmuson SK, Podivinsky E, Jordan BR, Bicknell R (2006) Deletion mapping of genetic regions associated with apomixis in *Hieracium*. Proc Natl Acad Sci 103: 18650–18655

Chaboudez P, Burdon JJ (1995) Frequency dependent selection in a wild plant-pathogen system. Oecologia 102: 490–493

Conner JA, Goel S, Gunawan G, Cordonnier-Pratt MM, Johnson VE, Liang C, Wang H, Pratt LH, Mullet JE, Debarry J, Yang L, Bennetzen JL, Klein PE, Ozias-Akins P (2008) Sequence

analysis of bacterial artificial chromosome clones from the apospory-specific genomic region of *Pennisetum*and *Cenchrus*. Plant Physiol 147: 1396–1411
Dawkins R (1989) The extended phenotype. Paperback edition. Oxford University Press, Oxford
Delmotte F, Leterme N, Bonhomme J, Rispe C, Simon JC (2001) Multiple routes to asexuality in an aphid species. Proc R Soc Lond B 268: 2291–2299
Docking TR, Saade FE, Elliott MC, Schoen DJ (2006) Retrotransposon sequence variation in four asexual plant species. J Mol Evol 62: 375–387
Dolgin ES, Charlesworth B (2006) The fate of transposable elements in asexual populations. Genetics 174: 817–827
Falque M, Keurentjes J , Bakx-Schotman JTM, Van Dijk PJ (1998) Development and characterization of microsatellite markers in the sexual-apomictic complex *Taraxacum officinale* (dandelion). Theor Appl Genet 97: 283–292
Goel S, Chen Z, Conner JA, Akiyama Y, Hanna WW, Ozias-Akins P (2003) Physical evidence that a single hemizygous chromosomal region is sufficient to confer aposporous embryo sac formation in *Pennisetum squamulatum* and *Cenchrus ciliaris*. Genetics 163: 1069–1082
Holm S, Ghatnekar L, Bengtsson BO (1997) Selfing and outcrossing but no apomixis in two natural populations of diploid *Potentilla argentea*. J Evol Biol 10: 343–352
Innes DJ, Hebert PDH (1988). The origin and genetic basis of obligate parthenogenesis in *Daphnia pulex*. Evolution 42: 1024–1035
Janzen DH (1977) What are dandelions and aphids? Am Nat 111: 586–589
Kondrashov AS (1982) Selection against harmful mutations in large sexual and asexual populations. Genet Res 40: 325–332
Li L, Jean M, Belzile F (2005) The impact of sequence divergence and DNA mismatch repair on homeologous recombination in *Arabidopsis*. Plant J 45: 908–916
Lynch M, Seyfert A, Eads B, Williams E (2008) Localization of the genetic determinants of meiosis suppression in *Daphnia pulex*. Genetics 180: 317–327
Marshall DR, Brown ADH (1981) The evolution of apomixis. Heredity 47: 1–15
Maynard Smith J (1978) The evolution of sex. Cambridge University Press, Cambridge, UK
Muller HJ (1964) The relation between recombination to mutational advance. Mutat Res 1: 2–9
Naumova TN, Van der Laak J, Osadtchiy J, Matzk F, Kravtchenko A, Bergervoet J, Ramulu KS, Boutilier K (2001) Reproductive development in apomictic populations of *Arabis holboellii* (Brassicaceae) Sex Plant Reprod 14: 195–200
Nogler GA (1984) Gametophytic apomixis. In: Johri BM (ed) Embryology of angiosperms. Springer, Berlin, pp. 475–518
Noyes RD, Baker R, Mai B (2007) Mendelian segregation for two-factor apomixis in *Erigeron annuus* (Asteraceae). Heredity 98: 92–98
Opperman R, Emmanuel E, Levy AA (2004) The effect of sequence divergence on recombination between direct repeats in *Arabidopsis*. Genetics 168: 2207–2215
Ozias-Akins P, Roche D, Hanna WW (1998) Tight clustering and hemizygosity of apomixis linked molecular markers in *Pennisetum squamulatum* implies genetic control of apospory by a divergent locus which may have no allelic form in sexual genotypes. Proc Natl Acad Sci USA 95: 5127–5132
Ozias-Akins P, van Dijk PJ (2007) Mendelian genetics of apomixis in plants. Annu Rev Genet 41: 509–537
Paland S, Colbourne JK, Lynch M (2005) Evolutionary history of contagious asexuality in *Daphnia pulex*. Evolution 59: 800–813
Rice WR (1987) Genetic hitchhiking and the evolution of reduced genetic activity of the Y sex chromosome. Genetics 116: 161–167
Richards AJ (1996) Why is gametophytic apomixis almost restricted to polyploids? The gametophyte-expressed lethal model. Apomixis Newslett 9: 1–3
Sørensen T (1958) Sexual chromosome-aberrants in triploid apomictic *Taraxaca* Bot Tidskr 54: 1–22
Sørensen T, Gudjonsson G (1946) Spontaneous chromosome-aberrants in apomictic *Taraxaca*. Kon Dansk Vidensk Selsk Biol Skrift 4: 1–48

Tas ICQ, Van Dijk PJ (1999) Crosses between sexual and apomictic dandelions (*Taraxacum*) I. The inheritance of apomixis. Heredity 83: 707–714

Van der Hulst RGM, Mes THM, Den Nijs JCM, Bachmann K (2000) Amplified fragment length polymorphism (AFLP) markers reveal that population structure of triploid dandelions (*Taraxacum officinale*) exhibits both clonality and recombination. Mol Ecol 9: 1–8

Van der Hulst RGM, Mes THM, Falque M, Stam P, Den Nijs JCM, Bachmann K (2003) Genetic structure of a population sample of apomictic dandelions. Heredity 90: 326–335

Van Dijk PJ (2003) Ecological and evolutionary opportunities of apomixis: insights from *Taraxacum* and *Chondrilla*. Philos Trans R Soc B Biol Sci 358: 1113–1121

Van Dijk PJ (2007) Potential and realized costs of sex in dandelions, *Taraxacum officinale*, s.*l.* In: Hörandl E, Grossniklaus U, van Dijk PJ, Sharbel TF (eds) Apomixis: Evolution, mechanisms and perspectives. ARG Gantner Verlag KG, Lichtenstein, pp. 215–234

Van Dijk PJ, Bakx-Schotman JMT (2004) Formation of unreduced megaspores (diplospory) in apomictic dandelions (*Taraxacum*) is controlled by a sex-specific dominant gene. Genetics 166: 483–492

Van Ooijen JW, Voorrips RE (2001) Joinmap® 3.0, software for the calculation of genetic maps. Plant Research International, Wageningen, The Netherlands

Verduijn MH, Van Dijk PJ, Van Damme JMM (2004) The role of tetraploids in the sexual-asexual cycle in dandelions (*Taraxacum*). Heredity 93: 390–398

Vijverberg K, Van der Hulst R, Lindhout P, Van Dijk PJ (2004) A genetic linkage map of the diplosporous chromosomal region in *Taraxacum* (common dandelion; Asteraceae). Theor Appl Genet 108: 725–732

West SA, Lively CM, Read AF (1999) A pluralist approach to sex and recombination. J Evol Biol 12: 1003–1012

Wu KK, Burnquist B,Sorrells ME, Tew TL, Moore PH, Tanksley SD (1992) The detection and estimation of linkage in polyploids using single dose restriction fragments. Theor Appl Genet 83: 94–300

Chapter 23
Allelic Sequence Divergence in the Apomictic *Boechera holboellii* Complex

Jose M. Corral, Marcin Piwczynski and Tim F. Sharbel

Abstract It has been suggested that the absence of meiosis in asexual lineages should lead to higher interallelic divergence at any given locus *within* an individual (i.e., allelic sequence divergence – ASD) compared to sexual populations (i.e., Meselson effect; Mark Welch and Meselson 2000). In the present study, ASD was investigated in 2 diploid sexual, 3 diploid apomictic and 3 triploid apomictic *Boechera* (Brassicaceae) using 8 microsatellite loci considering both repeat and flanking region polymorphism. A trend of higher ASD in apomictic versus sexual individuals, both in DNA sequence and repeat motif polymorphisms, was identified, although the pattern of polymorphism is complex and has likely resulted from a combination of mutation accumulation, gene duplication and hybridization. These data demonstrate that caution must be taken when using population genetic models to compare microsatellite variation between sexual and asexual taxa.

23.1 Introduction

Long-term accumulation of deleterious mutations in large sexual populations is prevented by natural selection and recombination (Kimura et al. 1963). In asexual eukaryotes, it has long been accepted that the absence of meiosis reduces rates of homologous recombination to extremely low levels and, in small obligate asexual populations, mutational decay seems unavoidable (Muller 1964; see also Chapter 5). In any generation, due to chance alone, there is a possibility that the class of individuals with the highest fitness in a population will not produce offspring. These genotypes are hence lost forever from the gene pool, and it is very improbable that this class of individuals can ever be reconstituted through meiosis and syngamy. This reduction in the effectiveness of natural selection is expected to increase the rate of deleterious mutation accumulation and reduce the rate of fixation of

J.M. Corral (✉)
Apomixis Research Group, Department of Cytogenetics and Genome Analysis, Institut für Pflanzengenetik und Kulturpflanzenforschung (IPK), D-06466 Gatersleben, Germany
e-mail: corral@ipk-gatersleben.de

I. Schön et al. (eds.), *Lost Sex*, DOI 10.1007/978-90-481-2770-2_23,

adaptive mutations. Under obligate asexuality, individuals can never generate offspring better than themselves, except in the rare case of a compensatory or back mutation. This idea has been taken to mean that the loss of sexual reproduction is a dead end in evolution, leading to early extinction (Maynard Smith 1978; Lynch et al. 1993).

Countering this generalization are a number of "ancient asexual scandals" (Judson and Normark 1996), including the bdelloid rotifers (Mark Welch and Meselson 2000; see also Chapter 13), oribatid mites (Norton and Palmer 1991; see also Chapter 12) and darwinulid ostracods (Martens et al. 2003; see also Chapter 11) which have apparently been devoid of any sexual reproduction for extremely long evolutionary periods. During the quest for an explanation of the evolutionary persistence of these asexual species, it has been suggested that the absence of meiosis in asexual lineages should lead to higher interallelic divergence at any given locus *within* an individual (i.e., allelic sequence divergence – ASD) compared to sexual populations (i.e., Meselson effect; Mark Welch and Meselson 2000). However, most studied asexual lineages fail to show high levels of neutral allelic divergence (Schön and Martens 2003; Schaefer et al. 2006), with the exception of the bdelloid rotifers (Mark Welch and Meselson 2000; Pouchkina-Stantcheva et al. 2007; but see Chapter 13) and *Meloidogyne* root knot nematodes (Lunt 2008).

Although allelic sequence divergence (ASD) has been accepted as a genetic test for the status of ancient asexuality, its absence does not necessarily imply sex (Butlin 2000). Homogenizing mechanisms such as gene conversion between alleles (Butlin 2000; Omilian et al. 2006), chromosomal modification (Forche et al. 2004), highly efficient DNA repair (Schön and Martens 1998), mitotic recombination (Omilian et al. 2006), reduction of ploidy, automixis (Liu et al. 2007), cyclical parthenogenesis (Halkett et al. 2005a) or clandestine sexual reproduction (Burt et al. 1996) can also be responsible for low levels of allelic divergence (Birky 1996; Schön and Martens 2003; Schaefer et al. 2006). Hybridization, a phenomenon which is frequently associated with asexual plants and animals, can alternatively give the impression that elevated intra-individual allelic divergence levels have resulted from ASD (Johnson 2006; Liu et al. 2007), although the spectrum of divergence should be reflected in the sexual populations from which the asexual lineages have arisen.

Whether the result of mutation accumulation and/or hybridization, some populations of asexual organisms are characterized by levels of genetic diversity that are higher than expected as predicted under models of strict clonal reproduction (Kashin et al. 2005; Johnson 2006). Understanding the origins, maintenance and spread of genetic diversity in asexual populations is further complicated by the fact that a number of taxa are characterized by both sexual and asexual members between which limited gene flow is possible, the consequences of which include (1) increased genetic diversity in the asexual gene pool and (2) the establishment of new asexual lineages in a contagious fashion (Halkett et al. 2005b).

Apomixis is a variable mode of reproduction in a number of plant taxa whereby seeds are produced asexually (Asker and Jerling 1992). Generally speaking (but see also Chapter 3 for more explanations), apomixis in plants is characterized by the production of a meiotically unreduced egg cell (apomeiosis), which undergoes parthenogenetic development into an embryo being genetically identical to the mother plant. Some aspects of sexuality are maintained, as fertilization (i.e., pseudogamy) is for the most part obligate for the production of a functional endosperm (i.e., the embryo's nourishing tissue) with a balanced maternal to paternal genome ratio. Apomictic plants are often facultative, and hence, a single individual can produce seeds both through sex and apomixis.

The *Boechera* (formerly *Arabis*) *holboellii* complex is composed of *B. holboellii*, *B. stricta* (formerly *B. drummondii*), and their hybrid *B.* × *divaricarpa* (Koch et al. 2003; Dobeš et al. 2004a, b). The breeding system of this complex is variable, consisting of both sexual and facultative apomictic forms (Böcher 1951; Naumova et al. 2001). Compounding this variability is the wide distribution of polyploidy (mostly $2n = 3x$) and aneuploidy ($2n = 2x+1$ or $2n = 3x+1$; Böcher 1951; Sharbel et al. 2004), where polyploidy has originated multiple times in geographically and genetically distinct populations (Sharbel and Mitchell-Olds 2001; Sharbel et al. 2005). *Boechera stricta* has been shown to be predominantly diploid and sexual, while *B. holboellii* and *B.* × *divaricarpa* are highly variable with respect to breeding system, ploidy, morphology, and genetic polymorphism (Roy and Rieseberg 1989; Roy 1995; Sharbel and Mitchell-Olds 2001; Koch et al. 2003; Dobeš et al. 2004a, b; Sharbel et al. 2004, 2005; Kantama et al. 2007).

Apomictic *Boechera* are characterized by *Taraxacum*-type diplospory whereby the megaspore mother cell (MMC) goes through meiosis I without completing the reductional phase (apomeiosis), followed by normal meiosis II leading to a nucleus which has the same ploidy as the mother plant (Böcher 1951; Naumova et al. 2001). As with many asexual taxa, microsporogenesis is typically disturbed in apomictic individuals, and normal reduced, non-reduced and aneuploid pollen can be found within and between different accessions (Böcher 1951; Dobeš et al. 2004b; Sharbel et al. 2005; Voigt et al. 2007). Analyses of meiosis in pollen cells have additionally demonstrated a range of chromosomal synapsis (univalent and multivalent) in apomictic accessions (Böcher 1951; Kantama et al. 2007), and both pseudogamous and autonomous endosperm formation have also been identified (Naumova et al. 2001; Voigt et al. 2007). Our recent comparisons of global gene expression patterns in sexual and apomeiotic ovules have identified many differentially expressed genes, whose expression levels are reflective of hybridization, heterochrony (shifts in gene expression timing) and gene duplication in apomicts (Sharbel et al. 2009).

Our ongoing analyses of microsatellite markers in natural *Boechera* populations have demonstrated high levels of variation in apomictic accessions (Sharbel et al. in preparation). Furthermore, a range of meiotic synapsis potential (i.e., fully synaptic, partially synaptic and fully asynaptic) in different apomictic accessions (Böcher 1951; Kantama et al. 2007), chromosomal heteromorphy (Kantama et al. 2007), in addition to aneuploidy and polyploidy (Sharbel and Mitchell-Olds 2001; Sharbel et al. 2004, 2005) together provide a context in which ASD could accumulate. As asexual *Boechera* are not part of an ancient asexual complex but rather pre-Pleistocene in origin (Dobeš et al. 2004b), it is unclear what the pattern and range of DNA sequence diversity in coding DNA regions on a genome-wide scale should resemble (Borevitz et al. 2007). We have thus chosen relatively rapidly evolving DNA loci, which are additionally assumed to undergo neutral evolutionary change, microsatellites and their flanking regions, in a first attempt to examine whether asexual *Boechera* genomes are characterized by ASD.

In most cases, microsatellites exhibit relatively high levels of polymorphism which are assumed to be selectively neutral, thus predisposing these markers to a number of analyses (e.g., Luty et al. 1990; Bruford and Wayne 1993; Roy et al. 1994). Microsatellite evolution is a complex mutational process driven mainly by DNA slippage (Levinson and Gutman 1987), but is additionally influenced by other factors including mismatch repair efficiency, length constraints, selection, repeat type, flanking sequence and the degree of perfection of the repeats (Schlötterer 2000; Huang et al. 2002). The evolution of these sequences has also been studied in asexual organisms (Sunnucks et al. 1996; Weetman et al. 2002), including apomictic plant lineages (Paun and Hörandl 2006). Interestingly, several studies have shown that the flanking regions of the microsatellites are highly variable (Mogg et al. 2002). The purpose of this chapter is to test whether we can detect the first evidence of ASD in the genomes of apomictic *Boechera*, using microsatellites and their flanking regions.

23.2 Microsatellite Variation

Two diploid sexual, 3 diploid apomictic and 3 triploid apomictic individuals were used in the sequence analysis (Table 23.1). The reproductive mode of all individuals was confirmed by using the flow cytometric seed screen (Matzk et al. 2000; Voigt et al. 2007). Ten microsatellite loci, which have been identified in previous studies, were selected for this present study (Table 23.2). Microsatellites BoechA10,

Table 23.1 Sexual and apomictic *Boechera* used in the microsatellite sequencing survey

Individual	Accession	Ploidy[a]	Reproduction	Origin
1	300-6-1	2C	Apomictic	21[b]
2	205-3-4	2C	Apomictic	6[b]
3	67-5-4	2C	Apomictic	21[b]
4	148-1-6	3C	Apomictic	20[b]
5	218-2-4	3C	Apomictic	4[b]
6	195-2-24	3C	Apomictic	4[b]
7	B06-1081 (*B. holboelli*)	2C	Sexual	ES 95 "Lower Hot Springs"[c]
8	B06-507 (*B. stricta*)	2C	Sexual	SAD 4 "Gold Creek"[d]

[a] Based on flow cytometric analyses of genome size (Sharbel and Mitchell-Olds 2001; Sharbel et al. 2005).
[b] Numbers correspond to map locations in Fig. 1 from Sharbel et al. (2005).
[c] Samples from Schranz et al. (2005).
[d] Samples collected by Bitty Roy.

Table 23.2 Microsatellite loci used for genotyping and sequencing survey

Locus[a]	Repeat motif	Primers (5' to 3')
ICE14	CAT	F-TCGAGGTGCTTTCTGAGGTT R-TACCTCACCCTTTTGACCCA
ADH1	A[b]	F-ACCACCGGACAGATTATTCG R-CCCAGAAGTAAACATCGGTGTG
ICE3	GA	F-GACTAATCATCACCGACTCAGCCAC R-ATTCTTCTTCACTTTTCTTGATCCCG
BF20	GA	F-TTCTCGGGAAAGTAATGAGGAG R-GCAAATCTGACCAATGCAAG
BoechA10	AT	F-GCAATTTTGGGAGGGAAAA R-GTTGTTTGGAGGGGACAGAA
D3	GAA	F-GGTTATGTGAGAGTTAAG R-ATTGTTGAATGCAACAGG
ICE8	CTT	F-GTGTTACCGATCTGGCTCTG R-TCAGCTTGAGCATTTCACAG
Bdru1220	AT	F-TCTATGCAAACAGCAAATCG R-TTCTTCTACGAAACATTCCTTGC
ATTS0392	CTT	F-TTTGGAGTTAGACACGGATCTG R-GTTGATCGCAGCTTGATAAGC
MBK21B3	CT	F-ATTGCTTCCGTTTTGCTCAT R-TTTCAATTCTCTCTGCCTCTCA

[a]Microsatellite primers taken from various sources (Clauss et al. 2002; Dobeš et al. 2004a; Song et al. 2006) T. Sharbel (unpublished data).
[b]We identified an "A" repeat motif, rather than the "GGT" repeat motif described in Clauss et al. (2002).

BF20, Bdru1220 and ICE3 have been recently mapped in *Boechera stricta* with the respective names A10, Bf_20, Bdru_12 and ICE3 (Schranz et al. 2007). An average of 16 clones per individual per locus was sequenced. Microsatellite and flanking regions were separated for various statistical analyses (Tables 23.3, 23.4 and 23.5). Despite repeated cloning attempts, reliable sequence data could not be obtained from loci ICE3 and ICE8, and thus these were not included in subsequent analyses.

Sequences were aligned using LASERGENE version 7.0 (DNASTAR), edited by BioEdit (version 7.0.5.2; Hall 1999) and imported into DnaSP (version 4.10; Rozas et al. 2003) and DAMBE (version 4.5.45 Xia and Xie 2001) in order to generate DNA statistics.

The published repeat motif of each microsatellite locus was confirmed by our sequencing data in all cases except for ADH1. This locus (ADH1) was characterized by a complex microsatellite, having both an insertion-deletion (indel) and a single "A" nucleotide repeat motif (Table 23.2, Fig. 23.1) rather than the "GGT" motif described in (Clauss et al. 2002). Variation at each locus was analyzed considering either (subsection 23.2.1) only the microsatellite repeat region, or (subsection 23.2.2) the complete cloned sequence, including both flanking and microsatellite repeat regions (Tables 23.3, 23.4 and 23.5). The sequences reported in this study have been entered into GenBank under the accession numbers EU214643 to EU214907.

Table 23.3 Nucleotide (π)/haplotype (H) diversities, as calculated from flanking region sequences and grouped for sexuals (2s), diploid apomicts (2a) and triploid apomicts (3a) for each locus

Class	ADH1	ATTS0392	Bdru1220	BF20	BoechA10	D3	ICE14	MBK21B3
2s	0/0	0.008/0.667	0.013/0.571	0.011/1.000	0.015/0.700	0.017/0.800	0/0	0.033/1.000
2a	0.011/0.900	0.016/0.857	0.005/0.600	0.005/0.618	0.022/0.722	0.021/1.000	0.005/0.644	0.006/0.583
3a	0.014/0.919	0.006/0.533	0.012/0.714	0.009/0.782	0.008/0.429	0.024/0.800	0.011/0.952	0.014/0.600
All	0.013/0.866	0.010/0.621	0.013/0.804	0.010/0.802	0.015/0.593	0.021/0.875	0.007/0.772	0.013/0.658

Table 23.4 Overview of microsatellite allele number and distribution in comparisons between grouped sexuals (2s), diploid apomicts (2a) and triploid apomicts (3a), including comparisons with single sexual individuals (e.g., Sex7 refers to individual 7, Table 23.1)

	Number of alleles				Shared alleles						
Locus	Total	2s	2a	3a	2a/2s	3a/2s	2a/3a	2a/Sex 7	2a/Sex 8	3a/Sex 7	3a/Sex 8
ADH1	21 (8)	2 (2)	4(4)	16(5)	0 (1)	1 (2)	0 (1)	0 (1)	0 (0)	0 (1)	1 (1)
ATTS0392	15 (5)	4 (3)	7(3)	6(3)	1 (2)	2 (2)	1 (2)	1 (1)	0 (1)	1 (1)	1 (1)
Bdru1220	27 (14)	7 (6)	8(5)	13(9)	0 (2)	1 (5)	0 (4)	0 (0)	0 (2)	0 (1)	1 (4)
BF20	26 (12)	4 (1)	11(9)	11(8)	0 (1)	0 (0)	2 (4)	0 (1)	0 (1)	0 (0)	0 (0)
BoechA10	18 (11)	5 (4)	8(4)	7(6)	0 (0)	0 (1)	2 (3)	0 (0)	0 (0)	0 (1)	0 (0)
D3	11 (5)	4 (4)	5(2)	5(2)	1 (1)	2 (2)	1 (1)	1 (1)	1 (1)	1 (1)	2 (2)
ICE14	11 (5)	2 (2)	8(3)	6(3)	1 (1)	0 (0)	2 (2)	1 (1)	0 (0)	0 (0)	0 (0)
MBK21B3	19 (9)	2 (2)	8(4)	10(7)	0 (0)	0 (1)	1 (1)	0 (0)	0 (0)	0 (0)	0 (1)

Allele numbers are calculated considering the complete microsatellite sequence (i.e., repeat motif and flanking region), or only the repeat motif (in brackets)

Table 23.5 Coefficients of variation (**CV** = σ / μ, where σ is the standard deviation of the number of repeated motifs of the different alleles of a microsatellite in one individual and μ is the mean of repeated motifs for the same microsatellite in the whole sample set of individuals) and number of alleles (in brackets) considering only the microsatellite repeated region

		2n apomict			3n apomict			2n sexual	
Locus	Repeat	1	2	3	4	5	6	7	8
ADH1	A	0.079 (2)	0 (1)	0 (1)	0.047 (5)	0.115 (4)	0.079 (7)	0 (1)	0 (1)
ATTS0392	CTT	0.096 (2)	0 (2)	0.078 (2)	0 (1)	0.078 (4)	0.130 (5)	0 (1)	0 (2)
Bdru1220	AT	0.146 (4)	0.073 (2)	0.196 (4)	0.239 (6)	0.331 (4)	0.371 (4)	0 (1)	0.124 (4)
BF20	GA	0.386 (4)	0.164 (3)	0.404 (4)	0.250 (3)	0.108 (5)	0.228 (5)	0 (2)	0 (2)
BoechA10	AT	0 (2)	0.056 (3)	0.279 (4)	0 (1)	0.279 (4)	0.056 (3)	0 (1)	0.214 (4)
D3	GAA	0 (1)	0 (1)	0 (3)	0 (2)	0 (2)	0.266 (2)	0.266 (2)	0.288 (3)
ICE14	CAT	0.263 (3)	0.088 (3)	0.076 (4)	0 (1)	0.152 (3)	0.152 (3)	0 (1)	0 (1)
MBK21B3	CT	0.061 (4)	0.054 (3)	0.131 (2)	0.322 (4)	0.107 (3)	0.267 (4)	0 (1)	0 (1)

```
         10         20         30         40
....|....| ....|....| ....|....| ....|....| ....
CCAC------ ---------- ---------- --AAAAACTA GTTA
....AAAAAC GACACCGTTC TTGTT----- AA........ ....
....AAAAAC GACACCGTTC TT-------- ---------- ....
....AAAAAC ---------- ---------- ---------- ....
....AAAAAC GACACCGTTC TTGTT---AA TA........ ....
....AAAAAC GACACCGTTC TTGTT----A GA........ ....
....AGAAAC GACACCGTTC TTGTTAAAAA AA........ ....
....AAAAAC GACACCGTTC TTGTT----A AA........ ....
....AGAAAC GACACCGTTC TTGTT-AAAA AA........ ....
....GAAAAC GACACCGTTC TTGTT----- AA........ ....
```

Fig. 23.1 Alignment of the repeat unit region (including 4 bp of non-variable sequence on each flank) of locus ADH1. Dots represent invariable sites (relative to first sequence), and dashes represent deletions

23.2.1 Variation in the Repeat Regions

Considering only length variation (i.e., PCR amplicon size) in the microsatellite region of each locus, 5 to 14 alleles per locus were observed for all samples grouped together (Table 23.4). Interestingly, the number of alleles counted in each reproductive class (i.e., diploid sexual, diploid apomict, triploid apomict) was occasionally higher than expected considering the number of samples and their ploidy. For example, 2 diploid sexual individuals were sequenced and thus a maximum of 4 alleles at any given locus was expected. This expectation was not violated, except in the case of locus Bdru1220, in which 6 different alleles (i.e., length variants) were identified, 4 of which were identified in the sexual *Boechera stricta* sample. Similarly, a maximum of 6 and 9 alleles was expected from the grouped diploid and triploid apomict samples, respectively, although 9 alleles in the diploid apomict samples were identified for locus BF20 (Table 23.4).

This trend was even clearer when examined on the individual level (Table 23.5). In the 2 sexual diploids, a higher, than expected, number of alleles was identified in individual 8 only (for loci Bdru1220, D3 and BoechA10; Table 23.5). The number of cases in which more alleles were sampled than expected was clearly greater at virtually all loci for all apomictic individuals (Table 23.5). These deviations from the expected number of alleles were more apparent when the complete amplicon sequence was examined (see below).

Relatively few alleles were shared between samples or groups of samples for any comparison. In comparisons between reproductive classes, the number of shared alleles was greatest between diploid and triploid apomictic individuals (Table 23.4). Alleles were shared across all 8 loci in the comparison between 2C and 3C apomictic individuals, while only 6 loci had shared alleles for the comparisons between the sexuals and apomicts (Table 23.4).

Intra-individual coefficients of variation were calculated per locus considering allelic variation in the repeat regions only (Table 23.5), and equaled zero in all but 4 out of 16 comparisons in the sexual diploids. The coefficients of variation were for the most part much higher in the apomicts than the sexuals, and higher in the triploid apomicts compared to the diploid apomicts (Table 23.5). The locus D3 exhibited a coefficient of variation pattern which was opposite to that characterized by the other 7 loci, having higher values in both diploid sexuals than in all but one triploid apomict (Table 23.5).

23.2.2 DNA Sequence Variation

DNA sequencing revealed more polymorphisms than were observed from amplicon length variation only. Referring to the microsatellite repeat regions, a number of phenomena other than changes in the repeat motif copy number were observed, including insertions or substitutions of single nucleotides (loci ADH1, Bdru1220, BF20, BoechA10, D3, MBK21B3) and complex repeat motifs composed of variable DNA sequence (loci ADH1, ATTS0392, BoechA10, D3, MBK21B3; Fig. 23.1). In a number of cases, the indel polymorphisms in the flanking regions led to homoplasious (i.e., convergence) changes in amplicon size (i.e., two microsatellite alleles having the same size but with different indel or microsatellite repeat variation).

If this information was included, it had the net effect of significantly increasing the number of observed alleles in all individuals, from 5 to 14 alleles (considering length variation only) to 11–26 alleles (considering sequence variation; Table 23.4). This effect was most obvious when comparing the pooled sexuals against both types of apomict (Table 23.4). For the two sexuals, combining the sequence data with the length variation had either: no effect on the number of observed alleles (4 loci), increased the number of observed alleles by 1 (3 loci), or increased them by 3 (1 locus; Table 23.4; Fig. 23.2). In contrast, the addition of sequence information to the apomictic groups increased the number of observed alleles at almost all loci, and doubled the observed number for 5 and 4 loci for diploid and triploid apomicts, respectively (Table 23.4; Fig. 23.2).

The number of alleles shared between reproductive classes and/or individuals did not change at loci D3 and ICE14, when DNA sequence variation was considered. At all other loci, the number of shared alleles either did not change or decreased when DNA sequence variation was considered (Table 23.4).

23.3 Allelic Sequence Divergence (ASD) in Microsatellite Flanking Regions

Pairwise DNA sequence divergence in the flanking regions of alleles for each locus was calculated using the DnaSP 4.10 (Rozas et al. 2003) and MEGA4 software packages (Tamura et al. 2007), and demonstrated three patterns of divergence in different comparisons between reproductive classes (Fig. 23.3). Loci ATTS0392 and BoechA10 demonstrated the highest levels of ASD when alleles were pooled and pairwise distances were calculated for the following comparisons: diploid

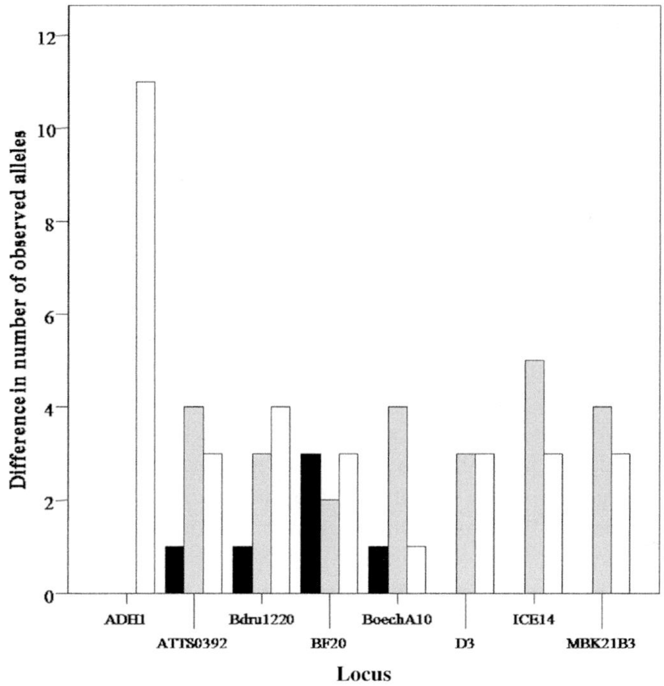

Fig. 23.2 Observed number of alleles as determined by sequencing minus the observed number of alleles as determined by length variation only, grouped for diploid sexual (*black*), diploid apomict (*grey*) and triploid apomict (*white*) per locus

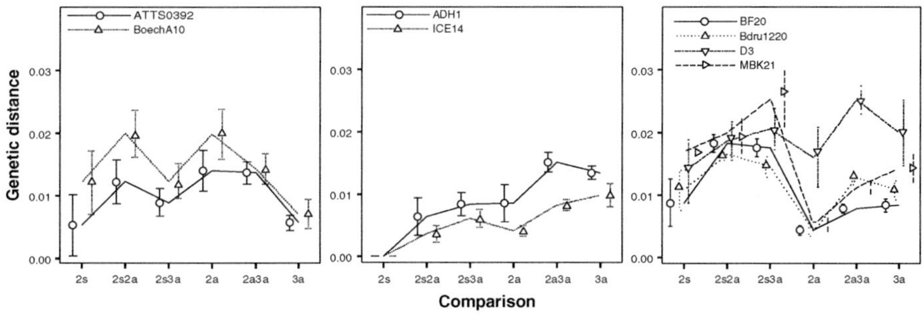

Fig. 23.3 Graphs of different patterns of pairwise genetic distance (Kimura 2-parameter) variation between alleles for comparisons between pooled alleles found in sexual (2s), diploid apomict (2a) and triploid apomict (3a). Genetic distances were calculated based upon sequence variation in the flanking regions only (i.e., microsatellite repeat regions were removed before analysis). Error bars demonstrate 95% confidence interval of the mean value

sexual-diploid apomict, diploid sexual-triploid apomict, diploid apomict, diploid apomict-triploid apomict (Fig. 23.3). Locus ATTS0392 showed low levels of divergence for diploid sexual and triploid apomicts, and locus BoechA10 displayed the lowest levels of divergence in triploid apomicts (Fig. 23.3).

In the second pattern of divergence, loci ADH1 and ICE14 both showed overall increasing levels of pairwise allelic sequence divergence in the following order: diploid sexual, diploid sexual-diploid apomict, diploid sexual-triploid apomict, diploid apomict, diploid apomict-triploid apomict, and triploid apomict (Fig. 23.3). The final pattern of divergence displayed the lowest allelic sequence divergence in the diploid apomict, and the highest levels in the diploid apomict-triploid apomict and diploid sexual-triploid apomict comparisons (Fig. 23.3).

Considering ASD in each reproductive class alone, diploid sexuals demonstrated the lowest ASD for loci ADH1 and ICE14, and the highest ASD for loci BF20, Bdru1220, D3 and MBK21B3 (Fig. 23.3). Loci BF20, Bdru1220 and MBK21B3 demonstrated the lowest ASD in diploid apomicts (Fig. 23.3).

ASD measured within individuals was for the most part lowest in the sexuals compared to the apomicts (Table 23.6). In three cases (locus ADH1, ICE14 and MBK21B3), the two sexuals showed a single allele, while most (but not all) apomicts had multiple alleles characterized by sequence divergence (Table 23.6). In one case (Bdru1220), both sexuals had multiple alleles as determined by length polymorphism, but no flanking region sequence variation was found (Table 23.6). The sexual *Boechera holboellii* accession had only a single allele compared to multiple alleles in the sexual *B. stricta* for loci ATTS0392 and BoechA10 (Table 23.6).

A split decomposition network analysis at different loci yielded 2 types of patterns with respect to allelic distribution among reproductive classes. The first pattern was characterized by little to no allelic variation between the 2 sexual individuals (Fig. 23.4). For example, locus ADH1 was characterized by a centrally-located allele (Hap_2), which was found in both sexuals (*B. holboellii* and *B. stricta*), as well as in members of both diploid and triploid apomicts (Fig. 23.4). This pattern was also characteristic of loci ATTS0392, D3 and ICE14 (data not shown).

The second pattern was characterized by allelic sequence divergence between both sexual individuals (i.e., species-specific alleles; Fig. 23.5). For example, locus Bdru1220 was characterized by alleles, which were specific to either *B. stricta* (Hap_5) or *B. holboellii* (Hap_8, 9 and 10). Those alleles are located peripherally in the split decomposition network (Fig. 23.5). This pattern was also characteristic of loci BF20, BoechA10 and MBK21B3 (data not shown), and was reflected in almost all nucleotide diversity values, which were highest in the grouped sexual diploids (Table 23.3).

Alleles, which were specific to the different reproductive classes were also apparent, as well as alleles which were shared between all classes, or shared only between apomicts (Table 23.4; Figs. 23.4 and 23.5). Finally, a pattern of allelic variation reflective of genome duplication and divergence was evident in some loci. In ADH1 for example, which is characterized by certain individuals having more cloned alleles than expected based upon ploidy (Table 23.4), certain alleles cluster together. Triploid apomictic individual 6 (Table 23.1) is characterized by 3 clusters of alleles: Hap_2 + Hap_11 + Hap_13; Hap_2 + Hap_9 + Hap_12; and Hap_14 + Hap_15 (Fig. 23.4). Similarly, triploid apomictic individual 4 (Table 23.1) is characterized by 2 clusters of alleles: Hap_2 + Hap_6 + Hap_7; and Hap_5 + Hap_8 (Fig. 23.4).

Table 23.6 Mean intra-individual genetic distance (Kimura 2-parameter) values considering microsatellite flanking regions only ("–" not calculated due to the presence of a single allele)

Individual	ADH1	ATTS0392	Bdru1220	BF20	BoechA10	D3	ICE14	MBK21B3
1	0.011	0	0.004	0.005	0.038	–	0.004	0.008
2	–	0.012	0	0.005	0.013	–	0.004	0.006
3	0.022	0.025	0.004	0.005	0.022	0.009	0.007	0
4	0.015	–	0.004	0.010	–	0.009	–	0.006
5	0.018	0.006	0.023	0.003	0.013	0	0.017	0.029
6	0.003	0.007	0.004	0.013	0	0.035	0.008	0.021
7	–	–	0	0.007	–	0.017	–	–
8	–	0.012	0	0.007	0.009	0.023	–	–

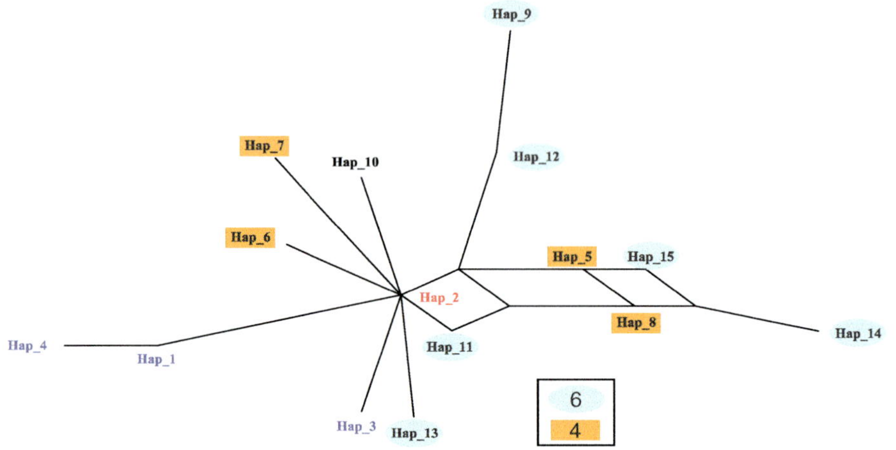

Fig. 23.4 Split decomposition network of 15 ADH1 alleles, considering flanking regions only. *Colored letters*: allele Hap_2 (*red*) is found in sexual (2s), diploid apomict (2a) and triploid apomict (3a), while blue and black alleles are found exclusively in diploid (2a) and triploid apomicts (3a) respectively. *Colored boxes* represent alleles cloned from same individual triploid apomictic individuals 4 and 6 (Hap_2 was also cloned in both 4 and 6)

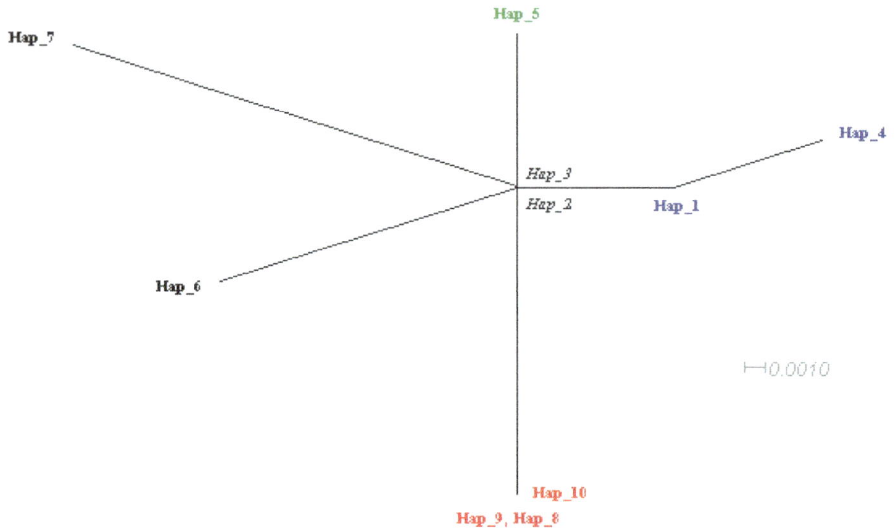

Fig. 23.5 Split decomposition network of 10 Bdru1220 alleles, considering flanking regions only. Red alleles are found only in sexual *B. holboellii*, the green allele is found only in sexual *B. stricta* and one triploid apomict, the 2 central alleles in italics are found in both diploid and triploid apomicts (2a, 3a), while blue and black alleles are found exclusively in diploid (2a) and triploid apomicts (3a), respectively

23.4 Microsatellite Evolution in Sexual and Apomictic Lineages

The complex pattern of both length variation and single nucleotide polymorphisms in flanking region confound interpretations of microsatellite evolution, at least with respect to apomictic *Boechera*. In concordance with previously-described genome plasticity in *Boechera*, including extensive chromosomal variation, disturbed meiosis and the presence of aberrant chromosomes in the apomictic lines (Böcher 1951; Sharbel and Mitchell-Olds 2001; Dobeš et al. 2007; Kantama et al. 2007), the data presented here are similarly reflective of a dynamic genome. Additionally, hybridization, which has played a significant role in the evolution of *Boechera* (Dobeš et al. 2004a, b; Kantama et al. 2007), is often characterized by large scale genomic and global gene expression changes (Shaked et al. 2001; Hegarty and Hiscock 2005).

Our initial intention was to analyze for ASD in rapidly evolving genomic regions, and in doing so we identified multiple phenomena which appeared to be associated with asexuality. A number of differences were detected between the three groups of samples, the most striking of which were between the sexual and apomictic groups. Understanding these phenomena requires first an overview of the hypothetical interrelationships between the different reproductive and ploidy classes:

1. Diploid sexual *Boechera* produce diploid apomictic *Boechera* via hybridization (Koch et al. 2003; Dobeš et al. 2004a, b; Schranz et al. 2005; Kantama et al. 2007).
2. Diploid apomictic *Boechera* produce, via apomeiosis, unreduced male and female gametes (Voigt et al. 2007). Fertilization of unreduced apomictic gametes with reduced sexual gametes leads to the production of polyploids (mostly 3C; Schranz et al. 2005; Voigt et al. 2007).
3. Polyploid (3C) *Boechera* are for the most part apomictic, and produce unreduced 3C gametes. 3C pollen can occasionally fertilize sexually-produced reduced (C) egg cells, and sexually-produced reduced (C) pollen can occasionally fertilize unreduced 3C egg cells to produce tetraploid offspring.
4. The production of reduced gametes (both male and female) by apomictic individuals is relatively rare (Voigt et al. 2007).

The dynamics of gene flow between sexual and apomictic *Boechera* lineages are thus complex, due to the production of reduced and unreduced gametes which vary in fertilization potential. Crosses between apomictic and sexual lineages are possible, although the offspring are not necessarily apomictic (Schranz et al. 2005). The following evolutionary pathway appears to be the major evolutionary trend, and will thus be considered for the purpose of interpreting the data presented here: 2C sexual → 2C apomict → 3C apomict. Exceptions to this trend (i.e., tetraploidy, fertilization of apomeiotically-derived gametes, etc.) have also been documented (Naumova et al. 2001; Voigt et al. 2007), but appear to be relatively rare phenomena (Voigt et al. 2007).

23.5 Allelic Variation and Genome Duplication

One of the most important observations of this study is the discrepancy between the numbers of alleles calculated considering amplicon length polymorphism versus calculations from DNA sequencing. Allelic homoplasy, the same size of microsatellite amplicons resulting from different types of DNA sequence variation, has been documented in other non-apomictic taxa (Grimaldi and Crouau-Roy 1997; Vowles and Amos 2004; Webster and Hagberg 2007). In all cases presented here, the number of observed alleles increased when DNA sequences were considered, due to the result of SNPs and indels in both the flanking and microsatellite repeat regions (Fig. 23.1). This discrepancy was accentuated in the apomictic genomes (Table 23.4, Fig. 23.2).

Part of this variation can be explained in the context of tandem duplication events, which also appear to be characteristic of the apomicts (Figs. 23.4 and 23.5). Every analyzed locus demonstrated evidence of tandem duplication (i.e., more than the expected number of alleles considering ploidy, Tables 23.3 and 23.4), although it was not possible to estimate the number of copies of each locus based upon the data presented here. The pattern of DNA sequence variation in flanking regions is concordant with multiple alleles (i.e., heterozygosity) existing at a single locus in an apomictic genome, followed by duplication of the locus and mutational divergence of the alleles in the flanking regions (Figs. 23.4 and 23.5).

The implications of this for population studies of asexual taxa are significant, as the estimation of allelic variation in terms of amplicon length only biases interpretations of clonal lineage origin and evolution. The incorrect determination of allelic polymorphism would lead to an underestimation of the genetic variability characteristic of different asexual lineages, and to an overestimation of the genetic relatedness between different lineages. Although it has not been empirically tested, we predict that reproductive and genetic isolation should act on a relatively smaller scale between apomictic lineages compared to sexual *Boechera*, a result of limited to no gene flow in the former. Measures of genetic isolation by distance (IBD) would therefore have a different meaning for sexual versus apomictic lineages, as homologous microsatellite alleles are expected to be lost between isolated populations via drift, whereas alleles with high frequencies may be generated by mutations in a convergent fashion (Estoup et al. 2002).

23.6 Allelic Sequence Divergence (ASD)

ASD in asexual bdellloid rotifers has been demonstrated by elevated synonymous and non-synonymous mutations in four genes in comparison to nearly identical allele copies in sexually-reproducing nonbdelloid rotifers (Mark Welch and Meselson 2000; but see Mark Welch et al. (2008) and Chapter 13 for a new interpretation of the results). Furthermore, it has recently been shown that adaptive functional divergence in allelic copies of a gene linked with desiccation tolerance in bdelloids has occurred (Pouchkina-Stantcheva et al. 2007), thereby demonstrating how ASD could provide beneficial genetic diversity in an asexual organism.

In contrast, most of the loci analyzed here are non-coding and hence assumed to be selectively neutral. Intra-individual ASD appears to be higher in apomicts, considering both microsatellite repeat polymorphism (Table 23.5) and flanking region sequence variation (Tables 23.3 and 23.6). Apomictic *Boechera* exhibit a range of chromosomal synapsis potential (i.e., fully synaptic, partially synaptic and fully asynaptic) during meiosis I (Böcher 1951; Kantama et al. 2007), and thus it is conceivable that ASD could accumulate at loci which no longer undergo synapsis and recombination. Alternatively, the observed ASD could also be explained or influenced by gene duplication and/or hybridization.

As described above, our sequencing data demonstrate that many of the examined microsatellite loci exist in multiple copies, as a result of hypothesized tandem duplications in the apomictic genomes (Table 23.4; Fig. 23.2). Furthermore, the pattern of genetic similarity between alleles is concordant with duplication of a heterozygous locus, followed by mutation accumulation in both the original and duplicated alleles (Figs. 23.4 and 23.5). One potential source of elevated ASD in apomictic lineages is thus the measurement of divergence between duplicated alleles rather than that between two alleles at the same locus. In fact, in asexual bdelloid rotifers, it has been found that for four genes located in the *hsp82* gene region, synonymous divergence between alleles of the same locus is almost 10-fold lower than the synonymous divergence comparing duplicated alleles for that locus (Mark Welch et al. 2008).

Of the 8 loci, Bdru1220 and BoechA10 demonstrate evidence for duplication in both sexuals and apomicts (Table 23.4). The remaining 6 loci imply duplication in the apomicts (i.e., more alleles than expected considering ploidy) but not in the sexuals (i.e., less than or equal to the number of alleles considering ploidy). Alternatively, they are also duplicated in the sexuals but with little to no sequence divergence between allele copies (Table 23.4). If ASD between alleles at duplicated loci is the reason behind the observed elevated ASD values, then it would be expected to similarly affect the two loci duplicated in both sexuals and apomicts (Table 23.4). This is clearly not the case for loci Bdru1220 and BoechA10 (duplicated in both sexuals and apomicts), which in the sexuals demonstrate little to no flanking sequence divergence in multiple alleles (Table 23.6) as determined by length polymorphism (Table 23.4).

Hybridization, the likely source of diploid apomictic *Boechera* (Dobeš et al. 2004a, b, 2007; Kantama et al. 2007), could also give the impression of ASD if the two parental taxa were characterized by species- (or population) specific alleles (Delmotte et al. 2003; Lunt 2008). The interallelic relationships among accessions (i.e., samples) was reflective of species-specific alleles for loci Bdru1220, BF20, BoechA10 and MBK21B3 (Table 23.3, Fig. 23.5), and thus, we cannot differentiate between hybridization and ASD as the cause of the elevated allelic variation in apomicts observed at these loci (Table 23.6). The remaining analyzed loci, on the other hand, were not characterized by species-specific alleles for the two sexual taxa (Table 23.4, Fig. 23.4). ASD could thus explain elevated allelic variation in the apomicts at these loci, although the limited number of analyzed sexual accessions could simply mean that the alleles characteristic of the parental taxa in the hybridization event were not sampled. It is thus unclear whether ASD has accumulated in the apomictic accessions by flanking sequence variation only.

Clearly, hybridization has contributed to much of the variation characteristic of the apomictic genomes in *Boechera*, both on the chromosomal (Kantama et al. 2007) and DNA sequence levels (Sharbel et al. 2009). A phylogenetic analysis based on non-coding chloroplast DNA sequences (*trn*L intron and *trn*L/F intergenic spacer) resolved *B. stricta* as a monophyletic taxon, but found *B. holboellii* to bear chloroplast haplotypes from highly diverged evolutionary lineages (Dobeš et al. 2004a, b). Pleistocene fragmentation, colonization events and subsequent radiations have likely played an important role in shaping the current distribution of genetic variation in *Boechera* populations (Dobeš et al. 2004a, b, 2007). In addition, initial or repeated hybridization with formation of allopolyploids (Kantama et al. 2007) and facultative apomixis (Schranz et al. 2005) appear to be ongoing and influencing the actual genetic diversity in apomictic lineages. As the taxonomy of *Boechera* is complex and under revision (Windham and Al-Shehbaz 2006, 2007), it cannot be determined whether hybridization has occurred between different species or differentiated populations which may have originated from different glacial refugia (Dobeš et al. 2004a, b, 2007).

Alternatively, our analyses of repeat length polymorphisms demonstrate higher CV's (i.e., coefficients of variation) in apomicts at all but one locus (D3; Table 23.5), and the picture becomes more complicated if both flanking sequence and repeat length polymorphisms are considered together (Table 23.4). Taken together, the higher number of alleles characterizing the apomicts is demonstrative of a complex pattern of locus duplication, intra-individual variation in repeat polymorphisms, and (perhaps) accumulated DNA sequence polymorphism in flanking regions. Whether or not this elevated variation in apomicts can be considered ASD *sensu stricto* is unclear, but what is apparent is that the selective constraints acting upon microsatellite evolution differ between the sexual and apomictic genomes.

Using variation measured from the chloroplast *trn*L intron, Dobeš et al. (2004a) estimated a divergence date between the most distantly related cpDNA lineages of *B. stricta* and *B. holboellii* to be 1–3 myr before present. We did not perform a similar estimation using the flanking region polymorphisms presented here due to the limited number of individuals sampled, the unknown rate and pattern of mutations of the studied sequences, and the lack of information regarding the structure, size, and dynamics of the sampled populations.

23.7 Conclusions

The current ploidy and reproductive diversity in *Boechera* is reflective of a complex biogeographic history involving genetic isolation and divergence, adaptation and post-glacial recolonization of North America. The recurrent origins of polyploidy and (likely) apomictic reproduction (Sharbel and Mitchell-Olds 2001; Sharbel et al.

2005) have led to the establishment of numerous asexual lineages, which differ in evolutionary age. Upon their origin, these asexual lineages have undergone independent evolution from one another, and have accumulated both small (SNP) and large-scale (chromosome duplication and deletion) mutations, some of which may have perturbed meiosis I to produce the spectrum of chromosomal synapsis potential seen across different accessions. The partial or complete absence of meiosis in asexual lineages is expected to lead to elevated ASD when compared to sexual populations (Mark Welch and Meselson 2000).

Our analyses of microsatellites and their flanking regions have demonstrated a trend of higher ASD in apomictic versus sexual individuals, both in DNA sequence and repeat motif polymorphisms. The pattern of polymorphism is nonetheless complex, and based upon the limited sampling of both accessions and loci, it is impossible to differentiate between the relative contributions of ASD versus other phenomena (e.g., gene duplication and hybridization). We nonetheless feel that our study has shown that caution must be taken when using population genetic models to compare microsatellite variation between sexual and asexual taxa.

Acknowledgments This study was supported by funding to TFS from a "Pakt für Forschung und Innovation 2007" grant from the Leibniz Association. TFS would like to warmly thank Tom Mitchell-Olds for his foresight, encouragement, and access to seed material. We would also like to thank all members of the apomixis research group for their support, ideas and friendship throughout this project.

References

Asker S, Jerling L (1992) Apomixis in plants. CRC press, Boca Raton

Birky W Jr. C (1996) Heterozygosity, heteromorphy, and phylogenetic trees in asexual eukaryotes. Genetics 144: 427–437

Böcher TW (1951) Cytological and embryological studies in the amphi-apomictic *Arabis holboellii* complex. Kong Dansk. Vidensk Selsk Biol Skr 6: 1–59

Borevitz JO, Hazen SP, Michael TP, Morris GP, Baxter IR et al. (2007) Genome-wide patterns of single-feature polymorphism in *Arabidopsis thaliana*. Proc Natl Acad Sci USA 104: 12057–12062

Bruford MW, Wayne RK (1993) Microsatellites and their application to population genetic studies. Curr Opin Genet Dev 3: 939–943

Burt A, Carter DA, Koenig GL, White TJ, Taylor JW (1996) Molecular markers reveal cryptic sex in the human pathogen *Coccidioides immitis*. Proc Natl Acad Sci USA 93: 770–773

Butlin RK (2000) Virgin rotifers. Trends Ecol Evol 15: 389–390

Clauss MJ, Cobban H, Mitchell-Olds T (2002) Cross-species microsatellite markers for elucidating population genetic structure in *Arabidopsis* and *Arabis* (Brassicaeae). Mol Ecol 11: 591–601

Delmotte F, Sabater-Munoz B, Prunier-Leterme N, Latorre A, Sunnucks P, Rispe C, Simon J-C (2003) Phylogenetic evidence for hybrid origins of asexual lineages in an aphid species. Evolution 57: 1291–1303

Dobeš C, Mitchell-Olds T, Koch MA (2004a) Extensive chloroplast haplotype variation indicates Pleistocene hybridization and radiation of North American *Arabis drummondii, A.* × *divaricarpa*, and *A. holboellii* (Brassicaceae). Mol Ecol 13: 349–370

Dobeš C, Mitchell-Olds T, Koch MA (2004b) Intraspecific diversification in North American *Boechera stricta* (= *Arabis drummondii*), *Boechera* x *divaricarpa*, and *Boechera holboellii*

(Brassicaceae) inferred from nuclear and chloroplast molecular markers – an integrative approach. Am J Bot 91: 2087–2101
Dobeš C, Sharbel TF, Koch M (2007) Towards understanding the dynamics of hybridization and apomixis in the evolution of the genus *Boechera* (Brassicaceae). System Biodivers 5: 321–331
Estoup A, Jarne P, Cornuet JM (2002) Homoplasy and mutation model at microsatellite loci and their consequences for population genetics analysis. Mol Ecol 11: 1591–1604
Forche A, Magee PT, Magee BB, May G (2004) Genome-wide single-nucleotide polymorphism map for *Candida albicans*. Eukaryot Cell 3: 705–714
Grimaldi M-C, Crouau-Roy B (1997) Microsatellite allelic homoplasy due to variable flanking sequences. J Mol Evol 44: 336–340
Halkett F, Simon JC, Balloux F (2005a) Tackling the population genetics of clonal and partially clonal organisms. Trends Ecol Evol 20: 194–201
Halkett F, Plantegenest M, Prunier-Leterme N, Mieuzet L, Delmotte F, Simon JC (2005b) Admixed sexual and facultatively asexual aphid lineages at mating sites. Mol Ecol 14: 325–336
Hall TA (1999) BioEdit: a user friendly biological sequence alignment editor and analysis program for Windows 95/98/NT. Nucl Acids Symp Ser 41: 95–98
Hegarty MJ, Hiscock SJ (2005) Hybrid speciation in plants: new insights from molecular studies. New Phytol 165: 411–423
Huang QY, Xu FH, Shen H, Deng HY, Liu YJ et al. (2002) Mutation patterns at dinucleotide microsatellite loci in humans. Am J Hum Genet 70: 625–634
Johnson SG (2006) Geographic ranges, population structure, and ages of sexual and parthenogenetic snail lineages. Evolution 60: 1417–1426
Judson PO, Normark BB (1996) Ancient asexual scandals. Trends Ecol Evol 11: 41–46
Kantama L, Sharbel TF, Schranz EM, Mitchell-Olds T, De Vries S, de Jong H (2007) Diploid apomicts of the *Boechera holboelli* complex display large scale chromosome substitutions and aberrant chromosomes. Proc Natl Acad Sci of USA 104: 14026–14031
Kashin AS, Anfalov VE, Demochko Iu A (2005) Allozyme variation in sexual and apomictic *Taraxacum* and *Pilosella* (Asteraceae) populations. Genetika 41: 203–215
Kimura M, Maruyama T, Crow JF (1963) The mutation load in small populations. Genetics 48: 1303–1312
Koch MA, Dobeš C, Mitchell-Olds T (2003) Multiple hybrid formation in natural populations: concerted evolution of the Internal Transcribed Spacer of nuclear ribosomal DNA (ITS) in North American *Arabis divaricarpa* (Brassicaceae). Mol Biol Evol 20: 338–350
Levinson G, Gutman GA (1987) Slipped-strand mispairing: a major mechanism for DNA sequence evolution. Mol Biol Evol 4: 203–221
Liu QL, Thomas VP, Williamson VM (2007) Meiotic parthenogenesis in a root-knot nematode results in rapid genomic homozygosity. Genetics 176: 1483–1490
Lunt DH (2008) Genetic tests of ancient asexuality in root knot nematodes reveal recent hybrid origins. BMC Evol Biol 8: 194
Luty JA, Guo Z, Willard HF, Ledbetter DH, Ledbetter S, Litt M (1990) Five polymorphic microsatellite VNTRs on the human X chromosome. Am J Hum Genet 46: 776–783
Lynch M, Burger R, Butcher D, Gabriel W (1993) The mutational meltdown in asexual populations. J Hered 84: 339–344
Mark Welch D, Meselson M (2000) Evidence for the evolution of bdelloid rotifers without sexual reproduction or genetic exchange. Science 288: 1211–1215
Mark Welch DB, Mark Welch JL, Meselson M (2008) Evidence for degenerate tetraploidy in bdelloid rotifers. Proc Natl Acad Sci USA 105: 5145–5149
Martens K, Rossetti G, Horne DJ (2003) How ancient are ancient asexuals? Proc R Soc Lond B 270: 723–729
Matzk F, Meister A, Schubert I (2000) An efficient screen for reproductive pathways using mature seeds of monocots and dicots. Plant J 21: 97–108
Maynard Smith J (1978) The evolution of sex. Cambridge University Press, Cambridge, UK

Mogg R, Batley J, Hanley S, Edwards D, O'Sullivan H, Edwards KJ (2002) Characterization of the flanking regions of *Zea mays* microsatellites reveals a large number of useful sequence polymorphisms. Theor Appl Genet 105: 532–543

Muller HJ (1964) The relation of recombination to mutational advance. Mutat Res 106: 2-9

Naumova TN, van der Laak J, Osadtchiy J, Matzk F, Kravtchenko A, Bergervoet J, Kamisetti SR, Boutilier K (2001) Reproductive development in apomictic populations of *Arabis holboellii* (Brassicaceae). Sex Plant Reprod 14: 195–200

Norton RA, Palmer SC (1991) The distribution, mechanisms and evolutionary significance of parthenogenesis in oribatid mites; In: Schuster R, Murphy PW (eds) The Acari: reproduction, development and life-history strategies. Chapman and Hall, London, pp. 107–136

Omilian AR, Cristescu ME, Dudycha JL, Lynch M (2006) Ameiotic recombination in asexual lineages of *Daphnia*. Proc Natl Acad Sci USA 103: 18638–18643

Paun O, Hörandl E (2006) Evolution of hypervariable microsatellites in apomictic polyploid lineages of *Ranunculus carpaticola*: directional bias at dinucleotide loci. Genetics 174: 387–398

Pouchkina-Stantcheva NN, McGee BM, Boschetti C, Tolleter D, Chakrabortee S, Popova AV, Meersman F, Macherel D, Hincha DK, Tunnacliffe A (2007) Functional divergence of former alleles in an ancient asexual invertebrate. Science 318: 268–271

Roy BA (1995) The breeding systems of six species of *Arabis* (Brassicaceae). Am J Bot 82: 869–877

Roy BA, Rieseberg LH (1989) Evidence for apomixis in *Arabis*. J Hered 80: 506–508

Roy MS, Geffen E, Smith D, Ostrander EA, Wayne RK (1994) Patterns of differentiation and hybridization in North American wolflike canids, revealed by analysis of microsatellite loci. Mol Biol Evol 11: 553–570

Rozas J, Sanchez-DelBarrio JC, Messeguer X, Rozas R (2003) DnaSP, DNA polymorphism analyses by the coalescent and other methods. Bioinformatics 19: 2496–2497

Schaefer I, Domes K, Heethoff M, Schneider K, Schön I, Norton RA, Scheu S, Maraun M (2006) No evidence for the "Meselson effect" in parthenogenetic oribatid mites (Oribatida, Acari). J Evol Biol 19: 184–193

Schlötterer C (2000) Evolutionary dynamics of microsatellite DNA. Chromosoma 109: 365–371

Schön I, Martens K (1998) DNA repair in an ancient asexual: a new solution for an old problem? J Nat Hist 32: 943–948

Schön I, Martens K (2003) No slave to sex. Proc R Soc London B 270: 827–833

Schranz ME, Dobeš C, Koch MA, Mitchell-Olds T (2005) Sexual reproduction, hybridization, apomixis, and polyploidization in the genus *Boechera* (Brassicaceae). Am J Bot 92: 1797–1810

Schranz ME, Windsor AJ, Song BH, Lawton-Rauh A, Mitchell-Olds T (2007) Comparative genetic mapping in *Boechera stricta*, a close relative of *Arabidopsis*. Plant Physiol 144: 286–298

Shaked H, Kashkush K, Ozkan H, Feldman M, Levy AA (2001) Sequence elimination and cytosine methylation are rapid and reproducible responses of the genome to wide hybridization and allopolyploidy in wheat. Plant Cell 13: 1749–1759

Sharbel TF, Mitchell-Olds T (2001) Recurrent polyploid origins and chloroplast phylogeography in the *Arabis holboellii* complex (Brassicaceae). Heredity 87: 59–68

Sharbel TF, Voigt ML, Mitchell-Olds T, Kantama L, de Jong H (2004) Is the aneuploid chromosome in an apomictic *Boechera holboellii* a genuine B chromosome? Cytogenet Genome Res 106: 173–183

Sharbel TF, Mitchell-Olds T, Dobes C, Kantama L, de Jong H (2005) Biogeographic distribution of polyploidy and B chromosomes in the apomictic *Boechera holboellii* complex. Cytogenet Genome Res 109: 283–292

Sharbel TF, Voigt M-L, Corral JM, Thiel T, Varshney A, Kumlehn J, Vogel H, Rotter B (2009) Molecular signatures of apomictic and sexual ovules in the *Boechera holboellii* complex. Plant J 58: 870–882

Song B-H, Clauss MJ, Pepper A, Mitchell-Olds T (2006) Geographic patterns of microsatellite variation in *Boechera stricta*, a close relative of *Arabidopsis*. Mol Ecol 15: 357–369

Sunnucks P, England PR, Taylor AC, Hales DF (1996) Microsatellite and chromosome evolution of parthenogenetic *Sitobion* aphids in Australia. Genetics 144: 747–756

Tamura K, Dudley J, Nei M, Kumar S (2007) MEGA4: Molecular evolutionary genetics analysis (MEGA) software version 4.0. Mol Biol Evol 24: 1596–1599

Voigt ML, Melzer M, Rutten T, Mitchell-Olds T, Sharbel TF (2007) Gametogenesis in the apomictic *Boechera holboellii* complex: the male perspective. In: Hörandl E, Grossniklaus U, Van Dijk P, Sharbel TF (eds) Apomixis: evolution, mechanisms and perspectives. ARG Gantner Verlag KG, Lichtenstein, pp. 235–258

Vowles EJ, Amos W (2004) Evidence for widespread convergent evolution around human microsatellites. PLoS Biol 2: e199

Webster MT, Hagberg J (2007) Is there evidence for convergent evolution around human microsatellites? Mol Biol Evol 24: 1097–1100

Weetman D, Hauser L, Carvalho GR (2002) Reconstruction of microsatellite mutation history reveals a strong and consistent deletion bias in invasive clonal snails, *Potamopyrgus antipodarum*. Genetics 162: 813–822

Windham MD, Al-Shehbaz IA (2006) New and noteworthy species of *Boechera* (Brassicaceae) I: sexual diploids. Harv Pap Bot 11: 61–88

Windham MD, Al-Shehbaz IA (2007) New and noteworthy species of *Boechera* (Brassicaceae) II: apomictic hybrids. Harv Pap Bot 11: 257–274

Xia X, Xie Z (2001) DAMBE: software package for data analysis in molecular biology and evolution. J Hered 92: 371–373

Chapter 24
Asexual Reproduction in Infectious Diseases

Thierry De Meeûs, Franck Prugnolle and Philip Agnew

Abstract Parasitic organisms in the strict sense (eukaryotes) represent a significant part of the general biodiversity which has been described and, with 179 species affecting people worldwide, are of relevance for mankind in particular. Contrary to the classical view, many of these species are clonal. For example, 72% of human parasites use this means of reproduction. Such parasites represent a major threat to human health. A cumulative inventory leads to an impressive total of 1339 million people being affected by clonal parasites worldwide. These clonal parasites can be classified into different groups depending on how and where asexuality takes place in the life cycle. The demography and population genetics of these groups differ, which is relevant for their treatment. Recent empirical studies have found that the sampling strategy used can dramatically influence how results are interpreted. Furthermore, the role of individual hosts and their gender has been identified as being important for some parasites and that these parasites harbour an unexpected amount of genetic diversity on a very local scale. These issues are discussed in terms of how they may influence the design of therapeutic strategies.

24.1 Introduction

Infectious diseases can be considered as predominant in the living world. Many of the species which have been described are infectious (De Meeûs and Renaud 2002), and all (or almost all) living organisms are exposed to the risk of infectious disease at some stage in their life (Poulin and Morand 2000). Asexual reproduction is omnipresent in infectious agents, especially so in prokaryotic organisms. Even when viruses and bacteria are excluded from consideration, many eukaryotic infectious agents (or parasites) undergo asexual reproduction. Asexual reproduction is challenging to population biologists because it drives the population genetics of

T. De Meeûs (✉)
IRD, UMR 177 IRD-CIRAD "Trypanosomoses" Centre International de Recherche-Développement sur l'Elevage en zone Subhumide (CIRDES), 01 BP 454, Bobo-Dioulasso 01, Burkina-Faso
e-mail: demeeus@mpl.ird.fr

I. Schön et al. (eds.), *Lost Sex*, DOI 10.1007/978-90-481-2770-2_24,

these organisms far away from classical expectations based on models with sexual reproduction. Furthermore, these departures from prediction also depend on how asexuality intervenes in the life cycle (see De Meeûs et al. 2006 for review). Recent studies have shown that different life cycles involving asexuality can strongly influence the distribution of genetic diversity and its maintenance in natural populations of these pathogenic agents, both within and among hosts (patients). It is thus relevant to study the consequences of clonality on the population genetic structure of such organisms. As advocated by many authors (cited in De Meeûs et al. 2007), this is not a pure academic matter as such studies can provide important clues as to the epidemiology of these organisms. Such findings are to be taken into account while evaluating the efficiency of drugs or vaccines against these parasites.

In this review, we try to assess how diverse clonal parasites are, how many human parasites use the clonal mode of reproduction in at least at one stage of their life cycle, which different kinds of life cycle are present and, using empirical data, how genetic diversity is distributed in these systems.

24.2 Infectious Diseases in the Living World

Parasitism, even when focusing only on eukaryotic pathogens, is widespread. All species are concerned by parasitism as being either a host or a parasite. Several attempts have been made to evaluate the proportion of organisms that are parasitic. Some estimates based on extrapolation provide approximations varying from 50% (Timm and Clauson 1988) to 70% (Price 1980) of all species being parasitic. Data based on species counts from published databases provide a more modest estimate of 30% of the ~2 million described eukaryotic species (De Meeûs and Renaud 2002). Nevertheless, a safe assumption is to consider each host species possessing at least one specific parasite (Poulin and Morand 2000). This would lead to an estimate with a lower bound close to Timm and Clauson's estimate. If we focus strictly on mankind, we can count 179 eukaryotic species that are recognised as pathogens or true parasites (obligate parasites) of humans (Table 24.1). Among these, 35 species are specific (Table 24.1), i.e., they are obligate parasites of *Homo sapiens* with no known reservoir host. If each free living species acts as a host as humans do, then parasitism is the most frequently used ecological niche in terms of species diversity. The picture is reinforced when one realises that many parasites are themselves hosts for other parasites. This phenomenon, called hyperparasitism, is widespread in parasitic and parasitoid arthropods (Samish and Rehacek 1999; Sullivan and Völkl 1999). It also occurs in many other groups, e.g., cestodes or nematodes are often parasitic towards other cestodes (Rego and Gibson 1989). Some fungi and microsporidia are also parasitic towards other parasitic protozoa (Sassuchin 1934). There is even a case of a nematode parasitizing a parasitic ciliate within the intestine of *Hyrax capensis* (Sassuchin 1934). Even if this last example is doubtful, it remains likely that parasites represent the most diverse and numerous ecological group in nature.

Table 24.1 The different species of eukaryotic parasites affecting humans, their type of reproductive cycle, specificity and numerical importance (at bottom of table)

Taxa	Species	Life
Excavata (12)		
Metamonadina (1)		
Diplomonadida (1)	*Giardia lamblia*	A, NSp
Trichomonadida (1)	*Trichomonas vaginalis*	A, Sp
Percolozoa (1)	*Naegleria fowleri*	A, NSp
Kinetoplastida (9)	*Trypanosoma brucei gambiense*	A, Sp
	T. brucei rhodesiense	A, NSp
	T. cruzi	A, NSp
	Leishmania braziliensis	A, NSp
	L. major	A, NSp
	L. tropica	A, NSp
	L. mexicana	A, NSp
	L. donovani	A, NSp
	L. peruviana	A, NSp
Chromoalveolata (12)		
Ciliata (1)	*Ballantidium coli*	A, NSp
Sporozoea (11)	*Toxoplasma gondii*	S, NSp
	Cryptosporidium parvum	S, NSp
	C. hominis	S, NSp
	C. canis	S, NSp
	C. felis	S, NSp
	C. maleagridis	S, NSp
	C. muris	S, NSp
	Plasmodium falciparum	S, Sp
	P. vivax	S, Sp
	P. ovale	S, Sp
	P. malariae	S, Sp
Unichonta (155)		
Amoebozoa (1)		
Archamoebae (1)	*Entamoeba histolytica*	A, Sp
Opistochonta (154)		
Platyhelmintha (21)		
Cestoidea (9)	*Echinococcus granulosus*	I, NSp
	E. multilocularis	I, NSp
	Diphyllobothrium latum	I, NSp
	Dipylidium caninum	Sex, NpS
	Hymenolepis nana	Sex, NSp
	H. diminuta	Sex, NSp
	Taenia multiceps	Sex, NSp
	T. saginata	Sex, Sp
	T. solium	Sex, Sp
Trematoda (12)	*Schistosoma japonicum*	I, NSp
	S. mekongi	I, NSp
	S. haematobium	I, NSp
	S. mansoni	I, NSp
	Fasciola hepatica	I, NSp
	F. gigantica	I, NSp
	Fasciolopsis buski	I, NSp

Table 24.1 (continued)

Taxa	Species	Life
	Paragonimus westermani	I, NSp
	Clonorchis sinensis	I, NSp
	Opistorchis Viverrini	I, NSp
	Heterophyes heterophyes	I, NSp
	Metagonimus yokogawai	I, NSp
Nematozoa (13)		
Nematoda (13)	*Capillaria philippensis*	Sex, NSp
	Trichinella spiralis	Sex, NSp
	Trichiuris trichiura	Sex, Sp
	Strongyloides stercoralis	Sex, NSp
	Necator americanus	Sex, Sp
	Ascaris lumbricoides	Sex, Sp
	Toxocara canis	Sex, NSp
	T. cati	Sex, NSp
	Enterobius vermicularis	Sex, Sp
	Dracunculus medinensis	Sex, Sp
	Wuchereria bancrofti	Sex, Sp
	Loa loa	Sex, Sp
	Onhocerca volvulus	Sex, Sp
Arthropoda (6)		
Arachnida (3)	*Sarcoptes scabiei*	Sex, Sp
	Demodex follicularum	Sex, Sp
	D. brevis	Sex, Sp
Hexapoda (3)	*Pediculus capitis*	Sex, Sp
	P. humanus	Sex, Sp
	Phtirius pubis	Sex, Sp
Eumyceta (114)		
Ascomycota (97)	*Aspergillus alliaceus*	A, NSp
	A. alutaceus	A, NSp
	A. atroviolaceus	A, NSp
	A. caesiellus	A, NSp
	A. candidus	A, NSp
	A. carneus	A, NSp
	A. chevalieri	A, NSp
	A. clavato-nanicus	A, NSp
	A. clavatus	A, NSp
	A. conicus	A, NSp
	A. deflectus	A, NSp
	A. fischerianus	A, NSp
	A. flavipes	A, NSp
	A. flavus	A, NSp
	A. fumigatus	A, NSp
	A. glaucus	A, NSp
	A. hollandicus	A, NSp
	A. janus	A, NSp
	A. japonicus	A, NSp
	A. nidulans	A, NSp
	A. niger	A, NSp
	A. niger var. awamorii	A, NSp

Table 24.1 (continued)

Taxa	Species	Life
	A. niveus	A, NSp
	A. ochraceus	A, NSp
	A. oryzae	A, NSp
	A. penicilloides	A, NSp
	A. reptans	A, NSp
	A. restrictus	A, NSp
	A. rubrobrunneus	A, NSp
	A. sejunctus	A, NSp
	A. spinosus	A, NSp
	A. sydowii	A, NSp
	A. tamarii	A, NSp
	A. terreus	A, NSp
	A. tetrazonus	A, NSp
	A. unguis	A, NSp
	A. ustus	A, NSp
	A. versicolor	A, NSp
	Blastomyces dermatitidis	A, NSp
	Candida albicans	A, NSp
	C. tropicalis	A, NSp
	C. glabrata	A, NSp
	C. parapsilosis	A, NSp
	C. krusei	A, NSp
	C. lusitaniae	A, NSp
	Coccidioides immitis	A, NSp
	C. posadasii	A, NSp
	Histoplasma capsulatum	A, NSp
	Paracoccidioides brasiliensis	A, NSp
	Sporothrix schenckii	A, NSp
	Epidermophyton floccosum	A, Sp
	Microsporum audouinii	A, Sp
	M. distortum	A, NSp
	M. ferrugineum	A, Sp
	M. vanbreuseghemii	A, NSp
	M. canis	A, NSp
	M. gypseum	A, NSp
	M. nanum	A, NSp
	Trichophyton concentricum	A, Sp
	T. interdigitale	A, Sp
	T. megnini	A, Sp
	T. mentagrophytes	A, NSp
	T. rubrum	A, Sp
	T. schoenleinii	A, Sp
	T. soudanense	A, Sp
	T. tonsurans	A, Sp
	T. verrucosum	A, NSp
	T. violaceum	A, Sp
	T. yaoundei	A, Sp
	Acremonium falciforme	A, NSp
	A. kiliense	A, NSp

Table 24.1 (continued)

Taxa	Species	Life
	A. recifei	A, NSp
	Fusarium solani	A, NSp
	Scopulariopsis brevicaulis	A, NSp
	Onychocola canadensis	A, NSp
	Scytalidium dimidiatum	A, NSp
	Hortaea werneckii	A, NSp
	Cladosporium elatum	A, NSp
	C. herbarum	A, NSp
	C. sphaerospermum	A, NSp
	C. cladosporioides	A, NSp
	Curvularia lunata	A, NSp
	Bipolaris spicifera	A, NSp
	B. australiensis	A, NSp
	B. hawaiiensis	A, NSp
	Exserohilum rostratum	A, NSp
	E. meginnisii	A, NSp
	E. longirostratum	A, NSp
	Exophiala jeanselmei	A, NSp
	E. moniliae	A, NSp
	E. spinifera	A, NSp
	Scedosporium prolificans	A, NSp
	Pseudallescheria boydii	A, NSp
	Ochroconis gallopavum	A, NSp
	Coniothyrium fuckelii	A, NSp
	Phialophora verrucosa	A, NSp
	Wangiella dermatitidis	A, NSp
Basidiomycota (11)	*Cryptococcus neoformans*	A, NSp
	Trichosporon cutaneum	A, NSp
	T. asteroides	A, NSp
	T. ovoides	A, NSp
	T. inkin	A, NSp
	T. asahii	A, NSp
	T. mucoides	A, NSp
	Malassezia furfur	A, NSp
	M. globosa	A, NSp
	M. sympodialis	A, NSp
	M. pachydermatis	A, NSp
Zygomycota (6)	*Absidia corymbifera*	A, NSp
	Rhizomucor pusillus	A, NSp
	R. miehei	A, NSp
	R. variabilis	A, NSp
	Rhizopus arrhizus	A, NSp
	R. microsporus	A, NSp

Total: 179 species, 35 specific (23%), 154 with clonality (72%), 128 A (83% of clonals), 15 I (10% of clonals), 11 S (7% of clonals)

In the column entitled "Life", the different kinds of life cycles correspond to Simple Life Cycle (A), Complex Life Cycle with one round of clonal propagation (I), Complex Life Cycle with Several rounds of clonal propagation (S) and pure sexual reproduction (Sex) (see text for details). The specificity is expressed by Sp (specific to humans) or NSp (not specific).

24.3 Clonality in Eukaryotic Infectious Agents

Assessing the importance of asexuality in eukaryote parasites is not a simple task. Not only because of the diversity and taxonomic distribution of parasitism, as described above, but because we need to agree on a definition of asexuality. First, reproduction is asexual if the offspring are produced by a single parent and are strictly genetically identical to it at the scale of the genome (with the exception of somatic mutations). Second, if to be asexual is to never reproduce sexually then it is nearly impossible to make a list of asexual parasites. To our knowledge, a lack of sexuality has only been demonstrated for few taxa, such as bdelloid rotifers (see Chapter 13; Mark-Welsh and Meselson 2000; strictly clonal), some oribatid mites (Maraun et al. 2004; see Chapter 12; strictly automictic) and darwinulid ostracods (see Chapter 11; Schön and Martens 2003; probably strictly clonal; see also Chapter 9 for clone definitions) and strict asexuality must indeed be rare. However, it may be interesting to know how many organisms use asexual reproduction (see Chapters 3 and Chapter 4). It will be seen that many parasites use asexual reproduction at least at one stage or another during their life cycle.

It can be useful to distinguish four different life cycles depending on when, how and where asexuality occurs (Fig. 24.1). These factors are important because they influence the demography and population genetics associated with these different life cycles. In the first life cycle, clonality may be absent in purely sexually reproducing organisms; note, this category does not exclude organisms reproducing by selfing. Clonality may be involved in life cycles where there is an obligate alternation between sexual and asexual reproduction. Here, two different categories are worth distinguishing. The first category involves a single (I) round of clonality where all sexually produced propagules that survive experience a single event of clonal propagation (with many propagules as in flukes or fewer as in armadillos). This kind of cycle is typical of trematodes (flukes). It can be contrasted with complex life cycles with several rounds of clonality (S) where the clonal phase corresponds to a succession of clonal generations, more or less numerous before the sexual phase. S is typical for numerous fungi and aphids. Finally, clonal and sexual reproduction can occur without having any particular or regular cycle (Acyclic = A). In these cases, some individuals reproduce sexually while others reproduce asexually, or each individual invests a particular amount of reproductive potential to one or the other reproductive mode at any time.

What proportion of parasitic eukaryotes fall into these four categories? To answer this question, we used the database developed by De Meeûs and Renaud (2002). The results are presented in Table 24.2. As shown, most parasitic species lack a clonal phase (90% of described parasite species). For the remaining species, most fall into the category S (6%), with the rest being I and A (2% each). This lack of asexual parasites probably also reflects a bias due to which species have been described, as has previously been stressed by De Meeûs and Renaud (2002). A good illustration of this bias is seen with the Cycliophora (all S). Only a single species has been described so far, but several thousand probably exist (De Meeûs and Renaud 2002).

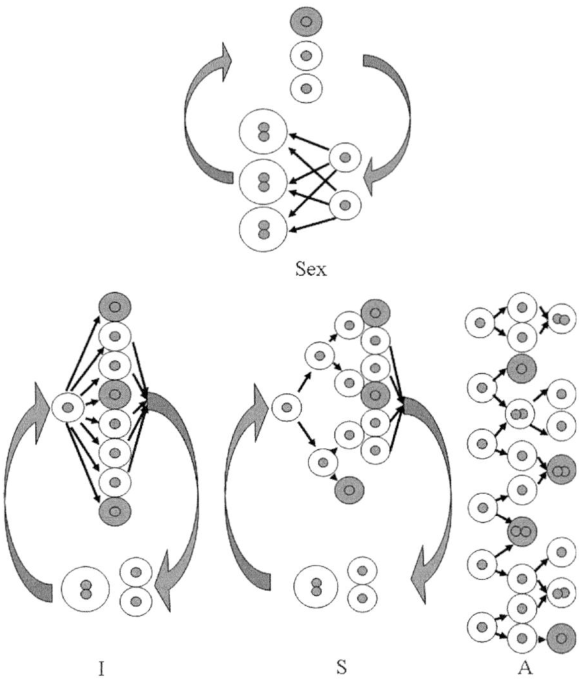

Fig. 24.1 The four categories of life cycles considered in this article. The purely sexual cycle (**Sex**) corresponds to organisms that can only reproduce through sexual reproduction. Complex life cycles with an instantaneous clonal phase with only one (**I**) clonal generation per cycle. Complex life cycles with Several generations of asexuality (**S**) where the clonal phase involves more than one clonal generation. In both cases, and for all surviving individuals, sexual reproduction (segregation and recombination) must intervene at one point in the cycle to form zygotes. Sexual reproduction may be more or less frequent (or even absent) with an acyclic pattern (**A**). In this case, the life cycle is not defined by a regular pattern of sexual or asexual reproduction. In the three cases, some individuals, symbolised in grey, die before they can reproduce

Parasite diversity is probably better described in humans and a glance at Table 24.1 provides an alternative view.

Most human parasites reproduce clonally at one stage or another in their life cycle (most being acyclic). As far as human health is concerned, clonal parasites are a major threat. From Table 1.1 in Bush et al. (2001), we can see that *Plasmodium* spp. (S), agents of Malaria, affect 300 million people, Distomatoses, provoked by trematode flatworms (flukes, schistosomes) (I) affect 221 million people, Amoebiasis (A) affects more than 500 million people, Giardiasis (A) affects 200 million people, and kinetoplastids (Chagas disease, African trypanosomiasis, leishmaniasis) (A) affect 118 million people worldwide, mostly in tropical and sub-tropical countries. Thus, the importance of clonal parasites is not purely anecdotic and finding ways to promote a better understanding of the population biology of these fairly unknown organisms is a worthy goal.

Table 24.2 Main parasitic eukaryotic taxa and the kind of life cycle (defined in the text and Fig. 24.1) they belong to (according to De Meeûs and Renaud 2002)

Main taxa	Species	Sex	I	S	A
Rhizaria	60				Most
Ciliata	1800				All
Sporozoa	5000			Most	
Dinoflagelata	72				Most
Parabasalia	2000				All
Metamonadina	300				All
Kinetoplastida	580				All
Plantae	1612	Most			
Oomycota	800			All	
Opalina	400			All	
Rhizopoda	40				All
Microsporidia	800		Some?		Some?
Eumycota	19731			Most	~100(1)
Myxozoa	1200		All		
Chordata	223	All			
Orthonectida	25		All		
Annelida	788	Most			
Eumollusca	6889	All			
Rhombozoa	75				All
Myzostomida	170	All			
Cycliophora	1			All	
Trematoda	5080		All		
Monogenea	5000	4598	402 (2)		
Cestoda	5014	Most	Few		
Rotifera	1152	All			
Nematozoa	16290	Most			
Arthropoda	396989	Most			
All	~500000	~450000	~10000	~30000	~10000

(1) Deuteromycota, (2) Gyrodactylids (from Bakke et al. 2002).

24.4 Studying Populations of Clonal Parasites

Direct observations of clonal parasites are generally rendered difficult by their small size. Indirect methods, using molecular tools can provide good alternatives to study their ecology (Criscione et al. 2005; De Meeûs et al. 2007). However, depending on the life cycle involved, different strategies must be used. The combination of drift, variance in reproductive success, and clonal propagation generate replicates of genetically identical individuals. This will translate into what we call repeated multilocus genotypes. Most authors recommend only considering one isolate per unit of sampling (typically the patient for medical surveys), or deleting repeated genotypes from the data set (e.g., Shaw et al. 1994; Boerlin et al. 1996; Sunnuck et al. 1997; Arnavielhe et al. 2000; Delmotte et al. 2002; Fundyga et al. 2002). Such strategies may not always be ideal (De Meeûs et al. 2006) and how we must deal with these replicated genotypes depends on which kind of cycle is present and when individuals were sampled during this cycle.

Most publications dealing with population genetics inferences on clonal organisms use linkage disequilibrium measures and/or tree construction to infer the extent of clonality of the studied populations. However, as underlined in recent papers (De Meeûs and Balloux 2004; De Meeûs et al. 2006), linkage disequilibrium hardly reflects the reproductive system because it is very sensitive to population size and migration rate (e.g., Bartley et al. 1992; Vitalis and Couvet 2001a, b). Moreover, these two methods require a full description of the genetic diversity present at different levels, which is rarely achieved (see a discussion of this in Nébavi et al. 2006 and De Meeûs et al. 2007). Without this information, inaccurate interpretations of the data are possible. For example, this was the case in the analysis of *Trypanosoma brucei* isolates by Maynard-Smith et al. (1993) where several species of trypanosomes were gathered in the same sample and lead to inappropriate inferences (MacLeod et al. 2000; De Meeûs and Balloux 2005). Because of these problems and because such methods were extensively reviewed elsewhere (Milgroom 1996; Taylor et al. 1999; Tibayrenc and Ayala 2002; Halkett et al. 2005; De Meeûs et al. 2006 and references therein) we will mainly focus on diploid organisms and on more recent methods mostly developed in our laboratory.

24.4.1 I Parasites

In the case of I parasites, it has been shown that a better representation of a population's genetic structure is obtained if the samples are analysed without repeated genotypes, i.e., keeping a single representative for each multilocus genotype (Prugnolle et al. 2005a, c). However, discarding the information contained in repeated genotypes is not the best solution either, as we will see from an extensive study made on the population genetics of *Schistosoma mansoni* in Guadeloupe (French West Indies).

Schistosomiasis is a disease affecting 200 million people of which 20 million suffer from severe illness (Chitsulo et al. 2000). It is caused by trematode flatworms of the genus *Schistosoma*. Schistosomes are dioecious (two sexes) blood parasites of mammals and birds. The sex is determined by a chromosome pair ZW (as in birds): females are ZW and males are ZZ. These worms display a complex life cycle of the I type. In the vertebrate definitive host, adults mate and females release thousands of eggs (hundreds per day). These eggs have to find water, where they hatch into free swimming miracidia that must find an aquatic mollusc to infect. Within the mollusc host the parasite clonally produces thousands of cercariae that must find a suitable definitive host to complete the cycle. *S. mansoni* is responsible for intestinal schistosomiasis in Africa and South America. In Guadeloupe it no longer infects the human population and is found in the marshy forests of Grande-Terre Island where it relies on black rats (*Rattus rattus*) as its definitive host. The population genetic structure of adult populations of *S. mansoni* collected from rats in five foci of infection were analysed using seven microsatellite loci. The smallest scale of investigation was the parasite infra-population, which corresponds to parasites from within individual hosts. The authors observed a sex-biased genetic structure in the parasite population, between parasite infra-populations. There was genetic differentiation between females and males from single hosts. Furthermore genetic differentiation among infra-populations arising from different hosts was significantly more pronounced for female schistosomes (Prugnolle et al. 2002). This strange pattern remained even when a single representative for each multilocus genotype was used. It was shown that this pattern of population structure is in fact mainly produced by genotypic (heterozygosity) and sex-dependent clonal success (Prugnolle et al. 2004, 2005c). Indeed, the most heterozygous clones propagate better and the variance of this clonal propagation is stronger for females than for males. In these studies, excluding repeated genotypes was useful for studying population differentiation and genetic diversity of parasites from male and female hosts (Prugnolle et al. 2005b; Caillaud et al. 2006). Indeed, including all sampled individuals tended to increase variance in parameter estimates, leading to a severe drop in statistical power. Taking into account the information provided by repeated genotypes may however allow a better resolution and understanding of clonal reproduction and its consequences for the distribution of genetic variability at different scales.

24.4.2 S parasites

In the case of S parasites, it is hard to propose a clear answer as to how they should be analysed. Sampling is critical and when during the cycle sampling occurs, is highly important. If individuals are sampled just after sexual reproduction has occurred, then there should be no need to manipulate data. If individuals are sampled during the clonal phase, then analysing data with and without repeated genotypes is recommended. In most cases, single repeats should provide the most accurate parameter estimates, as in the I case above.

Plasmodium falciparum is the agent of the most malignant form of malaria and accounts for more than one million deaths each year (Razakandrainibe et al. 2005). For this hermaphroditic protozoan, sexual reproduction occurs in the mosquito vector (*Anopheles* sp.). This is quickly followed by meiosis and a clonal phase that begins in the mosquito and is followed by many clonal generations in the human host, after injection by the mosquito vector. During most of its life cycle this parasite is asexual and haploid. It is commonly admitted that its rate of outbreeding varies across areas with different intensities of transmission. In sites with low transmission, the rate of outbreeding is believed to be low and, hence, the resulting population structure is sometimes awkwardly referred as a "clonal population structure" (e.g., Razakandrainibe et al. 2005). On the contrary, in sites of high transmission, the rate of outbreeding increases and is believed to be nearly panmictic (e.g., Razakandrainibe et al. 2005 and references therein). Such inferences are mainly based on data from parasites sampled from the blood of patients, and thus during the haploid phase. Such conclusions rely on linkage disequilibrium measures but, because of multi-infections, only major bands of the genetic marker used (the "dominant" allele at each locus) are kept or lots of data are excluded (Anderson et al. 2000; Leclerc et al. 2002), which drastically reduces the power to accurately infer the genetic structure of the population. Razakandrainibe et al. (2005) studied microsatellite variation in the infra-populations of *P. falciparum* inside the guts of its vector *Anopheles gambiae*. The study of the diploid phase of the parasite revealed a strong level of inbreeding. This was partially explained by selfing (F_{IS}≈0.15) however most was due to infra-population subdivision (F_{ST}≈0.36) among mosquitoes. Because most *A. gambiae* probably had a single blood meal (Koella et al. 1998) this strong partitioning also reflects infra-population structure in human hosts and thus the clonal and haploid phase of the cycle. Here, because individual parasites were sampled just after sexual reproduction as zygotes, there was no need to remove repeated genotypes. Consequently, because of a strong subdivision of *P. falciparum* into numerous hosts, together with a substantial amount of selfing (≈0.26), the strong inbreeding observed within individual hosts is accompanied by strong genetic diversity across hosts in areas of high transmission.

24.4.3 A Parasites

In the case of A parasites, theoretical studies show that keeping all genotyped individuals for analyses is the best possible strategy. This is particularly so for diploid species where F_{IS} measures appear very informative as to the structure of the population (Balloux et al. 2003; De Meeûs and Balloux 2005; De Meeûs et al 2006), when combined with an optimal sampling strategy (De Meeûs and Balloux 2005; De Meeûs et al. 2006). This was illustrated by the application of these concepts to study populations of the opportunistic yeast *Candida albicans* in AIDS patient from Abidjan (Côte d'Ivoire) (Nébavi et al. 2006).

Candida albicans is a diploid fungus of the gastrointestinal and genitourinary flora of most healthy humans and other mammals (e.g., Hull et al. 2000; Berman and Sudbery 2002). In immunocompromised patients, *C. albicans* may invade host tissues. For instance, HIV patients frequently suffer from recurrent oral candidiasis. This in itself is not life threatening, but if the parasite gains access to the blood stream it may cause severe damage (e.g., Hull et al. 2000; Berman and Sudbery 2002; Bougnoux et al. 2004). This yeast is the main cause of nosocomial fungal infections (e.g., Verduyn Lunel et al. 1999; Gupta et al. 2004; Correira et al. 2004). To date, the population biology of *C. albicans* remains unclear with respect to population size, transmission rate, and reproductive strategy.

In Nébavi et al. (2006), attention was given to sample within individual hosts (patients) as putative sub-populations. Because *C. albicans* is suspected to be strongly clonal (many repeated genotypes), it was helpful to sample an equal number of isolates per patient (here five). A balanced sampling design is the best way to generate a fair representation of the population without the need to use complex weighting. The work of Nébavi et al. (2006) concerned 42 HIV-positive patients with oral candidiasis from Abidjan and the periphery of the town (19 females and 23 males). All patients were sampled by buccal swabbing at arrival in hospital and then treated with anti-fungal drugs. After two weeks, 13 patients relapsed (four females and nine males) and were successfully treated again after another buccal swabbing. All isolates (five per swabbing) were genotyped with 14 enzymatic (allozyme) loci. The data analysis was complex (see Nébavi et al. 2006 for details) but clearly revealed a high degree, if not absolute, of clonality within patients and a strong genetic structure among patients. This was suggested by strong heterozygote excess within patients (as compared to Hardy-Weinberg expectations), a Wright's $F_{IS}=-0.85$ (-1 is the minimum possible value), and a differentiation between individual patients of $F_{Patient}=0.5$, which is the maximum possible value for pure clones (De Meeûs and Balloux 2005; De Meeûs et al. 2006). Interestingly, important differences in population structure both within and between patients were found for female and male patients. These differences would be compatible with different sub-population sizes and/or rare events of sexual reproduction occurring in female patients (yeasts maintain a higher genetic diversity within individual female patients and a higher variance in F_{IS} is found for females than for males). The assumption of pure clonality agrees with the data when some loci are removed: those loci are only polymorphic in one gender (six loci) and one locus with $F_{IS}=1$ in females (only homozygous) and $F_{IS}=-1$ in males (heterozygous in a single patient). Removing these loci provides data compatible with pure (or almost pure) clonality in *C. albicans* and a strongly structured population with many sub-populations of two kinds: (i) female patients maintaining more strains within individual patients than males, and (ii) male patients with smaller sub-populations. Heterozygote excess is so strong, even in female patients, that it is incompatible with a rate of sexual reproduction above 0.005 (1/total number of isolates). If indeed sex does exist, simulations (unpublished) suggest the rate would be much lower because rare sexual events leave a strong signature on population genetics parameters and these were not found in the study. In particular, F_{IT} should be above 0 while it is never different from 0, as expected in fully clonal and strongly subdivided populations. We cannot exclude, however, that some recombination events may occur so rarely that it is nearly impossible to detect them. Here, global genetic diversity is high as illustrated by the fact that there were almost as many multilocus genotypes (37) as patients (42) sampled. All these conclusions drawn from "old fashioned" allozymes were made possible because sampling was exceptional (several isolates per patient, sampling was balanced and involved many loci) and used methods based on heterozygosity. This kind of meticulous sampling is not reflected in the general behaviour of mycologists where the gold standard of sampling still consists of one isolate per patient, and comparing patients from different origins and date. This kind of sampling seriously tends to limit conclusions in terms of the population biology of the fungus under study (see Nébavi et al. 2006 and references therein for an extended discussion).

24.5 Discussion

The absence of sex, besides being difficult to prove (see also Chapter 11), seems to be extremely rare in eukaryotes. However, this does not mean that absence of clonality is frequent, far from it. Organisms practicing asexual reproduction (in the strict genetic sense), in at least at one stage of their life cycle are not rare, especially if one looks at parasite species (30% of all described species), 10% of which use clonal reproduction to multiply. A glance at the well documented guild of human parasites (179 species) provides a stronger picture as 72% of these species appear to be clonal and together they pose a health threat to more than a billion people worldwide.

The population biology of most, if not all, of these parasites can only be assessed by molecular markers and population genetic tools that need to be adapted to this particular reproductive mode. Hereby, the life cycle is important to consider and thus must be known, at least approximately, as data will not be handled in the same way for parasites with simple life cycles or complex life cycles. Another important factor to take into account is where in the life cycle individuals (or isolates) were sampled (just after sex or not). Obviously, a complex cycle will make data analysis more complex. In any case, sampling is critical and in the three examples provided in this review, individual hosts (or patients) played a significant (schistosomes in rats) or major (*Plasmodium* and *Candida*) role in the population biology and in shaping the distribution of genetic diversity in parasite populations that would have been overlooked if a classical sampling design had been used instead (of one parasite or isolate per host). These recent studies highlighted high levels of genetic diversity being maintained at local scales and revealed a combined role of drift, selection and sometimes of unexpected factors, such as host gender. These high levels of polymorphism maintained by a balance between subdivision, clonality and drift (and probably clonal selection) can only be beneficial to the evolutionary potential of these clonally propagating parasites. This may even partly explain how easy it is for them to develop adaptive mechanisms against therapeutic drugs or escape the effects of vaccines.

Clonality does not necessarily lead to an evolutionary "cul-de-sac" after all, at least for parasites.

Acknowledgments The authors are financed by the CNRS and IRD. We thank Anne-Laure Bañuls for very useful discussions.

References

Anderson TJC, Haubold B, Williams JT, Estrada-Franco JG, Richardson L, Mollinedo R, Bockarie M, Mokili J, Mharakurwa S, French N, Whitworth J, Velez ID, Brockman AH, Nosten F, Ferreira MU, Day KP (2000) Microsatellite markers reveal a spectrum of population structures in the malaria parasite *Plasmodium falciparum*. Mol Biol Evol 17: 1467–1482

Arnavielhe S, de Meeûs T, Blancart A, Mallié M, Renaud F, Bastide JM (2000) Multicentric study of *Candida albicans* isolates from non-neutropenic patients: population structure and mode of reproduction. Mycoses 43: 109–117

Bakke TA, Harris PD, Cable J (2002) Host specificity dynamics: observations on gyrodactylid monogeneans. Int J Parasitol 32: 281–308

Balloux F, Lehmann L, De Meeûs T (2003) The population genetics of clonal or partially clonal diploids. Genetics 164: 1635–1644

Bartley D, Bagley M, Gall G, Bentley B (1992) Use of linkage disequilibrium data to estimate effective size of hatchery and natural fish populations. Conserv Biol 6: 365–375

Berman J, Sudbery PE (2002) *Candida albicans*: a molecular revolution built on lessons from budding yeast. Nature Rev Genet 3: 918–928

Boerlin P, Boerlin-Petzold F, Goudet J, Durussel C, Pagani JL, Chave JP, Bille J (1996) Typing *Candida albicans* oral isolates from human immunodeficiency virus-infected patients by multilocus enzyme electrophoresis and DNA fingerprinting. J Clinical Microbiol 34: 1235–1248

Bougnoux ME, Aanensen DM, Morand S, Théraud M, Spratt BG, d'Enfert C (2004) Multilocus sequence typing of *Candida albicans*: strategies, data exchange and applications. Infect Genet Evol 4: 243–252

Bush AO, Fernández JC, Esch GW, Seed JR (2001) Parasitism: the diversity and ecology of animal parasites. Cambridge University Press, Cambridge, UK

Caillaud D, Prugnolle F, Durand P, Théron A, De Meeûs T (2006) Host sex and parasite genetic diversity. Microbes Infect 8: 2477–2483

Chitsulo L, Engels D, Montresor A, Savioli L (2000) The global status of schistosomiasis and its control. Acta Trop 77: 41–51

Correira A, Sampaio P, Almeida J, Pais C (2004) Study of molecular epidemiology of candidiasis in Portugal by PCR fingerprinting of *Candida* clinical isolates. J Clin Microbiol 42: 5899–5903

Criscione CD, Poulin R, Blouin MS (2005) Molecular ecology of parasites: elucidating ecological and microevolutionary processes. Mol Ecol 14: 2247–2257

Delmotte F, Leterme N, Gauthier JP, Rispe C, Simon JC (2002) Genetic architecture of sexual and asexual populations of the aphid *Rhopalosiphum padi* based on allozyme and microsatellite markers. Mol Ecol 11: 711–723

De Meeûs T, Balloux F (2004) Clonal reproduction and linkage disequilibrium in diploids: a simulation study. Infect Genet Evol 4: 345–351

De Meeûs T, Balloux F (2005) *F*-statistics of clonal diploids structured in numerous demes. Mol Ecol 14: 2695–2702

De Meeûs T, Lehmann L, Balloux F (2006) Molecular epidemiology of clonal diploids: a quick overview and a short DIY (Do It Yourself) notice. Infect Genet Evol 6: 163–170

De Meeûs T, McCoy KD, Prugnolle F, Chevillon C, Durand P, Hurtrez-Boussès S, Renaud F (2007) Population genetics and molecular epidemiology or how to "débusquer la bête". Infect Genet Evol 7: 308–332

De Meeûs T, Renaud F (2002) Parasites within the new phylogeny of eukaryotes. Trends Parasitol 18: 247–251

Fundyga RE, Lott TJ, Arnold J (2002). Population structure of *Candida albicans*, a member of the human flora, as determined by microsatellite loci. Infect Genet Evol 2: 57–68

Gupta N, Haque A, Lattif AA, Narayan RP, Mukhopadhyay G, Prasad R (2004) Epidemiology and molecular typing of *Candida* isolates from burn patients. Mycopathologia 158: 397–405

Halkett F, Simon JC, Balloux F (2005) Tackling the population genetics of clonal and partially clonal organisms. Trends Ecol Evol 20: 194–201

Hull CM, Raisner RM, Johnson AD (2000) Evidence for mating of the "Asexual" yeast *Candida albicans* in a mammalian host. Science 289: 307–310

Koella JC., Sørensen FL, Anderson RA (1998) The malaria parasite, *Plasmodium falciparum*, increases the frequency of multiple feeding of its mosquito vector, *Anopheles gambiae*. Proc R Soc Lond B 265: 763–768

Leclerc MC, Durand P, De Meeûs T, Robert V, Renaud F (2002) Genetic diversity and population structure of *Plasmodium falciparum* from Senegalense samples investigated by nineteen microsatellites and four genes coding for antigenic determinants. Microbes Infect 4: 685–692

MacLeod A, Tweedie A, Welburn SC, Maudlin I, Turner CMR, Tait A. (2000) Minisatellite marker analysis of *Trypanosoma brucei*: Reconciliation of clonal, panmictic, and epidemic population genetic structures. Proc Natl Acad Sci USA 97: 13442–13447
Maraun M, Heethoff M, Schneider K, Scheu S, Weigmann G, Cianciolo J, Thomas RH, Norton RA (2004) Molecular phylogeny of oribatid mites (Oribatida, Acari): evidence for multiple radiations of parthenogenetic lineages. Exp Appl Acarol 33: 183–201
Mark-Welsh D, Meselson M (2000) Evidence for the evolution of bdelloid rotifers without sexual reproduction or genetic exchange. Science 288: 1211–1215
Maynard-Smith J, Smith NH, Orourke M, Spratt BG (1993) How clonal are bacteria? Proc Natl Acad Sci USA 90: 4384–4388
Milgroom MG (1996) Recombination and the multilocus structure of fungal populations. Annu Rev Phytopathol 34: 457–477
Nébavi F, Ayala FJ, Renaud F, Bertout S, Eholié S, Koné M, Mallié M, de Meeûs T (2006) Clonal population structure and genetic diversity of *Candida albicans* in AIDS patients from Abidjan (Côte d'Ivoire). Proc Natl Acad Sci USA 103: 3663–3668
Poulin R, Morand S (2000) The diversity of parasites. Quart Rev Biol 75: 277–293
Price PW (1980) Evolutionary biology of parasites. Princeton University Press, Princeton
Prugnolle F, Choisy M, Théron A, Durand P, De Meeûs T (2004) Sex-specific correlation between heterozygosity and clone size in the trematode *Schistosoma mansoni*. Mol Ecol 13: 2859–2864
Prugnolle F, De Meeûs T, Durand P, Sire C, Théron A (2002) Sex specific genetic structure in *Schistosoma mansoni*: evolutionary and epidemiological implications. Mol Ecol 11: 1231–1238
Prugnolle F, Liu H, De Meeûs T, Balloux F (2005a) Population genetics of complex life cycle parasites: the case of monoecious trematodes. Int J Parasitol 35: 255–263
Prugnolle F, Théron A, Pointier JP, Jabbour-Zahad R, Jarne P, Durand P, De Meeûs T. (2005b) Dispersal in a parasitic worm and its two hosts and its consequence for local adaptation. Evolution 59: 296–303
Prugnolle F, Roze D, Théron A, De Meeûs T (2005c) F-statistics under alternation of sexual and asexual reproduction: a model and data from schistosomes. Mol Ecol 14: 1355–1365
Razakandrainibe FG, Durand P, Koella JC, De Meeûs T. Rousset F, Ayala FJ, Renaud F (2005) "Clonal" population structure of the malaria agent *Plasmodium falciparum* in high-infection regions. Proc Natl Acad Sci USA 102: 17388–17393
Rego AA, Gibson DI (1989) Hyperparasitism by helminths: new records of cestodes and nematodes in proteocephalid cestodes from South American siluriform fishes. Mem Instit Oswaldo Cruz, Rio de Janeiro 84: 371–376
Samish M, Rehacek J (1999) Pathogens and predators of ticks and their potential in biological control. Annu Rev Entomol 44: 159–182
Sassuchin DN (1934) Hyperparasitism in protozoa. Quart Rev Biol 9: 215–224
Schön I, Martens K (2003) No slave to sex. Proc R Soc Lond Ser B 270: 827–833
Shaw P, Ryland JS, Beardmore JA (1994) Population genetic parameters within a sea anemone family (Sagartiidae) encompassing clonal, semiclonal and aclonal modes of reproduction. In: Beaumont, A (ed) Genetics and evolution of aquatic organisms. Chapman and Hall, London, pp. 351–358
Sullivan DJ, Völkl W (1999) Hyperparasitism: multitrophic ecology and behaviour. Annu Rev Entomol 44: 291–315
Sunnuck P, De Barro PJ, Lushai G., MacLean N, Hales D (1997) Genetic structure of an aphid studied using microsatellites: cyclic parthenogenesis, differentiated lineages and specialization. Mol Ecol 6: 1059–1073
Taylor JW, Geiser DM, Burt A, Koufopanou V (1999) The evolutionary biology and population genetics underlying fungal strain typing. Clin Microbiol Rev 12: 126–146
Tibayrenc M, Ayala FJ (2002) The clonal theory of parasitic protozoa: 12 years on. Trends Parasitol 18: 405–410
Timm RM, Clauson BL (1988) Coevolution: Mammalia. In: 1988 McGraw-Hill yearbook of science & technology. McGraw-Hill Book Company, New York, pp. 212–214

Verduyn Lunel F, Meis JFGM,Voss A (1999) Nosocomial fungal infections: Candidemia. Diagn Microbiol Infect Dis 34: 213–220
Vitalis R, Couvet D (2001a) ESTIM 1.0: a computer program to infer population parameters from one- and two-locus gene identity probabilities. Mol Ecol Notes 1:354–356
Vitalis R, Couvet D (2001b) Estimation of effective population size and migration rate from one- and two-locus identity measures. Genetics 157:911–925

Chapter 25
What's in a Clone: The Rapid Evolution of Aphid Asexual Lineages in Relation to Geography, Host Plant Adaptation and Resistance to Pesticides

Hugh D. Loxdale

Abstract The term "clone", coined over a hundred years ago (see Chapter 9), is still in common parlance and widely used throughout the world, but its usage depends on definition, which is still obscure, especially if this involves the concept of genetic fidelity between clone mates rather than just the offspring from an asexual female founder (more correctly, an asexual lineage). To date, there have been no DNA sequencing studies proving such fidelity; on the contrary, the various DNA molecular marker studies performed on aphids display widespread genetic variation within and between different clonal lineages, as expected since mutation is a fundamental property of the DNA and hence the genome itself. In this overview, I use aphids as a model system to show that, rather than being an unchanging evolutionary "dead end", asexual aphid lineages show rapid and widespread adaptive changes to changing ecological conditions in the field, including in relation to geography, host plant factors, and to insecticide applications. This being so, the so-called clone cannot be a fixed entity in time and space, but like all other living organisms in the real world, is evolving in response to its environment.

25.1 Introduction

Aphids (Insecta: Hemiptera: Aphididae), like all other living organisms, are not immune from the effects of mutation, selection/genetic drift, adaptation, and extinction (Thompson 1994). With this in mind, it is thus strange that the members of such a lineage are still assumed by some scientists to be "genetically identical" to their stem mother, and thereby, with little genetic/functional plasticity to be able to adapt in the face of changing ecological circumstances (Lushai et al. 2003). If the aphid "clone" *sensu stricto* really did exist (see Chapter 9 for further discussions),

H.D. Loxdale (✉)
Institute of Ecology, Friedrich Schiller University, Dornburger Str. 159, 07743 Jena, Germany; Department of Entomology, Max Planck Institute for Chemical Ecology, Hans-Knoell-Strasse 8, D-07745 Jena, Germany
e-mail: Hugh.Loxdale@uni-jena.de; hloxdale@ice.mpg.de

I. Schön et al. (eds.), *Lost Sex*, DOI 10.1007/978-90-481-2770-2_25,

a natural population of these organisms would be represented by a population mean with no variance for any given trait/s in question, a bizarre state of affairs (see Fig. 1 in Loxdale and Lushai 2003a) and indeed, a Creationist's dream! This is surely an incorrect view of the real world that we know from studies of numerous other organisms (Loxdale and Lushai 2003b).

In reality, due to their fast rate of asexual reproduction (parthenogenetic, apomictic) and short generation time ($\sim$ 10 days), and involving "telescoping of generations" (Dixon 1998), a single virgin female of a given aphid species, e.g., the Peach-potato aphid, *Myzus persicae* (Sulzer), can typically give rise to 30–90 offspring (Blackman 1971). Under ideal conditions (climatic and with a dearth of predators, parasitoids and pathogens), this can potentially result in billions of individuals derived from one individual in a single growing season (Dixon 1989; Harrington 1994). Even at typical mutation rates of 10^{-9} to 10^{-6} per gene per generation, a large number of mutant aphids are likely to be produced with such an astronomical rate of reproduction. For example, in one large alfalfa growing region in California alone, it has been calculated that some 1.7×10^{11} Spotted Alfalfa aphids, *Therioaphis trifolii* form *maculata* (Buckton) were produced in a couple of growing seasons (Dickson 1962). At a conservative mutation rate of 10^{-7}, this means that at any given locus, around 17,000 mutations might arise. The large majority of these mutations are likely to be mildly deleterious and at non-coding regions of the genome; however, others are probably important, that is to say, at coding regions and are thus positively or negatively selected for (Lambert and Moran 1998; Lynch and Blanchard 1998; Fry et al. 1999; Korona 2004; Begin and Schoen 2006; Paland and Lynch 2006; Barraclough et al. 2007; see also Vorwerk and Forneck 2007). In the case of the Spotted Alfalfa aphid, some mutations were shown to confer resistance to organo-phosphates insecticides, so that the insect became resistant to these pesticides within a relatively short time after its introduction into the USA in the early 1950s (Dickson 1962; Blackman and Eastop 2000).

In 1837, Charles Darwin (1809–1882) stated in his "B" notebook, relating to the phenotype of animals and plants (since genes had yet to be identified) that "If all organisms merely replicated their kind by vegetative budding or splitting, history would show a succession of identical individuals holding no potential for alterations of any kind" (Browne 1996). In this century, Simon et al. (2003a) mainly relating to the genotype of organisms in a review on parthenogenesis, further argued that "It appears that most unisexual taxa occupy terminal nodes of phylogenetic trees. This suggests that, even if initially successful, they are evolutionary 'dead-ends'". Yet there are certainly exceptions to this pattern. For example, bdelloid rotifers and some ostracod species have existed totally asexually for aeons (40–200 myr), whilst bdelloids are very species "rich" with some 350 spp. and appear to be adaptively radiating and speciating (Mark Welch and Meselson 2000; Schön et al. 2003; Martens et al. 2003; Birky et al. 2005; see also Chapters 11 and 13).

In relation to aphids, Dan Janzen argued in 1977 that because both aphids and dandelions have asexual phases during which the offspring produced are "genetically identical", the members of such a clone are in effect a super-organism, i.e., a single "evolutionary individual", and are thereby able to exploit a much larger geographic region and its resources.

In the present chapter, I wish to show how the use of biochemical techniques, particularly protein-based genetic markers (allozymes) and more recently, DNA-based markers (reviewed by Loxdale and Lushai 1998; Loxdale 2001; Behura 2006) have transformed our understanding of what constitutes a clone (asexual lineage) as well as higher levels of evolutionary divergence, whilst even what constitutes an aphid population is now under scrutiny and re-assessment. This has occurred as new molecular knowledge has changed once well-established notions concerning morphology, genetics and evolution. Since I have spent most of my career studying cereal aphid pests, I shall tend to concentrate on these.

25.2 The Nature of the Clone

25.2.1 Evidence for Variation Within the Clonal Genotype

Some modern molecular evidence gives credence to clonal fidelity, other evidence does not. The use of microsatellite markers (Goldstein and Schlötterer 1999) has shown that aphids can have multilocus genotypes or MLG's (they are not strictly "genotypes" since microsatellite regions are non-coding), which remain consistent over a range of loci (12 tested; Haack et al. 2000). This suggests that such MLG's have descended from a common stem mother (Haack et al. 2000; Miller 2000). However, this may be an illusion of constancy. Mini- and microsatellites, essentially selectively neutral (but see Li et al. 2002), are known to be fast mutating and evolving, with mutation rates typically of the order 10^{-6} to 10^{-3} per gene per generation (Hancock 1999). There are thousands of such loci, perhaps tens of thousands, scattered throughout the genome (Loxdale and Lushai 2003b). In addition, even sampling 12 loci is but a tiny proportion of the total genome and it is not known what the rest of the genome is undergoing in terms of mutational changes. Even if particular aphid clones have arisen from a common foundress, there is empirical evidence from microsatellites for recent mutational changes in *Sitobion* aphids (Wilson et al. 1999), *Myzus persicae* (Kasprowicz 2006) and in tansy aphids, *Macrosiphoniella tanacetaria* (Kaltenbach) (Loxdale, unpublished).

Other regions of the aphid genome are seemingly changing within and between clonal lineages, including ribosomal DNA (rDNA) regions. Thus, some examined *Myzus persicae* clonal lineages have been found to show intra- and interclonal polymorphisms for the size and number of IGS (intergenic spacer) repeats (Fenton et al. 2003, 2005), whilst persistent selection with chemical pesticides (disulfoton) over 200 generations (four years) has been demonstrated to be associated with an alteration in the IGS genotype of Greenbug, *Schizaphis graminum* (Rondani) clones, involving the loss of specific bands (Shufran et al. 2003).

With regard to genotypic banding patterns as seen on gels using predominantly dominant markers (markers which usually do not provide heterozygous genotypes but are nevertheless useful in clonal studies; Loxdale and Lushai 1998), new mutated bands have been recognised using synthetic oligonucleotide probes, e.g., $[(GATA)_4]_n$, RAPDs (random amplified polymorphic DNA; De Barro et al. 1994; Lushai et al. 1998) and AFLPs (amplified fragment length polymorphisms; Forneck et al. 2001a, b; Vorwerk and Forneck 2007). Furthermore, such random (somatic) mutations were observed within 1–14 generations in the grain aphid, *Sitobion avenae* (F.) and Grape Rootstock Phylloxera, *Daktulosphaira vitifoliae* Fitch, and in one case (*S. avenae*), in the germ line (Lushai et al. 1998). Using RAPDs, Lushai et al. (1997) revealed banding pattern differences between both clones and morphs of some *S. avenae* asexual lineages. As yet, the mechanism for such changes remains unknown, but may involve transposons.

In terms of physiology, genetically-based intermorphic differences between lineages in colour and life cycle are known (Jenkins et al. 1999; Simon et al. 2002). Intraclonal variations in ovariole number have also been recorded in the Black Bean aphid, *Aphis fabae* Scopoli (Dixon, 1989), whilst highly insecticide resistant *M. persicae* clones (R_2 and R_3; see Section 25.7) are known to undergo a decline of gene expression, but not gene number, when chemical selection ceases (Sawicki et al. 1980; Hick et al. 1996; Field et al. 1999). Field and Blackman (2003) detail the changes observed in the highly resistant strains of *M. persicae* in relation to E4 expression.

25.2.2 Evidence for Variation Between Clonal Lineages

Using rDNA markers, Fenton et al. (1998a) have shown that certain clonal *Myzus persicae* lineages had two ITS (internal transcribed spacer) haplotypes, suggestive of an introgression event between *M. persicae sensu stricto* (*s.s.*) and another close relative, *M. certus* (Walker) By applying fluorescent in-situ hybridisation (FISH) techniques, Blackman et al. (2000) revealed interclonal variation in the number and position of rDNA arrays in the chromosomes of aphids belonging to the genus *Trama* (of which males are unknown[†], although there is other molecular evidence for sexual recombination events having occurred; e.g., Normark 1999). Since it is not known how such polymorphisms are generated between clones, it must be assumed they are also generated within clones. Large interclonal variations are also found in the number of males produced in different life cycle forms of *Sitobion avenae*, especially androcyclic clones (Helden and Dixon 2002).

Using microsatellite and mtDNA markers, evidence for introgression events between different lifecycle morphs *within* species as well as *between* closely-related species has been demonstrated. Thus, for example, the two species *Sitobion avenae* (predominantly anholocyclic, [1] i.e., obligate asexual), and the Blackberry-grain aphid, *S. fragariae* (Walker) (predominantly holocyclic, i.e., with annual sexual

[1]For further details on aphid life cycles, refer to Blackman and Eastop (2000), Carter et al. (1980) and Dixon (1998).

phase), both with chromosome numbers 2n = 18 (Blackman and Eastop 2000), clearly cross breed. Using microsatellites, Sunnucks et al. (1997) showed high levels of allelic/genotypic variation in *Sitobion avenae sensu lato* (*s.l.*) collected from wheat and cocksfoot grass, *Dactylis glomerata*. Three apparently almost non-interbreeding genotypic groups were identified, with high levels of sexual recombination within each. Host specialisation was apparent: thus there were wheat-specific lineages, lineages common to both wheat and *D. glomerata*, and lineages from *D. glomerata* only, which were found to bear many alleles from *S. fragariae*. However, the genotype class with *S. avenae*-like and *S. fragariae*-like alleles also carried *S. fragariae*-like mtDNA in 80% of cases. Such asymmetry suggests that *S. avenae* males are attracted via similar/identical sex pheromones (Goldansaz 2003) to *S. fragariae* females (Sunnucks et al. 1997; see also Vialatte et al. 2005).

In the Bird Cherry-Oat aphid, *Rhopalosiphum padi* (L.), studies have shown that the anholocyclic and holocyclic lineages are very divergent in terms of their mtDNA (by some 0.4–1.4 myr; Martinez-Torres 1994, cited in Simon et al. 1996). Newer investigations suggest, however, that a hybridisation event may have occurred in recent times between *R. padi s.s.* and a closely-related unknown species, causing asexuality (Delmotte et al. 2003).

Other evidence (Delmotte et al. 2001) points to "sexual leakage" (gene flow) between *R. padi* lifecycle morphs and to asexuals deriving from sexuals, probably by three main mechanisms: (1) repeated mutations at the gene/s controlling sexuality; (2) hybridisation between closely-related species; and (3) mating between sexual lineages, e.g., mitochondrial haplotype hII, and males from asexual lineages, e.g., haplotype hI (Delmotte et al. 2001). The DNA introgression data mentioned above in the case of *Sitobion* aphids (Sunnuck et al. 1997) and *M. persicae s.l.* (Fenton et al. 1998a) confirms that aphids of different species mix and mate sympatrically, probably commonly as in the case of *S. avenae* and *S. fragariae*, and that such pairings may indeed cause asexuality, as suggested by Loxdale and Brookes (1990; see below). It is well known in stick insects (Phasmatodea) that hybridisation can result in the genesis of asexual lineages (e.g., Scali et al. 2003; see also Chapter 16).

As mentioned earlier, while microsatellites reveal multi-locus genotypes to be identical at a range of loci (Haack et al. 2000), the rest of the genome is untested and may conceal widespread variations, both within and between asexual lineages. Hence, at present it is not possible to definitively say that clone mates of asexual lineages from particular aphid species are identical throughout their entire genomes (nuclear and mitochondrial), without sequencing these entirely and directly comparing them between individuals (Caillaud et al. 2004; see also Sabater-Muñoz et al. 2006 and Tagu et al. 2008).

25.2.3 Evidence of Higher Level Lineage Evolution, Especially in Relation to the Host Plant

Apparently "good" taxonomic species can show a range of polymorphisms associated with the host plant and where distributions overlap, this implies the beginnings of sympatric speciation. Such changes are seen in both the chromosomal karyotype as well as molecular DNA differences. As an example of the first, chromosomal differences have been observed in the Corn Leaf aphid, *Rhopalosiphum maidis* (Fitch), in relation to whether the insect feeds on barley *or* sorghum and maize in the northern hemisphere (2n=10 and usually 8, respectively; Brown and Blackman 1988; Blackman and Eastop 2000).

Biotypes are well known in the Aphididae (Eastop 1973). In the case of the Greenbug, *Schizaphis graminum*, host adapted forms, almost leading to the point of speciation, have been detected by mtDNA COI sequence analysis (Anstead et al. 2002). The three main clades found involve many wild species of Poaceae (grasses and cereals) strongly suggesting that the evolution of such host-based differences in this species pre-dates the development of agriculture in historical times, i.e., the last 5,000 years.

Using nuclear DNA markers, the existence of host-based stratification was first demonstrated in *Sitobion* species by De Barro et al. (1995a) employing RAPDs, a phenomenon that has been further explored by a number of researchers since (Loxdale and Lushai 2007). Clear evidence for host preference in *S. avenae* infesting wheat, wheat volunteers, barley and maize was also found in French populations using microsatellites (Haack et al. 2000). Thus, many genotypes were detected, some with apparent host preferences, whilst two genotypes from maize were also found on all other hosts. These data suggest specialist (s) and generalist (g) clones, "g" clones seemingly being able to colonise large geographical areas and persist for several years. Such a scenario could be favoured by agricultural practice. Besides these two "g" clones, a continual replacement of rare "s" genotypes was observed in maize during the two years of the study. It was hypothesized by the authors that selection occurs *via* aphid-plant genotype interactions and natural enemies.

Other field experiments involving the application of molecular markers (RAPDs) have revealed that host preference is shown by early immigrants into the crop. Thus, winged *S. avenae* foundresses showed such host preferences when landing on spring cereal and grass hosts and their genotypes could accordingly be split into four main clades (Lushai et al. 2002).

Regarding what constitutes a natural species population, early evidence using a single polymorphic allozyme locus (GOT, glutamate oxaloacetate transaminase) revealed that *S. fragariae* may be a complex of species/forms (Loxdale and Brookes 1990). Samples collected at various sites within a $\sim$ 65 km radius around Rothamsted Research, Hertfordshire in southern England, and tested for GOT variation showed that, at certain sites, high frequencies of a slow (S) allele occurred within large sub-samples from grass (*D. glomerata*). However, genotypes bearing this allele were largely absent from samples collected from blackberry, *Rubus fruticosus* agg., the primary host, as if such individuals were not completing the holocycle by returning in autumn to this host. These may constitute an anholocyclic strain or perhaps (less likely) a cryptic species (Loxdale and Brookes 1990).

The available evidence reveals convincingly that a "super-clonal" population (= "evolutionary individual") as proposed by Janzen (1977) is unlikely to be real, because of the clear molecular heterogeneity and levels of host adaptation shown to exist within and between natural aphid populations of given species, real or apparent. Even so, as shown by molecular markers, some generalist clones of certain species (so-called "Superclones") have a wide distribution and may persist for some years, e.g., the already mentioned "g" clone of *S. avenae* (Haack et al. 2000; see also Figueroa et al. 2005) and *M. persicae* in Australia (Vorburger et al. 2003a), as well as in Scotland, notably the "Braveheart" or "J" clone (Fenton et al. 1998b, 2003, 2005; Kasprowicz 2006; Kasprowicz et al. 2008a, b; see also Malloch et al. 2006) and certain clones of *R. padi* (Gilabert et al. 2009). Whether these clones or strains are successful in the long-term, has yet to be shown and perhaps never can be (but see Clonal Selection, Section 25.7), whilst the ecological reasons for their success have yet to be fully understood (Kasprowicz 2006; Kasprowicz et al. 2008a, b). That ecological scenarios change, for example by host switching, has been demonstrated in the alfalfa and red clover-associated host races (subspecies) of the pea aphid, *Acyrthosiphon pisum* (Harris) (Via 1999; Via et al. 2000; Frantz et al. 2006) and in other insect species, notably the tephritid fruit fly, *Rhagoletis pomonella* (Walsh), which has hawthorn and apple-preferring forms (Feder et al. 1998). In the case of both *A. pisum* and *R. pomonella*, the level of sympatric speciation, whilst strong, is still only partial, i.e., with no fixed allelic differences between host-preferring forms. In many aphids, the level of such speciation may be greater and pre-date agriculture. Certainly when chromosome polymorphisms are also involved, especially differences in number, population divergence is more likely, perhaps to the point of total differentiation (Blackman 1980; Blackman et al. 1989). Other examples of population divergence seem to have occurred between aphids on wild hosts and cultivated crops, such as cereals and wild grasses in the case of *Sitobion* aphids (Vialatte et al. 2005, 2007). The theory of, and empirical evidence for, sympatric speciation in aphids has been discussed by several authors, especially Guldemond and Mackenzie (1994), Mackenzie and Guldemond (1994) and Via (2001) and most recently, by Via and West (2008). These last authors show that in the initial speciating races of pea aphids, *A. pisum*, "extensive 'divergence hitchhiking' occurs [around quantitative trait (QTL) loci] because reduced inter-race mating and negative selection decreases the opportunity of recombination between chromosomes bearing different locally adapted QTL alleles" (see also Smadja et al. 2008). In *A. pisum*, biotypic host preference is associated with strain differences of the endosymbiotic bacteria of the aphid mycetome, comprising specialist cells (mycetocytes) in the aphid body cavity which they inhabit (Simon et al. 2003b).

25.3 Clonal Persistence

The question of whether aphid clones persist and for how long can only be answered – if it can yet at all – by providing some background of sexual versus asexual modes of reproduction, including the positive and negative aspects of both. Sex increases variance within populations and eliminates deleterious alleles (Felsenstein 1974; Normark et al. 2003). However, negatively, there are costs involved: (i) two organisms are needed to produce one offspring, unlike asexuals, and the clonal offspring from an asexual female multiplies at twice the rate of the progeny descended from a sexual female (half of all offspring are males, and only contribute to the next generation by fertilizing the females), whilst a sexual female has only 50% of the fitness of an asexual female (so-called "two fold cost of sex and cost of meiosis" Maynard Smith 1978; Ridley 1993); (ii) recombination breaks up favourable gene combinations that have increased in frequency under the action of natural selection (Barton and Charlesworth 1998; West et al. 1999), whilst maladapted genes may be incorporated into the genome ("cost of recombination"; Felsenstein, 1974) (iii) an individual has to find a mate, which may be rare and widely dispersed and requires much effort to find ("cost of rarity"; e.g., Schreiber 2003); and lastly (iv), there is an increased risk of infection by pathogens and transposons brought about by sex (Arkhipova and Meselson 2000).

In the case of asexuals, a major potential benefit includes a fast rate of reproduction and adaptation to favourable environmental conditions, involving *r*-type selection (Dixon 1998). But negatively, there is great potential for the accumulation of deleterious alleles within clonal lineages, leading to "mutational meltdown" (Muller's ratchet; Muller 1964; Lynch et al. 1993; see also Chapter 5). Recent molecular evidence using AFLPs of asexual aphids (Phylloxera) seems to cast some doubt on this theory, since most polymorphisms were characterised as random mutations that were not continuously detected in later generations (Vorwerk and Forneck 2007).

With aphids, it may be asked why sex is retained in so many species, even rare sex. In fact, even those species that are apparently obligately asexual may have males and oviparae, but as a consequence of very low frequency in the population, they have yet to be found (see below). For host alternating aphids (some 10% of recorded species, Eastop 1986), even finding the right host and a mate is difficult and there is much more than a two-fold cost involved. Ward et al. (1998) estimated that only around 0.6% of *Rhopalosiphum padi* winged autumn migrants (presexual females = gynoparae) were able to locate the widely-dispersed primary host, the Bird Cherry, *Prunus padus*, thereafter produce oviparae (which mate with the winged males) and hence reproduce sexually – assuming that the males are successful in "homing in" on the ovipara's sex pheromones in order to find the females. Even for non-host alternating species, if the plant host is rare or widely dispersed, finding it may still be problematic to winged migrants. Thus, the host plant acts effectively as a bottleneck, thereby causing loss of alleles and genotypes from the aphid population and leading to monomorphism at given loci (Loxdale and Brookes 1988). In addition, some aphid lineages have seemingly persisted asexually for long periods, i.e., many years (> 30) and generations (> 540) (van Emden 1988; see also Kasprowicz 2006). Lastly, some species do seem to be truly obligate asexuals, e.g., the Shallot aphid, *Myzus ascalonicus* Doncaster (Blackman and Eastop 2000) and aphids of the genus *Trama* (tribe Tramini)† (Normark 1999), although sexual forms may yet one day be found, as happened in the case of the Corn Leaf aphid, *Rhopalosiphum maidis* (Fitch) (Blackman and Eastop 2000).

For aphids that rarely indulge in sex, or the general absence of suitable hosts on which to complete the holocycle (e.g., *Myzus persicae* in Scotland where the primary host Peach, *Prunus persica* is very rare; Kasprowicz 2006), sex appears not to be *that* useful at a population level for either *ridding the germ line of mildly deleterious mutations* or for *increasing population genetic variation*, although these mechanisms clearly may have some longer-term importance. Why is it then maintained in most aphid species? Part of the reason is certainly the production of cold hardy overwintering eggs (Blackman 1980), although holocyclic aphids may still go through a sexual phase if the primary host is present and even if conditions are relatively mild, e.g., southern France in the case of *M. persicae* (Guillemaud et al. 2003). Another hypothesis is that sex has to occur to reset chromosomal telomere length (Loxdale and Lushai 2003c) which, if true, may have profound consequences on population genetic structure (Lushai and Loxdale 2007).

25.4 Aerial Displacements

Aphid aerial movement is important for the super-organism hypothesis (Janzen 1977) because it potentially spreads the genes, or genotypes, far and wide and allows asexual lineages to get established in new regions (Taylor 1965, 1986a, b; Loxdale et al. 1993). However, on arrival at the new destination, lineages are usually untested by abiotic and biotic selection, including pesticide treatment in the modern agro-ecosystem (see Section 25.7). As hosts senesce, it is of course also essential for many species to find new or alternative hosts and thereby at the same time, escape the effects of pathogens, predators and parasitoids – in effect, move into "enemy free space" (Jeffreys and Lawton 1984; Loxdale and Lushai 1999), although of course, the converse could also be true. Lastly, it is essential in other species to complete the holocycle by moving to an alternative primary or secondary host, including finding a mate and laying overwintering eggs.

In former times, ideas of the "aerial plankton" abounded to describe the effectively passive transport of aphids by winds above their flight speed in still air (Loxdale et al. 1993). Drake and Farrow (1989) in an article entitled "The 'aerial plankton' and atmospheric convergence" discuss the current ideas concerning this plankton, analogous to that in the sea, but also detail why the analogy is not that good, highlighting the fact that flying insects are not feeding or developing in the aerial medium, but merely being transported by it (unlike the zooplankton in water). With aphids, mostly only the post-teneral (newly emerged winged adult) forms disperse with a biphasic pattern of flight behaviour (Johnson1954); they settle on plant hosts in response to suitable visual and olfactory cues (Hardie 1993; Niemeyer

1990) as soon as possible in order to conserve fuel supplies (lipids) and maximise their potential for reproduction (Loxdale et al. 1993).

Taken as a group, aphids cannot be treated as a homogenous whole. Rather, they have to be considered as individual species, like the zooplankton, with different lifecycles and forms, which greatly impinge on the population structures observed empirically. After all, even some winged morphs of particular aphid species (e.g., virginoparae *vs.* gynoparae (pre-sexual females) of *R. padi* are known to have different flight heights and behaviours, since their ultimate targets are different. The former morph seeks Poaceae, which is usually abundant and widespread, whereas the latter morph locates the much less common and more isolated primary host *P. padus* and hence, on average, tends to fly at a greater height (Tatchell et al. 1988).

The use of molecular genetic markers has largely confirmed the aforementioned trend of species- and morph-dependent flight behaviour from comparative examination of aphids caught in 12.2 m high suction traps of the Rothamsted Insect Survey (RIS; Harrington et al. 2004). *Sitobion avenae* is very abundant there in June–July, whereas its congener, *S. fragariae*, is an order of magnitude rarer (Woiwod et al. 1988), perhaps pointing towards a more restricted flight behaviour and aerial displacement. Molecular ecological studies using allozymes and/or DNA markers show *S. avenae* to display rather similar allele frequencies over a wide geographical area in Britain (i.e., allelic homogeneity; Loxdale et al. 1985; Llewellyn et al. 2003), whereas *S. fragariae* shows patterns of restricted gene flow (i.e., allelic heterogeneity), even at relatively small spatial scales (Loxdale and Brookes 1990). In addition, samples of *S. fragariae* collected over a number of years show somewhat stable gene and genotype frequencies further supporting the lack of movement and gene flow (Loxdale and Brookes 1990).

In samples of *S. avenae* from one year, the distribution of genotypes appears as a function of latitude, governed by climatic factors which influence the proportion of holocyclic genotypes successfully overwintering as eggs versus obligate asexuals overwintering as live individuals (Llewellyn 2000; Llewellyn et al. 2003). Similar trends have also been documented in France (Simon et al. 1999). Other aphids clearly display genetic population patterns in relation to flight behaviour: Sycamore aphids, *Drepanosiphum platanoidis* (Schrank), very abundant in 12.2m suction traps, display homogeneous gene/genotypic patterns at all spatial scales from leaf, tree, to larger geographic scales (hundreds of kilometres; Wynne et al. 1994). In contrast, the Damson-hop aphid, *Phorodon humuli* (Schrank), resistant to a range of pesticides as expressed via elevated carboxylesterase activity (Devonshire et al. 1986), is much less migratory; its autumn migration is probably restricted to 15–20 km within the main hop growing centres in the UK, Herefordshire and Kent (Loxdale et al. 1998), a conclusion drawn from the use of allozyme markers (resistance and non-resistance), which supports earlier suction traps findings (Taylor et al. 1979). In the case of the tansy aphid, *Macrosiphoniella tanacetaria* (Kaltenbach), microsatellite markers show that the aphids display restricted gene flow (significant genetic heterogeneity at small spatial scales), with a metapopulation structure (Massonnet et al. 2002, Massonnet and Weisser 2004; see below).

Such species-specific molecular evidence lends support to the flight chamber (wind tunnel) experiments by Jim Hardie and colleagues at Silwood Park, Ascot, UK that show that the attraction of winged aphids to white lights (sky) or green targets (plant hosts on which to land) varies depending on the apparent "migratoriness" of the species or morphs concerned (Hardie 1993; Hardie and Campbell 1998). It seems certain that aphid migration has to be viewed in a case-specific manner, both in terms of the readiness of a given species to migrate and its ability to do so, a behaviour that is mirrored in population genetic patterns.

25.5 Geographic Populations

It is a well-established that many national aphid populations, especially those in the New World and Australasia, have been derived from introduced insects, probably very small founder populations, perhaps only a single individual in some instances. In most such cases, immigration probably results from human activities, i.e., aphids on imported plant material, root stocks, tubers, etc. However, instances appear to have occurred where winged aphids have travelled vast distances across landmasses and oceans to found new colonies (Bowden and Johnson 1976; Loxdale et al. 1993), although probably, most such insects never find land and a suitable host and die without trace (Loxdale and Lushai 1999).

With insects such as the grain aphid, *Sitobion avenae*, a lack of any clear relationship between genetic and geographic distance (Loxdale et al. 1985; Llewellyn 2000; Miller 2000; Llewellyn et al. 2003; Miller et al. 2003) hinders attempts to derive the migratory range or source of migrants, at least over the time scales studied (usually several years). Such a lack of correlation does not readily fit with expectations of "island" models of dispersal and population genetic structuring (Wright 1931, 1951; Slatkin 1985), nor with isolation by distance (Wright 1943) and "stepping stone" models (Slatkin 1985, 1993; see Llewellyn 2000 and Miller 2000 for discussions). With the tansy aphid, plotting measures of genetic *versus* geographical distance failed to show any sign of isolation by distance until populations were separated by *ca.* 470 km (Massonnet et al. 2002; Massonnet and Weisser 2004). Metapopulations showed significant local genetic heterogeneity, but there is clearly still sufficient local gene flow to offset the seasonal effects of drift and selection, with allopatric differentiation only becoming apparent at large geographic scales. Hence, interpretation of such data must be treated with caution since gene/genotype frequencies are probably rarely, if ever, in equilibrium in terms of the influences of selection and drift (Roderick 1996). Other data collected for *S. avenae* clearly show that some genes and multilocus genotypes are common across southern England and large parts of northern France (e.g., clone 53), thereby revealing that the Channel is in all probability not a geographical barrier to migration and gene flow (Llewellyn 2000).

At the very large geographical scale, the introduction of the Spotted Alfalfa aphid, *Therioaphis trifolii,* f. *maculata* into the USA in the 1950s (Blackman and Eastop 2000), the rose-grain aphid, *Metopolophium dirhodum* into New Zealand in the early 1980s (Nicol et al. 1997), and the Russian Wheat aphid, *Diuraphis noxia* (Mordvilko) into South Africa and the USA in the late 1970s and mid-1980s, respectively (Blackman and Eastop 2000), are all examples of the accidental importation of economic pests in recent historical times. With the aid of chromosomal and molecular markers, clues to the origins of such introductions have been forthcoming in recent years. Thus, for example, *Myzus persicae* in Australia probably derived from European stock (Wilson et al. 2002), *Sitobion* aphids in Australia and New Zealand came from Taiwan and Australia, respectively (Wilson et al. 1999), whilst *S. avenae* in Chile were probably also of European origin (Figuero et al. 2005). In applied entomology, markers may not allow determination of the exact source of an immigrant insect, but they can often exclude certain other sources. For example, the outbreak in California of the medfly, *Ceratitis capitata* (Wiedemann) (Diptera: Tethritidae) in 1989 and 1991 was shown by the use of mitochondrial DNA (mtDNA) restriction fragment length polymorphism (RFLP) analysis to probably *not* have derived from the Hawaiian islands, an assumed likely source (Sheppard et al. 1992).

25.6 Clonal Selection

Whatever the mode of transportation, aphid lineages are tested on arrival on their host plant/s by selection, be this abiotic or biotic. In fact, it may be disadvantageous for aphids acclimatised to certain ecological parameters, including photoperiods, to move long distances where they may be maladapted (Loxdale et al. 1993 and references therein; but see also Lushai et al. 1996 which contradicts this viewpoint). In the case of genotypes being resistant to insecticides, a susceptible individual moving to a sprayed area is clearly at a selective disadvantage compared with a resistant genotype (Loxdale et al. 1993). A single aphid can found an entire population (including sexual morphs in the case of holocyclic aphids) – indeed a national population as mentioned above. The problem with obligate asexuals or species with rare sex is that the genome is unrecombined, leading quickly to population linkage disequilibrium of loci, and whilst one or a few loci may be selected in one direction, the remainder of the genome is dragged with it in a kind of "mass hitchhiking effect" (Via and West, 2008 and Chapter 5). Scenarios can be imagined where this quickly leads to wild swings in gene/genotype frequency as asexual lineages are selected in one direction and then later in another (Smadja et al. 2008). By reshuffling the genome, sexual recombination may be beneficial in the longer term in reducing the worse effects of maladapted linkage of alleles. But even so, some loci will still react epistatically and other genes undoubtedly have negative pleiotropic influences, as is apparent in the highly-resistant forms of *Myzus persicae* (Foster et al. 2000).

In *M. persicae*, four main resistant genotypes occur in the field in the UK – S (susceptible), R_1, R_2, R_3, directly related to the level of amplification and expression of the carboxylesterase E4 and FE4[2] genes (80 x more genes in R_3 as compared with S), which confers resistance primarily to carbamates and organophosphates

[2]FE4 is a fast allelic variant of the typical and more frequent E4 gene product.

and to a lesser extent, pyrethroids (Field et al. 1999; Foster et al. 2000). These are assayed either using electrophoretic, immunoassay or PCR (polymerase chain reaction)-based methods, the last involving specific primers (Devonshire 1989; Field et al. 1999; Foster et al. 2000; Anstead et al. 2005). The highly E4 resistant genotypes (R_2 +) are also associated with a translocation of autosomes 1 and 3 (A1, 3; Blackman et al. 1995; Foster et al. 2000; Field and Blackman 2003). In addition, the various resistant genotypes may have other resistant mechanisms in association with the E4 genotype (Kasprowicz et al. 2008b). These include MACE (modified acetylcholinesterase), which confers resistance to chemicals blocking the normal hydrolytic function of the enzyme acetylcholinesterase in the nerve synaptic junction, and hence cause paralysis of the insect, and *kdr* ("knockdown resistance") as well as super-*kdr*. These last two are mutated forms of the gene which controls the expression of the allosteric binding site of the sodium gating channel in the insect nerve and which, as a consequence, confers resistance to pyrethroids, which would otherwise block these channels (Denholm et al. 1999; Anstead et al. 2005).

The possession of the *kdr* gene has several pleiotropic effects on fitness in resistant *Myzus persicae*, including response to alarm pheromone (Foster et al. 2000) and vulnerability to parasitoid attack (Foster et al. 2007a). While R_2 and R_3 genotypes may be highly selected for in the growing season as a consequence of chemical selective pressure, in the autumn and winter, they tend to be selected out of the population and decline in frequency, as seen from their reduced occurrence in 12.2 m high suction trap samples (Foster et al. 2002). Their frequency is thus generally reduced in the following spring (reviewed by Foster et al. 2007b). The frequency of the various S and R genotypes fluctuates at various sites in the UK during the growing season (Foster et al. 2000, 2002), a situation probably reflecting selection in relation to changing patterns of insecticide usage and fitness costs associated with insecticide resistance. It may also reflect the fact that highly resistant asexual lineages of *M. persicae* have linked genomes bringing about non-random associations between the resistance mechanisms and selection operating on asexual lineages (Foster et al. 2002). Hence, when pesticide selection pressure is reduced, the resistant geno-phenotypes may otherwise be maladapted (Foster et al. 1998). Furthermore, in Scottish *M. persicae* population samples, resistant genotypes (R, MACE and *kdr*), arbitrarily linked with microsatellites at a number of loci, are seen to fluctuate within natural populations but may be related to selection affecting other aspects of the geno-phenotype as earlier discussed (Fenton et al. 2005; Kasprowicz 2006; Kasprowicz et al. 2008a, b). Other evidence from population studies of *M. persicae* in various countries world-wide shows that both *kdr* and super-*kdr* homozygotes are apparently strongly selected against in the field, whilst in the UK, individuals bearing the super-*kdr* mutation are strongly selected against in any genotype in the absence of insecticide selective pressure (Anstead et al. 2007).

As also mentioned earlier, evidence has been found for latitudinally-dependent clinal changes in frequencies of aphid nuclear genes being related to climate and life cycle (e.g., Simon et al. 1999; Llewellyn et al. 2003). Similar clines have also been observed in populations of *Rhopalosiphum padi* from south-west to northern France in relation to non-recombinant mtDNA haplotypes (Martinez-Torres et al. 1997). At a smaller geographic scale, Llewellyn et al. (2004) found evidence for clonal/lineage selection of *S. avenae* multilocus microsatellite genotypes in wheat fields in southern England, possibly related to host plant cultivar, which is not surprising considering other evidence for host plant preference in *Sitobion* aphids (e.g., De Barro et al. 1995a, b; Haack et al. 2000; Lushai et al. 2002). Similarly, Vorburger (2006) has found evidence for some degree of host selection in predominantly asexual populations of *M. persicae* in Australia on some hosts (e.g., *Solanum physalifolium*), as has Kasprowicz (2006) for the same species in Scotland on *Brassicas*. Sunnucks et al. (1998) also provide evidence that chromosomal changes (rearrangements) may be associated with host plant preference/adaptation in aphids of the genus *Sitobion*. Whether general purpose genotypes or GPGs truly exist in aphids is debatable (see Chapter 6), but my view is that since the broad thrust of evolution is mainly towards specialisation, they probably do not (but see Van Doninck et al. (2002) and Chapter 11 for the case of ancient asexual ostracods). Empirical studies of host plant performance by Vorburger et al. (2003b), involving different life cycle forms, seemingly deny the existence of GPGs. For sure, linkage of the loci of the asexual aphid genome would, at least on theoretical grounds, make the idea of GPGs seem unlikely.

25.7 Concluding Remarks

I hope that I have demonstrated that even a single female aphid lineage is an evolutionary force to be reckoned with and because it apparently defies many of the current beliefs and concerns about the importance of inbreeding, mutational meltdown, and lack of sex in many species, causes us to rethink our position on these issues. Clearly, asexual aphid lineages do not fit the broad criteria of an "evolutionary individual" or unit as originally proposed by Janzen (1977). What can be said with certainty is that aphids continue to be a thriving and successful group of organisms that have conquered most of the biomes. These include the arctic (Strathdee et al. 1993) and sub-antarctic regions (Hullé et al. 2003), whilst they have been found far out at sea (Hardy and Cheng 1986), on mountain tops (Bauer 2002), and infest a huge range of plants, usually monophagously, although some species are polyphagous within plant genera or families (e.g., Brassicas), and in the case of the ubiquitous and highly polyphagous *Myzus persicae*, this species attacks over 40 plant families worldwide (Blackman and Eastop 2000). They are a truly ancient group (fossils go back to the Triassic, some 220-210 myr; Heie 1987, 1994; Grimaldi and Engel 2005) and from their past and current performance as pests, causing both direct feeding damage and transmission of pathogenic plant viruses (van Emden and Harrington 2007), are highly likely to be around for a very long time to come, despite our best efforts, and using our most sophisticated technology, to combat them. The recent development of insecticide resistant forms in several species worldwide, notably *M. persicae* and involving cross-resistance mechanisms, clearly shows that aphid evolution continues on apace as it has always done and presumably, always will. Evolution in certain pest species examined can, as outlined, involve a loss of sex, sometimes predominantly or seemingly absolutely. In the short term, and over a short evolutionary timescale, i.e., several growing seasons, this may have more benefits than disadvantages, especially in an intensive agricultural context (e.g., in pest species, *S. avenae*, *S. fragariae*, *R. padi*, *S. graminum*, *M. persicae*, *M. ascalonicus,* etc.). However, over a longer period, perhaps decades, periodic sexual recombination may well have importance and without it, asexual lineages may lose fitness and eventually die out; hence, its retention, even if at low frequency, in the majority of aphid species.

Acknowledgments I thank Professor Wolfgang W. Weisser, to whom this article is dedicated, and Drs. Brian Fenton, Steve P. Foster, Nicola von Mende-Loxdale and Isa Schön for their valuable comments on the manuscript.

Glossary

Androcyclic: asexual aphid lineage that produces mostly males and some asexual females.
Anholocyclic: permanently viviparous asexual aphid lineage reproducing by apomictic parthenogenesis. This term can include an obligate asexual lineage or species, or a "facultative" holocyclic lineage of a species in which temperature and light conditions are favourable to maintain all year round parthenogenetic propagation (i.e., 16 h L: 8 h D and > 15°C ambient temperature), rather than induce the production of the pre-sexual and sexual forms (i.e., 8 h L: 16 h D and ambient temperature < 15°C).
Cline: here used in the sense of a gradual change in allele or genotypic frequencies at a given locus/loci, across the distributional range of a species population, and correlated with an environmental/ geographic transition.
Gynoparae: winged pre-sexual aphid female morph produced under shortening day, cooler ambient temperature conditions from a holocyclic asexual lineage, along with males. It migrates to a new host where it produces sexual females (oviparae) which in turn mate with the males to produce cold hardy overwintering eggs. The new host could be a primary woody one in the case of host alternating aphid species.
Holocyclic: an aphid lineage and/or species with an annual sexual phase, usually involving an autumnal winged migration. In the case of species that have an alternation of plant hosts, this is from a secondary herbaceous host to the primary woody host and *vice versa*, but there are exceptions (e.g., *Sitobion avenae* which remains on Poaceae all year round). Such holocyclic forms can also be "facultative" anholocyclic.
Ovipara: egg laying sexual aphid female. This morph attracts the males (winged in most species, but there are exceptions) using sex pheromones.
Telescoping of generations: the phenomenon whereby asexual aphid females have both their offspring (children) and *their* offspring (grandchildren) within them.
Virginoparae: asexual viviparous female aphids which produce further such asexual females.

Wheat volunteers: cultivated wheat sprouting from the seeds of plants sown in previous sowings.

References

Anstead JA, Burd JD, Shufran KA (2002) Mitochondrial DNA sequence divergence among *Schizaphis graminum* (Hemiptera: Aphididae) clones from cultivated and non-cultivated hosts: haplotype and host associations. Bull Entomol Res 92: 17–24

Anstead JA, Mallet J, Denholm I (2007) Temporal and spatial incidence of alleles conferring knockdown resistance to pyrethroids in the peach-potato aphid *Myzus persicae* (Hemiptera: Aphididae) and their association with other insecticide resistance mechanisms. Bull Entomol Res 97: 243–252

Anstead JA, Williamson MS, Denholm I (2005) Evidence for multiple origins of identical insecticide resistance mutations in the aphid *Myzus persicae.* Insect Biochem Mol 35: 249–256

Arkhipova I, Meselson M (2000) Transposable elements in sexual and ancient asexual taxa. Proc Natl Acad Sci USA 97: 14473–14477

Barraclough TG, Fontaneto D, Ricci C, Herniou EA (2007) Evidence for inefficient selection against deleterious mutations in cytochrome oxidase I of asexual bdelloid rotifers. Mol Biol Evol 24: 1952–1962

Barton NH, Charlesworth, B (1998) Why sex and recombination? Science 281: 1986–1989

Bauer JA (2002) Research on the Roan (re: Woolly Balsam Adelgid) The Tennessee Conservationist wwwstatetnus/environment/tn_consv/archive/roanehtm

Begin M, Schoen DJ (2006) Low impact of germline transposition on the rate of mildly deleterious mutation in *Caenorhabditis elegans*. Genetics 174: 2129–2136

Behura SK (2006) Molecular marker systems in insects: current trends and future avenues. Mol Ecol 15: 3087–3113

Birky CW, Wolf C, Maughan H, Hebertson L, Henry E (2005) Speciation and selection without sex. Hydrobiologia 546: 29–45

Blackman RL (1971) Variation in the photoperiodic response within natural populations of *Myzus persicae* (Sulz). Bull Entomol Res 60: 533–546

Blackman RL (1980) Chromosomes and parthenogenesis in aphids. In: Blackman RL, Hewitt GM, Ashburner M (eds) Insect cytogenetics, 10th Symposium of the Royal Entomological Society. Blackwell Scientific Publ, Oxford, pp. 133–148

Blackman RL, Brown PA, Furk C, Seccombe AD, Watson GW (1989) Enzyme differences within species groups containing pest aphids In: Loxdale HD, den Hollander J (eds), Electrophoretic studies on agricultural pests. Systematics association special volume No. 39. Clarendon Press, Oxford, pp. 271–295

Blackman RL, Eastop VF (2000) Aphids on the world's crops: An identification and information guide (2nd edition). John Wiley and Sons Ltd, Chichester

Blackman RL, Spence JM, Field LM, Devonshire AL (1995) Chromosomal location of the amplified esterase genes conferring resistance to insecticides in *Myzus persicae* (Homoptera: Aphididae). Heredity 75: 297–302

Blackman RL, Spence JM, Normark BB (2000) High diversity of structurally heterozygous karyotypes and rDNA arrays in parthenogenetic aphids of the genus *Trama* (Aphididae: Lachninae). Heredity 84: 254–260

Bowden J, Johnson CG (1976) Migrating and other terrestrial insects at sea. In: Cheng L (ed) Marine insects. North-Holland Publ, Amsterdam, pp. 97–117

Brown PA, Blackman RL (1988) Karyotype variation in the corn leaf aphid *Rhopalosiphum maidis* (Fitch) species complex (Hemiptera: Aphididae) in relation to host-plant and morphology. Bull Entomol Res 78: 351–363

Browne J (1996) Charles Darwin: Voyaging. Volume 1 of a biography. Pimlico London, p. 363

Caillaud M, Edwards O, Field L, Giblot-Ducray D, Gray S, Hawthorne D, Hunter W, Jander G, Moran N, Moya A, Nakabachi A, Robertson H, Shufran K, Simon J-C, Stern, D, Tagu D (2004) Proposal to sequence the genome of the pea aphid (Acyrthosiphon pisum) www.genome.gov/Pages/Research/Sequencing/SeqProposals/AphidSeq.pdf

Carter N, McLean IFG, Watt AD, Dixon AFG (1980) Cereal aphids: a case study and review. Appl Biol 5: 271–348

De Barro PJ, Sherratt T, Wratten S, Maclean N (1994) DNA fingerprinting of cereal aphids using $(GATA)_4$. Eur J Entomol 91: 109–114

De Barro PJ, Sherratt TN, Brookes CP, David O, Maclean N (1995a) Spatial and temporal variation in British field populations of the grain aphid *Sitobion avenae* (F) (Hemiptera: Aphididae) studied using RAPD-PCR. Proc R Soc Lond B 262: 321–327

De Barro PJ, Sherratt TN, David O, Maclean N (1995b) An investigation of the differential performance of clones of the aphid *S. avenae* on two hosts. Oecologia 104: 379–385

Delmotte F, Leterme N, Bonhomme J, Rispe C, Simon J-C (2001) Multiple routes to asexuality in an aphid. Proc R Soc Lond B 268: 2291–2299

Delmotte F, Sabater-Munoz B, Prunier-Leterme N, Latorre A, Sunnucks P, Rispe C, Simon J-C (2003) Phylogenetic evidence for hybrid origins of asexual lineages in an aphid species. Evolution 57: 1291–1303

Denholm I, Pickett JA, Devonshire AL (eds) (1999) Insecticide resistance: From mechanisms to management. CABI Publishing, Wallingford, Oxford

Devonshire AL (1989) The role of electrophoresis in the biochemical detection of insecticide resistance In: Loxdale HD, den Hollander J (eds), Electrophoretic studies on agricultural pests. Clarendon Press, Oxford, pp. 363–374

Devonshire AL, Moores GD, ffrench-Constant RH (1986) Detection of insecticide resistance by immunological estimation of carboxylesterase activity in *Myzus persicae* (Sulzer) and cross reaction of the antiserum with *Phorodon humuli* (Schrank) (Hemiptera Aphididae). Bull Entomol Res 76: 97–107

Dickson RC (1962) Development of the spotted alfalfa aphid population in North America. Int Kongress für Entomol (Wien 1960) 2: 26–28

Dixon AFG (1989) Parthenogenetic reproduction and the rate of increase in aphids. In: Minks AK, Harrewijn P (eds) Aphids – their biology, natural enemies and control, Volume A. Elsevier, Amsterdam, pp. 269–287

Dixon AFG (1998) Aphid ecology (2nd edition). Chapman and Hall, London

Drake VA, Farrow RA (1989) The 'aerial plankton' and atmospheric convergence. Trends Ecol Evol 4: 381–385

Eastop VF (1973) Biotypes of aphids In: Lowe AD (ed) Perspectives in aphid biology. Entomological Soc of New Zealand, Auckland, pp. 40–41

Eastop VF (1986) Aphid-plant associations. In: Stone AR, Hawksworth DL (eds) Coevolution and systematics. Systematics Association Special Volume No. 32. Clarendon Press, Oxford, pp. 35–54

Feder JL, Berlocher SH, Opp SB (1998) Sympatric host-race formation and speciation in *Rhagoletis* (Diptera: Tephritidae): A tale of two species for Charles D. In: Mopper S Strauss SY (eds) Genetic structure and local adaptation in natural insect populations. Chapman and Hall, New York, pp. 408–441

Felsenstein, J (1974) The evolutionary advantage of recombination. Genetics 78: 737–756

Fenton B, Malloch G, Germa F (1998a) A study of variation in rDNA ITS regions shows that two haplotypes coexist within a single aphid genome. Genome 41: 337–345

Fenton B, Malloch G, Navajas M, Hillier J, Birch ANE (2003) Clonal composition of the peach–potato aphid *Myzus persicae* (Homoptera: Aphididae) in France and Scotland: Comparative analysis with IGS fingerprinting and microsatellite markers. Ann Appl Biol 142: 255–267

Fenton B, Malloch G, Woodford JAT, Foster SP, Anstead J, Denholm I, King L, Pickup J (2005) The attack of the clones: tracking the movement of insecticide resistant peach–potato aphids *Myzus persicae* (Hemiptera: Aphididae). Bull Entomol Res 95: 483–494

Fenton B, Woodford JAT, Malloch G (1998b) Analysis of clonal diversity of the peach-potato aphid *Myzus persicae* (Sulzer) in Scotland UK and evidence for the existence of a predominant clone. Mol Ecol 7: 1475–1487

Field LM, Blackman RL (2003) Insecticide resistance in the aphid *Myzus persicae* (Sulzer): chromosome location and epigenetic effects on esterase gene expression in clonal lineages. Biol J Linn Soc 79: 107–113

Field LM, Blackman RL, Tyler-Smith C, Devonshire AL (1999) Relationship between amount of esterase and gene copy number in insecticide-resistant *Myzus persicae* (Sulzer). Biochem J 339: 737–742

Figueroa CC, Simon J-C, Le Gallic JF, Prunier-Leterme N, Briones LM, Dedryver C-A, Niemeyer HM (2005) Genetic structure and clonal diversity of an introduced pest in Chile-the cereal aphid *Sitobion avenae*. Heredity 95: 24–33

Forneck A, Walker MA, Blaich R (2001a) An *in vitro* assessment of phylloxera (*Daktulosphaira vitifoliae* Fitch) (Hom., Phylloxeridae) life cycle. J Appl Entomol 125: 443–447

Forneck A, Walker MA, Blaich R (2001b) Ecological and genetic aspects of grape phylloxera (*Daktulosphaira vitifoliae* Fitch) performance on rootstock hosts. Bull Entomol Res 91: 445–451

Foster SP, Denholm I, Devonshire AL (2000) The ups and downs of insecticide resistance in peach-potato aphids (*Myzus persicae*) in the UK. Crop Prot 19: 873–879

Foster SP, Denholm I, Harling ZK, Moores GD, Devonshire AL (1998) Intensification of insecticide resistance in UK field populations of the peach-potato aphid *Myzus persicae* (Hemiptera: Aphididae) in 1996. Bull Entomol Res 88: 127–130

Foster SP, Devine G, Devonshire AL (2007b) Insecticide resistance. In: van Emden HF, Harrington R (eds) Aphids as crop pests. CABI, Wallingford, Oxford, pp. 261–286

Foster SP, Harrington R, Dewar AM, Denholm I, Devonshire AL (2002) Temporal and spatial dynamics of insecticide resistance in *Myzus persicae* (Sulzer). Pest Manag Sci 58: 895–907

Foster SP, Tomiczek M, Thompson R, Denholm I, Poppy G, Kraaijeveld AR, Powell W (2007a) Behavioural side-effects of insecticide resistance in aphids increase their vulnerability to parasitoid attack. Anim Behav 74: 621–632

Frantz A, Plantegenest M, Mieuzet L, Simon J-C (2006) Ecological specialisation correlates with genotypic differentiation in sympatric host populations of the pea aphid. J Evol Biol 19: 392–401

Fry JD, Keightley PD, Heinsohn SL, Nuzhdin SV (1999) New estimates of the rates and effects of mildly deleterious mutation in *Drosophila melanogaster*. Proc Natl Acad Sci USA 96: 574–579

Gilabert A, Simon J-C, Mieuzet L, Halkett F, Stoeckel S, Plantegenest M, Dedryver C-A (2009) Climate and agricultuarl context shape reproductive mode variation in an aphid crop pest. Mol Ecol, in press

Goldansaz SH (2003) Étude comportementale et écologie chimique de la recherche d'un partenaire sexuel chez le puceron de la pomme de terre *Macrosiphum euphorbiae* (Thomas) (Homoptera: Aphididae) PhD Thesis, Laval University, Quebec, p. 66

Goldstein DB, Schlötterer C (1999) (eds) Microsatellites: Evolution and applications. Oxford University Press, Oxford

Grimaldi D, Engel MS (2005) Evolution of the insects. Cambridge University Press, New York, pp. 293–297

Guldemond JA, Mackenzie A (1994) Sympatric speciation in aphids. I. Host race formation by escape from gene flow. In: Leather SR, Watt AD, Mill NJ, Walter KPA (eds) Individuals, populations and patterns in ecology. Intercept Ltd, Andover, Hampshire, pp. 367–378

Guillemaud T, Mieuzet L, Simon J-C (2003) Spatial and temporal genetic variability in French populations of the peach–potato aphid *Myzus persicae*. Heredity 91: 143–152

Haack L, Simon J-C, Gautheir J-P, Plantegenest M, Dedryver C-A (2000) Evidence for predominant clones in a cyclically parthenogenetic organism provided by combined demographic and genetic analyses. Mol Ecol 9: 2055–2066

Hancock JM (1999) Microsatellites and other simple sequences: genomic context and mutational mechanisms In: Goldstein DB, Schlötterer C (eds) Microsatellites: Evolution and applications. Oxford University Press, Oxford, pp. 1–9
Hardie J (1993) Flight behaviour in migrating insects. J Agric Entomol 10: 239–245
Hardie J, Campbell CAM (1998) The flight behaviour of spring and autumn forms of the damson–hop aphid *Phorodon humuli* in the laboratory In: Nieto Nafria JM, Dixon AFG (eds) Aphids in natural and managed ecosystems. Universidad de León, León, pp. 205–212
Hardy AC, Cheng L (1986) Studies in the distribution of insects by aerial currents. III. Insect drift over the sea. Ecol Entomol 11: 283–290
Harrington R (1994) Aphid layer (letter). Antenna 18: 50
Harrington R et al. (2004) 'EXAMINE' (EXploitation of Aphid Monitoring in Europe): An EU thematic network for the study of global change impacts on aphids. In: Simon J-C, Dedryver C-A, Rispe C, Hulle M (eds) Aphids in a new millennium. Proceedings of the 6th International Aphid Symposium, Rennes 3–7 September 2001. INRA Editions, Rennes, pp. 45–49
Heie OE (1987) Palaeontology and phylogeny In: Minks AK, Harrewijn P (eds) Aphids, their biology, natural enemies and control, Volume 2A. Elsevier, Amsterdam, pp. 367–391
Heie OE (1994) Aphid ecology in the past and a new view on the evolution of Macrosiphini. In: Leather SR, Watt AD, Mill NJ, Walter KPA (eds) Individuals, populations and patterns in ecology. Intercept Ltd, Andover, Hampshire, pp. 409–418
Helden A J, Dixon AFG (2002) Life-cycle variation in the aphid *Sitobion avenae*: costs and benefits of male production. Ecol Entomol 27: 692–701
Hick CA, Field LM, Devonshire AL (1996) Changes in methylation of amplified esterase DNA during loss and reselection of insecticide resistance in peach-potato aphids *Myzus persicae.* Insect Biochem Mol 26: 41–47
Hullé M, Pannetier D, Simon J-C, Vernon P, Frenot Y (2003) Aphids of sub-Antarctic Iles Crozet and Kerguelen: species diversity, host range and spatial distribution. Antarct Sci 15: 203–209
Janzen DH (1977) What are dandelions and aphids? Am Nat 111: 586–589
Jeffreys MJ, Lawton JH (1984) Enemy free space and the structure of ecological communities. Biol J Linn Soc 23: 269–286
Jenkins RL, Loxdale HD, Brookes CP, Dixon AFG (1999) The major carotenoid pigments of the grain aphid *Sitobion avenae* (F) (Hemiptera: Aphididae). Physiol Entomol 24: 171–178
Johnson CG (1954) Aphid migration in relation to weather. Biol Rev 29: 87–118
Kasprowicz L (2006) The molecular ecology and eco-physiology of the peach-potato aphid *Myzus persicae*. PhD Thesis, University of Dundee, Dundee
Kasprowicz L, Malloch G, Pickup J, Fenton B (2008a) Spatial and temporal dynamics of *Myzus persicae* clones in fields and suction traps. Agr Forest Entomol 10: 91–100
Kasprowicz L, Malloch G, Foster S, Pickup J, Zhan J, Fenton B (2008b) Clonal turnover of MACE-carrying peach-potato aphids (*Myzus persicae* (Sulzer), Homoptera: Aphididae) colonising Scotland. Bull Entomol Res 98: 115–124
Korona R (2004) Experimental studies of deleterious mutation in *Saccharomyces cerevisiae*. Res Microbiol 155: 301–310
Lambert JD, Moran NA (1998) Deleterious mutations destabilize ribosomal RNA in endosymbiotic bacteria. Proc Natl Acad Sci USA 95: 4458–4462
Li Y-C, Korol AB, Fahima T, Beiles A, Nevo E (2002) Microsatellites: genomic distribution putative functions and mutational mechanisms: a review. Mol Ecol 11: 2453–2465
Llewellyn KS (2000) Genetic structure and dispersal of cereal aphid populations. PhD Thesis, University of Nottingham, Nottingham
Llewellyn KS, Loxdale HD, Harrington R, Brookes CP, Clark SJ, Sunnucks P (2003) Migration and genetic structure of the grain aphid (*Sitobion avenae*) in Britain related to climate and clonal fluctuation as revealed using microsatellites. Mol Ecol 12: 21–34
Llewellyn KS, Loxdale HD, Harrington R, Clark SJ, Sunnucks P (2004) Evidence for gene flow and local clonal selection in field populations of the grain aphid (*Sitobion avenae*) in Britain revealed using microsatellites. Heredity 93: 143–153

Loxdale HD (2001) Tracking flying insects using molecular markers. Antenna 25: 242–250
Loxdale HD, Brookes CP (1988) Electrophoretic study of enzymes from cereal aphid populations. V. Spatial and temporal genetic similarity between holocyclic populations of the bird cherry-oat aphid *Rhopalosiphum padi* (L) (Hemiptera: Aphididae) in Britain. Bull Entomol Res 78: 241–249
Loxdale HD, Brookes CP (1990) Genetic stability within and restricted migration (gene flow) between local populations of the blackberry-grain aphid *Sitobion fragariae* in south-east England. J Anim Ecol 59: 495–512
Loxdale HD, Brookes CP, Wynne IR, Clark SJ (1998) Genetic variability within and between English populations of the damson-hop aphid *Phorodon humuli* (Hemiptera: Aphididae) with special reference to esterases associated with insecticide resistance. Bull Entomol Res 88: 513–526
Loxdale HD, Hardie J, Halbert S, Foottit R, Kidd NAC, Carter CI (1993) The relative importance of short- and long-range movement of flying aphids. Biol Rev 68: 291–311
Loxdale HD, Lushai G (1998) Molecular markers in entomology (Review). Bull Entomol Res 88: 577–600
Loxdale HD, Lushai G (1999) Slaves of the environment: the movement of insects in relation to their ecology and genotype. Philos Trans R Soc 354: 1479–1495
Loxdale HD, Lushai G (2003a) Rapid changes in clonal lines: the death of a 'sacred cow'. Biol J Linn Soc 79: 3–16
Loxdale HD, Lushai G (eds) (2003b) Intraclonal genetic variation: ecological and evolutionary aspects. Proceedings of the joint Royal Entomological Society-Linnean Society Symposium. Biol J Linn Soc 79: 1–208
Loxdale HD, Lushai G (2003c) Maintenance of aphid clonal lineages: images of immortality. Infect Genet Evol 3: 259–269
Loxdale HD, Lushai G (2007) Population genetic issues: the unfolding story revealed using molecular markers. In: van Emden HF, Harrington R (eds) Aphids as crop pests. CABI, Wallingford, Oxford, pp. 31–67
Loxdale HD, Tarr IJ, Weber CP, Brookes CP, Digby PGN, Castañera P (1985) Electrophoretic study of enzymes from cereal aphid populations. III. Spatial and temporal genetic variation of populations of *Sitobion avenae* (F) (Hemiptera: Aphididae). Bull Entomol Res 75: 121–141
Lushai G, De Barro PJ, David O, Sherratt TN, Maclean N (1998) Genetic variation within a parthenogenetic lineage. Insect Mol Biol 7: 337–344
Lushai G, Hardie J, Harrington R (1996) Inheritance of photoperiodic response in the bird cherry aphid *Rhopalosiphum padi.* Physiol Entomol 21: 297–303
Lushai G, Loxdale HD (2007) The potential role of chromosome telomere resetting consequent upon sex in the population dynamics of aphids: an hypothesis. Biol J Linn Soc 90: 719–728
Lushai G, Loxdale HD, Allen JA (2003) The dynamic clonal genome and its adaptive potential. Biol J Linn Soc 79: 193–208
Lushai G, Loxdale HD, Brookes CP, von Mende N, Harrington R, Hardie J (1997) Genotypic variation among different phenotypes within aphid clones. Proc R Soc Lond B 264: 725–730
Lushai G, Markovitch O, Loxdale HD (2002) Host-based genotype variation in insects revisited. B Entomol Res 92: 159–164
Lynch M, Blanchard JL (1998) Deleterious mutation accumulation in organelle genomes. Genetica 102–103: 29–39
Lynch M, Burger R, Butcher D, Gabriel W (1993) The mutational meltdown in asexual populations. J Hered 84: 339–344
Mackenzie A, Guldemond JA (1994) Sympatric speciation in aphids. II. Host race formation in the face of gene flow. In: Leather SR, Watt AD, Mill NJ, Walter KPA (eds) Individuals, populations and patterns in ecology. Intercept Ltd, Andover, Hampshire, pp. 379–393
Malloch G, Highet F, Kasprowicz L, Pickup J, Neilson R, Fenton B (2006) Microsatellite marker analysis of peach-potato aphids (*Myzus persicae*; Homoptera: Aphididae) from Scottish suction traps. Bull Entomol Res 96: 573–582

Mark Welch DB, Meselson M (2000) Evidence for the evolution of bdelloid rotifers without sexual reproduction or genetic exchange. Science 288: 1211–1215

Martínez-Torres DA (1994) Molecular characterisation and population analysis of mitochondrial variation in the aphid *Rhopalosiphum padi* (L) (Homoptera: Aphididae). PhD Thesis, University of Valencia, Valencia

Martínez-Torres D, Moya A, Hebert PDN, Simon J-C (1997) Geographic distribution and seasonal variation of mitochondrial DNA haplotypes in the aphid *Rhopalosiphum padi* (Hemiptera: Aphididae). Bull Entomol Res 87: 161–167

Martens K, Rossetti, G, Home DJ (2003) How ancient are ancient asexuals? Proc R Soc Lond B 270: 723–729

Massonnet B, Simon J-C, Weisser WW (2002) Metapopulation structure of the specialised herbivore *Macrosiphoniella tanacetaria* (Homoptera: Aphididae). Mol Ecol 11: 2511–2521

Massonnet B, Weisser WW (2004) Patterns of genetic differentiation between populations of the specialised herbivore *Macrosiphoniella tanacetaria* (Homoptera: Aphididae). Heredity 93: 577–584

Maynard-Smith J (1978) The evolution of sex. Cambridge University Press, Cambridge, UK

Miller NJ (2000) Population structure and gene flow in a host-alternating aphid *Pemphigus bursarius*. PhD Thesis, University of Birmingham, Birmingham

Miller NJ, Birley AJ, Overall ADJ, Tatchell, GM (2003) Population genetic structure of the lettuce root aphid *Pemphigus bursarius* (L.) in relation to geographic distance, gene flow and host plant usage. Heredity 91: 217–223

Muller HJ (1964) The relation of recombination to mutational advance. Mutat Res 1: 2–9

Nicol D, Armstrong KF, Wratten SD, Cameron CM, Frampton C, Fenton B (1997) Genetic variation in an introduced aphid pest (*Metopolophium dirhodum*) in New Zealand and relation to individuals from Europe. Mol Ecol 6: 255–265

Niemeyer HM (1990) The role of secondary plant compounds in aphid-host interactions. In: Campbell RK, Eikenbary RD (eds) Aphid-plant genotype interactions. Elsevier Amsterdam, pp. 187–205

Normark BB (1999) Evolution in a putatively ancient asexual aphid lineage: Recombination and rapid karyotype change. Evolution 53: 1458–1469

Normark BB, Judson OP, Moran NA (2003) Genomic signatures on ancient asexual lineages. Biol J Linn Soc 79: 69–84

Paland S, Lynch M (2006) Transitions to asexuality result in excess amino acid substitutions. Science 311: 990–992

Ridley M (1993) Evolution. Blackwell, Cambridge, Massachusetts, USA

Roderick G K (1996) Geographic structure of insect populations: gene flow, phylogeography and their uses. Annu Rev Entomol 41: 325–352

Sabater-Muñoz B, Legeai F, Rispe C, Bonhomme J, Dearden P, Dossat C, Duclert A, Gauthier J-P, Giblot Ducray D, Hunter W, Dang P, Kambhampati S, Martinez-Torres D, Cortes T, Moya A, Nakabachi A, Philippe C, Prunier-Leterme N, Rahbé Y, Simon J-C, Stern DL, Wincker P, Tagu D (2006) Large-scale gene discovery in the pea aphid *Acyrthosiphon pisum* (Hemiptera). Genome Biol 7: R21

Sawicki RM, Devonshire AL, Payne RW, Petzing SM (1980) Stability of insecticide resistance in the peach-potato aphid *Myzus persicae* (Sulzer). Pestic Sci 11: 33

Scali V, Passamonti M, Marescalchi O, Mantovani B (2003) Linkage between sexual and asexual lineages: genome evolution in *Bacillus* stick insects. Biol J Linn Soc 79: 137–150

Schön I, Martens K, van Doninck K, Butlin RK (2003) Evolution in the slow lane: molecular rates of evolution in sexual and asexual ostracods (Crustacea: Ostracoda). Biol J Linn Soc 79: 93–100

Schreiber SJ (2003) Allee effects, extinctions, and chaotic transients in simple population models. Theor Popul Biol 64: 201–209

Sheppard WS, Steck GS, McPheron BA (1992) Geographic populations of the medfly may be differentiated by mitochondrial DNA variation. Experientia 48: 1010–1013

Shufran KA, Mayo ZB, Crease TJ (2003) Genetic changes within an aphid clone: homogenization of rDNA intergenic spacers after insecticide selection. Biol J Lin Soc 79: 101–105

Simon J-C, Baumann S, Sunnucks P, Hebert PDN, Pierre JS, Le Gallic JF, Dedryver C-A (1999) Reproductive mode and population genetic structure of the cereal aphid *Sitobion avenae* studied using phenotypic and microsatellite markers. Mol Ecol 8: 531–545

Simon J-C, Delmotte F, Rispe C, Crease T (2003a) Phylogenetic relationships between parthenogens and their sexual relatives: the possible routes to parthenogenesis in animals. Biol J Lin Soc 79: 151–163

Simon J-C, Carre S, Boutin M, Prunier-Leterme N, Sabater-Munoz B, Latorre A, Bournoville R (2003b) Host-based divergence in populations of the pea aphid: insights from nuclear markers and the prevalence of facultative symbionts. Proc R Soc Lond B 270: 1703–1712

Simon J-C, Martinez-Torres D, Latorre A, Moya A, Hebert PDN (1996) Molecular characterization of cyclic and obligate parthenogens of the aphid *Rhopalosiphum padi* (L.). Proc R Soc Lond B 263: 481–486

Simon J-C, Rispe C, Sunnucks P (2002) Ecology and evolution of sex in aphids. Trends Ecol Evol 17: 34–39

Slatkin M (1985) Gene flow in natural populations. Annu Rev Ecol Syst 16: 393–430

Slatkin M (1993) Isolation by distance in equilibrium and non-equilibrium populations. Evolution 47: 264–279

Smadja C, Galindo J, Butlin R (2008) Hitching a lift on the road to speciation. Mol Ecol 17: 4177–4180

Strathdee AT, Bale JS, Block WC, Webb NR, Hodkinson ID, Coulson SJ (1993) Extreme adaptive life-cycle in a high arctic aphid *Acyrthosiphon svalbardicum.* Ecol Entomol 18: 254–258

Sunnucks P, Chisholm D, Turak E, Hales DF (1998) Evolution of an ecological trait in parthenogenetic *Sitobion* aphids. Heredity 81: 638–647

Sunnucks P, De Barro PJ, Lushai G, Maclean N, Hales D (1997) Genetic structure of an aphid studied using microsatellites: cyclic parthenogenesis, differentiated lineages and host specialization. Mol Ecol 6: 1059–1073

Tagu D, Klingler JP, Moya A, Simon J-C (2008) Early progress in aphid genomics and consequences for plant-aphid interactions studies. Mol Plant Microbe Interact 21: 701–708

Tatchell GM, Plumb RT, Carter N (1988) Migration of alate morphs of the bird cherry aphid (*Rhopalosiphum padi*) and implications for the epidemiology of barley yellow dwarf virus. Ann Appl Biol 112: 1–11

Taylor LR (1965) Flight behaviour and aphid migration. Proc North Central Branch Entomol Soc Am 20: 9–19

Taylor LR (1986a) The distribution of virus disease and the migrant vector aphid. In: McLean GD, Garrett RG, Ruesink WG (eds) Plant virus epidemics: monitoring, modelling and predicting outbreaks. Acad Press Australia, pp. 35–57

Taylor LR (1986b) Synoptic dynamics migration and the Rothamsted insect survey. J Anim Ecol 55: 1–38

Taylor LR, Woiwod IP, Taylor RAJ (1979) The migratory ambit of the hop aphid and its significance in aphid population dynamics. J Anim Ecol 48: 955–972

Thompson JN (1994) The Co-evolutionary process. University of Chicago Press, Chicago and London

Van Doninck K, Schön I, De Bruyn L, Martens K (2002) A general purpose genotype in an ancient asexual. Oecologia 132: 205–212

van Emden HF (1988) The peach-potato aphid *Myzus persicae* (Sulzer) (Hemiptera: Aphididae) – more than a decade on a fully defined chemical diet. Entomol 107: 4–10

van Emden HF, Harrington R (eds) (2007) Aphids as crop pests. CABI, Wallingford, Oxford

Via S (1999) Reproductive isolation between sympatric races of pea aphids. I. Gene flow restriction and habitat choice. Evolution 53: 1446–1457

Via S (2001) Sympatric speciation in animals: the ugly ducking grows up. Trends Ecol Evol 16: 381–390

Via S, Bouck AC, Skillman S (2000) Reproductive isolation between divergent races of pea aphids on two hosts. II. Selection against migrants and hybrids in the parental environments. Evolution 54: 1626–1637
Via S, West J (2008) The genetic mosaic suggests a new role for hitchhiking in ecological speciation. Mol Ecol 17: 4334–4345
Vialatte A, Dedryver C-A, Simon J-C, Galman M, Plantegenest M (2005) Limited genetic exchanges between populations of an insect pest living on uncultivated and related cultivated host plants. Proc R Soc Lond B 272: 1075–1082
Vialatte A, Plantegenest M, Simon J-C, Dedryver C-A (2007) Farm-scale assessment of movement patterns and colonization dynamics of the grain aphid in arable crops and hedgerows. Agric For Entomol 9: 337–346
Vorburger C (2006) Temporal dynamics of genotypic diversity reveal strong clonal selection in the aphid *Myzus persicae.* J Evol Biol 19: 97–107
Vorburger C, Lancaster M, Sunnucks P (2003a) Environmentally related patterns of reproductive modes in the aphid *Myzus persicae* and the predominance of two 'superclones' in Victoria, Australia. Mol Ecol 12: 3493–3504
Vorburger C, Sunnucks P, Ward SA (2003b) Explaining the coexistence of asexuals with their sexual progenitors: no evidence for general-purpose genotypes in obligate parthenogens of the peach-potato aphid *Myzus persicae*. Ecol Lett 6: 1091–1098
Vorwerk S, Forneck A (2007) Analysis of genetic variation within clonal lineages of grape phylloxera (*Daktulosphaira vitifoliae* Fitch) using AFLP fingerprinting and DNA sequencing. Genome 50: 660–667
Ward SA, Leather SR, Pickup J, Harrington R (1998) Mortality during dispersal and the cost of host specificity in parasites: how many aphids find hosts? J Anim Ecol 67: 763–773
West SA, Lively CM, Read AF (1999) A pluralist approach to sex and recombination. J Evol Biol 12: 1003–1012
Wilson ACC, Sunnucks P, Blackman RL, Hales DF (2002) Microsatellite variation in cyclically parthenogenetic populations of *Myzus persicae* in south-eastern Australia. Heredity 88: 258–266
Wilson ACC, Sunnucks P, Hales D F (1999) Microevolution, low clonal diversity and genetic affinities of parthenogenetic *Sitobion* aphids in New Zealand. Mol Ecol 8: 1655–1666
Woiwod IP, Tatchell GM, Dupuch MJ, Macaulay EDM, Parker SJ, Riley AM, Taylor MS (1988) Rothamsted Insect Survey: nineteenth annual summary: suction traps 1987. Annu Rep Rothamsted Experimental Station 1987 Part 2, pp. 195–229
Wright S (1931) Evolution in Mendelian populations. Genetics 16: 97–159
Wright S (1943) Isolation by distance. Genetics 28: 114–138
Wright S (1951) The genetical structure of populations. Ann Eugen 15: 323–354
Wynne IR, Howard JJ, Loxdale HD, Brookes CP (1994) Population genetic structure during aestivation in the sycamore aphid *Drepanosiphum platanoidis* (Hemiptera: Drepanosiphidae). Eur J Entomol 91: 375–383

† Since writing this article, I have found a reference to the occurrence of rare *Trama* males and oviparae, at least in *T. troglodytes,* which may possibly be sexually functional according to the authors. See: Blackman RL, De Boise E, Czylok A (2001) Occurrence of sexual morphs in *Trama troglodytes* von Heyden, 1837 (Hemiptera, Aphididae). *Journal of Natural History*, 35: 779–785

Chapter 26
Epigenetic Mechanisms in Mammals and Their Effects on Cloning Procedures

Pasqualino Loi, Grazyna Ptak and Robert Feil

Abstract Early embryonic development is characterized by dramatic modifications in chromatin organization. These contribute to the formation of the totipotent cells that will differentiate into the embryonic and extra-embryonic lineages. However, some groups of genes are not affected by these global changes. Amongst these, are the genes controlled by genomic imprinting, an epigenetic mechanism that makes the maternal and the paternal genomes functionally non-equivalent in development. Somatic cell nuclear transfer (SCNT), used to clone animals, mimics the chromatin reprogramming that occurs in the post-fertilization embryo. However, SNCT leads to aberrant development, particularly of the extra-embryonic tissues. As a consequence, live-born animals are obtained at very low frequency. Here, we review how different steps in the SCNT procedure may give rise to aberrant development. Imprinted genes are amongst the frequently affected genes, and their perturbed expression explains in part the abnormal developmental phenotypes. The future challenge will be to modify currently used cloning procedures, in order to increase the efficiency of nuclear reprogramming, without affecting the gene loci at which chromatin is normally not altered during development.

26.1 Introduction

The generation of viable animals by transfer of a nucleus from a somatic cell into an enucleated egg was first achieved in *Xenopus* frogs, fifthy years ago (Gurdon et al. 1975). In mammalian species, cloning from somatic cells was successfully accomplished about ten years ago, for the first time in the sheep (Wilmut et al. 1997). Through subsequent research in a rapidly expanding number of laboratories, it emerged that reprogramming of a somatic cell to a condition of full totipotency is not easily achieved. In different mammalian species, somatic cell nuclear

P. Loi (✉)
Faculty of Veterinary, Department of Comparative Biomedical Sciences, University of Teramo, Piazza Aldo Moro, 45, 64020, Teramo, Italy
e-mail: ploi@unite.it

I. Schön et al. (eds.), *Lost Sex*, DOI 10.1007/978-90-481-2770-2_26,

transfer (SCNT) technologies were found to be associated with a multitude of developmental defects in the resulting embryos and newborn animals. Below, we summarise the developmental complications that arise following cloning procedures in mammals, which have also been reviewed in detail elsewhere (Meissner and Jaenisch 2006). One of the emerging insights is that the development and functioning of the extra-embryonic membranes is particularly affected by SCNT. Defects in placentation have dramatic consequences on embryonic development and growth, and it is, therefore, important to understand the underlying mechanisms. One possibility, which will be discussed below, is that gene regulation in extra-embryonic membranes is less tightly controlled by covalent modifications on DNA and associated chromatin than in embryonic lineages. Consequently, even minor disruptions in chromatin regulatory mechanisms could affect gene expression more readily in the extra-embryonic tissues than in lineages that give rise to the embryo proper.

The aim of cloning is to achieve the best-possible remodelling of a somatic cell nucleus without affecting genes and chromosomal regions that should normally not be altered. Indeed, many genes and certain types of repeated DNA sequences do normally not change the organisation of their chromatin during pre-implantation development, and certain chromosomal regions need to be maintained in a repressed state throughout development. Importantly, these chromosomal regions should not become aberrantly altered as a consequence of cloning. This is particularly true for imprinted genes. These genes are developmentally essential and are expressed from one of the two alleles only, depending on the parental origin of the allele. As imprinted genes are expressed from only one of the two copies, either the maternally or the paternally inherited one, they are responsible for the functional non-equivalence of the maternal and paternal genomes (Kawahara et al. 2007). Their allele-specific expression is programmed by chromatin features that originate from either the egg or the sperm, and which are maintained throughout development. Several studies have shown that these rather unusual genes are particularly susceptible to becoming altered as a consequence of cloning procedures. In an ideal cloning experiment, imprinted genes should retain their germ-line derived epigenetic marks and not alter their allelic expression status. However, some imprinted genes have a strong tendency to become perturbed during the cloning procedures and this has consequences for the growth and development of the resulting conceptuses. Besides imprinted genes, also other groups of genes must not become aberrantly reprogrammed in the zygote. This raises the question of how to balance efficient global reprogramming whilst keeping different groups of essential genes unaltered.

Reprogramming of a somatic cell nucleus after its transfer into the enucleated egg should ideally lead to remodelling of chromatin at developmental genes, to enable them to acquire the potential of becoming expressed in the appropriate lineage. Neuronal genes, for instance, must re-acquire the potential of becoming expressed when the cloned embryo initiates the development of its neuronal lineages. Genes which are required to maintain cells undifferentiated and pluripotent, such as *OCT4* and *NODAL*, should have their chromatin reprogrammed to become expressed again in the nucleus of the donor somatic cells. Global chromatin remodelling occurs after the somatic cell nucleus has been introduced into the egg, and during the first cell divisions. The details of this process remain largely elusive. More is known about chromatin remodelling in normal post-fertilisation embryos. Particularly, recent work shows that dynamic, global changes in DNA methylation and modifications of the histone proteins are part of the normal remodelling process after fertilisation (Reik 2007). To which extent chromatin remodelling in the reconstituted zygote of cloned embryos is comparable to this natural process is subject to intense research efforts. One actively pursued idea is that the somatic cell nucleus could possibly be pre-conditioned by specific treatments, to make its remodelling after transfer into the egg more efficient.

A challenge for the future will be to devise novel strategies to improve the efficiency of cloning. Even in species for which cloning is relatively successful and with minor developmental defects only, the efficiency of the currently used procedures remains low. How can these procedures be improved to minimise adverse developmental and epigenetic effects? Various interesting approaches have been suggested by different groups. However, so far, none of these has led to a dramatic increase in cloning efficiencies. Possibly, a multi-step approach that includes a pre-treatment of donor cells to finely modify their chromatin may be the most promising.

There are challenging days ahead for developmental and reproductive biologists who are interested in SCNT. Our review discusses some of the issues at stake, with a particular emphasis on how epigenetic mechanisms, such as imprinting, may weigh on the outcome of cloning procedures in different mammalian species.

26.2 Is Cloning in Mammals Compatible with Normal Development?

This question was addressed by Wilmut and colleagues who reconstructed enucleated sheep oocytes with cultured epithelial cells established from the mammary gland of a ewe, and transferred the resulting embryos into the womb of surrogate mothers. A single lamb was obtained, with apparently normal growth rate and reproductive performance (Wilmut et al. 1997). Thus, cloning with somatic cells (SCNT) is indeed compatible with normal development in placental mammals. Not surprisingly, the removal of the major dogma in mammalian developmental biology has exerted a powerful influence on cutting edge research over the last 10 years. As a result, so far eleven mammalian species have been successfully cloned (Meissner and Jaenisch 2006), and SCNT micro-manipulation procedures have been simplified to allow the production of larger numbers of cloned embryos (Lagutina et al. 2005; Vajta 2007). However, despite the major efforts during the last years, the frequency of live offspring obtained by SCNT remains low, roughly about 1–5% of the transferred embryos.

26.3 Why is SCNT So Unsuccessful?

In the light of the very low frequency of normal development of clones across all tested species, the question might be easily reversed: how does it come that SCNT is sometimes successful? The reversal of cellular differentiation through SNCT is a rare and unpredictable event (Tsunoda and Kato 2002). The epigenetic formatting, which takes place during male and female germ cell development, and the maturation of female germ cells during weeks/months/years -depending on the species, are bypassed in SCNT (Morgan et al. 2005). The main factor responsible for the developmental failure of clones seems to be the oocyte's incapability to convey totipotency to the transplanted somatic cell nucleus, a largely unknown process referred to as "Nuclear Reprogramming" (Rideout et al. 2001). Our understanding of nuclear reprogramming remains essentially phenomenological. Epigenetic deregulation and abnormal gene expression in pre- and post-implantation embryos (Ogura et al. 2002; Latham 2005), in newborn animals and extra-embryonic tissues (Tamashiro et al. 2003; Kremenskoy et al. 2006; Loi et al. 2006) are commonly observed as consequences of "Nuclear Reprogramming" malfunction.

Epigenetic alterations in cloned animals lead to similar developmental phenotypes in all species studied, suggesting that common causal mechanisms are involved. Although abnormalities have been reported in cloned foetuses/offspring (Wells et al. 1998), these embryonic defects often result from placental fluid imbalances that affect kidney function and therefore general homeostasis (Li et al. 2005). The extra-embryonic tissues are severely affected at all stages of foetal development (Cezar et al. 2003), also when full-term clones are obtained (Loi et al. 2006). Owing to differences between species in the timing and type of implantation, there are distinct placental lesions in clones of different species. So far, functional and morphological abnormalities in placentation have been studied in detail only in mice, cattle and sheep.

In mice (Wakisaka-Saito et al. 2006), and to a certain extent also in cattle (Hashizume et al. 2002), placental hypertrophy, or placentomegaly, has been described, but not in sheep (Loi et al. 2006). The main histological lesions are increased numbers of glycogen cells and enlarged spongiotrophoblast cells (Wakisaka-Saito et al. 2006). Enlargement of the trophoblast giant cells and disorganisation of the labyrinth layer have also been reported (Wakisaka-Saito et al. 2006). SCNT placentas in cattle display fewer placentomes, and these are often larger than normal and irregular in size (Chavatte-Palmer et al. 2006; Constant et al. 2006). Histological examination revealed a hypotrophic trophoblastic epithelium associated with reduced vascularization (Hill et al. 2000). These histological alterations have also been found in placentas of cloned sheep (Loi et al. 2006). In addition, ultrastructural studies on cloned sheep revealed features indicative of placental ageing, including thickening of the trophoblast basement membrane (Palmieri et al. 2007). Ageing is one of the issues raised after Dolly's birth. Dolly was produced from a cell line that was established from a six years old sheep. Therefore, the biological clock of these cells was already midway the average sheep's life. There had been uncertainty about Dolly's biological age, and this question has been addressed by looking at the telomere length of its cells. Telomeres are short repetitive arrays at the end of chromosomes, which are recognised by the DNA replication machinery. Every cell replication cycle removes a telomere repeat. Therefore, telomeres shorten progressively throughout the finite proliferative life of a cell. Accordingly, shorter telomeres were found in Dolly's cells, despite its phenotypic similarity with age-matched ewes of her breed (Shiels et al. 1999). A later report in cattle challenged this observation, showing that telomere length is restored in bovine SCNT clones (Tian et al. 2000). Further reports in cattle confirmed the latter data (Lanza et al. 2000), adding novel insights into the regenerative power of the mammalian oocyte.

Our own observations are also indicative of placental senescence in ovine clones (Palmieri et al. 2007). We are not aware of reports on telomere length in placental tissue of clones, but this should be an interesting question to address in future studies.

Paradoxically, the first mammalian species to be cloned, the sheep, is the one which displays the most severe placental abnormalities. These not only account for foetal losses, but probably also for the peri- and post-natal demises of cloned lambs. The situation seems to be different in some other large animal species, where clones that survive the critical post-natal phase continue their development till adulthood, particularly in cattle (Tian et al. 2007).

26.4 Epigenetic Inequality Between the Parental Genomes

In mammals, the maternal and paternal genomes are functionally not equal, and they are therefore both required for normal development (Surani et al. 1984). This important knowledge had first emerged from the observation that parthenogenetic mouse embryos (which have two maternal genomes) do not develop to term (Graham 1974). Subsequent studies were performed, in which pronuclei were transplanted in fertilised eggs in such a way, that reconstituted zygotes contained either two maternal or two paternal pronuclei. These studies demonstrated that differences between the parental genomes were responsible for these developmental failures, rather than factors present in the cytoplasm (Surani et al. 1984). Gynogenetic embryos (with two maternal genomes, and no paternal genome) failed to develop beyond day 10 of gestation and displayed major developmental abnormalities. A similarly dramatic phenotype was observed in androgenetic embryos (with two paternal genomes, and no maternal genome). These studies clearly proved that the maternal and paternal genomes do not express their genetic information in the same way, which makes them functionally different (Dean et al. 2001). At about the same time, requirement of both a maternal and a paternal copy was shown for individual chromosomes as well, through extensive studies on uniparental disomic mice (Cattanach and Kirk 1985). Together, these studies established that the maternal and paternal genomes are marked in a different way in order to become functionally different and that both are required for normal development. This epigenetic phenomenon has been named "genomic imprinting", and has been observed through embryological studies in other mammalian species as well. In sheep, for instance, parthenogenetic conceptuses do not develop much beyond the embryonic stages, and are grossly abnormal, particularly in their extra-embryonic membranes (Feil et al. 1998; Hagemann et al.1998; Loi et al. 1998).

To date, more than eighty genes in mice and humans have been found to be expressed from only one of their two alleles, depending on the parental origin of the allele (Kono et al. 2004; Morison et al. 2005).

Genomic imprinting is regulated by heritable epigenetic modifications, which are put onto the DNA (methylation) and the chromatin (histone modifications) at imprinted gene loci. These epigenetic marks (imprints) are present at DNA sequence elements that regulate the allelic expression of imprinted genes, the "imprinting control regions" (ICRs). Their establishment occurs either in the egg or in sperm, depending on the imprinted locus in question, and their somatic maintenance from the zygote stage onwards is discussed in several recent reviews (Delaval and Feil 2004; Feil and Berger 2007; Edwards and Ferguson-Smith 2007). Importantly, if the allelic methylation status of ICRs is not maintained properly during development, this can lead to different growth-related diseases, such as the Beckwith-Wiedemann syndrome of foetal overgrowth (Arnaud and Feil 2005; Delaval et al. 2006). Also several steps of the cloning procedure may affect the methylation imprints at ICRs. This can have negative consequences for the development and growth of the cloned embryos and foetuses.

The parental imprints at imprinted gene loci are not the only differences in DNA methylation that exist between the sperm and egg derived genomes. In contrast to ICRs, the global methylation status of the maternal and paternal sets of chromosomes undergoes dramatic changes after fertilisation of the egg, such that both parental genomes become comparable in their levels of methylation at the blastocyst stage. Specifically, sperm DNA is overall highly methylated, but directly after fertilisation, most of this DNA methylation is actively removed by a yet-unknown demethylase activity in the egg (Mayer et al. 2000). This global DNA demethylation process does not affect the ICRs and the imprinted genes controlled by them. In many species, global DNA demethylation of the sperm genome is extensive, whereas in others, such as the sheep, it appeared to be more reduced. However, such apparent differences between species could be due to when and how the experimental staining procedures are performed. The maternal genome in the zygote is, on the contrary, not subject to an active demethylation process.

Possibly this could be related to the fact that its DNA is organised around nucleosomes, and thereby somehow protected. Thus, the maternal genome will be slowly passively demethylated in order for both genomes to be largely demethylated at blastocyst stage. Because of the dramatically different way the egg and the sperm-derived genomes are organised, they are being remodelled very differently in the zygote and during the early cleavage stages. As mentioned, the paternal genome undergoes global DNA demethylation and its protamines are replaced by nucleosomes (Rousseaux et al. 2005). Dynamic changes in histone methylation have also been described, many of which are different between the maternal and the paternal genomes at the early stages following fertilisation.

In SCNT, the somatic cell nucleus that is introduced into the enucleated egg has nucleosomally organised chromatin on both its genomes, and its DNA methylation levels are reflective of the somatic cell type the nucleus originated from. It remains poorly understood to which extent the chromatin remodelling machineries present in the egg are acting on the somatic cell-derived chromosomes, and whether the overall effect is comparable to that achieved in naturally fertilised eggs. This is one of the key questions related to cloning in mammals (Gurdon and Melton 2008).

26.5 Does Cloning Affect the Parental Epigenetic Information?

In the mouse, regardless of the donor cell type used, aberrant imprinted gene methylation and expression were reported in a large number of different cloning experiments (Oghane et al. 2004; Humpherys et al. 2001, 2002; Inoue et al. 2002; Young et al. 2003). Imprinted gene expression was found to be perturbed in the placenta in many cases, and the imprinted domain comprising the Insulin-like growth factor-2 gene was frequently affected (Fig. 26.1). This finding could be relevant to the observed placentomegaly, which is a recurrent phenotype in cloned mouse conceptuses (Wakayama et al. 1999; Ogura et al. 2002). Interestingly, the placental phenotype in cloned mice is highly reminiscent of the defects sometimes detected after in vitro manipulation and embryo/cell culture (Khosla et al. 2001a; Pannetier and Feil 2007), indicating that at least part of the phenotype could have a common origin. However, it is exclusively in cloned animals that expansion of the basal layer is observed with an increase in glycogen cells, suggesting that this could be specifically linked to SCNT. Although the current data are too limited to understand to which extent imprinted genes contribute to the hypertrophy of the placenta, altered expression levels have been reported for the imprinted *Mest*, *Grb10* and *Gtl2* genes.

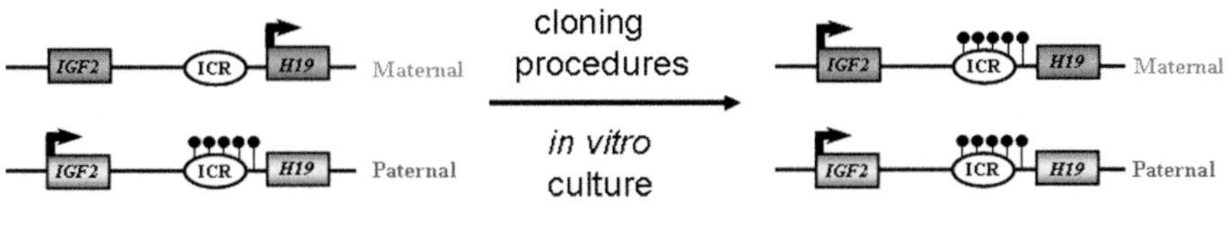

Fig. 26.1 Cloning-induced gain of DNA methylation at the *IGF2-H19* imprinted domain. The imprinting control region (ICR) of the *IGF2-H19* domain is DNA-methylated (filled lollipops) at its paternally inherited copy. This epigenetic imprint mediates the paternal expression of the *IGF2* growth factor gene, and the maternal expression of the neighbouring *H19* gene, which encodes a non-coding RNA of unknown function. In cells and embryos of different mammalian species, somatic cell nuclear transfer and in vitro culture can induce aberrant DNA methylation at this ICR, leading to biallelic expression of *IGF2* ("loss of imprinting") and loss of *H19* expression. Upon differentiation and during subsequent embryonic development, the epigenetic alteration and the aberrant imprinted expression persist, and affect foetal growth

Why is imprinted gene expression so often deregulated in the placenta? One explanation could be that the extra-embryonic tissues have a short functional life compared to that of the embryo proper. Therefore, the epigenetic marks securing the heritable expression or repression of genes do not have to be strictly regulated, and altered gene dosage could be more tolerable. For instance, aneuploid or polynucleated cells are more readily tolerated in extra-embryonic tissues (Weier et al. 2005) than in the foetus proper. Moreover, tetraploid embryos develop a functional placenta capable of nourishing a foetus derived from diploid embryonic stem cells till term (Nagy et al. 1990). Further evidence on the more relaxed gene regulation in the extra-embryonic lineage can be observed as early as the blastocyst stage. In all mammals analysed, the trophectoderm has significantly lower levels of repressive DNA methylation, including at CpG islands and gene promoters, compared to the inner cell mass of the blastocyst (Bird 2002; Reik et al. 2003). At later developmental stages, the trophoblast (placenta) also shows low levels of DNA methylation; therefore, it is thought that DNA methylation plays a less important role in gene regulation in the extra-embryonic membranes (Wagschal and Feil 2006; Wagschal et al. 2008).

Mammalian DNA methylation can be assessed by immuno-staining with antibodies directed against methylated 5-Cytosine (5 meC). In many of such studies, the mouse has been used as a reference model. It is in the mouse that global DNA methylation changes were first discovered to occur in early embryonic development (Haaf 2006; Santos et al. 2005). Specifically, following fertilization of the oocyte, active DNA demethylation takes place after the formation of the pronuclei leading to an almost complete loss of methylation in the sperm-derived pronucleus (Mayer et al. 2000). The maternal pronucleus is protected against this demethylation process which occurs directly after fertilisation. However, a progressive and passive demethylation process occurs during the early cleavages. This leads to the gradual reduction in DNA methylation in the maternally-derived genome (Santos et al. 2005). Consequently, at the blastocyst stage, both parental genomes have acquired a hypo-methylated state. Upon implantation, there is acquisition of novel DNA methylation at many regions of the genome, starting during the initial differentiation process that occurs in the ICM cells, while the trophoblast cells remain hypo-methylated at this stage (Haaf 2006). It has been observed that the global waves of DNA demethylation and the subsequent de novo methylation are not occurring as they should in cloned embryos, in which relatively high levels of DNA methylation persist in both the ICM and the trophectoderm (Santos et al. 2003, 2005).

DNA methylation is a landmark of epigenetic regulation, and one, therefore, expects its role in embryonic development to be the same in different mammals. To address this question, several groups have explored the dynamics of DNA methylation in pre-implantation embryos of different mammalian species. In rat, pig, and bovine embryos, global changes in DNA methylation were indeed found to be comparable to those in mouse embryos, with the zygote displaying active demethylation of the paternal pronucleus, followed by passive demethylation of the maternal genome (Haaf 2006). Also in these species, the asymmetric methylation between the trophectoderm and ICM cell lineages has been observed, although with some quantitative and temporal species differences (Haaf 2006). Several other studies, however, have suggested that sheep and rabbit zygotes do not undergo active demethylation of the paternal genome. In these species, also the passive DNA demethylation typical of the early cleavage stages seems to be less pronounced (Beaujean et al. 2004; Shi et al. 2004). It should be important to determine whether these two species are indeed different from other mammalian species, or whether the non-detection of DNA demethylation could be explained by the procedures used and the time after fertilisation at which these studies were performed.

In human zygotes, DNA demethylation was reported, albeit less pronouncedly than in the mouse and apparently not in all zygotes analysed (Fulka et al. 2004). Further work is required before firm conclusions can be drawn. In an interspecific ICSI study, we demonstrated that a species that appears to be resistant to the male genome's demethylation (sheep), showed demethylation of its male pronuclei when introduced into mouse oocytes. Conversely, sheep oocytes can only partially demethylate mouse sperm (Beaujean et al. 2004), but again, this could be linked to the timing of the experiment.

It is difficult to draw definite conclusions from the studies on pre-implantation cloned embryos that have been performed so far. The emerging picture, however, is that the oocyte is not able to completely reprogram the genome of the somatic cell by DNA demethylation. As mentioned above, in normal embryos, imprinted genes and their ICRs are fully protected against the global waves of DNA demethylation and de novo methylation. In fact, the methylation marks at imprinted gene loci originate from either the sperm or the egg, and are normally faithfully maintained throughout pre- and post-implantation development.

Another pitfall of SCNT concerns genes that are important for the maintenance of pluripotency. During normal development, genes that are essential for early differentiation and ontogenesis, like *OCT4* and *NANOG*, are also marked by de novo DNA methylation after implantation and they become repressed in differentiated somatic cells. Not surprisingly, many SCNT pre-implantation embryos display absence or reduction of expression of these genes, probably due to an incomplete removal of the repressive DNA methylation in the donor somatic cell nucleus (Boiani et al. 2002).

The small amount of genomic DNA that can be obtained from a single pre-implantation embryo has hindered high-throughput analysis of clones. Encouragingly, the miniaturization of gene expression detection methods makes it possible to monitor a reasonable panel of genes in single blastocysts (Smith et al. 2007). DNA micro-arrays have shed light on gene expression profiles of individual SCNT bovine embryos (Somers et al. 2006).

These finding suggest that SCNT does not much affect the patterns of gene expression monitored in high-throughput expression analyses (Smith et al. 2005). By paradox, major effects on gene expression are exerted by the culture system used, rather than the SCNT procedure itself (Smith et al. 2007). So far, the published data suggest that, while major alterations are detected in the global epigenetic reorganization of SCNT clones, specifically in DNA methylation (Beaujean et al. 2004), analyses carried out on transcription are indicative of a consistent reprogramming following nuclear transfer (Smith et al. 2005).

In conclusion, the gene expression data so far do not give insights into the severe phenotypes observed in cloned foetuses and their extra-embryonic tissues. Most likely, some locus-specific epigenetic changes are not detected by the current technologies, and/or these could exert their effects on gene expression later in development, as it has been shown in the mouse (Jouneau et al. 2006). This is particularly true for epigenetic alterations at genes, which are expressed only in specific differentiated cell lineages, and whose phenotypic consequences are detectable only later in development. Nevertheless, gene expression and epigenetic analyses remain important since they could unravel the relevant molecular mechanisms that need to be understood when aiming at improving nuclear reprogramming procedures (Wuensch et al. 2007).

26.6 Strategies to Improve Cloning

Dolly, the cloned sheep, was originally conceived to demonstrate that selected genotypes could be multiplied though nuclear transfer. However, despite 10 years of research, little progress has been made in the field of reproductive cloning. One reason has been the growing interest of laboratories in applications other than cloning. Reproductive aspects aside, the demonstration of the full reversibility of the differentiated state of a cell has opened the prospect of using nuclear transfer for the generation of patient-tailored cells for therapeutic purposes (Pomerantz and Blau 2004), an approach called therapeutic cloning. The potential of therapeutic cloning was documented a couple of years ago in a disease's animal model, in which nuclear transfer, gene therapy and stem cell biology were applied together (Rideout et al. 2002). Since then, major progress has been made in refining the process of cell dedifferentiation. Recent studies show that somatic cells can also be reprogrammed to become totipotent stem cells by transfecting them with combinations of pluripotency-associated genes (Takahashi and Yamanaka 2006; Okita et al. 2007; Wernig et al. 2007). This approach has been consolidated during the last few years and is becoming a promising alternative to therapeutic cloning.

Several recent reviews cover the advances made in therapeutic cloning (Hochedlinger and Jaenisch 2006; Carey et al. 2009). In our chapter, therefore, we have focused on the different strategies for reproductive cloning. SCNT has many potential applications in animal breeding (Wells 2003), production of transgenic animals (Prather et al. 2003; Trounson 2006; Robl et al. 2007; Niemann and Kues 2007), and as a tool in conservation efforts (Loi et al. 2001; Holt et al. 2004). To successfully achieve these goals, it is mandatory to develop efficient nuclear reprogramming approaches that lead to the production of normal animals.

Based on the published data, we assume that all mammalian species can be cloned using different types of somatic cells. However, the frequency of development of clones to term remains extremely low. Different solutions have been proposed to optimise nuclear reprogramming, but overall these have had only minor effects. The first paper where cloning was achieved with an adult cell line (Wilmut et al. 1997), and an earlier report on cloning from embryonic cell lines (Campbell et al. 1996), suggested that the induction of nuclear quiescence could be the trick to reset the memory of a somatic cell. The second mammalian species cloned, the mouse, was produced thanks to prolonged exposure of the transplanted cumulus cell nucleus to the cytoplasm of an enucleated, non-activated oocyte (Wakayama et al. 1998). Both these strategies have a solid scientific basis. Indeed, during G0, but also following chromosome condensation, cell specific transcription factors are displaced from the genome (Withfield et al. 1985; Martínez-Balbás et al. 1995), potentially facilitating the action of reprogramming factors. A later report, however, challenged these hypotheses by demonstrating that the "reprogrammability" of the nucleus was not strictly dependent on the cell cycle (Cibelli et al. 1998), although the compatibility of donor nuclei cell cycle status with the metaphase II oocyte plays still an important role (Campbell et al. 1996). Nevertheless, nuclear quiescence and delayed activation following nuclear transfer are widely used strategies in current SCNT protocols (Wakayama 2007).

As we mentioned before, absent or reduced DNA demethylation are commonly observed in cloned embryos (Morgan et al. 2005). Therefore, the use of molecules that can remove the repressive methyl marks from differentiated cells before nuclear transfer has been attempted. The drug 5-Azacytidine (5-Aza) has been used to globally remove DNA methylation in the donor cell nucleus prior to its injection into the egg, but the results have been somewhat disappointing. Very few cloned bovine embryos derived from 5-Aza-treated cells reached the blastocyst stage (Enright et al. 2003).

This finding is not surprising. It is known that removal of DNA methylation by 5-Aza treatment leads to massive DNA rearrangements and the formation of fragmented small nuclei (Kiziltepe et al. 2007), thus exacerbating the intrinsic drift of SCNT cloned embryos towards aneuploidy (Shi et al. 2004). Possibly, treatments at lower concentrations should be attempted to avoid/limit chromosomal abnormalities.

Nuclei of cleavage stage embryo are larger than those of somatic cells. Their chromatin also has an open structure to allow extensive transcriptional activity of a large proportion of genes, probably the most intense transcriptional burst that can be exerted. The chromatin histone composition is central for the establishment of such an "open" conformation, particularly the acetylation of histone 4 (H4) tails. Accordingly, both parental genomes (the paternal first) are heavily acetylated at H4 on lysine18 in the zygote (van der Heijden et al. 2005; Yoshida et al. 2007), and this is likely to facilitate chromatin remodelling. In *Drosophila*, for instance, hyper-acetylation of H4 on lysine-18 mediates the increased levels of transcription from the X chromosome in male flies (Mendjan and Akhtar 2007). One strategy, therefore, has been to boost global levels of histone acetylation by treatment with specific drugs that inhibit histone deacetylases (HDACs). A well-characterised HDAC inhibitor is Trichostatin-A (TSA), which, upon short treatments, confers to the chromatin an open configuration with significantly increased levels of histone acetylation. TSA has been used to treat the somatic donor cells before nuclear transfer (Enright et al. 2003), and to treat SCNT embryos shortly after nuclear transfer (Kishigami et al. 2006), at about ten hours post activation. TSA-treatment of donor cells in bovine experiments increased the frequency of development to the blastocyst stage (Enright et al. 2003), although subsequent in vivo development was not assessed. In two recent mouse studies, the protocol of treatment was slightly changed, and TSA was used after nuclear transfer. Both these studies reported an improved frequency of development to the blastocyst stage, and, remarkably, also to term (5% versus 1% in the controls), indicating a positive effect of the drug on the efficiency of nuclear reprogramming (Kishigami et al. 2006, 2007; Rybouchkin et al. 2006). These results are the most significant breakthrough so far accomplished in SCNT and they strengthen the idea that modification of the chromatin in differentiated cells can make them more prone to "remodelling" by factors present in the oocyte's cytoplasm.

Furthermore, the choice of donor cells for SNCT has also been evaluated in view of improving SCNT. The information gained so far is that there is an inverse relationship between the level of differentiation of the cell and its "reprogrammability" in mammals, as already observed in the amphibians (Gurdon 2006). Terminally differentiated cells, like B or T lymphocytes, or olfactory neurons, do not allow the production of live offspring following SCNT. This problem, however, can be overcome by using a "two-step nuclear transfer" procedure, in which embryonic stem (ES) cells established from cloned blastocysts are injected into tetraploid blastocysts (Hochedlinger and Jaenisch 2002; Eggan et al. 2004). The need for a two-step nuclear transfer procedure was recently circumvented by a group that managed to produce cloned pups from nuclear transfer of not-yet-fully differentiated B cells (Inoue et al. 2005). Therefore, one should select undifferentiated or partially differentiated cells for nuclear transfer, since these appear to be more easily reprogrammable by the oocyte's cytoplasm. This is indeed the case in mice, where undifferentiated ES cells have been used for nuclear transfer experiments with a high frequency of development to term (Wakayama et al. 1999; Rideout et al. 2000).

Different groups also compared the efficiency of nuclear reprogramming, assessed by early embryonic development, or development to term, of SCNT embryos in which stem cells isolated from adult tissue or differentiated cells were used as donor cells. All these different cell types were equally efficient (Mizutani et al. 2006; Inoue et al. 2006, 2001). However, it seems likely that efficiency differences are present among the different kinds of adult stem cells. Indeed, the published data as a whole indicate that tissue-specific stem cells exhibit marked variations in the ability to produce cloned offspring, probably as a result of the epigenetic status of the original chromatin (Oback and Wells 2006).

Only few of the solutions proposed (summarized in Fig. 26.2) to improve nuclear reprogramming are actually effective as so far only TSA (Wakayama 2007) and the use of embryonic stem cells as nuclear donors seem to increase efficiency (Wakayama et al. 1999; Rideout et al. 2000). This situation is however far from satisfactory for several reasons. First, the effect of TSA on development to term of clones has been tested only in mice, and no data have been published on other

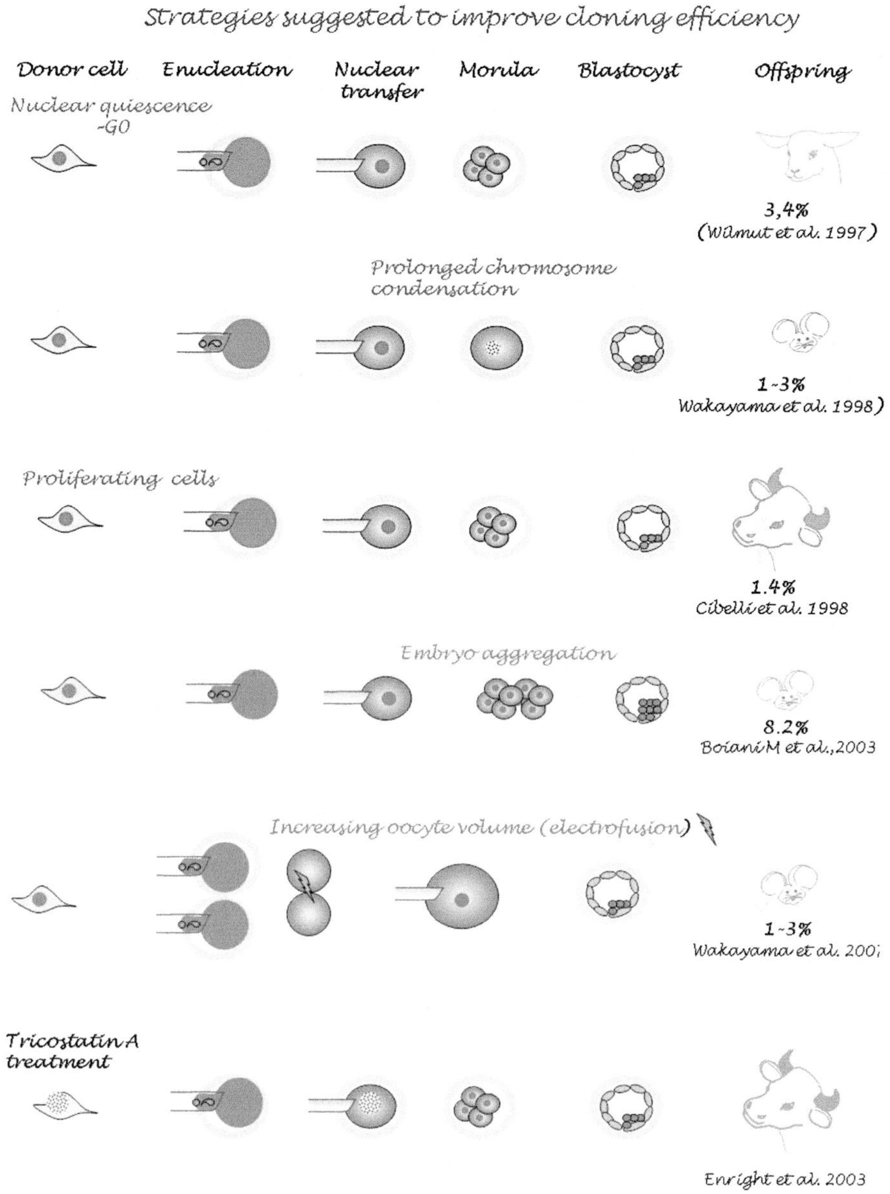

Fig. 26.2 The cartoon illustrates the main strategies suggested so far to improve cloning efficiency

species. Second, the use of ES cells as nuclear donor for SCNT is possible only in the mouse, because no ES cells have been isolated in other "clonable" mammals. Moreover, the advantages resulting from the use of TSA or ES cells are still very limited. As far as ES cells are concerned, development into viable pups is higher, but

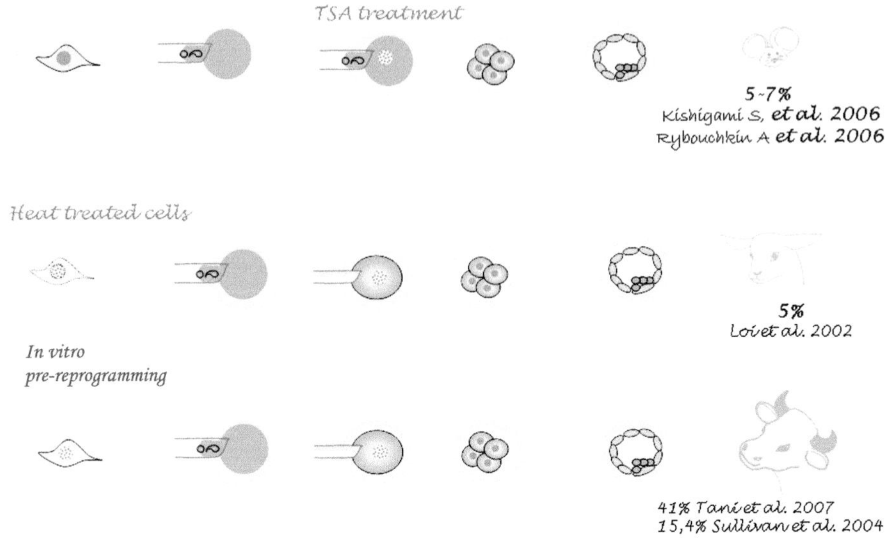

Fig. 26.2 (continued)

development to the blastocyst stage is lower than in control clones in which somatic cells were used; therefore, the overall efficiency remains comparable. Concerning TSA, 5% of the TSA treated clones developed into pups, compared to the 1% in the untreated group. At the moment, SCNT still misses the essential breakthroughs that could provide a vital leap forward, i.e., from 1–5% to about 20–25% of live offspring, with absence or drastic reduction of adverse phenotypes.

26.7 Outlook

While the nuclear reprogramming achieved in the mouse is sufficient to allow the derivation of ES cells from cloned blastocysts, it rarely results in the production of normal offspring. Several factors account for this low efficiency. SCNT is a complex multi-step procedure, which includes oocyte maturation, enucleation, cell fusion/injection, oocyte activation and embryo culture, and the efficiency achieved in each step accounts for the final success. Oocyte physiology, activation dynamics and pre-implantation embryo metabolism differ markedly among species. Therefore, it will be unlikely that a unique, standardized, protocol can be derived for all mammals. The fact that the cloning efficiency is highest in cattle, a species in which the most advanced embryo technologies are being used (Tian et al. 2007), underlines this point. Another critical factor affecting nuclear reprogramming is the timing of zygotic genome activation (ZGA). Species where ZGA is delayed until the morula stage may benefit more from the reprogramming machinery of the egg. Accordingly, in the mouse, in which ZGA starts late during the first cell cycle, SCNT is less efficient than in bovines, where ZGA takes place during the fourth cell cycle (Meissner and Jaenisch 2006). However, species differences have a minor influence on cloning outcomes, if we consider that even in the bovine, the species in which SCNT seems to be easiest, the efficiency remains disappointingly low. The main limiting factor to a satisfactory application of SCNT is the abnormal nuclear reprogramming (Latham 2005), complicated by the high frequency of aneuploidy in SCNT clones (Shi et al. 2004).

Nuclear reprogramming efficiency is still too low. The potential solutions tested so far, including DNA demethylation and TSA treatments (Enright et al. 2003; Kishigami et al. 2006) rely on bulk, non-specific effects that could therefore lead to positive as well as negative effects. A multi-step approach, which takes into consideration the gamete/oocyte biology to better control nuclear reprogramming, could be envisaged. A great deal of efforts should be put into the optimization of in vitro systems for mass production of fully competent recipient oocytes. Robust protocols are available for the maturation of ovine, bovine and pig oocytes, but nothing is known about other clonable species, particularly wild and rare animals. Pre-implantation development of cloned embryos is often carried out in vitro, using culture media formulated for normal embryos. There is evidence that SCNT embryos develop better in complex media, suggesting that some of the metabolic pathways of differentiated cell are still active after nuclear transfer (Chung et al. 2002). The formulation of such media should improve viability, and nuclear reprogramming in cloned embryos (Boiani et al. 2005; Cavaleri et al. 2006). From recent studies on human and mouse embryonic stem cells, however, it has become apparent that developmental and cancer-related genes as well as imprinted genes may acquire aberrant DNA methylation in certain media (Pannetier and Feil 2007). Particularly, the supplementation of media with serum can have dramatic developmental effects, and extreme care needs to be taken when choosing the culture medium (Khosla et al. 2001b; Rivera et al. 2007).

Far more complicated is nuclear reprogramming itself, although we must acknowledge that small ameliorations have been achieved (Wakayama 2007). An important tool will be the possibility to monitor in living embryos the expression of critical genes through coupling with fluorescent tags (Wuensch et al. 2007). Such systems are of precious value because they allow one to not only objectively assess the extent of reprogramming but also to evaluate in a short time the effectiveness of a specific cloning protocol/variant. These improvements, together with the continuous progress in our understanding of the biochemical essential reprogramming steps, like DNA methylation (Jaenisch and Bird, 2003; Reik 2007; Loi et al. 2008), are sure indicators on the bright future waiting ahead for mammalian cloning by SCNT.

Acknowledgments RF acknowledges grant funding from the ESF EUROCORES programme EuroSTELLS, ARC programme ARECA, and the Cancéropôle Grand Sud-Ouest, and thanks the 6th Framework NoE EPIGENOME (LSHG-CT-2004-503433) for support and encouragement. PL and GP acknowledge ESF EUROCORES programme EuroSTELLS, PRIN MIUR 2006, Interlink MIUR and Teramo University Funds.

The European Research Council has provided financial support under the European Community's Seventh Framework Programme (FP7/2007-2013)/ ERC grant agreement n° 210103.

Glossary

Dedifferentiation:the erasure of the differentiation memory of the cells (also described as Nuclear Reprogramming).
Embryo proper:referes to the fetus, which will give rise to the individual.
Disomic:embryo with a normal number of chromosomes (2n).
Labyrinth layer:the vascular structure of the mouse placenta.
Placentomes:functional units of the placenta in ruminants (resulting from the interdigitation of the fetal placenta with specific uterine structures, the cotiledons).
Spongitrophoblast cells:cells composing the external layer of the placenta.
Trophoblast basement membrane:the membrane, which separates the trophoblast from the other extraembryonic memebranes.
Trophoblastic cells:the cells that line the blastocele in early embryos (during blastulation, there are two diferent type of cells, the inner mass cells, which will produce the fetus and the trophoblast cells, which will produce the placenta, extraembryonic tissue).

References

Arnaud P, Feil R (2005) Epigenetic deregulation of genomic imprinting in human disorders and following assisted reproduction. Birth Defects Res (Part C) 75: 81–97

Beaujean N, Taylor J, Gardner J, Wilmut I, Meehan R, Young L (2004) Effect of limited DNA methylation reprogramming in the normal sheep embryo on somatic cell nuclear transfer. Biol Reprod 71: 185–193

Beaujean N, Taylor JE, McGarry M, Gardner JO, Wilmut I, Loi P, Ptak G, Galli C, Lazzari G, Bird A, Young LE, Meehan RR (2004) The effect of interspecific oocytes on demethylation of sperm DNA. Proc Natl Acad Sci USA 101: 7636–7440

Bird A (2002) DNA methylation patterns and epigenetic memory. Genes Dev 16: 6–21

Boiani M, Eckardt S, Scholer HR, McLaughlin KJ (2002) Oct4 distribution and level in mouse clones: consequences for pluripotency. Genes Dev 16: 1209–1219

Boiani M, Gentile L, Gambles VV, Cavaleri F, Redi CA, Scholer HR (2005) Variable reprogramming of the pluripotent stem cell marker Oct4 in mouse clones: distinct developmental potentials in different culture environments. Stem Cells 23: 1089–1094

Campbell KH, Loi P, Otaegui PJ, Wilmut I (1996) Cell cycle co-ordination in embryo cloning by nuclear transfer. Rev Reprod 1: 40–46

Campbell KH, McWhir J, Ritchie WA, Wilmut I (1996) Sheep cloned by nuclear transfer from a cultured cell line. Nature 380: 64–66

Carey BW, Markoulaki S, Hanna J, Saha K, Gao Q, Mitalipova M, Jaenisch R (2009) Reprogramming of murine and human somatic cells using a single polycistronic vector. Proc Natl Acad Sci USA 106: 157–162

Cattanach BM, Kirk M (1985) Differential activity of maternally and paternally derived chromosome regions in mice. Nature 315: 496–498

Cavaleri F, Gentile L, Scholer HR, Boiani M (2006) Recombinant human albumin supports development of somatic cell nuclear transfer embryos in mice: toward the establishment of a chemically defined cloning protocol. Cloning Stem Cells 8: 24–40

Cezar GG, Bartolomei MS, Forsberg EJ, First NL, Bishop MD, Eilertsen KJ (2003) Genome-wide epigenetic alterations in cloned bovine fetuses. Biol Reprod 68: 1009–1014.

Chavatte-Palmer P, de Sousa N, Laigre P, Camous S, Ponter AA, Beckers JF, Heyman Y (2006) Ultrasound fetal measurements and pregnancy associated glycoprotein secretion in early pregnancy in cattle recipients carrying somatic clones. Theriogenology 66: 829–840

Chung YG, Mann MR, Bartolomei MS, Latham KE (2002) Nuclear-cytoplasmic "tug of war" during cloning: effects of somatic cell nuclei on culture medium preferences of pre-implantation cloned mouse embryos. Biol Reprod 66: 1178–1184
Cibelli JB, Stice SL, Golueke PJ, Kane JJ, Jerry J, Blackwell C, Ponce de Leon FA, Robl JM (1998) Cloned transgenic calves produced from nonquiescent fetal fibroblasts. Science 280: 1256–1258
Constant F, Guillomot M, Heyman Y, Vignon X, Laigre P, Servely JL, Renard JP, Chavatte-Palmer P (2006) Large offspring or large placenta syndrome? Morphometric analysis of late gestation bovine placentomes from somatic nuclear transfer pregnancies complicated by hydrallantois. Biol Reprod 75: 122–130
Dean WL, Kelsey G, Reik W (2001) Generation of monoparental embryos for investigation into genomic imprinting. Methods Mol Biol 181: 1–19.
Delaval K, Feil R (2004). Epigenetic regulation of mammalian genomic imprinting. Curr Opin Genet Dev 14: 188–195
Delaval K, Wagschal A, Feil R (2006). Epigenetic deregulation of imprinting in congenital diseases of aberrant growth. BioEssays 28: 453–459
Edwards CA, Ferguson-Smith AC (2007) Mechanisms regulating imprinted genes in clusters. Curr Opin Cell Biol 19: 281–289
Eggan K, Baldwin K, Tackett M, Osborne J, Gogos J, Chess A, Axel R, Jaenisch R (2004) Mice cloned from olfactory sensory neurons. Nature 428: 44–49
Enright BP, Kubota C, Yang X, Tian XC (2003) Epigenetic characteristics and development of embryos cloned from donor cells treated by trichostatin A or 5-aza-2′-deoxycytidine. Biol Reprod 69: 896–901
Feil R, Berger F (2007) Convergent evolution of genomic imprinting in plants and mammals. Trends Genet 23: 192–199
Feil R, Khosla S, Cappai P, Loi P (1998) Genomic imprinting in Ruminants: Allele-specific gene expression in parthenogenetic sheep. Mamm Genome 9: 831–834
Fulka H, Mrazek M, Tepla O, Fulka J Jr (2004) DNA methylation pattern in human zygotes and developing embryos. Reproduction 128: 703–708
Graham CF (1974) Production of parthenogenetic mammalian embryos and their use in biological-research. Biol Rev Cambridge Philos Soc 49: 399–401
Gurdon JB (2006) From nuclear transfer to nuclear reprogramming: the reversal of cell differentiation. Annu Rev Cell Dev Biol 22: 1–22
Gurdon JB, Laskey RA, Reeves OR (1975) The developmental capacity of nuclei transplanted from keratinized skin cells of adult frogs. J Embryol Exp Morphol 34: 93–112
Gurdon JB, Melton DA (2008) Nuclear reprogramming in cells. Science 322:1811–1815
Haaf T (2006) Methylation dynamics in the early mammalian embryo: implications of genome reprogramming defects for development. Curr Top Microbiol Immunol 310: 13–22
Hagemann LJ, Peterson AJ, Weilert LL, Lee RS, Tervit HR (1998) In vitro and early in vivo development of sheep gynogenones and putative androgenones. Mol Reprod Dev 50: 154–162
Hashizume K, Ishiwata H, Kizaki K, Yamada O, Takahashi T, Imai K, Patel OV, Akagi S, Shimizu M, Takahashi S, Katsuma S, Shiojima S, Hirasawa A, Tsujimoto G, Todoroki J, Izaike Y (2002) Implantation and placental development in somatic cell clone recipient cows. Cloning Stem Cells 4: 197–209
Hill JR, Burghardt RC, Jones K, Long CR, Looney CR, Shin T, Spencer TE, Thompson JA, Winger QA, Westhusin ME (2000) Evidence for placental abnormality as the major cause of mortality in first-trimester somatic cell cloned bovine fetuses. Biol Reprod 63: 1787–1794
Hochedlinger K, Jaenisch R (2002) Monoclonal mice generated by nuclear transfer from mature B and T donor cells. Nature 415: 1035–1038
Hochedlinger K, Jaenisch R (2006) Nuclear reprogramming and pluripotency. Nature 441: 1061–1067
Holt WV, Pickard AR, Prather RS (2004) Wildlife conservation and reproductive cloning. Reproduction 127: 317–324

Humpherys D, Eggan K, Akutsu H, Friedman A, Hochedlinger K, Yanagimachi R, Lander ES, Golub TR, Jaenisch R (2002) Abnormal gene expression in cloned mice derived from embryonic stem cell and cumulus cell nuclei. Proc Natl Acad Sci USA 99: 12889–12894

Humpherys D, Eggan K, Akutsu H, Hochedlinger K, Rideout WM 3rd, Biniszkiewicz D, Yanagimachi R, Jaenisch R (2001) Epigenetic instability in ES cells and cloned mice. Science 293: 95–97

Inoue K, Kohda T, Lee J, Ogonuki N, Mochida K, Noguchi Y, Tanemura K, Kaneko-Ishino T, Ishino F, Ogura A (2002) Faithful expression of imprinted genes in cloned mice. Science 295: 297

Inoue K, Noda S, Ogonuki N, Miki H, Inoue S, Katayama K, Mekada K, Miyoshi H, Ogura A (2007) Differential developmental ability of embryos cloned from tissue-specific stem cells. Stem Cells 25: 1279–1285

Inoue K, Ogonuki N, Miki H, Hirose M, Noda S, Kim JM, Aoki F, Miyoshi H (2006) Inefficient reprogramming of the hematopoietic stem cell genome following nuclear transfer. J Cell Sci 119: 1985–1991

Inoue K, Wakao H, Ogonuki N, Miki H, Seino K, Nambu-Wakao R, Noda S, Miyoshi H, Koseki H, Taniguchi M, Ogura A (2005) Generation of cloned mice by direct nuclear transfer from natural killer T cells. Curr Biol 21: 1114–1118

Jaenisch R, Bird A (2003) Epigenetic regulation of gene expression: how the genome integrates intrinsic and environmental signals. Nat Genet 33 Suppl: 245–254

Jouneau A, Zhou Q, Camus A, Brochard V, Maulny L, Collignon J, Renard JP (2006) Developmental abnormalities of NT mouse embryos appear early after implantation. Development 133: 1597–1607

Kawahara M, Wu Q, Takahashi N, Morita S, Yamada K, Ito M, Ferguson-Smith AC, Kono T (2007) High frequency generation of viable mice from engineered bi-maternal embryos. Nat Biotechnol 25: 1045–1050

Khosla S, Dean W, Brown D, Reik W, Feil R (2001a) Culture of preimplantation mouse embryos affects fetal development and the expression of imprinted genes. Biol Reprod 64: 918–926

Khosla, S, Dean W, Reik W, Fei R (2001b) Culture of pre-implantation embryos and its long-term effects on gene expression and phenotype. Human Reprod Update 7: 419–427

Kishigami S, Bui HT, Wakayama S, Tokunaga K, Van Thuan N, Hikichi T, Mizutani E, Ohta H, Suetsugu R, Sata T, Wakayama T (2007) Successful mouse cloning of an outbred strain by trichostatin A treatment after somatic nuclear transfer. J Reprod Dev 53: 165–170

Kishigami S, Mizutani E, Ohta H, Hikichi T, Thuan NV, Wakayama S, Bui HT, Wakayama T (2006) Significant improvement of mouse cloning technique by treatment with trichostatin A after somatic nuclear transfer. Biochem Biophys Res Commun 340: 183–189

Kiziltepe T, Hideshima T, Catley L, Raje N, Yasui H, Shiraishi N, Okawa Y, Ikeda H, Vallet S, Pozzi S, Ishitsuka K, Ocio EM, Chauhan D, Anderson KC (2007) 5-Azacytidine, a DNA methyltransferase inhibitor, induces ATR-mediated DNA double-strand break responses, apoptosis, and synergistic cytotoxicity with doxorubicin and bortezomib against multiple myeloma cells. Mol Cancer Ther 6: 1718–1727

Kono T, Obata Y, Wu QL, Niwa K, Ono Y, Yamamoto Y, Park ES, Seo JS, Ogawa H (2004) Birth of parthenogenetic mice that can develop to adulthood. Nature 428: 860–864

Kremenskoy M, Kremenska Y, Suzuki M, Imai K, Takahashi S, Hashizume K, Yagi S, Shiota K (2006) DNA methylation profiles of donor nuclei cells and tissues of cloned bovine fetuses. J Reprod Dev 52: 259–266

Lagutina I, Lazzari G, Duchi R, Colleoni S, Ponderato N, Turini P, Crotti G, Galli C (2005) Somatic cell nuclear transfer in horses: effect of oocyte morphology, embryo reconstruction method and donor cell type. Reproduction 130: 559–567

Lanza RP, Cibelli JB, Blackwell C, Cristofalo VJ, Francis MK, Baerlocher GM, Mak J, Schertzer M, Chavez EA, Sawyer N, Lansdorp PM, West MD (2000) Extension of cell life-span and telomere length in animals cloned from senescent somatic cells. Science 288: 665–669

Latham KE (2005) Early and delayed aspects of nuclear reprogramming during cloning. Biol Cell 97: 119–132

Li N, Wells DN, Peterson AJ, Lee RS (2005) Perturbations in the biochemical composition of fetal fluids are apparent in surviving bovine somatic cell nuclear transfer pregnancies in the first half of gestation. Biol Reprod 73: 139–148

Loi P, Beaujean N, Khochbin S, Fulka J Jr, Ptak G (2008) Asymmetric nuclear reprogramming in somatic cell nuclear transfer? Bioessays 30: 66–74

Loi P, Clinton M, Barboni B, Fulka J Jr, Cappai P, Feil R, Moor RM, Ptak G (2002) Nuclei of nonviable ovine somatic cells develop into lambs after nuclear transplantation. Biol Reprod 67: 126–132

Loi P, Clinton M, Vackova I, Fulka J Jr, Feil R, Palmieri C, Della Salda L, Ptak G (2006) Placental abnormalities associated with post-natal mortality in sheep somatic cell clones. Theriogenology 65: 1110–1121

Loi P, Ledda S, Fulka J, Cappai P, Moore RM (1998) Development of parthenogenetic and cloned ovine embryos: effect of activation protocols. Biol Reprod 58: 1177–1187

Loi P, Ptak G, Barboni B, Fulka J Jr, Cappai P, Clinton M (2001) Genetic rescue of an endangered mammal by cross-species nuclear transfer using post-mortem somatic cells. Nat Biotechnol 19: 962–964

Martínez-Balbás MA, Dey A, Rabindran SK, Ozato K, Wu C (1995) Displacement of sequence-specific transcription factors from mitotic chromatin. Cell 83: 29–38

Mayer W, Niveleau A, Walter J, Fundele R, Haaf T (2000) Demethylation of the zygotic paternal genome. Nature 403: 501–502

Meissner A, Jaenisch R (2006) Mammalian nuclear transfer. Dev Dyn 235: 2460–2469

Mendjan S, Akhtar A (2007) The right dose for every sex. Chromosoma 116: 95–106

Mizutani E, Ohta H, Kishigami S, Van Thuan N, Hikichi T, Wakayama S, Kosaka M, Sato E, Wakayama T (2006) Developmental ability of cloned embryos from neural stem cells. Reproduction 132: 849–857

Morgan HD, Santos F, Green K, Dean W, Reik W (2005) Epigenetic reprogramming in mammals. Hum Mol Genet 14: R47–58

Morison IM, Ramsay JP, Spencer HG (2005) A census of mammalian imprinting. Trends Genet 21: 457–465

Nagy A, Gocza E, Diaz EM, Prideaux VR, Ivanyi E, Markkula M, Rossant J (1990) Embryonic stem cells alone are able to support fetal development in the mouse. Development 110: 815–821

Niemann H, Kues WA (2007) Transgenic farm animals: an update. Reprod Fertil Dev 19: 762–770

Oback B, Wells DN (2006) Donor cell differentiation, reprogramming, and cloning efficiency: Elusive or illusive correlation? Mol Reprod Dev 74: 646–654

Oghane J, Wakayama T, Senda S, Yamazaki Y, Inoue K, Ogura A, Marh J, Tanaka S, Yanagimachi R, Shiota K (2004) The Sall3 locus is an epigenetic hotspot of aberrant DNA methylation associated with placentomegaly of cloned mice. Genes Cells 9: 253–260

Ogura A, Inoue K, Ogonuki N, Lee J, Kohda T, Ishino F (2002) Phenotypic effects of somatic cell cloning in the mouse. Cloning Stem Cells 4: 397–405

Okita K, Ichisaka T, Yamanaka S (2007) Generation of germline-competent induced pluripotent stem cells. Nature 48: 313–317

Palmieri C, Loi P, Reynolds LP, Ptak G, Della Salda L (2007) Placental abnormalities in ovine somatic cell clones at term: a light and electron microscopic investigation. Placenta 28: 577–384

Pannetier M, Feil R (2007) Epigenetic stability of embryonic stem cells and developmental potential. Trends Biotechnol 25: 556–562

Pomerantz J, Blau HM (2004) Nuclear reprogramming: a key to stem cell function in regenerative medicine. Nat Cell Biol 6: 810–816

Prather RS, Hawley RJ, Carter DB, Lai L, Greenstein JL (2003) Transgenic swine for biomedicine and agriculture. Theriogenology 59: 115–123

Reik W. (2007) Stability and flexibility of epigenetic gene regulation in mammalian development. Nature 447: 425–432

Reik W, Santos F, Mitsuya K, Morgan H, Dean W (2003) Epigenetic asymmetry in the mammalian zygote and early embryo: relationship to lineage commitment? Philos Trans R Soc Lond B 358: 1403–1409

Rideout WM 3rd, Hochedlinger K, Kyba M, Daley GQ, Jaenisch R (2002) Correction of a genetic defect by nuclear transplantation and combined cell and gene therapy. Cell 109: 17–27

Rideout WM 3rd, Wakayama T, Wutz A, Eggan K, Jackson-Grusby L, Dausman J, Yanagimachi R, Jaenisch R (2000) Generation of mice from wild-type and targeted ES cells by nuclear cloning. Nat Genet 24: 109–110

Rideout WM III, Eggan K, Jaenisch R (2001) Nuclear cloning and epigenetic reprogramming of the genome. Science 293: 1093–1098

Rivera RM, Stein P, Weaver JR, Mager J, Schultz RM, Bartolomei MS (2007) Manipulations of mouse embryos prior to implantation result in aberrant expression of imprinted genes on day 9.5 of development. Hum Mol Gen 17: 1–14.

Robl JM, Wang Z, Kasinathan P, Kuroiwa Y (2007) Transgenic animal production and animal biotechnology. Theriogenology 67: 127–133

Rousseaux S, Caron C, Govin J, Lestrat C, Faure AK, Khochbin S (2005) Establishment of male-specific epigenetic information. Gene 345: 139–153

Rybouchkin A, Kato Y, Tsunoda Y (2006) Role of histone acetylation in reprogramming of somatic nuclei following nuclear transfer. Biol Reprod 74: 1083–1089

Santos F, Peters AH, Otte AP, Reik W, Dean W (2005) Dynamic chromatin modifications characterise the first cell cycle in mouse embryos. Dev Biol 280: 225–236

Santos F, Zakhartchenko V, Stojkovic M, Peters A, Jenuwein T, Wolf E, Reik W, Dean W (2003) Epigenetic marking correlates with developmental potential in cloned bovine preimplantation embryos. Curr Biol 13: 1116–1121

Shi W, Dirim F, Wolf E, Zakhartchenko V, Haaf T (2004) Methylation reprogramming and chromosomal aneuploidy in in vivo fertilized and cloned rabbit preimplantation embryos. Biol Reprod 71: 340–347

Shiels PG, Kind AJ, Campbell KH, Wilmut I, Waddington D, Colman A, Schnieke AE (1999) Analysis of telomere length in Dolly, a sheep derived by nuclear transfer. Cloning 1: 119–125

Smith C, Berg D, Beaumont S, Standley NT, Wells DN, Pfeffer PL (2007) Simultaneous gene quantitation of multiple genes in individual bovine nuclear transfer blastocysts. Reproduction 133: 231–242

Smith SL, Everts RE, Tian XC, Du F, Sung LY, Rodriguez-Zas SL, Jeong BS, Renard JP, Lewin HA, Yang X (2005) Global gene expression profiles reveal significant nuclear reprogramming by the blastocyst stage after cloning. Proc Natl Acad Sci USA 102: 17582–17587

Somers J, Smith C, Donnison M, Wells DN, Henderson H, McLeay L, Pfeffer PL (2006) Gene expression profiling of individual bovine nuclear transfer blastocysts. Reproduction 131: 1073–1084

Surani MA, Barton SC, Norris ML (1984) Development of reconstituted mouse eggs suggests imprinting of the genome during gametogenesis. Nature 308: 548–550

Takahashi K, Yamanaka S (2006) Induction of pluripotent stem cells from mouse embryonic and adult fibroblast cultures by defined factors. Cell 126: 663–676

Tamashiro KL, Sakai RR, Yamazaki Y, Wakayama T, Yanagimachi R (2003) Phenotype of cloned mice: development, behaviour, and physiology. Exp Biol Med 228: 1193–1200

Tani T, Shimada H, Kato Y, Tsunoda Y (2007) Bovine oocytes with the potential to reprogram somatic cell nuclei have a unique 23-kDa protein, phosphorylated transcriptionally controlled tumor protein (TCTP). Cloning Stem Cells 9: 267–280

Tian XC, Smith SL, Zhang SQ, Kubota C, Curchoe C, Xue F, Yang L, Du F, Sung LY, Yang X (2007) Nuclear reprogramming by somatic cell nuclear transfer: the cattle story. Soc Reprod Fertil Suppl 64: 327–339

Tian XC, Xu J, Yang X (2000) Normal telomere lengths found in cloned cattle. Nat Genet 26: 272–273

Trounson AO (2006) Future and applications of cloning. Methods Mol Biol 348: 319–332

Tsunoda Y, Kato Y (2002) Recent progress and problems in animal cloning. Differentiation 69: 158–161

Vajta G (2007) Handmade cloning: the future way of nuclear transfer? Trends Biotechnol 25: 250–253

van der Heijden GW, Dieker JW, Derijck AA, Muller S, Berden JH, Braat DD, van der Vlag J, de Boer P (2005) Asymmetry in histone H3 variants and lysine methylation between paternal and maternal chromatin of the early mouse zygote. Mech Dev 122: 1008–1022

Wagschal A, Sutherland HG, Woodfine K, Henckel A, Chebli K, Schultz R, Oakey, RJ, Bickmore WA, Feil R (2008) G9a histone methyltransferase contributes to imprinting in the mouse placenta. Mol Cell Biol 28: 1104–1113

Wagschal A, Feil R (2006) Genomic imprinting in the placenta. Cytogenet Genome Res 113: 90–98

Wakayama T (2007) Production of cloned mice and ES cells from adult somatic cells by nuclear transfer: how to improve cloning efficiency? J Reprod Dev 53: 13–26

Wakayama T, Perry AC, Zuccotti M, Johnson KR, Yanagimachi R (1998) Full-term development of mice from enucleated oocytes injected with cumulus cell nuclei. Nature 394: 369–374

Wakayama T, Rodriguez I, Perry AC, Yanagimachi R, Mombaerts P (1999) Mice cloned from embryonic stem cells. Proc Natl Acad Sci USA 21: 14984–14989

Wakisaka-Saito N, Kohda T, Inoue K, Ogonuki N, Miki H, Hikichi T, Mizutani E, Wakayama T, Kaneko-Ishino T, Ogura A, Ishino F (2006) Chorioallantoic placenta defects in cloned mice. Biochem Biophys Res Commun 349: 106–114

Weier JF, Weier HU, Jung CJ, Gormley M, Zhou Y, Chu LW, Genbacev O, Wright AA, Fisher SJ (2005) Human cytotrophoblasts acquire aneuploidies as they differentiate to an invasive phenotype. Dev Biol 279: 420–432

Wells DN (2003) Cloning in livestock agriculture. Reprod Suppl 61:131–150

Wells DN, Misica PM, Day AM, Peterson AJ, Tervit HR (1998) Cloning sheep from cultured embryonic cells, Reprod Fertil Dev 10: 615–626

Wernig M, Meissner A, Foreman R, Brambrink T, Ku M, Hochedlinger K, Bernstein BE, Jaenisch R (2007) In vitro reprogramming of fibroblasts into a pluripotent ES-cell-like state. Nature 448: 318–324

Wilmut I, Schnieke AE, McWhir J, Kind AJ, Campbell KKH (1997) Viable offspring derived from fetal and adult mammalian cells. Nature 385: 810–813

Withfield JF, Boynton AL, Rixon RH, Youdale T (1985) The control of animal cell proliferation by calcium, Ca^{2+}-calmodulin, and cyclic AMP. In: Boynton AL, Leffert HL (eds) Control of Animal Cell Proliferation, Vol. 1. Academic Press, London, pp. 331–365

Wuensch A, Habermann FA, Kurosaka S, Klose R, Zakhartchenko V, Reichenbach HD, Sinowatz F, McLaughlin KJ, Wolf E (2007) Quantitative monitoring of pluripotency gene activation after somatic cloning in cattle. Biol Reprod 76: 983–991

Yoshida N, Brahmajosyula M, Shoji S, Amanai M, Perry AC (2007) Epigenetic discrimination by mouse metaphase II oocytes mediates asymmetric chromatin remodeling independently of meiotic exit. Dev Biol 301: 464–477

Young LE, Schnieke AE, McCreath KJ, Wieckowski S, Konfortova G, Fernandes K, Ptak G, Kind AJ, Wilmut I, Loi P, Feil R (2003) Conservation of IGF2-H19 and IGF2R imprinting in sheep: effects of somatic cell nuclear transfer. Mech Dev 120: 1433–1442

Chapter 27
Grapevine (*Vitis* ssp.): Example of Clonal Reproduction in Agricultural Important Plants

Astrid Forneck, Andrej Benjak and Ernst Rühl

Abstract This review approaches the concept of clonality (asexual reproduction) and its implications for phenotypic and genetic stability. Grapevine cultivars (*Vitis vinifera* L.) are composed of clones showing homogeneous ampelographic characteristics with minor differences. The concept of clonal selection (through vegetative propagation) implies very low genetic variation within a "population of genotypes identical to the ancient progenitor except of mutations". Yet, the accumulated genetic variation is higher than expected in cultivated grapevine clones. This variation is combined with numerous mechanisms of asexual strategies to enhance variation and to provide an open system for adaptation and selection processes. This chapter presents insights into the clonal selection of grapevine exemplifying the cultivar Pinot noir (*V. vinifera* L.) to elucidate the potential sources of its phenotypic and genotypic variation. The impact of clonal propagation of this agricultural important crop will also be discussed.

27.1 Introduction

Grapevine (*Vitis vinifera* L.), economically one of the most important crop plants, comprises multiple cultivars and clones. The identification and characterization of grape varieties has always been an intrinsic concern for agriculture as well as for breeding research programs. Traditionally, ampelographic and ampelometric methods based on morphological characteristics have been implemented for distinguishing grapevine cultivars and often resulted in insufficient or even unsuccessful differentiation. Grapevine is commercially propagated vegetatively and the cultivars existing today result from the selection of advanced genotypes with an ancient origin, mostly generated by spontaneous crosses centuries ago (Mullins and Meredith 1989). Each ancient cultivar expresses distinct phenotypes, resulting

A. Forneck (✉)
Department of Applied Plant Sciences and Plant Biotechnology, University of Natural Resources and Applied Life Sciences, Vienna, Peter-Jordan Str. 82, A-1190, Vienna, Austria
e-mail: astrid.forneck@boku.ac.at

I. Schön et al. (eds.), *Lost Sex*, DOI 10.1007/978-90-481-2770-2_27,

in sets of morphologically diverging clones. These clones are spread worldwide adjusting to different environments and cultivation techniques. Currently, almost 16.000 prime named grapevine cultivars are listed in the International *Vitis* Variety Catalogue (http://www.genres.de).

Grapevine domestication from *Vitis sylvestris* Gmel. or *Vitis caucasica* Vav. dates back to the Neolithic period (Negrul 1946; Levadoux 1956; Ambrosi and Becker 1978). Apart from their various usages as fresh fruit, dried fruit, jam, wine or vinegar, the simple vegetative multiplication was a key reason for early domestication. Vegetative propagation of grapevines is straightforward: the long and flexible wooden canes are used for "natural" layering and single, selected dormant cuttings can be easily rooted. In this way, clonal selection may have additionally triggered grapevine domestication, since promising phenotypes were multiplied, spread according to human transportation and conserved over centuries. Vegetative propagation has consequently been advantageous and used from the very beginning of grapevine domestication (Billiard 1913) and also for many fruit species (Zohary and Spiegel-Roy 1975). But sexual reproduction was not completely evaded either, though generative propagation of grapevine is difficult: seeds germinate erratically and plants grown from seeds vary extremely, due to high levels of heterozygosity (Bowers et al. 1999). Parentage studies show that natural crosses must have happened. Table grapes were eaten and the seeds spat out or exuded, wine grapes were pressed and the pomace dumped in the vicinity of the winery, leading -though rarely- to superior varieties. Some varieties appear to be immediate selections from wild types, e.g., the variety Traminer (Regner 1999), while others are crosses between existing cultured varieties, e.g., Cabernet Sauvignon, a cross between Sauvignon blanc and Cabernet franc (Bowers and Meredith 1997; Regner et al. 1998). There are also crosses between wild types and cultured varieties, e.g., Riesling, a cross between Gouais and most likely a Traminer *V. silvestris* hybrid (Regner 1999). In many cases, parent varieties were at their time important varieties, but have virtually disappeared from modern viticulture, like the variety Gouais also called Heunisch in Germany, which is a parent of more than 70 different cultivars (Boursiquot et al. 2004), e.g., Chardonnay or Gamay (Bowers et al. 1999).

For grapevine clones, the concept of individuality is straightforward and relies on propagation records and morphological features leading to the breeding concept of clonal selection. The first descriptions of the need for clonal selection and useful methods are found in Roman sources as indicated by Columella (ca. 60), emphasizing regular visual evaluations and positive mass selection according to quality-related traits (yield, fruit set). Since grapevine is a highly priced crop, significant viticultural research in describing and analyzing phenotypes has been performed since the 19th century. Subsequently, with the onset of genomic research, *Vitis* ssp. has been focus of many studies using elaborate tissue culture, transformation and molecular genetic techniques. In general, investigations on clonal variation within grapevine cultivars have shown that the degree of detected genetic divergence usually depends on the applied marker system and on the scope and type of plant samples (Forneck 2005; see also Chapter 9). The retrotransposon-based marker systems SSAP or ISTR have shown higher levels of polymorphism (Sensi et al. 1996; Labra et al. 2004) than the standard AFLP or SSR techniques.

In this chapter, we intend to explain the importance of asexual = vegetative propagation and clones for an agricultural crop leading to the breeding strategy of clonal selection (see Chapter 3 for other definitions of asexual plants). We give a short review on the successes of clonal selection and discuss the mechanisms behind the clonal variation in grapevine introducing the variety Pinot noir as a well-analysed example, revealing high genetic similarity and also indicating the origin of asexual reproduction (Ye et al. 1998; Regner et al. 2000). Compared to other grapevine cultivars, Pinot noir clones are characterised by high phenotypic diversity, originating by spontaneous mutations of different kinds. Several pale coloured mutants have emerged from the red grape Pinot noir. Pinot gris has been identified as periclinal chimera resulting from somatic mutation at the berry colour locus (Walker et al. 2006). The white-skinned Pinot blanc is considered to have also arisen from Pinot noir. The insertion of a retrotransposon into one as well as the deletion of the other allele of the *VvmybA1* gene has blocked the production of anthocyanin in the white grape (Yakushiji et al. 2006). To further contribute to the understanding of sexual propagation in grapevine, we add a short summary of the current *status quo* on grapevine genetics and breeding to highlight that sex has not been lost in grapevine and conclude with the advantages of grapevine clones for viticulture.

27.2 Clonality in Grapevine

27.2.1 Clonal Selection – the Art of Bringing Clonal Variation to the Fields

A grapevine clone is the vegetative progeny of a single plant. In the absence of mutations, all descendants of a clone have identical phenotypes and genotypes (but see Chapter 9). Modern clonal selection started in 1876, when a wine grower did research on quality traits (yield) of single Grapevine plants over a period of 20 years. The result of this work contributed to the first registered "grapevine clone" of the variety *V. vinifera* cv. Silvaner with an average yield of 6.637 kg per vine (Froelich 1900). The success of Froelich's approach resulted in numerous activities in clonal selection of grapevines, first in Germany, then in most vine-growing countries. Today, clonal propagation material is available from almost all important varieties and used world-wide. Currently in Germany, ca. 600 clones are registered (Bechers, pers.commun.) and in France more than 1000 (Boidron et al. 1997). Phenotypic differences among clones of the numerous grapevine cultivars have been reported by many authors (e.g., Sievers 1971; Silvestroni et al. 1995; Boidron et al. 1997; Rühl et al. 2000). Long-living grapevine is prone to adapt to environmental and pathogen effects, thus leading to both phenotypic and genotypic alterations that may mimic clonal variation. Clonal variation here is defined as changes in genomes other than sexually derived, which will be transmitted asexually to the descendants.

Virus infection can significantly alter the performance of vines. Inoculating *V. vinifera* cv. Albana and cv. Trebbiano Romagnolo vines with Grapevine Fanleaf or Grapevine Leafroll virus, reduced yield – depending on the virus type – by up to

72.9% and 80.4%, respectively (Credi and Babini 1997). Consequently, virus elimination can affect vine performance significantly (e.g., Mannini et al. 1994, 1998) and virus freedom is an essential prerequisite for the production of grapevine clones to be used as propagation material (Walter and Martelli 1996, 1997).

For clonal selection, phenotypic variation within a cultivar is to be clearly identified. While qualitative traits (e.g., number of bunches per shoot or bunch architecture) may be recognised on a single vine, quantitative traits (e.g., yield, sugar production, acidity) are only noticeable in larger plantings in experimental designs set up for clonal selection. Aims in clonal selection largely depend on the cultivar and its use. The emphasis in rootstocks will certainly be on virus freedom to stop virus transmission to the scion after grafting with detrimental effects. In tablegrapes, the major trait in focus will be appearance, which is of no importance at all in winegrapes.

27.2.2 Clonal Selection: An Example from V. vinifera cv. Pinot Noir

To illustrate clonal variation in grapevine, we exemplify a study on *V. vinifera* cv. Pinot noir. Forty-four Pinot noir clones were tested for viruses in spring 1988, grafted on Börner rootstocks and planted in a field trial in spring of 1989 in a fertile sandy loam at Geisenheim, Germany. The results are means of six years. Large clonal variation was found in three phenotypic traits: yield, acid content (titratable acidity) and susceptibility to botrytis bunch rot, the major disease of ripening berries caused by *Botrytis cinerea*. The sugar content, measured as total soluble solids, showed only small variation between clones (Fig. 27.1). One clone had an average yield of less than 700 g m^{-2}, while one produced more than 1600 g m^{-2}. Titratable acidity of different clones ranged from less than 9.5 g L^{-1} to more than 13.5 g L^{-1} and bunch rot susceptibility varied between less than 2% to more than 26% of botrytis-infected berries. Titratable acidity and bunch rot incident did not resemble a normal distribution. Plotting berry sugar content of a clone as a function of its corresponding yield revealed a trend to lower sugar levels with increasing yields (Fig. 27.2a).

Clones with higher yield have lower sugar levels and *vice versa*. Clones may be grouped by a system published by Oustric (1994). A vertical line at the average yield and a horizontal line at the mean total soluble solids value divide the graph in 4 quadrants. Quadrant A holds clones with generally low yield and high sugar level, B clones with both a high yield and high sugar level, C high yielding clones with low sugar level and D clones with both low yield and low sugar level. A-clones are well suited for the production of premium wines, B-clones, depending on the cropping level, may be used for both premium and bulk wine production and C-clones mostly for bulk wines. Variation becomes even more obvious, if bunch rot incidence is plotted as a function of titratable acidity (Fig. 27.2b). Four major, phenotypically diverging groups of clones emerge: clones with compact bunches,

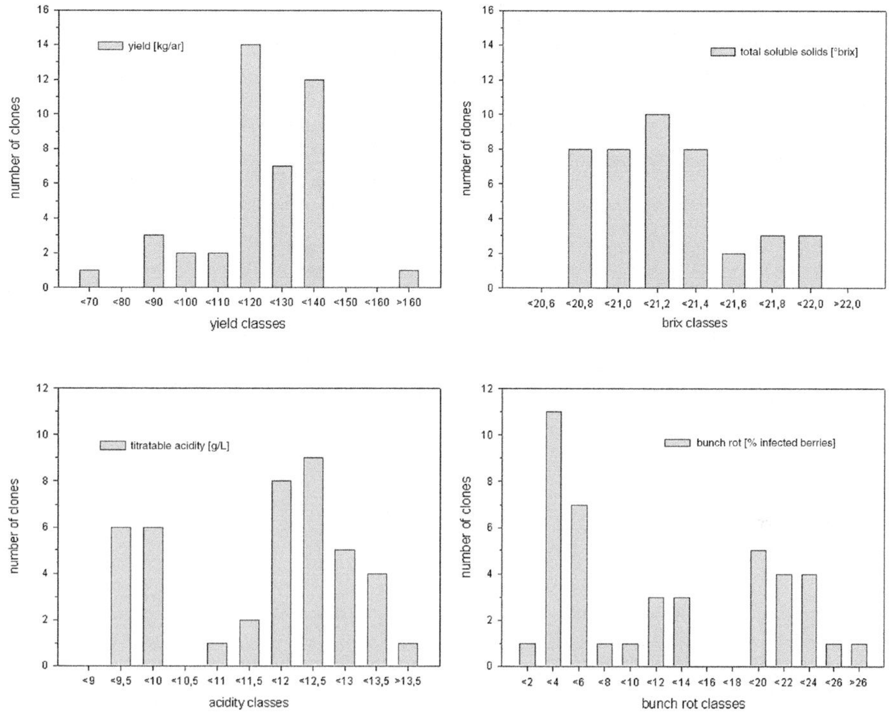

Fig. 27.1 Yield, total soluble solids (sugar content), titratable acidity and susceptibility to botrytis bunch rot of 44 different Pinot noir clones. Results were collected between 1993 and 1995 ($n = 6$) in the experimental vineyards of the Geisenheim Research Centre at Geisenheim. Clones were grouped in classes. Bars indicate the number of clones in this class

high risk of bunch rot and high acidity; clones with upright growing shoots, average incidence of botrytis and average acidity; clones with loose clusters, low acidity and low botrytis risk and clones with small berries, high acidity and low incidence of botrytis (Rühl et al. 2000). With the choice of the right clone, a grower largely determines the plant performance and fruit quality in the vineyards.

27.3 Sources of Clonal Variation in Grapevine Explored in Viticulture

Knowledge about grapevine genetics is still rare and new discoveries will allow to better understand how naturally occurring mutations influence the phenotype of different cultivars and their clones and to adequately modify some of their genomic properties for a more successful breeding. The grapevine genome consists of various confirmed sources of genetic variation relying on mutation events. Along with the mutability in random soma and germinal cells, there is an interaction driving force

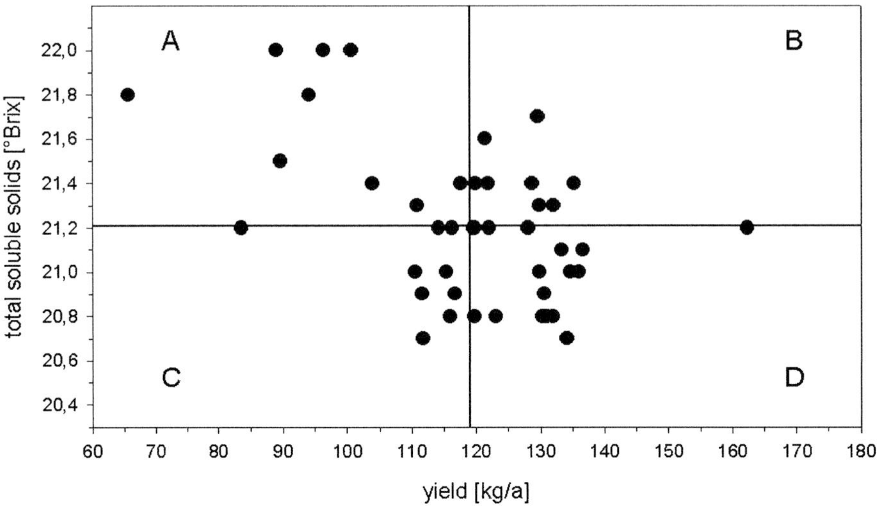

Fig. 27.2 Correlation between yield and total soluble solids and titratable acidity and botrytis bunch rot, respectively, of 44 Pinot noir clones. Data are means of the harvests 1993–1998 ($n = 6$)

among tissues resulting in chimeric structures. Each of these layers contributes to an individual genotype and thus may display polymorphic AFLP-patterns. In the case of sectorial chimerism or mosaicims, this would lead to a pooled mix of genotypes (of varying quantities) within one DNA template. However, there is abundant evidence for the occurrence of somatic mutations in plants. Within-individual variability for polygenic and cytogenetic traits has been documented by several authors (e.g., Klekowski and Godfrey 1989). In addition, molecular genetic variation has been found among naturally occurring clones in several plant species (Caposella et al. 1992; Tuskan et al. 1996), though little is known about mutation rates during somatic development of plants (Gill et al. 1995). One explanation for variation within the progeny is the occurrence of spontaneous mutations (Forneck 2005) which may be traced by genetic fingerprinting techniques.

27.3.1 Clonal Variation by Random Mutation

The objective of genetically fingerprinting of grapevine clones is to confirm genetic similarities and to search and identifying reproducible sequence mutations. Clonal fingerprinting based on DNA sequence alterations has been performed with various PCR-assisted marker systems and is based on the assumption that distinct, individual DNA exists in individual plants. Experimental evidence shows that this assumption is not generally fulfilled (Blaich et al. 2008). Chimerism, tissue-specific and time-specific methylation and stress-related dynamic transposition events (Benjak et al. 2006) exemplify the multitude of processes resulting in genomic expansion. As to date we cannot deny that rapidly micro-evolving genomes may exist, reflecting dynamic sequence mutations that are not soundly transmitted in subsequent vegetation cycles, but can be traced by highly sensitive genetic marker techniques. Recently, two articles describing the sequence of the Vitis genome (*V. vinifera* cv. Pinot noir) have been published (Jaillon et al. 2007; Velasco et al. 2007) and a shotgun sequence of grapevine genome has been made available, opening the possibility for a genome-wide bioinformatical analysis for clonal variation. "Measuring" clonal variation among grapevine clones of a given cultivar remains difficult, since the original motherplant and its genotype are rarely known or existent. One way to approach the nature of mutation events among agronomically cultivated grapevine clone may be the identification of a "most common clonal genotype". This has been conducted by Hocquigny et al. (2004) analysing 145 accessions, belonging to five Pinot cultivars (*V. vinifera* L.) at 50 loci. A Pinot "genotype I" has been proposed due to the facts that (1) 65% of all samples shared this genotype, (2) the remaining variant clones shared a minimum of 95% of all loci and (3) most of the loci showed fixed heterozygosity. This study postulates that genotype I is likely the most common ancestor of five Pinot cultivars, which implies that these were generated by asexual propagation from a single unique zygote. A study analysing the clonal variation among seventy Pinot noir (*V. vinifera* L.) clones deriving from a single cultivar but displaying various phenotypes (cluster architecture, maturity, canopy growth) implemented 178 AFLP-markers in a replicated, stringent design (Blaich et al. 2008). A "most common" genotype comprises eighteen (25.7%) identically fingerprinted clones. The biggest group of 48 clones (68.6%) was analyzed within the range of 99% genetic similarity compared to the main identical group. A group of 24 clones could be situated further away than 1% differentiation with 99.1–94.0% genetic similarity. The identification of clonal variation depends on the sample size and the selected molecular marker system (see also Chapter 9 on a more extensive discussion of this topic). Furthermore, the selected clone samples play a major role. Given the possible existence of one common zygote for all existing Pinot clones (approximately 500), one would have to identify the major selection lines prior to search for random mutations, since otherwise, these may mimic the genetic variation existing in world-wide Pinot selections.

27.3.2 Clonal Variation by Transposition

As earlier studies on other organisms (reviewed in Bennetzen 2000) as well as grapevine itself showed (Verries et al. 2000; Kobayashi et al. 2004), genome modifications induced by transposable elements are one key to our understanding of grapevine genetics. Transposable elements (TEs) are DNA segments possessing the ability to move or multiply within genomes, thereby generating self-copies interspersed with non-repetitive DNA (reviewed in Feschotte et al. 2002). Many of them encode protein(s) required for their mobility and solely use the host cellular machinery for their transcription and translation. They are called autonomous elements. On the other hand, the mobility of non-autonomous elements relies on proteins encoded by related autonomous TEs. TEs are classified into two classes based on their mechanism of transposition (Finnegan 1989): the class I elements, also called retrotransposons, use a RNA intermediate and a reverse transcriptase whereas the class II elements, or DNA transposons, use a DNA intermediate and a transposase. Retrotransposons are divided into two principal groups, the LTR (Long Terminal Repeat) and the non-LTR retrotransposons. LTR-retrotransposons themselves are further divided into Ty1/*copia*, Ty3/*gypsy* and the non-autonomous LARDs (Large Retrotransposon Derivates) and TRIMs (Terminal-repeat Retrotransposons In Miniature; Witte et al. 2001; Kalendar et al. 2004). Non-LTR retrotransposons consist of LINEs (Long INterspersed Elements) and the non-autonomous SINEs (Short INterspersed Elements). DNA transposons have Terminal Inverted Repeats (TIRs) flanking the gene for a transposase (in the autonomous elements). Eukaryotic DNA transposons are classified into 5 to 7 different superfamilies (Feschotte et al. 2002; Robertson 2002).

RAPD amplification products have provided the first indication for the presence of retroelements and remnants thereof in the grapevine genome. Cloned repetitive sequences showed high similarities to retrotransposons of higher plants, and were found to be dispersed throughout the genome (Böhm and Zyprian 1998). So far, only few TEs (of class I) have been described in grapevine, namely *Tvv1* (Pelsy and Merdinoglu 2002), *Vine-1* (Verries et al. 2000), and *Gret1* (Kobayashi et al. 2004). The last two were found to be inserted in the *Adhr* and the *VvmybA1* genes, respectively, confirming that TEs have the potential to alter genes in grapevine. Recent studies on the skin colour mutation of grapevine have been conducted on the black-skinned Pinot Noir and the white-skinned Pinot Blanc (Yakushiji et al. 2006). Pinot noir, heterozygous for *VvmybA1*, comprises a functional allele, capable of anthocyanin expression, and a non-functional allele, which has lost its capability by the insertion of *Gret1*. Pinot blanc considered to be arisen from the dark-skinned Pinot noir, possesses only a non-functional allele lacking the functional part of the gene. Studies on the identification and isolation of this null-allele in Pinot blanc are in progress (Yakushiji et al. 2006). Moreover, it has been observed that recombination between the LTRs of *Gret1* have lead to solo LTRs or even to a total loss of the retrotransposon in coloured cultivars. This excision event resulted in new dark-skinned varieties originating from white progenitors, such as Red Chardonnay which is derived from Chardonnay (Kobayashi et al. 2004; Yakushiji et al. 2006).

Retrotransposons have repeatedly been used for studying polymorphisms among grapevine cultivars and clones, and have revealed promising results. Relying on the presence of retrotransposon reverse transcriptase sequences, inverse sequence-tagged repeat analyses (ISTR) have been conducted for investigating genetic diversity among closely related *Sangiovese* accessions. ISTR fingerprints provided a high level of polymorphism whereby clonal distinction was successful (Sensi et al. 1996). S-SAP analyses, implementing primers based on the LTRs of *Vine-1* could distinguish particular clones such as Traminer clones. But the distinction of *Pinot* clones failed, indicating different clonal variability in different cultivars (Imazio et al. 2002; Labra et al. 2003). Pereira et al. (2005) utilised molecular markers based on LTRs of *Gret1* for REMAP and IRAP analyses to reveal polymorphism among Portuguese cultivars, while the techniques failed in finding polymorphisms between clones of the same cultivars. Pelsy et al. (2003) have assessed the discriminative power of S-SAPs relying on the LTRs of grapevine retrotransposons within 12 *Vitis vinifera* varieties. The authors confirmed their efficiency in distinguishing each variety from one another (Pelsy et al. 2003).

Up to date, the information on abundance and transcriptional activities of TEs and RTEs in grapevine is scarce. First global and detailed results on the abundance of TEs have been presented based on *in silico* analysis of the publicly available sequences of the Vitis genome (Jaillon et al. 2007; Velasco et al. 2007). Over 1150 potentially complete grapevine transposons as well as more than 3500 defective copies were characterized, representing approximately 2.0% of the grapevine genome (Benjak et al.2008) including a number of potentially domesticated transposases. TEs of four superfamilies (CACTA, hAT, Mutator, PIF/Harbinger) were identified; however, no elements representing the Tc1-Mariner superfamily could be found. The two recent reports on the draft sequence of the genome of *Vitis vinifera* (Jaillon et al. 2007; Velasco et al. 2007) predict a higher copy number of transposon-related sequences (6,344 and 9,562%) but with lower transposon content in terms of genome fraction (0.43 and 1.6%, respectively). According to the study of Benjak et al. (2008), the mean length of the described copies is 2.3 Kb/ element, which is more than three times bigger when compared to previous reports, most possible due to a more stringent *in silico* analysis. The hAT is the most prevalent superfamily of transposons in grapevine with more than 1500 hAT-related elements in the grapevine genome, which can be grouped into different families by phylogenetic analysis. Some hAT families are expressed in grapevine and could have retained the ability to transpose since they present conserved TIRs and TSD of 8 nucleotides as typical for hAT transposons. Nevertheless, some of these families also contain a high number of defective copies suggesting that some of the hAT subfamilies are relatively old in grapevine. TEs seem to be involved in phenotypic grapevine variation as reported by Benjak et al. (2008), because many grapevine TEs match EST (expressed sequence tags) sequences, which point towards their potential activity. Various insertion polymorphisms are found within genic sequences and give rise to different transcribed TE species suggesting that insertion polymorphisms are likely linked to phenotypic variation. *In silico* analysis furthermore confirmed that expression of TEs in *Vitis* spp. is induced by stress (Benjak et al. 2008).

Transposition seems to play a significant role in the generation of clonal variation. Results applying a SSAP – approach combining *Mse*-primers with degenerated transposase primers analyzing six Pinot clones (20, 20–13, 1–84, 1–44, 18 Gm) in four biological replications each provide evidence that the similarity among these clones (despite the lower number of genotypes employed) is lower than in comparable AFLP-studies (Wegscheider 2007).

27.3.3 Clonal Variation by Chimerism

Chimeric grapevines, in particular periclinal chimeras, have been observed in the past and implemented in clonal selection programs. By convention, periclinal chimera have a two-layered-tunica above a corpus with one or more genetically different apical cell layers. Each of these cell layers remains developmentally independent from the adjacent layers. Because of the stratified meristem morphology, most somatic mutations are not fetal (Hocquiny et al. 2004). This structure is a stabilized chimeric form and can be maintained and amplified through vegetative propagation. Somatic mutation events may be induced within these meristematic layers through either mutated cells deriving from the initial shoot meristem, or mutated soma cells that may be incorporated into an adventitious meristem, which then develops into a shoot for the mutant phenotype. First descriptions of chimeric grapevine phenotypes were reported in the middle of the 19th century describing red and white coloured Pinot clusters occurring on one vine (e.g., Breider 1967). Molecular analysis has added proof through recent genetic studies covering periclinal cytochimeras for Gamay (Thompson and Olmo 1963) to grapevine bud sports resulting in multiple colored grapes (e.g., Hocquigny 2004; Walker et al. 2006). The presence of chimera has been confirmed by studies employing microsatellites within the Pinot group (Hocquigny et al. 2004), P. meunier (Franks et al. 2002), Chardonnay (Riaz et al. 2002) and Greco di Tufo (Crespan et al. 2006) through the presence of a third or fourth microsatellite allele. There is evidence that chimeric clones arise from many grapevine cultivars (e.g., Riaz et al. 2002). Rarely, a grapevine cultivar has been identified to be periclinal chimeric, such as *V. vinifera* cv. Pinot meunier (Skene and Barlass 1983). The P. Meunier phenotype has hairy leaves but eventually exhibits mutations (loss of trichomes on leaf surfaces). Studies found P. meunier to be tri-allelic at several loci instead of the usual di-allelic genotypes in grapevines. At locus VVS2, Pinot meunier shares two alleles with Pinot noir (138:153 bp) plus one additional allele (129 bp) (Franks et al. 2002). The underlying genetic mutation is not yet clearly identified. Stenkamp et al. (2009) studied clonal variation of chimeric Pinot meunier in 11 Pinot meunier wild type clones of various origins by comparing mutated genotypes of various ages by AFLP-PCR. Eighteen primer combinations generated a total of 670 reproducible bands, of which 161 (24.02%) were polymorphic. Variation (presented as the rate of interclonal polymorphisms in percent, Stenkamp et al. 2009) among all samples of both groups of wildtype and mutated P. meunier clones was with a mean 1.5% per sample similar to other Pinot varieties (Blaich et al. 2008). Interestingly, the variation among clones were higher in the non-chimeric mutations (1.3%) than in the chimeric wild type (0.6%) confirming the stability of a periclinal chimera (Stenkamp et al. 2009). However, this study also showed that chimeric forces contribute only partly to the overall clonal variation in grapevine. This is in accordance with an extensive genetic variation experiment on five "Pinot" cultivars (P. noir, P. gris, P.blanc, P. meunier, P. moure) performed by Hocquigny et al. (2004) who propose a common, diallelic ancestor (genotype I) for all five Pinot cultivars and provide experimental evidence that divergent genotypes have arisen from this genotype through differential mutation accumulations and cell layer arrangements driven by yet unknown chimeric forces.

27.4 Sex is Not Lost in Grapevine, But Rare

Sexual reproduction is employed for breeding purposes and genetic research. A short current *status quo* on grapevine genetics and molecular breeding is introduced here. The grape cultivars currently in worldwide cultivation are mostly the centuries-old progeny of vines that mated promiscuously in vineyards and are highly heterozygous (Bowers et al. 1999). Selfing seems to be a rare mechanisms of parentage, although grapevine cultivars are hermaphrodite and do self-pollinate (di Gaspero et al. 2005). At the molecular level, heterozygosity manifests itself in DNA sequence divergence among the different species and between cultivars and clones of *V. vinifera* as evidenced by results from molecular based genotyping and from sequencing of allelic variants of genes (e.g., Salamaso et al. 2004; Adam-Blondon et al. 2004; Hoffmann et al. 2006). Molecular maps have been developed (e.g., Doligez et al. 2002; Fischer et al. 2004; Grando et al. 2003; Riaz et al. 2004) covering more than 425 Mbp of the grapevine genome. Grapevine heterozygosity is also expected to be existing in the assortment of genes expressed and in the level at which they are transcribed (Fung et al. 2007). Grape ESTs libraries have been constructed and assembled in a combined effort to facilitate gene discovery, transcription profiling and SNP marker development (e.g., Moser et al. 2005 for a review), providing insights into organ-specific expression of berry, root, leaf, bud, shoot and inflorescences of several grape cultivars. Several EST-banks are open to the scientific community and are used for both applied and molecular breeding efforts (e.g., Goes da Silva et al. 2005). The nucleotide sequence of *V. vinifera* cv. Pinot noir via a shot gun approach has been recently released by Velasco et al. (2006) and will greatly boost further research on grapevine genomics and allow further insights into effects on both sexual and asexual propagation.

27.5 Advantages of Grapevine Clones

Apart from the already mentioned advantages of clones in grapevine breeding, what are the advantages of clones for growers? The major benefit is certainly the identical genotype of every plant in a vineyard and consequently, identical behaviour and growth stages. Plants of a clone will have their bud burst at the same time; their shoots will grow at the same speed and direction, which makes canopy management much easier. All plants of a clonal vineyard will require crop protection at the same time and at the same dosage, which increases efficiency, reduces costs, the amount of pesticides used and their impact on the environment. At the end of the growing season, all plants of a clonal vineyard will commence ripening simultaneously and will all be ready for harvesting at the same time. So, all grapes can be harvested at the right time with a maximum in quality. Therefore, the use of clonal material has many economical and ecological advantages.

Are there also disadvantages in the use of clones? Looking at a clonal vineyard from an ecological point of view, it is an extreme form of monoculture. Identical genotypes are growing side by side throughout the field, a pest or disease specialised in this genotype could wipe out the whole planting (see also Chapters 7 and 25). But so far, there is no evidence that clonal plantings are more threatened by pests and diseases than other plantings. The obvious reason for this is that a vineyard is largely a monoculture anyhow and the genetic differences between clones regarding resistance to pests and diseases are – apart from bunch rot (see Fig. 27.1) – very small. Therefore, clones do not contribute much to the monocultural character of a vineyard and do not increase the pest and disease risk. In the case of botrytis, a tolerant clone is far better than a mixed clonal planting.

27.5.1 Do Grapevine Clones Make Wines Better?

This is a very difficult question to answer, as quality and wine quality in particular can not be measured. Quality is a subjective term. Consequently, wine drinkers very often completely disagree on the quality of a wine. While we can not measure quality itself, we can measure quality parameter, e.g., acidity, sugar, alcohol, colour or aromas. That leads us to the question whether clones can influence quality parameters. As we have already seen (Fig. 27.1), different clones can produce wine with differences in acidity or sugar content. So, when a wine-grower chooses a clone he/she can decide on a particular wine type by selecting a clone with the required quality parameters. If the vineyard is in a humid area, it might be a good idea to choose a clone with loose clusters rather than one with tight compact bunches, being highly susceptible to bunch rot. In these cases, clones are certainly a measure to improve quality. But what about other parameters like flavour and complexity? Do wines from single clones not lack flavour and complexity? Particularly in countries, where clones are fairly new, the wines of often newly planted clones are reported as lower in quality. If a high yielding clone was planted, this is quite obvious. With other clones, the reasoning for these reports is that new, clonal vineyards are compared to old, non-clonal vineyards and that the wine quality of a vineyard usually increases with age. The reason for quality differences is the different age of the vineyards and not the use of clones. Consequently, apart from the economic and ecologic advantages mentioned above, clones can also be used to increase wine quality. Therefore, clones are increasingly used in viticulture world-wide.

27.6 Conclusions

Continuous asexual propagation is the basis for the clonal selection of superior clones in woody perennial crops such as grapevine. A term, describing clones in woody perennial crops could be "a clone is the assemblage of biotypes deriving from a single zygote through somatic mutations of various kinds and thus displays genetic and phenotypic variation".

Selection of grapevine clones combines the search for somatically derived genomic variation to produce new clones with new traits as well as the elimination of less favourable mutations of existing clones. Clonal selection uses both phenotypic and genotypic markers to select for new and to sustain existing grapevine clones. Research on asexually propagated grapevine clones will progress. New innovative techniques will facilitate closer studies of variation-inducing mechanisms of the grape's genome. In our view, these will also be expanded to the other plant genomes (mitochondrial, chloroplast) with new techniques. To successfully select for superior grapevine clones, long-term field studies and both fruit and wine analysis is required and need to be combined with molecular marker studies or metabolic profiling to reveal the molecular mechanisms involved in creating clonal variation. Though asexual propagation is not applied in viticulture, the breeder's efforts are essential for our understanding of the grapevine genome. The position and timing of the mutations will be studied in order to find ways to manipulate the variation-inducing events. Furthermore, quantification of such variation will be of great interest.

Acknowledgments We greatly appreciate the ongoing interest and discussion on grapevine clonality with our colleagues and students from University of Natural Resources and Applied Life Science, Vienna, Austria, and the Geisenheim Research Centre, Geisenheim, Germany. This work has been supported by a scholarship awarded to A.F. from the ESF- Parthenogenesis Network.

27.7 Glossary

Adventitious meristem: forms along the surfaces of stems and leaves.
Botrytis: (*Botrytis cinerea*) botrytis bunch rot.
Bud sports: shoot bearing a genetic mutation developed from a single bud of a plant, which differs from the rest of the plant, and which can also be grafted to grow new plants retaining this genetic difference as a new cultivar.
Cultivars: classification category (rank) of cultigens, and as a distinguishable group of cultigens (taxon).
Periclinal: cell division in a layer of cells that occurs parallel to an adjacent layer of cells.
Stratified meristems: tunica-corpus meristems.
Periclinal cytochimera: a chimera with periclinal layer structure and manifested in different tissues.
Trichomes: fine outgrowths or appendages on plants and certain protists. These are of diverse structure and function. Examples are hairs, glandular hairs, scales, and papillae.
Two-layered-tunica: the tunica layer(s) covers a multicellular, dome-shaped region in which anticlinal, periclinal and oblique divisions occur. The latter is the **corpus** consisting of the cells not displaying restricted divisions and making up the bulk of the meristem.

References

Adam-Blondon AF, Roux C, Claux D, Butterlin G, Merdinoglu D, This P (2004) Mapping 245 SSR markers on the *Vitis vinifera* genome: a tool for grape genetics. Theor Appl Genet 109: 1017–1027

Ambrosi H, Becker H (1978) Der deutsche Wein. Gräfe und Unzer Verlag, München

Benjak A, Konradi J, Blaich R, Forneck A (2006) Different DNA extraction methods can cause different AFLP profiles in grapevine (*Vitis vinifera* L.). Vitis 45: 15–21

Benjak A, Forneck A, Casacuberta JM (2008) Genome-wide analysis of the "cut-and paste" transposons of grapevine. PLoS One 3: e3107

Bennetzen JL (2000) Transposable element contributions to plant gene and genome evolution. Plant Mol Biol 42: 251–269

Billiard R (1913) La vigne dans l'Antiquité. Laffitte, Lyon

Blaich R, Konradi J, Rühl E, Forneck A (2008) Assessing genetic variation among Pinto noir (*Vitis vinifera* L.) clones with AFLP markers. Am J Enol Vitic 58: 526–529

Böhm A, Zyprian E (1998) RAPD marker in grapevine (*Vitis* spp.) similar to plant retrotransposons. Plant Cell Rep 17: 415–421

Boidron R, Boursiquot J-M, Doazan J-P, Leclair P, Leguay M, Walter B (1997) Catalogue of selected wine grape varieties and clones cultivated in France. ENTAV, Le Grau du Roi

Boursiquot JM, Lacombe T, Bowers J, Meredith C (2004) Le Gouais, un cépage clé du partimoine viticola européen. Bull de l'OIV 77: 5–19

Bowers JE, Meredith CP (1997) The parentage of a classic wine grape, Cabernet Sauvignon. Nature Genet 16: 84–87

Bowers J, Boursiquot JM, This P, Chu K, Johansson H, Meredith C (1999) Historical genetics: The parentage of Chardonnay, Gamay, and other wine grapes of northeastern France. Science 285: 1462–1565

Breider H (1967) Über eine Mutation bei der Sorte Würzburg B 48-21-4. Rebe Wein 20: 231–232

Caposella A, Silander JA, Jansen RK, Bergen, B, Talbot, DR (1992) Nuclear ribosomal DNA variation among ramets and genets of white clover. Evolution 46: 1240–1247

Credi R, Babini AR (1997) Effects of virus and virus-like infections on growth, yield, and fruit quality of Albana and Trebbiano Romagnolo grapevines. Am J Enol Vitic 48: 7–12

Crespan M, Cabello F, Giannetto S, Ibanez J, Karoglan Kontic J, Maletic E, Pejic I, Rodriguez-Torres I, Antonacci D (2006) Malvasia delle Lipari, Malvasia di Sardegna, Greco di Gerace, Malvasia de Sitges and Malvasia dubrovacka – synomyms of an old and famous grape cultivar. Vitis 45: 69–73

Columella LJM (ca. 60) De Re Rustica, Vol III. (German Translation:Ahrens K (1972) Columella über Landwirtschaft. Akademie Verlag, Berlin

Di Gaspero G, Cipriani G, Marrazzo MT, Andreetta D, Prado Castro MJ, Peterlunger E, Testolin R (2005) Isolation of (AC) n-microsatellites in *Vitis vinifera* L. and analysis of genetic background in grapevines under marker assisted selection. Mol Breed 15:11–20

Doligez A, Bouquet A, Danglot Y, Lahogue F, Riaz S, Meredith CP, Edwards KJ, This P (2002) Genetic mapping of grapevine (*Vitis vinifera* L.) applied to the detection of QTLs for seedlessness and berry weight. Theor Appl Genet 105:780–795

Feschotte C, Jiang N, Wessler SR (2002) Plant transposable elements: where genetics meets genomics. Nat Rev Genet 3: 329–341

Finnegan DJ (1989) Eukaryotic transposable elements and genome evolution. Trends Genet 5: 103–107

Fischer BM, Salakhutdinov I, Akkurt M, Eibach R, Edwards KJ, Töpfer R, Zyprian EM (2004) Quantitative trait locus analysis of fungal disease resistance factors on a molecular map of grapevine. Theor Appl Genet 108:501–515

Forneck A (2005) Plant Breeding: Clonality – a concept for stability and variability during vegetative propagation. In: Esser ULK, Beyschlag W, Murata J (eds) Progress in botany. Springer, Berlin, pp. 165–183

Franks T, Botta R, Thomas MR (2002) Chimerism in grapevines: implications for cultivar identity, ancestry and genetic improvement. Theor Appl Genet 104: 192–199

Froelich GA (1900) Zur Hybridisation der Reben und der Auswahl der Zuchtreben. Weinbau Weinhandel 18: 230–231

Fung RW, Gonzalo M, Fekete C, Kovacs LG, He Y, Marsh E, McIntyre LM, Schachtman DP, Qiu W (2007) Powdery mildew induces defense-oriented reprogramming of the transcriptome in a susceptible but not in a resistant grapevine. Plant Physiol. doi:10.1104/pp.107.108712

Gill DE, Chao L, Perkins SL, Wolf JB (1995) Genetic mosaicism in plants and clonal animals. Ann Rev Ecol Syst 26:423–444

Goes da Silva F, Iandolino A, Al-Kayal F, Bohlmann MC, Cushman MA, Lim H, Ergul A, Figueroa R, Kabuloglu EK, Osborne C, Rowe J, Tattersall E, Leslie A, Xu J, Baek J, Cramer GR, Cushman JC, Cook DR (2005) Characterizing the grape transcriptome. Analysis of expressed sequence tags from multiple vitis species and development of a compendium of gene expression during berry development. Plant Physiol 139: 574–597.

Grando MS, Bellin D, Edwards KJ, Pozzi C, Stefanini M, Velasso R (2003) Molecular linkage maps of *Vitis vinifera* L. and *Vitis riparia* Mchx. Theor Appl Genet 106: 1213–1224

Hocquigny S, Pelsy F, Dumas V, Kindt S, Heloir M-C, Merdinoglu D (2004) Diversification within grapevine cultivars goes through chimeric states. Genome 47: 579–589

Hoffmann P, Vaclavicek A, Blaich R, Forneck A (2006) A SNP Marker can differentiate loose clustered clones of Pinot noir (*V.vinifera* L.). In: Peterlunger E (ed) Intern Conference on grape genetics and breeding, 2–6th July 2006, Udine, Italy

Imazio S, Labra M, Grassi F, Winfield M, Bardini M, Scienza A (2002) Molecular tools for clone identification: the case of the grapevine cultivar 'Traminer'. Plant Breed 121: 531–535

Jaillon O, Aury J-M, Noel B, Policriti A, Clepet C, Casagrande A, Choisne N, Aubourg S, Vitulo N, Jubin C, Vezzi A, Legeai F, Hugueney P, Dasilva C, Horner D, Mica E, Jublot D, Poulain J, Bruyére C, Billault A, Segurens B, Gouyvenoux M, Ugarte E, Cattonaro F, Anthouard V, Vico V, Del Fabbro C, Alaux M, di Gaspero G, Dumas V, Felice N, Paillard S, Juman I, Moroldo M, Scalabrin S, Canaguier A, Le Clainche I, Malacrida G, Durand

E, Pesole G, Laucou V, Chatelet P, Merdinoglu D, Delledonne M, Pezzotti M, Lecharny A, Scarpelli C, Artiguenave F, Pé ME, Valle G, Morgante M, Caboche M, Adam-Blondon A-F, Weissenbach J, Quétier F, Wincker P (2007) The grapevine genome sequence suggests ancestral hexaploidization in major angiosperm phyla. Nature 449: 463–467.

Kalendar R, Vicient CM, Peleg O, Anamthawat-Jonsson K, Bolshoy A, Schulman AH (2004) Large retrotransposon derivatives: abundant, conserved but nonautonomous retroelements of barley and related genomes. Genetics, 166: 1437–1450.

Klekowski EJ, Godfrey PJ (1989) Aging and mutation in plants. Nature 340: 389–391

Kobayashi S, Goto-Yamamoto N, Hirochika H (2004) Retrotransposon-induced mutations in grape skin color. Science 304: 982

Labra M, Imazio F, Grassi F, Rossoni M, Sala F (2004) Vine-1 retrotransposon-based sequence-specific amplified polymorphism for *Vitis vinifera* L. genotyping. Plant Breed 123: 180–185

Labra M, Imazio S, Grassi F, Rossoni M, Citterio S, Sgorbati S, Scienza F, Failla O (2003) Molecular approach to assess the origin of cv. Marzemino. Vitis 42: 137–140

Levadoux L (1956) Les populations sauvages et cultivées de *Vitis vinifera* L. Ann Amélior Plant (Paris) 6: 59–118

Mannini F, Credi R, Argaimante N (1994) Changes in field performance of clones of grapevine cv. Nebbiolo after virus elimination by heat therapy. In: 6th International Symposium on Grape Breeding, Yalta, Crimea, Ukraine, 4–10 September 1994

Mannini F, Gerbi V, Credi R (1998) Heat-treated virus-infected grapevine clones: Agronomical and enological modifications. Acta Hortic 473: 155–163

Moser C, Segala C, Fontana P, Salakhudtinov I, Gatto P, Pindo M, Zyprian E, Toepfer R, Grando MS, Velasco R (2005) Comparative analysis of expressed sequence tags from different organs of *Vitis vinifera* L. Funct Integr Genomics 5: 208–217

Mullins MG, Meredith CP (1989) The nature of clonal vairaiton in winegrapes: a review. Proc 7th Aust Wine Ind Conference Adelaide, Australia: 14–16

Negrul AM (1946) Origine de la vigne cultivée et sa classification. Tome 1. Ampélographie d'URSS

Oustric J (1994) "Cahiers Techniques". Le Vigneron des Côtes-du-Rhône Nov 1994

Pelsy F, Merdinoglu D (2002) Complete sequence of Tvv1, a family of Ty1 copia-like retrotransposons of *Vitis vinifera* L., reconstituted by chromosome walking. Theor Appl Genet 105: 614–621

Pelsy F, Schehrer L, Merdinoglu D (2003) Development of grapevine retrotransposon-based molecular markers (S-SAP). Acta Hortic: 83–87

Pereira HS, Barao A, Delgado M, Morais-Cecilio L, Viegas W (2005) Genomic analysis of grapevine retrotransposon 1 (Gret1) in *Vitis vinifera*. Theor Appl Genet 111: 871–878

Regner F (1999) Genetische Analyse von Rebsorten. Deutsches Weinbau-Jahrbuch 2000 (51. Jahrgang). Waldkircher Verlag, Waldkirch, pp. 125–132

Regner F, Stadlbauer A, Eisenheld C (1998) Heunisch x Fränkisch, ein wichtiger Genpool europäischer Rebsorten (*Vitis vinifera* L *sativa*). Wein-Wissenschaft 53: 114–118

Regner F, Stadlbauer A, Eisenheld C, Kaserer H (2000) Genetic relationships among Pinots and related cultivars. Am J Enol Vitic 51: 7–14

Riaz S, Garrison KE, Dangl GS, Boursiquot J-M, Meredith CP (2002) Genetic divergence and chimerism within ancient asexually propagated winegrape cultivars. J Am Soc Hortic Sci 127: 508–514

Riaz S, Dangl GS, Edwards KJ, Meredith CP (2004) A microsatellite marker based framework linkage map of *Vitis vinifera* L. Theor Appl Genet 108: 864–872

Robertson HM (2002) Evolution of DNA transposons in eukaryotes. In: Craig NL, Craigie R, Gellert M, Lambowitz AM (eds) Mobile DNA II. Am Soc for Microbiol Press, Washington DC, pp. 1093–1110

Rühl EH, Konrad H, Schmitt K, Pfeiffer W (2000) Clonal variation in Pinot Noir. 5th Intern Symposium on cool climate viticulture and oenology, 16–20 January, 2000

Salmaso M, Faes G, Segala C, Stefanini M, Salakhutdinov I, Zyprian E, Toepfer R, Stella Grand OM, Velasco R (2004) Genome diversity and gene haplotypes in the grapevine (*Vitis vinifera* L.), as revealed by single nucleotide polymorphisms. Mol Breed 14: 385–395

Sensi E, Vignani R, Rohde W, Biricolti S (1996) Characterization of genetic biodiversity with *Vitis vinifera* L. Sangiovese and Colorino genotypes by AFLP and ISTR DNA marker technology. Vitis 35: 183–188

Skene KGM, Barlass M (1983) Studies on the fragmented shoot apex of grapevine. IV. Separation of phenotypes in a periclinal chimera in vitro. J Exp Bot 34: 1271–1280

Sievers E (1971) Anatomical differences in clones of the White Riesling cv. (Anatomische Unterschiede bei Klonen der Rebsorte Weisser Riesling). Mitt Klosterneuburg 21: 63–64

Silvestroni O, Intrieri C, Di Domizio N (1995) Valutazione della variabilità clonale in cultivar di *Vitis vinifera* mediante metodi fillometrici. Academ Ital Vite Vino 47: 111–125

Stenkamp S, Becker M, Hill B, Blaich R, Forneck A (2009) Clonal variation and stability assay of chimeric Pinot Muenier (*Vitis vinifera* L.) and descending sports. Euphytica 165: 197–209

Thompson MM, Olmo HP (1963) Cytohistological studies of cytochimeric and tetraploid grapes. Am J Bot 50: 901–906

Tuskan GA, Francis KE, Russ SL, Romme WH, Turner MG (1996) RAPD markers reveal diversity within and among clonal and seedling stands of aspen in Yellowstone National Park, USA. Can J For Res 26: 2088–2098

Velasco R, Zharkikh A, Troggio M, Bhatnagar S, Pindo M, Cartwright D, Coppola G, Eldredge G, Vezzulli S, Malacarne G, Mitchell J, Stefanini M, Segala C, Gutin N, Grando MS, Pruss D, Dematte L, Cestaro A, Toppo S, Fontana P, Skolnick M. Gutin A, Salamini F, Viola R (2006) The first genome sequence of an elite grapevine cultivar (Pinot noir *Vitis vinifera* L.): coping with a highly heterozygous genome. NBCI – database http://www.ncbi.nlm.nih.gov/

Velasco R, Zharkikh A, Troggio M, Cartwright DA, Cestaro A, Pruss D, Pindo M, FitzGerald LM, Vezzulli S, Reid J, Malacarne G, Iliev D, Coppola G, Wardell B, Micheletti D, Macalma T, Facci M, Mitchell JT, Perazzolli M, Eldredge G, Gatto P, Oyzerski R, Moretto M, Gutin N, Stefanini M, Chen Y, Segala C, Davenport C, Dematte L, Mraz A, Battilana J, Stormo K, Costa F, Tao Q, Si-Ammour A, Harkins T, Lackey A, Perbost C, Taillon B, Stella A, Solovyev V, Fawcett JA, Sterck L, Vandepoele K, Grando SM, Toppo S, Moser C, Lanchbury J, Bogden R, Skolnick M, Sgaramella V, Bhatnagar SK, Fontana P, Gutin A, Van de Peer Y, Salamini F, Viola R (2007) A high quality draft consensus sequence of the genome of a heterozygous grapevine variety, PLoS ONE 2: e1326.

Verries C, Bes C, This P, Tesniere C (2000) Cloning and characterization of Vine-1, a LTR-retrotransposon-like element in *Vitis vinifera* L., and other *Vitis* species. Genome 43: 366–376

Walker AR, Lee E, Robinson SP (2006) Two new grape cultivars, bud sports of Cabernet Sauvignon bearing pale-coloured berries, are the result of deletion of two regulatory genes of the berry colour locus. Plant Mol Biol 62: 623–635

Walter B, Martelli GP (1996) Clonal selection of the vine: Sanitary and pomological selection. Influence of viruses and quality. Part 1: Effects of viruses on the culture of the vine and its products. Bulletin de l'OIV 68: 945–971

Walter B, Martelli GP (1997) Clonal selection of the vine: Sanitary and pomological selection. Influence of viruses and quality. Part 2: Sanitary selection – pomological selection. Bulletin de l'OIV 70: 5–23

Wegscheider E (2007) Clonal variability in Pinot noir revealed by S-SAP involving universal retrotransposon-based sequences. Master Thesis. University of Natural Resources and applied Life Sciences, Vienna

Witte CP, Le QH, Bureau T, Kumar A (2001) Terminal-repeat retrotransposons in miniature (TRIM) are involved in restructuring plant genomes. Proc Natl. Acad Sci USA 98: 13778–13783

Yakushiji H, Kobayashi, S, Goto-Yamamoto N, Tae Jeong S, Sueta T, Mitani N, Azuma A (2006) A skin color mutation of grapevine, from black-skinned Pinot Noir to white-skinned Pinot

Blanc, is caused by deletion of the functional VvmybA1 allele. Biosci Biotechnol Biochem 70: 1506–1508

Ye GN, Soylemezoglu G, Weeden NF, Lamboy WF, Pool RM, Reisch BI (1998) Analysis of the relationship between grapevine cultivars, sports and clones via DNA fingerprinting. Vitis 37: 33–38

Zohary D, Spiegel-Roy P (1975) Beginnings of fruit growings in the old world. Science 187: 319–327

Index

I. Schön et al. (eds.), *Lost Sex*, DOI 10.1007/978-90-481-2770-2_BM2,

B

I

Q

R

GPSR Compliance

The European Union's (EU) General Product Safety Regulation (GPSR) is a set of rules that requires consumer products to be safe and our obligations to ensure this.

If you have any concerns about our products, you can contact us on ProductSafety@springernature.com

In case Publisher is established outside the EU, the EU authorized representative is:

Springer Nature Customer Service Center GmbH
Europaplatz 3
69115 Heidelberg, Germany

Batch number: 10822618

Printed by Printforce, the Netherlands